*Edited by*
*Andrew B. Hughes*

**Amino Acids, Peptides and Proteins in Organic Chemistry**

## *Further Reading*

Drauz, K., Gröger, H., May, O. (eds.)

**Enzyme Catalysis in Organic Synthesis**

**Third completely revised and enlarged edition**

3 Volumes

2011

ISBN: 978-3-527-32547-4

Fessner, W.-D., Anthonsen, T.

**Modern Biocatalysis**

**Stereoselective and Environmentally Friendly Reactions**

2009

ISBN: 978-3-527-32071-4

Pignataro, B. (ed.)

**Ideas in Chemistry and Molecular Sciences**

**Advances in Synthetic Chemistry**

2010

ISBN: 978-3-527-32539-9

Reek, J. N. H., Otto, S.

**Dynamic Combinatorial Chemistry**

2010

ISBN: 978-3-527-32122-3

Lutz, S., Bornscheuer, U. T. (eds.)

**Protein Engineering Handbook**

**2 Volume Set**

2009

ISBN: 978-3-527-31850-6

Sewald, N., Jakubke, H.-D.

**Peptides: Chemistry and Biology**

2009

ISBN: 978-3-527-31867-4

Jakubke, H.-D., Sewald, N.

**Peptides from A to Z**

**A Concise Encyclopedia**

2008

ISBN: 978-3-527-31722-6

Royer, J. (ed.)

**Asymmetric Synthesis of Nitrogen Heterocycles**

2009

ISBN: 978-3-527-32036-3

Hecht, S., Huc, I. (eds.)

**Foldamers**

**Structure, Properties, and Applications**

2007

ISBN: 978-3-527-31563-5

*Edited by*
*Andrew B. Hughes*

# Amino Acids, Peptides and Proteins in Organic Chemistry

## Volume 3 - Building Blocks, Catalysis and Coupling Chemistry

WILEY-VCH Verlag GmbH & Co. KGaA

**The Editor**

***Andrew B. Hughes***
La Trobe University
Department of Chemistry
Victoria 3086
Australia



**Library of Congress Card No.:** applied for

**British Library Cataloguing-in-Publication Data**
A catalogue record for this book is available from the British Library.

**Bibliographic information published by the Deutsche Nationalbibliothek**
The Deutsche Nationalbibliothek lists this publication in the Deutsche Nationalbibliografie; detailed bibliographic data are available on the Internet at http://dnb.d-nb.de.



**Composition** Thomson Digital, Noida, India
**Printing and Bookbinding** betz-druck GmbH, Darmstadt
**Cover Design** Schulz Grafik Design, Fußgönheim

Printed in the Federal Republic of Germany
Printed on acid-free paper

**ISBN:** 978-3-527-32102-5

# Contents



*Amino Acids, Peptides and Proteins in Organic Chemistry.*
*Vol.3 – Building Blocks, Catalysis and Coupling Chemistry.* Edited by Andrew B. Hughes

ISBN: 978-3-527-32102-5

# List of Contributors

***Knut Adermann***
Pharis Biotec GmbH
Feodor-Lynen-Strasse 31
30625 Hannover
Germany

***José Luis Adrio***
Neuron BPh
Avda. de la Innovación 1, Edificio BIC
Parque Tecnológico de ciencias de la Salud
18100 Armilla
Granada
Spain

***Fernando Albericio***
Institute for Research in Biomedicine
Barcelona Science Park
Baldiri Reixac 10
08028 Barcelona
Spain

and

Networking Centre on Bioengineering, Biomaterials and Nanomedicine (CIBER-BBN)
Barcelona Science Park
Baldiri Reixac 10
08028 Barcelona
Spain

***University of Barcelona***
Department of Organic Chemistry
Martí i Franqués 1–11
08028 Barcelona
Spain

***Andrea Bagno***
University of Padova
Department of Chemical Process Engineering
Via Marzolo 9
35131 Padova
Italy

***Kleomenis Barlos***
University of Patras
Department of Chemistry
Rion-Patras
Greece

***José Luis Barredo***
I+D Biologia
Antibióticos S.A.
Avda. Antibióticos, 59–61
24009 León
Spain

***Julie Bletz***
Institute for Systems Biology
1441 North 34th Street
Seattle, WA 98103-8904
USA

*Amino Acids, Peptides and Proteins in Organic Chemistry.*
*Vol.3 – Building Blocks, Catalysis and Coupling Chemistry.* Edited by Andrew B. Hughes

ISBN: 978-3-527-32102-5

*Jeffrey W. Bode*
Eidgenössische Technische
Hochschule Zürich
Laboratorium für Organische Chemie
Wolfgang Pauli Strasse 10
8093 Zürich
Switzerland

*M. Isabel Calaza*
Universidad de Zaragoza – CSIC
Instituto de Ciencia de Materiales
de Aragón
Departamento de Química Orgánica
50009 Zaragoza
Spain

*Stefano Carganico*
University of Firenze
Polo Scientifico e Tecnologico
Laboratory of Peptide and Protein
Chemistry and Biology
Via della Lastruccia 13
50019 Sesto Fiorentino
Italy

*PRES University of Cergy-Pontoise*
Laboratoire SOSCO-UMR 8123
5 mail Gay-Lussac, Neuville sur Oise
95031 Cergy-Pontoise
France

*Carlos Cativiela*
Universidad de Zaragoza – CSIC
Instituto de Ciencia de Materiales
de Aragón
Departamento de Química Orgánica
50009 Zaragoza
Spain

*Arnold L. Demain*
Drew University
RISE
HS-330
Madison, NJ 07940
USA

*Jan Deska*
Stockholm University
Arrhenius Laboratory
Department of Organic Chemistry
106 91 Stockholm
Sweden

*Monica Dettin*
University of Padova
Department of Chemical Process
Engineering
Via Marzolo 9
35131 Padova
Italy

*Carlo Di Bello*
University of Padova
Department of Chemical Process
Engineering
Via Marzolo 9
35131 Padova
Italy

*Ayman El-Faham*
King Saud University
College of Science
Department of Chemistry
PO Box 2455
1451 Riyadh
Kingdom of Saudi Arabia

*Alexandria University*
Faculty of Science
Department of Chemistry
Horria Street, PO Box 246, Ibrahimia
21321 Alexandria
Egypt

*Institute for Research in Biomedicine*
Barcelona Science Park
Baldiri Reixac 10
08028 Barcelona
Spain

***Christian P.R. Hackenberger***
Freie Universität Berlin
Institut für Chemie und Biochemie
Takustrasse 3
14195 Berlin
Germany

***Jan Hlaváč***
Palacky University
Department of Organic Chemistry
Trida 17, Listopadu 12
771 46 Olomouc
Czech Republic

***Steven A. Kates***
Ischemix
63 Great Road
Maynard, MA 01754
USA

***Viktor Krchňák***
University of Notre Dame
Department of Chemistry and Biochemistry
251 Nieuwland Science Center
Notre Dame, IN 46556
USA

***Ulrike Kusebauch***
Institute for Systems Biology
1441 North 34th Street
Seattle, WA 98103-8904
USA

***Cleber W. Liria***
University of São Paulo
Institute of Chemistry
Department of Biochemistry
Peptide Chemistry Laboratory
Av. Prof. Lineu Prestes, 748
05508-900 São Paulo
Brazil

***Garland R. Marshall***
Washington University
Center for Computational Biology
Departments of Biochemistry and Molecular Biophysics and Biomedical Engineering
700 S. Euclid Avenue
St. Louis, MO 63110
USA

***Joshua McBee***
Institute for Systems Biology
1441 North 34th Street
Seattle, WA 98103-8904
USA

***Maria Terêsa Machini Miranda***
University of São Paulo
Institute of Chemistry
Department of Biochemistry
Peptide Chemistry Laboratory
Av. Prof. Lineu Prestes, 748
05508-900 São Paulo
Brazil

***Robert L. Moritz***
Institute for Systems Biology
1441 North 34th Street
Seattle, WA 98103-8904
USA

***Yoshio Okada***
Kobe Gakuin University
Faculty of Pharmaceutical Sciences
Arise 518, Ikawadani-cho, Nishi-ku
651-2180 Kobe
Japan

***Michael I. Page***
University of Huddersfield
Department of Chemical and Biological Sciences
Queensgate
Huddersfield HD1 3DH
UK

***Anna Maria Papini***
University of Firenze
Polo Scientifico e Tecnologico
Laboratory of Peptide and Protein
Chemistry and Biology
Via della Lastruccia 13
50019 Sesto Fiorentino
Italy

***PRES University of Cergy-Pontoise***
Laboratoire SOSCO-UMR 8123
5 mail Gay-Lussac, Neuville sur Oise
95031 Cergy-Pontoise
France

***Emily J. Parker***
University of Canterbury
Department of Chemistry
PO Box 4800
Christchurch
New Zealand

***Andrew J. Pratt***
University of Canterbury
Department of Chemistry
PO Box 4800
Christchurch
New Zealand

***Cesar Remuzgo***
University of São Paulo
Institute of Chemistry
Department of Biochemistry
Peptide Chemistry Laboratory
Av. Prof. Lineu Prestes, 748
05508-900 São Paulo
Brazil

***Marta Rodriguez-Sáiz***
Antibióticos S.A.
Avda. Antibióticos, 59–61
24009 León
Spain

***Dirk Schwarzer***
Leibniz-Institut für Molekulare
Pharmakologie (FMP)
Chemical Biology Section
Robert-Rössle-Strasse 10
13125 Berlin
Germany

***Richard J. Simpson***
Ludwig Institute For Cancer Research
Joint Proteomics Laboratory
Royal Melbourne Hospital
Parkville, Victoria 3050
Australia

***Miroslav Soural***
Palacky University
Department of Organic Chemistry
Trida 17, Listopadu 12
771 46 Olomouc
Czech Republic

***Yuko Tsuda***
Kobe Gakuin University
Faculty of Pharmaceutical Sciences
Minatojima 1-1-3, Chuo-ku
650-8586 Kobe
Japan

***Judit Tulla-Puche***
Institute for Research in Biomedicine
Barcelona Science Park
Baldiri Reixac 10
08028 Barcelona
Spain

and

Networking Centre on Bioengineering,
Biomaterials and Nanomedicine
(CIBER-BBN)
Barcelona Science Park
Baldiri Reixac 10
08028 Barcelona
Spain

# Part One
# Amino Acids as Building Blocks

*Amino Acids, Peptides and Proteins in Organic Chemistry.*
*Vol.3 – Building Blocks, Catalysis and Coupling Chemistry.* Edited by Andrew B. Hughes

ISBN: 978-3-527-32102-5

# 1
# Amino Acid Biosynthesis

*Emily J. Parker and Andrew J. Pratt*

## 1.1
## Introduction

The ribosomal synthesis of proteins utilizes a family of 20 α-amino acids that are universally coded by the translation machinery; in addition, two further α-amino acids, selenocysteine and pyrrolysine, are now believed to be incorporated into proteins via ribosomal synthesis in some organisms. More than 300 other amino acid residues have been identified in proteins, but most are of restricted distribution and produced via post-translational modification of the ubiquitous protein amino acids [1]. The ribosomally encoded α-amino acids described here ultimately derive from α-keto acids by a process corresponding to reductive amination. The most important biosynthetic distinction relates to whether appropriate carbon skeletons are pre-existing in basic metabolism or whether they have to be synthesized *de novo* and this division underpins the structure of this chapter.

There are a small number of α-keto acids ubiquitously found in core metabolism, notably pyruvate (and a related 3-phosphoglycerate derivative from glycolysis), together with two components of the tricarboxylic acid cycle (TCA), oxaloacetate and α-ketoglutarate (α-KG). These building blocks ultimately provide the carbon skeletons for unbranched α-amino acids of three, four, and five carbons, respectively. α-Amino acids with shorter (glycine) or longer (lysine and pyrrolysine) straight chains are made by alternative pathways depending on the available raw materials. The strategic challenge for the biosynthesis of most straight-chain amino acids centers around two issues: how is the α-amino function introduced into the carbon skeleton and what functional group manipulations are required to generate the diversity of side-chain functionality required for the protein function?

The core family of straight-chain amino acids does not provide all the functionality required for proteins. α-Amino acids with branched side-chains are used for two purposes; the primary need is related to protein structural issues. Proteins fold into well-defined three-dimensional shapes by virtue of their amphipathic nature: a significant fraction of the amino acid side-chains are of low polarity and the hydrophobic effect drives the formation of ordered structures in which these side-chains are buried away from water. In contrast to the straight-chain amino

*Amino Acids, Peptides and Proteins in Organic Chemistry.*
*Vol.3 – Building Blocks, Catalysis and Coupling Chemistry.* Edited by Andrew B. Hughes

ISBN: 978-3-527-32102-5

acids, the hydrophobic residues have large nonpolar surface areas by virtue of their branched hydrocarbon side-chains. The other role of branched amino acids is to provide two useful functional groups: an imidazole (histidine) and a phenol (tyrosine) that exploit aromatic functional group chemistry.

This chapter provides an overview of amino acid biosynthesis from a chemical perspective and focuses on recent developments in the field. It highlights a few overarching themes, including the following:

i) The chemical logic of the biosynthetic pathways that underpin amino acid biosynthesis. This chemical foundation is critical because of the evolutionary mechanisms that have shaped these pathways. In particular, the way in which gene duplication and functional divergence (via mutation and selection) can generate new substrate specificity and enzyme activities from existing catalysts [2].
ii) The contemporary use of modern multidisciplinary methodology, including chemistry, enzymology, and genomics, to characterize new biosynthetic pathways.
iii) Potential practical implications of understanding the diverse metabolism of amino acid biosynthesis, especially medicinal and agrichemical applications.
iv) The higher-level molecular architectures that control the fate of metabolites, especially the channeling of metabolites between active sites for efficient utilization of reactive intermediates.

**Box 1.1: Nitrogen and Redox in Amino Acid Biosynthesis**

Ammonia is toxic and the levels of ammonia available for the biosynthesis of amino acids in most biochemical situations is low. There are a limited number of entry points of ammonia into amino acid biosynthesis, notably related to glutamate and glutamine. Once incorporated into key amino acids, nitrogen is transferred between metabolites either directly or via *in situ* liberation of ammonia by a multifunctional complex incorporating the target biosynthetic enzyme. The main source of *in situ* generated ammonia for biosynthesis is the hydrolysis of glutamine by glutaminases. *De novo* biosynthesis of amino acids, like element fixation pathways in general, is primarily reductive in nature. This may reflect the origins of these pathways in an anaerobic world more than 3 billion years ago.

**Box 1.2: The Study of Biosynthetic Enzymes and Pathways**

The source of an enzyme for biochemical study has important implications. Most core metabolism has been elaborated by studying a small number of organisms that were chosen for a variety of reasons, including availability, ease of manipulation, ethical concerns, scientific characterization, and so on. These exemplar organisms include the bacterium *Escherichia coli*, the yeast *Saccharomyces cerevisiae*, the plant *Arabidopsis thaliana*, and the rat as a typical mammal. Much of the detailed characterization of amino acid biosynthesis commenced with studies on these organisms. With the rise of genetic engineering techniques, biosynthetic enzymes from a wide variety of sources are available for scientific investigation, and there has been increasing emphasis on working with enzymes and pathways from alternative organisms.

Metabolic diversity is greatest among prokaryotes. One fundamental change in the underlying microbiology that has affected our understanding of pathway diversity has been the appreciation of the deep biochemical distinctions between what are now recognized to be two fundamental domains of prokaryotes: eubacteria and Archaea [3]. The former bacteria include those well known to be associated with disease and fermentation processes; while the latter include many methanogens and extremophiles (prokaryotes that grow in extreme conditions, such as hyperthermophiles that grow at temperatures above 60 °C or halophiles that grow in high ionic strength environments). Bioinformatics approaches are complementing conventional enzymological studies in identifying and characterizing interesting alternative biosynthetic pathways [4]. The greater understanding of microbial and biosynthetic diversity is presenting exciting opportunities for novel discoveries in biosynthesis.

Much of the focus of biosynthetic enzymology now focuses on enzymes from pathogens and hyperthermophiles. The focus on the study of enzymes from pathogens is predicated on the possibility that inhibitors of such enzymes may be useful as pesticides and therapeutic agents. Since humans have access to many amino acids in their food, they have lost the ability to make "dietary essential" amino acids that typically require extended dedicated biosynthetic pathways [5]. The biosynthetic enzymes of the corresponding pathways are essential for many pathogens and plants, but not for humans; hence, selective inhibitors of these biosynthetic enzymes are potentially nontoxic to humans, but toxic to undesirable organisms. Enzymes from hyperthermophilic organisms, produced by genetic engineering, are scrutinized mainly because of their ease of structural characterization. These enzymes retain their native structures at temperatures that denature most other proteins, including those of the host organism. These proteins are of high thermal stability and simple heat treatment can be used to effect high levels of purification of the desired protein.

## 1.2
## Glutamate and Glutamine: Gateways to Amino Acid Biosynthesis

Glutamate and the corresponding amide derivative, glutamine, are critical metabolites in amino acid metabolism. The biochemistry of these two amino acids also illustrates the distinct chemistry associated with the α-amino and side-chain functional groups, each of which is exploited in the biosynthesis of other amino acids. These amino acids derive from ammonia and α-KG. Glutamate dehydrogenase (GDH) interconverts α-KG and glutamate (Figure 1.1) [6]. Although glutamate is formed in this way by reductive amination, this enzyme is generally not dedicated to biosynthesis; the reverse reaction, an oxidative deamination to regenerate α-KG, is

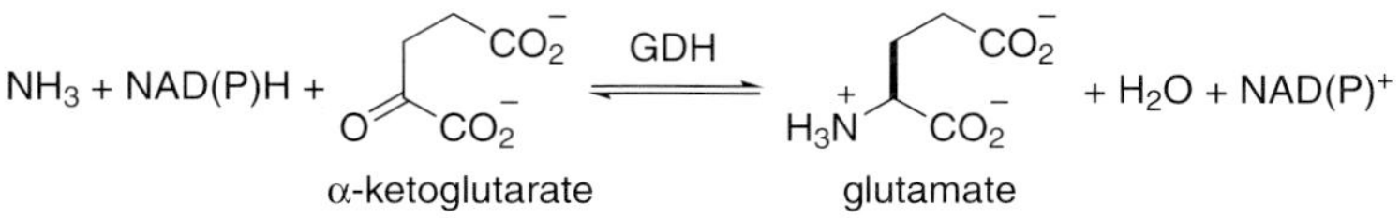

$$NH_3 + NAD(P)H + \alpha\text{-ketoglutarate} \xrightleftharpoons{GDH} \text{glutamate} + H_2O + NAD(P)^+$$

Figure 1.1 Interconversion of α-KG and glutamate catalyzed by GDH.

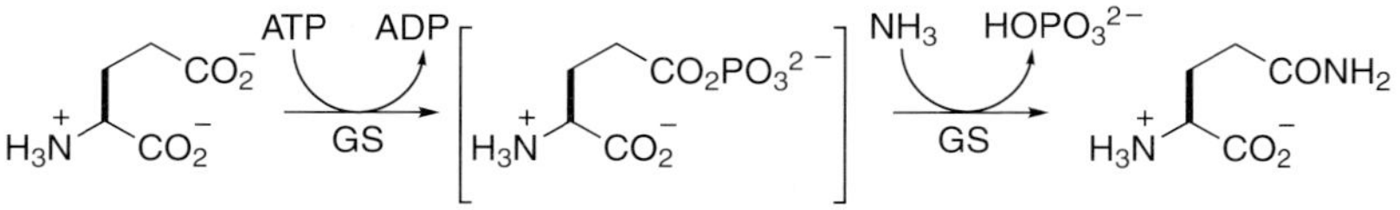

$$\text{glutamate} \xrightarrow[GS]{ATP \;\; ADP} \left[ \gamma\text{-glutamyl } CO_2PO_3^{2-} \right] \xrightarrow[GS]{NH_3 \;\; HOPO_3^{2-}} \text{glutamine } (CONH_2)$$

Figure 1.2 Conversion of glutamate to glutamine catalyzed by GS.

often an important *in vivo* role for this enzyme [7]. This deamination chemistry might be a factor in the relatively weak binding of ammonia (e.g., $K_M(NH_3)$ is 3 mM for the *E. coli* enzyme – above normal environmental concentrations). In many organisms there is an additional enzyme, glutamate synthase (GOGAT), dedicated to the biosynthesis of glutamate [8]. GOGAT utilizes ammonia generated *in situ* by the hydrolysis of glutamine and this enzyme will be described after a discussion of the biosynthesis of glutamine.

The conversion of glutamate to glutamine, catalyzed by glutamine synthetase (GS), requires the activation of the side-chain carboxylate as an acyl phosphate, prior to nucleophilic substitution of the resulting good leaving group by ammonia (Figure 1.2). The use of ATP, to produce γ-glutamyl phosphate, assists both the kinetics and the thermodynamics of amide formation: by producing a more reactive carboxylic acid derivative and overturning the intrinsically favorable nature of amide hydrolysis in water.

GS from enteric bacteria, such as *E. coli* and *Salmonella typhimurium*, is an exemplar of an amino acid biosynthetic enzyme; the overall reaction it catalyses is effectively irreversible *in vivo* ($K = 1200$). Being dedicated to biosynthesis, it has evolved tight binding of ammonia ($K_M(NH_3) < 200\,\mu M$) which allows efficient synthesis of glutamine under the low ammonia conditions (much less than 1 mM) found *in vivo*. Its *in vivo* role as an entry point for the biosynthesis of a wide range of nitrogen metabolites is eloquently communicated by the extensive feedback regulation of this enzyme by a range of nitrogen-containing metabolites, including glycine, serine, alanine, and histidine [9–12]. Glutamine is the primary store of ammonia in many cells; the side-chain amide is chemically unreactive, but its favorable hydrolysis can be catalyzed on demand by glutamine amidotransferase (GAT) enzymes [13].

### 1.2.1
### Case Study: GOGAT: GATs and Multifunctional Enzymes in Amino Acid Biosynthesis

In contrast to GDH, GOGAT is a dedicated biosynthetic enzyme. It is the primary source of glutamate in plants, eubacteria and lower animals [14, 15]. These iron–sulfur flavoproteins carry out the reductive amination of α-KG to glutamate via a five-

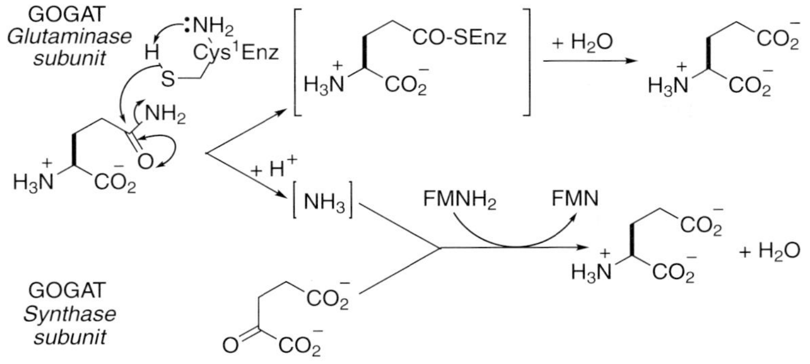

Figure 1.3 The biosynthesis of glutamate mediated by GOGAT.

step process that utilizes the *in situ* hydrolysis of glutamine as the source of ammonia for this reaction (Figure 1.3) [16].

As with many reductive biosynthetic enzymes, there are variants of the enzyme adapted to different electron sources; for example, both ferredoxin- and nicotinamide-dependent enzymes are known, and examples of both of these classes of GOGAT have been studied in detail [15]. They reveal many of the key features of metabolite channeling observed in biosynthetic enzymes utilizing glutamine as a nitrogen donor.

The NADPH-dependent GOGAT from *Azospirillum brasilense* is an α,β-heterodimer [17]. The β-subunit supplies the electrons for the reductive amination process: NADPH reduces FAD and the electrons are passed, in turn, to a 3Fe–4S center on the α-subunit and then on to the FMN cofactor at the active site for glutamate formation. The α-subunit consists of four domains. The N-terminal domain is a type II GAT, in which the N-terminal cysteine attacks glutamine releasing ammonia and generating an enzyme-bound thioester, which is subsequently hydrolyzed (Figure 1.3) (type I GATs, the other variant, utilize a combination of an internal cysteine and a histidine as catalytic residues [18]). When a class II GAT is active, a conserved Q-loop closes over the active site and prevents release of ammonia to the solution; instead the nascent ammonia travels through a hydrophilic internal tunnel approximately 30 Å in length to the third domain which is a $(\beta\alpha)_8$ barrel containing the 3Fe–4S cluster and the FMN active site. The latter site binds the substrate α-KG and carries out the synthesis of glutamate, presumably via reduction of an α-iminoglutarate intermediate. There is a gating mechanism for synchronization of GAT and reductive amination active sites: the glutaminase activity is dependent on the binding of both α-KG and reduced cofactor at the second site (Figure 1.4) [19].

The ferredoxin-dependent GOGAT from the cyanobacterium, *Synechocystis* sp. PCC 6803, has a similar structure to the *A. brasilense* enzyme, possessing a type II GAT domain and a synthase site linked by a 30-Å tunnel, which is gated in an analogous way [21]. The GAT domain exists in an inactive conformation, which can bind glutamine but not hydrolyze it. This is converted to the active conformation on binding of α-KG and reduced cofactor, $FMNH_2$, to the synthase site; this

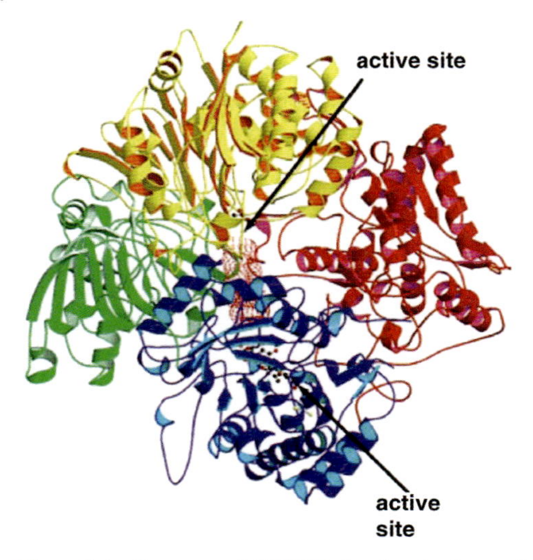


**Figure 1.4** Structure of GOGAT showing the internal tunnel for ammonia transfer between the GAT (gold) and synthase (blue) active sites. (Picture taken from [20].)

conformational switch also serves to open the entry point to the ammonia tunnel. A conserved glutamate residue (Glu1013 in the *Synechocystis* enzyme), present at the tunnel constriction, has been shown to be the key residue controlling the cross-regulation mechanism. This glutamate interacts with the N-terminal amino group of the protein, which is the active-site base of the glutaminase, as well as affecting the geometry of the tunnel entry point. Mutation of this residue to aspartate, asparagine, or alanine affected glutaminase activity and the sensitivity of glutaminase action to the binding of α-KG at the synthase site [22].

GOGAT exemplifies our growing awareness of details of glutamine-dependent enzymes, in particular, and biosynthetic pathways, in general. By exploiting the higher-level organization of multifunctional enzyme systems, metabolites can be channeled to the next enzyme of a pathway; thereby controlling their fate. Together with the potential for subtle levels of regulation, this organization ensures the efficient use of biosynthetic intermediates [20].

## 1.3
## Other Amino Acids from Ubiquitous Metabolites: Pyridoxal Phosphate-Dependent Routes to Aspartate, Alanine, and Glycine

### 1.3.1
### Pyridoxal Phosphate: A Critical Cofactor of Amino Acid Metabolism

Once glutamate is available, the α-amino function can be transferred to other α-keto acids via amino acid aminotransferase enzymes (Figure 1.5) [23]. This family of

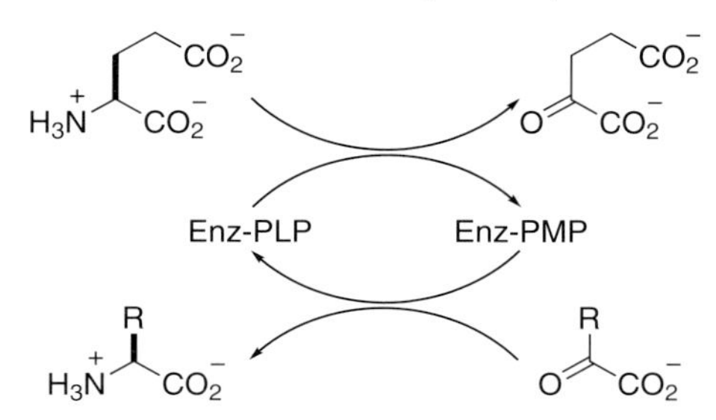


**Figure 1.5** Overall interconversion mediated by most aminotransferase enzymes (e.g., for AATases $R = CH_2CO_2^-$).

enzymes exploit the catalytic versatility of the cofactor pyridoxal 5′-phosphate (pyridoxal phosphate PLP), one of the active forms of vitamin $B_6$, which is interconverted with pyridoxamine phosphate (PMP) during the overall transformation [24, 25].

The aldehyde of PLP readily forms Schiff bases with amines and this cofactor is generally tethered to the active site of enzymes via a link to a lysine side-chain. Amino acid substrates bind to the cofactor by Schiff base exchange with the enzyme lysine, which is thereby liberated as a potential active-site base. The critical feature exploited in amino acid metabolism is the ability of PLP to act as an electron sink, stabilizing negative charge build up at $C^\alpha$ of the substrate (Figure 1.6). By delocalizing negative charge at this center PLP is able to mediate chemistry at the α-, β-, and γ-centers of appropriately functionalized amino acids (see Box 1.3).

Pyridoxal-dependent enzymes have been classified into five fold-types and the aspartate aminotransferase (AATase) family of enzymes belong to Fold-Type I [26, 27]. The cytosolic and mitochondrial AATases were the first PLP-dependent enzymes for which detailed structural information was obtained [28–30]. These enzymes interconvert glutamate and oxaloacetate with α-KG and aspartate (Figure 1.7). Glutamate is activated by binding to the PLP and the displaced Schiff base Lys258 acts as an acid–base catalyst to transfer a proton between $C^\alpha$ and C4′ of the PLP [31]. An aspartate residue (Asp222) interacts with the protonated nitrogen of the cofactor,

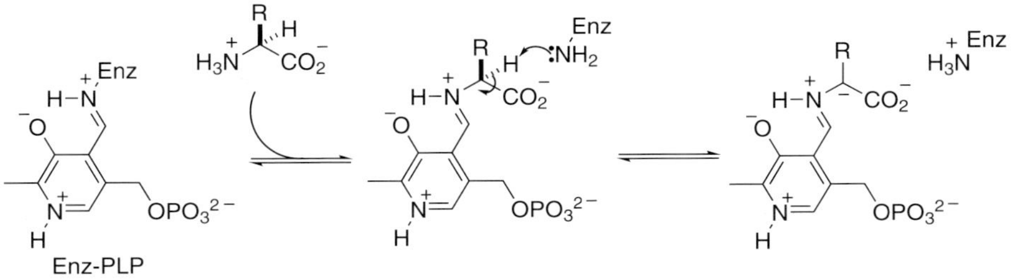

**Figure 1.6** Schiff base formation and anion stabilization by PLP-dependent enzymes.

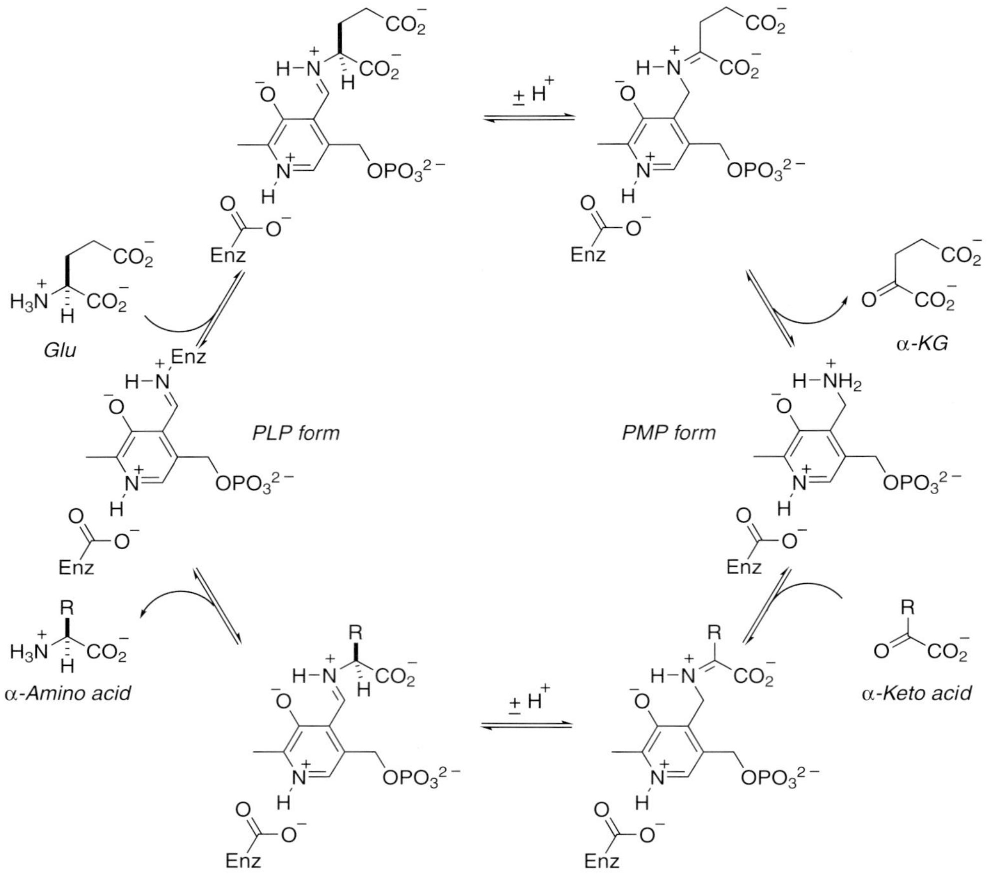

Figure 1.7 Mechanism of aminotransferase catalysis (for AATases R = $CH_2CO_2^-$).

stabilizing the pyridinium form and facilitating deprotonation of the substrate. Once a proton has been transferred from $C^{\alpha}$ to C4′, hydrolytic cleavage of the ketimine linkage liberates α-KG and leaves the PMP form of the cofactor. Binding of oxaloacetate and running the reaction in reverse leads to regeneration of the original enzyme and production of aspartate. Aminotransferase enzymes provide a general mechanism for interconverting α-amino acids and α-keto acids, illustrating a second route by which nitrogen is transferred between metabolites.

### 1.3.2
### Case Study: Dual Substrate Specificity of Families of Aminotransferase Enzymes

Aminotransferase enzymes pose an intriguing challenge for substrate specificity since they bind two different substrates successively at the same site and must

**Box 1.3: The Mechanistic Versatility of PLP: A Biochemical Electron Sink**

Amino acids bind to PLP by forming a Schiff base. Once bound, the ability of PLP to stabilize a negative charge at the α-center of bound amino acids has been harnessed by a range of amino acid biosynthetic enzymes to mediate chemistry at the α-, β- and γ-centers of suitably functionalized amino acids.

**α-Center Reactivity**

Cleavage of any of the three substituent bonds to the α-center can lead to a carbanionic species (Figure 1.8). Deprotonation of the α-proton, by the lysine liberated on Schiff base exchange, is used in transamination chemistry where the α-proton is relocated to the benzylic position of PLP *en route* to PMP as described above (and in some amino acid racemases). Decarboxylation provides a related anion, which can be protonated; this is the source of biological amines and is exploited in the biosynthesis of lysine via decarboxylation of the D-amino acid center of *meso*-diaminopimelate (DAP). Finally, when the amino acid side-chain contains a β-hydroxyl function, retro-aldol chemistry provides a way of cleaving this C–C bond. This is exploited in the biosynthesis of glycine, for example, by

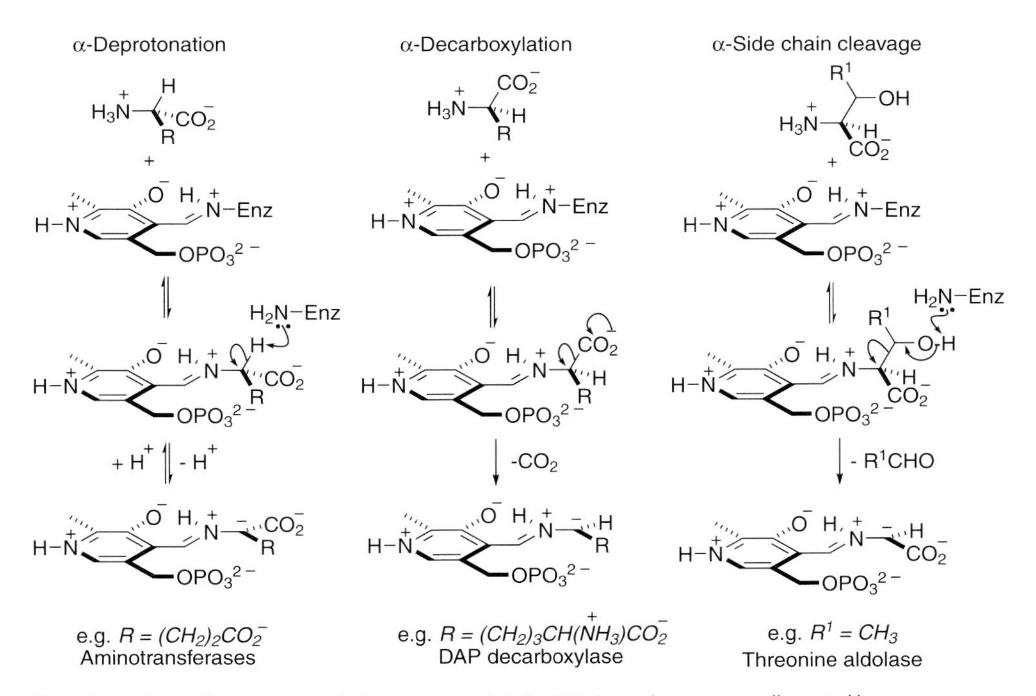


**Figure 1.8** Stereoelectronic control of α-center reactivity by PLP-dependent enzymes illustrated by enzymes involved in amino acid biosynthesis. As noted in the text, the decarboxylation example, DAP decarboxylase, utilizes a D-amino acid substrate.

threonine aldolase. Enzymes control the identity of the bond that is cleaved by exploiting stereoelectronic factors as originally proposed by Dunathan [32]. The cleaved bond must align with the delocalized π-orbitals of the PLP cofactor. By specific recognition of the α-amino acid functionalities, the enzyme can control the orientation of the substrate and hence its fate [33].

**β,γ-Center Reactivity**

Amino acids that contain a leaving group at the β-position can undergo elimination chemistry from the α-deprotonated intermediate. Nucleophilic attack on the aminoacryloyl intermediate leads to overall nucleophilic substitution, via an elimination–addition mechanism (Figure 1.9). This is exploited in the biosynthesis of cysteine and related amino acids. More extended proton relays can extend this chemistry to the γ-center (Figure 1.10) as observed in γ-cystathionine synthase.

recognize these substrates but not others. AATases selectively bind glutamate and aspartate. Two active-site arginine residues (Arg292 and Arg386) bind to the two carboxylates of these substrates, one of these, Arg292, controls the specificity forming an ion pair with the carboxylate side-chain of each substrate (Figure 1.11). Mutation of this arginine to an anionic aspartate depresses the activity ($k_{cat}/K_M$) of the enzyme with respect to anionic substrates by a factor of more than 100 000 [34].

Other families of aminotransferases face greater challenges with the dual substrate specificity that is a general feature of all these enzymes. Since glutamate is a common amino donor in these systems, these enzymes must accommodate the negatively charged γ-carboxylate of glutamate while also accepting side-chains of the alternative substrate with different sizes, polarities, and charges. Two different strategies are employed to deal with the issue: an "arginine switch," whereby the key arginine undergoes a conformational shift to accommodate the new side-chain, and the use of an extended hydrogen bond network to mediate substrate recognition, rather than the cationic charge of arginine (Figure 1.11) [35].

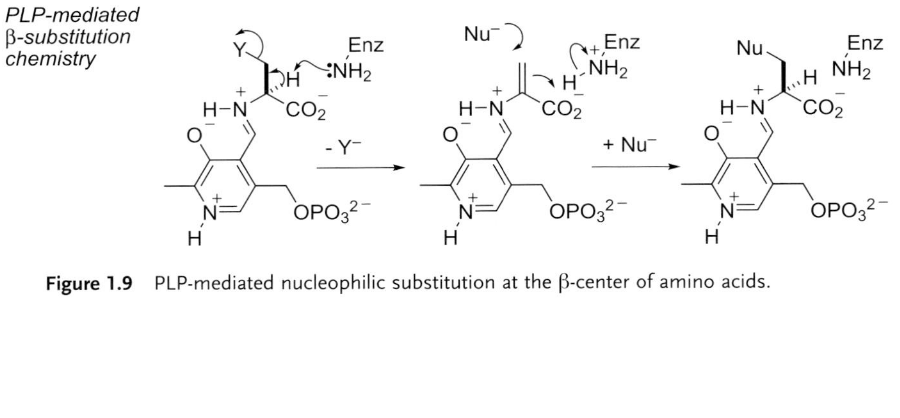

*PLP-mediated β-substitution chemistry*

**Figure 1.9** PLP-mediated nucleophilic substitution at the β-center of amino acids.

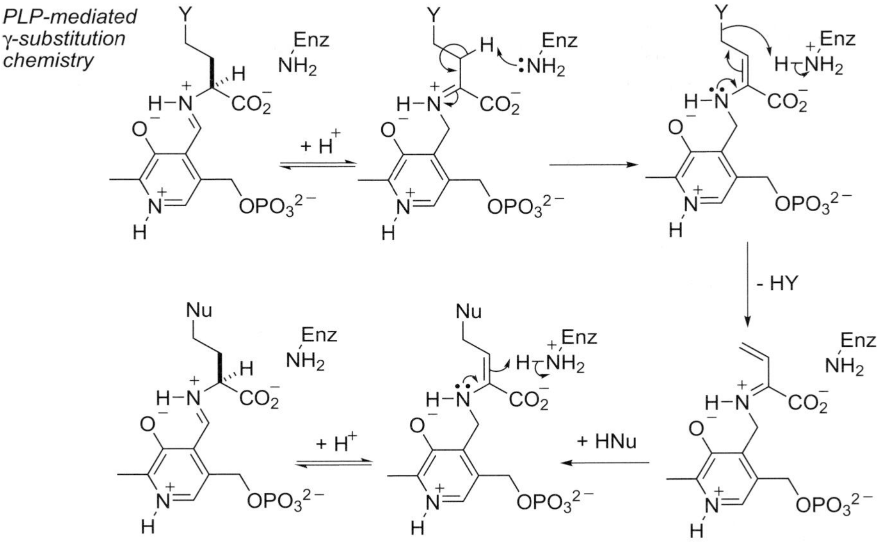

Figure 1.10 PLP-mediated nucleophilic substitution at the γ-center of amino acids.

Tyrosine aromatic amino transferases (TATases) utilize glutamate or aspartate as amino donors to produce the aromatic amino acids tyrosine, phenylalanine, and tryptophan. The TATase from *Paracococcus denitrificans* provides a clear example of an arginine switch [36]. The binding of a series of inhibitors to this enzyme shows that the active site utilizes Arg386 for specific recognition of the α-carboxylate and the

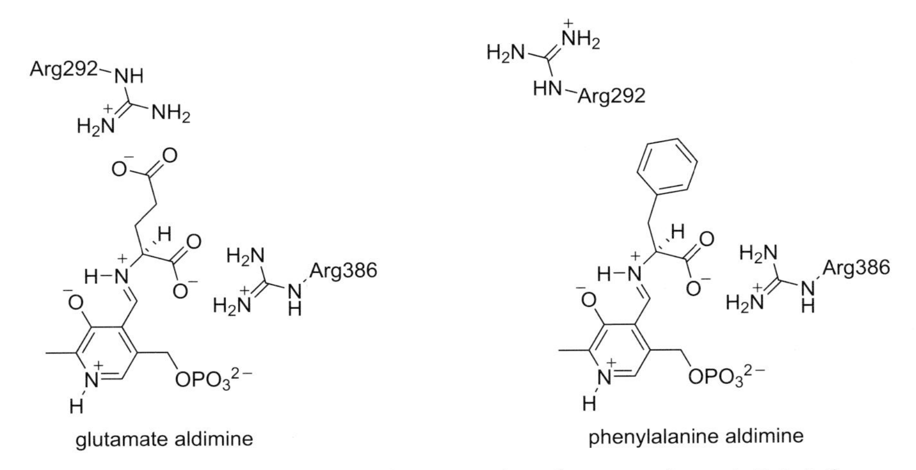


Figure 1.11 The arginine switch in the substrate specificity of aminotransferases: in *P. denitrificans* TATase Arg292 forms an electrostatic attraction to the glutamate γ-carboxylate; reorientation of Arg292 away from the active site allows binding of a nonpolar side-chain. (Adapted from [31].)

surrounding region, in the vicinity of the α- and β-centers of the substrate, is rigid. However, active-site residues that bind the large hydrophobic substituent are conformationally flexible and Arg292 moves out of the active site to accommodate bulky uncharged substrates [37]. The arginine switch has been engineered into AATase by site-directed mutation of six residues, thereby allowing transamination of large aromatic substrates [38]. The crystal structure of the resulting mutant provided the first structural evidence for the arginine switch [39].

Aspartate aminotransferase and tyrosine aminotransferase from *E. coli* are paralogs that share 43% sequence identity. It is likely that they evolved by gene duplication of an ancestral AATase gene. The role of gene duplication and evolution of new substrate specificities is an area of general interest [40]. Directed evolution, which mimics the action of natural selection, is a powerful strategy for tailoring protein properties [41]. It has been used to test these ideas. Repeated mutation of AATase, with selection for aromatic aminotransferase activity, leads to mutants with broadened substrate specificity [42], validating this evolutionary analysis. The first reports on the directed evolution of aminotransferases with modified substrate specificity were of the conversion of AATases to branched-chain aminotransferases [43]. In this case a mutant with 17 amino acid changes, remote from the active site, resulted in an arginine switch that allowed Arg292 to switch out of the active site. This change accommodates bulky hydrophobic side-chains (e.g., the catalytic efficiency ($k_{cat}/K_M$) of valine is increased by $2.1 \times 10^6$) [44, 45]. It appears that the arginine switch is readily accessible to evolution and that directed evolution strategies may provide a general tool for the development of new enzymes with tailored specificities.

The other mechanism for dual substrate specificity is the employment of an extended hydrogen bond network (Figure 1.12). The AATase [46] and TATase [47] from *Pyrococcus horikoshii* both use this strategy, as does the branched-chain aminotransferase from *E. coli* [48]. Binding glutamate at the active site without the

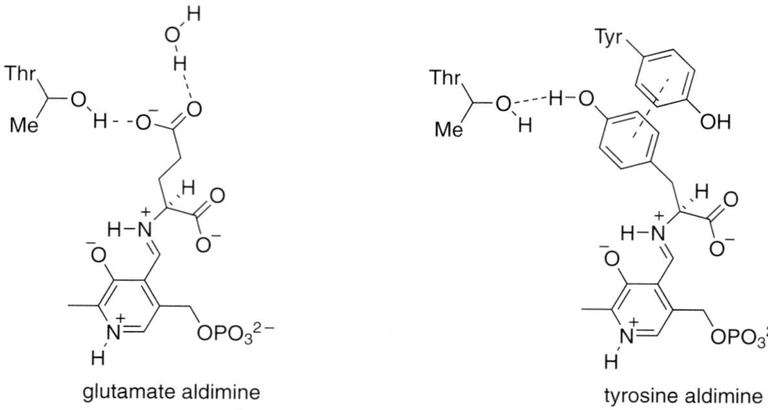

glutamate aldimine

tyrosine aldimine

**Figure 1.12** Extended hydrogen bond and π-stacking interactions in side-chain recognition of TATase from *P. horikoshii*. (Adapted from [31].)

presence of a cationic residue to recognize the side-chain reduces the electrostatic complexities for dual specificity. Interestingly, by using smaller, less flexible, residues than arginine for recognition, the branched-chain aminotransferase can more accurately distinguish between aspartate and glutamate.

### 1.3.3 PLP and the Biosynthesis of Alanine and Glycine

Two more of the protein amino acids, alanine and glycine, are biosynthesized by direct exploitation of α-center PLP chemistry. Essentially any α-amino acid can be created from the corresponding α-keto acid if an appropriate aminotransferase is available. Pyruvate is a ubiquitous metabolite and the corresponding amino acid, alanine, is readily available by transamination using aminotransferases of appropriate specificity (Figure 1.13).

There are three biosynthetic routes to glycine (Figure 1.14). Some organisms, such as the yeast *S. cerevisiae*, utilize all three. In organisms, such as *S. cerevisiae*, that have access to glyoxalate, transamination provides glycine directly. In this case the amino donor is alanine [49, 50].

The other two routes to glycine involve PLP-mediated cleavage of the protein β-hydroxy amino acids serine and threonine by the enzymes serine hydroxymethyltransferase (SHMT) and threonine aldolase. Enzymes of this class often have relaxed substrate specificity and can cleave the side-chain from a number of β-hydroxy-α-amino acids. Threonine aldolase is an important source of glycine in

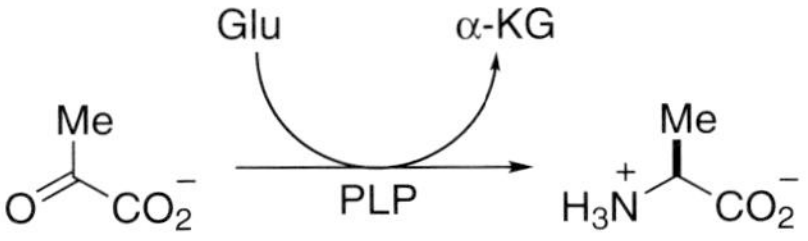

Glu α-KG Me $O\ CO_2^-$ PLP Me $H_3\overset{+}{N}\ CO_2^-$

**Figure 1.13** Biosynthesis of alanine from pyruvate.

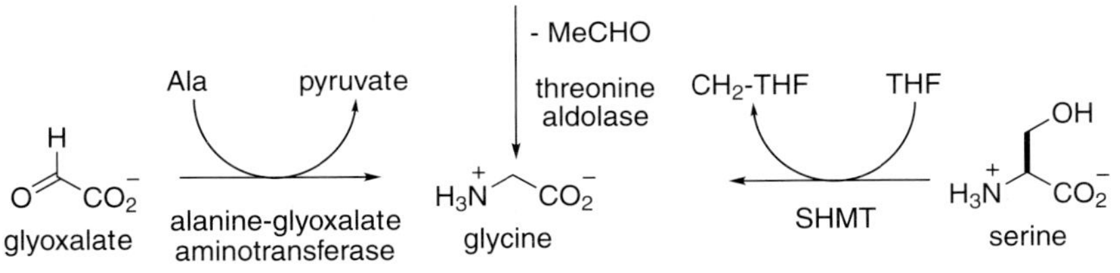

H OH $H_3\overset{+}{N}\ CO_2^-$ threonine - MeCHO threonine aldolase Ala pyruvate H $O\ CO_2^-$ glyoxalate alanine-glyoxalate aminotransferase $H_3\overset{+}{N}\ CO_2^-$ glycine $CH_2$-THF THF SHMT OH $H_3\overset{+}{N}\ CO_2^-$ serine

**Figure 1.14** Three PLP-dependent biosynthetic routes to glycine.

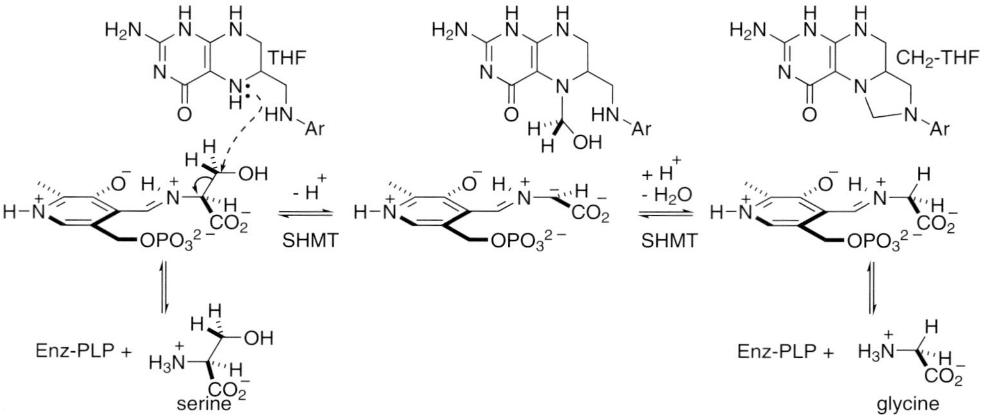

Figure 1.15 Proposed mechanism for SHMT.

*S. cerevisiae* [51]. Threonine forms a Schiff base with PLP which then catalyses a retro-aldol reaction to remove the side-chain as ethanal (see Box 1.3) [52, 53].

The biosynthesis of glycine in humans occurs primarily via the action of SHMT [49]. This enzyme is a critical source of both glycine and one-carbon units for metabolism [33]. Like threonine aldolase, the enzyme carries out a PLP-mediated side-chain cleavage reaction of a β-hydroxy-amino acid. However, in the case of serine the side-chain of the amino acid is a reactive aldehyde, methanal, and is not produced as a free intermediate. Instead it becomes attached to an essential cofactor as methylene-tetrahydrofolate ($CH_2$-THF). In this case, the THF cofactor is required in order to bring about the reaction. Extensive studies with isotopically labeled substrates and mutated enzymes, together with X-ray structural information, have attempted to resolve the question of whether the folate cofactor assists the cleavage reaction directly or simply reacts with methanal as soon as it is formed via retro-aldol chemistry (Figure 1.15) [54–58].

## 1.4 Routes to Functionalized Three-Carbon Amino Acids: Serine, Cysteine, and Selenocysteine

3-Phosphoglycerate is a key metabolite of glycolysis and is the precursor to the three-carbon protein amino acids with β-functional groups: serine, cysteine, and the 21st amino acid of the genetic code, selenocysteine.

### 1.4.1 Serine Biosynthesis

In Gram-negative bacteria, serine is biosynthesized in three steps from 3-phosphoglycerate [59]. The first step is oxidation to 3-phosphohydroxypyruvate and, as the

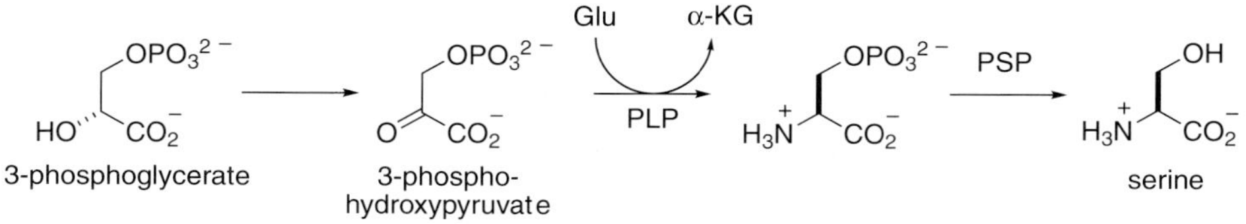

Figure 1.16 Biosynthesis of serine.

point of commitment to the biosynthetic pathway, it is feedback regulated by the end product, serine [60]. The resulting α-keto acid is a substrate for transamination with glutamate acting as the amino donor. Hydrolysis of the resulting serine-β-phosphate catalyzed by phosphoserine phosphatase (PSP) provides the free amino acid (Figure 1.16).

Systematic protein crystallography, exploiting the use of reactive intermediate analogs, has provided a detailed series of "snapshots" of intermediates in the catalytic cycle of the PSP from *Methanococcus jannaschii*, allowing the reaction to be visualized in three-dimensional detail (Figure 1.17) [61]. A conserved aspartate residue at the end of the active-site tunnel is a nucleophilic catalyst, attacking the serine-β-phosphate to generate an acyl phosphate intermediate. Release of serine allows the binding of a water molecule to mediate hydrolysis of the labile aspartate-β-phosphate to regenerate the starting enzyme.

The PSP from *Pseudomonas aeruginosa* has evolved the ability to bind homoserine rather than water in the second half of the reaction and transfer the activated phosphate of the aspartate-β-phosphate species to this amino acid providing access to homoserine-γ-phosphate, which is a biosynthetic precursor to threonine (Figure 1.18), as will be described later. This circumvents the need to expend ATP in phosphorylating this alcohol and is a rare example of an enzyme that transfers phosphoryl groups directly between non-nucleotide metabolites. This illustrates again the role of changed substrate specificity in generating new enzyme activities [62].

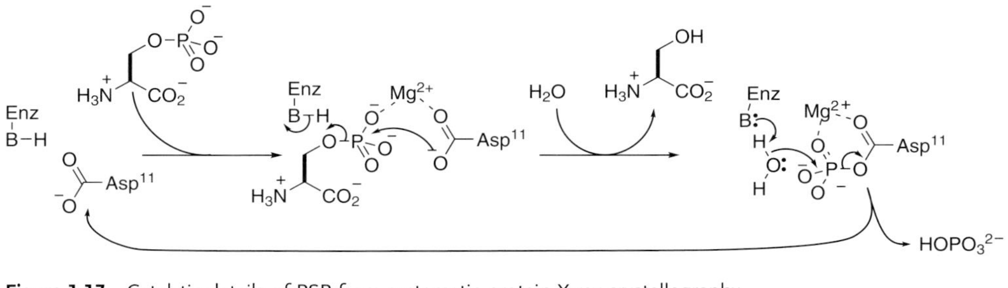

Figure 1.17 Catalytic details of PSP from systematic protein X-ray crystallography.

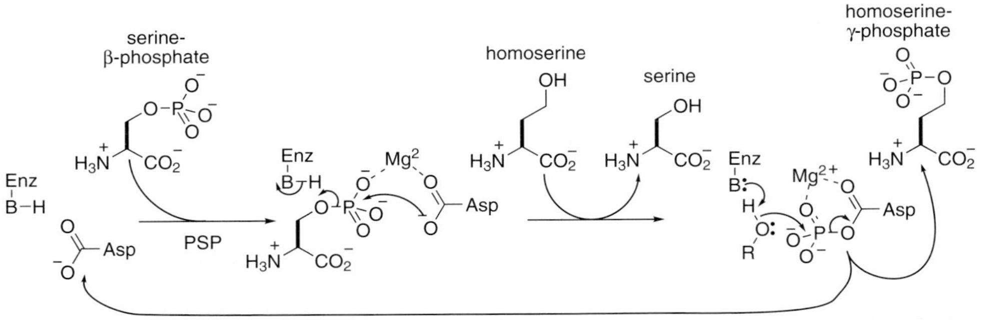

Figure 1.18 *P. aeruginosa* PSP-catalyzed biosynthesis of homoserine-γ-phosphate by phosphoryl transfer.

## 1.4.2
## Cysteine Biosynthesis

Serine is the starting material for the synthesis of the other three-carbon protein α-amino acids. There are two common pathways to cysteine: the sulfur assimilation pathway and the trans-sulfuration pathway. Vertebrates use the latter pathway, which interconverts homocysteine and cysteine. This latter pathway is discussed separately in the section on methionine biosynthesis.

The sulfur assimilation pathway to cysteine is found in plants, eubacteria and some Archaea. The two key steps in this synthesis are mediated by a bifunctional cysteine synthase complex [63]. Serine acetyltransferase activates the side-chain hydroxyl group of serine by derivatization with acetyl-CoA and the resulting *O*-acetylserine (OAS) reacts with a sulfur nucleophile, catalyzed by a PLP-dependent enzyme OAS sulfhydrylase (*O*-acetylserine sulfhydrylase, OASS) (Figure 1.19).

In enteric bacteria there are two isozymes of OASS that utilize different sulfur nucleophiles as substrates [64]. One isozyme, produced under aerobic conditions,

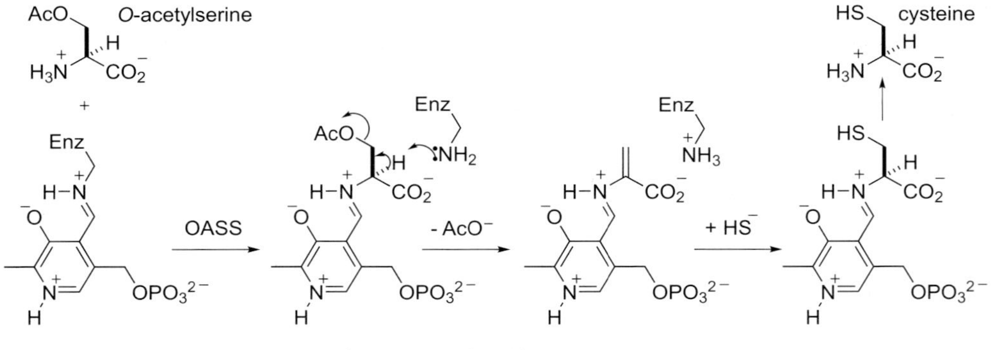

Figure 1.19 Cysteine formation catalyzed by OASS.

utilizes hydrosulfide (formed by a multistep reduction of sulfate) [65]. Under anaerobic conditions a second isozyme is produced which utilizes thiosulfate and produces *S*-sulfo-cysteine, which is transformed to cysteine by reaction with thiols.

The mechanism of OASS from *Salmonella typhimurium* has been studied in detail [65, 66]. This enzyme is a homodimer with an active-site PLP cofactor bound to Lys41. The initial stages of the reaction parallel those of aminotransferase enzymes. The monoanion form of OAS forms a Schiff base with PLP by amino exchange with Lys41, which is thereby liberated to act as an active-site acid–base catalyst. In this case, deprotonation of the α-center of PLP-linked OAS by Lys41 eliminates acetate and forms of a bound aminoacrylyl intermediate. After loss of acetate, hydrosulfide binds, in the second step of this ordered Ping Pong Bi Bi mechanism, and reacts with the aminoacrylate intermediate to produce cysteine. This mechanism is illustrative of a general class of PLP-dependent enzymes that facilitate reaction at the β-center of amino acids by facilitating the loss of a leaving group at that position (see Box 1.3).

### 1.4.3
### Case Study: Genome Information as a Starting Point for Uncovering New Biosynthetic Pathways

With the availability of a large number of genome sequences it is possible to identify the likely biosynthetic pathways operating in particular organisms based on the presence or absence of particular genes for biosynthetic enzymes. This has proved a powerful tool in expanding our understanding of the diversity and distribution of metabolic pathways. Genome analysis of the biosynthesis of cysteine and its incorporation into cysteinyl-tRNA have led to the discovery of two new pathways for the biosynthesis of this amino acid. These findings, in turn, have led to developments in our understanding of the biosynthesis of selenocysteine in humans [67]. This area presents a nice case study in the emerging use of genome analysis to identify new variants in biosynthetic pathways.

#### 1.4.3.1 Cysteine Biosynthesis in *Mycobacterium Tuberculosis*

Amino acid biosynthesis in *Mycobacterium tuberculosis* is under active investigation because of the growing health threat posed by tuberculosis. Inhibitors of distinctive essential metabolic pathways in this organism may be useful as antibiotics. The complete genome sequence of *M. tuberculosis* is known [68]. *M. tuberculosis* carries out cysteine biosynthesis via the sulfur assimilation pathway and adjacent genes, *cysE* and *cysK1*, encode the serine acetyltransferase and OASS activities of the cysteine synthase complex [69]. However, genome analysis revealed the presence of two other genes homologous to OASS, *cysK1* and *cysM*. Furthermore, *cysM* was found clustered with two other genes related to sulfur metabolism. One of these genes, now called *cysO*, is homologous to a family of small sulfide carrier proteins, such as ThiS, which play a role in thiamine pyrophosphate biosynthesis [70]. A thiocarboxylate derivative of the C-terminal group of these proteins is the sulfide carrier. The protein is activated by ATP, to form an acyl phosphate, and then converted to the corresponding

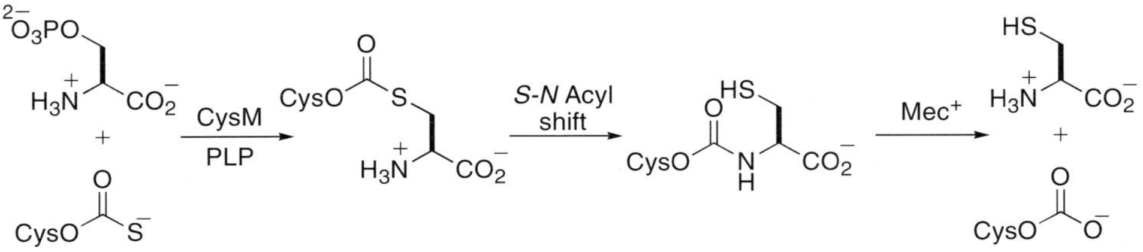

**Figure 1.20** Biosynthesis of cysteine in *M. tuberculosis*.

thiocarboxylate via nucleophilic substitution. Reaction of this thiocarboxylate, and hydrolysis of the resulting acyl derivative leads to overall transfer of sulfide. A second gene in this cluster, *mec*$^+$, encodes a potential hydrolase and this gene had previously been linked to sulfur amino acid metabolism in a *Streptomyces* species. This genome analysis led Begley *et al.* to investigate CysO as a potential sulfur source for cysteine biosynthesis (Figure 1.20). *In vitro* assays, making extensive use of protein mass spectrometry, confirmed this role [71]. CysO reacts with a suitably activated serine derivative to form a thioester, which rearranges to generate the corresponding peptide bond. Mec$^+$ is a zinc-dependent carboxypeptidase that removes the newly created cysteine from the temporarily homologated protein.

Subsequent studies have shown that CysO is part of a fully independent pathway to cysteine in this organism (Figure 1.20) [72]. The activated form of the serine substrate for CysM is *O*-phosphoserine rather than the *O*-acetylserine utilized by the sulfur assimilation pathway. This is the biosynthetic precursor to serine as described previously.

Cysteine plays a key role in responding to oxidative stress encountered by *M. tuberculosis* in its dormant phase. The CysO-dependent route to cysteine may be particularly important under these conditions because thiocarboxylate may be used as it is more resistant to oxidation that other sulfide sources. The absence of this biosynthetic route in other organisms, including humans, make inhibitors of these biosynthetic enzymes of great interest for the treatment of the persistent phase of tuberculosis.

#### 1.4.3.2 **Cysteine Biosynthesis in Archaea**

A similar genomics-based approach has uncovered an alternative cysteine biosynthetic pathway in Archaea. When the genome sequences of some methanogenic Archaea were sequenced they were found to lack the gene, *cysS* for the appropriate cysteinyl-tRNA synthetase. In one of these organisms, *Methanocaldococcus jannaschii*, it was found that Cys-tRNA$^{Cys}$ was generated via an alternative pathway (Figure 1.21) [73]. First, the relevant tRNA, tRNA$^{Cys}$, is ligated to phosphoserine by the enzyme *O*-phosphoseryl-tRNA synthetase (SepRS) which then undergoes a PLP-mediated exchange of the β-phosphate for thiol to generate Cys-tRNA$^{Cys}$, catalyzed by Sep-tRNA: Cys-tRNA synthase (SepCysS) – a type I PLP-dependent enzyme. When the gene for SepCysS is deleted in the related methanogen, *Methanococcus*

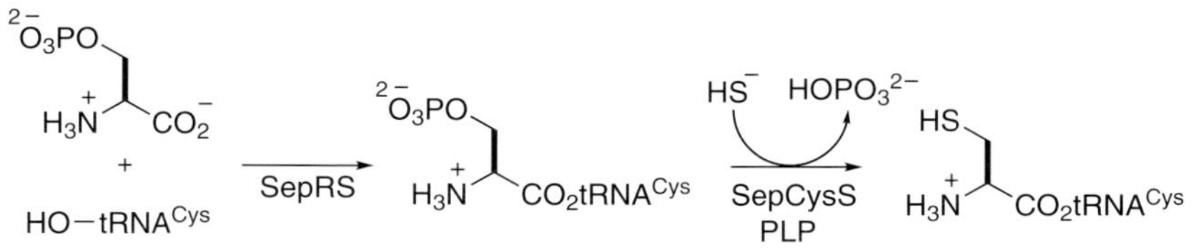

**Figure 1.21** Biosynthesis of Cys-tRNA$^{Cys}$ in the methanogenic archaeon *M. jannaschii*.

*maripaludis*, the organism is a cysteine auxotroph, indicating that this is the sole pathway to cysteine in this organism.

#### 1.4.3.3 **RNA-Dependent Biosynthesis of Selenocysteine and Other Amino Acids**

Developments in cysteine biosynthesis research have underpinned our understanding of the biosynthesis of the 21st protein amino acid, selenocysteine. Selenocysteine has been known to be an important residue for a range of enzymes since 1976 [74]. This amino acid is incorporated into proteins by the ribosome using a tRNA$^{Sec}$ – a suppressor tRNA that corresponds to a stop codon in the genetic code [75]. The utilization of this suppressor tRNA allows the expansion of the genetic code, but requires an additional elongation factor for the ribosome to insert the amino acid in the growing chain.

The biosynthesis of selenocysteine was first elucidated in *E. coli* and, like the archaeal route to cysteine, it is based on modification of aminoacyl-tRNAs (Figure 1.22). The pathway starts with the ligation of serine to tRNA$^{Sec}$, catalyzed by SerRS. The resulting ester undergoes PLP-mediated nucleophilic substitution of the side-chain hydroxyl group with a selenium-based nucleophile, selenophosphate, that is produced from selenide and ATP. The mechanism of selenophosphate synthetase from *E. coli* has been established using positional isotope exchange methodology [76, 77]. The reaction of Ser-tRNA$^{Sec}$ with selenophosphate is catalyzed by SelA [78]. The nucleophilic substitution reaction is assumed to follow a mechanism analogous to that of OASS involving an initial elimination of water to form an aminoacrylyl-tRNA$^{[Ser]Sec}$ intermediate that reacts with the selenophosphate and the

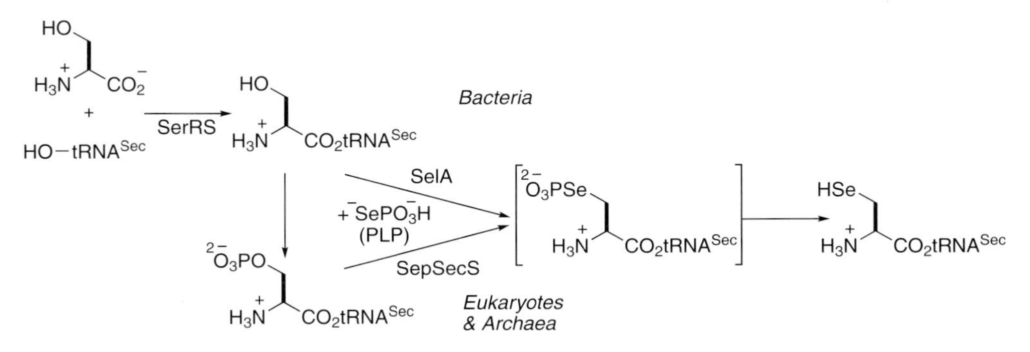


**Figure 1.22** Biosynthesis of the tRNA adduct of selenocysteine.

resulting phosphoselenocysteyl-tRNA$^{Sec}$ undergoes hydrolysis to generate selenocysteyl-tRNA$^{Sec}$.

This biosynthetic pathway was assumed to be common to all selenocysteine utilizing enzymes, but studies in eukaryotes failed to uncover the requisite biosynthetic enzymes. Subsequent studies have shown that selenocysteine biosynthesis occurs by a common pathway in Archaea and eukaryotes that is distinct, but related to that in bacteria. A protein believed to be associated with selenocysteine synthesis coprecipitated with the loaded selenocysteyl-tRNA$^{Sec}$ and bioinformatics analysis showed that the enzyme was a PLP-dependent enzyme [79]. The chemical similarity of cysteine and selenocysteine provided the clue to unraveling the biosynthetic pathway to the latter amino acid in eukaryotes: Sep-tRNA$^{Sec}$ was shown to be a substrate for the RNA-dependent biosynthesis of selenocysteine.

In the eukaryotic and archaeal version of the biosynthetic pathway, the β-hydroxyl group of Ser-tRNA$^{Sec}$ is activated by phosphorylation to form the phosphoserine derivative, Sep-tRNA$^{Sec}$, which then undergoes PLP-mediated nucleophilic substitution of the β-phosphate leaving group with selenophosphate catalyzed by selenocysteine synthase, SepSecS [80]. Selenocysteine synthase is homologous to OASS both in structure [81] and sequence [82, 83] and the catalytic mechanism is analogous, involving an initial elimination of phosphate to form an aminoacrylyl-tRNA$^{[Ser]Sec}$ intermediate which reacts with the selenophosphate. Despite selenocysteine being an addition to the 20 amino acids found ubiquitously in proteins, the phylogenetic data suggest that its biosynthesis is a primordial process and that selenocysteine has played a role in metabolism since before the divergence of the ancestors to the three kingdoms of life (bacteria, Archaea and eukaryotes) more than 3 billion years ago.

The synthesis of selenocysteine on a specialized tRNA scaffold assists in distinguishing the otherwise similar chemistry of selenocysteine and cysteine; the biosynthetic enzymes recognize structural features of tRNA$^{Sec}$. Selenocysteine is not the only amino acid synthesized by modification of an aminoacyl-tRNA. *N*-Formyl methionine is the N-terminal residue of proteins in eubacteria and eukaryotic organelles (mitochondria and chloroplasts). It is synthesized via formylation of Met-tRNA$^{fMet}$ in a process that also depends on binding to the tRNA$^{fMet}$ and is specific to this aminoacyl-tRNA species [84]. Interestingly, it has also been found that many organisms produce aminoacylated tRNAs for asparagine and glutamine by amidating aspartyl and glutamyl precursors. Again, genome analysis is proving useful in identifying the pathway(s) present in particular organisms [85]. For example, whereas Gln-tRNA$^{Gln}$ is synthesized from glutamine in the cytoplasm of eukaryotes, the majority of eubacteria and all Archaea make it by the transamidation route [86].

## 1.5
## Other Amino Acids from Aspartate and Glutamate: Asparagine and Side Chain Functional Group Manipulation

Glutamate and aspartate are the parents of six further amino acids that are ubiquitously found in proteins: asparagine, methionine, and threonine are produced from aspartate; and glutamine, proline, and arginine are derived from glutamate. The conversion of glutamate to glutamine has already been described and illustrates the strategy by which the remaining members of this group of protein amino acids are made. In each case the side-chain carboxylate undergoes functional group manipulation starting with activation to a short-lived acyl phosphate intermediate and then subsequent nucleophilic substitution. The two nucleophiles that are utilized are ammonia and hydride ion (delivered by redox cofactors), and these will be discussed in turn.

### 1.5.1 Asparagine Biosynthesis

There are two isozymes of asparagine synthetase in *E. coli* [87]. Each enzyme exploits the cleavage of ATP to AMP and pyrophosphate as a means of activating the β-carboxylate of aspartate. The two isozymes differ in their nitrogen source: AsnA utilizes ammonia [88], whereas AsnB uses glutamine. *E. coli* asparagine synthetase B is a multifunctional enzyme. The N-terminal domain is a class II GAT. The C-terminal domain binds aspartate and ATP, and generates β-aspartyl-adenylate as the reactive nucleophile (Figure 1.23). A kinetic model for the multistep reaction has been developed [89].

By producing an inactive mutant with the glutaminase active-site nucleophile, the N-terminal cysteine, changed to an alanine it has been possible to crystallize the enzyme with both glutamine and AMP bound, thus clearly revealing the relative locations of the two active sites (Figure 1.24) [90]. The two active sites are connected by a tunnel that is 19 Å long and lined primarily with low polarity functional groups. Ammonia traverses this tunnel and combines with the β-aspartyl-adenylate that is

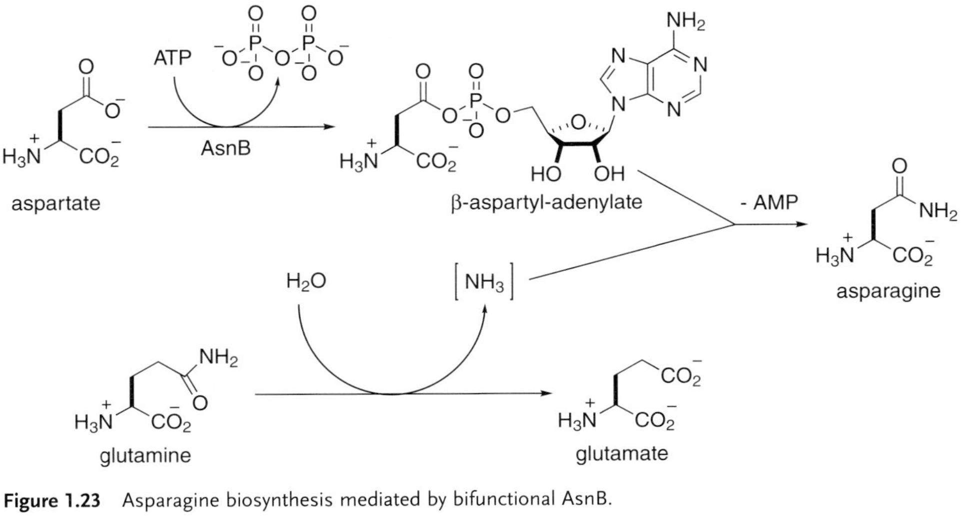

**Figure 1.23** Asparagine biosynthesis mediated by bifunctional AsnB.

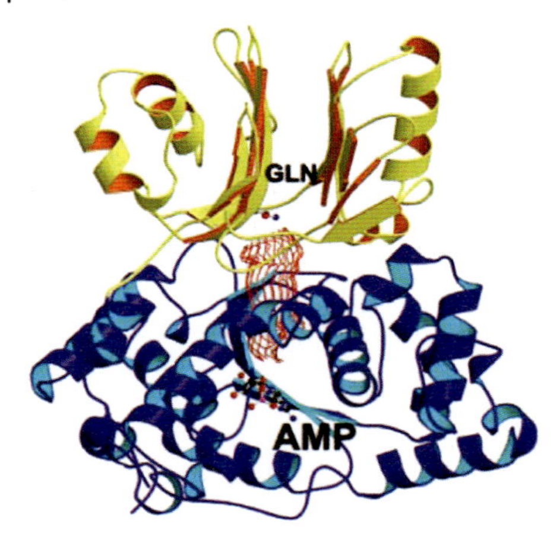


**Figure 1.24** Glutaminase (gold) and synthetase (purple) sites of AsnB are linked via an internal tunnel. (Picture taken from [20].)

formed at the second active site. This situation echoes that described for GOGAT and reinforces the pattern of multifunctional GAT-dependent enzymes with internal molecular tunnels for efficient delivery of nascent ammonia [20].

Like AsnB, asparagine synthetases from plants and animals are glutamine-dependent – illustrating the importance of glutamine, rather than ammonia, as a nitrogen carrier in higher organisms. Some leukemia cells have diminished levels of asparagine synthetase. When asparagine levels are reduced further by side-chain hydrolysis, mediated by the enzyme L-asparaginase, these cells become especially sensitive to chemotherapy Hence, L-asparaginase is a component of chemotherapeutic protocols for treating some acute childhood leukemias. Some of these leukemias develop resistance to chemotherapy by increasing their production of asparagine synthetase. For this reason, inhibitors of this enzyme are of potential significance as antileukemia agents and these are being developed based on the mechanistic studies of *E. coli* AsnB [91]. Mimics of the tetrahedral intermediate associated with ammonolysis of the aspartyl-β-adenylate intermediate inhibit the enzyme at submicromolar concentrations and inhibitors of this type (Figure 1.25) are being tested for their efficacy as chemotherapeutic agents [92].

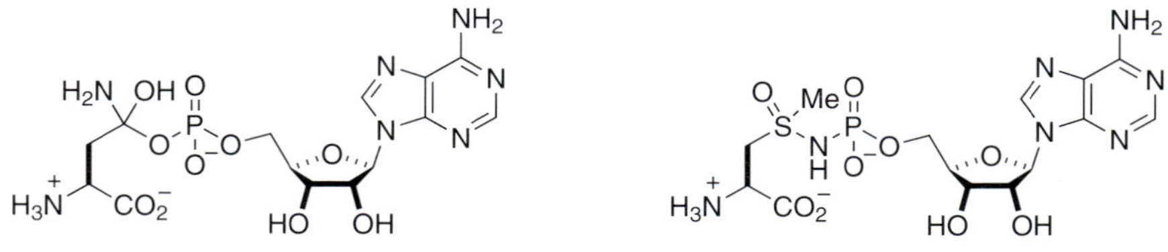

tetrahedral intermediate in AsnB catalysis

AsnB inhibitor

**Figure 1.25** Inhibitor of AsnB designed as a tetrahedral intermediate mimic.

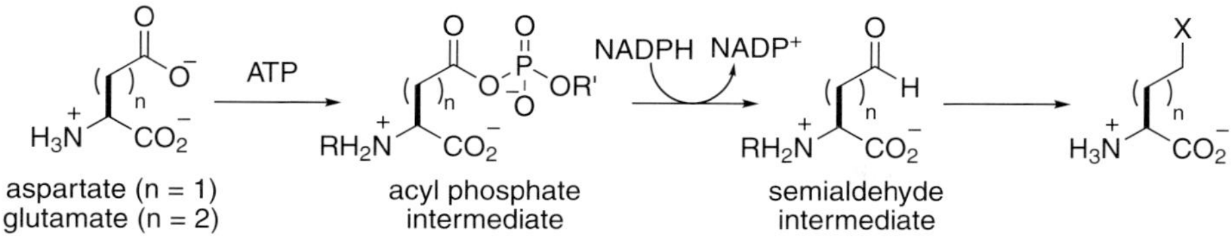

aspartate (n = 1)
glutamate (n = 2)
ATP
acyl phosphate intermediate
NADPH NADP+
semialdehyde intermediate

**Figure 1.26** Strategy of aspartate and glutamate family amino acid biosynthesis.

## 1.6
## Aspartate and Glutamate Families of Amino Acids

### 1.6.1
### Overview

The remaining four- and five-carbon amino acids are prepared by pathways based around a common chemical strategy (Figure 1.26): ATP-dependent activation of the side-chain carboxylate of the parent amino acid generates a reactive acyl phosphate intermediate that is reduced to the corresponding aldehyde by hydride transfer; the product amino aldehydes are also labile species and these are converted to more durable reduced analogs in the next stage.

These pathways illustrate several common features of amino acid biosynthetic pathways. Enzymes catalyzing analogous reactions in parallel pathways are often homologous in structure. Once enzymes are available that can catalyze a particular set of reactions then gene duplication and substrate specificity modification, via mutation and selection, can generate parallel pathways [93]. This evolutionary mechanism highlights the important role of the underlying chemical logic of the pathways that underpin this organization. A second general feature is that pathways involving reactive intermediates benefit from multifunctional enzyme systems that can efficiently channel metabolites to the active site that catalyzes the next stage in the pathway. This not only increases the yield of the reaction, but also controls the fate of the metabolite when there are competing metabolic uses of the product. For this reason there are often multiple isozymes to catalyze reactions that occur in multiple pathways and these are generally independently regulated [94]. For isozymes that catalyze reactions at branch-points of pathways, where a commitment to one or other final product is made, the pattern of feedback regulation provides a direct confirmation of the *in vivo* role of the specific form of the enzyme.

### 1.6.2
### Aspartate Family Amino Acids: Threonine and Methionine

The first three steps of the biosynthesis of threonine and methionine in plants and microbes are common to both pathways (Figure 1.27) [95]. Aspartokinase (AK) catalyzes the ATP-dependent β-phosphorylation of aspartate, which creates the

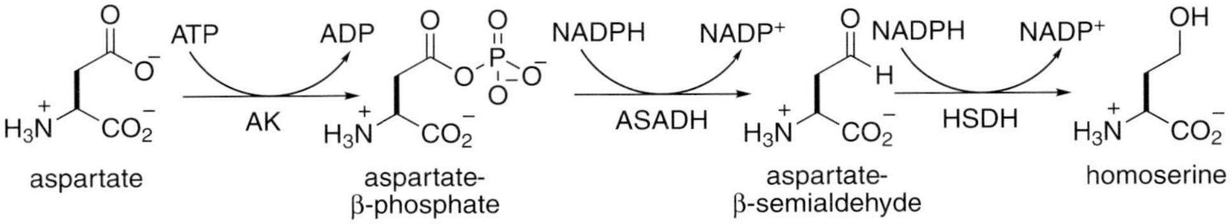

Figure 1.27 Biosynthesis of homoserine.

requisite leaving group for subsequent transformation [96]. The resulting aspartate-β-phosphate is reduced to the corresponding aldehyde by aspartate β-semialdehyde dehydrogenase (ASADH). Homoserine dehydrogenase (HSDH) catalyzes the further reduction of aspartate-β-semialdehyde to homoserine – a key intermediate for the biosynthesis of both threonine and methionine. NADPH is a hydride source for the reduction chemistry. In *E. coli* and other bacteria there are independently regulated isozymes of AK-HSDH for threonine and methionine biosynthesis.

ASADH from several sources has been characterized [97–99]. The mechanism of ASADH (Figure 1.28) is analogous to the oxidation of glyceraldehyde-3-phosphate to glycerate-3-phosphate – one of the key oxidation steps in glycolysis [100]. Aspartate-β-phosphate undergoes initial nucleophilic substitution of phosphate with the active-site thiol of cysteine-136. The resulting thioester is then reduced by a nicotinamide cofactor, NADPH, to aspartate-β-semialdehyde ASA. The resulting aldehyde, ASA, is sufficiently reactive with nucleophiles that the three-dimensional structure of a

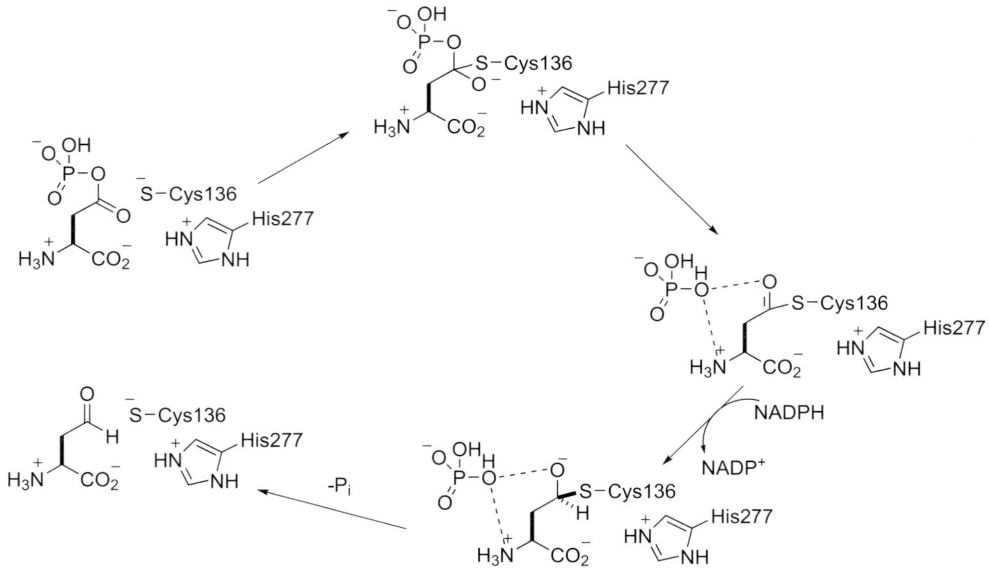

Figure 1.28 Mechanism of ASADH.

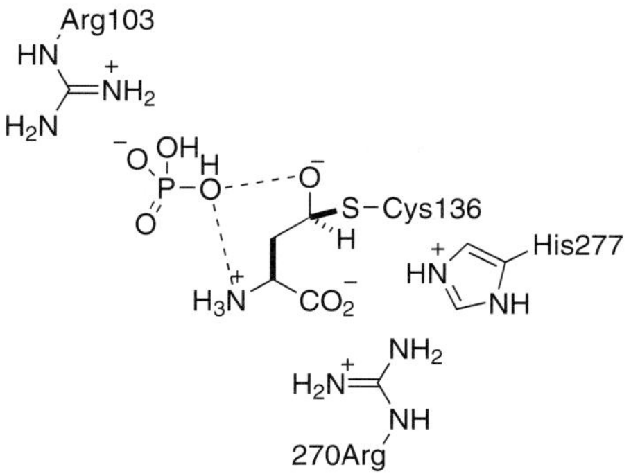

**Figure 1.29** Structure of the hemithioacetal intermediate in the mechanism of ASADH.

tetrahedral intermediate of the reduction step has been determined: when ASA and phosphate is incubated with the enzyme a hemithioacetal intermediate accumulates at the active site.

The structure of this intermediate provides a detailed snapshot of the catalytic machinery of the enzyme in action (Figure 1.29) [98]. The α-carboxylate of the substrate is bound to Arg270; a catalytic histidine (His277) is suitably placed for deprotonation of the thiol and one of two bound phosphates occupies the site of the displaced leaving group. As is expected from the Pauling view of enzyme catalysis, ASADH stabilizes a reactive intermediate on the pathway; in this case by hydrogen bonding to the positively charged side-chain of His277, the backbone peptide NH of Asp135 and the phosphate leaving group from the first half of the enzyme-catalyzed reaction.

With two labile species in the pathway, metabolite channeling is a feature of this biosynthetic chemistry. Interestingly, in a number of organisms, including *E. coli*, the first and third reactions are mediated by bifunctional AK-HSDH enzymes but the intervening ASADH reaction is carried out by a separate enzyme. It has proved difficult to provide direct kinetic evidence for the channeling of intermediates through a trifunctional enzyme complex AK-HSDH/ASADH; however, evidence for the presence of such a complex has accrued from a competition experiment [101]. An inactive ASADH mutant was generated by changing the essential active-site cysteine to alanine. When increasing amounts of this mutant were added to mixtures of wild-type AK-HSDH and ASADH the synthesis of homoserine was reduced. The inactive ASADH mutant binds to AK-HSDH in competition with wild-type ASADH and, when bound, prevents the direct flux of metabolites [102]. Channeling of aspartate-β-phosphate increases the efficiency of the pathway by minimizing possible losses from hydrolysis that might occur if this intermediate was freely exchanged with solution.

Homoserine dehydrogenase from *S. cerevisiae* has also been studied in detail and follows an ordered Bi Bi kinetic mechanism. The redox cofactor NADPH binds prior to ASA and homoserine is released before loss of the oxidized $NADP^+$ cofactor. The

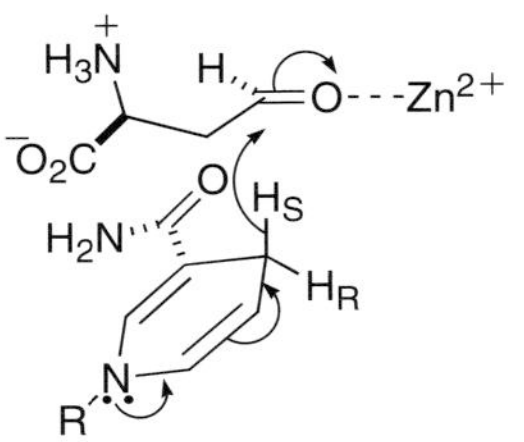

Figure 1.30 Stereospecificity of hydride transfer in the formation of homoserine.

*pro-(S)* hydride of stereospecifically deuterated NADP[$^2$H] is transferred and the reduction is catalyzed by carbonyl polarization by a protonated active-site lysine residue (Lys223) [103, 104] (Figure 1.30).

#### 1.6.2.1 Case Study: Evolution of Leaving Group Specificity in Methionine Biosynthesis

For processing to either methionine or threonine, the hydroxyl group of homoserine is converted into a good leaving group. Primary metabolism provides two main alternatives for hydroxyl activation: polyphosphates like ATP can generate phosphate leaving groups; alternatively thioesters (notably TCA cycle metabolites acetyl-CoA and succinyl-CoA) generate carboxylate leaving groups (Figure 1.31). There are variations in the pathway at this point depending on this choice. Phosphorylation of homoserine by homoserine kinase (HSK) to produce homoserine-γ-phosphate is ubiquitously used for threonine biosynthesis [105]. An interesting alternative phosphorylation route, based on the evolution of a novel bifunctional PSP, was described above (Figure 1.18) [62]. Homoserine-γ-phosphate is also the biosynthetic precursor to methionine biosynthesis in plants [95]. Other organisms use homoserine transacylases to activate homoserine for methionine biosynthesis and two different acyl groups are employed: Gram-negative bacteria make *O*-succinyl-homoserine, while yeasts and many clinically important bacteria (e.g., *M. tuberculosis* and *P. aeruginosa*) use an *O*-acetyl-homoserine as a precursor to methionine. Although the choice of leaving group does not fundamentally change the chemistry, it does have implications for the specificity of inhibitors and for the control of the pathways since, with distinct building blocks, the two pathways can be controlled independently.

All homoserine transacylases have a catalytic triad of residues situated at the end of a tunnel. The α-carboxylate of the substrate is recognized by an arginine and an active-site nucleophile (serine or cysteine) is assisted in catalysis by a histidine in

homoserine

$R = PO_3^{2-}$
$R = Ac$
$R = CO(CH_2)_2CO_2^-$

Figure 1.31 Three different activated forms of homoserine.

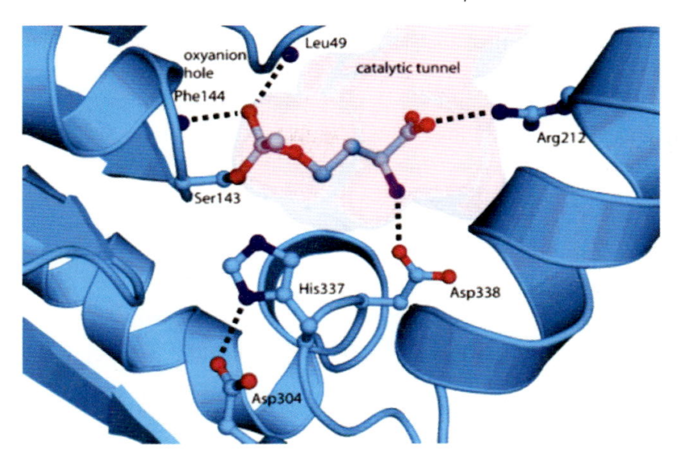


**Figure 1.32** Homoserine transacetylase from *H. influenzae*. (From [102].)

conjunction with either an aspartate or glutamate. The structures and sequences of homoserine transacylases group into two families related to the active-site nucleophile. Homoserine transacetylase from *Haemophilus influenzae* is typical of one class [106]. A conserved serine (Ser143) is in a strained conformation and acts as a reactive nucleophile to accept the acetyl group from acetyl-CoA (Figure 1.32). His337 is an adjacent acid–base catalyst and, like the active site of serine proteases, there is an oxyanion hole to stabilize the tetrahedral intermediate. The residues of the tunnel are well placed to direct homoserine to the acetylated active site and thereby assure transesterification outcompetes hydrolysis.

It had been believed that transsuccinylases comprised the cysteine-dependent family of transacylases. However, when one of this family of enzymes from *Bacillus cereus* was fully characterized it was found to be a transacetylase. The structure of this enzyme illustrates both the details of the active-site architecture, with the catalytic triad of Cys142, His235, and Glu237, and the basis for the substrate specificity of the enzyme (Figure 1.33a) [107]. A glutamate residue, Glu111, protrudes into the active

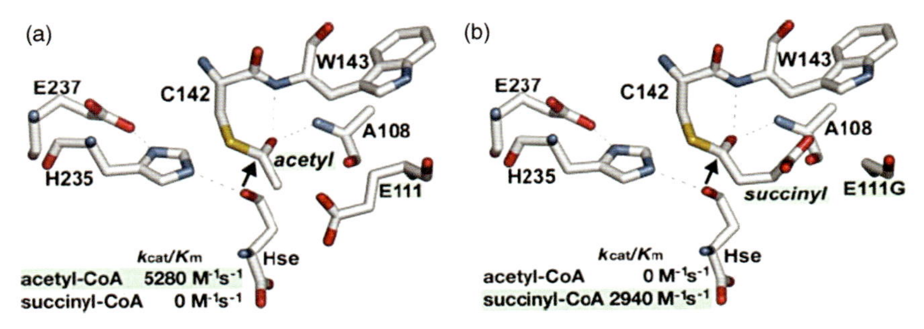


**Figure 1.33** Point mutagenesis (E111G) of *B. cereus* homoserine transacylase changes its substrate specificity from that of a transacetylase (a) to a transsuccinylase (b). (From [107].)

site limiting its size, allowing binding of acetyl species but excluding succinyl species on steric and electrostatic grounds. In many other enzymes of this class the corresponding residue is glycine, which presents no such impediments to succinyl derivatives. Making a single point mutation of this glutamate to glycine was sufficient to convert a specific transacetylase into a specific transsuccinylase (Figure 1.33b). This shows again the power of point mutations to engineer modifications to enzyme substrate specificity.

#### 1.6.2.2 **Threonine, Homocysteine, and PLP**

The manipulations of these activated homoserine derivatives to form threonine and homocysteine (*en route* to methionine) both involve catalysis of the loss of the side-chain leaving group. In each case this is catalyzed by a PLP-dependent enzyme. The ability to mediate the chemistry at β-, and γ-centers of amino acids, in addition to the α-center chemistry previously described, illustrates the catalytic versatility of PLP that makes it an indispensible cofactor in amino acid metabolism (see Box 1.3).

#### 1.6.2.3 **Threonine Synthase**

Threonine synthase (TS) catalyzes the conversion of homoserine-γ-phosphate to threonine. Threonine synthases are PLP-dependent enzymes, of type II fold, with a complex mechanism that utilizes the full capacity of PLP to stabilize reactive intermediates [108]. Sequence alignments have identified two subfamilies – class I and class II [109]. Class I enzymes are found in plants and some bacteria and Archaea, and are allosterically activated by *S*-adenosylmethionine [110]. Three-dimensional structures for both classes of enzymes are available [111–113].

Owing to its potential significance for the development of antibiotics to treat tuberculosis [114], the class I TS from *M. tuberculosis* has been studied in detail [115]. In the resting state, the PLP cofactor forms a Schiff base with Lys69. When homoserine-γ-phosphate binds to PLP it displaces Lys69, which then acts as a proton relay in a sequence of acid–base reactions that are shown in Figure 1.34, which gives an overview of the mechanism of the reaction.

Deprotonation of the α-proton of the substrate by Lys69 produces an aza-allyl anion. Reprotonation at the benzylic position of the PLP, by the conjugate acid of Lys69, generates a new iminium ion. Deprotonation at the β-position by the regenerated basic Lys69 leads to elimination of phosphate and the formation of a conjugated iminium ion. Now a proton transfer is required from the benzylic position of the PLP and the terminus of the conjugated system to produce a α,β-unsaturated ketimine ready for reaction with water at the β-position to generate threonine. While Lys69 is well placed to undertake the initial proton transfers, the molecular gymnastics required to also mediate the latter proton transfer is beyond its reach and there has been debate in the literature about the catalytic group required for this chemistry. Recent detailed structural studies show that the most likely acid–base catalyst is the 5′-phosphate of the PLP cofactor (Figure 1.35) [115]. This phosphate moiety is less than 5 Å away from both the benzylic position of the

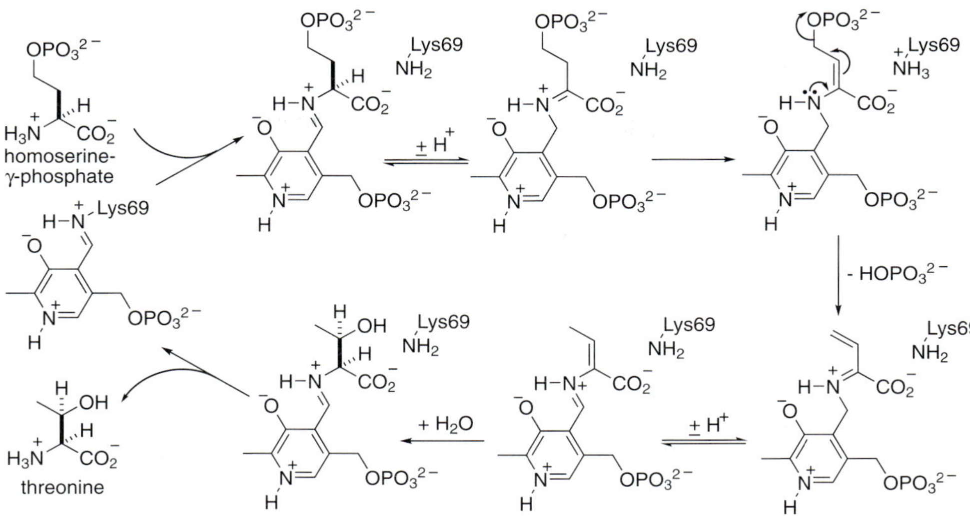

**Figure 1.34** Mechanism of threonine synthase.

PLP and the γ-carbon of the substrate. In the final step, Lys69 deprotonates a bound water molecule to generate hydroxide ion that attacks the α,β-unsaturated ketimine and produces threonine, which is released by hydrolysis of the Schiff base linkage to PLP.

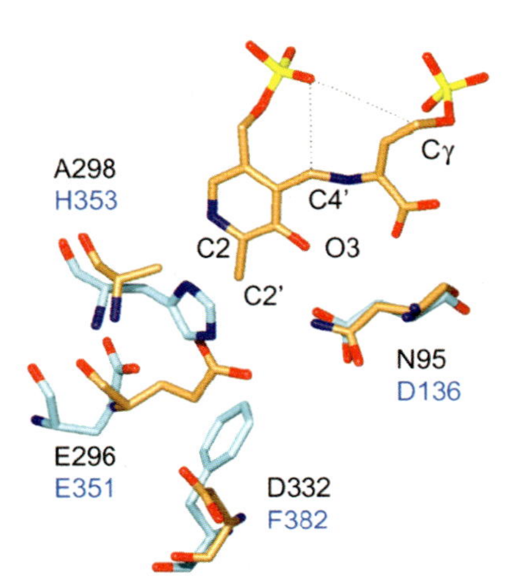


**Figure 1.35** Proximity of PLP phosphate to C4′ and Cγ sites at the active site of threonine synthase. (From [110].)

#### 1.6.2.4 Methionine, Cysteine, and Cystathionine

Methionine and cysteine are the two sulfur-containing protein amino acids. The side-chain sulfur is transferred between the two via the adduct, cystathionine. Most plants and microbes transfer the sulfur from cysteine to an activated homoserine derivative to make homocysteine, which is then methylated to produce methionine. In mammals only the reverse process is carried out. Some fungi undertake trans-sulfuration in both directions [116, 117]. The interconversion of cysteine and homocysteine involves facilitating leaving group chemistry at β- and γ-positions in two sequential PLP-dependent processes: cystathionine γ-synthase and cystathionine β-lyase (forward) or cystathionine β-synthase then cystathionine γ-lyase (reverse). The combined actions of cystathionine γ-synthase and cystathionine β-lyase from *E. coli* have been well characterized, and provide an example of the microbial route to homocysteine.

Cystathionine γ-synthase facilitates the loss of succinate from the γ-position of *O*-succinylhomoserine by a mechanism analogous to the first stage of threonine synthase catalysis. Cysteine attacks the bound vinylglycine intermediate in the reverse of the elimination chemistry and the overall result is nucleophilic substitution to produce cystathionine (Figure 1.36) [116].

Cystathionine β-lyase acts at the cysteine end of cystathionine to labilize homocysteine as a leaving group from the β-position. Hydrolysis results in the release of pyruvate as the other reaction product (Figure 1.37) [118].

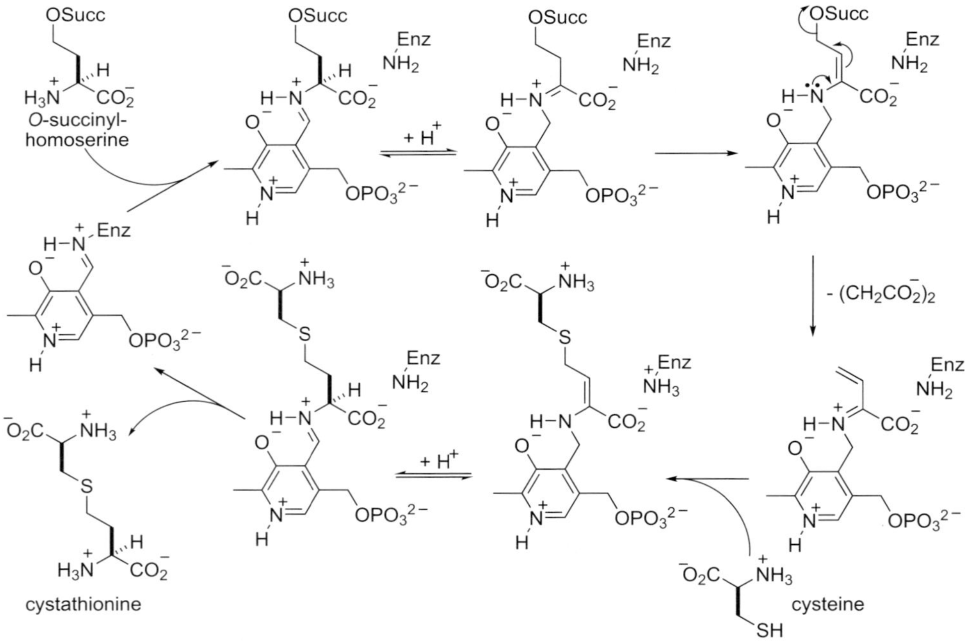

**Figure 1.36** Mechanism of cystathionine-γ-synthase.

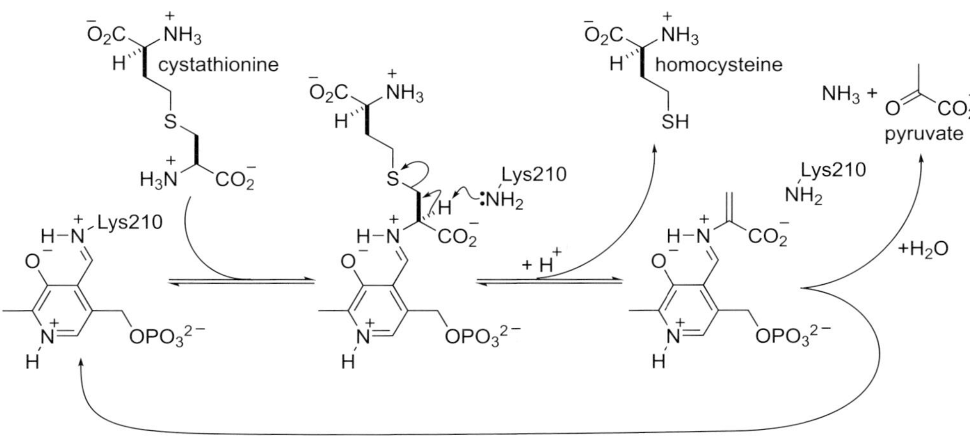

**Figure 1.37** Mechanism of cystathionine-β-lyase.

#### 1.6.2.5 **Methionine Synthase**

The biosynthesis of methionine is completed by methylation of homocysteine, catalyzed by methionine synthase (MS). The ultimate source of the methyl group is $N^5$-methyl-THF (Me-THF) where the methyl group can originate in the cleaved side-chain of serine (as seen in the biosynthesis of glycine). There are two distinct versions of this enzyme depending on the immediate source of the methyl group, either Me-THF or methyl cobalamin. In organisms that biosynthesize vitamin $B_{12}$, or obtain it from their environment, this is often carried out by a multifunctional MS that uses methyl cobalamin as the alkylating agent (Figure 1.38) [119]. After methionine synthesis, the methyl cobalamin is reconstituted by methyl transfer from Me-THF [120]. There are two MS enzymes in *E. coli* – one uses methyl cobalamin as methyl donor, whereas another uses Me-THF directly as the alkylating agent in a reaction that is dependent on a zinc ion for catalysis [121]. It is believed that homocysteine coordinates to the zinc as the thiolate and is thereby activated as a nucleophile to react with Me-THF.

### 1.6.3
### Glutamate Family Amino Acids: Proline and Arginine

Proline and arginine are made from glutamate by routes that utilize the same chemical strategy that is seen in methionine and threonine biosynthesis. This involves activation of the γ-carboxylate and biosynthetic reduction (Figure 1.39). A bacterial pathway has been characterized: glutamate-5-kinase [122] phosphorylates the γ-carboxylate of glutamate and NADPH-dependent reduction leads to glutamate-γ-semialdehyde, which undergoes spontaneous cyclization to the corresponding imine, $\Delta^1$-pyrroline-5-carboxylate. In plants a bifunctional enzyme mediates both of these steps and ensures efficient use of the reactive acyl phosphate intermediate [123, 124]. Two alternative pathways to this cyclic imine via oxidative deamination of ornithine have been reported [125, 126]. Reduction of the imine to proline is catalyzed

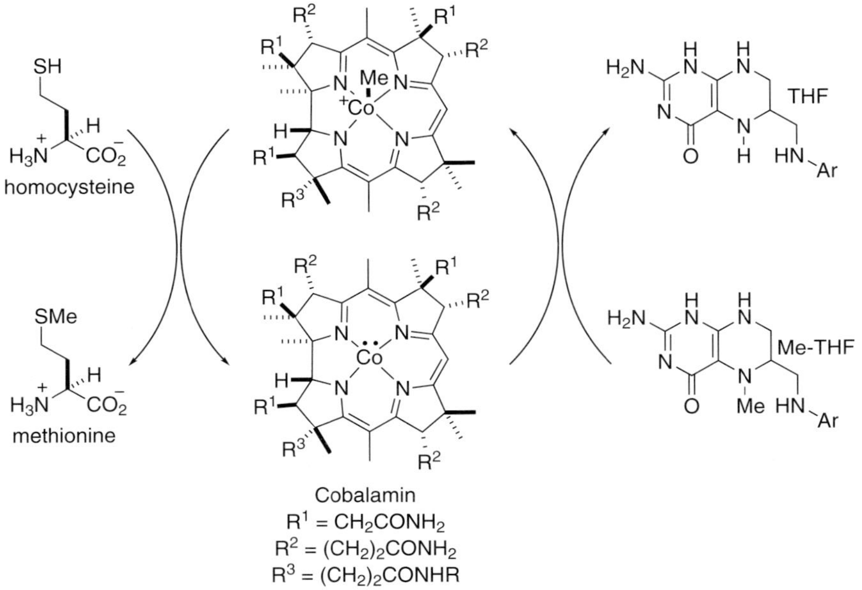
SH
H
$H_3N^+$ $CO_2^-$
homocysteine
SMe
H
$H_3N^+$ $CO_2^-$
methionine
Me
Co
Cobalamin
$R^1 = CH_2CONH_2$
$R^2 = (CH_2)_2CONH_2$
$R^3 = (CH_2)_2CONHR$
THF
Me-THF
HN
Ar

**Figure 1.38** Methyl transfer chemistry of methionine synthase.

by $\Delta^1$-pyrroline-5-carboxylate reductase and the structures of this enzyme from human pathogens (*Neisseria meningitidis* and *Streptococcus pyogenes*) have recently been characterized [127].

The chemical challenge in making the arginine precursor, ornithine, is in retaining an acyclic structure. This requires the prevention of cyclization of the α-amino group onto electrophilic functional groups in side-chain intermediates. A protecting group strategy is utilized for this purpose (Figure 1.39). The α-amino group of glutamate is first made non-nucleophilic by acetylation [128]. Phosphorylation [129, 130] and reduction [131] then produces *N*-acetylglutamate-γ-semialdehyde. In bacteria these two reactions (ArgB and ArgC) are mediated by enzymes that are homologous in structure to the corresponding enzymes, AK and ASADH, from the biosynthesis of ASA, illustrating the recurring pattern of pathway creation by gene duplication and mutation as noted above [93]. This aldehyde is converted to the corresponding amine by transamination with *N*-acetylornithine aminotransferase (Figure 1.40) [132]. This PLP-dependent enzyme utilizes glutamate as the amino donor. *N*-Acetylornithine aminotransferase has evolved an interesting multiple-substrate specificity and is also used to mediate the synthesis of *N*-succinyl-L,L-diaminopimelate in the bacterial biosynthesis of lysine [133]. This is a rare example of PLP-mediated transamination of a simple aldehyde, rather than an α-amino acid, illustrating substrate specificity divergence.

After the introduction of the γ-amino group the acetyl group is removed by hydrolysis or trans-acetylation to produce ornithine, and subsequent intermediates, *en route* to arginine (Figure 1.39) [134].

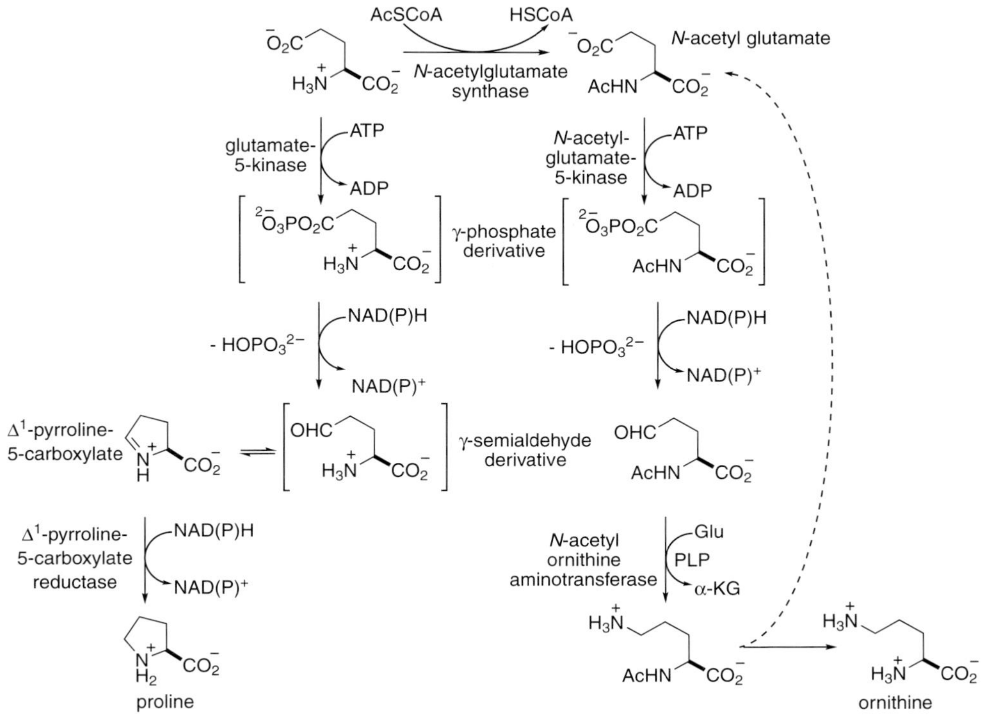

Figure 1.39 Biosynthesis of proline and ornithine.

Although the general chemical strategy for the biosynthesis of ornithine is common to many organisms, contemporary studies have revealed a diversity of detail in the pathways, associated with different acyl-transfer chemistry [135]. The first step of the biosynthesis in *E. coli*, which served as the initial exemplar for

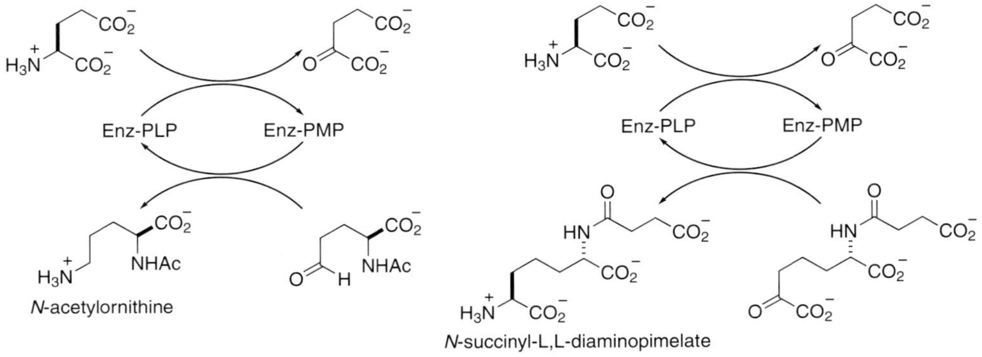

Figure 1.40 Multiple substrate specificity of *N*-acetylornithine aminotransferase.

bacterial, fungal and vertebrate pathways, is carried out by an *N*-acetylglutamate synthase [136, 137] that transfers an acetyl group directly from acetyl-CoA to glutamate. Subsequently, in other bacteria, a different family of smaller DITTOs were identified; these are sometimes fused to a second domain catalyzing the last step of arginine biosynthesis, argininosuccinate lyase – a hint of higher-level order in the biosynthetic apparatus. Other bacteria use a completely different enzyme for the acetylation chemistry: an ornithine acetyltransferase [138]. This enzyme acts via a sequential Bi Bi kinetic mechanism whereby it accepts an acetyl group from ornithine onto an active-site threonine, generating an acetyl-enzyme intermediate, and then, in a second step, transfers it to glutamate [139]. This route has the advantage of adding and removing the acetyl protecting group using a single enzyme and without paying the metabolic price of acetyl-CoA hydrolysis (although some of the ornithine acetyltransferases can be primed for acetyl donation by using acetyl-CoA as an alternative substrate) [140]. Two other biosynthetic variants have recently come to light: although ornithine is the normal product of this first section of the biosynthesis, there is evidence that some organisms only remove the *N*-acetyl group later in the pathway [141, 142]; mutation of substrate specificity, akin to that described for homoserine acyltransferases, has also underpinned the utilization of succinyl- rather than acetyltransferases in ornithine biosynthesis in *Bacteroides fragilis* [143].

The biosynthesis of arginine is completed by a three-step modification of the side-chain amino function of ornithine (Figure 1.41). Ornithine transcarbamoylase (OTCase) mediates carbamoylation with carbamoyl phosphate to produce citrulline [144]. It is one of a general class of transcarbamoylases [145] and is of clinical significance since OTCase deficiency is a relatively common genetic disease in

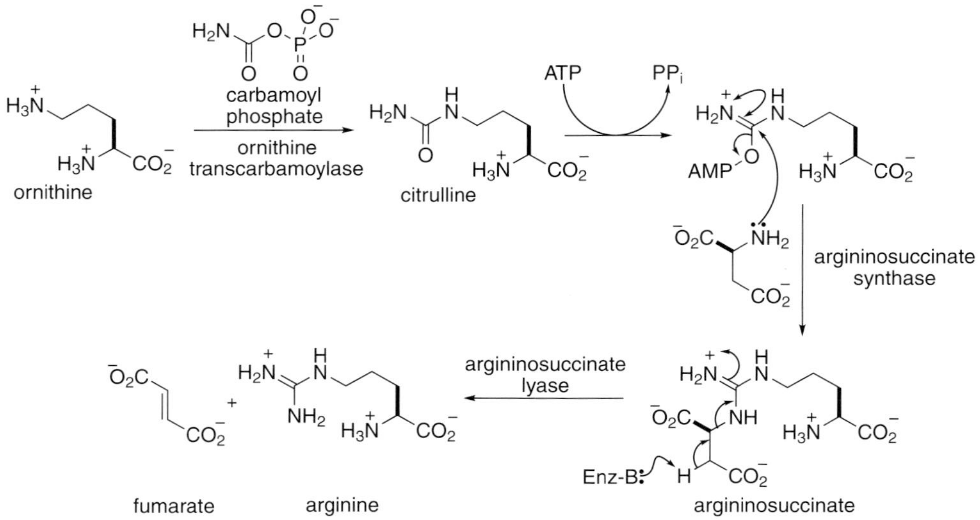

**Figure 1.41** Conversion of ornithine to arginine.

humans. The enzyme mechanism is now believed to involve acid–base catalysis by the phosphate group of the substrate, carbamoyl phosphate [146].

The urea group of citrulline is converted to an amidine by the introduction of the final amino group. This transformation occurs in a two-step process that exemplifies the third general mechanism of amine transfer in metabolism (after GAT and PLP action) – ligation of the α-amino group of aspartate and then elimination of fumarate, which is recycled via the TCA cycle [147]. Argininosuccinate synthetase is homologous to asparagine synthetase [148] and uses ATP to activate the urea by adenylation [149]; nucleophilic substitution by aspartate produces argininosuccinate. Argininosuccinate lyase, the terminal enzyme of the pathway, is a member of a family of aspartase enzymes that catalyze the elimination of fumarate (which feeds into the TCA cycle) from aspartate and *N*-aspartyl derivatives [150, 151], demonstrating the generality of this two step synthetase/lyase route for the indirect transfer of ammonia – the third general process for transfer of nitrogen between metabolites.

## 1.7
## Biosynthesis of Aliphatic Amino Acids with Modified Carbon Skeletons: Branched-Chain Amino Acids, Lysine, and Pyrrolysine

### 1.7.1
### Overview

Four of the ubiquitous aliphatic protein amino acids, valine, leucine, isoleucine, and lysine, are made by the synthesis of new carbon skeletons. Lysine is now known to be the precursor of the 22nd ribosomally encoded amino acid, pyrrolysine. The biosynthesis of these amino acids involves parallel pathways in which a chemical strategy is duplicated and homologous or promiscuous enzymes are used for the biosynthesis of more than one amino acid. Valine and isoleucine are made from precursor α-keto acids by a common route, which also provides a key biosynthetic precursor for leucine. The carbon skeleton of leucine is made from an α-keto acid intermediate in valine biosynthesis by a pathway homologous to the biosynthesis of α-KG in the TCA cycle. This route, in turn is utilized for the biosynthesis of another amino acid, α-aminoadipic acid, which is not incorporated into proteins but is a biosynthetic precursor to lysine in one of the pathways to that amino acid. Lysine is unusual in being made in different organisms by two completely different pathways – from α-aminoadipic acid in some fungi and bacteria, and from aspartate-β-semialdehyde in plants and other bacteria.

### 1.7.2
### Valine and Isoleucine

Three enzymes, acetohydroxyacid synthase (AHAS) [152], acetohydroxyacid isomeroreductase (AHIR) [153], and dihydroxyacid dehydratase [154], mediate the four key steps common to the biosynthesis of all the branched-chain amino acids [155]. Being specific to branched-chain amino acid biosynthesis and not found in animals, these enzymes are key targets for herbicide research [156]. AHAS, in particular, is the target of a number of commercial herbicides. All three enzymes have relaxed substrate specificity and process metabolites on two parallel pathways to produce two α-keto acid products (Figure 1.42).

Pyruvate reacts with the anion of the thiamine pyrophosphate cofactor of AHAS (Figure 1.43). Decarboxylation of the initial adduct leads to an acetyl α-anion equivalent, illustrating the role of thiamine pyrophosphate as an umpolung reagent in biochemistry. Reaction of this nucleophile with a second α-keto acid provides a branched carbon skeleton [157]. Two α-keto acids can participate as electrophiles in the second stage of the reaction: pyruvate or α-ketobutyrate; these lead to valine and isoleucine, respectively. The α-ketobutyrate can be biosynthesized in one step from threonine (see below).

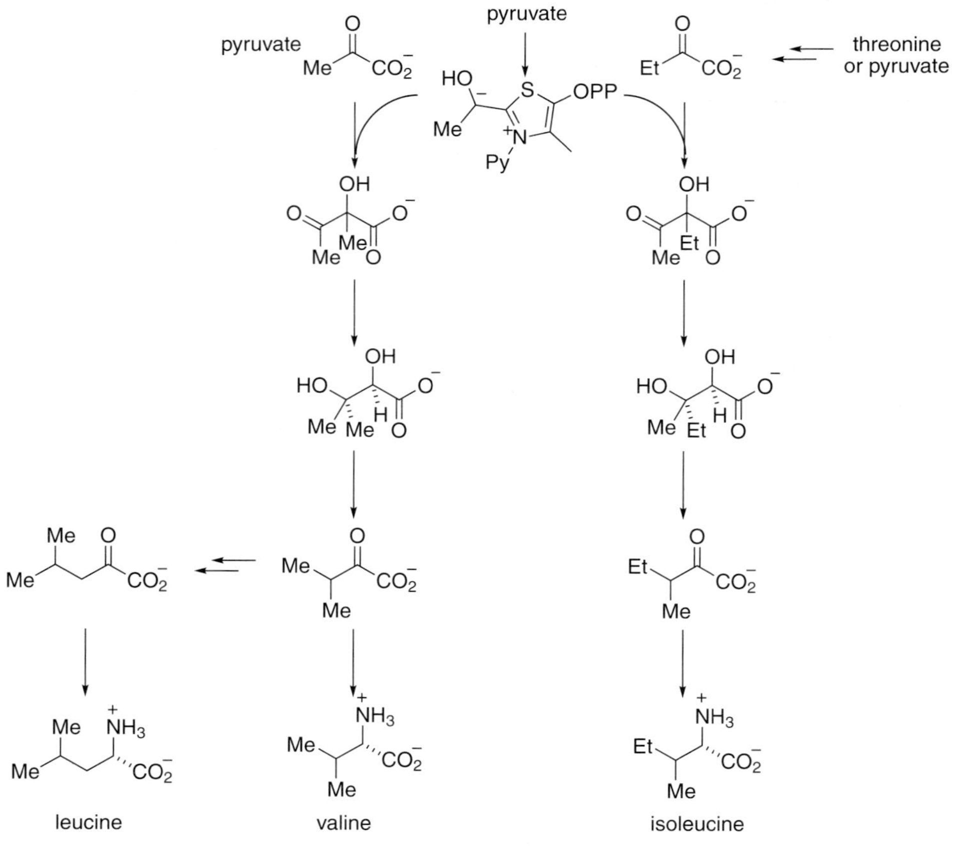

Figure 1.42 Overview of the parallel pathways to branched-chain amino acids.

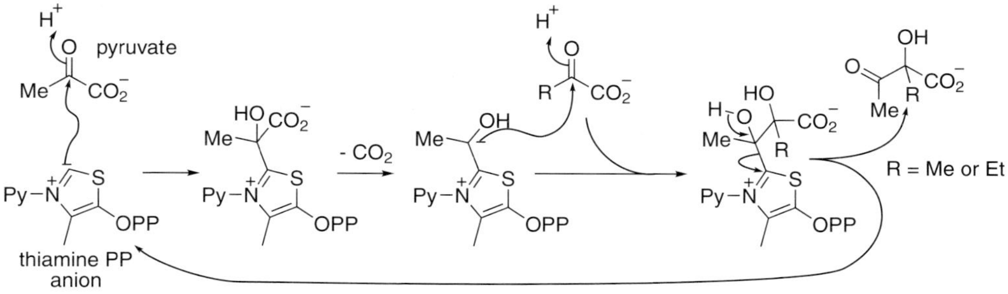

Figure 1.43 Mechanism of AHAS.

The structures of AHAS with bound sulfonylurea or imidazolinone show the mechanism of inhibition involves obstruction of the channel leading to the active site. The details of this inhibition provide an understanding of the basis of evolving herbicide resistance and will be useful in the generation of alternative herbicides [158].

Threonine deaminase (also known as threonine dehydratase) is a PLP-dependent enzyme that is the usual source of α-ketobutyrate needed for isoleucine biosynthesis [159, 160]. It is the point of commitment to isoleucine biosynthesis and the enzyme is subject to feedback regulation by isoleucine [161, 162]. Schiff base formation between threonine and the active-site PLP facilitates deprotonation at the α-center and consequent elimination of water, in a reaction characteristic of PLP-facilitated transformations of amino acids with β-leaving groups (Box 1.3). Cleavage of the link between PLP and the amino acid portion and hydrolysis chemistry produces α-ketobutyrate and ammonium ions as the product (Figure 1.44). An alternative route to α-ketobutyrate [163] based on the homologation of pyruvate is outlined later (Figure 1.48).

The two products of AHAS are processed by a bifunctional enzyme – AHIR [153]. While the enzyme acts on two different hydroxy acids, it has an absolute requirement

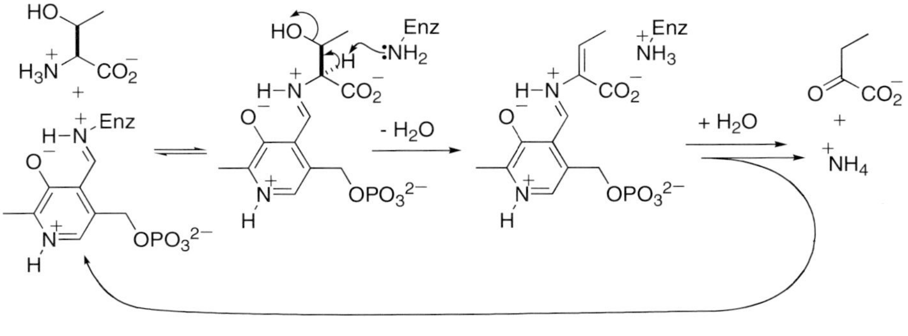

Figure 1.44 Conversion of threonine into α-ketobutyrate and ammonium ions.

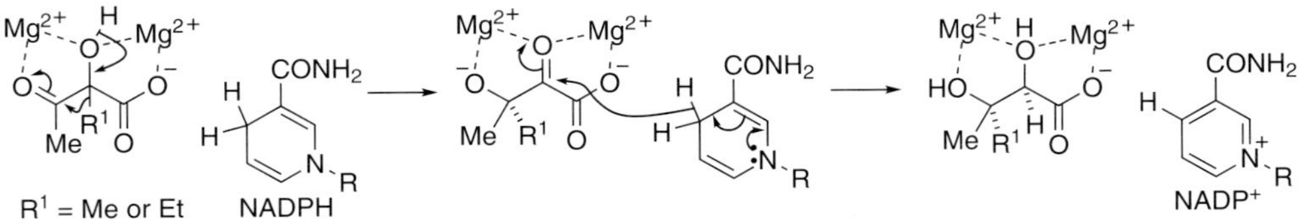

Figure 1.45 Two-stage mechanism of AHIR.

for the (S)-enantiomer of each substrate. Although no intermediate has been isolated, it is presumed that the reaction occurs in two stages – an alkyl migration and then a reduction step, mediated by NADPH (Figure 1.45).

The structure of plant AHIR with a herbicidal inhibitor bound [164] and, in another case, with the product and a modified cofactor bound, has allowed a mechanism to be proposed (Figure 1.45) [111]. Both steps of the reaction require $Mg^{2+}$ for both structural and catalytic reasons. The two $Mg^{2+}$ ions are suitably placed to polarize the starting keto group and facilitate deprotonation of the adjacent hydroxyl to facilitate push–pull catalysis of the alkyl transfer. The newly created ketone functionality is proximal to the nicotinamide cofactor and reduction traps the rearranged product.

Dehydration of the dihydroxy acid products by a dual-purpose enzyme completes the biosynthesis of the two key α-keto acids: α-ketoisovalerate and α-keto-β-methylvalerate (the precursors of valine and isoleucine, respectively). A [4Fe–4S] cluster of dihydroxyacid dehydratase catalyzes the elimination of water [154], in a mechanism believed to be analogous to that of aconitase (ACN) (Figure 1.46) [165]: coordination of the hydroxy acid to an iron center of the [4Fe–4S] cluster activates the hydroxyl as a leaving group and allows enzyme-induced elimination. Tautomerism of the resulting enol generates an α-keto acid product. The [4Fe–4S] cluster is labile to oxidation and one of the most significant antibacterial effects of nitric oxide, produced by the immune system, is now believed to be the inactivation of this enzyme [166], which leads to a diminished ability to make branched-chain amino acids [167]. The [4Fe–4S] center of dehydratases also appears to be a significant target for copper toxicity in bacteria [168].

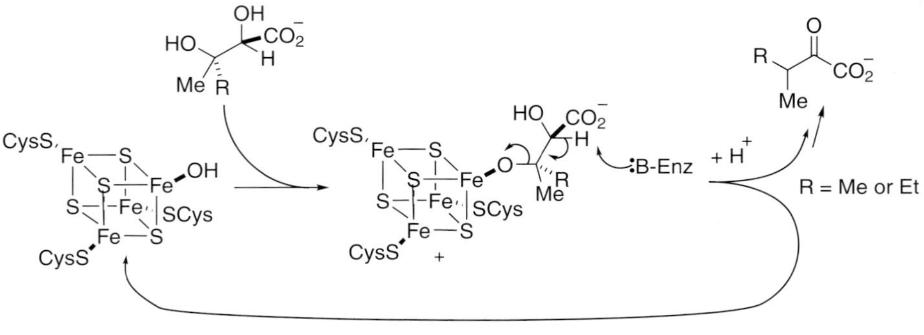

Figure 1.46 Proposed mechanism for dihydroxyacid dehydratase.

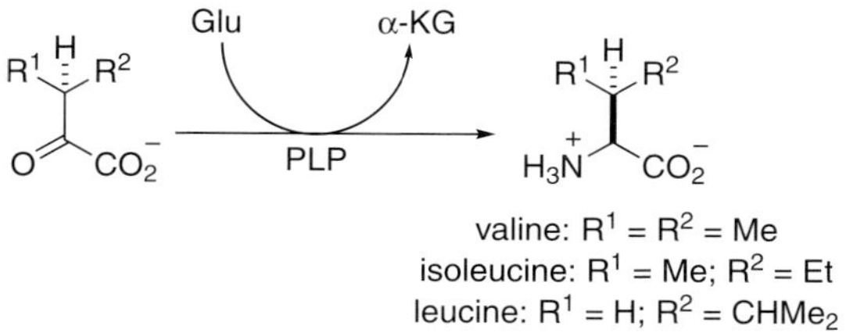

valine: $R^1 = R^2 = Me$
isoleucine: $R^1 = Me$; $R^2 = Et$
leucine: $R^1 = H$; $R^2 = CHMe_2$

**Figure 1.47** Transamination to form branched-chain amino acids.

Glutamate acts as amino donor in the transamination of α-ketoisovalerate and α-keto-β-methylvalerate catalyzed by the branched-chain aminotransferase (Figure 1.47). This produces valine and isoleucine, respectively. In addition to this fate, α-ketoisovalerate is also homologated to α-ketoisocaproate *en route* to leucine.

### 1.7.3
### Homologation of α-Keto Acids, and the Biosynthesis of Leucine and α-Aminoadipic Acid

The aliphatic amino acids described so far provide a wide range of chemical diversity for proteins. However, there are two significant omissions that are rectified in the universal genetic code. The nonpolar amino acids, valine and isoleucine, are branched at the β-position, which causes steric congestion when multiply incorporated within α-helical structures. Homologation of valine to leucine relieves this steric crowding by placing the branching at the γ-center of the amino acid, thus increasing the opportunity for forming stable hydrophobic cores in α-helical proteins. The other omission is the lack of an amino acid with a side-chain amino substituent. Ornithine is not incorporated into proteins, possibly because the amino side-chain can cleave an adjacent peptide link via lactamization to form a six-membered ring. The only amine side-chain among the protein amino acids is that of lysine. This is a homologated version of ornithine that would produce a less favorable seven-membered ring species on lactamization. Access to a homologation pathway for α-keto acids is used to address both these issues.

In the TCA cycle, α-KG is synthesized from oxaloacetate by a three enzyme homologation process. Citrate synthase (CS) condenses oxaloacetate with acetyl-CoA. Hydrolysis of the thioester adduct ensures that the condensation reaction is favorable. Hydroxyl migration occurs via a dehydration-rehydration sequence catalyzed by an iron–sulfur-dependent enzyme, ACN, which is named in honor of the intermediate alkene, aconitate. Isocitrate dehydrogenase (IDH) catalyses the oxidation of isocitrate to a β-keto acid that spontaneously decarboxylates, in the presence of an enzymic divalent cation, to produce α-KG. Gene duplication and functional divergence has generated variants of this three-step process that are used to homologate the other two common α-keto acids of primary metabolism (pyruvate and α-KG) and a range of other α-keto acids including α-ketobutyrate (the biosynthetic precursor to

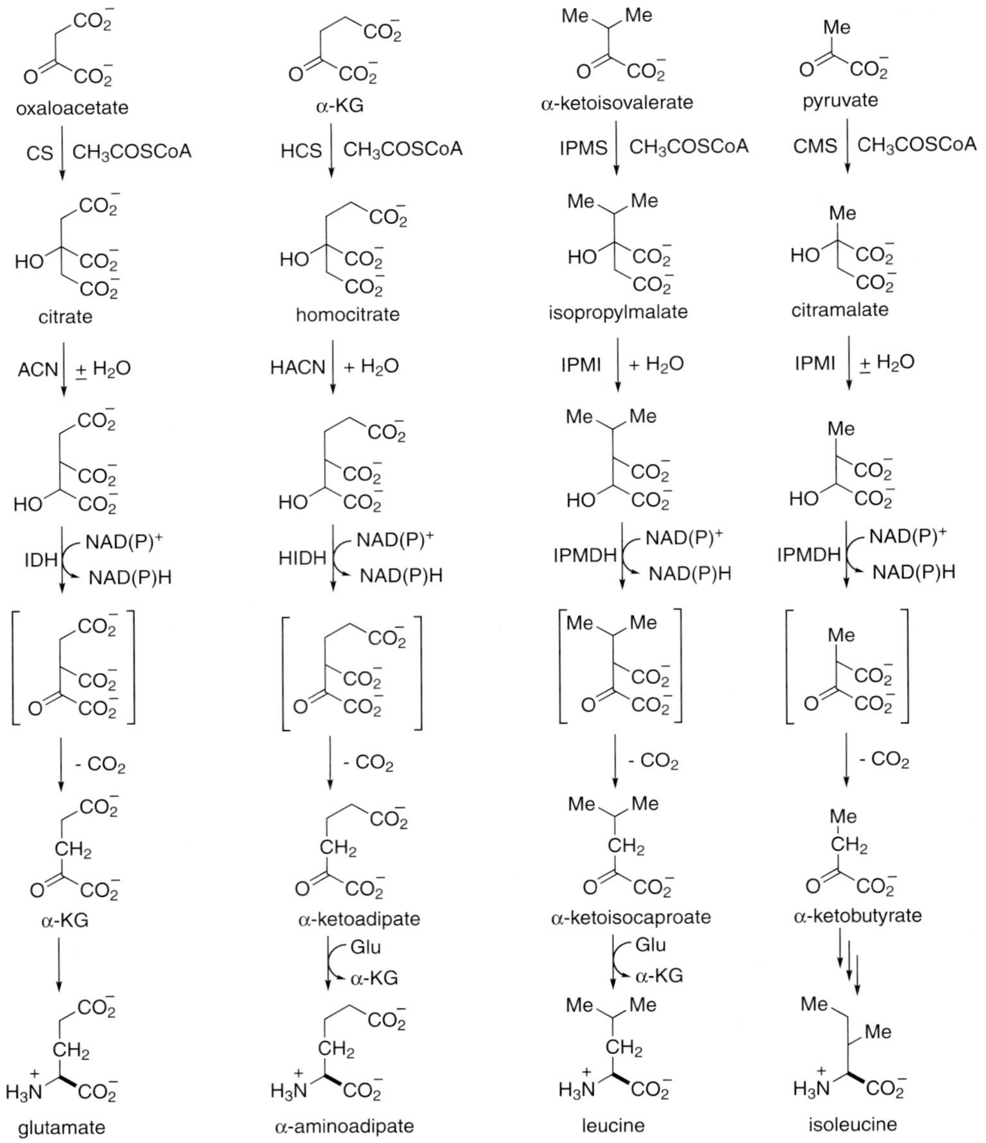

**Figure 1.48** Homologation of α-keto acids to produce amino acids.

valine) (Figure 1.48) [163]. Transamination of α-ketoisovalerate and of α-ketoadipic acid makes leucine and α-aminoadipic acid, respectively [169], while the homologation of pyruvate to α-ketobutyrate provides an alternative route to the starting material for leucine biosynthesis in some Archaea and bacteria. Many of these enzymes have not been studied comprehensively and mechanistic proposals are often drawn by analogy from the action of homologous enzymes that have been studied in more detail.

Homocitrate synthase (HCS) [170], isopropylmalate synthase (IPMS) [171], and citramalate synthase (CMS) [172] are homologous enzymes that catalyze the Claisen condensation of acetyl-CoA with α-KG, α-ketoisovalerate, and pyruvate, respectively. They are also homologous to one of the classes of CS found in anaerobic bacteria [173]. Hydrolysis of the initially formed thioester adduct makes the otherwise reversible Claisen condensation, favorable. The structure of IPMS has been determined [174] and that structure was used to shed light on the catalytic machinery of these enzymes [175]. Carbonyl groups of both substrates are polarized by the enzyme: acetyl-CoA by an arginine to facilitate enol formation and α-KG by chelation to an essential $Zn^{2+}$. On the basis of modeling, kinetic studies and site-directed mutagenesis, the general base for deprotonation of acetyl-CoA is believed to be a histidine, acting in concert with a glutamate as a catalytic dyad.

A variety of homologous isomerases catalyze the next step in the homologation process. These include homoaconitase (HACN) and isopropylmalate isomerase (IPMI), and are all members of the ACN superfamily [176]. As an example of the relatedness of these enzymes, in *P. horikoshii* a single enzyme, with dual substrate specificity, acts as both an IPMI and a HACN [177]. These enzymes are not fully characterized, but it is assumed that they exploit the dehydration–rehydration chemistry of a [4Fe–4S] cluster that has been characterized in ACN and was described for dihydroxyacid dehydratases above [165 (Figure 1.4–6)].

The final step in the homologation process is the oxidative decarboxylation of the β-hydroxy acids. In this step an oxidized nicotinamide cofactor ($NAD(P)^+$) accepts a hydride from the secondary alcohol center and generates a β-keto acid. This intermediate spontaneously decarboxylates, in the presence of an enzyme-bound divalent metal ion, to leave the homologated α-keto acids. This step is catalyzed by a superfamily of homologous enzymes, ICDH [178], isopropylmalate dehydrogenase (IPMDH) [179], and homoisocitrate dehydrogenase (HIDH) [180]. Dynamic X-ray crystallography has provided detailed structural information about the catalytic mechanism of ICDH which is the best characterized enzyme of the family [181]. Again, in some organisms dual substrate enzymes act in parallel pathways; for example, in both *Thermus thermophilus* [182] and *P. horikoshii* [183] a single enzyme catalyses two oxidative decarboxylation reactions producing both α-KG and α-ketoadipic acid. A loop at the active site of these enzymes appears to control specificity and mutations within this loop can modify the substrate specificity to change the pathway that an enzyme acts on. The conversion of *E. coli* ICDH to a bifunctional ICDH/IPMDH by mutagenesis has been reported by Koshland *et al.* Ser113 in the substrate recognition loop forms a hydrogen bond to the δ-carboxylate of isocitrate. Mutation of this residue to glutamate, which introduces electrostatic repulsion to the normal substrate, together with systematic mutation of two other local residues, Asp115 and Val116, produced several mutants with a preference for isopropylmalate as substrate. In *T. thermophilus* HICDH, Arg85 of the corresponding loop is involved in recognition of the δ/γ-carboxylate of isocitrate/homoisocitrate. When this residue is mutated to a large nonpolar residue, valine, the enzyme is converted to an IPMDH, retaining some HICDH activity, but losing its ability to oxidatively decarboxylate isocitrate [182]. Other substrate specificity changes have been described for these enzymes; for example, the 100-fold preference of the *T. thermophilus* IPMDH for $NAD^+$ as coenzyme has been reversed to a 1000-fold preference for $NADP^+$ by mutagenesis in which seven residues involved in a β-turn were exchanged for a 13-residue sequence of α-helix and a loop modeled on the corresponding specificity determinant in *E. coli* ICDH [184].

Finally, the resulting α-keto acids are transaminated to the corresponding amino acids by appropriate aminotransferases: either the branched-chain aminotransferase described earlier (to produce leucine) or a glutamate-α-ketoadipate transaminase to generate the latter amino acid (Figure 1.48). Both enzymes utilize glutamate as nitrogen donor [185–187] and the structural basis for the multiple substrate specificity of the *T. thermophilus* aminotransferase has been studied in detail by X-ray crystallography [188].

## 1.7.4
## Biosynthesis of Lysine: A Special Case

There are two distinct metabolic strategies for the biosynthesis of the six-carbon chain of lysine [169, 189]: the diaminopimelate pathway, which is found in plants, bacteria, and lower fungi utilizes pyruvate, and ASA (also a biosynthetic precursor to methionine and threonine as described above) as building blocks. The α-aminoadipate pathway, found in some bacteria, fungi, and euglenoids (eukaryotic single-celled flagellates), creates the carbon skeleton by a homologated version of the biosynthesis of α-KG (as found in the TCA cycle and described in the previous section). Both of these pathways are now known to occur in multiple variants [190, 191].

### 1.7.4.1 Diaminopimelate Pathway to Lysine

The DAP pathway produces both lysine and *meso*-DAP – an important bacterial cell wall building block (Figure 1.49). It starts with ASA, prepared by the same chemistry as the first two steps of the methionine/threonine pathway [95, 192]. As with other aspartate family amino acids, the biosynthetic enzymes are potential therapeutic and pesticide targets [193]. Dihydrodipicolinate synthase (DHDPS) forms a Schiff base with pyruvate to provide access to an enamine that condenses with ASA [194–196]. The resulting dihydrodipicolinate is reduced to tetrahydrodipicolinate [197, 198], after which the pathway diverges in different organisms.

In bacteria, *N*-acylation facilitates ring opening of tetrahydrodipicolinate; acetyl- and succinyl-CoA are used as acylating agents in different organisms. The resulting α-keto acid is transaminated, using glutamate as nitrogen donor; the dual role of this *E. coli* aminotransferase for both lysine and ornithine biosynthesis has been noted above (Figure 1.40). Removal of the *N*-acyl group gives L,L-DAP, which undergoes epimerization to *meso*-DAP catalyzed by an enzyme, DAP epimerase, that uses a pair of active-site cysteine residues to effect deprotonation and reprotonation at one of the α-amino acid centers [199, 200]. The D-amino acid

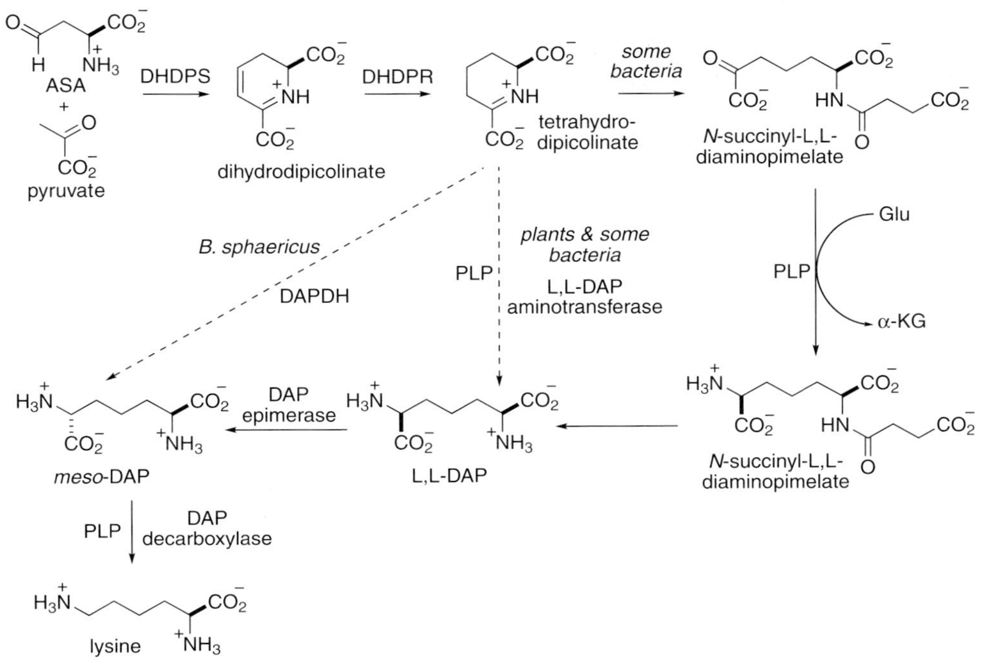

**Figure 1.49** DAP pathways for the biosynthesis of lysine.

center of *meso*-DAP is decarboxylated by a PLP-dependent enzyme [201–203] to generate lysine (see Box 1.3).

In plants and some bacteria the latter stages of the DAP pathway are truncated by omission of the acylation and deacylation steps [191]; in these organisms a L,L-DAP aminotransferase converts tetrahydrodipicolinate directly to L,L-DAP [204]. *Corynebacterium glutamicum*, used in the industrial production of lysine, can run either the full or truncated pathway, depending on conditions [160]. *Bacillus sphaericus* truncates this section of the pathway still further by directly converting tetrahydrodipicolinate to *meso*-DAP using an NADPH-dependent DAP dehydrogenase [205], presumably via a reductive amination reaction on the ring-opened form of tetrahydrodipicolinate.

### 1.7.4.2 α-Aminoadipic Acid Pathways to Lysine

There are now known to be two separate pathways to lysine from α-aminoadipic acid (Figure 1.50). Both utilize the strategy, anticipated from the biosynthesis of aspartate and glutamate family amino acids, of activation of the δ-carboxylate, reduction to a semialdehyde, and then an amination process. Interestingly, both pathways are now believed to involve ligation to carrier proteins. The most recently discovered pathway was identified in *T. thermophilus* and *P. horikoshii* by a bioinformatic approach [190]. A cluster of three genes homologous to the central three steps of ornithine biosynthesis (phosphorylation of the side-chain carboxylate, reduction of the acyl phosphate to a

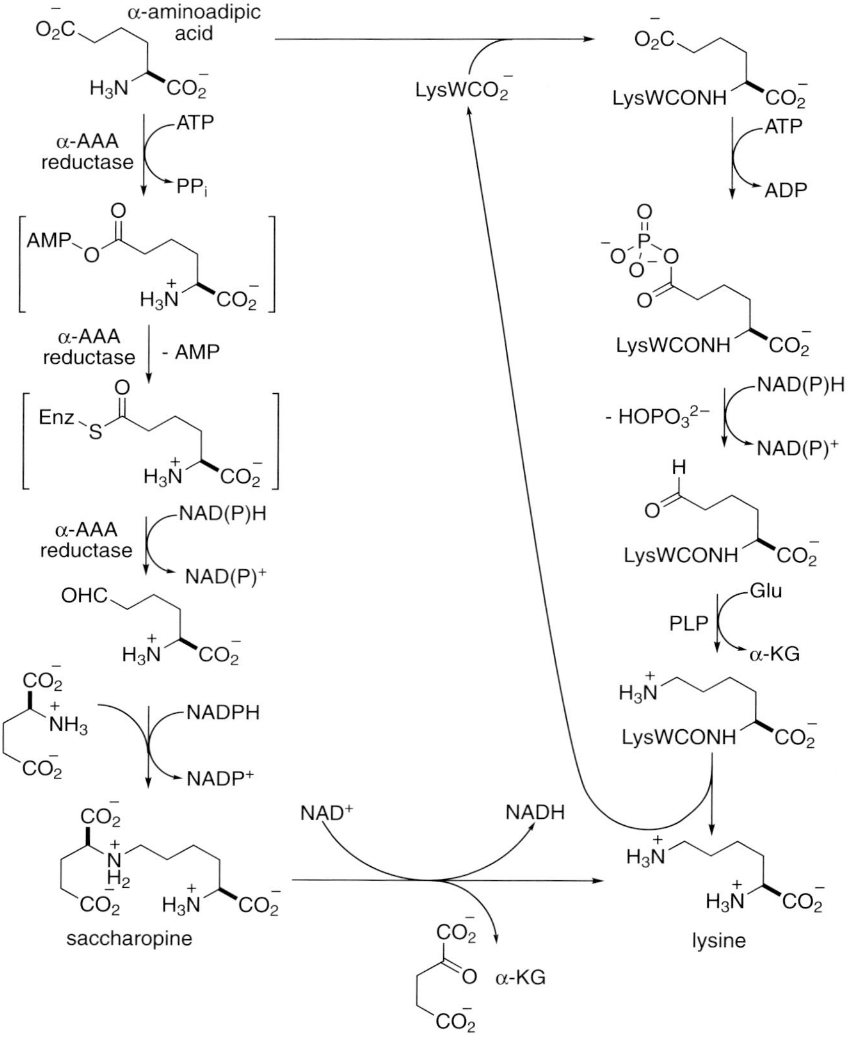

Figure 1.50 Biosynthetic routes to lysine from α-aminoadipic acid.

semialdehyde, and transamination to the amine) were identified in *T. thermophilus*. Genes for the first and last steps of arginine biosynthesis (for the protecting group chemistry of *N*-acylation and deacylation) were absent. Disruption of these putative arginine biosynthetic genes led to organisms auxotrophic for lysine. This demonstrated a biosynthetic route to lysine that is analogous to the biosynthesis of ornithine from glutamate [206]. In reconstituting the α-aminoadipic acid pathway from *T. thermophilus* it was discovered that an alternative *N*-acylation pathway operates in this bacterium. In the first step of the pathway the amino group of α-aminoadipic acid is protected by acylation by the side-chain carboxylate of the C-terminal glutamate of a small protein, LysW. The functional group modification of the α-aminoadipic acid is mediated on the LysW adduct which acts as a carrier protein for the biosynthesis. Following completion of the side-chain manipulation, the LysW adduct is cleaved to release lysine [207].

Although the discovery of a bacterial route to lysine from α-aminoadipic acid is a recent discovery, this amino acid has been known for some time to have two metabolic roles in fungi – both as an alternative precursor to lysine [169], and as a key precursor in the biosynthesis of penicillin and cephalosporin antibiotics [208]. For penicillin biosynthesis the δ-carboxylate of α-aminoadipic acid is incorporated into a tripeptide by a nonribosomal peptide synthetase, δ-(α-L-aminoadipyl)-L-cysteinyl-D-valine synthetase [209]. The peptide bonds, including that to the side-chain of α-aminoadipic acid, are made by ligating the carboxyl group of each component to a multifunctional enzyme via thioester links to 4′-phosphopantetheine groups which then undergo successive nucleophilic substitution reactions with the amine of the adjacent component [210].

Walsh *et al.* have proposed that the activation and reduction of the δ-carboxylate of α-aminoadipic acid occurs in an analogous fashion to nonribosomal peptide synthesis (Figure 1.50). α-Aminoadipate reductase comprises two proteins – Lys2 and Lys5. Lys2 is a large (155 kDa) multidomain protein. Sequence homologies with nonribosomal peptide synthetases suggest that residues 225–808 constitute a 60-kDa adenylation domain (A) that activates α-aminoadipic acid as the aminoacyl-δ-AMP derivative [211]. Hydrolysis of the pyrophosphate side-product makes the reaction favorable. Lys5 primes the PCP domain (residues 809–924) of Lys2 by adding a 4′-phosphopantatheine unit to Ser880. The aminoacyl-δ-AMP derivative is ligated to the 4′-phosphopantatheine unit as a thioester. NADPH-dependent reduction to α-aminoadipic acid-δ-semialdehyde, mediated by the reductase domain (residues 925–1392), cleaves the link to the carrier protein.

The biosynthesis of lysine in this pathway is completed by an unusual two-step transamination process that does not involve a PLP-dependent enzyme. Typical transamination chemistry involves the reductive amination of PLP to PMP by glutamate and subsequent reductive amination of the target carbonyl compound by PMP. In this case, however, a saccharopine dehydrogenase reductively aminates α-aminoadipic acid-δ-semialdehyde with glutamate directly to produce saccharopine [212–214]. A second saccharopine dehydrogenase, structurally related to alanine dehydrogenase, mediates the $NAD^+$-dependent oxidative cleavage of α-KG from the adduct to leave lysine [215]. Kinetic and structural studies have led to the proposal for a proton relay mechanism for this final step of the pathway [216]. $NAD^+$ abstracts a hydride to generate an imine that is converted to a carbinolamine by base-catalyzed addition of water. Further acid–base chemistry leads to fragmentation of the carbinolamine and formation of the two products (Figure 1.51).

#### 1.7.4.3 **Pyrrolysine**

Pyrrolysine is an uncommon amino acid found in some Archaea and eubacteria; only 1% of genomes thus far sequenced contain evidence for its presence. It was first

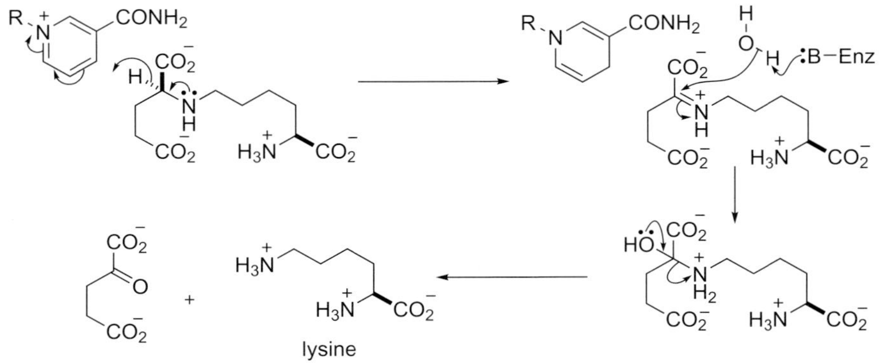

Figure 1.51 Formation of lysine by saccharopine dehydrogenase (lysine forming).

observed as an active-site residue in the crystal structure of monomethylamine methyltransferase from a methanogen, *Methanosarcina barkeri* [217], and it has now been found in other methylamine methyltransferases [218] associated with the metabolism of methylamines to methane. It is believed that the electrophilic imine of pyrrolysine is the functional group that activates methylamines for transfer of a methyl group to an adjacent cobalt corrin center [219].

In contrast to selenocysteine, this amino acid addition to the genetic code is biosynthesized from the free amino acid precursor, lysine, and then incorporated into an amber suppressor tRNA by a specific pyrrolysyl-tRNA synthetase [220–223]. The biosynthetic pathway to pyrrolysine is as yet undetermined, however a gene cluster from *Methanosarcina acetivorans pylBCD* encodes for the biosynthesis of this amino acid and recombinant *E. coli* containing these genes, together with the *pylTS* genes that encode $tRNA^{Pyl}$ and pyrrolysyl-tRNA synthetase, respectively, biosynthesize pyrrolysine and incorporate it into reading frames containing the amber codon [224]. The *pylBCD* genes show homology to genes that encode radical (*S*)-adenosylmethionine-dependent proteins, proteins forming amides and amino acid dehydrogenases, respectively; this combination of catalytic functionalities has led to the proposal of a possible biosynthetic route (Figure 1.52) [219]: manipulation of (3*R*)-3-methyl-D-glutamate, by analogy with the chemical strategy of the first stages of proline biosynthesis (reduction to the γ-aldehyde and spontaneous cyclization), could

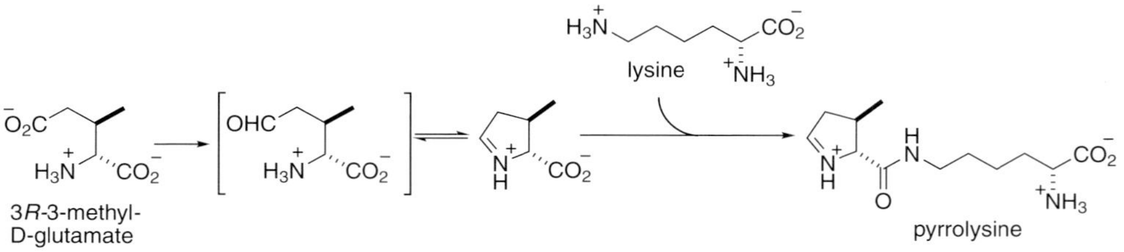

Figure 1.52 Possible biosynthetic route to pyrrolysine.

generate the required imino acid which could then be activated and used to acylate lysine to form pyrrolysine.

## 1.8 Biosynthesis of the Aromatic Amino Acids

### 1.8.1 Shikimate Pathway

The biosynthesis of the aromatic amino acids tryptophan, phenylalanine, and tyrosine begins with the shikimate pathway (Figure 1.53). This pathway consists of seven enzyme-catalyzed reactions, which convert erythrose 4-phosphate (E4P) and phosphoenol pyruvate (PEP), both originally derived from glucose, into the prearomatic compound chorismate [225, 226]. From chorismate, the pathway branches to produce tryptophan, or phenylalanine and tyrosine. Chorismate is also the precursor for other important aromatic compounds.

The first step of the shikimate pathway is catalyzed by 3-deoxy-D-*arabino*-heptulosonate-phosphate (DAH7P) synthase. Feedback regulation of this enzyme by

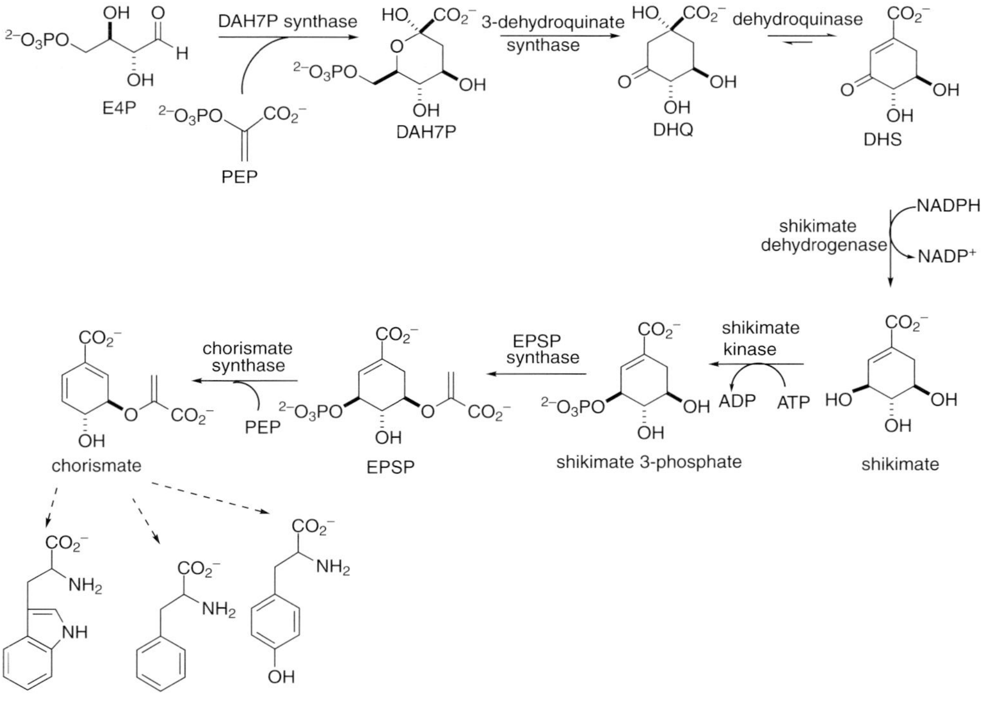

**Figure 1.53** Shikimate pathway.

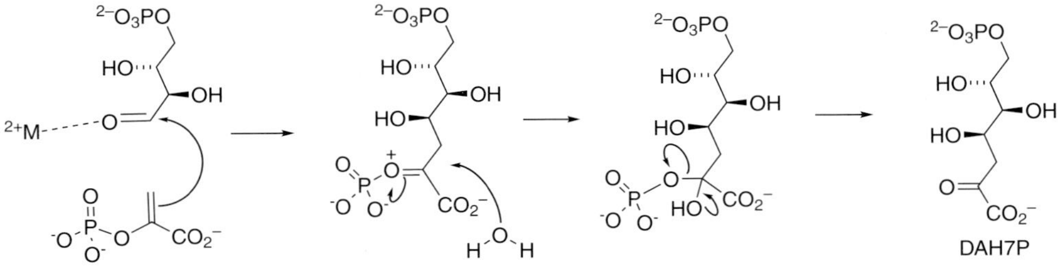

Figure 1.54 Proposed mechanism for DAH7P synthase.

aromatic amino acids is an important mechanism for control of flux through the shikimate pathway. DAH7P synthase catalyses an aldol-like reaction in which the three-carbon unit PEP condenses with the four-carbon aldehyde, E4P. It has been shown that the C–O bond of PEP is cleaved during the reaction, requiring water to act as a nucleophile on C2 of PEP, generating a phosphorylated hemiacetal intermediate (Figure 1.54). This intermediate then loses phosphate to produce the linear form of DAH7P. All DAH7P synthases have been shown to require a divalent metal ion for activity [227]. It is likely that the metal ion plays a role in activating the aldehydic carbon to nucleophilic attack (Figure 1.54) [228].

Directed evolution experiments have been used to replace DAH7P synthase with a pyruvate-utilizing aldolase, 2-keto-3-deoxy-6-phosphogalactonate (KDPGal) synthase (Figure 1.55) [229, 230]. Frost *et al.* showed that both the *E. coli* enzyme and that from *Klebsiella pneumoniae* could support biosynthesis of aromatic metabolites in a cell line lacking all isozymes of DAH7P synthase. Gene shuffling and modifications of the

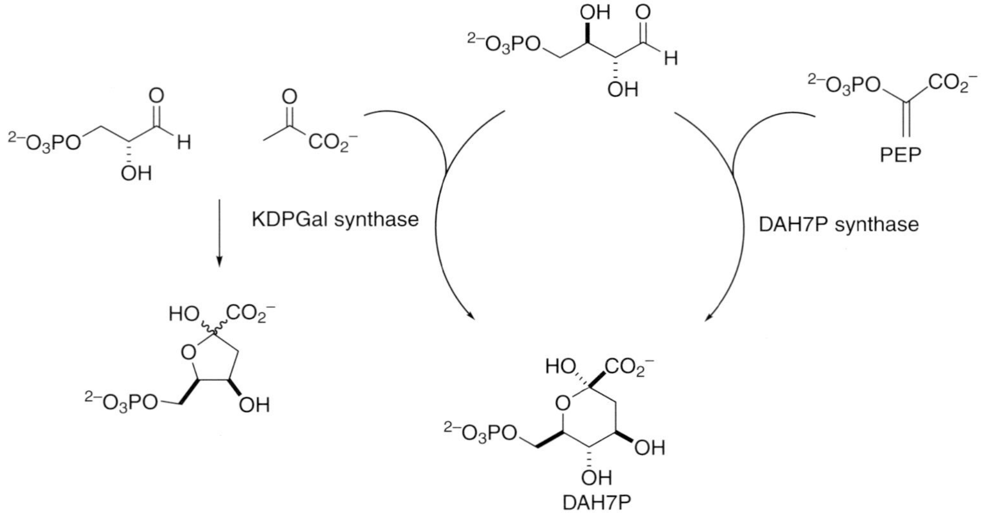

Figure 1.55 Directed evolution of the pyruvate-utilizing aldolase KDPGal synthase to replace DAH7P synthase.

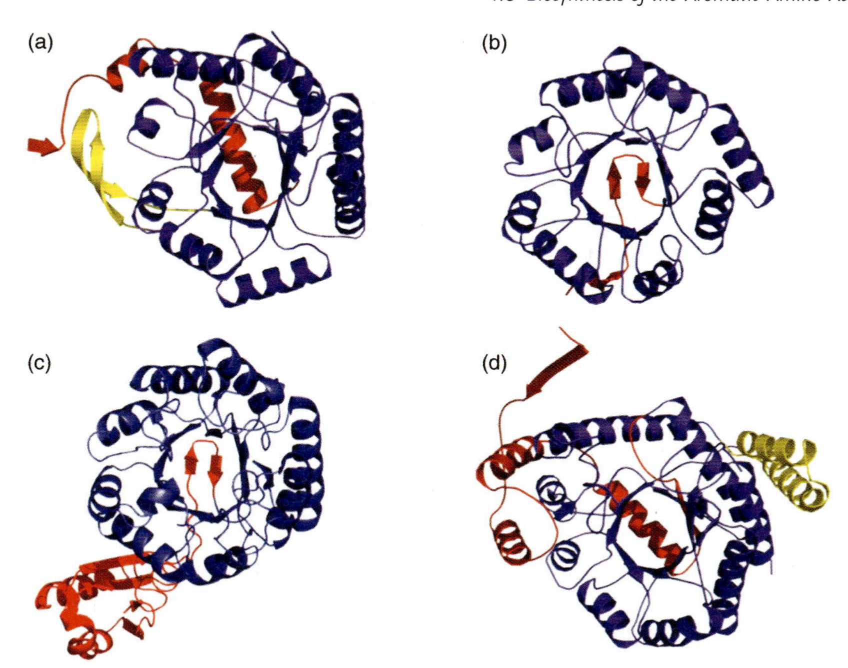


**Figure 1.56** Different forms of DAH7P synthase showing basic catalytic barrel in blue, and N-terminal and loop extensions in red and yellow, respectively, from (a) *E. coli* type I$\alpha$ (Phe sensitive), (b) *P. furiosus* type I$\beta$ (unregulated), (c) *T. maritima* I$\beta$ (Phe, Tyr inhibited) [235], and (d) *M. tuberculosis* (Phe + Tyr inhibited) [231].

*E. coli* enzyme by site directed mutagenesis produced mutant forms of the aldolase that exhibited a 60-fold increase of $k_{cat}/K_m$ over wild-type activity.

One of the most interesting aspects of DAH7P synthase is the variation in structures and regulation patterns that have been observed for this enzyme [231]. Structures of DAH7P synthases have been determined from a variety of sources including *E. coli, Pyrococcus furiosus, S. cerevisiae, Thermotoga maritima,* and *M. tuberculosis* (Figure 1.56) [231–234]. Despite large primary sequence variation, the core catalytic unit for each of these proteins is a standard $(\beta\alpha)_8$ barrel. Variable accessory domains have been shown to augment this barrel structure and different quaternary structures are observed for enzymes from different species. The variable extra-barrel elements are associated with feedback regulation. *E. coli* expresses three DAH7P synthase isozymes each with a binding site for a single aromatic amino acid. On the other hand, the *M. tuberculosis* genome encodes a single DAH7P synthase. This enzyme shows synergistic inhibition by a combination of tryptophan and phenylalanine, with both amino acids binding simultaneously to independent allosteric sites formed extra-barrel subdomains.

3-Dehydroquinate (DHQ) synthase, which catalyzes the second step of the shikimate pathway, is responsible for production of the first carbocycle of the pathway. This enzyme has attracted a great deal of attention due to the number of

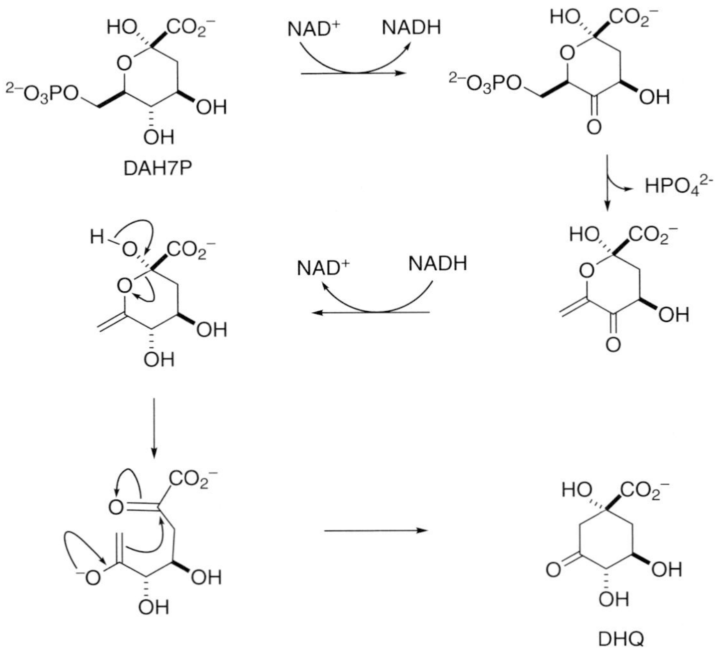

Figure 1.57 Reaction mechanism for dehydroquinate synthase.

different chemical steps catalyzed by a single enzyme (Figure 1.57) [236–238]. DHQ synthase uses the pyranose form of DAH7P. In the first step it undergoes temporary oxidation at C5 to facilitate *syn* elimination by an $E1_{cb}$ mechanism to produce an enol pyranose intermediate. The C5 keto group of the product is then reduced. These redox steps use $NAD^+$/NADH as a cofactor to accept and deliver the hydride. Following reduction, the enol pyranose intermediate ring opens and cyclizes via an intramolecular aldol reaction. There has been some debate about the role of the enzyme in these final two steps, as the enol pyranose intermediate, released *in situ* via photolysis of an *o*-nitrobenzyl protected form, was able to form dehydroquinate without assistance from the enzyme (Figure 1.58). However, as some epimer is produced in the nonenzymatic reaction (produced by the enol attacking the wrong

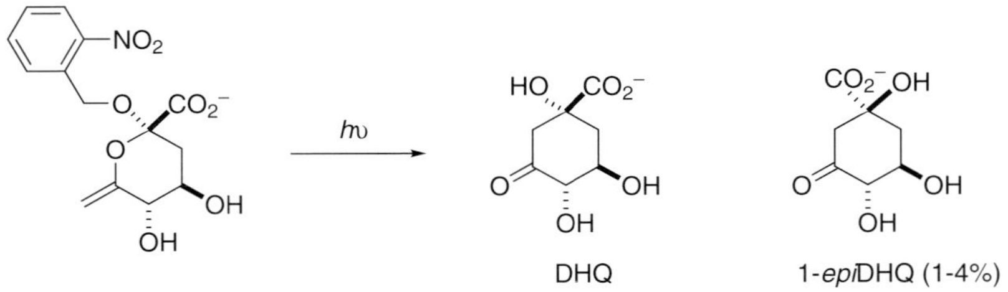

Figure 1.58 Nonenzymatic generation of DHQ.

face of the carbonyl), the enzyme appears to be required at least as a template to ensure the correct stereochemistry of the final product dehydroquinate is generated (Figure 1.58). DHQ synthase is one of a handful of enzymes that use transient oxidation to achieve its overall chemistry [239].

## 1.8.2
## Case Study: Alternative Synthesis of Dehydroquinate in Archaea

Euryarchaea have a different strategy for the biosynthesis of dehydroquinate (Figure 1.59) [240, 241]. Recent work on *Methanocaldococcus jannaschii* has shown that dehydroquinate production occurs through two alternative enzyme-catalyzed

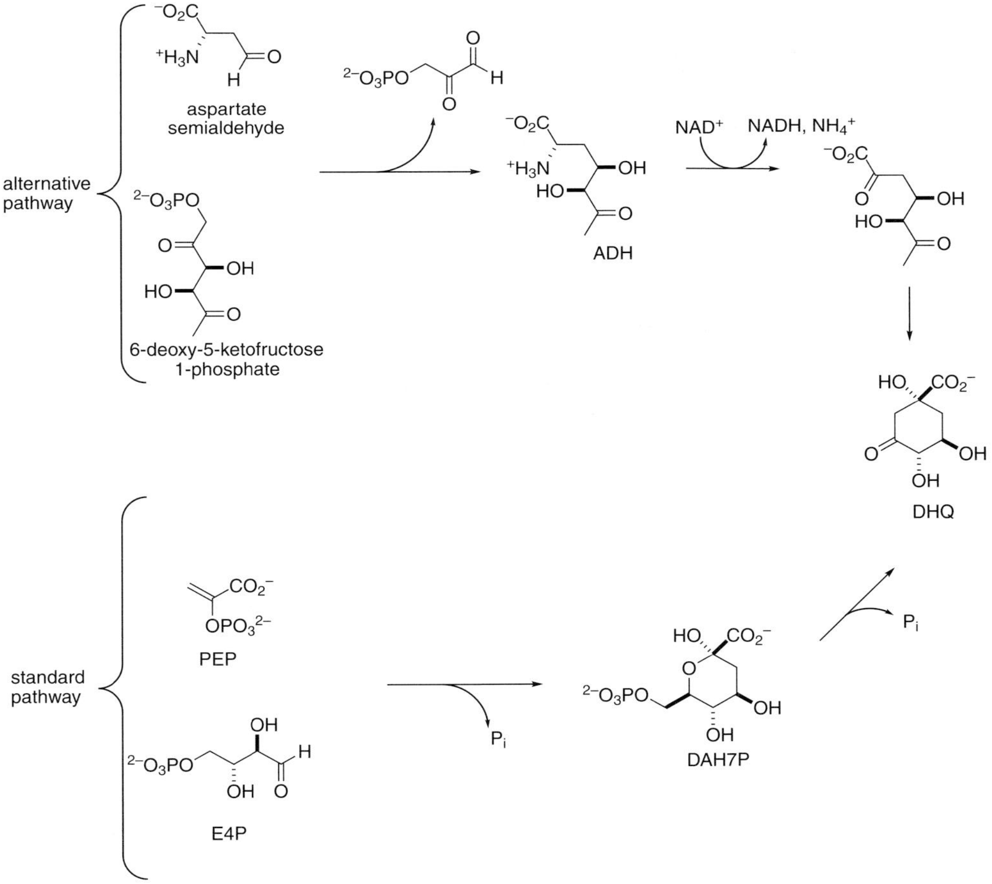

**Figure 1.59** Alternative routes for aromatic amino acid biosynthesis. Dehydroquinate synthase II catalyzes the oxidative deamination of ADH and its cyclization to form DHQ, which then feeds into the standard shikimate pathway.

steps. The first of these is catalyzed by a lysine-dependent, Schiff base-utilizing, type I aldolase. This enzyme, 2-amino-3,7-dideoxy-D-*threo*-hept-6-ulosonic acid (ADH) synthase, catalyses a transaldolase reaction between 6-deoxy-5-ketofructose 1-phosphate and ASA (a biosynthetic precursor to threonine, methionine, and lysine). The structure of this enzyme has revealed the likely active site residues and allowed the mechanism of this reaction to be predicted [242].

Following the production of ADH, the DHQ is produced in an oxidative deamination reaction that requires $NAD^+$. In contrast to DHQ synthase where oxidation is transient, an equivalent of $NAD^+$ is required in this transformation. Analogous to the DHQ synthase reaction that takes place as part of the canonical shikimate pathway, stereospecific intramolecular cyclization is required to produce DHQ. The discovery of this alternate route to DHQ in these organisms highlights how different strategies can be employed and different chemistries can evolve to achieve the same overall solution.

The third step in the pathway towards chorismate is catalyzed by dehydroquinase. There are two distinct types of dehydroquinase that operate in shikimate pathway metabolism that have evolved convergently and utilize different mechanisms for the dehydration reaction [243]. Whereas the type I enzyme catalyzes a reaction with overall *syn* elimination, the type II enzymes catalyze *anti* elimination in which the more acidic hydrogen is lost (Figure 1.60) [244]. Type I enzymes proceed via a Schiff base mechanism [245, 246]. The covalent attachment of substrate to the enzymes enables conformational change promoting the *syn* elimination to occur by increasing the acidity of the *pro-(S)* hydrogen. The reaction catalyzed by the type II enzyme proceeds via a standard $E1_{cb}$ mechanism, with the loss of the more acidic *pro-(R)* hydrogen. Type II enzymes operate in the shikimate pathway of some significant human pathogens and there has been considerable attention on the development of inhibitors. Several inhibitors with nanomolar inhibition constants have been described [247].

The fourth step of the shikimate pathway is the NADPH-mediated reduction of dehydroshikimate catalyzed by shikimate dehydrogenase [248, 249]. Structures of several shikimate dehydrogenases have been determined with the ternary shikimate/NADPH/enzyme complex of the shikimate dehydrogenase from *Staphylococcus epidermidis* revealing clearly the interaction of substrates with the enzyme, and

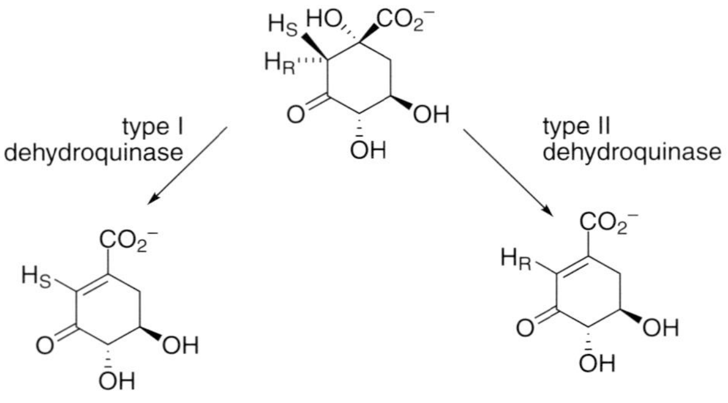

**Figure 1.60** Opposite stereochemical pathways catalyzed by the two types of dehydroquinase.

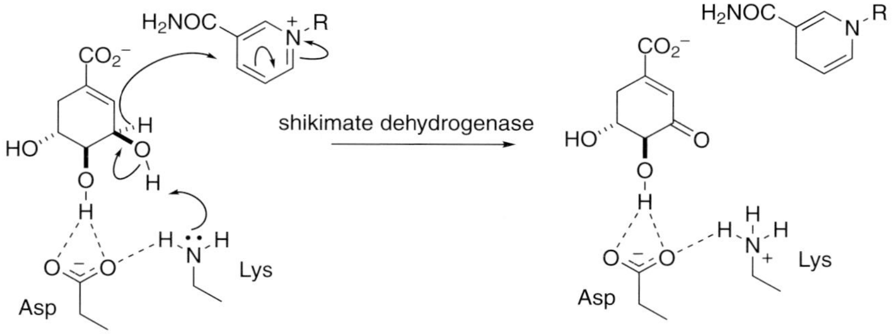

Figure 1.61 Reaction catalyzed by shikimate dehydrogenase.

identifying active-site lysine and aspartate residues as a key catalytic diad (Figure 1.61) [250–252].

In the fifth step of the pathway, shikimate becomes selectively phosphorylated on the C3 hydroxyl group to give shikimate 3-phosphate. This phosphoryl transfer reaction uses ATP as a cosubstrate and is catalyzed by shikimate kinase. Some organisms such as *E. coli* express two shikimate kinases that have quite different substrate affinities, while other organisms such as *M. tuberculosis* and *Helicobacter pylori* have a single enzyme responsible for catalyzing this transformation [253, 254].

The penultimate step of the pathway, catalyzed by enolpyruvylshikimate 3-phosphate synthase (EPSP synthase), has attracted a large amount of interest, as this enzyme is the target of the herbicide glyphosate [*N*-(phosphonomethyl)glycine] (Figure 1.62). The mechanism of the reaction, a stereospecific addition–elimination,

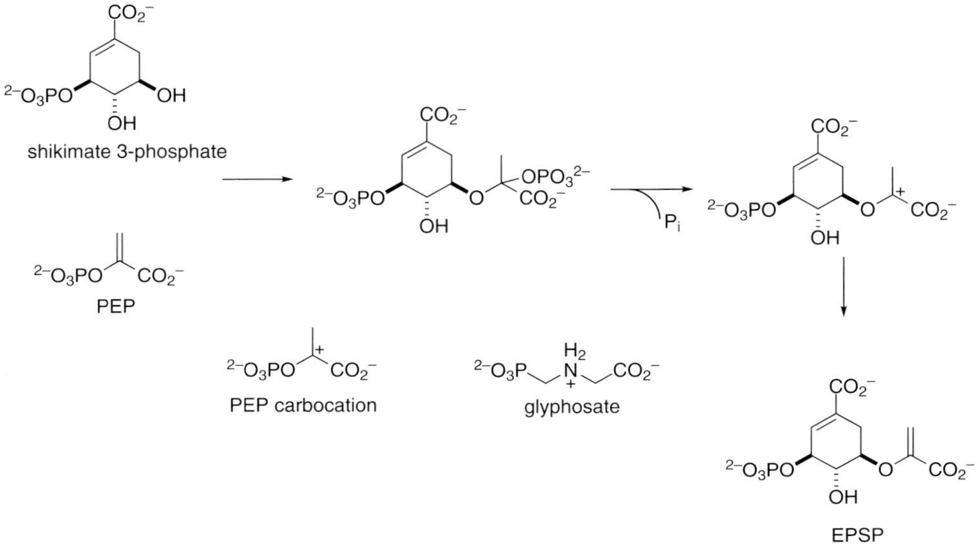

Figure 1.62 Reaction catalyzed by EPSP synthase.

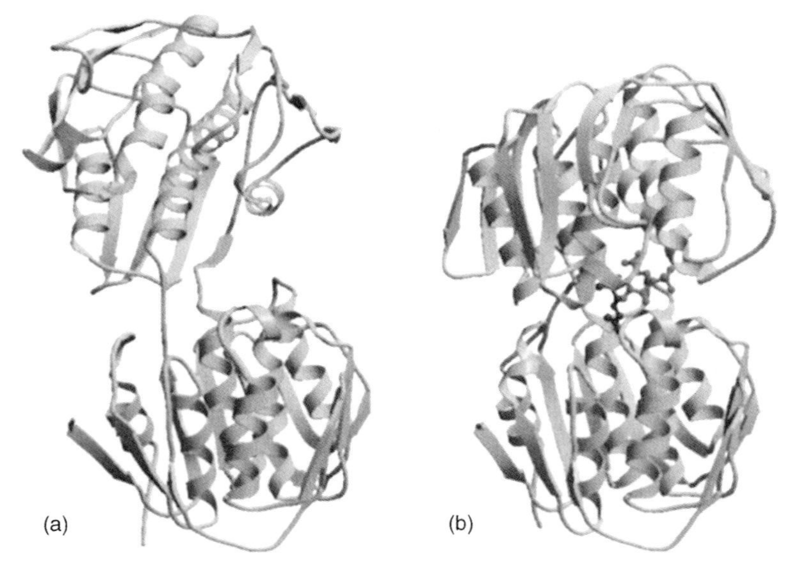


**Figure 1.63** EPSP synthase in open (unliganded, PDB ID: 1EPS) and closed forms (liganded). (From [257].)

has been well studied [255, 256]. The active site is found in a cleft between two domains and during the catalytic cycle there is extensive domain movement (Figure 1.63). Glyphosate inhibits by mimicking the cationic transition state associated with the protonated PEP substrate (Figure 1.64) [257]. Transition state analogs that include both PEP and shikimate 3-phosphate functionality are particularly

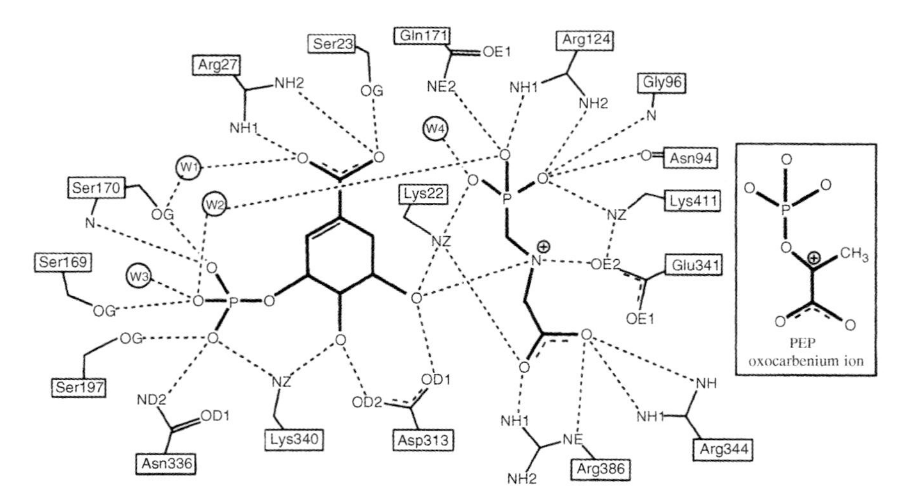


**Figure 1.64** Interactions of the glyphosate and shikimate 3-phosphate with EPSP synthase. (From [257].)

potent, but intriguingly show different abilities to inhibit enzymes that are glyphosate-tolerant to those which are sensitive to this inhibitor [256].

Chorismate synthase catalyzes the *anti* 1,4-elimination of phosphate and the C6 *pro-*(*R*) hydrogen, in the last step of the shikimate pathway. The mechanism of this reaction is intriguing, as the reaction, which is overall redox neutral, requires a reduced flavin mononucleotide – a cofactor associated with redox chemistry [258–260]. Some enzymes such as the chorismate synthases from *Neurospora crassa* and *S. cerevisiae* are bifunctional with both chorismate synthase and flavin reductase activity. The reduced flavin is predicted to help catalyze the reaction by a transient transfer of an electron, and by acting as a base to remove the *pro-*(*R*) hydrogen from EPSP (Figure 1.65). Various active site residues, notably histidine, are available to support this role [261, 262].

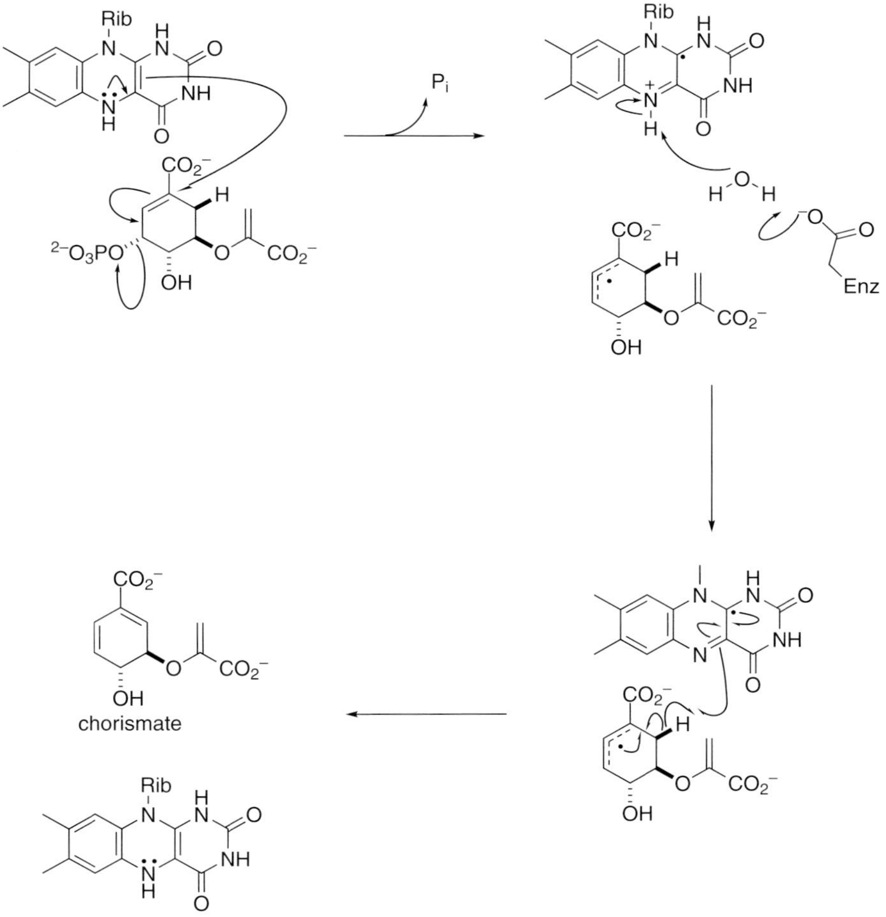

**Figure 1.65** Proposed reaction mechanism for chorismate synthase.

### 1.8.3
### Biosynthesis of Tryptophan, Phenylalanine, and Tyrosine from Chorismate

From chorismate the pathway for the synthesis of aromatic amino acids branches to produce either anthranilate for tryptophan biosynthesis or prephenate, the precursor to both phenylalanine and tyrosine. Chorismate is also the precursor to a range of other aromatic metabolites (Figure 1.66).

#### 1.8.3.1 **Tryptophan Biosynthesis**

Anthranilate synthase catalyses the first committed step in the biosynthesis of tryptophan from chorismate [263–265]. This enzyme is one of a group of chorismate utilizing enzymes that are predicted to share many mechanistic features and are likely to have common evolutionary origins (Figure 1.67) [266–268]. The first step is 1,4-nucleophilic substitution by ammonia. This is followed by elimination of pyruvate. The ammonia used in this reaction is generated *in situ* by a glutamine amidotransferase.

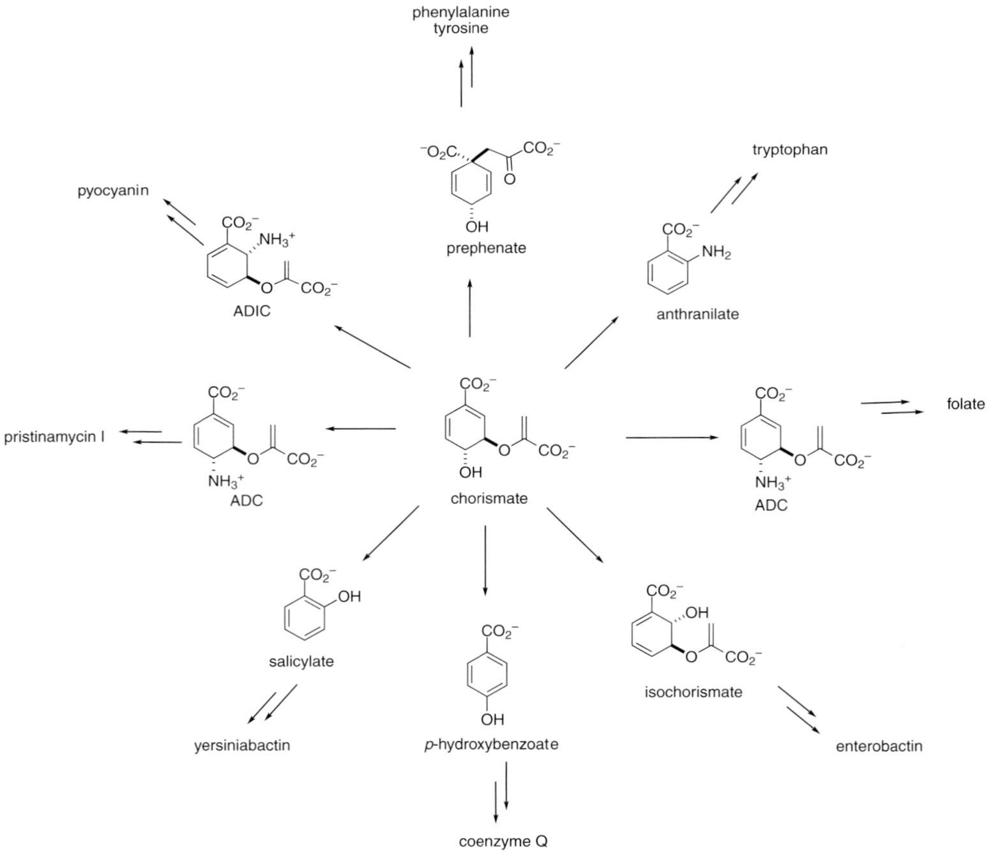

**Figure 1.66** Multiple end-products generated from chorismate.

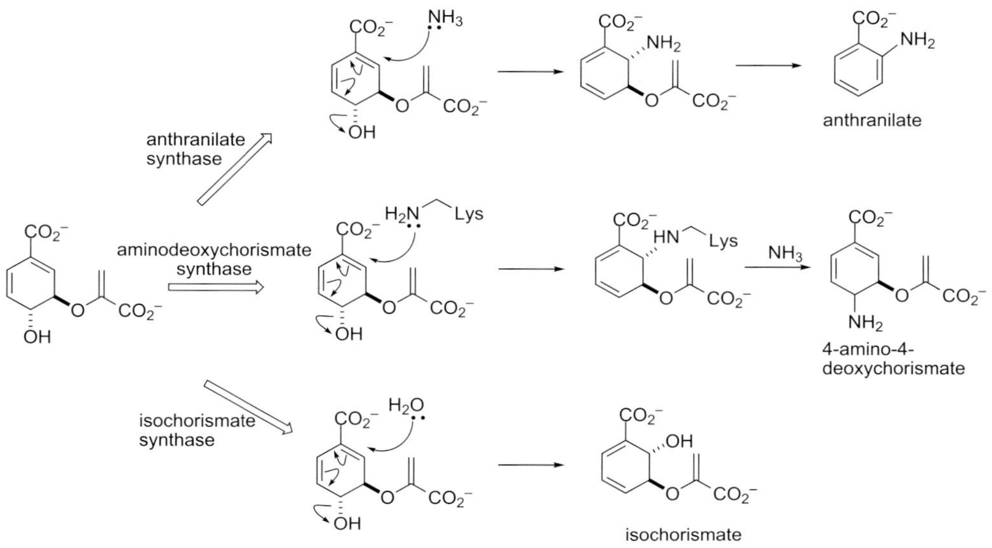

**Figure 1.67** Related enzymatic reactions using chorismate as a substrate.

The indole ring of tryptophan is created by the next three steps of the pathway (Figure 1.68). A ribose moiety is first tethered to anthranilate, in a reaction catalyzed by phosphoribosyl anthranilate synthase [269, 270]. Following isomerization of phosphoribosyl anthranilate involving an Amadori rearrangement (Figure 1.69), the indole ring is constructed by intramolecular electrophilic substitution. This reaction chemistry is analogous to that observed for histidine biosynthesis, as described below [271, 272].

The final step in tryptophan production is catalyzed by the tryptophan synthase complex [273]. This enzyme has attracted a great deal of attention and been extensively reviewed as the one of the first enzymes to be discovered that catalyzes two distinct reactions at independent active sites (in the α- and β-domains) that are connected via an internal tunnel [274]. This was the first well-documented example of substrate channeling in amino acid biosynthesis. The reaction chemistry is relatively straightforward; glyceraldehyde 3-phosphate is eliminated in the α-tryptophan synthase reaction, producing indole, which is channeled to the second active site (Figure 1.70). In the PLP-dependent β-reaction, serine undergoes nucleophilic displacement of water by indole via an elimination–addition mechanism typical of β-substitution reactions of PLP (Box 1.3). The nett result is the generation of tryptophan (Figure 1.68).

#### 1.8.3.2 **Phenylalanine and Tyrosine Biosynthesis**

Prephenate is formed by the reaction catalyzed by chorimate mutase (Figure 1.71) [275]. This is the only known example of a Claisen rearrangement

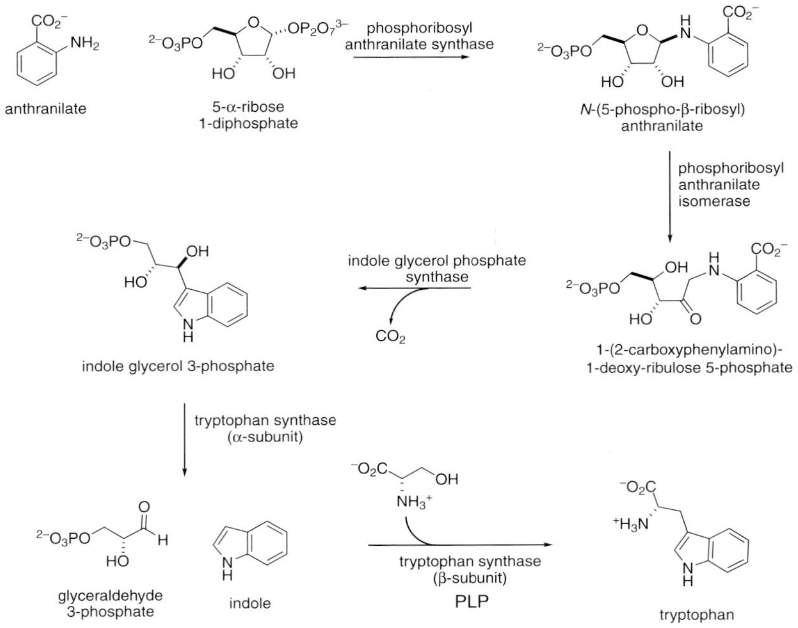

**Figure 1.68** Biosynthesis of tryptophan from anthranilate.

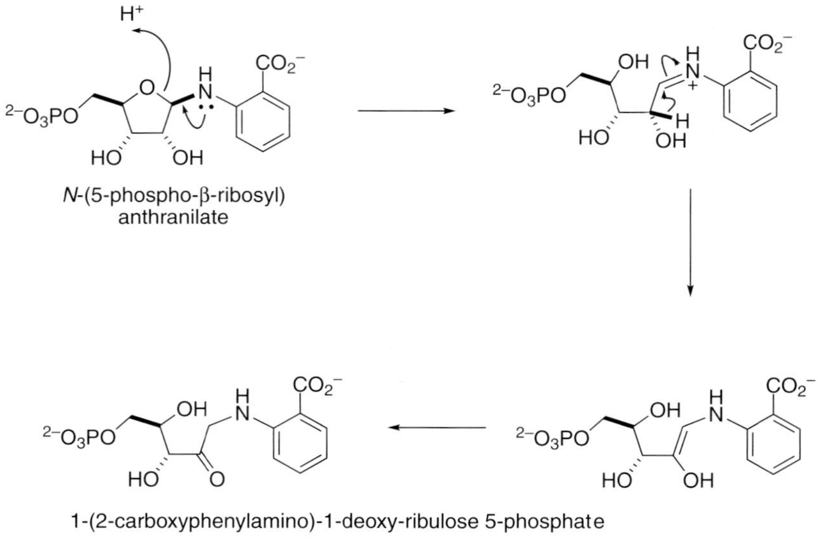

**Figure 1.69** Amadori rearrangement catalyzed by indole glycerol phosphate synthase.

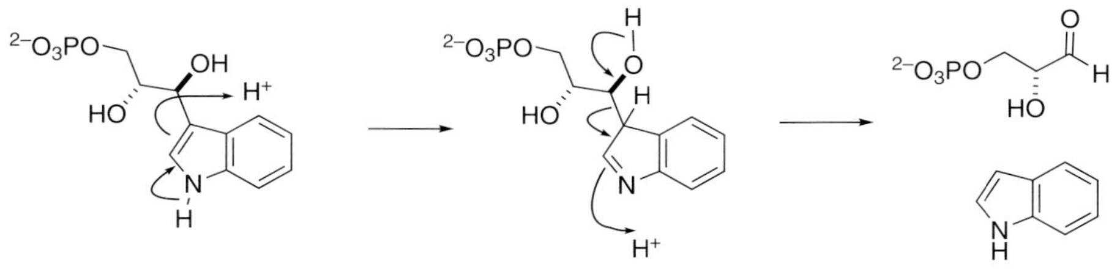

Figure 1.70 Generation of indole via a retro-aldol reaction catalyzed by the α-subunit of tryptophan synthase.

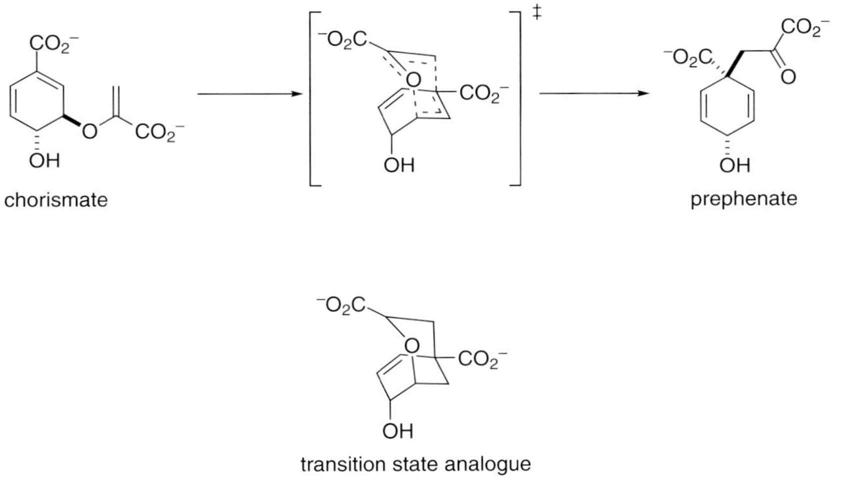

chorismate

prephenate

transition state analogue

Figure 1.71 Reaction catalyzed by chorismate mutase.

reaction in primary metabolism and one of the few enzyme-catalyzed pericyclic reactions. Chorismate mutase can be effectively inhibited with a transition state inhibitor, and the activity of the enzyme is often subject to inhibition by tyrosine and/or phenylalanine as part of regulation of the pathway.

From prephenate there are a number of alternative pathways to aromatic metabolites (Figure 1.72). Tyrosine is produced by a combination of oxidation and transamination reactions, whereas phenylalanine is produced by dehydration and transamination. As the order of these reactions can vary there are usually two pathways for tyrosine and phenylalanine generation [276–279].

### 1.8.4 Histidine Biosynthesis

Histidine is synthesized in eight steps from precursors phosphoribosyl pyrophosphate (PRPP) and ATP [280, 281]. The reaction chemistry for the formation of

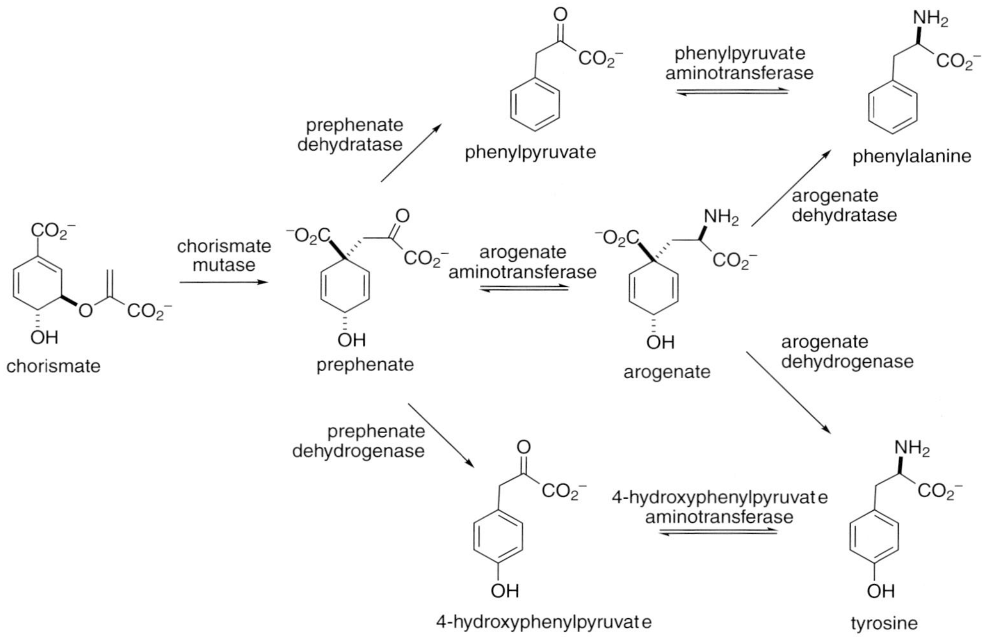

Figure 1.72 Pathways to phenylalanine and tyrosine.

histidine has distinct parallels with that of tryptophan biosynthesis. This relationship is apparent from the first step of the pathway with the reaction catalyzed by ATP-PRPP transferase (Figure 1.73) being analogous to the first step in tryptophan biosynthesis.

In many organisms the next two steps of the pathway – loss of diphosphate and hydrolysis of the adenine ring – are catalyzed by a single bifunctional protein, HisI

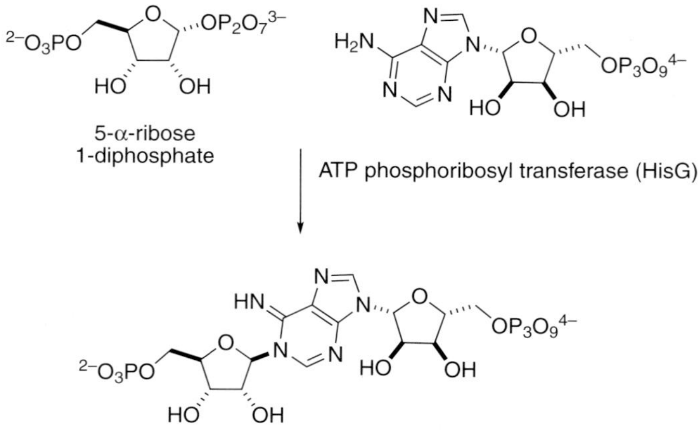

Figure 1.73 First step of histidine biosynthesis.

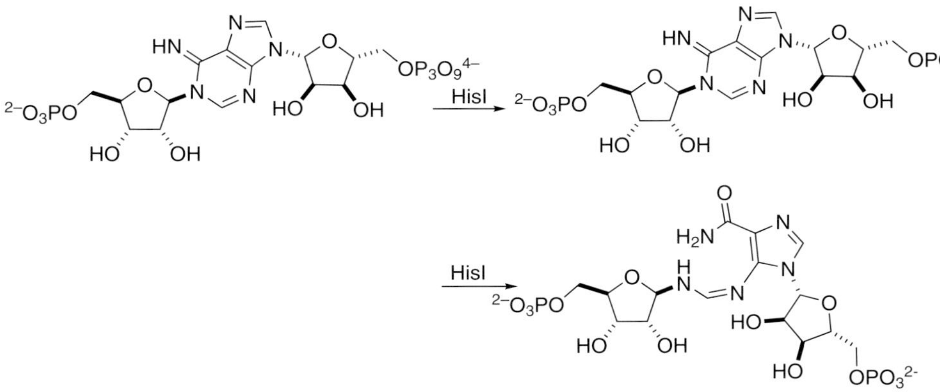

**Figure 1.74** Two sequential reactions catalyzed by HisI.

(Figure 1.74). However, there are cases in Archaea and proteobacteria where there are two distinct proteins responsible for these transformations. It has been speculated that the genes giving rise to these enzyme activities were originally separate entities and gene fusion was part of their evolution [282–285].

Following hydrolysis of the adenine ring, the ribofuranose ring is opened by way of the Amadori rearrangement (Figure 1.75), again in a step analogous to the reaction catalyzed by phosphoribosyl anthranilate isomerase as part of tryptophan biosynthesis [286–288]. The overlap between these reactions of tryptophan and histidine biosynthesis is more than just common reaction chemistry. Gene complementation studies in *Streptomyces coelicolor* have shown that a single isomerase operates in both tryptophan and histidine pathways, with dual substrate specificity.

Following the Amadori rearrangement the indole ring is formed [289, 290]. This step also is particularly interesting from a molecular evolution perspective. Ammonia is required for this transformation and this is provided by the glutaminase, HisF. This step also leads to the generation of the purine intermediate aminoimidazole carboxamide ribonucleotide.

The final steps of histidine biosynthesis involve transformation of the glycerol phosphate tail of the intermediate imidazole glycerol 3-phosphate (Figure 1.76). The first step is a dehydration reaction catalyzed by imidazole glycerol phosphate dehydratase (HisB) [291, 292]. As with other enzyme activities in this pathway, this reaction is catalyzed by an enzyme that forms one part of a bifunctional enzyme in many organisms. The other function of the protein is to catalyze the dephosphorylation reaction in a later step in the pathway. Biochemical studies indicate that these two enzyme activities operate independently and therefore that the bifunctional protein is likely to have arisen by gene fusion [285].

The intervening step between dehydration and dephosphorylation of the glycerol phosphate side-chain is catalyzed by the PLP-dependent histidinol phosphate aminotransferase, HisC [293–295]. The final steps for histidine biosynthesis are the oxidation of the primary alcohol functionality to the carboxylate, in two sequential $NAD^+$-dependent oxidations.

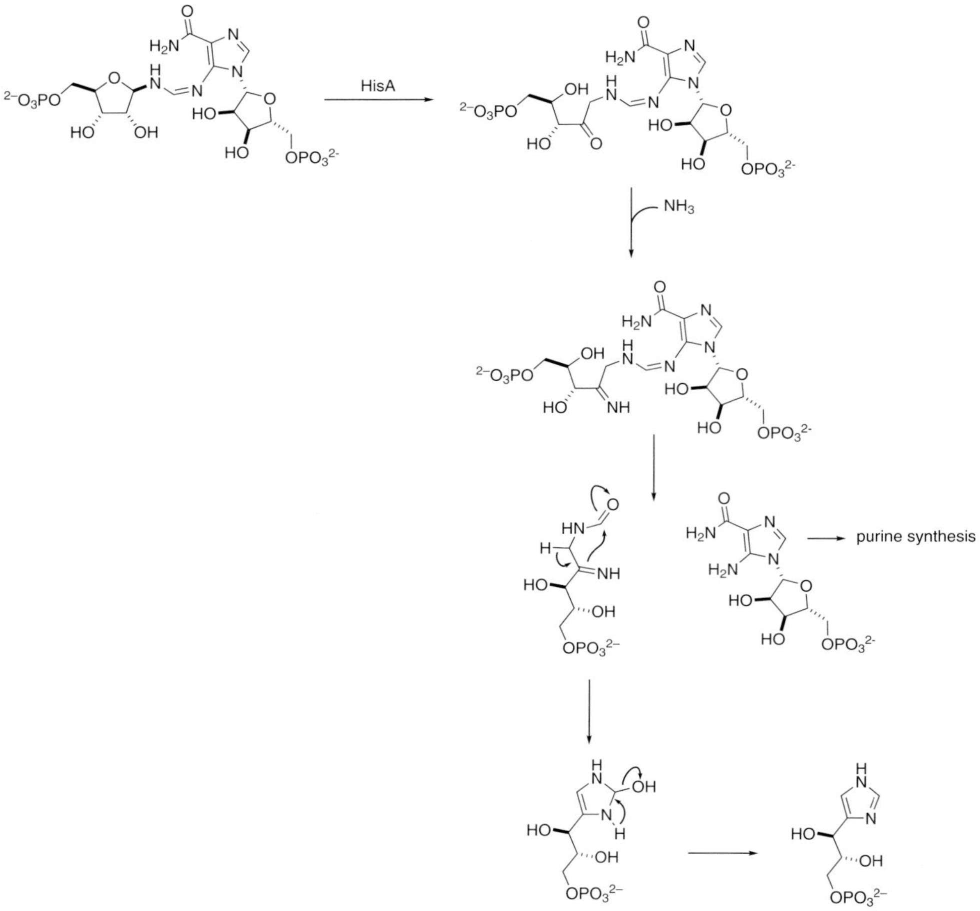

Figure 1.75 Amadori rearrangement and indole ring formation (catalyzed by HisA) reactions of histidine biosynthesis. This step requires ammonia generated by glutaminase, HisF.

## 1.9 Conclusions

This is an exciting time for research on amino acid biosynthesis. Genome projects and bioinformatics are providing rich details on metabolic diversity and the tailoring of metabolism to particular contexts. Structural biology is providing ever more three-dimensional detail on biosynthetic enzymes that underpins continued insights into enzyme catalysis and provides critical information for the design of synthetic inhibitors of critical biosynthetic enzymes of target organisms. Furthermore, evolutionary studies are providing new catalysts for interesting chemical transforma-

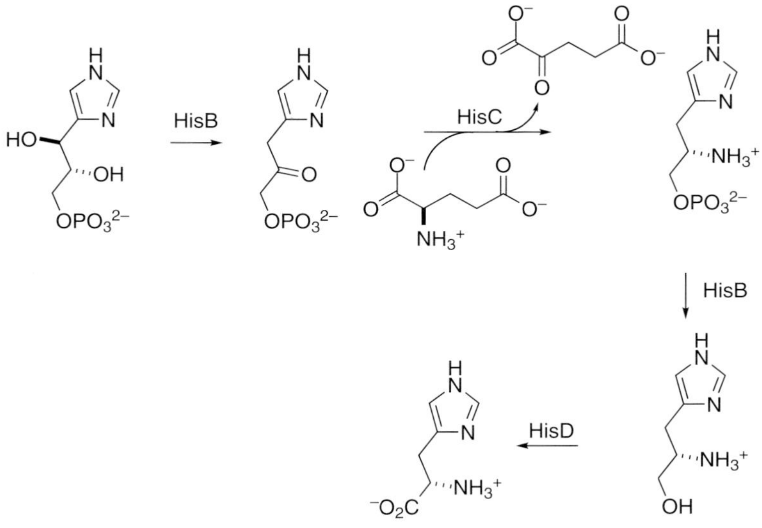

Figure 1.76 Final steps in the synthesis of histidine.

tions as well as new insights into the origins and development of this fascinating area of metabolism.

## References


1 Garavelli, J.S. (2004) The RESID database of protein modifications as a resource and annotation tool. *Proteomics*, 4, 1527–1533.

2 Copley, R.R. and Bork, P. (2000) Homology among (beta alpha)$_8$ barrels: implications for the evolution of metabolic pathways. *Journal of Molecular Biology*, **303**, 627–640.

3 Woese, C.R., Kandler, O., and Wheelis, M.L. (1990) Towards a natural system of organisms: proposal for the domains Archaea, Bacteria, and Eucarya. *Proceedings of the National Academy of Sciences of the United States of America*, **87**, 4576–4579.

4 Nishida, H. (2001) Evolution of amino acid biosynthesis and enzymes with broad substrate specificity. *Bioinformatics*, **17**, 1224–1225.

5 Shinivasan, V., Morowitz, H., and Smith, E. (2008) Essential amino acids, from LUCA to LUCY. *Complexity*, **13**, 8–9.

6 Hudson, R.C. and Daniel, R.M. (1993) l-Glutamate dehydrogenases – distribution, properties and mechanism. *Comparative Biochemistry and Physiology B*, **106**, 767–792.

7 Smith, T.J., Peterson, P.E., Schmidt, T., Fang, J., and Stanley, C.A. (2001) Structures of bovine glutamate dehydrogenase complexes elucidate the mechanism of purine regulation. *Journal of Molecular Biology*, **307**, 707–720.

8 Helling, R.B. (1994) Why does *Escherichia coli* have two primary pathways for synthesis of glutamate. *Journal of Bacteriology*, **176**, 4664–4668.

9 Woods, D.R. and Reid, S.J. (1993) Recent developments on the regulation and structure of glutamine-synthetase enzymes from selected bacterial groups. *FEMS Microbiology Reviews*, **11**, 273–283.

10 Stadtman, E.R. (2001) The story of glutamine synthetase regulation. *The*

*Journal of Biological Chemistry*, **276**, 44357–44364.

11 Reitzer, L. (2003) Nitrogen assimilation and global regulation in *Escherichia coli*. *Annual Review of Microbiology*, **57**, 155–176.

12 Liaw, S.H., Pan, C., and Eisenberg, D. (1993) Feedback inhibition of fully unadenylylated glutamine-synthetase from *Salmonella typhimurium* by glycine, alanine, and serine. *Proceedings of the National Academy of Sciences of the United States of America*, **90**, 4996–5000.

13 Zalkin, H. and Smith, J.L. (1998) Enzymes utilizing glutamine as an amide donor. *Advances in Enzymology*, **72**, 87–144.

14 van den Heuvel, R.H.H., Curti, B., Vanoni, M.A., and Mattevi, A. (2004) Glutamate synthase: a fascinating pathway from L-glutamine to L-glutamate. *Cellular and Molecular Life Sciences*, **61**, 669–681.

15 Lea, P.J. and Miflin, B.J. (2003) Glutamate synthase and the synthesis of glutamate in plants. *Plant Physiology and Biochemistry*, **41**, 555–564.

16 Vanoni, M.A. and Curti, B. (1999) Glutamate synthase: a complex iron–sulfur flavoprotein. *Cellular and Molecular Life Sciences*, **55**, 617–638.

17 Stabile, H., Curti, B., and Vanoni, M.A. (2000) Functional properties of recombinant *Azospirillum brasilense* glutamate synthase, a complex iron–sulfur flavoprotein. *European Journal of Biochemistry*, **267**, 2720–2730.

18 Massiere, F. and Badet-Denisot, M.-A. (1998) The mechanism of glutamine-dependent amidotransferases. *Cellular and Molecular Life Sciences*, **54**, 205–222.

19 Binda, C., Bossi, R.T., Wakatsuki, S., Arzt, S., Coda, A., Curti, B., Vanoni, M.A., and Mattevi, A. (2000) Cross-talk and ammonia channeling between active centers in the unexpected domain arrangement of glutamate synthase. *Structure*, **8**, 1299–1308.

20 Raushel, F.M., Thoden, J.B., and Holden, H.M. (2003) Enzymes with molecular tunnels. *Accounts of Chemical Research*, **36**, 539–548.

21 van den Heuvel, R.H.H., Ferrari, D., Bossi, R.T., Ravasio, S., Curti, B., Vanoni, M.A., Florencio, F.J., and Mattevi, A. (2002) Structural studies on the synchronization of catalytic centers in glutamate synthase. *The Journal of Biological Chemistry*, **277**, 24579–24583.

22 Dossena, L., Curti, B., and Vanoni, M.A. (2007) Activation and coupling of the glutaminase and synthase reaction of glutamate synthase is mediated by E1013 of the ferredoxin-dependent enzyme, belonging to loop 4 of the synthase domain. *Biochemistry*, **46**, 4473–4485.

23 Christen, P. and Mehta, P.K. (2001) From cofactor to enzymes. The molecular evolution of pyridoxal-5′-phosphate-dependent enzymes. *Chemical Record*, **1**, 436–447.

24 Hayashi, H. (1995) Pyridoxal enzymes: mechanistic diversity and uniformity. *Journal of Biochemistry*, **118**, 463–473.

25 John, R.A. (1995) Pyridoxal phosphate-dependent enzymes. *Biochimica et Biophysica Acta – Protein Structure and Molecular Enzymology*, **1248**, 81–96.

26 Jansonius, J.N. (1998) Structure, evolution and action of vitamin $B_6$-dependent enzymes. *Current Opinion in Structural Biology*, **8**, 759–769.

27 Schneider, G., Käck, H., and Lindqvist, Y. (2000) The manifold of vitamin B6 dependent enzymes. *Structure*, **8**, R1–R6.

28 Ford, G.C., Eichele, G., and Jansonius, J.N. (1980) Three-dimensional structure of a pyridoxal-phosphate-dependent enzyme, mitochondrial aspartate-aminotransferase. *Proceedings of the National Academy of Sciences of the United States of America*, **77**, 2559–2563.

29 Borisov, V.V., Borisova, S.N., Sosfenov, N.I., and Vainshtein, B.K. (1980) Electron density map of chicken heart cytosol aspartate transaminase at 3.5 Å resolution. *Nature*, **284**, 189–190.

30 Kirsch, J.F., Eichele, G., Ford, G.C., Vincent, M.G., Jansonius, J.N., Gehring, H., and Christen, P. (1984) Mechanism of action of aspartate-aminotransferase proposed on the basis of its spatial structure. *Journal of Molecular Biology*, **174**, 497–525.

31 Kochhar, S., Finlayson, W.L., Kirsch, J.F., and Christen, P. (1987) The stereospecific labilization of the C-4′ pro-*S* hydrogen of pyridoxamine 5′-phosphate is abolished

in (Lys258Ala) aspartate-aminotransferase. *The Journal of Biological Chemistry*, **262**, 11446–11448.

32 Dunathan, H.C. (1966) Conformation and reaction specificity in pyridoxal phosphate enzymes. *Proceedings of the National Academy of Sciences of the United States of America*, **55**, 712–716.

33 Matthews, R.G. and Drummond, J.T. (1990) Providing one-carbon units for biological methylations: mechanistic studies on serine hydroxymethyltransferase, methylenetetrahydrofolate reductase, and methyltetrahydrofolate-homocysteine methyltransferase. *Chemical Reviews*, **90**, 1275–1290.

34 Cronin, C.N. and Kirsch, J.F. (1988) Role of arginine-292 in the substrate-specificity of aspartate-aminotransferase as examined by site-directed mutagenesis. *Biochemistry*, **27**, 4572–4579.

35 Eliot, A.C. and Kirsch, J.F. (2004) Pyridoxal phosphate enzymes: Mechanistic, structural, and evolutionary considerations. *Annual Review of Biochemistry*, **73**, 383–415.

36 Okamoto, A., Nakai, Y., Hayashi, H., Hirotsu, K., and Kagamiyama, H. (1998) Crystal structures of *Paracoccus denitrificans* aromatic amino acid aminotransferase: a substrate recognition site constructed by rearrangement of hydrogen bond network. *Journal of Molecular Biology*, **280**, 443–461.

37 Okamoto, A., Ishii, S., Hirotsu, K., and Kagamiyama, H. (1999) The active site of *Paracoccus denitrificans* aromatic amino acid aminotransferase has contrary properties: flexibility and rigidity. *Biochemistry*, **38**, 1176–1184.

38 Onuffer, J.J. and Kirsch, J.F. (1995) Redesign of the substrate specificity of *Escherichia coli* aspartate aminotransferase to that of *Escherichia coli* tyrosine aminotransferase by homology modeling and site-directed mutagenesis. *Protein Science*, **4**, 1750–1757.

39 Malashkevich, V.N., Onuffer, J.J., Kirsch, J.F., and Jansonius, J.N. (1995) Alternating arginine-modulated substrate specificity in an engineered tyrosine aminotransferase. *Nature Structural Biology*, **2**, 548–553.

40 Penning, T.M. and Jez, J.M. (2001) Enzyme redesign. *Chemical Reviews*, **101**, 3027–3046.

41 Jäckel, C., Kast, P., and Hilvert, D. (2008) Protein design by directed evolution. *Annual Review of Biophysics*, **37**, 153–173.

42 Rothman, S.C. and Kirsch, J.F. (2003) How does an enzyme evolved *in vitro* compare to naturally occurring homologs possessing the targeted function? Tyrosine aminotransferase from aspartate aminotransferase. *Journal of Molecular Biology*, **327**, 593–608.

43 Yano, T., Oue, S., and Kagamiyama, H. (1998) Directed evolution of an aspartate aminotransferase with new substrate specificities. *Proceedings of the National Academy of Sciences of the United States of America*, **95**, 5511–5515.

44 Oue, S., Okamoto, A., Yano, T., and Kagamiyama, H. (1999) Redesigning the substrate specificity of an enzyme by cumulative effects of the mutations of non-active site residues. *The Journal of Biological Chemistry*, **274**, 2344–2349.

45 Oue, S., Okamoto, A., Yano, T., and Kagamiyama, H. (2000) Cocrystallization of a mutant aspartate aminotransferase with a C5-dicarboxylic substrate analog: structural comparison with the enzyme–C4-dicarboxylic analog complex. *Journal of Biochemistry*, **127**, 337–343.

46 Ura, H., Harata, K., Matsui, I., and Kuramitsu, S. (2001) Temperature dependence of the enzyme–substrate recognition mechanism. *Journal of Biochemistry*, **129**, 173–178.

47 Matsui, I., Matsui, E., Sakai, Y., Kikuchi, H., Kawarabayasi, Y., Ura, H., Kawaguchi, S., Kuramitsu, S., and Harata, K. (2000) The molecular structure of hyperthermostable aromatic aminotransferase with novel substrate specificity from *Pyrococcus horikoshii*. *The Journal of Biological Chemistry*, **275**, 4871–4879.

48 Goto, M., Miyahara, I., Hayashi, H., Kagamiyama, H., and Hirotsu, K. (2003) Crystal structures of branched-chain amino acid aminotransferase complexed with glutamate and glutarate: true reaction intermediate and double substrate recognition of the enzyme. *Biochemistry*, **42**, 3725–3733.

49 Schlösser, T., Gätgens, C., Weber, U., and Stahmann, K.-P. (2004) Alanine: glyoxylate aminotransferase of *Saccharomyces cerevisiae*-encoding gene AGX1 and metabolic significance. *Yeast*, **21**, 63–73.

50 Takada, Y. and Noguchi, T. (1985) Characteristics of alanine glyoxylate aminotransferase from *Saccharomyces cerevisiae*, a regulatory enzyme in the glyoxylate pathway of glycine and serine biosynthesis from tricarboxylic acid-cycle intermediates. *Biochemical Journal*, **231**, 157–163.

51 Monschau, N., Stahmann, K.-P., Sahm, H., McNeil, J.B., and Bognar, A.L. (1997) Identification of *Saccharomyces cerevisiae* GLY1 as a threonine aldolase: a key enzyme in glycine biosynthesis. *FEMS Microbiology Letters*, **150**, 55–60.

52 Liu, J.-Q., Nagata, S., Dairi, T., Misono, H., Shimizu, S., and Yamada, H. (1997) The GLY1 gene of *Saccharomyces cerevisiae* encodes a low-specific L-threonine aldolase that catalyzes cleavage of L-allo-threonine and L-threonine to glycine. *European Journal of Biochemistry*, **245**, 289–293.

53 Melendez-Hevia, E., de Paz-Lugo, P., Cornish-Bowden, A., and Cardenas, M.L. (2009) A weak link in metabolism: the metabolic capacity for glycine biosynthesis does not satisfy the need for collagen synthesis. *Journal of Biosciences*, **34**, 853–872.

54 Renwick, S.B., Snell, K., and Baumann, U. (1998) The crystal structure of human cytosolic serine hydroxymethyltransferase: a target for cancer chemotherapy. *Structure*, **6**, 1105–1116.

55 Rajaram, V., Bhavan, B.S., Kaul, P., Prakash, V., Rao, N.A., Savithri, H.S., and Murthyl, M.R.N. (2007) Structure determination and biochemical studies on *Bacillus stearothermophilus* E53Q serine hydroxymethyltransferase and its complexes provide insights on function and enzyme memory. *FEBS Journal*, **274**, 4148–4160.

56 Bhavani, S., Trivedi, V., Jala, V.R., Subramanya, H.S., Kaul, P., Prakash, V., Rao, N.A., and Savithri, H.S. (2005) Role of Lys-226 in the catalytic mechanism of *Bacillus stearothermophilus* serine hydroxymethyltransferase – crystal structure and kinetic studies. *Biochemistry*, **44**, 6929–6937.

57 Szebenyi, D.M.E., Musayev, F.N., di Salvo, M.L., Safo, M.K., and Schirch, V. (2004) Serine hydroxymethyltransferase: role of Glu75 and evidence that serine is cleaved by a retroaldol mechanism. *Biochemistry*, **43**, 6865–6876.

58 Trivedi, V., Gupta, A., Jala, V.R., Saravanan, P., Rao, G.S.J., Rao, N.A., Savithri, H.S., and Subramanya, H.S. (2002) Crystal structure of binary and ternary complexes of serine hydroxymethyltransferase from *Bacillus stearothermophilus* – insights into the catalytic mechanism. *The Journal of Biological Chemistry*, **277**, 17161–17169.

59 Stauffer, G.V. (1996) Biosynthesis of serine, glycine, and one-carbon unit, in *Escherichia coli and Salmonella: Cellular and Molecular Biology*, vol. I (eds F.C. Neidhardt, R. CurtissIII, J.L. Ingraham, E.C.C. Lin, K.B. Low, B. Magasanik, W.S. Reznikoff, M. Riley, M. Schaechter, and H.E. Umbarger), 2nd edn, American Society for Microbiology, Washington, DC, pp. 506–513.

60 Burton, R.L., Chen, S., Xu, X.L., and Grant, G.A. (2009) Transient kinetic analysis of the interaction of L-serine with *Escherichia coli*D-3-phosphoglycerate dehydrogenase reveals the mechanism of V-type regulation and the order of effector binding. *Biochemistry*, **48**, 12242–12251.

61 Wang, W., Cho, H.S., Kim, R., Jancarik, J., Yokota, H., Nguyen, H.H., Grigoriev, I.V., Wemmer, D.E., and Kim, S.-H. (2002) Structural characterization of the reaction pathway in phosphoserine phosphatase: crystallographic "snapshots" of intermediate states. *Journal of Molecular Biology*, **319**, 421–431.

62 Singh, S.K., Yang, K., Karthikeyan, S., Huynh, T., Zhang, X.J., Phillips, M.A., and Zhang, H. (2004) The *thrH* gene product of *Pseudomonas aeruginosa* is a dual activity enzyme with a novel phosphoserine: homoserine phosphotransferase activity. *The Journal of Biological Chemistry*, **279**, 13166–13173.

63 Feldman-Salit, A., Wirtz, M., Hell, R., and Wade, R.C. (2009) A mechanistic model of

the cysteine synthase complex. *Journal of Molecular Biology*, **386**, 37–59.

64 Mino, K. and Ishikawa, K. (2003) Characterization of a novel thermostable *O*-acetylserine sulfhydrylase from *Aeropyrum pernix* K1. *Journal of Bacteriology*, **185**, 2277–2284.

65 Rabeh, W.M. and Cook, P.F. (2004) Structure and mechanism of *O*-acetylserine sulfhydrylase. *The Journal of Biological Chemistry*, **279**, 26803–26806.

66 Tai, C.-H. and Cook, P.F. (2001) Pyridoxal 5′-phosphate-dependent $\alpha,\beta$-elimination reactions: mechanism of *O*-acetylserine sulfhydrylase. *Accounts of Chemical Research*, **34**, 49–59.

67 Su, D., Hohn, M.J., Palioura, S., Sherrer, R.L., Yuan, J., Söll, D., and O'Donoghue, P. (2009) How an obscure archaeal gene inspired the discovery of selenocysteine biosynthesis in humans. *IUBMB Life*, **61**, 35–39.

68 Cole, S.T., Brosch, R., Parkhill, J., Garnier, T., Churcher, C., Harris, D., Gordon, S.V., Eiglmeier, K., Gas, S., Barry, C.E., Tekaia, F., Badcock, K., Basham, D., Brown, D., Chillingworth, T., Connor, R., Davies, R., Devlin, K., Feltwell, T., Gentles, S., Hamlin, N., Holroyd, S., Hornsby, T., Jagels, K., Krogh, A., McLean, J., Moule, S., Murphy, L., Oliver, K., Osborne, J., Quail, M.A., Rajandream, M.-A., Rogers, J., Rutter, S., Seeger, K., Skelton, J., Squares, R., Squares, S., Sulston, J.E., Taylor, K., Whitehead, S., and Barrell, B.G. (1998) Deciphering the biology of *Mycobacterium tuberculosis* from the complete genome sequence. *Nature*, **393** 537–544.

69 Schnell, R., Oehlmann, W., Singh, M., and Schneider, G. (2007) Structural insights into catalysis and inhibition of *O*-acetylserine sulfhydrylase from *Mycobacterium tuberculosis*:crystal structures of the enzyme $\alpha$-aminoacrylate intermediate and an enzyme–inhibitor complex. *The Journal of Biological Chemistry*, **282**, 23473–23481.

70 Park, J.-H., Dorrestein, P.C., Zhai, H., Kinsland, C., McLafferty, F.W., and Begley, T.P. (2003) Biosynthesis of the thiazole moiety of thiamin pyrophosphate (vitamin $B_1$). *Biochemistry*, **42**, 12430–12438.

71 Burns, K.E., Baumgart, S., Dorrestein, P.C., Zhai, H., McLafferty, F.W., and Begley, T.P. (2005) Reconstitution of a new cysteine biosynthetic pathway in *Mycobacterium tuberculosis*. *Journal of the American Chemical Society*, **127**, 11602–11603.

72 Ågren, D., Schnell, R., Oehlmann, W., Singh, M., and Schneider, G. (2008) Cysteine synthase (CysM) of *Mycobacterium tuberculosis* is an *O*-phosphoserine sulfhydrylase: evidence for an alternative cysteine biosynthesis pathway in Mycobacteria. *The Journal of Biological Chemistry*, **283**, 31567–31574.

73 Sauerwald, A., Zhu, W.H., Major, T.A., Roy, H., Palioura, S., Jahn, D., Whitman, W.B., Yates, J.R., Ibba, M., and Söll, D. (2005) RNA-dependent cysteine biosynthesis in archaea. *Science*, **307**, 1969–1972.

74 Cone, J.E., Del Rio, M.R., Davis, J.N., and Stadtman, T.C. (1976) Chemical characterization of selenoprotein component of clostridial glycine reductase: identification of selenocysteine as the organoselenium moiety. *Proceedings of the National Academy of Sciences of the United States of America*, **73**, 2659–2663.

75 Leinfelder, W., Zehelein, E., Mandrandberthelot, M.A., and Bock, A. (1988) Gene for a novel transfer-RNA species that accepts L-serine and cotranslationally inserts selenocysteine. *Nature*, **331**, 723–725.

76 Mullins, L.S., Hong, S.-B., Gibson, G.E., Walker, H., Stadtman, T.C., and Raushel, F.M. (1997) Identification of a phosphorylated enzyme intermediate in the catalytic mechanism for selenophosphate synthetase. *Journal of the American Chemical Society*, **119**, 6684–6685.

77 Midelfort, C.F. and Rose, I.A. (1976) A stereochemical method for detection of ATP terminal phosphate transfer in enzymatic reactions. Glutamine synthetase. *The Journal of Biological Chemistry*, **251**, 5881–5887.

**78** Forchhammer, K. and Böck, A. (1991) Selenocysteine synthase from *Escherichia coli*. Analysis of the reaction sequence. *The Journal of Biological Chemistry*, **266**, 6324–6328.

**79** Kernebeck, T., Lohse, A.W., and Grötzinger, J. (2001) A bioinformatical approach suggests the function of the autoimmune hepatitis target antigen soluble liver antigen/liver pancreas. *Hepatology*, **34**, 230–233.

**80** Ganichkin, O.M., Xu, X.-M., Carlson, B.A., Mix, H., Hatfield, D.L., Gladyshev, V.N., and Wahl, M.C. (2008) Structure and catalytic mechanism of eukaryotic selenocysteine synthase. *The Journal of Biological Chemistry*, **283**, 5849–5865.

**81** Araiso, Y., Palioura, S., Ishitani, R., Sherrer, R.L., O'Donoghue, P., Yuan, J., Oshikane, H., Domae, N., DeFranco, J., Söll, D., and Nureki, O. (2008) Structural insights into RNA-dependent eukaryal and archaeal selenocysteine formation. *Nucleic Acids Research*, **36**, 1187–1199.

**82** Xu, X.-M., Carlson, B.A., Mix, H., Zhang, Y., Saira, K., Glass, R.S., Berry, M.J., Gladyshev, V.N., and Hatfield, D.L. (2007) Biosynthesis of selenocysteine on its tRNA in eukaryotes. *PLoS Biology*, **5**, 96–105.

**83** Yuan, J., Palioura, S., Salazar, J.C., Su, D., O'Donoghue, P., Hohn, M.J., Cardoso, A.M., Whitman, W.B., and Söll, D. (2006) RNA-dependent conversion of phosphoserine forms selenocysteine in eukaryotes and archaea. *Proceedings of the National Academy of Sciences of the United States of America*, **103**, 18923–18927.

**84** Schmitt, E., Panvert, M., Blanquet, S., and Mechulam, Y. (1998) Crystal structure of methionyl-$tRNA_f^{Met}$ transformylase complexed with the initiator formylmethionyl-$tRNA_f^{Met}$. *The EMBO Journal*, **17**, 6819–6826.

**85** Curnow, A.W., Hong, K.-W., Yuan, R., Kim, S.-I., Martins, O., Winkler, W., Henkin, T.M., and Söll, D. (1997) Glu-$tRNA^{Gln}$ amidotransferase: a novel heterotrimeric enzyme required for correct decoding of glutamine codons during translation. *Proceedings of the National Academy of Sciences of the United States of America*, **94**, 11819–11826.

**86** Ibba, M. and Soll, D. (2000) Aminoacyl-tRNA synthesis. *Annual Review of Biochemistry*, **69**, 617–650.

**87** Humbert, R. and Simoni, R.D. (1980) Genetic and biochemical studies demonstrating a second gene coding for asparagine synthetase in *Escherichia coli*. *Journal of Bacteriology*, **142**, 212–220.

**88** Cedar, H. and Schwartz, J.H. (1969) Asparagine synthetase of *Escherichia coli* I. Biosynthetic role of the enzyme, purification, and characterization of the reaction products. *The Journal of Biological Chemistry*, **244**, 4112–4121.

**89** Tesson, A.R., Soper, T.S., Ciustea, M., and Richards, N.G.J. (2003) Revisiting the steady state kinetic mechanism of glutamine-dependent asparagine synthetase from *Escherichia coli*. *Archives of Biochemistry and Biophysics*, **413**, 23–31.

**90** Larsen, T.M., Boehlein, S.K., Schuster, S.M., Richards, N.G.J., Thoden, J.B., Holden, H.M., and Rayment, I. (1999) Three-dimensional structure of *Escherichia coli* asparagine synthetase B: a short journey from substrate to product. *Biochemistry*, **38**, 16146–16157.

**91** Richards, N.G.J. and Kilberg, M.S. (2006) Asparagine synthetase chemotherapy. *Annual Review of Biochemistry*, **75**, 629–654.

**92** Koizumi, M., Hiratake, J., Nakatsu, T., Kato, H., and Oda, J. (1999) A potent transition-state analogue inhibitor of *Escherichia coli* asparagine synthetase A. *Journal of the American Chemical Society*, **121**, 5799–5800.

**93** Zhang, Y., Thiele, I., Weekes, D., Li, Z., Jaroszewski, L., Ginalski, K., Deacon, A.M., Wooley, J., Lesley, S.A., Wilson, I.A., Palsson, B., Osterman, A., and Godzik, A. (2009) Three-dimensional structural view of the central metabolic network of *Thermotoga maritima*. *Science*, **325**, 1544–1549.

**94** Curien, G., Biou, V., Mas-Droux, C., Robert-Genthon, M., Ferrer, J.-L., and Dumas, R. (2008) Amino acid biosynthesis: new architectures in allosteric enzymes. *Plant Physiology and Biochemistry*, **46**, 325–339.

**95** Azevedo, R.A., Lancien, M., and Lea, P.J. (2006) The aspartic acid metabolic pathway, an exciting and essential pathway in plants. *Amino Acids*, **30**, 143–162.

96 Liu, X., Pavlovsky, A.G., and Viola, R.E. (2008) The structural basis for allosteric inhibition of a threonine-sensitive aspartokinase. *The Journal of Biological Chemistry*, **283**, 16216–16225.

97 Hadfield, A., Shammas, C., Kryger, G., Ringe, D., Petsko, G.A., Ouyang, J., and Viola, R.E. (2001) Active site analysis of the potential antimicrobial target aspartate semialdehyde dehydrogenase. *Biochemistry*, **40**, 14475–14483.

98 Blanco, J., Moore, R.A., and Viola, R.E. (2003) Capture of an intermediate in the catalytic cycle of L-aspartate-β-semialdehyde dehydrogenase. *Proceedings of the National Academy of Sciences of the United States of America*, **100**, 12613–12617.

99 Blanco, J., Moore, R.A., Kabaleeswaran, V., and Viola, R.E. (2003) A structural basis for the mechanism of aspartate-β-semialdehyde dehydrogenase from *Vibrio cholerae*. *Protein Science*, **12**, 27–33.

100 Holland, M.J. and Westhead, E.W. (1973) Purification and characterization of aspartic beta-semialdehyde dehydrogenase from yeast and purification of an isoenzyme of glyceraldehyde-3-phosphate dehydrogenase. *Biochemistry*, **12**, 2264–2270.

101 Geck, M.K. and Kirsch, J.F. (1999) A novel, definitive test for substrate channeling illustrated with the aspartate aminotransferase/malate dehydrogenase system. *Biochemistry*, **38**, 8032–8037.

102 James, C.L. and Viola, R.E. (2002) Production and characterization of bifunctional enzymes. Substrate channeling in the aspartate pathway. *Biochemistry*, **41**, 3726–3731.

103 DeLaBarre, B., Thompson, P.R., Wright, G.D., and Berghuis, A.M. (2000) Crystal structures of homoserine dehydrogenase suggest a novel catalytic mechanism for oxidoreductases. *Nature Structural Biology*, **7**, 238–244.

104 Jacques, S.L., Ejim, L.J., and Wright, G.D. (2001) Homoserine dehydrogenase from *Saccharomyces cerevisiae*: kinetic mechanism and stereochemistry of hydride transfer. *Biochimica et Biophysica Acta – Protein Structure and Molecular Enzymology*, **1544**, 42–54.

105 Huo, X. and Viola, R.E. (1996) Substrate specificity and identification of functional groups of homoserine kinase from *Escherichia coli*. *Biochemistry*, **35**, 16180–16185.

106 Mirza, I.A., Nazi, I., Korczynska, M., Wright, G.D., and Berghuis, A.M. (2005) Crystal structure of homoserine transacetylase from *Haemophilus influenzae* reveals a new family of α/β-hydrolases. *Biochemistry*, **44**, 15768–15773.

107 Zubieta, C., Arkus, K.A.J., Cahoon, R.E., and Jez, J.M. (2008) A single amino acid change is responsible for evolution of acyltransferase specificity in bacterial methionine biosynthesis. *The Journal of Biological Chemistry*, **283**, 7561–7567.

108 Laber, B., Gerbling, K.-P., Harde, C., Neff, K.-H., Nordhoff, E., and Pohlenz, H.-D. (1994) Mechanisms of interaction of *Escherichia coli* threonine synthase with substrates and inhibitors. *Biochemistry*, **33**, 3413–3423.

109 Laber, B., Maurer, W., Hanke, C., Gräfe, S., Ehlert, S., Messerschmidt, A., and Clausen, T. (1999) Characterization of recombinant *Arabidopsis thaliana* threonine synthase. *European Journal of Biochemistry*, **263**, 212–221.

110 Curien, G., Job, D., Douce, R., and Dumas, R. (1998) Allosteric activation of *Arabidopsis* threonine synthase by S-adenosylmethionine. *Biochemistry*, **37**, 13212–13221.

111 Thomazeau, K., Curien, G., Dumas, R., and Biou, V. (2001) Crystal structure of threonine synthase from *Arabidopsis thaliana*. *Protein Science*, **10**, 638–648.

112 Omi, R., Goto, M., Miyahara, I., Mizuguchi, H., Hayashi, H., Kagamiyama, H., and Hirotsu, K. (2003) Crystal structures of threonine synthase from *Thermus thermophilus* HB8 – conformational change, substrate recognition, and mechanism. *The Journal of Biological Chemistry*, **278**, 46035–46045.

113 Garrido-Franco, M., Ehlert, S., Messerschmidt, A., Marinkovic, S., Huber, R., Laber, B., Bourenkov, G.P., and

Clausen, T. (2002) Structure and function of threonine synthase from yeast. *The Journal of Biological Chemistry*, **277**, 12396–12405.
**114** Sassetti, C.M. and Rubin, E.J. (2003) Genetic requirements for mycobacterial survival during infection. *Proceedings of the National Academy of Sciences of the United States of America*, **100**, 12989–12994.
**115** Covarrubias, A.S., Högbom, M., Bergfors, T., Carroll, P., Mannerstedt, K., Oscarson, S., Parish, T., Jones, T.A., and Mowbray, S.L. (2008) Structural, biochemical, and *in vivo* investigations of the threonine synthase from *Mycobacterium tuberculosis*. *Journal of Molecular Biology*, **381**, 622–633.
**116** Clausen, T., Huber, R., Prade, L., Wahl, M.C., and Messerschmidt, A. (1998) Crystal structure of *Escherichia coli* cystathionine γ-synthase at 1.5 Å resolution. *The EMBO Journal*, **17**, 6827–6838.
**117** Kong, Y., Wu, D., Bai, H., Han, C., Chen, J., Chen, L., Hu, L., Jiang, H., and Shen, X. (2008) Enzymatic characterization and inhibitor discovery of a new cystathionine γ-synthase from *Helicobacter pylori*. *Journal of Biochemistry*, **143**, 59–68.
**118** Clausen, T., Huber, R., Laber, B., Pohlenz, H.-D., and Messerschmidt, A. (1996) Crystal structure of the pyridoxal-5′-phosphate dependent cystathionine β-lyase from *Escherichia coli* at 1.83 Å. *Journal of Molecular Biology*, **262**, 202–224.
**119** Goulding, C.W., Postigo, D., and Matthews, R.G. (1997) Cobalamin-dependent methionine synthase is a modular protein with distinct regions for binding homocysteine, methyltetrahydrofolate, cobalamin, and adenosylmethionine. *Biochemistry*, **36**, 8082–8091.
**120** Matthews, R.G., Koutmos, M., and Datta, S. (2008) Cobalamin-dependent and cobamide-dependent methyltransferases. *Current Opinion in Structural Biology*, **18**, 658–666.
**121** Gonzalez, J.C., Peariso, K., PennerHahn, J.E., and Matthews, R.G. (1996) Cobalamin-independent methionine synthase from *Escherichia coli*: a zinc metalloenzyme. *Biochemistry*, **35**, 12228–12234.
**122** Marco-Marin, C., Gil-Ortiz, F., Perez-Arellano, I., Cervera, J., Fita, I., and Rubio, V. (2007) A novel two-domain architecture within the amino acid kinase enzyme family revealed by the crystal structure of *Escherichia coli* glutamate 5-kinase. *Journal of Molecular Biology*, **367**, 1431–1446.
**123** Hu, C.A., Delauney, A.J., and Verma, D.P. (1992) A bifunctional enzyme (δ-1-pyrroline-5-carboxylate synthetase) catalyzes the first two steps in proline biosynthesis in plants. *Proceedings of the National Academy of Sciences of the United States of America*, **89**, 9354–9358.
**124** Turchetto-Zolet, A.C., Margis-Pinheiro, M., and Margis, R. (2009) The evolution of pyrroline-5-carboxylate synthase in plants: a key enzyme in proline synthesis. *Molecular Genetics and Genomics*, **281**, 87–97.
**125** Costilow, R.N. and Laycock, L. (1971) Ornithine cyclase (deaminating) – purification of a protein that converts ornithine to proline and definition of optimal assay conditions. *The Journal of Biological Chemistry*, **246**, 6655–6660.
**126** Graupner, M. and White, R.H. (2001) Methanococcus *jannaschii* generates L-proline by cyclization of L-ornithine. *Journal of Bacteriology*, **183**, 5203–5205.
**127** Nocek, B., Chang, C., Li, H., Lezondra, L., Holzle, D., Collart, F., and Joachimiak, A. (2005) Crystal structures of δ-1-pyrroline-5-carboxylate reductase from human pathogens *Neisseria meningitides* and *Streptococcus pyogenes*. *Journal of Molecular Biology*, **354**, 91–106.
**128** Caldovic, L. and Tuchman, M. (2003) N-Acetylglutamate and its changing role through evolution. *Biochemical Journal*, **372**, 279–290.
**129** Fernandez-Murga, M.L., Gil-Ortiz, F., Llacer, J.L., and Rubio, V. (2004) Arginine biosynthesis in *Thermotoga maritima*: characterization of the arginine-sensitive *N*-acetyl-L-glutamate kinase. *Journal of Bacteriology*, **186**, 6142–6149.
**130** Ramon-Maiques, S., Marina, A., Gil-Ortiz, F., Fita, I., and Rubio, V. (2002) Structure of acetylglutamate kinase, a key enzyme for arginine biosynthesis and a prototype for the amino acid kinase enzyme family, during catalysis. *Structure*, **10**, 329–342.

131 Cherney, L.T., Cherney, M.M., Garen, C.R., Niu, C., Moradian, F., and James, M.N.G. (2007) Crystal structure of *N*-acetyl-γ-glutamyl-phosphate reductase from *Mycobacterium tuberculosis* in complex with NADP$^+$. *Journal of Molecular Biology*, **367**, 1357–1369.

132 Rajaram, V., Prasuna, P.R., Savithri, H.S., and Murthy, M.R.N. (2008) Structure of biosynthetic *N*-acetylornithine aminotransferase from *Salmonella typhimurium*: studies on substrate specificity and inhibitor binding. *Proteins – Structure Function and Bioinformatics*, **70**, 429–441.

133 Ledwidge, R. and Blanchard, J.S. (1999) The dual biosynthetic capability of *N*-acetylornithine aminotransferase in arginine and lysine biosynthesis. *Biochemistry*, **38**, 3019–3024.

134 Slocum, R.D. (2005) Genes, enzymes and regulation of arginine biosynthesis in plants. *Plant Physiology and Biochemistry*, **43**, 729–745.

135 Xu, Y., Labedan, B., and Glansdorff, N. (2007) Surprising arginine biosynthesis: a reappraisal of the enzymology and evolution of the pathway in microorganisms. *Microbiology and Molecular Biology Reviews*, **71**, 36–47.

136 Bachmann, C., Krähenbühl, S., and Colombo, J.P. (1982) Purification and properties of acetyl-CoA: L-glutamate *N*-acetyltransferase from human liver. *Biochemical Journal*, **205**, 123–127.

137 Shi, D., Sagar, V., Jin, Z.M., Yu, X.L., Caldovic, L., Morizono, H., Allewell, N.M., and Tuchman, M. (2008) The crystal structure of *N*-acetyl-L-glutamate synthase from *Neisseria gonorrhoeae* provides insights into mechanisms of catalysis and regulation. *The Journal of Biological Chemistry*, **283**, 7176–7184.

138 Alonso, E. and Rubio, V. (1989) Participation of ornithine aminotransferase in the synthesis and catabolism of ornithine in mice. Studies using gabaculine and arginine deprivation. *Biochemical Journal*, **259**, 131–138.

139 Lqbal, A., Clifton, I.J., Bagonis, M., Kershaw, N.J., Domene, C., Claridge, T.D.W., Wharton, C.W., and Schofield, C.J. (2009) Anatomy of a simple acyl intermediate in enzyme catalysis: combined biophysical and modeling studies on ornithine acetyl transferase. *Journal of the American Chemical Society*, **131**, 749–757.

140 Marc, F., Weigel, P., Legrain, C., Almeras, Y., Santrot, M., Glansdorff, N., and Sakanyan, V. (2000) Characterization and kinetic mechanism of mono- and bifunctional ornithine acetyltransferases from thermophilic microorganisms. *European Journal of Biochemistry*, **267**, 5217–5226.

141 Shi, D., Yu, X., Roth, L., Morizono, H., Hathout, Y., Allewell, N.M., and Tuchman, M. (2005) Expression, purification, crystallization and preliminary X-ray crystallographic studies of a novel acetylcitrulline deacetylase from *Xanthomonas campestris*. *Acta Crystallographica F*, **61**, 676–679.

142 Shi, D., Morizono, H., Yu, X., Roth, L., Caldovic, L., Allewell, N.M., Malamy, M.H., and Tuchman, M. (2005) Crystal structure of *N*-acetylornithine transcarbamylase from *Xanthomonas campestris*: a novel enzyme in a new arginine biosynthetic pathway found in several eubacteria. *The Journal of Biological Chemistry*, **280**, 14366–14369.

143 Shi, D., Morizono, H., Cabrera-Luque, J., Yu, X.L., Roth, L., Malamy, M.H., Allewell, N.M., and Tuchman, M. (2006) Structure and catalytic mechanism of a novel *N*-succinyl-L-ornithine transcarbamylase in arginine biosynthesis of *Bacteroides fragilis*. *The Journal of Biological Chemistry*, **281**, 20623–20631.

144 Allewell, N.M., Shi, D., Morizono, H., and Tuchman, M. (1999) Molecular recognition by ornithine and aspartate transcarbamylases. *Accounts of Chemical Research*, **32**, 885–894.

145 Labedan, B., Boyen, A., Baetens, M., Charlier, D., Chen, P., Cunin, R., Durbeco, V., Glansdorff, N., Herve, G., Legrain, C., Liang, Z., Purcarea, C., Roovers, M., Sanchez, R., Toong, T.-L., Van de Casteele, M., van Vliet, F., Xu, Y., and Zhang, Y.-F. (1999) The evolutionary history of carbamoyltransferases: a complex set of paralogous genes was already present in the last universal

common ancestor. *Journal of Molecular Evolution*, **49**, 461–473.

**146** Sankaranarayanan, R., Cherney, M.M., Cherney, L.T., Garen, C.R., Moradian, F., and James, M.N.G. (2008) The crystal structures of ornithine carbamoyltransferase from *Mycobacterium tuberculosis* and its ternary complex with carbamoyl phosphate and L-norvaline reveal the enzyme's catalytic mechanism. *Journal of Molecular Biology*, **375**, 1052–1063.

**147** Yu, Y., Terada, K., Nagasaki, A., Takiguchi, M., and Mori, M. (1995) Preparation of recombinant argininosuccinate synthetase and argininosuccinate lyase – expression of the enzymes in rat tissues. *Journal of Biochemistry*, **117**, 952–957.

**148** Bork, P. and Koonin, E.V. (1994) A P-loop-like motif in a widespread ATP pyrophosphatase domain: implications for the evolution of sequence motifs and enzyme activity. *Proteins – Structure Function and Genetics*, **20**, 347–355.

**149** Goto, M., Omi, R., Miyahara, I., Sugahara, M., and Hirotsu, K. (2003) Structures of argininosuccinate synthetase in enzyme-ATP substrates and enzyme-AMP product forms: stereochemistry of the catalytic reaction. *The Journal of Biological Chemistry*, **278**, 22964–22971.

**150** Tsai, M., Koo, J., and Howell, P.L. (2005) Recovery of argininosuccinate lyase activity in duck δ1 crystallin. *Biochemistry*, **44**, 9034–9044.

**151** Fujii, T., Sakai, H., Kawata, Y., and Hata, Y. (2003) Crystal structure of thermostable aspartase from *Bacillus* sp YM55-1: structure-based exploration of functional sites in the aspartase family. *Journal of Molecular Biology*, **328**, 635–654.

**152** Dailey, F.E. and Cronan, J.E. (1986) Acetohydroxy acid synthase I, a required enzyme for isoleucine and valine biosynthesis in *Escherichia coli* K-12 during growth on acetate as the sole carbon source. *Journal of Bacteriology*, **165**, 453–460.

**153** Dumas, R., Biou, V., Halgand, F., Douce, R., and Duggleby, R.G. (2001) Enzymology, structure, and dynamics of acetohydroxy acid isomeroreductase. *Accounts of Chemical Research*, **34**, 399–408.

**154** Flint, D.H., Emptage, M.H., Finnegan, M.G., Fu, W., and Johnson, M.K. (1993) The role and properties of the iron–sulfur cluster in *Escherichia coli* dihydroxy-acid dehydratase. *The Journal of Biological Chemistry*, **268**, 14732–14742.

**155** Chipman, D., Barak, Z., and Schloss, J.V. (1998) Biosynthesis of 2-aceto-2-hydroxy acids: acetolactate synthases and acetohydroxyacid synthases. *Biochimica et Biophysica Acta – Protein Structure and Molecular Enzymology*, **1385**, 401–419.

**156** Kishore, G.M. and Shah, D.M. (1988) Amino acid biosynthesis inhibitors as herbicides. *Annual Review of Biochemistry*, **57**, 627–663.

**157** Engel, S., Vyazmensky, M., Vinogradov, M., Berkovich, D., Bar-Ilan, A., Qimron, U., Rosiansky, Y., Barak, Z., and Chipman, D.M. (2004) Role of a conserved arginine in the mechanism of acetohydroxyacid synthase: catalysis of condensation with a specific ketoacid substrate. *The Journal of Biological Chemistry*, **279**, 24803–24812.

**158** McCourt, J.A., Pang, S.S., King-Scott, J., Guddat, L.W., and Duggleby, R.G. (2006) Herbicide-binding sites revealed in the structure of plant acetohydroxyacid synthase. *Proceedings of the National Academy of Sciences of the United States of America*, **103**, 569–573.

**159** Scarselli, M., Padula, M.G., Bernini, A., Spiga, O., Ciutti, A., Leoncini, R., Vannoni, D., Marinello, E., and Niccolai, N. (2003) Structure and function correlations between the rat liver threonine deaminase and aminotransferases. *Biochimica et Biophysica Acta – Proteins and Proteomics*, **1645**, 40–48.

**160** Eikmanns, B.J., Eggeling, L., and Sahm, H. (1993) Molecular aspects of lysine, threonine, and isoleucine biosynthesis in *Corynebacterium glutamicum*. *Antonie Van Leeuwenhoek International Journal of General and Molecular Microbiology*, **64**, 145–163.

**161** Shulman, A., Zalyapin, E., Vyazmensky, M., Yifrach, O., Barak, Z., and Chipman, D.M. (2008) Allosteric regulation of *Bacillus subtilis* threonine deaminase, a biosynthetic threonine deaminase with a single regulatory domain. *Biochemistry*, **47**, 11783–11792.

**162** Eisenstein, E. (1991) Cloning, expression, purification, and characterization of biosynthetic threonine deaminase from *Escherichia coli. The Journal of Biological Chemistry*, **266**, 5801–5807.

**163** Drevland, R.M., Waheed, A., and Graham, D.E. (2007) Enzymology and evolution of the pyruvate pathway to 2-oxobutyrate in *Methanocaldococcus jannaschii. Journal of Bacteriology*, **189**, 4391–4400.

**164** Biou, V., Dumas, R., CohenAddad, C., Douce, R., Job, D., and PebayPeyroula, E. (1997) The crystal structure of plant acetohydroxy acid isomeroreductase complexed with NADPH, two magnesium ions and a herbicidal transition state analog determined at 1.65 Å resolution. *The EMBO Journal*, **16**, 3405–3415.

**165** Beinert, H., Kennedy, M.C., and Stout, C.D. (1996) Aconitase as iron–sulfur protein, enzyme, and iron-regulatory protein. *Chemical Reviews*, **96**, 2335–2374.

**166** Duan, X., Yang, J., Ren, B., Tan, G., and Ding, H. (2009) Reactivity of nitric oxide with the [4Fe–4S] cluster of dihydroxyacid dehydratase from *Escherichia coli. Biochemical Journal*, **417**, 783–789.

**167** Hyduke, D.R., Jarboe, L.R., Tran, L.M., Chou, K.J.Y., and Liao, J.C. (2007) Integrated network analysis identifies nitric oxide response networks and dihydroxyacid dehydratase as a crucial target in *Escherichia coli. Proceedings of the National Academy of Sciences of the United States of America*, **104**, 8484–8489.

**168** Macomber, L. and Imlay, J.A. (2009) The iron–sulfur clusters of dehydratases are primary intracellular targets of copper toxicity. *Proceedings of the National Academy of Sciences of the United States of America*, **106**, 8344–8349.

**169** Xu, H., Andi, B., Qian, J., West, A.H., and Cook, P.F. (2006) The α-aminoadipate pathway for lysine biosynthesis in fungi. *Cell Biochemistry and Biophysics*, **46**, 43–64.

**170** Wulandari, A.P., Miyazaki, J., Kobashi, N., Nishiyama, M., Hoshino, T., and Yamane, H. (2002) Characterization of bacterial homocitrate synthase involved in lysine biosynthesis. *FEBS Letters*, **522**, 35–40.

**171** de Carvalho, L.P.S. and Blanchard, J.S. (2006) Kinetic and chemical mechanism of alpha-isopropylmalate synthase from *Mycobacterium tuberculosis. Biochemistry*, **45**, 8988–8999.

**172** Howell, D.M., Xu, H.M., and White, R.H. (1999) (*R*)-Citramalate synthase in methanogenic archaea. *Journal of Bacteriology*, **181**, 331–333.

**173** Li, F., Hagemeier, C.H., Seedorf, H., Gottschalk, G., and Thauer, R.K. (2007) Re-citrate synthase from *Clostridium kluyveri* is phylogenetically related to homocitrate synthase and isopropylmalate synthase rather than to *Si*-citrate synthase. *Journal of Bacteriology*, **189**, 4299–4304.

**174** Koon, N., Squire, C.J., and Baker, E.N. (2004) Crystal structure of LeuA from *Mycobacterium tuberculosis*, a key enzyme in leucine biosynthesis. *Proceedings of the National Academy of Sciences of the United States of America*, **101**, 8295–8300.

**175** Qian, J., Khandogin, J., West, A.H., and Cook, P.F. (2008) Evidence for a catalytic dyad in the active site of homocitrate synthase from *Saccharomyces cerevisiae. Biochemistry*, **47**, 6851–6858.

**176** Gruer, M.J., Artymiuk, P.J., and Guest, J.R. (1997) The aconitase family: three structural variations on a common theme. *Trends in Biochemical Sciences*, **22**, 3–6.

**177** Yasutake, Y., Yao, M., Sakai, N., Kirita, T., and Tanaka, I. (2004) Crystal structure of the *Pyrococcus horikoshii* isopropylmalate isomerase small subunit provides insight into the dual substrate specificity of the enzyme. *Journal of Molecular Biology*, **344**, 325–333.

**178** Hurley, J.H., Thorsness, P.E., Ramalingam, V., Helmers, N.H., Koshland, D.E., and Stroud, R.M. (1989) Structure of a bacterial enzyme regulated by phosphorylation, isocitrate dehydrogenase. *Proceedings of the National Academy of Sciences of the United States of America*, **86**, 8635–8639.

**179** Imada, K., Sato, M., Tanaka, N., Katsube, Y., Matsuura, Y., and Oshima, T. (1991) 3-Dimensional structure of a highly thermostable enzyme, 3-isopropylmalate dehydrogenase of *Thermus thermophilus* at 2.2 Å resolution. *Journal of Molecular Biology*, **222**, 725–738.

**180** Howell, D.M., Graupner, M., Xu, H., and White, R.H. (2000) Identification of enzymes homologous to isocitrate dehydrogenase that are involved in coenzyme B and leucine biosynthesis in methanoarchaea. *Journal of Bacteriology*, **182**, 5013–5016.

181 Bolduc, J.M., Dyer, D.H., Scott, W.G., Singer, P., Sweet, R.M., Koshland, D.E., and Stoddard, B.L. (1995) Mutagenesis and Laue structures of enzyme intermediates: isocitrate dehydrogenase. *Science*, **268**, 1312–1318.

182 Miyazaki, J., Kobashi, N., Nishiyama, M., and Yamane, H. (2003) Characterization of homoisocitrate dehydrogenase involved in lysine biosynthesis of an extremely thermophilic bacterium, *Thermus thermophilus* HB27, and evolutionary implication of beta-decarboxylating dehydrogenase. *The Journal of Biological Chemistry*, **278**, 1864–1871.

183 Miyazaki, K. (2005) Bifunctional isocitrate-homoisocitrate dehydrogenase: a missing link in the evolution of β-decarboxylating dehydrogenase. *Biochemical and Biophysical Research Communications*, **331**, 341–346.

184 Chen, R., Greer, A., and Dean, A.M. (1996) Redesigning secondary structure to invert coenzyme specificity in isopropylmalate dehydrogenase. *Proceedings of the National Academy of Sciences of the United States of America*, **93**, 12171–12176.

185 Miyazaki, T., Miyazaki, J., Yamane, H., and Nishiyama, M. (2004) α-Aminoadipate aminotransferase from an extremely thermophilic bacterium, *Thermus thermophilus*. *Microbiology*, **150**, 2327–2334.

186 Matsuda, M. and Ogur, M. (1969) Enzymatic and physiological properties of yeast glutamate-α-ketoadipate transaminase. *The Journal of Biological Chemistry*, **244**, 5153–5158.

187 Matsuda, M. and Ogur, M. (1969) Separation and specificity of yeast glutamate-α-ketoadipate transaminase. *The Journal of Biological Chemistry*, **244**, 3352–3358.

188 Tomita, T., Miyagawa, T., Miyazaki, T., Fushinobu, S., Kuzuyama, T., and Nishiyama, M. (2009) Mechanism for multiple-substrates recognition of α-aminoadipate aminotransferase from *Thermus thermophilus*. *Proteins – Structure Function and Bioinformatics*, **75**, 348–359.

189 Velasco, A.M., Leguina, J.I., and Lazcano, A. (2002) Molecular evolution of the lysine biosynthetic pathways. *Journal of Molecular Evolution*, **55**, 445–449.

190 Nishida, H., Nishiyama, M., Kobashi, N., Kosuge, T., Hoshino, T., and Yamane, H. (1999) A prokaryotic gene cluster involved in synthesis of lysine through the amino adipate pathway: a key to the evolution of amino acid biosynthesis. *Genome Research*, **9**, 1175–1183.

191 Hudson, A.O., Gilvarg, C., and Leustek, T. (2008) Biochemical and phylogenetic characterization of a novel diaminopimelate biosynthesis pathway in prokaryotes identifies a diverged form of LL-diaminopimelate aminotransferase. *Journal of Bacteriology*, **190**, 3256–3263.

192 Viola, R.E. (2001) The central enzymes of the aspartate family of amino acid biosynthesis. *Accounts of Chemical Research*, **34**, 339–349.

193 Coulter, C.V., Gerrard, J.A., Kraunsoe, J.A.E., and Pratt, A.J. (1999) Escherichia *coli* dihydrodipicolinate synthase and dihydrodipicolinate reductase: kinetic and inhibition studies of two putative herbicide targets. *Pesticide Science*, **55**, 887–895.

194 Dobson, R.C.J., Valegard, K., and Gerrard, J.A. (2004) The crystal structure of three site-directed mutants of *Escherichia coli* dihydrodipicolinate synthase: further evidence for a catalytic triad. *Journal of Molecular Biology*, **338**, 329–339.

195 Devenish, S.R.A., Huisman, F.H.A., Parker, E.J., Hadfield, A.T., and Gerrard, J.A. (2009) Cloning and characterisation of dihydrodipicolinate synthase from the pathogen *Neisseria meningitidis*. *Biochimica et Biophysica Acta – Proteins and Proteomics*, **1794**, 1168–1174.

196 Dobson, R.C.J., Griffin, M.D.W., Devenish, S.R.A., Pearce, F.G., Hutton, C.A., Gerrard, J.A., Jameson, G.B., and Perugini, M.A. (2008) Conserved main-chain peptide distortions: a proposed role for Ile203 in catalysis by dihydrodipicolinate synthase. *Protein Science*, **17**, 2080–2090.

**197** Ge, X., Olson, A., Cai, S., and Sem, D.S. (2008) Binding synergy and cooperativity in dihydrodipicolinate reductase: implications for mechanism and the design of biligand inhibitors. *Biochemistry*, **47**, 9966–9980.

**198** Pearce, F.G., Sprissler, C., and Gerrard, J.A. (2008) Characterization of dihydrodipicolinate reductase from *Thermotoga maritima* reveals evolution of substrate binding kinetics. *Journal of Biochemistry*, **143**, 617–623.

**199** Pillai, B., Moorthie, V.A., van Belkum, M.J., Marcus, S.L., Cherney, M.M., Diaper, C.M., Vederas, J.C., and James, M.N.G. (2009) Crystal structure of diaminopimelate epimerase from *Arabidopsis thaliana*, an amino acid racemase critical for L-lysine biosynthesis. *Journal of Molecular Biology*, **385**, 580–594.

**200** Usha, V., Dover, L.G., Roper, D.L., and Besra, G.S. (2008) Characterization of *Mycobacterium tuberculosis* diaminopimelic acid epimerase: paired cysteine residues are crucial for racemization. *FEMS Microbiology Letters*, **280**, 57–63.

**201** Hu, T., Wu, D., Chen, J., Ding, J., Jiang, H., and Shen, X. (2008) The catalytic intermediate stabilized by a "Down" active site loop for diaminopimelate decarboxylase from *Helicobacter pylori*: enzymatic characterization with crystal structure analysis. *The Journal of Biological Chemistry*, **283**, 21284–21293.

**202** Lee, J., Michael, A.J., Martynowski, D., Goldsmith, E.J., and Phillips, M.A. (2007) Phylogenetic diversity and the structural basis of substrate specificity in the β/α-barrel fold basic amino acid decarboxylases. *The Journal of Biological Chemistry*, **282**, 27115–27125.

**203** Gokulan, K., Rupp, B., Pavelka, M.S., Jacobs, W.R., and Sacchettini, J.C. (2003) Crystal structure of *Mycobacterium tuberculosis* diaminopimelate decarboxylase, an essential enzyme in bacterial lysine biosynthesis. *The Journal of Biological Chemistry*, **278**, 18588–18596.

**204** Watanabe, N., Cherney, M.M., van Belkum, M.J., Marcus, S.L., Flegel, M.D., Clay, M.D., Deyholos, M.K., Vederas, J.C., and James, M.N.G. (2007) Crystal structure of LL-diaminopimelate aminotransferase from *Arabidopsis thaliana*: a recently discovered enzyme in the biosynthesis of L-lysine by plants and *Chlamydia*. *Journal of Molecular Biology*, **371**, 685–702.

**205** White, P.J. (1983) The essential role of diaminopimelate dehydrogenase in the biosynthesis of lysine by *Bacillus sphaericus*. *Journal of General Microbiology*, **129**, 739–749.

**206** Miyazaki, J., Kobashi, N., Nishiyama, M., and Yamane, H. (2001) Functional and evolutionary relationship between arginine biosynthesis and prokaryotic lysine biosynthesis through alpha-aminoadipate. *Journal of Bacteriology*, **183**, 5067–5073.

**207** Horie, A., Tomita, T., Saiki, A., Kono, H., Taka, H., Mineki, R., Fujimura, T., Nishiyama, C., Kuzuyama, T., and Nishiyama, M. (2009) Discovery of proteinaceous N-modification in lysine biosynthesis of *Thermus thermophilus*. *Nature Chemical Biology*, **5**, 673–679.

**208** Baldwin, J.E., Shiau, C.Y., Byford, M.F., and Schofield, C.J. (1994) Substrate-specificity of L-δ-(α-aminoadipoyl)-L-cysteinyl-D-valine synthetase from *Cephalosporium acremonium*: demonstration of the structure of several unnatural tripeptide products. *Biochemical Journal*, **301**, 367–372.

**209** Aharonowitz, Y., Bergmeyer, J., Cantoral, J.M., Cohen, G., Demain, A.L., Fink, U., Kinghorn, J., Kleinkauf, H., MacCabe, A., Palissa, H., Pfeifer, E., Schwecke, T., Vanliempt, H., Vondohren, H., Wolfe, S., and Zhang, J.Y. (1993) δ-(L-α-Aminoadipyl)-L-cysteinyl-D-valine synthetase, the multienzyme integrating the four primary reactions in β-lactam biosynthesis, as a model peptide synthetase. *Bio/Technology*, **11**, 807–810.

**210** Byford, M.F., Baldwin, J.E., Shiau, C.-Y., and Schofield, C.J. (1997) The mechanism of ACV synthetase. *Chemical Reviews*, **97**, 2631–2650.

**211** An, K.-D., Nishida, H., Miura, Y., and Yokota, A. (2003) Molecular evolution of adenylating domain of aminoadipate reductase. *BMC Evolutionary Biology*, **3**, 9.

**212** Storts, D.R. and Bhattacharjee, J.K. (1987) Purification and properties of saccharopine dehydrogenase (glutamate forming) in the *Saccharomyces cerevisiae* lysine biosynthetic pathway. *Journal of Bacteriology*, **169**, 416–418.

**213** Johansson, E., Steffens, J.J., Lindqvist, Y., and Schneider, G. (2000) Crystal structure of saccharopine reductase from *Magnaporthe grisea*, an enzyme of the alpha-aminoadipate pathway of lysine biosynthesis. *Structure*, **8**, 1037–1047.

**214** Andi, B., Cook, P.F., and West, A.H. (2006) Crystal structure of the His-tagged saccharopine reductase from *Saccharomyces cerevisiae* at 1.7 Å resolution. *Cell Biochemistry and Biophysics*, **46**, 17–26.

**215** Burk, D.L., Hwang, J., Kwok, E., Marrone, L., Goodfellow, V., Dmitrienko, G.I., and Berghuis, A.M. (2007) Structural studies of the final enzyme in the α-aminoadipate pathway – saccharopine dehydrogenase from *Saccharomyces cerevisiae*. *Journal of Molecular Biology*, **373**, 745–754.

**216** Xu, H.Y., Alguindigue, S.S., West, A.H., and Cook, P.F. (2007) A proposed proton shuttle mechanism for saccharopine dehydrogenase from *Saccharomyces cerevisiae*. *Biochemistry*, **46**, 871–882.

**217** Hao, B., Gong, W., Ferguson, T.K., James, C.M., Krzycki, J.A., and Chan, M.K. (2002) A new UAG-encoded residue in the structure of a methanogen methyltransferase. *Science*, **296**, 1462–1466.

**218** Soares, J.A., Zhang, L., Pitsch, R.L., Kleinholz, N.M., Jones, R.B., Wolff, J.J., Amster, J., Green-Church, K.B., and Krzycki, J.A. (2005) The residue mass of L-pyrrolysine in three distinct methylamine methyltransferases. *The Journal of Biological Chemistry*, **280**, 36962–36969.

**219** Krzycki, J.A. (2004) Function of genetically encoded pyrrolysine in corrinoid-dependent methylamine methyltransferases. *Current Opinion in Chemical Biology*, **8**, 484–491.

**220** Blight, S.K., Larue, R.C., Mahapatra, A., Longstaff, D.G., Chang, E., Zhao, G., Kang, P.T., Green-Church, K.B., Chan, M.K., and Krzycki, J.A. (2004) Direct charging of $tRNA^{CUA}$ with pyrrolysine *in vitro* and *in vivo*. *Nature*, **431**, 333–335.

**221** Polycarpo, C., Ambrogelly, A., Berube, A., Winbush, S.A.M., McCloskey, J.A., Crain, P.F., Wood, J.L., and Soll, D. (2004) An aminoacyl-tRNA synthetase that specifically activates pyrrolysine. *Proceedings of the National Academy of Sciences of the United States of America*, **101**, 12450–12454.

**222** Yuan, J., O'Donoghue, P., Ambrogelly, A., Gundllapalli, S., Sherrer, R.L., Palioura, S., Simonovic, M., and Söll, D. (2010) Distinct genetic code expansion strategies for selenocysteine and pyrrolysine are reflected in different aminoacyl-tRNA formation systems. *FEBS Letters*, **584**, 342–349.

**223** Kavran, J.M., Gundlliapalli, S., O'Donoghue, P., Englert, M., Söll, D., and Steitz, T.A. (2007) Structure of pyrrolysyl-tRNA synthetase, an archaeal enzyme for genetic code innovation. *Proceedings of the National Academy of Sciences of the United States of America*, **104**, 11268–11273.

**224** Longstaff, D.G., Larue, R.C., Faust, J.E., Mahapatra, A., Zhang, L., Green-Church, K.B., and Krzycki, J.A. (2007) A natural genetic code expansion cassette enables transmissible biosynthesis and genetic encoding of pyrrolysine. *Proceedings of the National Academy of Sciences of the United States of America*, **104**, 1021–1026.

**225** Abell, C. (1999) Enzymology and molecular biology of the shikimate pathway. *Comprehensive Natural Products Chemistry*, **1**, 573–607.

**226** Herrmann, K.M. and Weaver, L.M. (1999) The shikimate pathway. *Annual Review of Plant Physiology and Plant Molecular Biology*, **50**, 473–503.

**227** Stephens, C.M. and Bauerle, R. (1991) Analysis of the metal requirement of 3-deoxy-D-*arabino*-heptulosonate-7-phosphate synthase from *Escherichia coli*. *The Journal of Biological Chemistry*, **266**, 20810–20817.

**228** Shumilin, I.A., Bauerle, R., Wu, J., Woodard, R.W., and Kretsinger, R.H. (2004) Crystal structure of the reaction complex of 3-deoxy-D-*arabino*-heptulosonate-7-phosphate synthase from *Thermotoga maritima* refines the catalytic mechanism and indicates a new mechanism of allosteric regulation. *Journal of Molecular Biology*, **341**, 455–466.

**229** Ran, N., Draths, K.M., and Frost, J.W. (2004) Creation of a shikimate pathway variant. *Journal of the American Chemical Society*, **126**, 6856–6857.

**230** Ran, N. and Frost, J.W. (2007) Directed evolution of 2-keto-3-deoxy-6-phosphogalactonate aldolase to replace 3-deoxy-D-*arabino*-heptulosonic acid 7-phosphate synthase. *Journal of the American Chemical Society*, **129**, 6130–6139.

**231** Webby, C.J., Baker, H.M., Lott, J.S., Baker, E.N., and Parker, E.J. (2005) The structure of 3-deoxy-D-*arabino*-heptulosonate 7-phosphate synthase from *Mycobacterium tuberculosis* reveals a common catalytic scaffold and ancestry for type I and type II enzymes. *Journal of Molecular Biology*, **354**, 927–939.

**232** Wagner, T., Shumilin, I.A., Bauerle, R., and Kretsinger, R.H. (2000) Structure of 3-deoxy-D-*arabino*-heptulosonate-7-phosphate synthase from *Escherichia coli*: Comparison of the $Mn^{2+}$*2-phosphoglycolate and the $Pb^{2+}$*2-phosphoenolpyruvate complexes and implications for catalysis. *Journal of Molecular Biology*, **301**, 389–399.

**233** Schofield, L.R., Anderson, B.F., Patchett, M.L., Norris, G.E., Jameson, G.B., and Parker, E.J. (2005) Substrate ambiguity and structure of *Pyrococcus furiosus* 3-deoxy-D-*arabino*-heptulosonate-7-phosphate synthase: an ancestral 3-deoxyald-2-ulosonate phosphate synthase? *Biochemistry*, **44**, 11950–11962.

**234** König, V., Pfeil, A., Braus, G.H., and Schneider, T.R. (2004) Substrate and metal complexes of 3-deoxy-D-*arabino*-heptulosonate-7-phosphate synthase from *Saccharomyces cerevisiae* provide new insights into the catalytic mechanism. *Journal of Molecular Biology*, **337**, 675–690.

**235** Wu, J., Howe, D.L., and Woodard, R.W. (2003) Thermotoga *maritima* 3-deoxy-D-*arabino*-heptulosonate 7-phosphate (DAHP) synthase: the ancestral eubacterial DAHP synthase? *The Journal of Biological Chemistry*, **278**, 27525–31.

**236** Widlanski, T., Bender, S.L., and Knowles, J.R. (1989) Dehydroquinate synthase: a sheep in wolf's clothing? *Journal of the American Chemical Society*, **111**, 2299–2300.

**237** Carpenter, E.P., Hawkins, A.R., Frost, J.W., and Brown, K.A. (1998) Structure of dehydroquinate synthase reveals an active site capable of multistep catalysis. *Nature*, **394**, 299–302.

**238** Bartlett, P.A. and Satake, K. (1988) Does dehydroquinate synthase synthesize dehydroquinate? *Journal of the American Chemical Society*, **110**, 1628–1630.

**239** Tanner, M.E. (2008) Transient oxidation as a mechanistic strategy in enzymatic catalysis. *Current Opinion in Chemical Biology*, **12**, 532–538.

**240** Porat, I., Sieprawska-Lupa, M., Teng, Q., Bohanon, F.J., White, R.H., and Whitman, W.B. (2006) Biochemical and genetic characterization of an early step in a novel pathway for the biosynthesis of aromatic amino acids and *p*-aminobenzoic acid in the archaeon *Methanococcus maripaludis*. *Molecular Microbiology*, **62**, 1117–1131.

**241** White, R.H. (2004) l-Aspartate semialdehyde and a 6-deoxy-5-ketohexose 1-phosphate are the precursors to the aromatic amino acids in *Methanocaldococcus jannaschii*. *Biochemistry*, **43**, 7618–7627.

**242** Morar, M., White, R.H., and Ealick, S.E. (2007) Structure of 2-amino-3,7-dideoxy-D-threo-hept-6-ulosonic acid synthase, a catalyst in the archaeal pathway for the biosynthesis of aromatic amino acids. *Biochemistry*, **46**, 10562–10571.

**243** Gourley, D.G., Shrive, A.K., Polikarpov, I., Krell, T., Coggins, J.R., Hawkins, A.R., Isaacs, N.W., and Sawyer, L. (1999) The two types of 3-dehydroquinase have distinct structures but catalyze the same overall reaction. *Nature Structural Biology*, **6**, 521–525.

**244** Shneier, A., Harris, J., Kleanthous, C., Coggins, J.R., Hawkins, A.R., and Abell, C. (1993) Evidence for opposite stereochemical courses for the reactions catalyzed by type I and type II dehydroquinases. *Bioorganic & Medicinal Chemistry Letters*, **3**, 1399–402.

**245** Shneier, A., Kleanthous, C., Deka, R., Coggins, J.R., and Abell, C. (1991) Observation of an imine intermediate on dehydroquinase by electrospray mass spectrometry. *Journal of the American Chemical Society*, **113**, 9416–9418.

**246** Kleanthous, C., Reilly, M., Cooper, A., Kelly, S., Price, N.C., and Coggins, J.R.

(1991) Stabilization of the shikimate pathway enzyme dehydroquinase by covalently bound ligand. *The Journal of Biological Chemistry*, **266**, 10893–10898.

**247** Gonzalez-Bello, C. and Castedo, L. (2007) Progress in type II dehydroquinase inhibitors: from concept to practice. *Medicinal Research Reviews*, **27**, 177–208.

**248** Chaudhuri, S. and Coggins, J.R. (1985) The purification of shikimate dehydrogenase from *Escherichia coli*. *Biochemical Journal*, **226**, 217–223.

**249** Chaudhuri, S., Anton, I.A., and Coggins, J.R. (1987) Shikimate dehydrogenase from *Escherichia coli*. *Methods in Enzymology*, **142**, 315–320.

**250** Han, C., Hu, T., Wu, D., Qu, S., Zhou, J., Ding, J., Shen, X., Qu, D., and Jiang, H. (2009) X-ray crystallographic and enzymatic analyses of shikimate dehydrogenase from *Staphylococcus epidermidis*. *FEBS Journal*, **276**, 1125–1139.

**251** Singh, S., Korolev, S., Koroleva, O., Zarembinski, T., Collart, F., Joachimiak, A., and Christendat, D. (2005) Crystal structure of a novel shikimate dehydrogenase from *Haemophilus influenzae*. *The Journal of Biological Chemistry*, **280**, 17101–17108.

**252** Michel, G., Roszak, A.W., Sauve, V., Maclean, J., Matte, A., Coggins, J.R., Cygler, M., and Lapthorn, A.J. (2003) Structures of shikimate dehydrogenase AroE and its paralog YdiB: a common structural framework for different activities. *The Journal of Biological Chemistry*, **278**, 19463–19472.

**253** Segura-Cabrera, A. and Rodriguez-Perez, M.A. (2008) Structure-based prediction of *Mycobacterium tuberculosis* shikimate kinase inhibitors by high-throughput virtual screening. *Bioorganic & Medicinal Chemistry Letters*, **18**, 3152–3157.

**254** Pauli, I., Caceres, R.A., and Filgueira de Azevedo, W. (2008) Molecular modeling and dynamics studies of shikimate kinase from *Bacillus anthracis*. *Bioorganic and Medicinal Chemistry*, **16**, 8098–8108.

**255** Berti, P.J. and Chindemi, P. (2009) Catalytic residues and an electrostatic sandwich that promote enolpyruvyl shikimate 3-phosphate synthase (AroA) catalysis. *Biochemistry*, **48**, 3699–3707.

**256** Funke, T., Healy-Fried, M.L., Han, H., Alberg, D.G., Bartlett, P.A., and Schönbrunn, E. (2007) Differential inhibition of class I and class II 5-enolpyruvylshikimate-3-phosphate synthases by tetrahedral reaction intermediate analogues. *Biochemistry*, **46**, 13344–13351.

**257** Schonbrunn, E., Eschenburg, S., Shuttleworth, W.A., Schloss, J.V., Amrhein, N., Evans, J.N.S., and Kabsch, W. (2001) Interaction of the herbicide glyphosate with its target enzyme 5-enolpyruvylshikimate 3-phosphate synthase in atomic detail. *Proceedings of the National Academy of Sciences of the United States of America*, **98**, 1376–1380.

**258** Fernandes, C.L., Breda, A., Santos, D.S., Basso, L.A., and de Souza, O.N. (2007) A structural model for chorismate synthase from *Mycobacterium tuberculosis* in complex with coenzyme and substrate. *Computers in Biology and Medicine*, **37**, 149–158.

**259** Bornemann, S., Balasubramanian, S., Coggins, J.R., Abell, C., Lowe, D.J., and Thorneley, R.N. (1995) Escherichia *coli* chorismate synthase: a deuterium kinetic-isotope effect under single-turnover and steady-state conditions shows that a flavin intermediate forms before the C-(6*proR*)-H bond is cleaved. *Biochemical Journal*, **305**, 707–710.

**260** Balasubramanian, S., Coggins, J.R., and Abell, C. (1995) Observation of a secondary tritium isotope effect in the chorismate synthase reaction. *Biochemistry*, **34**, 341–348.

**261** Rauch, G., Ehammer, H., Bornemann, S., and Macheroux, P. (2008) Replacement of two invariant serine residues in chorismate synthase provides evidence that a proton relay system is essential for intermediate formation and catalytic activity. *FEBS Journal*, **275**, 1464–1473.

**262** Rauch, G., Ehammer, H., Bornemann, S., and Macheroux, P. (2007) Mutagenic analysis of an invariant aspartate residue in chorismate synthase supports its role as an active site base. *Biochemistry*, **46**, 3768–74.

**263** Lin, X., Xu, S., Yang, Y., Wu, J., Wang, H., Shen, H., and Wang, H. (2009) Purification and characterization of anthranilate synthase component I (TrpE) from *Mycobacterium tuberculosis* H37Rv. *Protein Expression and Purification*, **64**, 8–15.

**264** Morollo, A.A. and Eck, M.J. (2001) Structure of the cooperative allosteric anthranilate synthase from *Salmonella typhimurium*. *Nature Structural Biology*, **8**, 243–247.

**265** Tang, X.-F., Ezaki, S., Atomi, H., and Imanaka, T. (2001) Anthranilate synthase without an LLES motif from a hyperthermophilic archaeon is inhibited by tryptophan. *Biochemical and Biophysical Research Communications*, **281**, 858–865.

**266** Bulloch, E.M.M., Jones, M.A., Parker, E.J., Osborne, A.P., Stephens, E., Davies, G.M., Coggins, J.R., and Abell, C. (2004) Identification of 4-amino-4-deoxychorismate synthase as the molecular target for the antimicrobial action of (6*S*)-6-fluoroshikimate. *Journal of the American Chemical Society*, **126**, 9912–9913.

**267** He, Z., Lavoie, K.D.S., Bartlett, P.A., and Toney, M.D. (2004) Conservation of mechanism in three chorismate-utilizing enzymes. *Journal of the American Chemical Society*, **126**, 2378–2385.

**268** He, Z. and Toney, M.D. (2006) Direct detection and kinetic analysis of covalent intermediate formation in the 4-amino-4-deoxychorismate synthase catalyzed reaction. *Biochemistry*, **45**, 5019–5028.

**269** Kim, C., Xuong, N.-H., Edwards, S., Madhusudan; Yee, M.-C., Spraggon, G., and Mills, S.E. (2002) The crystal structure of anthranilate phosphoribosyltransferase from the enterobacterium *Pectobacterium carotovorum*. *FEBS Letters*, **523**, 239–246.

**270** Schlee, S., Deuss, M., Bruning, M., Ivens, A., Schwab, T., Hellmann, N., Mayans, O., and Sterner, R. (2009) Activation of anthranilate phosphoribosyltransferase from *Sulfolobus solfataricus* by removal of magnesium inhibition and acceleration of product release. *Biochemistry*, **48**, 5199–5209.

**271** Sterner, R., Merz, A., Thoma, R., and Kirschner, K. (2001) Phosphoribosylanthranilate isomerase and indoleglycerol-phosphate synthase: tryptophan biosynthetic enzymes from *Thermotoga maritima*. *Methods in Enzymology*, **331**, 270–280.

**272** Sterner, R., Dahm, A., Darimont, B., Ivens, A., Liebl, W., and Kirschner, K. (1995) $(\beta\alpha)_8$-Barrel proteins of tryptophan biosynthesis in the hyperthermophile *Thermotoga maritima*. *The EMBO Journal*, **14**, 4395–4402.

**273** Raboni, S., Bettati, S., and Mozzarelli, A. (2009) Tryptophan synthase: a mine for enzymologists. *Cellular and Molecular Life Sciences*, **66**, 2391–2403.

**274** Leopoldseder, S., Hettwer, S., and Sterner, R. (2006) Evolution of multi-enzyme complexes: the case of tryptophan synthase. *Biochemistry*, **45**, 14111–14119.

**275** Sasso, S., Ökvist, M., Roderer, K., Gamper, M., Codoni, G., Krengel, U., and Kast, P. (2009) Structure and function of a complex between chorismate mutase and DAHP synthase: efficiency boost for the junior partner. *The EMBO Journal*, **28**, 2128–2142.

**276** Song, J., Bonner, C.A., Wolinsky, M., and Jensen, R.A. (2005) The TyrA family of aromatic-pathway dehydrogenases in phylogenetic context. *BMC Biology*, **3**, 13.

**277** Bonner, C.A., Jensen, R.A., Gander, J.E., and Keyhani, N.O. (2004) A core catalytic domain of the TyrA protein family: arogenate dehydrogenase from *Synechocystis*. *The Biochemical Journal*, **382**, 2792–91.

**278** Bonner, C.A., Disz, T., Hwang, K., Song, J., Vonstein, V., Overbeek, R., and Jensen, R.A. (2008) Cohesion group approach for evolutionary analysis of TyrA, a protein family with wide-ranging substrate specificities. *Microbiology and Molecular Biology Reviews*, **72**, 13–53.

**279** Sun, W., Shahinas, D., Bonvin, J., Hou, W., Kimber, M.S., Turnbull, J., and Christendat, D. (2009) The crystal structure of *Aquifex aeolicus* prephenate dehydrogenase reveals the mode of tyrosine inhibition. *The Journal of Biological Chemistry*, **284**, 13223–13232.

**280** Alifano, P., Fani, R., Lio, P., Lazcano, A., Bazzicalupo, M., Carlomagno, M.S., and Bruni, C.B. (1996) Histidine biosynthetic pathway and genes: structure, regulation, and evolution. *Microbiological Reviews*, **60**, 44–69.

281 Martin, R.G., Berberich, M.A., Ames, B.N., Davis, W.W., Goldberger, R.F., and Yourno, J.D. (1971) Enzymes and intermediates of histidine biosynthesis in *Salmonella typhimurium*. *Methods in Enzymology*, **17**, 3–44.

282 Papaleo, M.C., Russo, E., Fondi, M., Emiliani, G., Frandi, A., Brilli, M., Pastorelli, R., and Fani, R. (2009) Structural, evolutionary and genetic analysis of the histidine biosynthetic "core" in the genus *Burkholderia*. *Gene*, **448**, 16–28.

283 Fondi, M., Emiliani, G., Lio, P., Gribaldo, S., and Fani, R. (2009) The evolution of histidine biosynthesis in archaea: insights into the his genes structure and organization in LUCA. *Journal of Molecular Evolution*, **69**, 512–526.

284 Jung, S., Chun, J.-Y., Yim, S.-H., Cheon, C.-I., Song, E., Lee, S.-S., and Lee, M.-S. (2009) Organization and analysis of the histidine biosynthetic genes from *Corynebacterium Glutamicum*. *Genes Genomics*, **31**, 315–323.

285 Fani, R., Brilli, M., Fondi, M., and Lio, P. (2007) The role of gene fusions in the evolution of metabolic pathways: the histidine biosynthesis case. *BMC Evolutionary Biology*, **7** (Suppl. 2), S12.

286 Henn-Sax, M., Thoma, R., Schmidt, S., Hennig, M., Kirschner, K., and Sterner, R. (2002) Two $(\beta\alpha)_8$-barrel enzymes of histidine and tryptophan biosynthesis have similar reaction mechanisms and common strategies for protecting their labile substrates. *Biochemistry*, **41**, 12032–12042.

287 Leopoldseder, S., Claren, J., Juergens, C., and Sterner, R. (2004) Interconverting the catalytic activities of (beta alpha)8-barrel enzymes from different metabolic pathways: sequence requirements and molecular analysis. *Journal of Molecular Biology*, **337**, 871–879.

288 Mayans, O., Ivens, A., Nissen, L.J., Kirschner, K., and Wilmanns, M. (2002) Structural analysis of two enzymes catalysing reverse metabolic reactions implies common ancestry. *The EMBO Journal*, **21**, 3245–54.

289 Lipchock, J.M. and Loria, J.P. (2008) 1H, 15N and $^{13}$C resonance assignment of imidazole glycerol phosphate (IGP) synthase protein HisF from *Thermotoga maritima*. *Biomolecular NMR Assignments*, **2**, 219–221.

290 Chaudhuri, B.N., Lange, S.C., Myers, R.S., Davisson, V.J., and Smith, J.L. (2003) Toward understanding the mechanism of the complex cyclization reaction catalyzed by imidazole glycerolphosphate synthase: crystal structures of a ternary complex and the free enzyme. *Biochemistry*, **42**, 7003–7012.

291 Glynn, S.E., Baker, P.J., Sedelnikova, S.E., Davies, C.L., Eadsforth, T.C., Levy, C.W., Rodgers, H.F., Blackburn, G.M., Hawkes, T.R., Viner, R., and Rice, D.W. (2005) Structure and mechanism of imidazoleglycerol-phosphate dehydratase. *Structure*, **13**, 1809–1817.

292 Sinha, S.C., Chaudhuri, B.N., Burgner, J.W., Yakovleva, G., Davisson, V.J., and Smith, J.L. (2004) Crystal structure of imidazole glycerol-phosphate dehydratase: duplication of an unusual fold. *The Journal of Biological Chemistry*, **279**, 15491–15498.

293 Marineo, S., Cusimano, M.G., Limauro, D., Coticchio, G., and Puglia, A.M. (2008) The histidinol phosphate phosphatase involved in histidine biosynthetic pathway is encoded by SCO5208 (hisN) in *Streptomyces coelicolor* $A3_2$. *Current Microbiology*, **56**, 6–13.

294 Fernandez, F.J., Vega, M.C., Lehmann, F., Sandmeier, E., Gehring, H., Christen, P., and Wilmanns, M. (2004) Structural studies of the catalytic reaction pathway of a hyperthermophilic histidinol-phosphate aminotransferase. *The Journal of Biological Chemistry*, **279**, 21478–21488.

295 le Coq, D., Fillinger, S., and Aymerich, S. (1999) Histidinol phosphate phosphatase, catalyzing the penultimate step of the histidine biosynthesis pathway, is encoded by *ytvP* (*hisJ*) in *Bacillus subtilis*. *Journal of Bacteriology*, **181**, 3277–3280.

# 2
# Heterocycles from Amino Acids

*M. Isabel Calaza and Carlos Cativiela*

## 2.1
## Introduction

Heterocycles make up an extraordinarily important class of compounds. Many natural products, such as hormones, antibiotics, or alkaloids, as well as pharmaceuticals and products of industrial significance (dyes, luminophores, herbicides, pesticides, etc.) are heterocyclic in nature. The great diversity of heterocycles alongside their chemical, biological, and technological relevance justifies a permanent effort to devise efficient synthetic procedures for their preparation. Special attention has been devoted to the development of methodologies for the asymmetric synthesis of heterocyclic compounds, due to their biological importance. In this context, proteinogenic α-amino acids are the most extensively used precursors for the synthesis of enantiomerically pure heterocycles, because they constitute a natural source of chirality that is readily available and display a multifunctional character that facilitates synthetic transformations [1]. In view of that, this chapter describes the use of proteinogenic α-amino acids as enantiomerically pure starting materials for the synthesis of heterocyclic compounds, provides additional information on the utility of the products in selected synthetic transformations, and focuses mainly on synthetic methods of general applicability to different amino acids. The diverse synthetic approaches have been organized according to the nature of the reactions that the α-amino acid functional groups undergo to generate the heterocyclic skeleton (i.e., intramolecular or intermolecular cyclizations and cycloaddition reactions).

## 2.2
## Heterocycles Generated by Intramolecular Cyclizations

### 2.2.1
### α-Lactones and α-Lactams

The multifunctional character of α-amino acids allows the preparation of a diverse range of heterocycles just through intramolecular ring-closure reactions. In this

*Amino Acids, Peptides and Proteins in Organic Chemistry.*
*Vol.3 – Building Blocks, Catalysis and Coupling Chemistry.* Edited by Andrew B. Hughes

ISBN: 978-3-527-32102-5

context, two cyclization modes that exclusively involve the carboxylic and amino functionalities of the α-amino acid **1** could be envisioned. The ring-closure could be accomplished either by turning the amino function into a leaving group and subsequent nucleophilic attack of the carboxylate onto the α-carbon or by nucleophilic displacement of the amino group to a suitably activated carboxylic moiety. Such cyclization sequences would furnish α-lactones (**2**) and α-lactams (**5**), respectively (Scheme 2.1). These three-membered ring heterocycles are endowed with a certain strain energy that influences their stability and reactivity. In fact, α-lactones **2** have mostly been proposed as intermediates in various transformations such as the diazotization of α-amino acids to produce α-hydroxy or α-halo acids **3** with retention of configuration [2]. Conversely, α-lactams **5** have been synthesized and have proven useful synthons to regioselectively react with nucleophiles [3]. The treatment of *N*-benzyloxycarbonyl α-amino acids with a dehydrating agent and a base was initially reported for the synthesis of aziridinones [4]; however, this report was later shown to be incorrect since oxazol-5(4*H*)-ones were identified as the reaction products [5]. Instead, one of the methods currently employed to synthesize α-lactams **5** is the dehydrohalogenation of α-halo amides **4** [6], which are often generated from carboxylic acids rather than α-amino acids because isolable α-lactams need to have at least one tertiary alkyl or aryl group attached at C3.

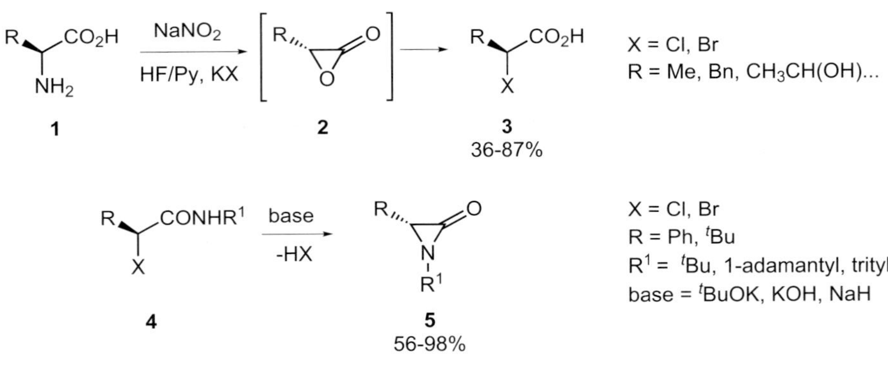

$NaNO_2$, HF/Py, KX: **1** → [**2**] → **3** (36-87%); X = Cl, Br; R = Me, Bn, $CH_3CH(OH)$...

base, -HX: **4** → **5** (56-98%); X = Cl, Br; R = Ph, $^tBu$; $R^1$ = $^tBu$, 1-adamantyl, trityl; base = $^tBuOK$, KOH, NaH

**Scheme 2.1**

### 2.2.2 Indolines

Intramolecular ring-closure reactions may also involve functionalities present in the α-amino acid side-chains. In particular, α-amino acids with aromatic groups may afford a straightforward entry to indolines and pyrroloindolines, which are key structural elements in a wide selection of natural products with a diverse range of biological activities [7]. The connection of the nitrogen moiety and the benzene ring requires activation either by external oxidants (such as the two-electron oxidant $F^+$) or functionalities incorporated into the product (halogen substituent at the *ortho* positions) to undergo transition metal-catalyzed aryl amination reactions. In that manner, the preparation of enantiomerically pure indolines **7** and **8**, from phenylalanine and tyrosine, has been achieved either by palladium-catalyzed amination using a $F^+$ source as the oxidant [8] or through a tandem C–H bond iodination/amination using Pd(II)/Cu(I) catalysts (Scheme 2.2) [9].

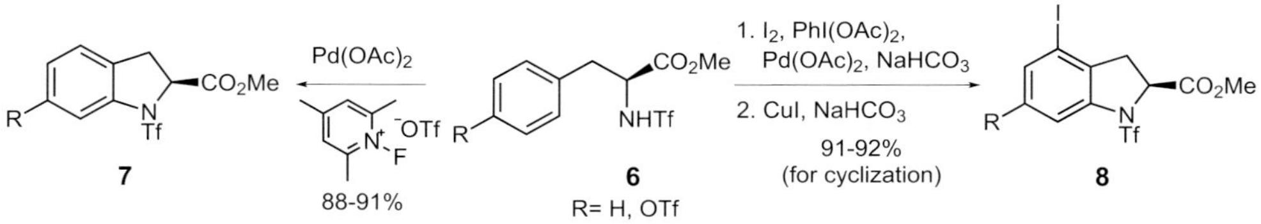

Scheme 2.2

Correspondingly, the hexahydropyrrolo[2,3-*b*]indole [10] skeleton can be generated by the ring closure of tryptophan. In particular, the treatment of **9** with $H_3PO_4$ followed by *N*-tosylation (Scheme 2.3) affords a single diastereoisomer (**10**), which has been used as a chiral building block for the enantiospecific synthesis of α-alkylated tryptophans. In addition, 3*a*-hydroxy- and 3*a*-(methylthiomethyl)hexahydropyrrolo[2,3-*b*]indoline structural motifs (**12** and **13**), of high pharmaceutical interest, have been generated via benzylic oxidations [11] and alkylative cyclizations [12] of suitably protected tryptophan derivatives, respectively. On the other hand, synthetic approaches towards tetrahydropyrrolo[2,3-*b*]indoles **11** rely on oxidative cycloaromatizations [13, 14] or intramolecular copper-catalyzed couplings [15] of suitable iodinated tryptophan derivatives.

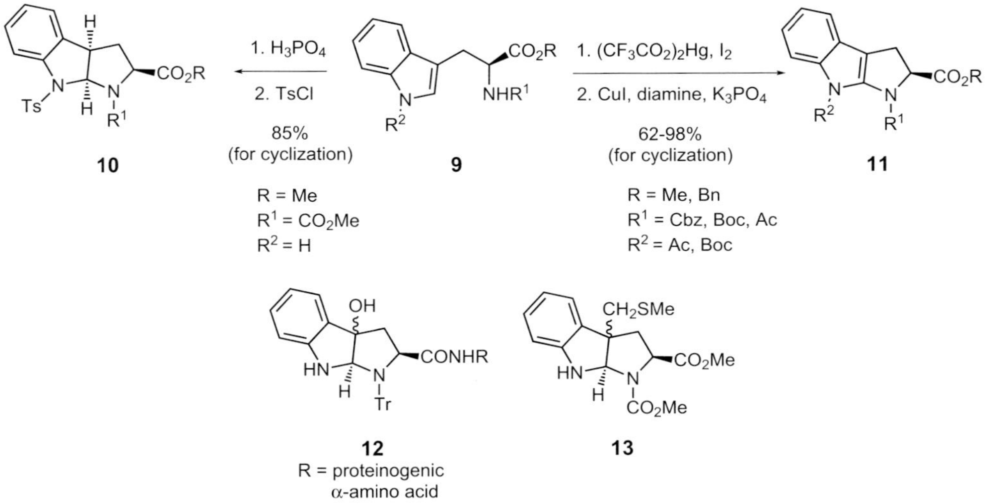

Scheme 2.3

### 2.2.3
### Aziridinecarboxylic Acids and Oxetanones

The hydroxyl functional group in serine (and threonine), acting as a nucleofuge, may provide chiral aziridine-2-carboxylic esters (**16**) through an intramolecular nucleophilic displacement by an amide anion or, alternatively, optically pure α-amino-β-lactones (**18**) [16] if the carboxylate acts as the nucleophile (Scheme 2.4). Both, aziridine-2-carboxylic acids and α-amino-β-lactones (oxetanones), have proven highly valuable heterocycles for the synthesis of a large variety of unnatural amino acids and natural products [17]. Furthermore, aziridine-2-carboxylic acids find utility as ligands for catalytic purposes, synthons, and to construct four- and five-membered ring heterocycles through ring-expansion reactions, even in solid-phase synthesis [18]. In particular, the preparation of *N*-Boc aziridines has been reported to require a temporary amine protection (i.e., trityl protecting group) to avoid the formation of a five-membered oxazolonium through the nucleophilic attack of the carbonyl oxygen (Scheme 2.4) [19, 20]. *N*-Trityl-protected aziridines **15** have been generated either by the conversion of the hydroxyl group of **14** into a leaving group and subsequent cyclization under basic conditions or by direct treatment of **14** with sulfuryl chloride [21]. Alternatively, diethoxytriphenylphosphine (DTPP) effects a one-pot activation and cyclization of free serine, although with variable yields due to the fragile nature of DTPP. On the other hand, the treatment of *N*-Cbz- or *N*-Boc-serine under modified Mitsunobu [22] reaction conditions (Scheme 2.4) produces oxetanones **18** via hydroxyl group activation with triphenylphosphine and methyl azodicarboxylate [23].

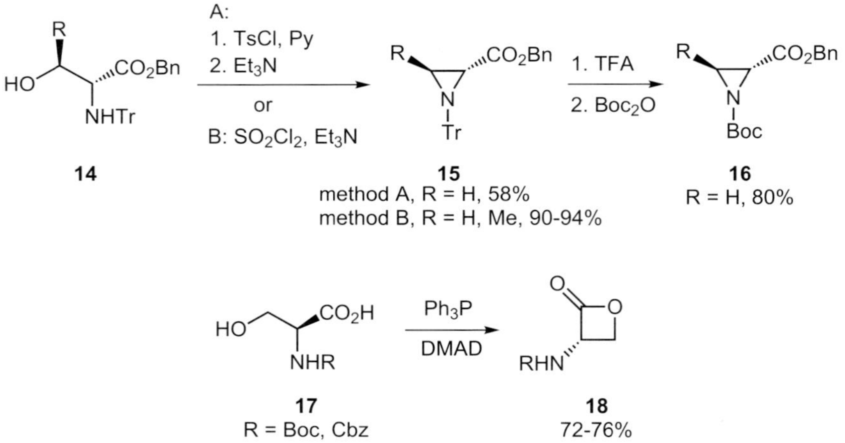

R
HO
$CO_2Bn$
NHTr
**14**
A:
1. TsCl, Py
2. $Et_3N$
or
B: $SO_2Cl_2$, $Et_3N$
R
$CO_2Bn$
N
Tr
**15**
method A, R = H, 58%
method B, R = H, Me, 90-94%
1. TFA
2. $Boc_2O$
R
$CO_2Bn$
N
Boc
**16**
R = H, 80%

HO
$CO_2H$
NHR
**17**
R = Boc, Cbz
$Ph_3P$
DMAD
O
O
RHN
**18**
72-76%

**Scheme 2.4**

An additional useful method to α-amino-protected β-lactones involves the use of asparagine as the starting amino acid (Scheme 2.5). In this case, the side-chain

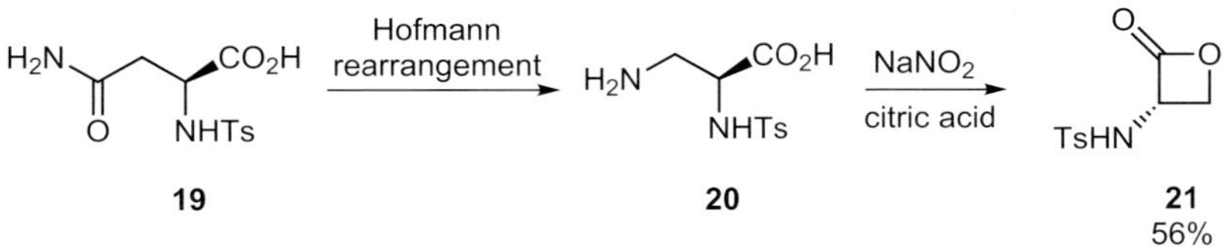

Scheme 2.5

functional group is transformed into a leaving group via Hofmann rearrangement and diazotization followed by an *in situ* cyclization [24].

### 2.2.4
### β-Lactams and Pyroglutamic Acid Derivatives

Aspartic and glutamic acid, the naturally occurring α-amino acids that possess a carboxylic acid group at the side-chain, are attractive precursors for the preparation of β- and γ-lactams, respectively. The intramolecular nucleophilic attack at the side-chain carboxyl moiety by the α-amino acid nitrogen group is one of many methodologies developed for the preparation of these heterocycles of great synthetic utility. The β-lactam moiety is critical to the efficacy of a large class of broad-spectrum antibiotics against bacterial infections, thus being a heterocycle of great pharmacological importance. In particular, β-lactam **24** has been obtained from L-aspartic acid via a Grignard-mediated ring closure – a procedure that occurs without racemization (Scheme 2.6) [25]. The synthetic route involves the preparation of **23** by reacting

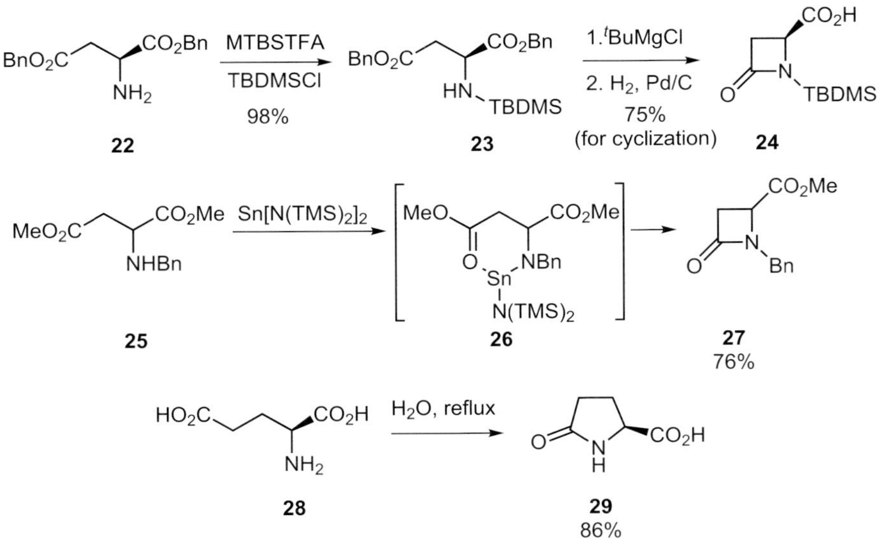

Scheme 2.6

the free amine **22** with *N*-methyl-*N*-*tert*-butyldimethylsilyltrifluoroacetamide (MTBSTFA). The subsequent treatment with *tert*-butylmagnesium chloride yields β-lactam which affords the free acid **24** by hydrogenolysis. Alternatively, the addition of bis(bis(trimethylsilyl)amino)tin(II) over *rac*-**25** has been reported to produce a tin (II) amide that cyclizes to give a β-lactam when the amino group lacks a sterically demanding substituent [26]. On the other hand, the synthetic applications of pyroglutamic acid as a chiral building block or auxiliary in asymmetric synthesis are enormous [27] since this γ-lactam may undergo transformations on the carboxylic group, the ring, the carbonyl group of the lactam, and ring-opening reactions. This heterocycle is synthesized by direct dehydration and ring closure of L-glutamic acid [28].

### 2.2.5 Amino Lactams and Amino Anhydrides

In the same way to the aforementioned intramolecular ring closures, the α-amino acid lysine may afford an ε-lactam by nucleophilic attack of the moiety that is present on the side-chain to the α-carboxylic group. Thus, amino caprolactam **31** has been conveniently made by ring closure of trimethylsilyl amino ester derived from **30** [29] or, alternatively, the amidation of the Fmoc-protected amino acid with 1-(3-dimethylaminopropyl)-3-ethylcarbodiimide hydrochloride (EDCI) in the presence of *N*-hydroxybenzotriazole (HOBt) (Scheme 2.7) [30].

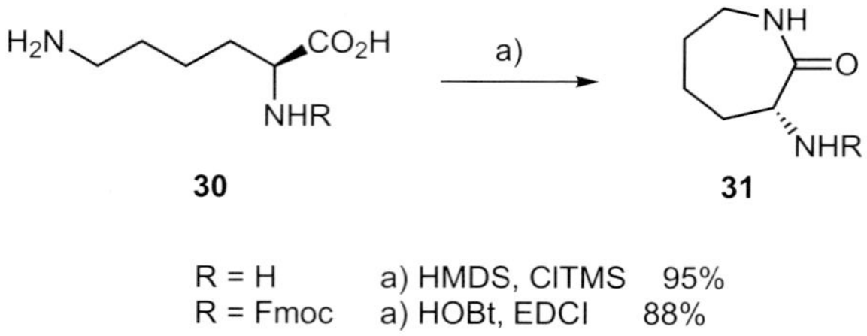

$H_2N$ ... $CO_2H$, NHR **30** —a)→ **31** (NH, =O, NHR)

R = H a) HMDS, CITMS 95%
R = Fmoc a) HOBt, EDCI 88%

Scheme 2.7

Conversely, the condensation of both carboxylic moieties in aspartic and glutamic acid gives access to amino anhydrides **33** in enantiomerically pure form (Scheme 2.8) [31].

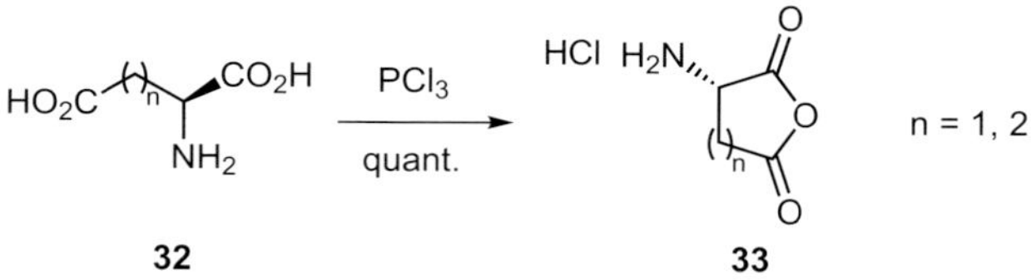

$HO_2C$ ($)_n$ $CO_2H$, $NH_2$ **32** —$PCl_3$, quant.→ HCl $H_2N$ **33** n = 1, 2

Scheme 2.8

### 2.2.6
### Azacycloalkanecarboxylic Acids

The construction of a variety of chiral nitrogen heterocycles can be achieved by means of intramolecular ring-closure reactions via the nucleophilic attack of an enolate, generated at the α-carbon of the α-amino acid, to a leaving group present in an alkyl chain attached to the nitrogen moiety rather than in the α-amino acid side-chain (Scheme 2.9) [32, 33]. The key insights of the methodology are (i) the preservation of the chirality of a starting *N*-alkylated amino ester, such as **34**, in the form of transient conformational chirality of a reactive enolate intermediate and (ii) a high stereoselectivity during the subsequent cyclization procedure. In general, this protocol constitutes a direct method for the asymmetric alkylation of α-amino acid derivatives without the aid of external chiral sources such as chiral auxiliaries or chiral catalysts, and it is referred to as "memory of chirality" [34]. In the case of intramolecular alkylations, this method has been reported to be very useful for the enantiodivergent synthesis of α-substituted prolines. With potassium or sodium amide bases in dimethylformamide (DMF) or tetrahydrofuran (THF), cyclizations proceed with retention of configuration, while inversion of configuration is observed with lithium amide bases in THF or toluene. By varying the nature of the *N*-alkyl chain attached to the α-amino acid precursor, azetidine and piperidine 2-carboxylic acids or α-tetrasubstituted tetrahydroisoquinoline derivatives, such as **38** and **39**, have also been prepared [32, 33, 35].

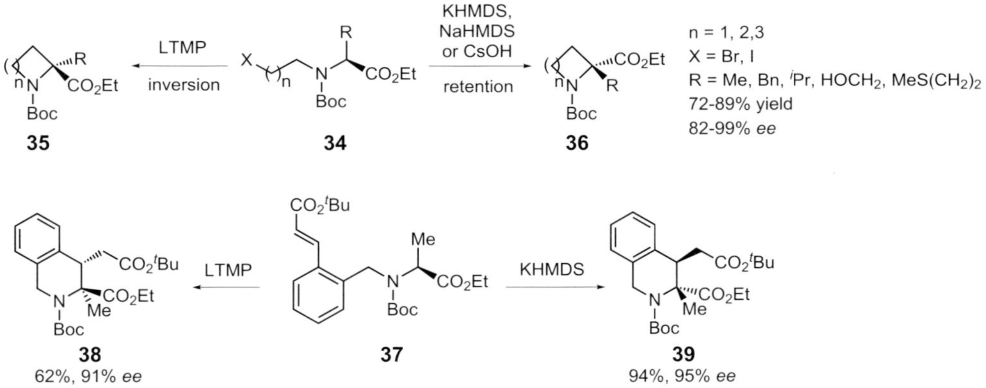

LTMP inversion ← **34** → KHMDS, NaHMDS or CsOH retention

**35** **34** **36**

n = 1, 2,3
X = Br, I
R = Me, Bn, $^i$Pr, $HOCH_2$, $MeS(CH_2)_2$
72-89% yield
82-99% *ee*

**38** 62%, 91% *ee* ← LTMP — **37** — KHMDS → **39** 94%, 95% *ee*

**Scheme 2.9**

## 2.3
## Heterocycles Generated by Intermolecular Cyclizations

### 2.3.1
### Metal Complexes

α-Amino acids form stable five-membered chelate rings with various metal ions through the amine and carboxylate moieties (*N,O*-chelation). What is more, amino acids with coordinating side-chains (histidine, tyrosine, cysteine, etc.) may act as a tridentate ligands. As a consequence of this multifunctional character, amino acids form organometallic complexes [36] with a variety of structures that, in turn, display a great diversity of applications according to their particular physicochemical properties. α-Amino acids cannot only be stabilized by organometallic complexes, they can also be activated. In fact, such heterocyclic complexes have been shown to be useful precursors for the synthesis of substituted amino acids, the controlled synthesis of peptides, the development of new stereoselective reactions, and for labeling and catalytic functions. Methodologically, the α-amino acid *N,O*-chelates can be prepared by substitution reactions of α-amino acid anions on chloro-bridged complexes (Scheme 2.10). Cyclic carbonyl (**43**), $\eta^1$-alkyl (**44**), $\eta^2$-olefin (**45**), $\eta^3$-allyl, $\eta^5$-cyclopentadienyl (**46**), or $\eta^6$-arene (**47**) complexes with a variety of metals have been synthesized and their applications reviewed (some representative examples appear in Scheme 2.10).

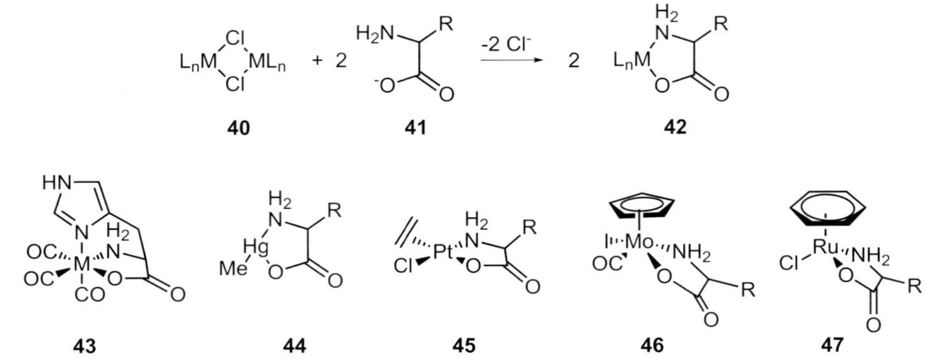

M = Cr, Mo, W, Mn

**Scheme 2.10**

On the other hand, complexation of α-amino acids with boron affords boroxazolidinones, which have found application as a simultaneous protection of the α-amino acid moiety to perform side-chain modifications [37] and, when boron is a stereogenic center, as precursors of optically pure α-alkyl α-amino acids via the generation of chiral enolates that react with asymmetric memory [38].

### 2.3.2
### α-Amino Acid *N*-Carboxyanhydrides and Hydantoins

Other five-membered ring heterocycles of great synthetic value may be prepared through intermolecular cyclizations which involve the reaction of the amine and carboxylate moieties of the α-amino acid with a one-carbon atom dielectrophile. In

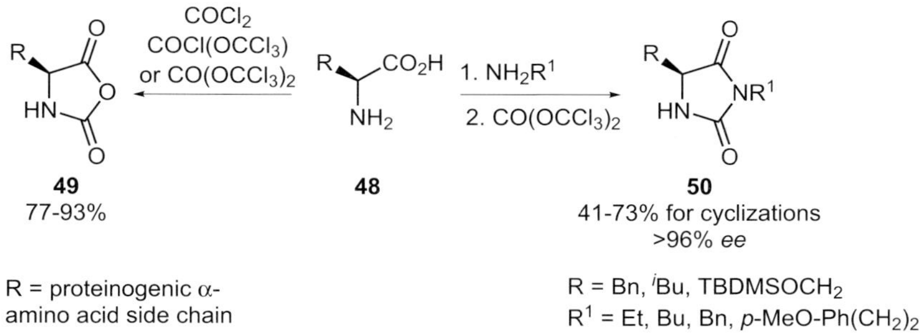

$COCl_2$
$COCl(OCCl_3)$
or $CO(OCCl_3)_2$
1. $NH_2R^1$
2. $CO(OCCl_3)_2$

**49** 77-93%  **48**  **50** 41-73% for cyclizations >96% *ee*

R = proteinogenic α-amino acid side chain

R = Bn, $^i$Bu, $TBDMSOCH_2$
$R^1$ = Et, Bu, Bn, *p*-MeO-$Ph(CH_2)_2$

**Scheme 2.11**

particular, the treatment of α-amino acids with phosgene, diphosgene, or triphosgene gives access to *N*-carboxyanhydrides **49** (Scheme 2.11) [39]. The *N*-carbamoyl intermediates spontaneously cyclize to produce the anhydrides and only in the case of proline the use of a non-nucleophilic base is required for the cyclization to occur [40]. These protected and activated amino acid derivatives are extensively used for stepwise peptide synthesis; in particular, for the preparation of polypeptides by ring-opening polymerizations [41]. In fact, synthetic polypeptides are emerging as very promising compounds in research fields such as drug delivery or material science [42]. In addition, enantiomerically pure hydantoins **50** [43] can be prepared when optically pure α-amino amides are treated with triphosgene. These heterocycles are also useful intermediates in the synthesis of noncoded amino acids, and, additionally, constitute an important structural moiety found in several natural products and therapeutically useful compounds.

## 2.3.3 Oxazolidinones and Imidazolidinones

The condensation of aldehydes with α-amino acids or *N*-protected amide derivatives gives access to oxazolidinones and imidazolidinones, respectively [44]. The use of formaldehyde as dielectrophile provides oxazolidinones with a single chiral center that are useful for the enantiospecific synthesis of α-amino ketones [45], site-selective transformations of α-amino acids [46], or the synthesis of *N*-methyl amino acids [47]. More substituted aldehydes afford chiral heterocycles, differing in the nature of the substituents on the ring and their relative configuration, by means of diastereoselective procedures (Scheme 2.12). The *cis*- and *trans*-oxazolidinone ratios are highly dependent upon several factors including the C2-substituent of the α-amino acid, the aldehyde employed to promote the cyclization, the acylating agent, and the reactions conditions. In example, using *N*-benzoyl protection and pivalaldehyde, and varying the alkyl stereodirecting group of the amino acid, *cis*-oxazolidinones predominate with stereoselectivities ranging from 71 : 29 to 83 : 17, while using benzaldehyde, the *trans*-diastereoisomer is favored with stereoselectivities ranging from 25 : 75 to 12 : 88. In addition, highly stereoselective

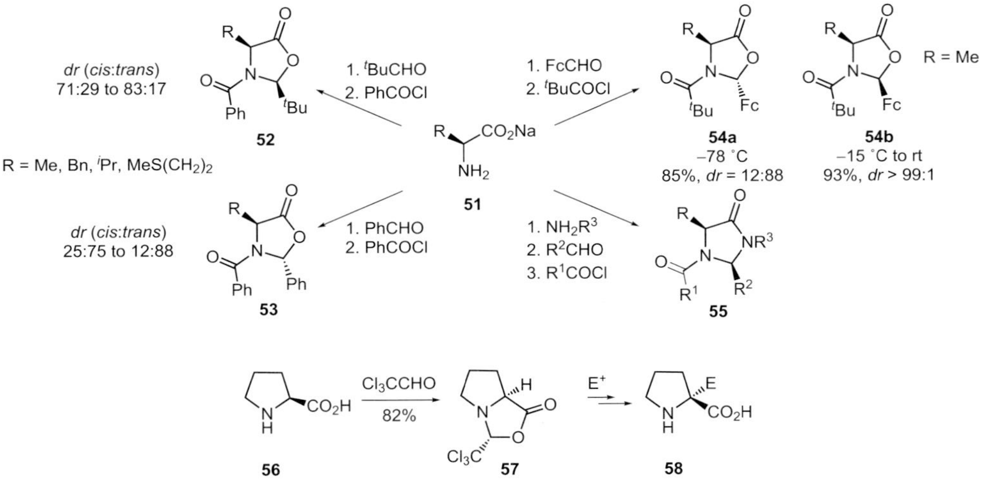

Scheme 2.12

cyclizations of a range of ferrocenyl imines, derived from ferrocenecarboxaldehyde and α-amino acids, have been reported [48]. In these cases, the careful choice of the reaction temperature during the treatment with pivaloyl chloride determines the generation of the *trans* isomer (kinetic control) or the *cis* isomer (thermodynamic control). In general, these compounds have been shown to be extremely useful for the synthesis of enantiopure α-alkyl α-amino acids through the use of the "self-reproduction of chirality" concept [49], by which the stereogenic center of the amino acid generates a temporary center of chirality that is used to diastereoselectively introduce a new substituent at the original chiral center. Among them, glycine- and proline-derived [50] oxazolidinones and imidazolidinones have proved to be particularly useful chiral auxiliaries for the stereoselective synthesis of α-substituted analogs (Scheme 2.12). In particular, chiral **52** (R = H) can be dialkylated in a single procedure and it is the order of addition of the two different electrophiles that determines the absolute configuration of the α-alkyl amino acid isolated after heterocycle hydrolysis. On the other hand, the condensation of L-proline with trichloroacetaldehyde is completely stereoselective and provides **57**, which undergoes α-alkylations with electrophiles with retention of the configuration.

Hexafluoroacetone undergoes heterocyclization with the amino and carboxylic groups of α-amino acids, and provides access to cyclic oxazolidinones **60**. The hexafluoroacetone protection occurs site selectively even in the presence of unprotected side-chain functionalities like the β-carboxy group of aspartic acid. The carboxy-derivatization of these oxazolidinones by nucleophiles is accompanied by *N*-deprotection and, as a result, these bidentate reagents find extensive applications in the synthesis of peptides (Scheme 2.13) [51].

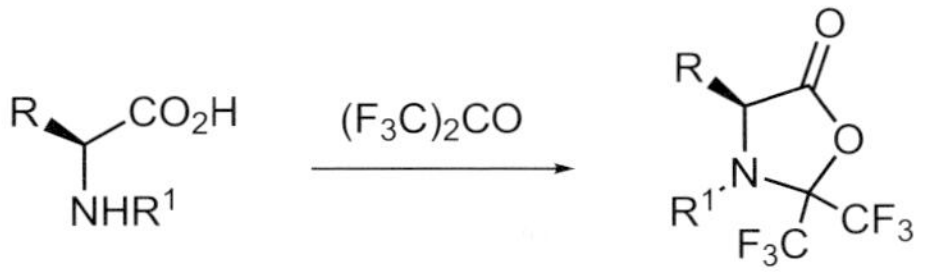


59

60
64-100%

61

R = proteinogenic α-amino acid side chain
$R^1$ = H, alkyl

**Scheme 2.13**

## 2.3.4
## Oxazolones

Saturated 5(4*H*)-oxazolones **63**, also known as 2-oxazolin-5-ones, are easily obtained from *N*-acylamino acids in the presence of cyclodehydrating agents such as anhydrides or carbodiimides, acting as the dielectrophiles (Scheme 2.14) [52]. Similarly, the treatment of *N*-alkoxycarbonyl-α-amino acids generates 2-alkoxy-5(4*H*)-oxazolones. Although 5(4*H*)-oxazolones are obtained in good enantiomeric excesses, the tautomeric equilibrium towards saturated 5(2*H*)-oxazolones (3-oxazolin-5-ones), by a 1,3-prototropic shift, can be a problem due to the racemization associated with such a process. In this context, the cyclizations with perfluoroacylating agents seem to be quite general for the synthesis of 5(2*H*)-oxazolones **66** with aromatic substituents directly bonded to the heterocyclic ring. Additionally, unsaturated 5(4*H*)-oxazolones **64** are obtained by treatment of *N*-acetyl- or *N*-benzoylglycine and an aldehyde

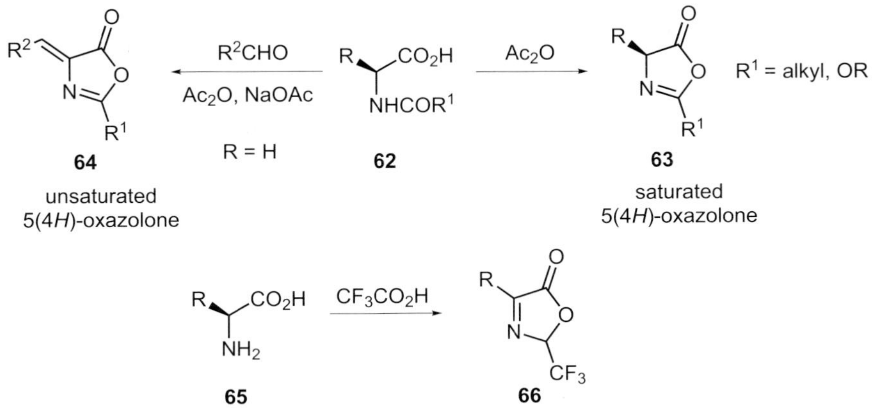


64
unsaturated
5(4*H*)-oxazolone

62

63
saturated
5(4*H*)-oxazolone

65

66
saturated
5(2*H*)-oxazolone

**Scheme 2.14**

in the presence of a cyclodehydrating agent. The synthesis usually proceeds with a high degree of stereoselectivity to favor the thermodynamically more stable (*Z*) isomer, which can be isolated by recrystallization. In general, saturated 5(4*H*)-oxazolones are extensively used in coupling reactions as synthetic equivalents of amino acids in the synthesis of peptides. On the other hand, unsaturated 5(4*H*)-oxazolones are very versatile heterocycles as they undergo ring-opening reactions with a range of nucleophiles and a plethora of reactions on the exocyclic double bond, and exhibit many different applications such as dyes, fungicides, and organic luminophores.

### 2.3.5
### Oxazinones and Morpholinones, Pyrazinones and Diketopiperazines

Intermolecular cyclizations that involve the reaction of the amine and carboxylate moieties of the α-amino acid with compounds that possess two adjacent electrophilic carbon centers gives access to six-membered heterocycles. In particular, oxazinone **69** is a chiral alanine [53] synthetic equivalent of great utility for the asymmetric synthesis of α-methylated amino acids that are important compounds in medicinal chemistry (Scheme 2.15) [54]. The condensation of the potassium salt of *N-tert*-butoxycarbonyl alanine with an aromatic bromoketone affords diastereomeric esters, which can be separated by recrystallization and/or flash chromatography. After *N*-Boc deprotection and cyclization, *trans*-oxazinone **69** is obtained and only slight epimerization at C3 is detected. Conversely, the *cis*-oxazinone suffers epimerization at C6.

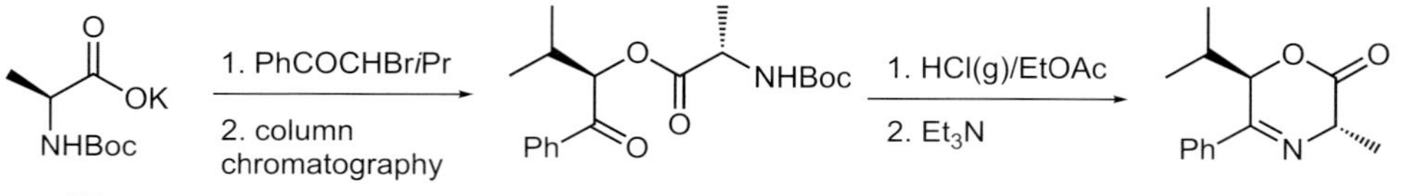

67 → (1. PhCOCHBr*i*Pr; 2. column chromatography) → 68 → (1. HCl(g)/EtOAc; 2. $Et_3N$) → 69, 70%, *dr* 20:1

**Scheme 2.15**

Alternatively, oxazinones **71** and **72** can be prepared through intermolecular cyclizations, which involve the carboxylic group of the α-amino acid acting as an electrophilic carbon center and the amine moiety as a nucleophile (Scheme 2.16). In that manner, glycine synthetic equivalent **71** [55] has been obtained and used as a template for the asymmetric synthesis of dialkylated α-amino acids. The enolates generated from chiral oxazinones are very soft and can be alkylated in a highly diastereoselective manner even under phase-transfer catalysis conditions. Additionally, **71** and glycine imine **72** have been employed for the synthesis of α,β-didehydro-α-amino acids that can be hydrogenated, cyclopropanated, or submitted to cycloaddition reactions [56]. Moreover, the catalytic hydrogenation of oxazinones **73** affords morpholin-2-ones **74** in good yields and high diastereoselectivities.

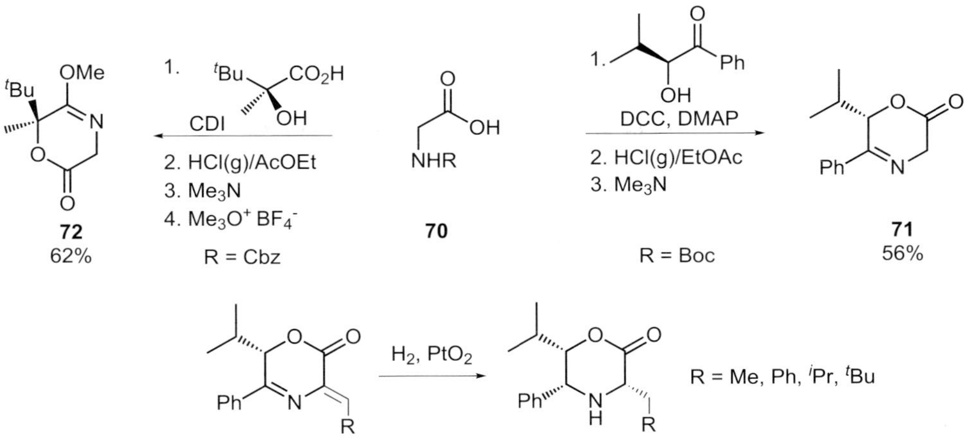


**72** 62% **70** **71** 56% **73** **74** 62-95% *dr* (*cis*:*trans*) = 91:9 to 98:2

**Scheme 2.16**

In a similar manner to the abovementioned preparation of oxazinone **71**, the condensation of two substrates with nucleophilic nitrogens adjacent to electrophilic carbon centers renders pyrazinones. In particular, pyrazinone **78** has been accessed by reaction of α-amino ketone **75** with mixed *N*-Boc-L-alanine-pivalic anhydride followed by cyclization (Scheme 2.17) [57]. This heterocycle is obtained in the same diastereomeric ratio as oxazinone **69** (Scheme 2.15), which means that both compounds have similar acidity at C3. In general, pyrazinones are less sensitive to aqueous acidic and basic media than oxazinones, thus increasing the yields for purified compounds.

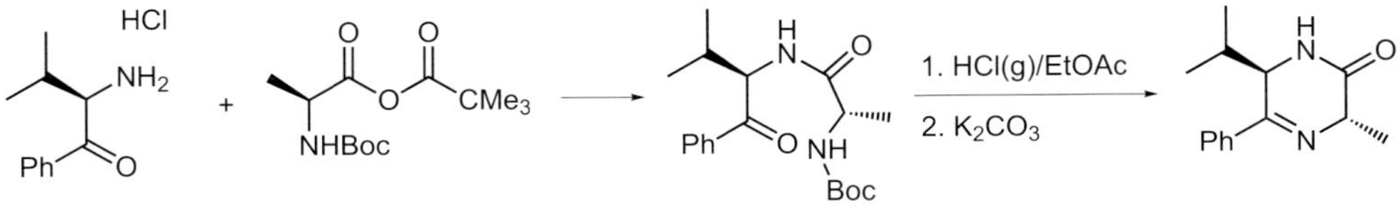


**75** **76** **77** 95% **78** 86%

**Scheme 2.17**

On the other hand, the coupling of *N*-protected α-amino esters **79** to a compound with an electrophilic carboxylic group, followed by nitrogen displacement of the leaving group and intramolecular cyclization of the N1–C2 bond, gives access to unsymmetrical 2,5-diketopiperazines **81** (Scheme 2.18) [58], which are among the most important backbones in today's drug discovery because they may confer more

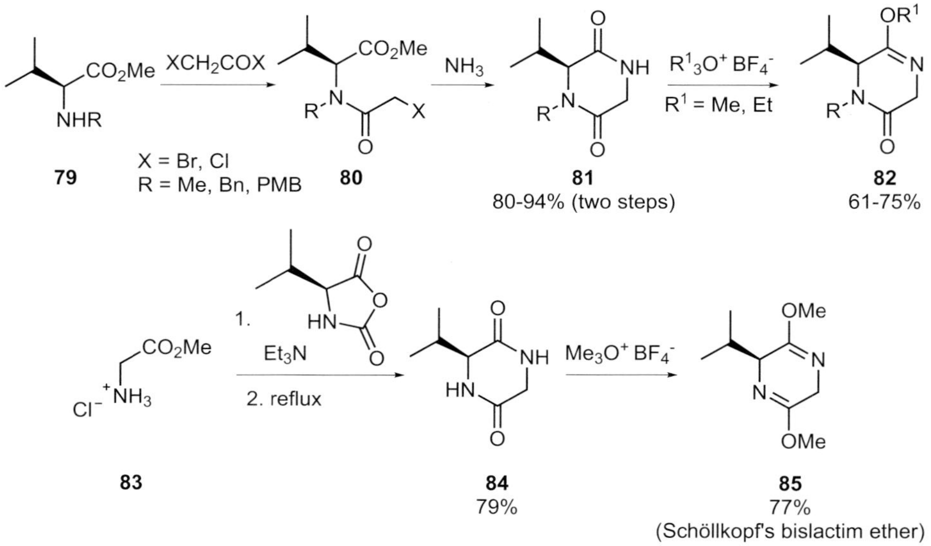

X = Br, Cl
R = Me, Bn, PMB

79 → ($XCH_2COX$) 80 → ($NH_3$) 81 80-94% (two steps) → ($R^1{}_3O^+ BF_4^-$, $R^1$ = Me, Et) 82 61-75%

83 → (1. Et$_3$N, 2. reflux) 84 79% → ($Me_3O^+ BF_4^-$) 85 77% (Schöllkopf's bislactim ether)

Scheme 2.18

drug-like properties to peptide molecules by constraining the nitrogen atoms of an α-amino amide into a ring. Alternatively, analogous unprotected 2,5-diketopiperazine **84** has been obtained, in a 50-g scale, by reaction of glycine methyl ester and the anhydride derived from valine. Such mixed heterocycles can be further converted into monolactim **82** or bislactim ethers **85** [59] that are very useful templates for the preparation of α-substituted amino acids through regiospecific deprotonations followed by stereoselective alkylations.

### 2.3.6
### Tetrahydroisoquinolines and β-Carbolines

Intermolecular cyclizations may also involve functionalities present at the α-amino acid side-chain. Those α-amino acids with aromatic side-chains are ideally suited for the construction of tetrahydroisoquinolines and β-carbolines scaffolds that appear in a diverse array of biologically active compounds of both natural and synthetic origin [60]. The aromatic moieties most often become incorporated into the structure of these heterocycles by means of Pictet–Spengler [61] cyclizations of an appropriately substituted α-amino acid and an aldehyde or aldehyde precursor (Scheme 2.19). In that manner, L-phenylalanine **86** gives access to tetrahydroisoquinoline **87**, which is a conformationally constrained α-amino acid (Tic) of utility, when incorporated into bioactive peptidic structures, for structure–activity relationship analysis and for the generation of compounds with improved biological activity. Correspondingly, the synthetic potential of the Pictet–Spengler cyclizations in aprotic media with either tryptophan methyl ester **88** or its $N_b$-benzyl derivative (**88**, R = Bn) is quite general

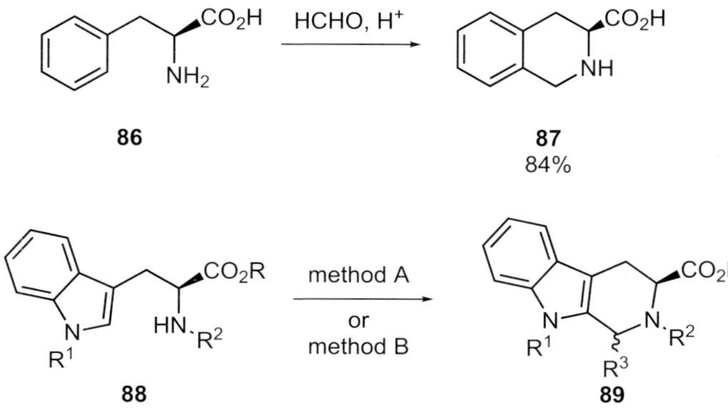

$CO_2H$ HCHO, $H^+$ → $CO_2H$

$NH_2$ NH

**86** **87** 84%

$CO_2R$ method A or method B → $CO_2R$

N HN $R^2$ N N $R^2$

$R^1$ $R^1$ $R^3$

**88** **89**

method A: $R^3CHO$, aprotic solvent

| | | |
|---|---|---|
| R = Me | R = Me | R = Me |
| $R^1$ = H | $R^1$ = Me | $R^1$ = H |
| $R^2$ = H | $R^2$ = H | $R^2$ = Bn |
| $R^3$ = Ph, Et, Pr, $Ph(CH_2)_2$ | $R^3$ = Ph, $C_6H_{11}$, Et | $R^3$ = Ph, $HC(OEt)_2$, Pr |
| *dr* (*cis*:*trans*) = 80:20 | *dr* (*cis*:*trans*) = 0:100 | *dr* (*cis*:*trans*) = 0:100 |

method B: 1. $(CH_2O)n$, camphorsulfonic acid 2. $Et_3SiH$, $F_3CCO_2H$

R = H
$R^1$ = CHO
$R^2$ = Cbz
$R^3$ = H

**Scheme 2.19**

and can be employed with a variety of aldehydes, which can even contain acid-labile functional groups. In particular, the treatment of tryptophan methyl ester **88** with aldehydes at low temperature affords *cis*-carbolines **89** with moderate diastereoselectivity and complete retention of the absolute configuration. Such stereoselectivities have been recently improved by means of crystallization-induced asymmetric transformations in appropriate solvents [62]. On the other hand, $N_b$-benzyl tryptophans afford mainly the *trans* products, possibly to avoid steric interactions among contiguous equatorial substituents in the intermediate carbenium ions. Additionally, non-aromatic α-amino esters may be employed in catalytic enantioselective syntheses of tetrahydroisoquinolines through phase-transfer alkylation–cyclization processes [63].

### 2.3.7
### Oxazo/Thiazolidinones, Oxazo/Thiazolidines, and Oxazo/Thiazolines

The side-chain functionalities of serine, threonine, and cysteine have been employed to generate a range of five-membered ring heterocycles (oxazo/thiazolidinones, oxazo/thiazolidines, and oxazo/thiazolines), which are constituents of numerous bioactive natural products with antitumor, antiviral, and antibiotic properties. In particular, oxazolidinone derivatives of serine and threonine **92** can be prepared by treatment with phosgene under basic conditions, followed by esterification and *N*-acetylation (Scheme 2.20) [64]. After such a protection scheme, these heterocycles may undergo acylations under basic conditions since the β-elimination at the side-chain is strongly retarded on a stereoelectronic basis. Correspondingly, 2-thiazolidinone derivative **94** has been shown to be an effective protective surrogate of the thiol group in cysteine derivatives [65]. It is noteworthy that the preparation of **94**, by treatment with phenylchloroformate under basic conditions is successfully attained in a one-pot procedure and, after subsequent elaborations at the C4 substituent, the thiol group can be liberated by simple heating in DMF.

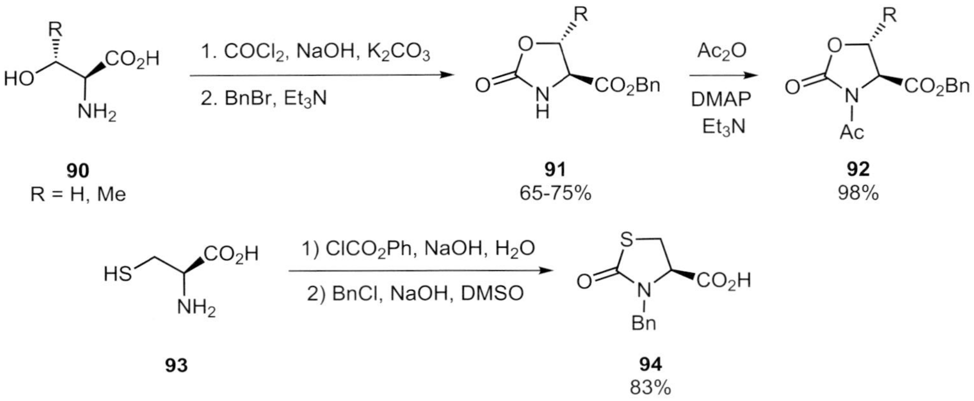

1. $COCl_2$, NaOH, $K_2CO_3$; 2. BnBr, $Et_3N$

**90** R = H, Me → **91** 65-75% → ($Ac_2O$, DMAP, $Et_3N$) → **92** 98%

1) $ClCO_2Ph$, NaOH, $H_2O$; 2) BnCl, NaOH, DMSO

**93** → **94** 83%

Scheme 2.20

In a similar way, the preparation of oxazo- and thiazolidines was also conceived by means of simultaneous protection of the hydroxyl/thiol and nitrogen moieties in serine, threonine, and cysteine (Scheme 2.21). Oxazolidines **96** have been generated

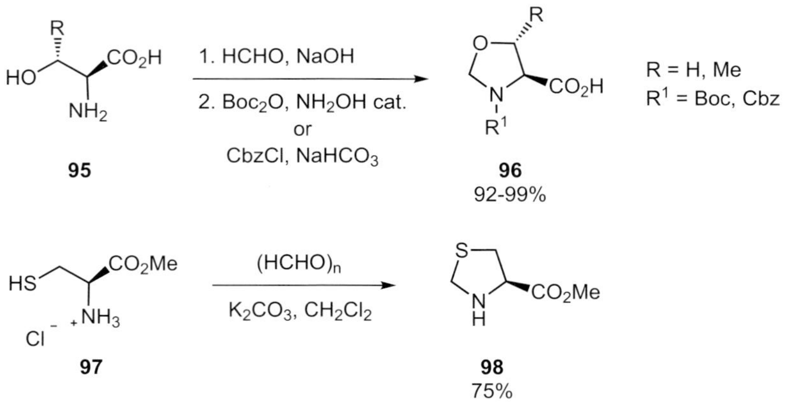

1. HCHO, NaOH; 2. $Boc_2O$, $NH_2OH$ cat. or CbzCl, $NaHCO_3$

**95** → **96** 92-99%; R = H, Me; $R^1$ = Boc, Cbz

$(HCHO)_n$; $K_2CO_3$, $CH_2Cl_2$

**97** → **98** 75%

Scheme 2.21

in optically pure form by the treatment of the α-amino acids serine and threonine with aqueous formaldehyde, followed by nitrogen protection and trapping of the adduct at a pH value greater than 7 [66]. On the other hand, thiazolidine **98** has been synthesized in good yield by the condensation of cysteine ester **97** with paraformaldehyde in nonaqueous conditions [67].

Substituted oxazo- and thiazolidines can be prepared, in a diastereoselective manner, by using bulky aldehydes as electrophiles [68]. In particular (Scheme 2.22), oxazo- and thiazolidine **100** are generated as *cis/trans* mixtures but, owing to facile ring–opening/ring–closure reactions, the *N*-acylation of the mixtures gives rise to a single *cis* isomer [69]. The *t*Bu derivatives **100** have proven very useful to generate enolates, stable with regard to β-elimination, which undergo alkylation from the side opposite to that shielded by the *t*Bu group. Such oxazo- and thiazolidines may afford **101** via the introduction of a leaving group and subsequent elimination. The resulting compounds are useful synthetic equivalents to undergo stereoselective nucleophilic additions to the double bond. On the other hand, L-cysteine reacts with aromatic aldehydes to produce thiazolidines **103**, while serine and threonine give rise to ring-chain tautomeric mixtures due to the lower nucleophilicity and the higher steric strain on the oxygen atom [70]. Additionally, the cyclocondensation of L-cysteine with sugars generates thiazolidines with antioxidative properties and, therefore, their applications in nutrition or drug design (oxidative stress xenobiotics) are being extensively studied [71].

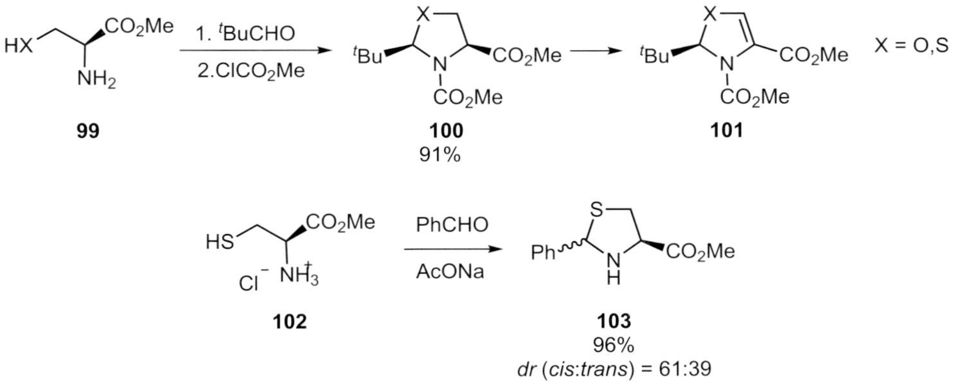

HX–CH₂–CH(NH₂)–CO₂Me (**99**) → 1. $^t$BuCHO, 2. $ClCO_2Me$ → **100** (91%) → **101**; X = O,S

**102** ($HS$, $Cl^-$ $NH_3^+$, $CO_2Me$) → PhCHO, AcONa → **103** (96%, *dr* (*cis:trans*) = 61:39)

**Scheme 2.22**

On the other hand, the use of 2,2-dimethoxypropane or acetone as electrophiles affords dimethyloxazo- **106** and thiazolidines **113** [72]. In particular, oxazolidines **106** have been extensively applied to the synthesis of difficult peptides as backbone protectors for the Fmoc/*t*Bu solid-phase strategy. Such application is limited to serine and threonine derivatives due to their major acid lability (removed by trifluoroacetic acid (TFA) within minutes) in comparison to thiazolidines (removed by TFA within hours) [73]. Among dimethyloxazolidine derivatives generated from α-amino acids, Garner's aldehyde **110** constitutes a very remarkable compound, because it is one of the most valuable chiral building blocks in asymmetric synthesis in recent years [74]. Its value is due to its simple structure that may undergo diastereoselective nucleophilic additions to aldehydes and participate in Diels–Alder reactions. Garner's original synthesis [75] implies Boc protection of serine, methyl esterification, isopropylidenation, and ester reduction. However, the overall procedure has been subject of several improvements that have led to better yields and high optical purities, such as the reversal of the first two steps [76], the use of $BF_3{\cdot}OEt_2$ to catalyze the oxazolidine formation [77] or the use of a $LiAlH_4$-Swern protocol [78] for the ester reduction and subsequent oxidation to the aldehyde (Scheme 2.23).

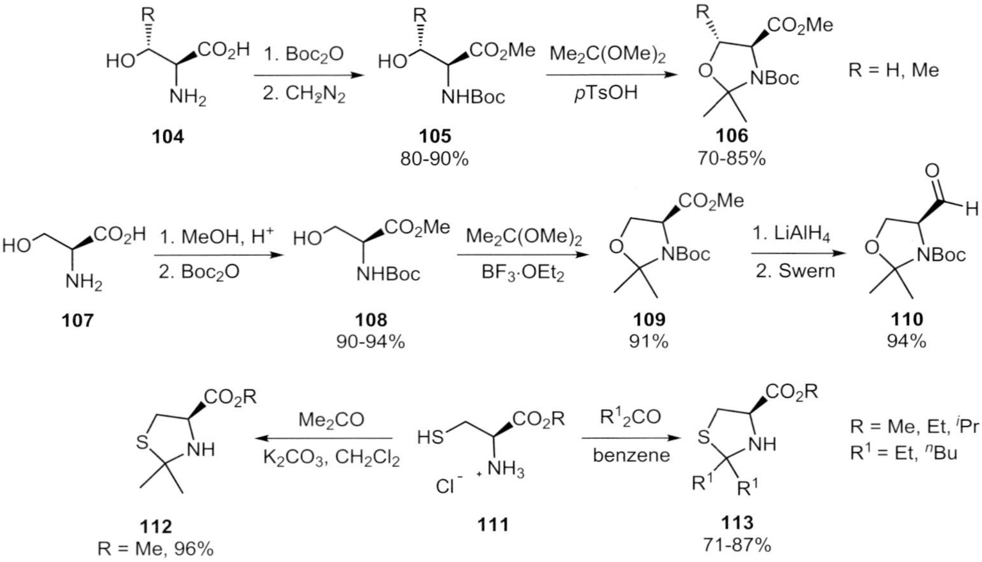

**Scheme 2.23**

The chemical synthesis of 2-oxazolines and 2-thiazolines usually involves the dehydrative cyclization of serine, threonine and cysteine residues. In particular, Mo (VI) oxides efficiently catalyze this process and produce oxazolines **115** in good yields and with complete retention of configuration at the β-position [79]. In the case of cysteine derivatives the use of bis(quinolinolato)dioxomolybdenum(VI) complexes is required to suppress the loss of stereochemical integrity at the C2-exomethine position. Additionally, 2-thiazolines **116** have been prepared by ruthenium-catalyzed oxidation of thiazolidines, condensation of nitriles iminoethers, iminium triflates, or α,α-difluoroalkylamines with cysteine derivatives [80], as well as by a two-step sequence which involves the thioacylation of serine followed by intramolecular cyclization via a mesylate. Conversely, the coupling of aromatic acids and serine *tert*-butyl ester by *N*-(3-dimethylaminopropyl)-*N′*-ethylcarbodiimide (EDC), followed by the cyclization using (diethylamino)sulfur trifluoride (DAST), gives oxazolines **120** in high yields (Scheme 2.24) [81]. These racemic compounds have been

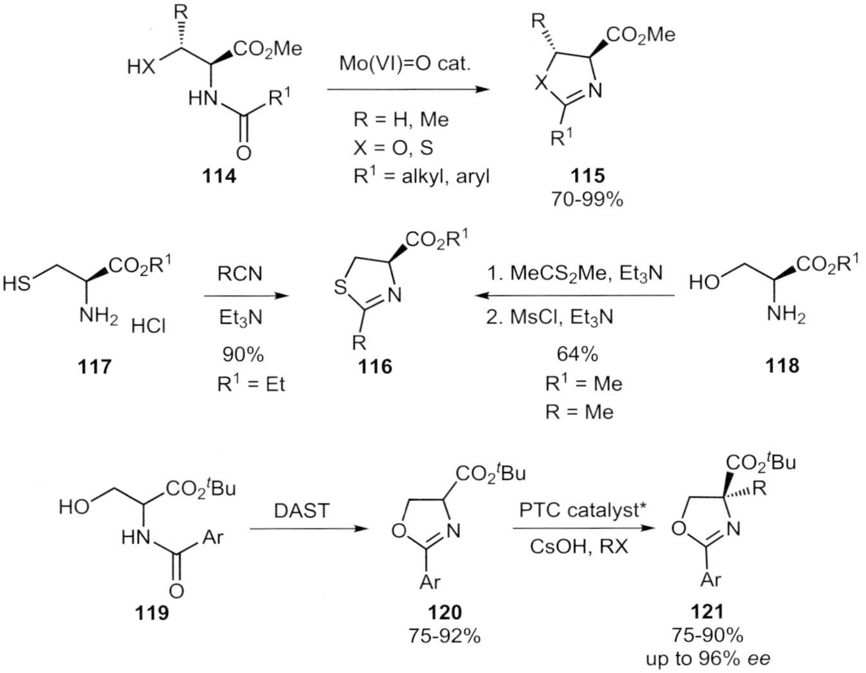

Scheme 2.24

employed for the enantioselective synthesis of α-alkylserines under phase-transfer catalysis conditions.

## 2.3.8
## Sulfamidates

Sulfamidate derivatives of serine and threonine permit protection of the nitrogen moiety and conversion of the hydroxyl group into a leaving group. The most direct approach for their construction is the treatment of esters **122** with sulfuryl chloride; however, this process leads to the corresponding aziridines in excellent yields (Scheme 2.25). Owing to this, cyclic sulfamidates are prepared by oxidation of

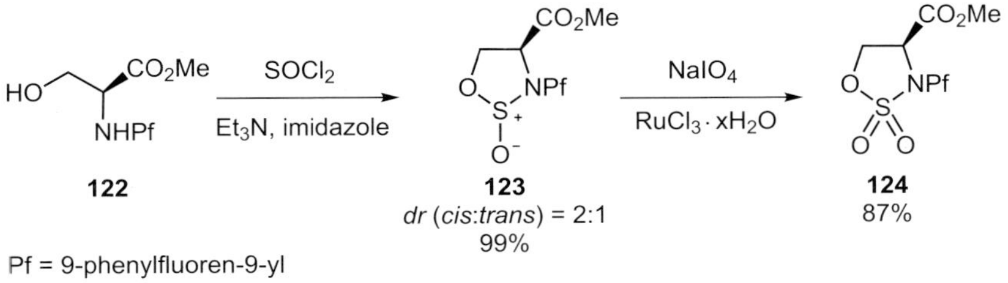

Pf = 9-phenylfluoren-9-yl

Scheme 2.25

the sulfamidites **123** that, in turn, are obtained by the treatment of serine and threonine esters **122** with thionyl chloride, triethylamine, and imidazole [82]. Both heterocyclic compounds can be ring-opened by several heteroatomic and carbon nucleophiles, giving access to valuable β-functionalized α-amino acids and glycopeptides [83].

### 2.3.9 Tetrahydropyrimidinones

The α-amino acid asparagine is the precursor for the preparation of enantiomerically pure tetrahydropyrimidinones **126**, which can also be converted in dihydropyrimidinones **127** (Scheme 2.26) [84]. One-pot syntheses, amenable for the production of large quantities of tetrahydropyrimidinones **126**, have been accomplished when pivalaldehyde or *p*-chlorobenzaldehyde are employed as dielectrophiles. Following these procedures, the newly formed heterocycles exhibit a *cis* relationship of the substituents at C2 and C6. Tetrahydropyrimidinone **126** has proven an efficient chiral auxiliary for the synthesis of enantiomerically pure α-substituted derivatives, following the principle of "self-reproduction of chirality" pioneered by Seebach in that the C2 substituent directs the alkylation stereochemistry at C6 [85].

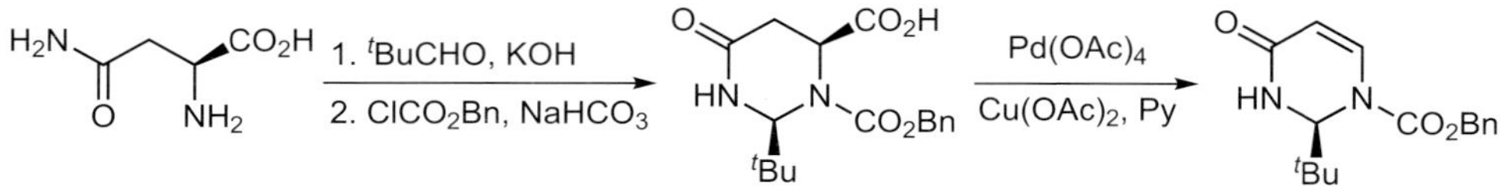

$H_2N$ $CO_2H$ O $NH_2$ **125** — 1. $^tBuCHO$, KOH; 2. $ClCO_2Bn$, $NaHCO_3$ → O $CO_2H$ HN N $CO_2Bn$ $^tBu$ **126** 80% — $Pd(OAc)_4$, $Cu(OAc)_2$, Py → O HN N $CO_2Bn$ $^tBu$ **127** 60%

Scheme 2.26

## 2.4 Heterocycles Generated by Cycloadditions

The manipulation of α-amino acids allows the generation of azomethine ylides and imino dienophiles which have been extensively utilized in 1,3-dipolar cycloadditions and aza-Diels–Alder reactions, respectively. Such methodologies are high yielding, efficient, and regio- and stereocontrolled reactions that find extensive use for the synthesis of heterocyclic compounds. In general, azomethine ylides, generated upon treatment of aromatic and aliphatic imines of α-amino acid esters **128** with a base and a metal salt, undergo highly stereoselective intermolecular 1,3-cycloadditions to electron-deficient olefins and furnish polysubstituted pyrrolidines **129**, which occur in many pharmacologically active compounds. The stereocontrol can be achieved either by means of a chiral auxiliary approach, based on the presence of a chiral

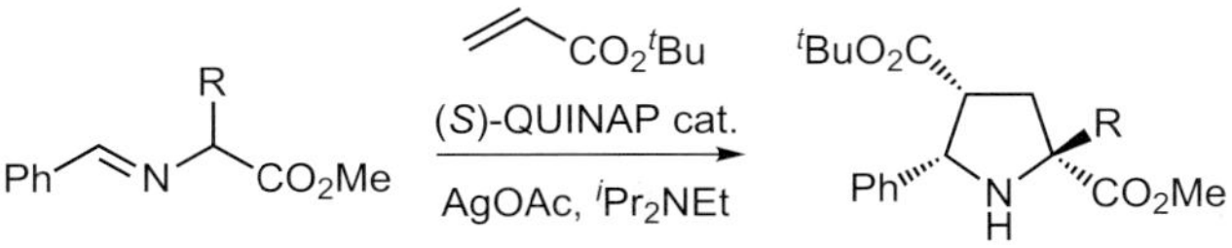


**128**
R = H, Me...

**129**
84%, 80% *ee*

**Scheme 2.27**

moiety at the azomethine ylide or the dipolarophile, or by employing chiral catalysts (Scheme 2.27) [86]. Additionally, azomethine ylides generated from secondary α-amino esters or α-amino acids have been shown to be very useful for the synthesis of polycyclic proline and pyrrolidine derivatives through cycloadditions carried out intramolecularly [87], and also on solid-phase supports [88].

On the other hand, the reaction of dienophiles **131**, derived from α-amino acid esters, with cyclic and open-chain 1,3-dienes renders bicyclic aza compounds **132** and substituted dehydropiperidines in acceptable yields, and with useful diastereomer ratios that depend on the steric demand of the α-amino acid side-chain (Scheme 2.28). The iminium ion generated *in situ* by the reaction of the α-amino acid with the aldehyde is assumed to be the reactive intermediate for the cycloaddition reactions. In this way, the observed diastereoselectivities can be explained by an antiparallel orientation of dipoles given by the carboxyl group and the iminium function on the dienophile, and a preferential attack of the diene from the side opposite to the α-amino acid side-chain [89].

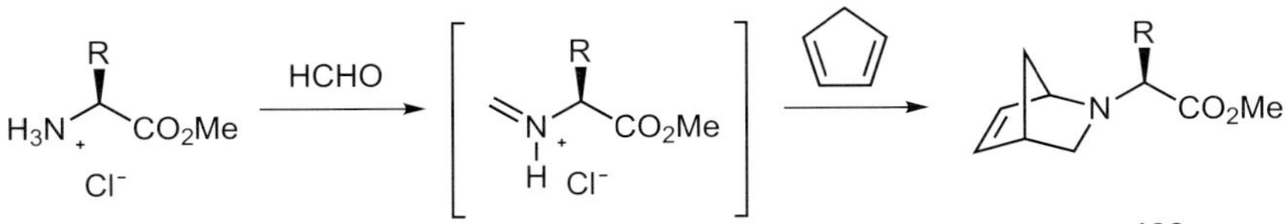


**130**

**131**

**132**

R = *i*Bu, 57%, *dr* 93:7
R = Me, 99%, *dr* 73:27

**Scheme 2.28**

When Danishefsky's diene **134** is combined with aminoester imines **133** in the presence of a Lewis acid, a tandem Mannich–Michael reaction takes places, and enaminones **135** are formed in high yields and with high diastereoisomeric ratios (Scheme 2.29) [89]. In particular, the reaction of electron-rich siloxydienes with imines derived from tryptophan has opened new routes to the polycyclic framework of naturally occuring alkaloids of both reserpine and yohimbine in enantiomerically pure form.

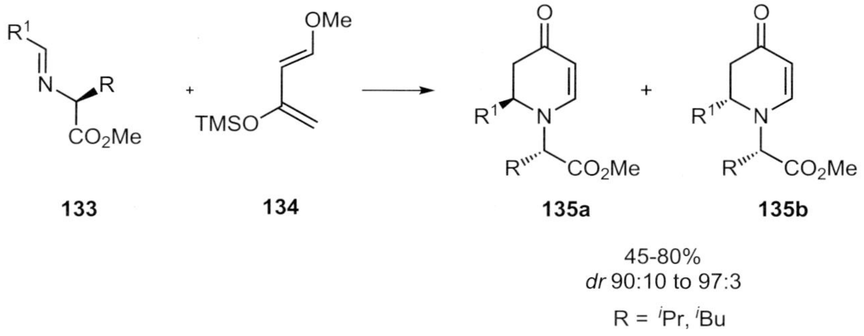
R1
N
R
CO2Me
OMe
TMSO
O
R1
N
R
CO2Me
+
O
R1
N
R
CO2Me
133
134
135a
135b
45-80%
dr 90:10 to 97:3
R = iPr, iBu

Scheme 2.29

## 2.5 Conclusions

The ease of access to α-amino acids in both enantiomerically pure forms, their diversity, and their multifunctional character makes them extremely valuable starting materials for the asymmetric synthesis of complex molecules by means of stereoselective transformations. In this context, we have tried to provide a general perspective, rather than an exhaustive revision due to the vast literature in the field, of the applicability of α-amino acids for the asymmetric synthesis of heterocycles. The synthetic methodologies have been organized according to the type of cyclization that the α-amino acid undergoes to assemble the heterocycle (intramolecular or intermolecular cyclizations and cycloadditions). In this way, the reader may easily realize the potential behind the α-amino acid functional moieties to generate a great diversity of heterocyclic systems.

## 2.6 Experimental Procedures

### 2.6.1 Synthesis of 1-*tert*-Butyl-3-phenylaziridinone (5)

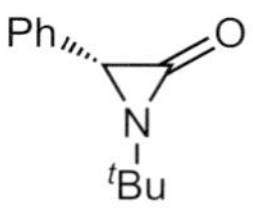
Ph
O
N
tBu

To a well-stirred solution of *N*-*tert*-butyl-2-bromo-2-phenylacetamide (6.75 g, 0.025 mol) in diethyl ether (200 ml) at 0 °C, a suspension of potassium *tert*-butoxide (3.70 g, 0.033 mol) in anhydrous ether (150 ml) was added over a period of 1 h. The reaction mixture was filtered and the filtrate was evaporated under reduced pressure to afford an oily residue. Recrystallization of this residue from dry pentane gave aziridinone **5** (2.60 g, 0.014 mol) in 56% yield.

### 2.6.2
### Synthesis of Dimethyl (2*S*,3a*R*,8a*S*)-1,2,3,3a,8,8a-Hexahydropyrrolo[2,3-*b*]indole-1,2-dicarboxylate (Precursor of 10)

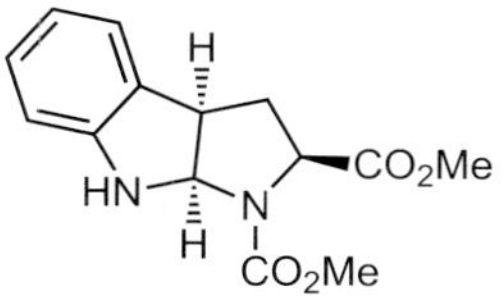

85% Phosphoric acid (43 ml) was added to **9** (4.00 g, 14.46 mmol) at room temperature and the mixture was stirred for 3 h until a clear solution was obtained. This solution was added dropwise to a vigorously stirred two-phase system consisting of 15% aqueous sodium carbonate (90 ml/mmol) and dichloromethane (400 ml) in such a manner as to prevent build-up of local regions of acid pH (the pH of the mixture was greater than 8 throughout the addition). The two phases were separated and the aqueous layer was extracted with chloroform (2 × 20 ml). The combined organic phases were dried with calcium chloride and concentrated under reduced pressure to give dimethyl (2*S*,3a*R*,8a*S*)-1,2,3,3a,8,8a-hexahydropyrrolo[2,3-*b*]indole-1,2-dicarboxylate as an oil (3.40 g, 12.30 mmol, 85% yield) which solidified when set aside at 0 °C.

### 2.6.3
### Synthesis of Benzyl (*R*)-1-Tritylaziridine-2-carboxylate (15)

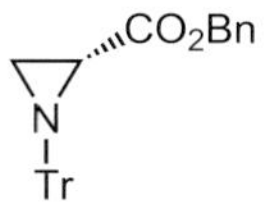

To a solution of **14** (R = H) (65.6 g, 0.15 mol) in dry pyridine (200 ml) at −10 °C a solution of toluenesulfonyl chloride (85.8 g, 0.45 mol) in dry pyridine (50 ml) was added over a period of 1 h. After 24 h of stirring at 0 °C, the solvent was removed *in vacuo* and the residual oil was partitioned between ethyl acetate and water. The organic layer was washed with 10% citric acid solution and water, dried over sodium sulfate and filtered. The solvent was removed *in vacuo* and the oily residue was dissolved in dry THF (100 ml). The solution was treated with triethylamine (39.2 ml, 0.28 mol) and it was then heated at reflux (75 °C) for 24 h. After that, the reaction mixture was concentrated *in vacuo* to give an oily product which was dissolved in ethyl acetate. The organic layer was washed with 10% citric acid solution, 1 M sodium hydrogen carbonate, and water, dried over sodium sulfate, and filtered. The solvent was evaporated under reduced pressure and the product was crystallized from ethyl acetate/ether/hexane (37.4 g, 0.09 mol, 58% yield).

### 2.6.4
### Synthesis of (*S*)-*N*-*tert*-Butoxycarbonyl-3-aminooxetan-2-one (18)

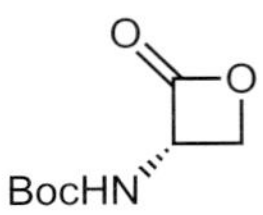


To a solution of dry triphenylphosphine (6.43 g, 24.5 mmol) in anhydrous THF (100 ml) at −78 °C, distilled methyl azodicarboxylate (2.70 ml, 24.5 mmol) was added dropwise over 10 min. After an additional 10 min, a solution of dried *N*-*tert*-butoxycarbonyl-L-serine (5.00 g, 24.4 mmol) in THF (100 ml) was added dropwise over 15 min. The mixture was stirred for 20 min at −78 °C and then 2.5 h at 20 °C. The solvent was removed *in vacuo* at 35 °C and the residue was purified by chromatography to afford **18** as a white crystalline solid (3.29 g, 0.017 mol, 72% yield).

### 2.6.5
### Synthesis of (*S*)-1-(*tert*-Butyldimethylsilyl)-4-oxoazetidine-2-carboxylic Acid (24)

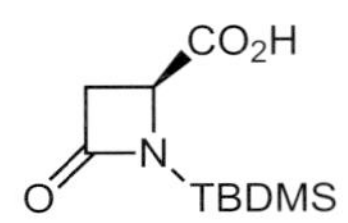


A solution of dibenzyl aspartate **22** (660 mg, 2.10 mmol), *N*-methyl-*N*-*tert*-butyldimethylsilyltrifluoroacetamide (1.5 ml, 6.30 mmol) and *tert*-butylchlorodimethylsilane (32.0 mg, 0.21 mmol) in acetonitrile (1 ml) was stirred for 30 min. The elimination of the solvent under reduced pressure and the excess of silylating reagent under high vacuum (1 mmHg) for 18 h, furnished **23** (875 mg, 2.05 mmol) as a pale yellow oil. This crude product was dissolved in diethyl ether (20 ml) and cooled to 0 °C. *tert*-Butylmagnesium chloride (1.13 ml, 2.26 mmol) was added dropwise and the resulting yellow suspension was stirred overnight while warming to room temperature. After careful addition of saturated ammonium chloride (20 ml), the stirring was continued for 10 min. The organic layer was washed with water (20 ml) and brine (30 ml), dried over magnesium sulfate, and filtered. The solvent was concentrated *in vacuo* to afford a yellow oil which was purified by column chromatography (hexanes/diethyl ether 1: 1) to yield **24** (356 mg, 75% yield) as a colorless oil.

### 2.6.6
### Synthesis of 9*H*-Fluoren-9-ylmethyl (*R*)-Hexahydro-2-oxo-1*H*-azepin-3-yl Carbamate (31)

A solution of **30** (R = Fmoc), as TFA salt, (719 mg, 1.49 mmol) in dichloromethane (57 ml) containing DMF (14 ml) at 18 °C, under an inert atmosphere, was treated with

Hünig's base (290 μl, 1.10 mmol). After 10 min of stirring, the white suspension was cooled to 0 °C and it was then treated with HOBt (244 mg, 1.79 mmol) and EDCI (344 mg, 1.79 mmol). After 16 h at 18 °C, a 1 M aqueous solution of tartaric acid (20 ml) was added to the reaction mixture and then it was extracted with ethyl acetate (3 × 30 ml). The combined organic layers were washed with sodium bicarbonate (50 ml), water (50 ml) and brine (50 ml), dried over magnesium sulfate and filtered. The solvent was concentrated *in vacuo* to afford a light yellow oil which was purified by column chromatography eluting with methanol/chloroform/triethylamine 1: 98: 1 to yield **31** (460 mg, 1.31 mmol, 88%) as a white powder.

### 2.6.7
### Synthesis of Ethyl (*S*)-*N*-(*tert*-Butoxycarbonyl)-α-(*tert*-butoxymethyl)proline Ester (36)

A mixture of powdered cesium hydroxide (56 mg, 0.38 mmol) in dry dimethyl sulfoxide (water <0.005%, 1.0 ml) was vigorously stirred (1000–1100 rpm) under an argon atmosphere at 20 °C for 5 min. A solution of **34** (R = $CH_2OtBu$, X = I, n = 2) (96 mg, 0.25 mmol) in dry dimethyl sulfoxide (1.5 ml) was added to the suspension at 20 °C. After vigorously stirring (1000–1100 rpm) for 30 min, the reaction mixture was poured into saturated aqueous ammonium chloride and extracted with ethyl acetate (30 ml). The organic phase was washed with saturated aqueous sodium bicarbonate and brine, dried over sodium sulfate, filtered, and concentrated *in vacuo*. The residue was purified by preparative thin-layer chromatography (hexanes/ethyl acetate 3: 1) and furnished **36** (57 mg, 0.22 mmol) in 89% yield and 93% e.e.

### 2.6.8
### Synthesis of Proline *N*-Carboxyanhydride (49)

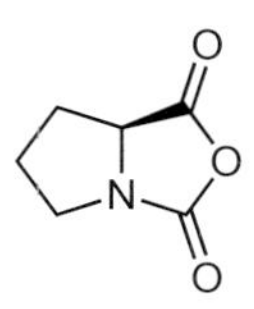

L-Proline (0.5 g, 4.34 mmol) was suspended in dry THF (5 ml) in a round-bottomed flask provided with a calcium chloride tube. The mixture was heated at 50 °C and triphosgene (0.64 g, 2.16 mmol) was added in one portion. The reaction mixture was stirred for 1 h and then evaporated under reduced pressure. The oily residue was dissolved in dry THF (5 ml) and added onto polymer-supported amine DEAM-PS (3 equiv.) which had been previously swollen in dry THF into a polypropylene syringe fitted with a polyethylene filter disk and kept (and purged) under argon atmosphere for 3 h (at room temperature). Proline *N*-carboxyanhydride was obtained by filtration, washing with additional dry THF. The solvent was removed under reduced pressure and the crude was recrystallized from ethyl acetate/hexanes, cooling it to −20 °C, to furnish **49** (0.57 g, 4.03 mmol) in 93% yield and 99% purity.

### 2.6.9
### Synthesis of (2*S*,4*S*)-2-Ferrocenyl-3-pivaloyl-4-methyl-1,3-oxazolidin-5-one (54b)

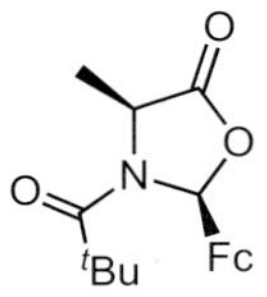

O N O O $^t$Bu Fc

Molecular sieves (4 Å) and ferrocene carboxaldehyde (2.3 g, 10.8 mmol) were added to a solution of L-alanine sodium salt (0.92 g, 10.3 mmol) in ethanol and the resulting mixture was stirred at room temperature for 5 h. The sieves were removed by filtration and the filtrate was concentrated *in vacuo*. The residue was suspended in pentane, filtered, and dried *in vacuo* to yield an imine (3.00 g, 9.77 mmol), which was suspended in dichloromethane with 4 Å molecular sieves and the mixture was cooled to −15 °C. After dropwise addition of pivaloyl chloride (1.2 ml, 9.77 mmol) in dichloromethane, the reaction mixture was allowed to warm to room temperature over 16 h, and then it was filtered. The filtrate was concentrated *in vacuo* (avoiding strong heating in order to prevent possible racemization). Several portions of diethyl ether were added, and the mixture was then passed through a short plug of celite and silica. The filtrate was removed under reduced pressure and the residue was recrystallized from diethyl ether/pentane to afford **54b** (3.42 g, 9.28 mmol, 90% yield, greater than 98% d.e.).

### 2.6.10
### Synthesis of (6*S*)-6-Isopropyl-5-phenyl-3,6-dihydro-2*H*-1,4-oxazin-2-one (71)

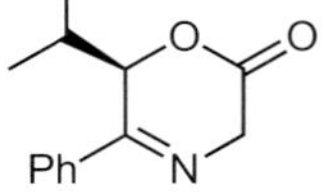

O O Ph N

To a solution of *N*,*N*′-dicyclohexylcarbodiimide (2.27 g, 11 mmol), *N*-Boc-glycine (1.89 g, 10 mmol) and a catalytic amount of *N*,*N*-(dimethylamino)pyridine in dichloromethane (25 ml) was added a solution of (*R*)-2-hydroxy-3-methyl-1-phenylbutanone (1.78 g, 10 mmol) in dichloromethane (25 ml), and the mixture was stirred overnight at room temperature. The resulting precipitate was filtered off, the filtrate was concentrated *in vacuo*, and the residue was purified by column chromatography to afford an ester (2.68 g, 8 mmol), which was treated with a saturated solution of hydrogen chloride in ethyl acetate (25 ml) for 1 h. The solvent was evaporated *in vacuo*, and the solid was washed with ether and filtered. The solid was dissolved in dichloromethane (3 ml) and a solution of trimethylamine in dichloromethane (5 ml) (obtained by extracting a sodium chloride saturated 45% solution of trimethylamine in water (3 ml) with dichloromethane (6 ml)) was added. The resulting mixture was stirred for 1 h and then cooled to 0 °C, hexane (8 ml) was added, and the precipitate was filtered off and washed with a 1 : 1 hexanes/dichloromethane mixture (2 × 5 ml). The combined filtrates were evaporated *in vacuo*, affording oxazinone **71** (1.21 g, 5.60 mmol, 56% yield).

### 2.6.11
### Synthesis of (3*S*,6*R*)-6-Isopropyl-3-methyl-5-phenyl-1,2,3,6-tetrahydro-2-pyrazinone (78)

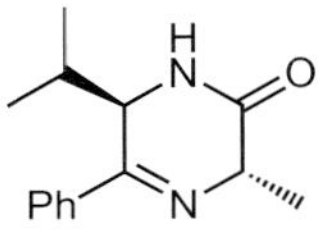

A solution of ketone **77** (3.48 g, 10 mmol) in a 3 M solution of hydrogen chloride in ethyl acetate (50 ml) was stirred for 1 h. The solvent was evaporated, the residue dissolved in hydrochloric acid (0.1 M, 10 ml), and the solution was washed with ether (10 ml). The organic phase was discarded, and the aqueous phase was treated with saturated solution of potassium carbonate (50 ml) and extracted with ethyl acetate (3 × 15 ml). The organic layer was dried over sodium sulfate, filtered, and the solvent was removed under reduced pressure to afford pyrazinone **78** (1.98 g, 8.6 mmol) as a pale yellow oil in 86% yield.

### 2.6.12
### Synthesis of (3*S*)-3,6-Dihydro-2,5-dimethoxy-3-isopropylpyrazine (85)

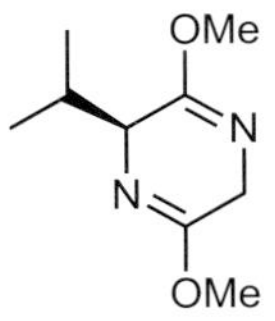

Triethylamine (87.55 g, 0.87 mol) was added to a suspension of glycine methyl ester hydrochloride (55.44 g, 0.43 mol) in chloroform (500 ml) under nitrogen and the reaction mixture was cooled to −78 °C. A solution of the anhydride derived from L-valine (58.62 g, 0.41 mol) in THF (300 ml) was added dropwise over a period of 4 h resulting in the precipitation of triethylamine hydrochloride as a gelatinous white solid. After stirring for 1 h at −78 °C, the reaction mixture was stored at −20 °C for 12 h. The precipitate was removed by filtration through Celite and the filtrate was concentrated *in vacuo* to afford a crude oil which was redissolved in THF (300 ml).

Filtration through Celite and evaporation of the solvent gave a colorless unstable oil that was refluxed in toluene (300 ml) for 24 h. The solvent was removed under reduced pressure to provide a pink solid that was decolorized by treatment with charcoal in water (300 ml), under reflux, for 1 h. After filtering off the charcoal the solvent was removed *in vacuo* with the aid of ethanol (300 ml). The resulting solid was dried at 90 °C for 24 h to provide **84** as a white powder (53.1 g, 79% yield). Trimethyloxonium tetrafluoroborate (135 g, 915 mmol) and **84** (50 g, 320 mmol) were efficiently mixed together, as solids, and then covered with dichloromethane (500 ml) under an atmosphere of nitrogen. The heterogeneous reaction mixture was stirred at room temperature for 72 h. The biphasic mixture was then poured slowly, with rapid stirring and cooling, into a saturated solution of sodium bicarbonate (750 ml) ensuring that the solution remained above pH 7.5. The resulting solution was extracted with dichloromethane (2 × 200 ml), filtered through Celite, the organic layer dried over magnesium sulfate, and concentrated *in vacuo* to afford **85** (45 g, 245 mmol, 77% yield) as a pale yellow oil.

### 2.6.13
### Synthesis of (3*S*)-1,2,3,4-Tetrahydroisoquinoline-3-carboxylic Acid (87)

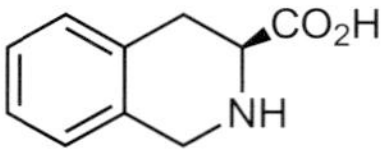

To a suspension of L-phenylalanine (5.0 g, 0.03 mol) in chloroform (50 ml), formaldehyde (27 ml) and concentrated hydrochloric acid (45 ml) were added dropwise, and the resulting mixture was stirred at 80–90 °C for 10 h. After cooling the reaction to room temperature, the precipitate was collected by filtration and washed with water and acetone to afford **87** as a colorless powder (4.5 g, 0.025 mol, 84%).

### 2.6.14
### Synthesis of Methyl (*S*)-*N*-*tert*-Butoxycarbonyl-2,2-dimethyloxazolidine-4-carboxylate (109)

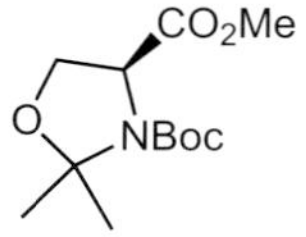

Ester **108** (3.05 g, 13.9 mmol) was dissolved in a mixture of acetone (50 ml) and 2,2,-dimethoxypropane (15 ml) to which boron trifluoride etherate (0.1 ml) was added. The resulting solution was stirred at room temperature for 2 h. The solvent was removed under reduced pressure, the residual oil was taken up in dichloromethane (50 ml), and washed with a mixture of saturated sodium bicarbonate and water (1 : 1, 30 ml) and brine (30 ml). The organic layer was dried over magnesium sulfate, filtered, and the solvent was removed *in vacuo* to give **109** as a pale yellow oil (3.28 g, 12.6 mmol, 91% yield).

2.6.15
### Synthesis of (2*S*,6*S*)-2-*tert*-Butyl-1-carbobenzoxy-4-oxopyrimidin-6-carboxylic Acid (126)

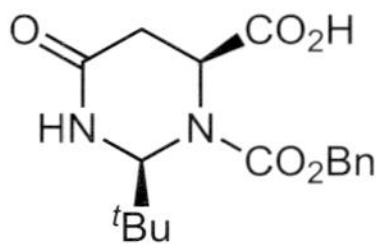

A slurry of L-asparagine monohydrate (6.17 g, 41.1 mmol) in water (4 ml) was treated with KOH (2.30 g, 41.1 mmol) dissolved in water (6 ml). The slightly yellow solution was heated to 60 °C and evaporated overnight under high vacuum (0.1 Torr), producing a slightly yellow glassy solid. The solid was dissolved with absolute methanol (63 ml), treated at room temperature with pivalaldehyde (5.1 ml, 46 mmol), and heated at reflux with vigorous stirring for 6–8 h. Evaporation of solvent under reduced pressure (50 Torr) and heat (60 °C) left behind an off-white lumpy solid that was dissolved in aqueous solution of sodium bicarbonate (83 ml, 3.19 g, 41.1 mmol). Benzyl chloroformate (7.28 ml, 49.3 mmol) was added dropwise to the ice-cooled solution from a pressure equalizing dropping funnel. After stirring for 8 h at room temperature, the mixture was quenched with 5% HCl (32 ml, 46.0 mmol) and evaporated to one-third volume, whereupon a white precipitate was produced. The slurry was extracted with ethyl acetate (100 ml), and the aqueous layer was back-extracted with 20% methanolic ethyl acetate (3 × 200 ml). The organic layers were combined and dried over anhydrous magnesium sulfate. Filtration, evaporation of the solvent and prolonged evacuation (1 Torr; 48 h) gave **126** as a white solid (11.0 g, 32.9 mmol, 80% yield).

## Acknowledgments

Financial support from the Ministerio de Educación y Ciencia – FEDER (project CTQ2007-62245) and Gobierno de Aragón (research group E40) is gratefully acknowledged.

## References

1 Sardina, F.J. and Rapoport, H. (1996) *Chemical Reviews*, **96**, 1825–1872.
2 (a) Olah, G.A., Shih, J., and Prakash, G.K.S. (1983) *Helvetica Chimica Acta*, **66**, 1028–1030; (b) Quast, H. and Leybach, H. (1991) *Chemische Berichte*, **124**, 849–859.
3 Hoffman, R.V. and Cesare, V. (2005) *Science of Synthesis*, **21**, 591–608 and references therein.
4 Miyoshi, M. (1973) *Bulletin of the Chemical Society of Japan*, **46**, 212–218.
5 Jones, J.H. and Witty, M.J. (1977) *Journal of the Chemical Society. Chemical Communications*, 281–282.
6 Cesare, V., Lyons, T.M., and Lengyel, I. (2002) *Synthesis*, 1716–1720.
7 Anthoni, U., Christophersen, C., and Nielsen, P.H. (1999) Naturally Occurring

Cyclotryptophans and Cyclotryptamines, in *Alkaloids: Chemical and Biological Perspectives*, vol. 13 (ed. S.W. Pelletier) Pergamon: New York, pp. 163–236.
8 Mei, T.-S., Wang, X., and Yu, J.-Q. (2009) *Journal of the American Chemical Society*, **131**, 10806–10807.
9 Li, J.-J., Mei, T.-S., and Yu, J.-Q. (2008) *Angewandte Chemie International Edition*, **47**, 6452–6455.
10 Crich, D. and Banerjee, A. (2007) *Accounts of Chemical Research*, **40**, 151–161 and references therein.
11 May, J.P., Fournier, P., Pellicelli, J., Patrick, B.O., and Perrin, D.M. (2005) *The Journal of Organic Chemistry*, **70**, 8424–8430 and references therein.
12 Kawahara, M., Nishida, A., and Nakagawa, M. (2000) *Organic Letters*, **2**, 675–678.
13 Bailey, P.D., Cochrane, P.J., Irvine, F., Morgan, K.M., Pearson, D.P.J., and Veal, K.T. (1999) *Tetrahedron Letters*, **40**, 4593–4596.
14 Ohno, M., Spande, T.F., and Witkop, B. (1970) *Journal of the American Chemical Society*, **92**, 343–348.
15 Coste, A., Toumi, M., Wright, K., Razafimahaléo, V., Couty, F., Marrot, J., and Evano, G. (2008) *Organic Letters*, **10**, 3841–3844.
16 Yang, H.W. and Romo, D. (1999) *Tetrahedron*, **55**, 6403–6434 and references therein.
17 Sweeney, J.B. (2002) *Chemical Society Reviews*, **31**, 247–258.
18 Olsen, C.A., Christensen, C., Nielsen, B., Mohamed, F.M., Witt, M., Clausen, R.P., Kristensen, J.L., Franzyk, H., and Jaroszewski, J.W. (2006) *Organic Letters*, **8**, 3371–3374.
19 Osborn, H.M.I. and Sweeney, J. (1997) *Tetrahedron: Asymmetry*, **8**, 1693–1715 and references therein.
20 Beresford, K.J.M., Church, N.J., and Young, D.W. (2006) *Organic & Biomolecular Chemistry*, **4**, 2888–2897.
21 Kuyl-Yeheskiely, E., Lodder, M., van der Marel, G.A., and van Boom, J.H. (1992) *Tetrahedron Letters*, **33**, 3013–3016.
22 Swamy, K.C.K., Kumar, N.N.B., Balaraman, E., and Kumar, K.V.P.P. (2009) *Chemical Reviews*, **109**, 2551–2651.
23 Arnold, L.D., Kalantar, T.H., and Vederas, J.C. (1985) *Journal of the American Chemical Society*, **107**, 7105–7109.
24 Miyoshi, M., Fujii, T., Yoneda, N., and Okumura, K. (1969) *Chemical & Pharmaceutical Bulletin*, **17**, 1617–1622.
25 Baldwin, J.E., Adlington, R.M., Gollins, D.W., and Schofield, C.J. (1990) *Tetrahedron*, **46**, 4733–4748.
26 Wang, W.-B. and Roskamp, E.J. (1993) *Journal of the American Chemical Society*, **115**, 9417–9420.
27 (a) Nájera, C. and Yus, M. (1999) *Tetrahedron: Asymmetry*, **10**, 2245–2303; (b) Panday, S.K., Prasad, J., and Dikshit, D.K. (2009) *Tetrahedron: Asymmetry*, **20**, 1581–1632.
28 (a) Pellegata, R., Pinza, M., and Pifferi, G. (1978) *Synthesis*, 614–616; (b) Martin, L.L., Scott, S.J., Agnew, M.N., and Setescak, L.L. (1986) *The Journal of Organic Chemistry*, **51**, 3697–3700.
29 Parker, M.F., Bronson, J.J., Barten, D.M., Corsa, J.A., Du, W., Felsenstein, K.M., Guss, V.L., Izzarelli, D., Loo, A., McElhone, K.E., Marcin, L.R., Padmanabha, R., Pak, R., Polson, C.T., Toyn, J.H., Varma, S., Wang, J., Wong, V., Zheng, M., and Roberts, S.B. (2007) *Bioorganic & Medicinal Chemistry Letters*, **17**, 5790–5795.
30 Banwell, M.G. and McRae, K.J. (2001) *The Journal of Organic Chemistry*, **66**, 6768–6774.
31 (a) Yan, G., Wu, Y., Lin, W., and Zhang, X. (2007) *Tetrahedron: Asymmetry*, **18**, 2643–2648; (b) Lin, W., He, Z., Zhang, H., Zhang, X., Mi, A., and Jiang, Y. (2001) *Synthesis*, 1007–1009.
32 (a) Kawabata, T., Matsuda, S., Kawakami, S., Monguchi, D., and Moriyama, K. (2006) *Journal of the American Chemical Society*, **128**, 15394–15395; (b) Kawabata, T., Moriyama, K., Kawakami, S., and Tsubaki, K. (2008) *Journal of the American Chemical Society*, **130**, 4153–4157 and references therein; (c) Moriyama, K., Sakai, H., and Kawabata, T. (2008) *Organic Letters*, **10**, 3883–3886.
33 Kolaczkowski, L. and Barnes, D.M. (2007) *Organic Letters*, **9**, 3029–3032.
34 (a) Fuji, K. and Kawabata, T. (1998) *Chemistry – A European Journal*, **4**, 373–376; (b) Kawabata, T. and Fuji, K.

(2003) Memory of chirality: asymmetric induction based on the dynamic chirality of enolates, in *Topics in Stereochemistry* (ed. S.E. Denmark), John Wiley & Sons, Inc., New York, pp. 175–205.

35 Bonache, M.A., Cativiela, C., García-López, M.T., and González-Muñiz, R. (2006) *Tetrahedron Letters*, **47**, 5883–5887.

36 (a) Shimazaki, Y., Takani, M., and Yamauchi, O. (2009) *Dalton Transactions*, 7854–7869; (b) Severin, K., Bergs, R., and Beck, W. (1998) *Angewandte Chemie (International Edition in English)*, **37**, 1634–1654.

37 (a) Gong, B. and Lynn, D.G. (1990) *The Journal of Organic Chemistry*, **55**, 4763–4765; (b) Schade, D., Töpker-Lehmann, K., Kotthaus, J., and Clement, B. (2008) *The Journal of Organic Chemistry*, **73**, 1025–1030.

38 (a) Vedejs, E., Fields, S.C., and Schrimpf, M.R. (1993) *Pure and Applied Chemistry*, **65**, 723–728; (b) Vedejs, E., Fields, S.C., and Schrimpf, M.R. (1993) *Journal of the American Chemical Society*, **115**, 11612–11613; (c) Ferey, V., Toupet, L., Le Gall, T., and Mioskowski, C. (1996) *Angewandte Chemie (International Edition in English)*, **35**, 430–432.

39 Smeets, N.M.B., van der Weide, P.L.J., Meuldijk, J., Vekemans, J.A.J.M., and Hulshof, L.A. (2005) *Organic Process Research & Development*, **9**, 757–763 and references therein.

40 Gulín, O.P., Rabanal, F., and Giralt, E. (2006) *Organic Letters*, **8**, 5385–5388.

41 Isidro-Llobet, A., Álvarez, M., and Albericio, F. (2009) *Chemical Reviews*, **109**, 2455–2504.

42 (a) Kricheldorf, H.R. (2006) *Angewandte Chemie International Edition*, **45**, 5752–5784; (b) Matsusaki, M., Waku, T., Kaneko, T., Kida, T., and Akashi, M. (2006) *Langmuir*, **22**, 1396–1399.

43 Zhang, D., Xing, X., and Cuny, G.D. (2006) *The Journal of Organic Chemistry*, **71**, 1750–1753.

44 (a) Cativiela, C. and Ordóñez, M. (2009) *Tetrahedron: Asymmetry*, **20**, 1–63; (b) Cativiela, C. and Díaz-de-Villegas, M.D. (2007) *Tetrahedron: Asymmetry*, **18**, 569–623.

45 Paleo, M.R., Calaza, M.I., and Sardina, F.J. (1997) *The Journal of Organic Chemistry*, **62**, 6862–6869.

46 Hoffmann, M.G. and Zeiss, H.-J. (1992) *Tetrahedron Letters*, **33**, 2669–2672.

47 (a) Aurelio, L. and Hughes, A.B. (2009) Synthesis of *N*-alkyl amino acids, in *Amino Acids, Peptides and Proteins in Organic Chemistry. Origins and Synthesis of Amino Acids*, vol. 1 (ed. A.B. Hughes), Wiley-VCH Verlag GmbH, Weinheim, pp. 245–289; (b) Hoveyda, H.R. and Pinault, J.-F. (2006) *Organic Letters*, **8**, 5849–5852; (c) Zhang, S., Govender, T., Norström, T., and Arvidsson, P.I. (2005) *The Journal of Organic Chemistry*, **70**, 6918–6920; (d) Aurelio, L., Brownlee, R.T.C., and Hughes, A.B. (2004) *Chemical Reviews*, **104**, 5823–5846.

48 (a) Alonso, F., Davies, S.G., Elend, A.S., and Smith, A.D. (2009) *Organic & Biomolecular Chemistry*, **7**, 518–526; (b) Alonso, F., Davies, S.G., Elend, A.S., Leech, M.A., Roberts, P.M., Smith, A.D., and Thomson, J.E. (2009) *Organic & Biomolecular Chemistry*, **7**, 527–536.

49 Seebach, D., Sting, A.R., and Hoffmann, M. (1996) *Angewandte Chemie (International Edition in English)*, **35**, 2708–2748.

50 Calaza, M.I. and Cativiela, C. (2008) *European Journal of Organic Chemistry*, 3427–3448.

51 Spengler, J., Böttcher, C., Albericio, F., and Burger, K. (2006) *Chemical Reviews*, **106**, 4728–4746.

52 Cativiela, C. and Díaz-de-Villegas, M.D. (2004) 5(2*H*)-Oxazolones and 5(4*H*)-oxazolones, in *The Chemistry of Heterocyclic Compounds. Oxazoles: Synthesis, Reactions, and Spectroscopy, Part B*, vol. 60 (ed. D.C. Palmer), John Wiley & Sons, Inc., New York, pp. 129–330.

53 Chinchilla, R., Falvello, L.R., Galindo, N., and Nájera, C. (1997) *Angewandte Chemie (International Edition in English)*, **36**, 995–997.

54 (a) Nájera, C., Abellán, T., and Sansano, J.M. (2000) *European Journal of Organic Chemistry*, 2809–2820; (b) Abellán, T., Chinchilla, R., Galindo, N., Guillena, G., Nájera, C., and Sansano, J.M. (2000) *European Journal of Organic Chemistry*, 2689–2697.

55 Chinchilla, R., Falvello, L.R., Galindo, N., and Nájera, C. (2000) *The Journal of Organic Chemistry*, **65**, 3034–3041.

56 Koch, C.-J., Simonyiová, S., Pabel, J., Kärtner, A., Polborn, K., and Wanner, K.T. (2003) *European Journal of Organic Chemistry*, 1244–1263.

57 Abellán, T., Nájera, C., and Sansano, J.M. (1998) *Tetrahedron: Asymmetry*, **9**, 2211–2214.

58 (a) Paradisi, F., Porzi, G., and Sandri, S. (2001) *Tetrahedron: Asymmetry*, **12**, 3319–3324; (b) Davies, S.G., Garner, A.C., Ouzman, J.V.A., Roberts, P.M., Smith, A.D., Snow, E.J., Thomson, J.E., Tamayo, J.A., and Vickers, R.J. (2007) *Organic & Biomolecular Chemistry*, **5**, 2138–2147; (c) Dinsmore, C.J. and Beshore, D.C. (2002) *Tetrahedron*, **58**, 3297–3312.

59 (a) Schöllkopf, U. (1983) *Pure and Applied Chemistry*, **55**, 1799–1806; (b) Bull, S.D., Davies, S.G., and Moss, W.O. (1998) *Tetrahedron: Asymmetry*, **9**, 321–327.

60 (a) Vicario, J.L., Badía, D., Carrillo, L., and Etxebarria, J. (2003) *Current Organic Chemistry*, **7**, 1775–1792; (b) Aurelio, L., Brownlee, R.T.C., and Hughes, A.B. (2002) *Organic Letters*, **4**, 3767–3769.

61 Cox, E.D. and Cook, J.M. (1995) *Chemical Reviews*, **95**, 1797–1842.

62 Xiao, S., Lu, X., Shi, X.-X., Sun, Y., Liang, L.-L., Yu, X.-H., and Dong, J. (2009) *Tetrahedron: Asymmetry*, **20**, 430–439.

63 Ooi, T., Takeuchi, M., and Maruoka, K. (2001) *Synthesis*, 1716–1718.

64 Xi, N. and Ciufolini, M.A. (1995) *Tetrahedron Letters*, **36**, 6595–6598.

65 Seki, M., Kimura, M., Hatsuda, M., Yoshida, S.-I., and Shimizu, T. (2003) *Tetrahedron Letters*, **44**, 8905–8907.

66 Falorni, M., Conti, S., Giacomelli, G., Cossu, S., and Soccolini, F. (1995) *Tetrahedron: Asymmetry*, **6**, 287–294.

67 Braga, A.L., Milani, P., Vargas, F., Paixao, M.W., and Sehnem, J.A. (2006) *Tetrahedron: Asymmetry*, **17**, 2793–2797.

68 (a) Calmes, M., Escale, F., and Paolini, F. (1997) *Tetrahedron: Asymmetry*, **8**, 3691–3697; (b) Pinho e Melo, T.M.V.D. (2004) 1,3-Thiazolidine-4-carboxylic acids as building blocks in organic synthesis, in *Targets in Heterocyclic Systems*, vol. 8 (eds O.A. Attanasi and D. Spinelli,) Italian Society of Chemistry, Rome, pp. 288–329.

69 Jeanguenat, A. and Seebach, D. (1991) *Journal of the Chemical Society, Perkin Transactions 1*, 2291–2298.

70 Fülöp, F. and Pihlaja, K. (1993) *Tetrahedron*, **49**, 6701–6706.

71 Yan, Y., Wan-Shun, L., Bao-Qin, H., and Hai-Zhou, S. (2006) *Nutrition Research*, **26**, 369–377.

72 Garner, P. and Park, J.M. (1987) *The Journal of Organic Chemistry*, **52**, 2361–2364.

73 Kemp, D.S. and Carey, R.I. (1989) *The Journal of Organic Chemistry*, **54**, 3640–3646.

74 Liang, X., Andersch, J., and Bols, M. (2001) *Journal of the Chemical Society, Perkin Transactions 1*, 2136–2157 and references therein.

75 Garner, P. (1984) *Tetrahedron Letters*, **25**, 5855–5858.

76 McKillop, A., Taylor, R.J.K., Watson, R.J., and Lewis, N. (1994) *Synthesis*, 31–33.

77 Moriwake, T., Hamano, S., Saito, S., and Torii, S. (1987) *Chemistry Letters*, 2085–2088.

78 Dondoni, A. and Perrone, D. (1997) *Synthesis*, 527–529.

79 Sakakura, A., Kondo, R., Umemura, S., and Ishihara, K. (2009) *Tetrahedron*, **65**, 2102–2109.

80 Gaumont, A.-C., Gulea, M., and Levillain, J. (2009) *Chemical Reviews*, **109**, 1371–1401 and references therein.

81 Lee, Y.-J., Lee, J., Kim, M.-J., Kim, T.-S., Park, H.-G., and Jew, S.-S. (2005) *Organic Letters*, **7**, 1557–1560.

82 Wei, L. and Lubell, W.D. (2001) *Canadian Journal of Chemistry*, **79**, 94–104.

83 Meléndez, R.E. and Lubell, W.D. (2003) *Tetrahedron*, **59**, 2581–2616.

84 Chu, K.S., Negrete, G.R., Konopelski, J.P., Lakner, F.J., Woo, N.-T., and Olmstead, M.M. (1992) *Journal of the American Chemical Society*, **114**, 1800–1812.

85 Hopkins, S.A., Ritsema, T.A., and Konopelski, J.P. (1999) *The Journal of Organic Chemistry*, **64**, 7885–7889.

86 Pandey, G., Banerjee, P., and Gadre, S.R. (2006) *Chemical Reviews*, **106**, 4484–4517.

87 Coldham, I. and Hufton, R. (2005) *Chemical Reviews*, **105**, 2765–2809.

88 Harju, K. and Yli-Kauhaluoma, J. (2005) *Molecular Diversity*, **9**, 187–207.

89 Waldmann, H. (1995) *Synlett*, 133–141.

# 3
# Radical-Mediated Synthesis of α-Amino Acids and Peptides

*Jan Deska*

## 3.1
## Introduction

The development of efficient and flexible methods for the preparation of α-amino acids and related peptides has been attracting considerable attention over the past decades, and numerous different approaches have been explored, meeting the challenges of regioselectivity issues, stereocontrol, and functional group tolerance. Apart from the extensively exploited ionic reactions employing nucleophilic amino acid enolates or electrophilic amino acid cation equivalents, the number of investigations towards radical-mediated procedures has steadily increased over the years. Offering the benefits of strictly neutral conditions such as absence of racemization hazards and compatibility with a wide range of functionalities, a variety of free radical processes have been developed. The different approaches can be categorized in three major strategies according to the nature of the amino acid precursors (Figure 3.1). One field of research deals with the free radical formation of amino acid or peptide building blocks, either at the α-carbon or the amino acid side-chain, and their modification through appropriate trapping reactions. A second pathway includes the radical additions to glyoxylate imine derivatives as radicalophilic glycine precursors. In a third concept, α,β-dehydroamino acids are used as precursors, which are modified via radical conjugate addition. The different theoretical considerations, practical procedures, stereoselective solutions, and synthetic applications of the respective strategies are discussed in the following sections.

## 3.2
## Free Radical Reactions

One major strategy for the radical-mediated synthesis or modification of amino acids deals with the generation of carbon-centered free radicals at either the α-carbon or the amino acid side-chain, which are subsequently trapped by functional radical acceptors. The radical formation can be accomplished by either hydrogen atom abstraction

*Amino Acids, Peptides and Proteins in Organic Chemistry.*
*Vol.3 – Building Blocks, Catalysis and Coupling Chemistry.* Edited by Andrew B. Hughes

ISBN: 978-3-527-32102-5

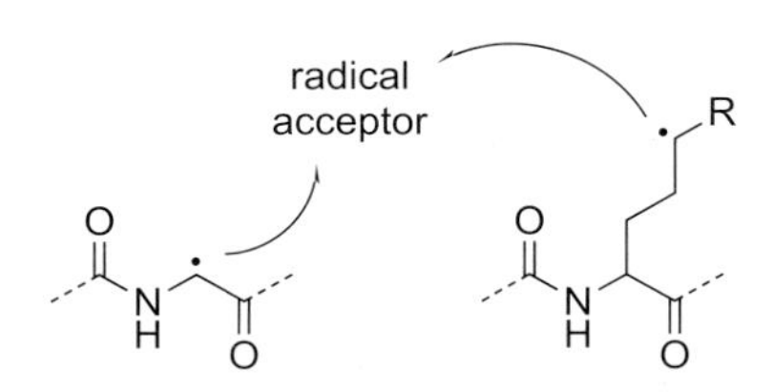


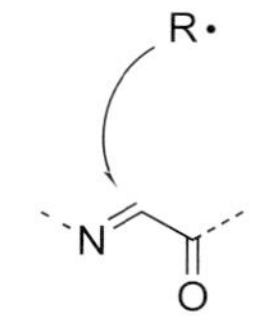


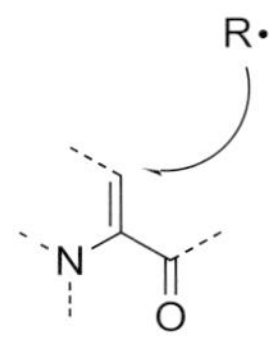


**Figure 3.1** Strategies for the radical-mediated synthesis or modification of α-amino acids.

or through radical degradation of functional groups present in the amino acid or peptide building block.

### 3.2.1
### Hydrogen Atom Transfer Reactions

Hydrogen transfer reactions offer certain advantages as they allow for the direct substitution of nonactivated hydrogens both in carbon–carbon and carbon–heteroatom coupling reactions. Although amino acids, and in particular peptides, can exhibit a broad number of carbon–hydrogen bonds capable of undergoing homolytic cleavage, a high degree of regio- and chemoselectivity is often observed. In general, hydrogen atom transfer reactions show high selectivities to form the most stabilized radical, which is in most cases the captodative α-carbon-centered radical. Both with the amino function as the free base or as *N*-amido derivatives, the unpaired spin density of α-carbon-centered amino acid radicals is extensively delocalized as a combined resonance effect from the electron-withdrawing (capto) carboxy group and the electron-donating (dative) amino or amido group, respectively. In contrast, α-radical formation of protonated or quaternized amino acid derivatives is less favored due to the lack of dative resonance stabilization (Figure 3.2) [1].

Easton *et al.* studied the radical formation behavior of different *N*-acyl amino acid methyl esters and their relative stabilities [2]. Contrary to the expectation that tertiary radicals are more stable than secondary ones, in the α-bromination reaction using *N*-bromosuccinimide, the reaction of the glycine derivative was faster than the bromination of corresponding alanine and valine derivatives. This phenomenon has been attributed to the destabilizing effects of nonbonding interactions in the different radical species. The stabilization of the captodative α-radical results from overlap of the semioccupied p-orbital with the π-orbitals of the amido and the methoxycarbonyl groups, which is maximal in a planar conformation. In comparison to the glycinyl radical, the alaninyl radical will thereby be destabilized due to the nonbonding interaction between the methyl side-chain and the amido oxygen; even stronger destabilizing effects result from nonbonding interactions with the sterically more demanding isopropyl group in the valinyl radical. By this means, *N*-methylation causes also destabilization, so that both alanine and sarcosine derivatives show comparable reaction rates; on the other, hand methyl pyroglutaminate, relatively free

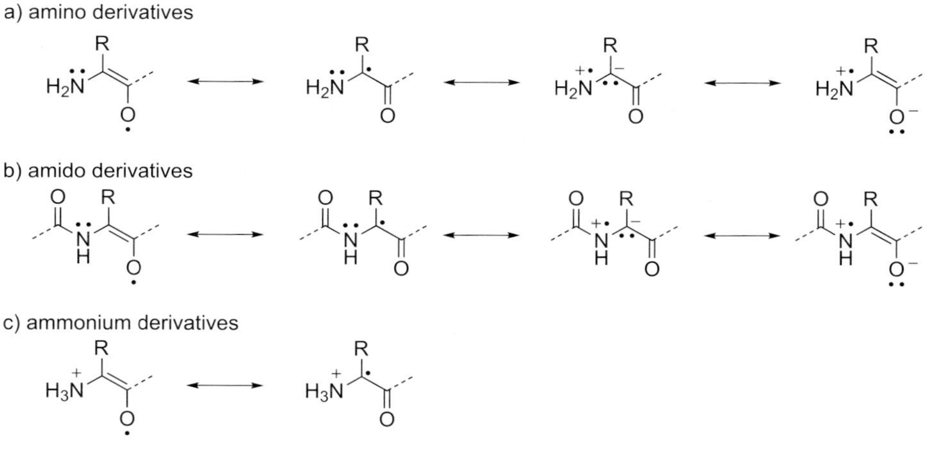

**Figure 3.2** Resonance stabilization of amino, amido, and ammonium α-radicals.

of nonbonding interactions in the radical intermediate, reacts faster than the glycine ester. See Figure 3.3.

These differences in reaction rates can be exploited in the selective modification of glycine-containing peptide derivatives. In the pioneering work of Elad, the photochemical alkylation of amino acids [3], peptides [4], and proteins [5] has been investigated. Using an α-diketone together with di-*tert*-butylperoxide as the photoinitiating system, *tert*-butoxy radicals are formed, which act as hydrogen-abstracting species, leading to α-radical generation in both the peptide as well as in the alkylating agent toluene and recombination of the peptide radical with the tolyl radical results in the formation of a phenylalanine–valine dipeptide (**2**), although without significant asymmetric induction from the adjacent valine moiety (Figure 3.4). Various alkylating agents were used such as methoxytoluene, fluorotoluene, 1-butene, and acetic anhydride, yielding the corresponding tyrosine, fluorophenylalanine, norleucine, or aspartic acid derivatives.

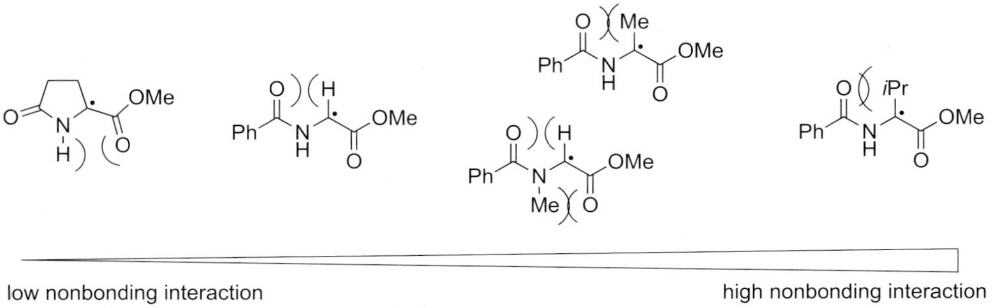

**Figure 3.3** Nonbonding interactions in different amino acid α-carbon-centered radicals.

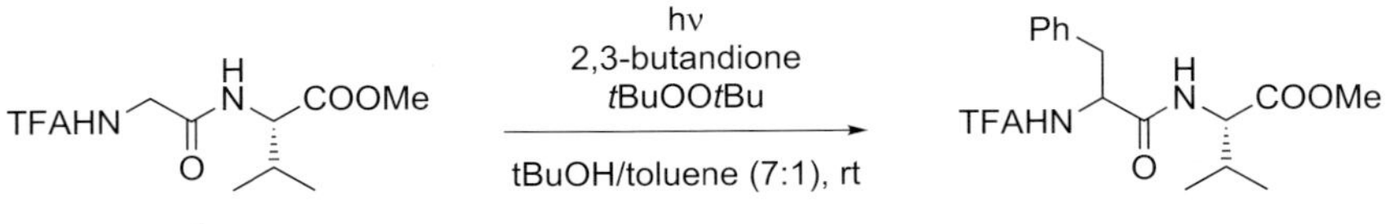


**1**

**2**, 50% yield, dr = 1:1

**Figure 3.4** Photochemical peptide alkylation.

Easton *et al.* elaborated a two-step procedure for the preparation of allylated peptides consisting of a regioselective radical bromination of glycine moieties in dipeptides followed by a radical allylation employing allylstannanes [6]. Skrydstrup *et al.* used potassium *O*-ethyl dithiocarbonate to trap the unstable α-bromoglycine peptide intermediates forming peptidic α-xanthates, which served as α-radical precursors in C–C coupling reactions such as radical allylations and radical additions to olefins [7]. As in the case of Elad's photoalkylation, the reactions with linear peptides only reached modest diastereoselectivities. This limitation was overcome by the use of *N,N′*-diacetyl-diketopiperazines [8]. The cyclic dipeptide **3**, derived from (*S*)-valine and glycine, reacted highly regio- and diastereoselective both in the bromination and the allylation step to give the (*R*)-allylglycine derivative **5** in good yield as a single stereoisomer (Figure 3.5).

Apart from the diastereocontrol in peptides, there have also been efforts to develop stereoselective radical-mediated protocols for the enantioselective synthesis of single amino acids by α-hydrogen atom transfer of glycine derivatives bearing chiral auxiliaries attached either to the amine or the carboxyl group. In a fully radical-mediated approach presented by Hamon *et al.*, chiral α-brominated 8-phenylmenthyl glycinate, derived via radical bromination of **6** [9], was treated with various allylstannanes and the corresponding (*S*)-allylglycine derivatives were obtained in high yield and excellent stereoselectivity (Figure 3.6) [10]. Furthermore propargyl- and

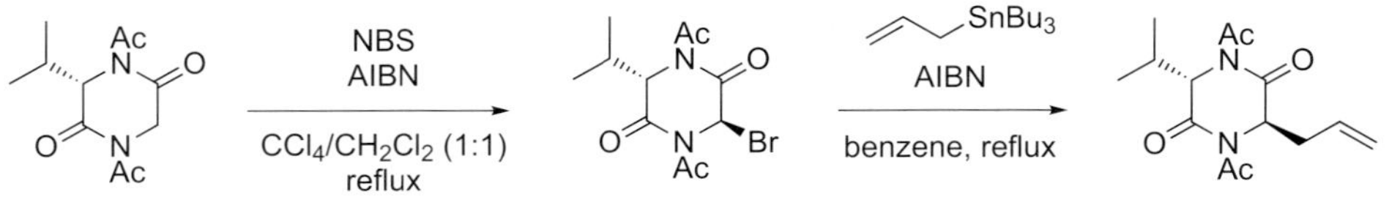


**3**

**4**, 86% yield

**5**, 60% yield, dr > 99:1

**Figure 3.5** Radical bromination-allylation sequence of diketopiperazines.

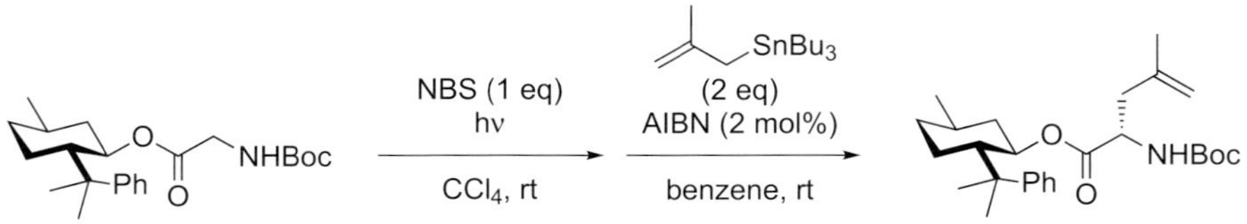


**6**

**7**, 81% yield, dr = 24:1

**Figure 3.6** 8-Phenylmenthyl-controlled glycine allylation.

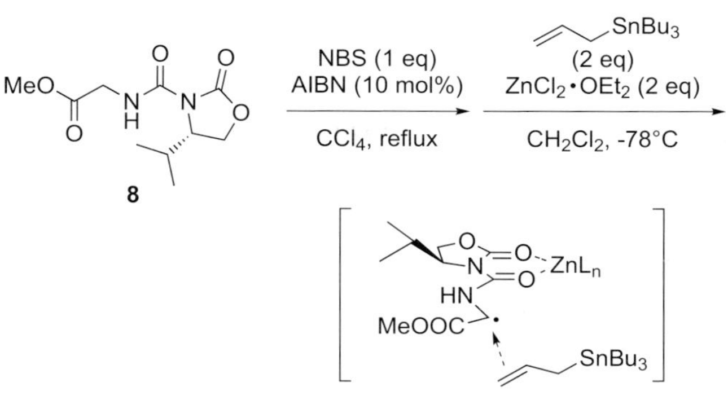

Figure 3.7 Oxazolidinone-urea-controlled glycine allylation.

allenylstannanes were employed in this reaction, yielding the corresponding alkyne- and allene-substituted amino acids, respectively.

Yamamoto *et al.* have used a chiral oxazolidinone-urea modified glycine ester in a similar bromination–allylation approach [11]. Here, $ZnCl_2$ acts as radical initiator as well as a chelating agent, indispensable for a fixed geometry and efficient shielding of one face of the intermediate captodative glycine radical (Figure 3.7). Also in this example, the allylglycine **9** was obtained in high yield, although with a slightly lower degree of stereoselectivity.

Intramolecular α-alkylation procedures have been presented by Giese *et al.* [12] in a series of stereoselective photocyclizations yielding cyclic amino acid derivatives. A phenylketone moiety linked to the N- or C-terminal nitrogen of a glycine acts as an internal radical initiator, which, after being excited, abstracts a hydrogen atom via 1,6-hydrogen atom transfer from the glycine α-carbon. The biradical formed recombines to give the cyclic amino acid. In an auxiliary controlled approach, a C2-symmetrical pyrrolidine at the C-terminus forces the biradical into a conformation minimizing the steric repulsion between the phenyl group and the auxiliary, resulting in a highly selective recombination from the less hindered *Re*-face, leading to 3-hydroxyproline derivative **11**, which was isolated in good yield as a single diastereoisomer (Figure 3.8) [12].

A related photocyclization could also be performed with dipeptides, in which the adjacent side-chain of the C-terminal amino acid caused the asymmetric induction [13]. Radical formation occurred regioselectively at the glycine moiety, not only due to the higher stability of the glycine radical (see above), but also determined by the preferred conformation of the substrate, leading to the formation of conformationally restricted peptide species **14** and **15** in good yield and high diastereoselectivity (Figure 3.9).

Following the idea of α-carbon radical formation by intramolecular hydrogen atom transfer, Wood *et al.* investigated the use of the 2-iodobenzoyl moiety as "protecting-radical-translocating" group [14]. In an asymmetric approach exploiting Seebach's concept of "self-regenerating of stereocenters" [15], *N*-2-iodobenzoylated oxazoli-

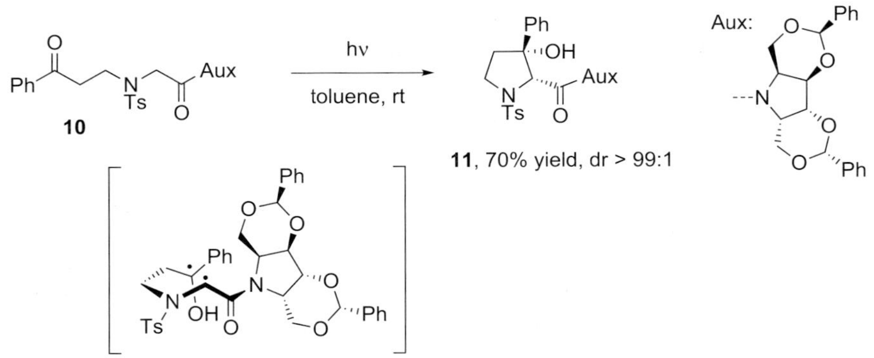

Figure 3.8 Stereocontrolled photocyclization.

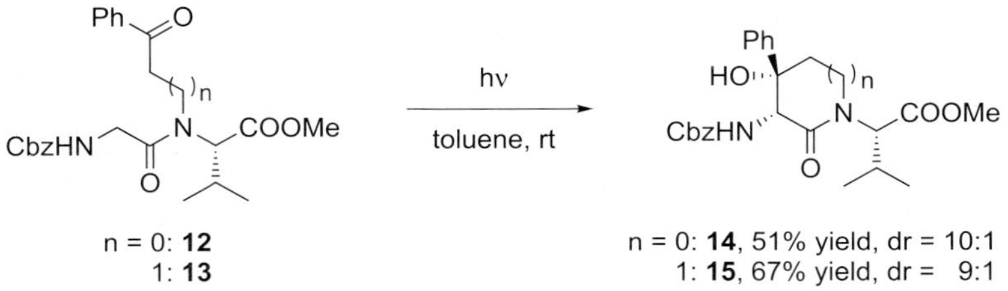

Figure 3.9 Constrained peptides via photocyclization.

dines, derived from enantiomerically pure β-amino alcohols and pivalaldehyde, were reacted with *in situ* generated $Bu_3SnH$ and acrylonitrile forming diastereomerically pure *N*-benzoyl oxazolidines, which after hydrolysis and oxidation of the alcohol yielded enantiomerically pure α,α-disubstituted amino acid derivatives (**18**) (Figure 3.10). Hereby, the initially formed tributyltin radical abstracts iodine leaving a phenyl radical moiety that undergoes 1,5-hydrogen atom transfer forming a carbon-centered radical α to the nitrogen, which finally adds to acrylonitrile. Along with the α,α-dialkylated product, reduced *N*-benzoyloxazolidine was obtained, caused by undesired quenching of either the phenyl or the oxazolidine radical.

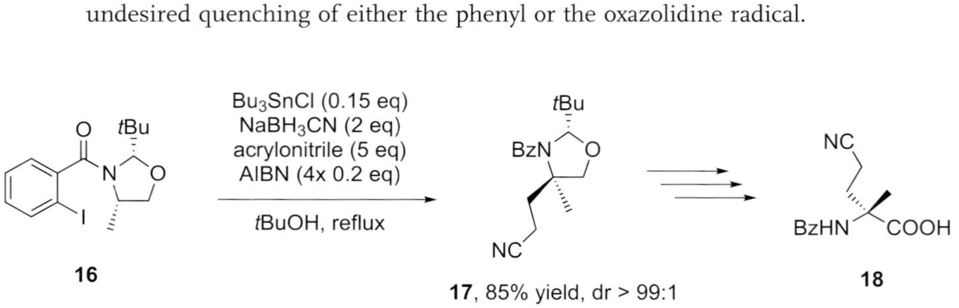

Figure 3.10 Iodobenzoyl as "radical-translocating" protective group.

### 3.2.2 Functional Group Transformations

In contrast to hydrogen atom transfer reactions that usually generate prochiral α-radicals and therefore require stereoselective solutions in order to obtain enantiomerically pure amino acids, radical transformations can also be performed at the amino acid side-chain, using side-chain functionalities for the radical generation without affecting the stereochemical information present at the α-carbon.

In particular, the side-chain carboxylic acid function of aspartic and glutamic acid derivatives has been the target for numerous investigations in the field of radical side-chain modifications. For example, decarboxylative radical transformations of corresponding Barton esters have been used to introduce heteroatoms into the amino acid side-chain such as halides [16] or selenides [17], which subsequently were employed as a handle for a variety of following reactions [18]. Further, it was possible to add the intermediate carbon-centered radical to activated olefins such as acrylates, vinylsulfones, or nitroethylene forming new carbon–carbon bonds (Figure 3.11) [19].

A different kind of C–C coupling reaction of aspartic and glutamic acid derivatives can be accomplished either by Kolbe electrolysis of the free acids or by photolytic decomposition of related diacyl peroxides, both in a symmetrical or unsymmetrical fashion. In both cases, carboxyl radicals are formed, which subsequently undergo decarboxylation resulting in the formation of alkyl radicals. Recombination finally yields the modified amino acid (Figure 3.12).

Kolbe electrolysis has been applied in the synthesis of the cystine isostere (*S*,*S*)-diaminosuberic acid as well as for the preparation of (*S*,*S*)-diaminoadipic acid [20]. In these homocoupling procedures, *N*-protected α-esters have been subjected to anodic oxidation on a platinum electrode in the presence of catalytic amounts of sodium methoxide giving rise to the amino acid dimers in moderate yield (Figure 3.13). Attempts to perform electrolytic dimerization between two different glutaminic acid derivatives led to the formation of mixtures of homo- and heterocoupling products [21].

More successful heterocouplings were performed by Vederas *et al.* by photochemical decomposition of diacyl peroxides (Figure 3.14) [22]. UV irradiation under solvent-free conditions at low temperatures generated the heterodimers in acceptable yield without the formation of cross-over products. This method allowed for the decarboxylative coupling between two amino acid derivatives, but also between aspartic/glutamic acid and other kinds of acyl derivatives. Interestingly, even sub-

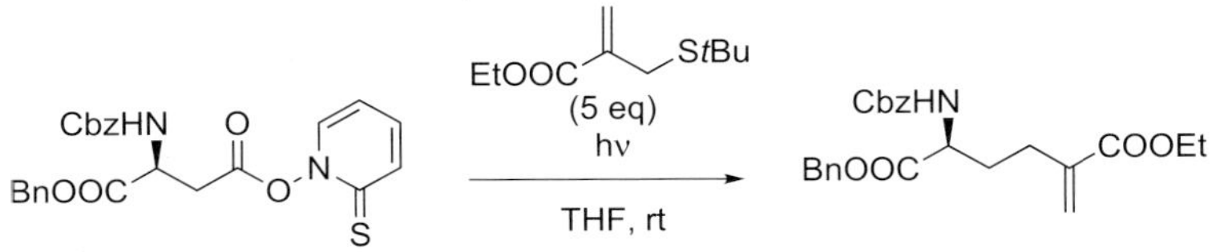

CbzHN O BnOOC O N S **19** — EtOOC S*t*Bu (5 eq) hν, THF, rt → CbzHN BnOOC COOEt **20**, 34% yield

**Figure 3.11** Decarboxylative side-chain allylation.

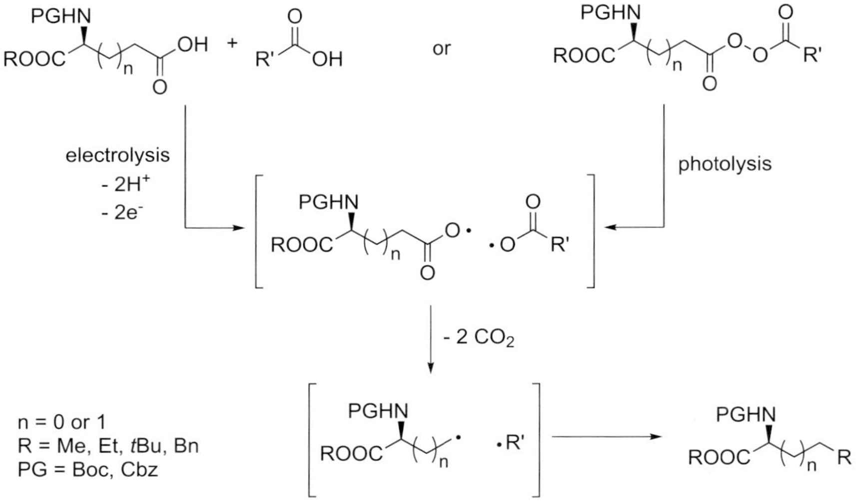

electrolysis
$-2H^+$
$-2e^-$

photolysis

$-2\,CO_2$

n = 0 or 1
R = Me, Et, *t*Bu, Bn
PG = Boc, Cbz

**Figure 3.12** Radical decarboxylative coupling of aspartic and glutamic acids.

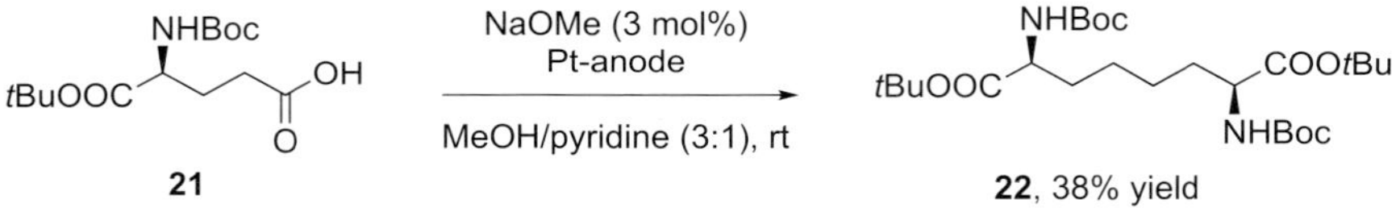

NaOMe (3 mol%)
Pt-anode
MeOH/pyridine (3:1), rt

**21**

**22**, 38% yield

**Figure 3.13** Electrolytic decarboxylative dimerization.

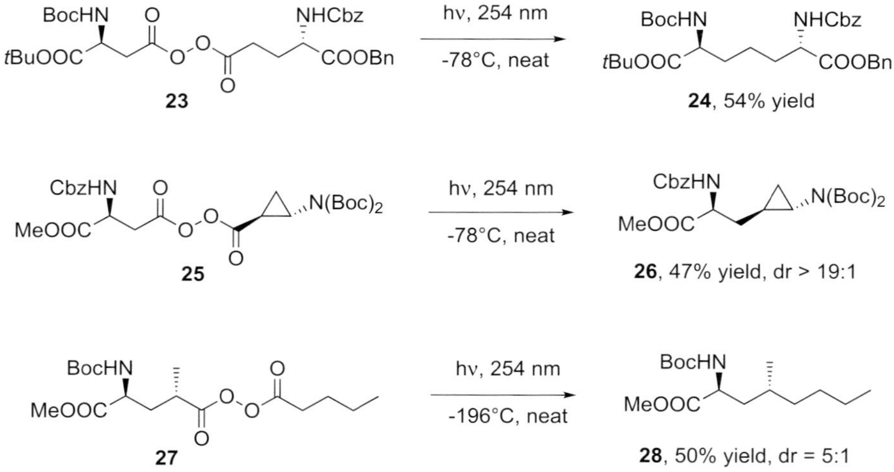

hν, 254 nm
−78°C, neat

**23**

**24**, 54% yield

hν, 254 nm
−78°C, neat

**25**

**26**, 47% yield, dr > 19:1

hν, 254 nm
−196°C, neat

**27**

**28**, 50% yield, dr = 5:1

**Figure 3.14** Photolytic heterocoupling of diacyl peroxides.

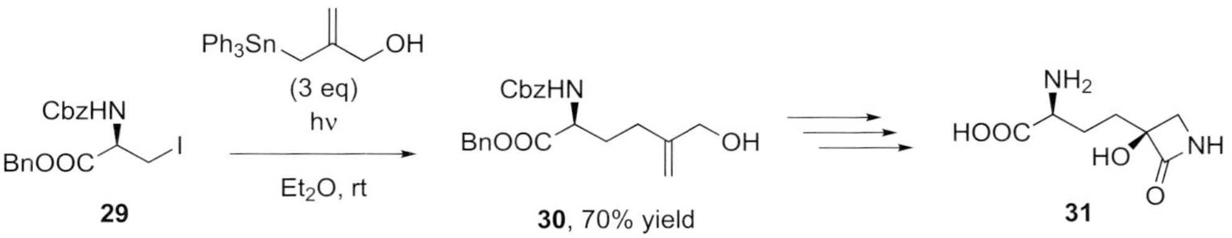

**Figure 3.15** Iodine atom transfer radical coupling.

strates with chiral centers next to the decarboxylating acyl peroxide, forming prochiral radical intermediates, could be coupled with significant retention of configuration, which was attributed to the marginal mobility of the reactive partners at low temperature.

Another amino acid that proved to be suitable for radical functional group manipulations is serine. One possibility is the transformation of the side-chain alcohol function into an iodide, which allows for subsequent iodine atom transfer reactions. Both the groups of Kiyota [23] and Baldwin [24] employed serine-derived 3-iodoalanine derivatives in the radical coupling with various allylic stannanes in the synthesis of the unnatural amino acid tabtoxinine β-lactam (**31**) (Figure 3.15).

Skrydstrup *et al.* extensively studied $SmI_2$-promoted *C*-alkylations of small to medium-sized peptides. Herein, serine acted as a handle to regioselectively introduce the reducible pyridyl sulfide group by degradation to the α-acetoxyglycine using Pb $(OAc)_4$ and subsequent nucleophilic displacement with mercaptopyridine [25]. Alternatively this kind of α-activation was also achieved via Easton's selective *N*-bromosuccinimide bromination of glycine moieties [26]. In the C–C coupling step, the peptidic α-pyridyl sulfide reacts with $SmI_2$, forming a captodative glycine radical, which is further reduced by a second equivalent of the lanthanide reagent. The resulting chelated samarium enolate finally undergoes an aldol addition with carbonyl compounds (Figure 3.16).

This method was applied to various linear as well as cyclic peptides and the β-hydroxy amino acid-containing derivatives were obtained in generally good yield. Surprisingly, and in contrast to related aldol reactions employing base-generated peptide enolates [27], only a few examples showed significant stereochemical induction caused by adjacent amino acids (Figure 3.17); more frequently, diastereomeric mixtures were obtained, which was attributed to the high flexibility in the proposed seven-membered samarium enolate chelate.

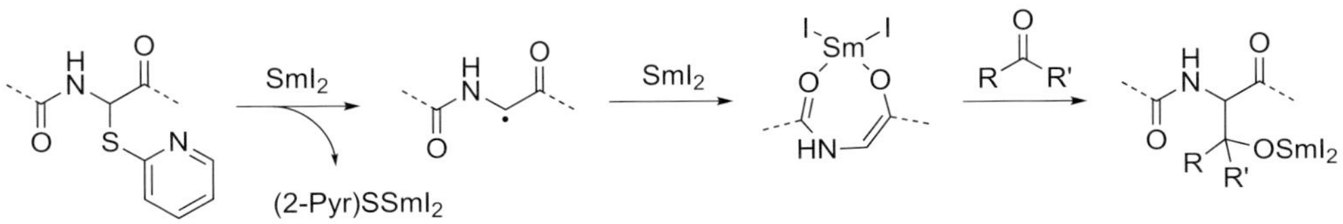

**Figure 3.16** Mechanism for the $SmI_2$-mediated reductive aldol addition.

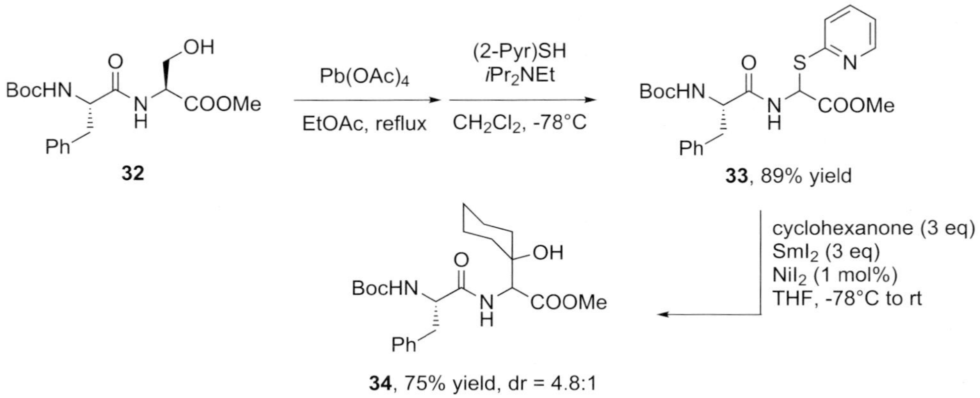

**Figure 3.17** $SmI_2$-promoted dipeptide alkylation via serine degradation.

### 3.3
### Radical Addition to Imine Derivatives

As a second general pathway for the radical mediated synthesis of α-amino acids and related compounds, the field of carbon-centered radical addition to C=N double bonds, mostly imine derivatives, as radical acceptors has emerged rapidly in the past years [28]. Though radical additions to aldimines often suffer from the low reactivity of the C=N double bond, glyoxylate imine derivatives react already under very mild conditions with nucleophilic carbon radicals to give the corresponding amino acid, due to the activating, electron-withdrawing properties of the neighboring carboxyl substituent. Further enhancement of the reactivity can be achieved by the addition of Lewis acids, which coordinate to the imine nitrogen resulting in a lower lowest unoccupied molecular orbital energy of the radical acceptor and a decreased electron density at the iminyl carbon atom [29]. Thereby, different kinds of glyoxylate imines such as oximes, hydrazones, nitrones, or simple imines can be applied.

Radical formation is usually accomplished by atom transfer processes using alkyl halides together with an appropriate initiator. A variety of different initiator systems such as tributyltin hydride [30] and $Mo_2(CO)_{10}$ [31], as well as metallic zinc [32] and indium [33], have been investigated, but the most widespread practice involves the use of triethylborane or diethylzinc in the presence of air. Hereby, triethylborane (alike diethylzinc) plays, simultaneously, the role of an initiator, activating complexing reagent and chain transfer agent [34]. Initially, the borane reacts with oxygen to form an ethyl radical, which undergoes iodine atom transfer with the alkyl iodide. The so-provided alkyl radical adds to a preactivated imine–borane complex generating a *N*-boryl iminyl radical that finally liberates another ethyl radical (Figure 3.18). Nevertheless, sometimes the formation of undesired ethylated side-products detract from the benefits in using triethylborane (i.e., mild reaction conditions and the absence of toxic metals).

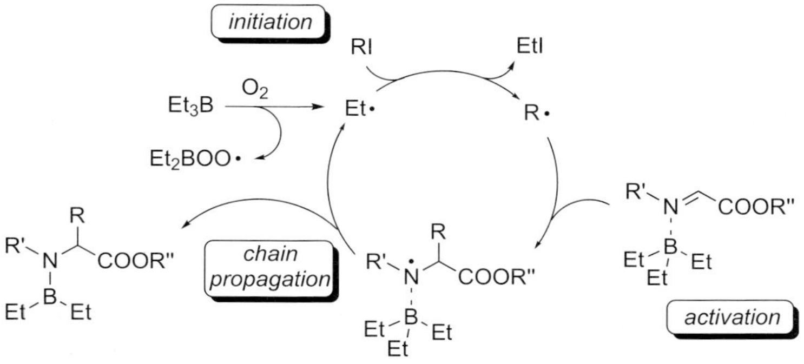

**Figure 3.18** Multiple functions of triethylborane in the radical addition to glyoxylate imines.

## 3.3.1
## Glyoxylate Imines as Radical Acceptors

Bertrand *et al.* studied the use of glyoxylate imines, prepared from chiral amines, in the stereocontrolled, radical-mediated alkylation with alkyl halides. Different initiator systems ($Bu_3SnH$/2,2′-azobisisobutyronitrile (AIBN) [35], $Et_3B$, or $Et_2Zn$ [36]) have been investigated. In particular, diethylzinc turned out to be not only an appropriate initiator, but also due to its ability to form bidentate substrate–zinc complexes it allowed for high diastereoselectivities in the reaction with acyclic imine derivatives. Several chiral methyl glyoxylate imines, derived from phenylethylamine, valinol, or norephedrine, were reacted with secondary and tertiary alkyl iodides and diethylzinc, yielding the desired amino acids in moderate-to-good diastereoselectivities. The best stereocontrol was observed in the reaction with the *O*-methyl-norephedrine-derived imine **35** and *tert*-butyl iodide giving rise to the (*R*)-*tert*-leucine derivative **36** in 69% yield and 86% d.e. The origin for the high selectivity observed has been attributed to an efficient *Si*-face shielding by formation of an *N*,*O*-chelate (Figure 3.19).

In a different strategy, not introducing the amino acid side-chain but the carbonyl moiety in a radical addition step, *in situ* prepared imines were used as acceptors by

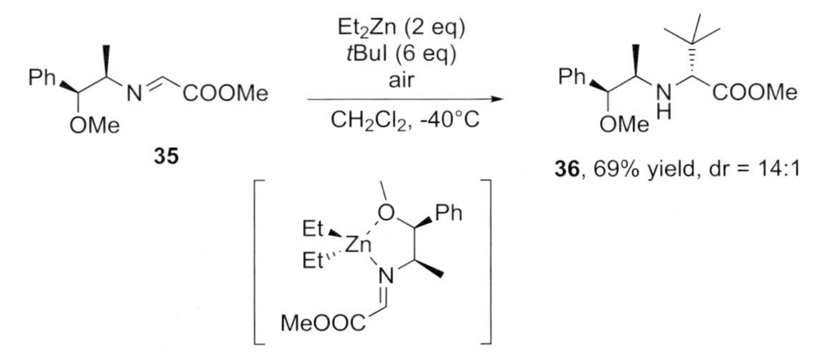


**Figure 3.19** Diethylzinc-mediated alkylation of imines.

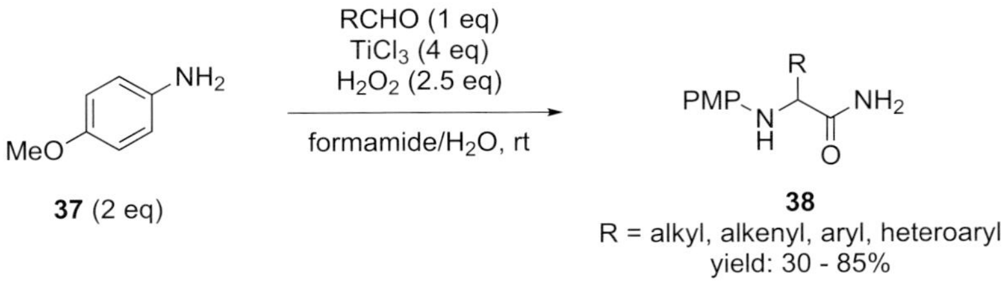

**Figure 3.20** Titanium-mediated three-component radical addition.

Porta *et al.* in a free-radical Strecker-related multicomponent reaction [37]. Initially, Ti (III) reduces hydrogen peroxide to hydroxy radicals, abstracting hydrogen atoms from formamide and generating carbamoyl radicals. The Lewis acidic Ti(IV) formed initiates the condensation of an aldehyde with *p*-methoxyaniline and activates the resulting *p*-methoxyphenyl imine, which is finally attacked by the nucleophilic carbamoyl radical. This one-pot multicomponent reaction was applied to various aliphatic, vinylic and aromatic aldehydes yielding the respective racemic α-amino acid amides in moderate-to-good yields (Figure 3.20).

### 3.3.2
### Oximes and Hydrazones as Radical Acceptors

Among the different types of glyoxylate imine derivatives used as radical acceptors, oxime ethers and hydrazones are the most commonly applied ones. They provide an additional stabilization of the intermediate aminyl radical due to the π-bonding interaction of the adjacent heteroatom's lone pair (Figure 3.21) [38]. Furthermore, oxime ethers and hydrazones benefit from enhanced hydrolytic stability. However, both acceptors require an additional step for the cleavage of the nitrogen–heteroatom bond to yield the desired amino acid derivative.

Both glyoxylate oximes and hydrazones have been the target of extensive studies by Naito *et al.* During the course of their investigations they explored different aspects such as Lewis acid influence [39], solid-phase applications [40], and *in situ* generation of the oxime ethers [41]. Asymmetric induction was achieved by using Oppolzer's camphorsultam as chiral auxiliary and high diastereoselectivities were obtained, both with oxime ether **39** and hydrazone **41** [32, 38, 42]. The stereochemical outcome can

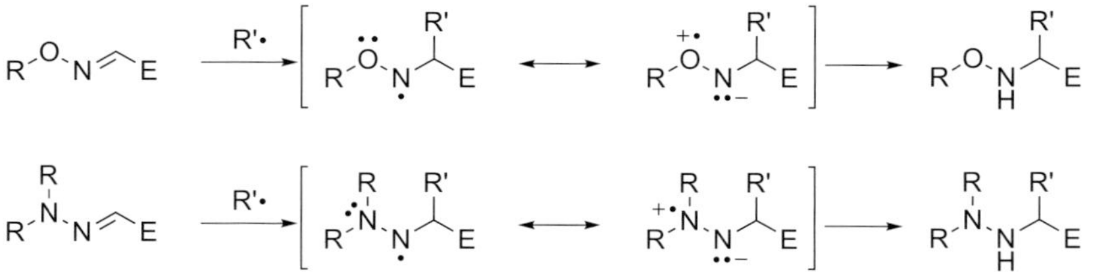

**Figure 3.21** Aminyl radical stabilization of oxime ethers and hydrazones.

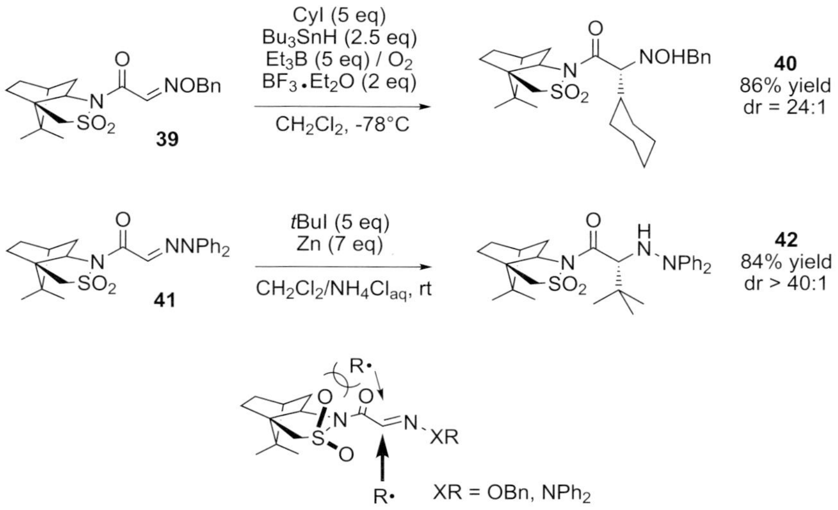

Figure 3.22 Radical additions to camphorsultam-derived oxime ethers and hydrazones.

be rationalized by a preferred conformation, shown below, in which dipolar interactions between functional groups are minimized and radical attack from the less hindered *Re*-face is favored (Figure 3.22). Scherrmann *et al.* successfully applied this approach with more complex, sugar-derived alkyl iodides in the stereoselective synthesis of *C*-glycosyl amino acids [43].

Instead of glyoxylic acceptors, Friestad *et al.* employed a pyruvate-derived ketimine (**43**) bearing a chiral oxazolidinone moiety at the N-terminus, which can be used in the stereoselective preparation of α,α-disubstituted amino acids [31]. $Mo(CO)_5$-radicals, formed by photolytic homolysis of $Mn_2(CO)_{10}$, acted as initiator providing alkyl radicals via iodine atom transfer. $InCl_3$ was used as the activating agent for the hydrazone, among the different Lewis acids tested, and provided the best results both in terms of yield and stereoselectivity. Primary and secondary iodoalkanes could be employed, whereby especially secondary iodides gave the quaternary α-methylamino acid derivatives such as the α-methylvaline ester **44** not only in high yields, but also good stereoselectivities (Figure 3.23).

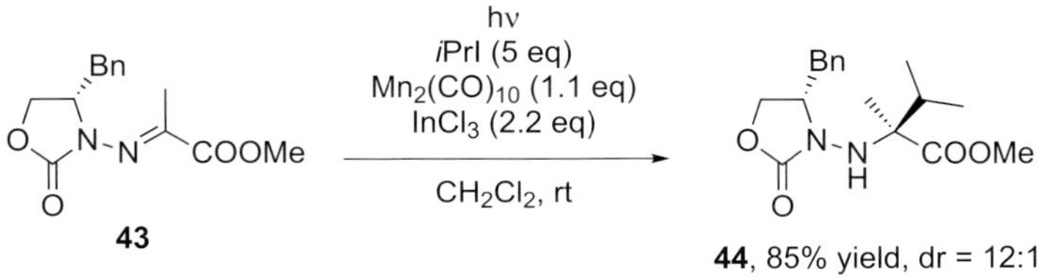

Figure 3.23 Auxiliary-controlled stereoselective synthesis of α,α-disubstituted amino acids.

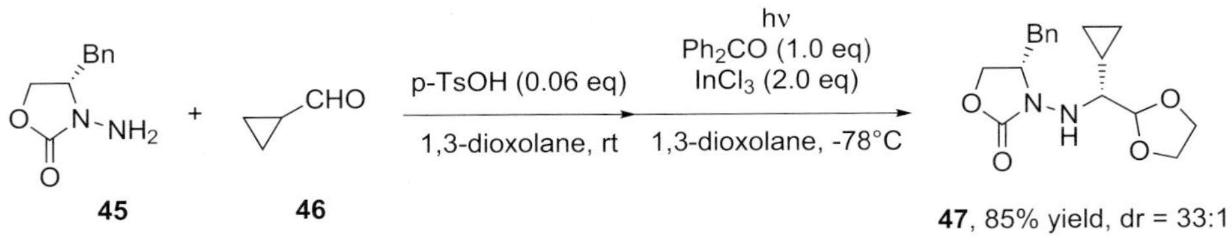

**Figure 3.24** Three-component condensation-radical addition of 1,3-dioxolanyl radicals.

Fernandez and Alonso applied the same auxiliary (**45**) in a different approach, not introducing the amino acid side-chain in the radical addition step, but using the photolytically generated 1,3-dioxolanyl radical as a synthon for the carboxyl group (Figure 3.24) [44]. In a one-pot procedure, different alkyl- and arylaldehydes were condensed with aminooxazolanone **45** and the hydrazones formed reacted subsequently under irradiation, in the presence of benzophenone as photoinitiator and $InCl_3$ as activating Lewis acid, with the dioxolanyl radical, generated from the solvent 1,3-dioxolane. Thus, protected α-amino aldehyde derivatives were obtained in high yield and moderate to high diastereoselectivities, which delivered the (*R*)-amino acids after reductive cleavage of the N–N bond and oxidation of the dioxolane. The photolytic generation of α-oxo carbon-centered radicals and their addition to imine derivatives has also been used in a substrate-controlled approach in a formal total synthesis of the polyhydroxylated amino acid (+)-myriocin [45].

So far, asymmetric catalysis in radical additions remains an unsolved problem. Attempts to conduct reactions in the presence of chiral Lewis acids suffer from the lack of catalytic turnover, due to the fact that the amine formed during the reaction is not released from the Lewis acid, since the imine represents a weaker ligand than the product amino acid. However, Naito *et al.* took up the concept of asymmetric radical additions controlled by chiral Lewis acids in a stoichiometric approach using a bisoxazoline-modified magnesium complex as activating agent in a tin-mediated alkylation of methyl glyoxylate oxime ether **48** [39]. In the presence of equimolar amounts of $MgBr_2$ and the C2-symmetric bisoxazoline **49** the reaction to the (*R*)-valine derivative **50** proceeded in excellent yield and moderate enantioselectivity (Figure 3.25).

In this context, Cho and Jang reported the enantioselective radical addition to glyoxylic oxime ethers mediated by quaternary ammonium hypophosphites derived from cinchona alkaloids [46]. Under biphasic conditions, the reactions with secondary and tertiary alkyl iodides gave high yields and good to excellent enantioselectivities and even the sterically highly demanding adamantyl radical reacted smoothly, yielding the corresponding amino acid derivative in enantiomerically pure form, depending on the cinchona species used as either (*S*)- or (*R*)-enantiomer (Figure 3.26). However, in some cases the formation of substantial amounts of ethylated side-product was observed. The stereoselectivity of the reaction is explained by hydrogen bonding between the ammonium proton and the oxime nitrogen. Together with a π-stacking interaction between benzyl ether and the quinoline moiety, this results in an efficient shielding of one face of the oxime ether.

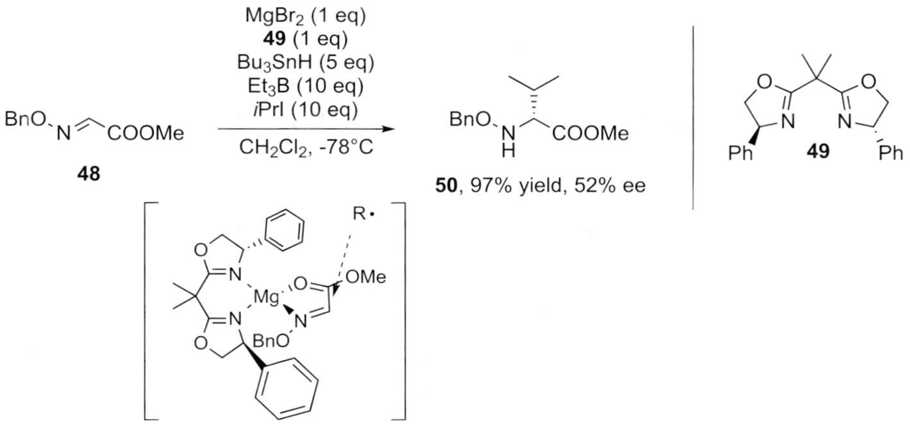

Figure 3.25 Mg-bisoxazoline-mediated enantioselective radical addition to glyoxylate oxime ethers.

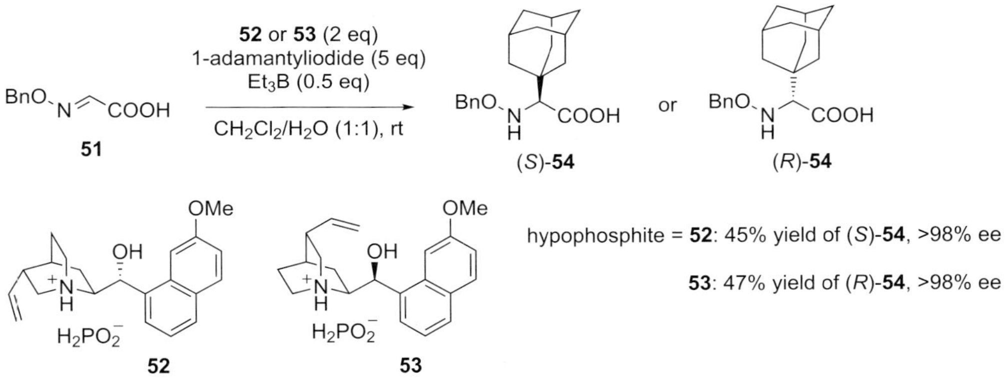

Figure 3.26 Cinchona-controlled enantioselective radical alkylation.

### 3.3.3
### Nitrones as Radical Acceptors

Like oxime ethers and hydrazones, nitrones also offer the additional stabilization of the aminyl radical by the adjacent oxygen atom; however, there are only a few examples for amino acid synthesis using nitrones as radical acceptors. Naito *et al.* studied asymmetric $Et_3B$-mediated iodine atom transfer reactions between alkyliodides and chiral glyoxylic nitrones [47]. Both acyclic camphorsultam-modified and cyclic 1,2-diphenylaminoethanol-derived radical acceptors have been employed and in either case a high degree of stereocontrol was achieved (Figure 3.27). Nevertheless, the formation of *O*-alkylated side-products, low stability (in the case of the acyclic

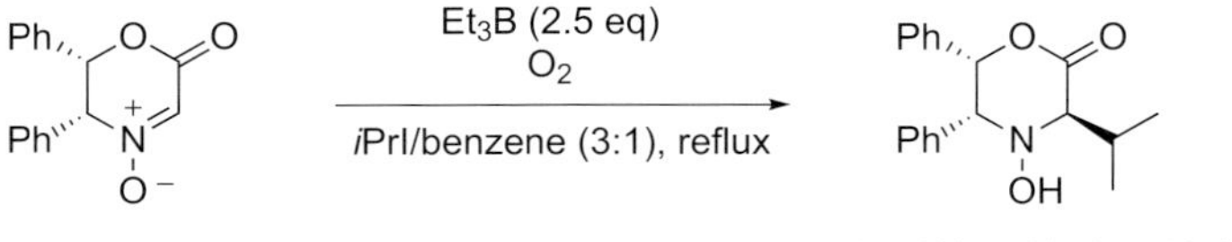

Figure 3.27 Auxiliary-controlled addition to nitrones.

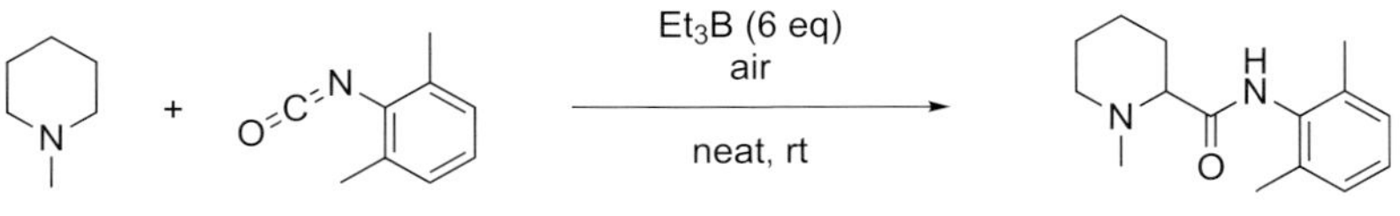

Figure 3.28 Amine α-carbamoylation.

derivative), and the need for high excesses of the alkylating agent, up to its use as solvent, make this method less attractive.

### 3.3.4
### Isocyanates as Radical Acceptors

Yoshimitsu *et al.* reported on a method employing the addition of carbon centered radicals to the C=N double bond of isocyanates as a related approach to the addition to imine derivatives [48]. In contrast to the previous strategies, instead of introduction of the amino acid side-chain the isocyanate serves as precursor for a carboxamide group. $Et_3B$ is used as initiator abstracting hydrogen atoms from the α-carbon of tertiary amines and the generated α-amino carbon radicals add to aromatic isocyanates giving rise to amino acid anilides in moderate yield. This method was used for the one-step synthesis of the anesthetic pipecolinic acid derivative mepivacaine (**59**) from readily available materials (Figure 3.28).

## 3.4
## Radical Conjugate Addition

Dehydroamino acids and related derivatives can act as radical acceptors using alkyl halides or selenides as alkylating agents. The conjugate addition of nucleophilic alkyl radicals proceeds generally with a high degree of regioselectivity at the β-carbon, forming a captodatively stabilized α-radical [49]. The trapping of the α-radical is a crucial step since it represents the stereodetermining factor in asymmetric amino acid syntheses of this kind and is commonly accomplished by hydrogen atom transfer using tin hydrides. $Bu_3SnH$ acts both as a quenching reagent for the amino

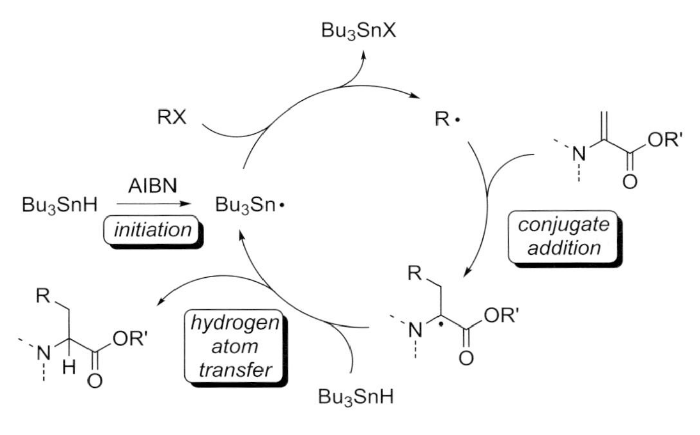


**Figure 3.29** Bu$_3$SnH-mediated radical conjugate addition.

acid radical as well as a chain propagating agent, where the *in situ* formed tin radical facilitates the generation of the reactive alkyl radical by atom transfer from alkyl halides or selenides (Figure 3.29). Alternatively, alkylmercury halides can be used as a direct alkylating agent via *in situ* generation of alkylmercury hydride species by treatment with $NaBH_4$ [50].

The latter method has been applied by Crich *et al.* for the alkylation of dehydroalanines [51] and dehydroalanine-containing di- and tripeptides [52], and a variety of primary, secondary, and tertiary alkylmercury halides could be reacted in generally high yields. Unfortunately, in the case of peptidic substrates, the stereochemical induction from adjacent amino acids in the hydrogen atom transfer step was relatively low (Figure 3.30).

High stereoselectivities could be obtained in intramolecular radical conjugate additions of peptide derivatives. Colombo *et al.* described the diastereoselective synthesis of conformationally restricted peptides via 7-*endo-trig* cyclization of substituted dehydroalanine–proline dipeptides (Figure 3.31) [53]. Selenoethyl or iodoethyl side-chains in the 5-position of the proline served as a handle for the generation of the carbon-centered radical using $Bu_3SnH$ and AIBN yielding 7,5-fused bicyclic lactams in good yield and excellent stereoselectivity. This method has also been used for the synthesis of 8,5- [54] and 6,5-fused bicyclic systems; however, in the latter case lower selectivities were observed.

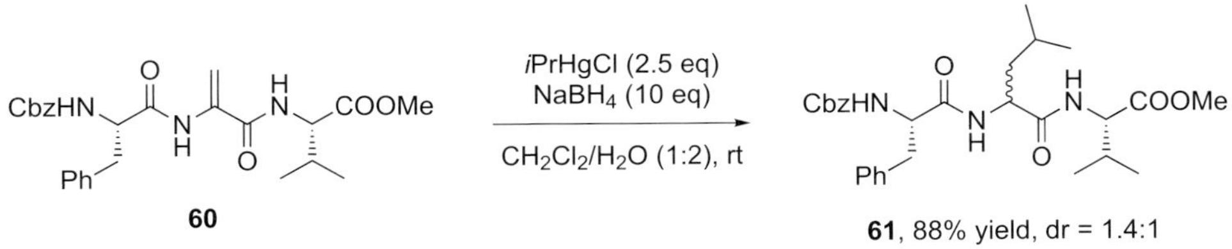

CbzHN … COOMe, Ph — **60**

*i*PrHgCl (2.5 eq), $NaBH_4$ (10 eq)

$CH_2Cl_2/H_2O$ (1:2), rt

CbzHN … COOMe, Ph — **61**, 88% yield, dr = 1.4:1

**Figure 3.30** Conjugate addition to peptidic dehydroalanines.

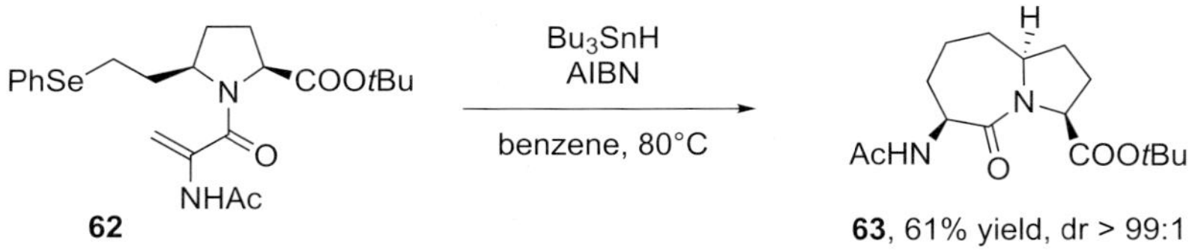

Figure 3.31 Synthesis of conformationally restricted dipeptides via 7-*endo-trig* cyclization.

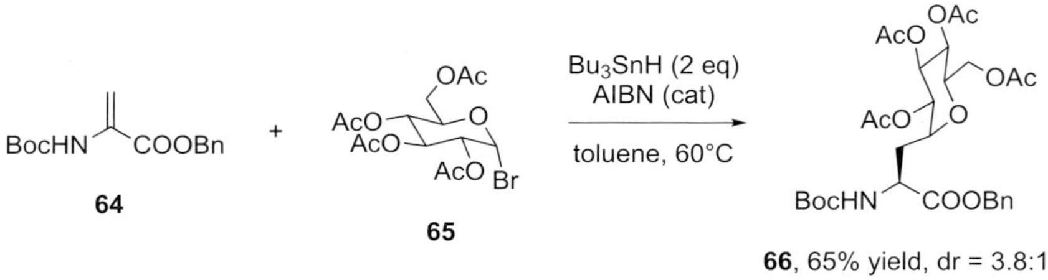

Figure 3.32 Substrate-controlled synthesis of *C*-glycoamino acids.

Kessler *et al.* also used a tin-mediated method for the synthesis of *C*-glycoamino acids and *C*-glycopeptides [55]. Herein, glycosyl bromides, derived from glucose, galactose, and lactose, were reacted with dehydroalanine derivatives yielding the α-configured *C*-glycosides in good yield. In this case, the newly introduced chiral glycosyl group controlled the hydrogen transfer and moderate-to-good diastereoselectivity was observed (Figure 3.32).

With the objective of attaining a better control over the stereochemical outcome of radical conjugate addition reactions, different kinds of auxiliary controlled diastereoselective methods have been employed. Beckwith *et al.* investigated the radical addition to chiral 4-methyleneoxazolidin-5-ones [56]. Applying either tin-mediated alkylations with alkyl iodides or reductive alkylmercury additions, in both cases good yields and high diastereoselectivities were obtained. Great benefit in this method arises from the fact that different nitrogen protecting groups can give rise to opposite diastereomers, as a change as small as switching from benzoyl to phenylacetyl led to a complete reversal in stereoselectivity from the *trans* (**69**) to *cis* product (**70**) (Figure 3.33). Furthermore, this procedure has been extended to the application of

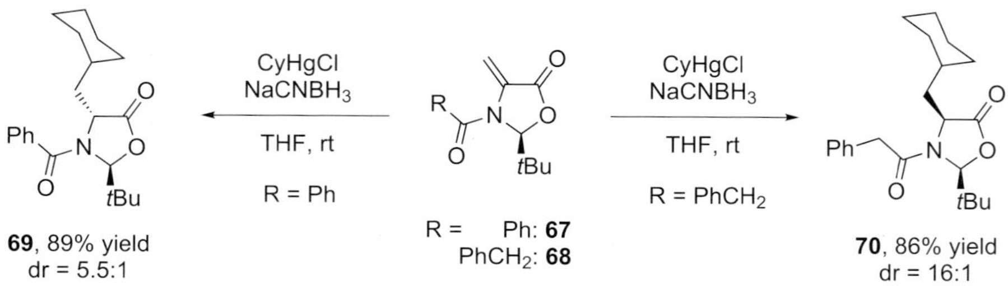

Figure 3.33 Complementary protective groups in additions to methyleneoxazolidinones.

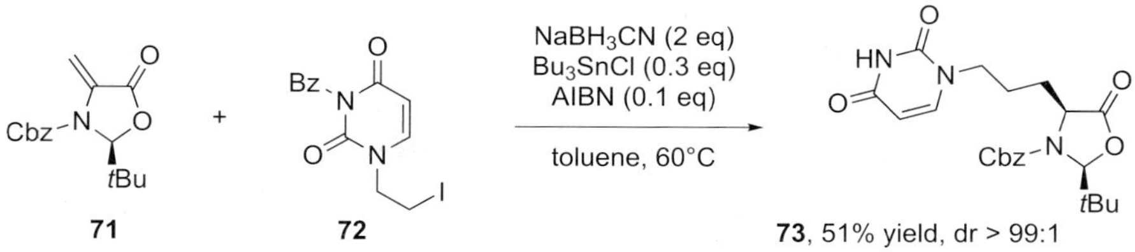

**Figure 3.34** Nucleobase-modified amino acids via radical conjugate addition.

methyleneimidazolidinones as precursors for the enantioselective synthesis of amino acid amides [57].

Enantiopure 4-methyleneoxazolidin-5-ones were later applied by other groups, for example, as acceptors in the photolytic C–C coupling with alcohols and ethers [58] or in the tin-mediated enantioselective synthesis of *C*-glycoamino acids [59]. Jones *et al.* used this approach for the synthesis of nucleobase-modified amino acids by coupling of iodoalkyl-substituted benzoyl-protected purine- and pyrimidine derivatives, giving rise to the (debenzoylated) products in moderate yield but excellent stereoselectivity (Figure 3.34) [60].

In another auxiliary-controlled approach, Chai and King described the use of methylenediketopiperazines as radical acceptors [61]. Interestingly, even with the relatively small methyl group as the stereoinducing element in this cyclic dipeptide derivative (**74**), diastereopure radical addition products could be obtained in moderate-to-good yields after reacting with either alkylmercury halides and $NaBH_4$ or alkylhalides in presence of $Bu_3SnH$ (Figure 3.35).

Belokon *et al.* reported on the tin-mediated radical conjugate addition to dehydroalanine–nickel complexes, modified by a chiral proline-derived benzophenone imine auxiliary [62]. Especially with bulky alkyl radicals, very high diastereomeric excesses were observed, rationalized by an efficient *Si*-face shielding of the intermediate α-radical in the square-planar nickel complex (Figure 3.36). The lower selectivity observed for smaller radicals such as ethyl or benzyl could easily be increased (up to 98% d.e.) by a subsequent NaOMe-catalyzed epimerization of the diastereomeric mixture.

As an alternative to the use of auxiliaries, Sibi *et al.* developed an enantioselective method employing stoichiometric amounts of a chiral Lewis acid derived from

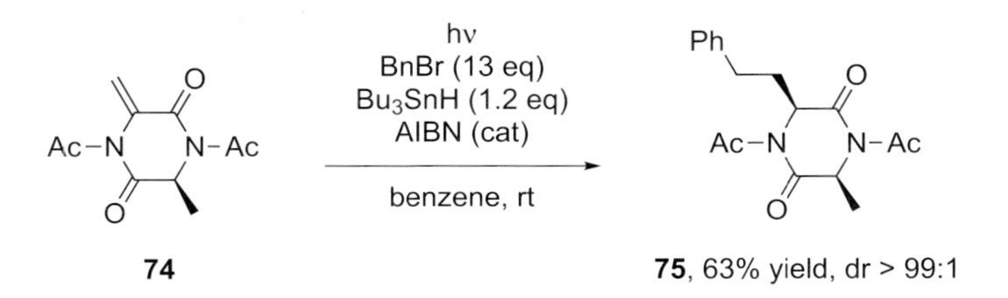


**Figure 3.35** Methylenediketopiperazines as chiral radical acceptors.

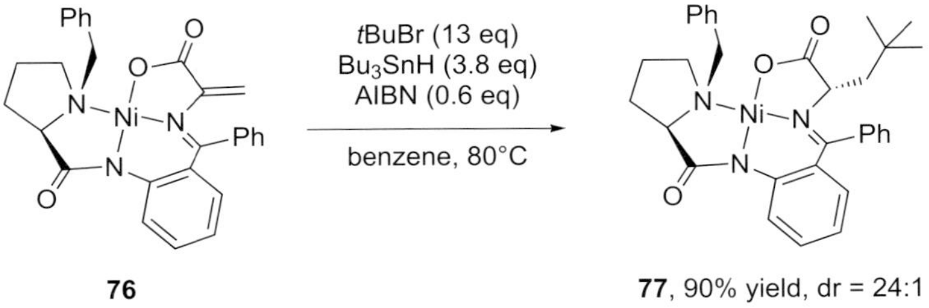

Figure 3.36 Auxiliary-controlled conjugate addition using dehydroalanine–nickel complexes.

$Mg(ClO_4)_2$ and an appropriate bisoxazoline ligand (**79**), which acts as an activating agent forming a seven-membered chelate with *N*-acyl dehydroalanine esters and controls the stereoselectivity in the hydrogen atom transfer by shielding the *Re*-face of the captodative radical [63]. Thus, alkylation with a series of alkyl halides using $Et_3B/O_2$ as radical initiator and $Bu_3SnH$ as hydrogen atom donor could be performed in good yields and moderate-to-good enantioselectivities (Figure 3.37). The 2-naphthoyl protective group turned out to be crucial for high enantiomeric excesses.

In radical conjugate additions, β-substituted dehydroamino acids, yielding β-branched amino acids, have been rarely employed [64]. Castle and Srikanth addressed this task by using α-nitroacrylamides and -esters as radical acceptors, which were afterwards reduced to the corresponding amino acid derivatives [65]. In an enantioselective approach related to Sibi's bisoxazoline-magnesium procedure, β-branched α-amino acids could be obtained in good yield with very high enantioselectivities [66]. Unfortunately the *syn*/*anti* selectivity remained an unsolved problem (Figure 3.38). Recently, the radical conjugate addition to α-nitroacrylamides has been extended to peptidic acceptors, where moderate substrate stereocontrol was achieved using $Zn(OTf)_2$ as Lewis acid. This protocol has recently been applied in the total synthesis of the bicyclic octapeptide celogentin C [67].

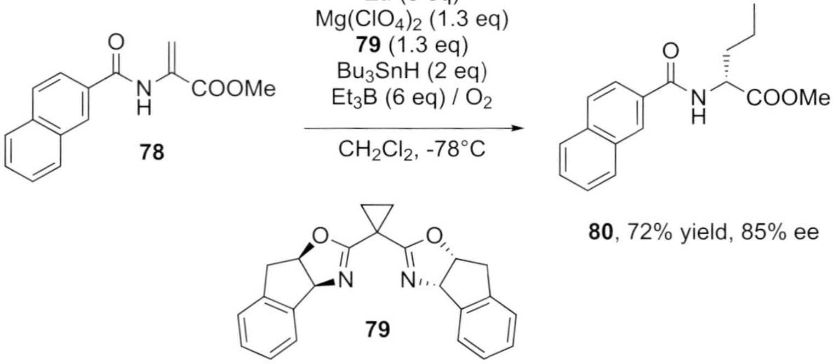

Figure 3.37 Mg-bisoxazoline-mediated enantioselective radical conjugate addition.

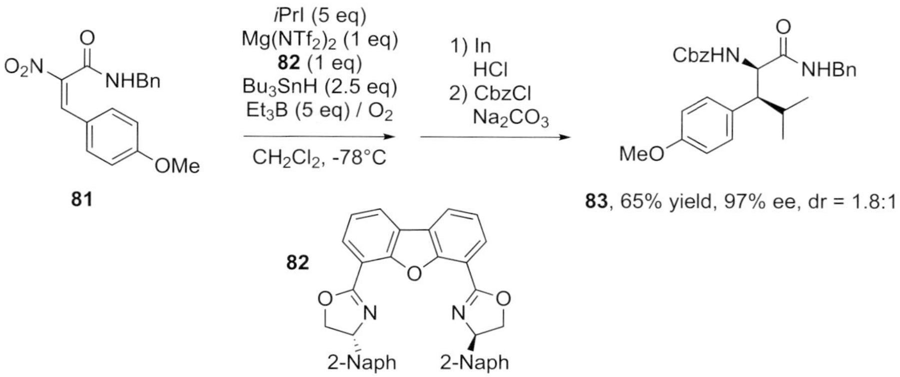

Figure 3.38 Mg-dibenzofuradiylphenyloxazoline-mediated addition to α-nitroacrylamides.

## 3.5
## Conclusions

The field of radical-mediated methods for the synthesis and modification of amino acids and peptides has steadily evolved during the past decades. A variety of different complementary approaches has been established, offering versatile protocols dealing with α-alkylation, side-chain modification, or *de novo* synthesis. Some of these methods provide flexible and selective solutions for the construction of complex amino acid building blocks with high synthetic value, and which have found their way into natural product synthesis or application in the preparation of biologically active compounds.

## 3.6
## Experimental Protocols

### 3.6.1
### Preparation of ((1*R*,2*S*,5*R*)-5-methyl-2-(1-methyl-l-phenylethyl)cyclohexyl 2-[(*tert*-butoxycarbonyl)amino]-4-methyl-pent-4-enoate) (7)

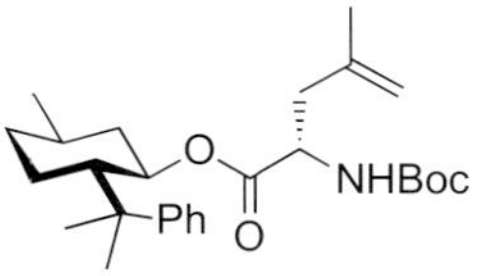

A solution of the bromoglycine derivative (91 mg, 0.193 mmol), tri-*n*-butyl(2-methyl-2-propenyl)stannane (133 mg, 0.386 mmol) and AIBN (about 1 mg) in dry benzene (0.68 ml) was let stand at 20 °C for 16 h. The benzene was evaporated at reduced pressure and the residue was chromatographed on silica to give **7** (69 mg, 81%). High-performance liquid chromatography (HPLC) analysis indicated the presence of two components in a 95: 5 ratio. $^{1}$H-nuclear magnetic resonance (NMR): δ = 7.15–7.35 (m, 5H), 4.84 (dt, $J$ = 4.3, 10.7 Hz, 1H), 4.75 (s, 1H), 1.20 (s, 3H), 4.63 (s, 1H), 4.41 (d, $J$ = 8.3 Hz, 1H), 3.77 (m, 1H), 1.64 (s, 3H), 1.45 (s, 9H), 1.30 (s, 3H), 0.86 (d, $J$ = 6.4 Hz, 3H); $^{13}$C-NMR: δ = 171.6, 154.9, 151.7, 140.9, 128.1, 125.3, 125.2, 113.9, 79.2, 75.7, 51.7, 50.2, 41.4, 40.4, 39.5, 34.5, 31.2, 29.0, 28.3, 26.4, 23.9, 21.9, 21.8; MS m/z: 443 ($M^+$), 387 ($M^+$-$C_4H_8$); high-resolution mass spectrometry (HRMS) calculated for $C_{27}H_{42}NO_4$ ($[M + H]^+$): 444.311, found: 444.313; analytically calculated for $C_{27}H_{41}NO_4$: C 73.10, H 9.32, found: C 72.70, H 9.03.

### 3.6.2
### Synthesis of (2*S*)-3-{(1*R*,2*S*)-2-[(*N*-bis-Boc)amino]-1-cyclopropyl}-2-benzyloxycarbonylamino-propionic Acid Methyl Ester (26)

A solution of **25** (40 mg, 0.07 mmol) in $CHCl_3$ was transferred to a crystallization dish (150 mm × 75 mm) and the solvent was evaporated with the aid of dry air, and finally purged with argon gas in order to generate a thin film of compound at the bottom of the dish. The dish was then covered with a quartz plate, cooled to −78 °C, and photolyzed with 254 nm UV lamp (0.9 A) for 36 h. Purification of the crude product by flash chromatography on silica gel (4: 1 hexanes/ethyl acetate) provided **26** (16 mg, 47%) as clear colorless oil. $[\alpha]_{26}^{D} = +11.1$ ($c$ 1, $CHCl_3$); IR ($CHCl_3$ cast) 3321, 1731, 1706, 1530 cm$^{-1}$; $^{1}$H-NMR ($CDCl_3$, 500 MHz): δ = 7.36–7.26 (m, 5H), 7.01 (d, $J$ = 9.0 Hz, 1H), 5.14 (d, 1H, $J$ = 12.5 Hz), 5.08 (d, 1H, $J$ = 12.5 Hz), 4.53 (dt, 1H, $J$ = 4.5, 9.0 Hz), 3.70 (s, 3H), 2.41–2.38 (m, 1H), 2.06 (dt, 1H, $J$ = 4.5, 14.5), 1.61-1.56 (m, 1H), 1.45 (s, 18H), 0.96–0.90 (m, 1H), 0.77–0.73 (m, 1H), 0.66 (q, 1H, $J$ = 6.5 Hz); $^{13}$C-NMR ($CDCl_3$, 100 MHz): δ = 172.3, 156.4, 153.4, 136.7, 128.3, 127.8, 127.7, 82.7, 66.5, 53.4, 52.1, 34.5, 34.4, 27.9, 18.2, 14.5; HRMS (electrospray) calculated for $C_{25}H_{36}N_2O_8Na$: 515.2363, found 515.2361.

### 3.6.3
### Synthesis of (3a*R*,6*S*,7a*S*)-hexahydro-8,8-dimethyl-1-[(2*R*)-3,3-dimethyl-1-oxo-2-(2,2-diphenylhydrazino)butyl]-3*H*-3a,6-methano-2,1-benzisothiazole 2,2-dioxide (42)

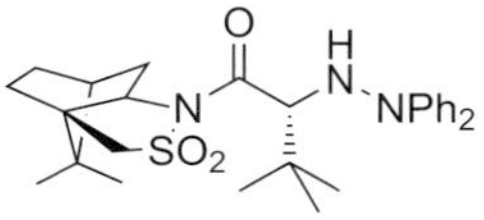

To a micro tube containing **41** (50 mg, 0.11 mmol), *tert*-butyl iodide (101 mg, 0.55 mmol), zinc (50.3 mg, 0.77 mmol), and $CH_2Cl_2$ (0.1 ml) was added dropwise aqueous saturated $NH_4Cl$ (0.4 ml) at 20 °C over 5 min. After being stirred at the same temperature for 22 h, the reaction mixture was diluted with 36% potassium sodium (+)-tartrate and then extracted with $CH_2Cl_2$. The organic phase was washed, dried over $MgSO_4$, and concentrated at reduced pressure. Purification of the residue by preparative thin-layer chromatography (hexane/EtOAc, 3: 1) afforded **42** as colorless oil. $[\alpha]^{D}_{24}$ +50.6 (*c* 1.11, $CHCl_3$); IR ($CHCl_3$): 2958, 1685, 1589, 1467 $cm^{-1}$; $^1H$ NMR ($CDCl_3$): δ = 7.29–6.97 (m, 10H), 4.63 (brs, 1H), 4.11 (s, 1H), 3.66 (t, *J* = 6.3 Hz, 1H), 3.43 (d, *J* = 13.5 Hz, 1H), 3.37 (d, *J* = 13.5 Hz, 1H), 2.05–2.01 (brm, 2H), 1.93–1.81 (brm, 3H), 1.35–1.20 (brm, 3H), 1.08 (m, 12H), 0.93 (s, 3H); $^{13}C$-NMR ($CDCl_3$): δ = 174.1, 149.3, 128.7, 122.6, 121.4, 69.3, 65.9, 53.2, 47.6, 47.5, 44.5, 38.6, 35.8, 33.2, 27.1, 26.4, 20.6, 20.0. HRMS calculated for $C_{28}H_{37}N_3O_3S$ ($M^+$): 495.2554, found: 495.2546.

### 3.6.4
### Synthesis of *N*-(2,6-diphenyl-methylpiperidine-2-carboxamide (59)

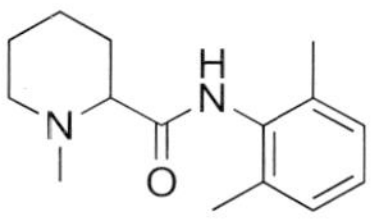

To a solution of 2,6-dimethylphenylisocyanate (**58**) (420 μl, 3.0 mmol) in *N*-methylpiperidine (**57**) (12.8 ml, 105 mmol) was added $Et_3B$ (2.6 ml, 18 mmol) at room temperature under an argon atmosphere. After removal of the argon balloon, the reaction mixture was stirred at the same temperature with continuous bubbling of dry air through a syringe needle with a balloon (flow rate: about 10–20 ml/h/mmol isocyanate) for 42 h. MeOH (5 ml) was added and the mixture was stirred at 60 °C for an additional 12 h without air admission. The reaction mixture was concentrated under reduced pressure, diluted with $CH_2Cl_2$, and transferred into a separatory funnel where it was acidified with 3 N HCl and extracted with $CH_2Cl_2$. The aqueous layer was carefully basified by portionwise addition of solid $Na_2CO_3$ and extracted with $CH_2Cl_2$. The combined organic extracts were dried over $K_2CO_3$, filtered, and concentrated. Silica gel chromatography ($MeOH/CH_2Cl_2$ 1: 60) yielded the pure **59** (381 mg, 52%) as colorless prisms. Melting point: 149–151 °C (Et2O/hexane); IR (KBr): ν 3159, 1655 $cm^{-1}$; $^1H$-NMR (300 MHz, $CDCl_3$): δ = 8.04 (brs, 1H), 7.12–7.02 (m, 3H), 3.00 (brd, 1H, *J* = 11.5 Hz), 2.66 (dd, 1H, *J* = 3.3, 11.4 Hz), 2.42 (s, 3H), 2.25 (s, 6H), 2.20–2.07 (m, 2H), 1.83–1.51 (m, 4H), 1.36–1.21 (m, 1H); $^{13}C$-NMR (75 MHz, CDCl3): δ = 172.3, 134.9, 133.3, 128.0, 126.7, 69.8, 55.3, 45.1, 31.2, 25.2, 23.1, 18.6; HRMS (electron ionization) calculated for $C_{15}H_{22}N_2O$ ($M^+$): 246.1732, found: 246.1725.

### 3.6.5
### Synthesis of Methyl 2-(2-naphthylcarbonylamino)pentanoate (80)

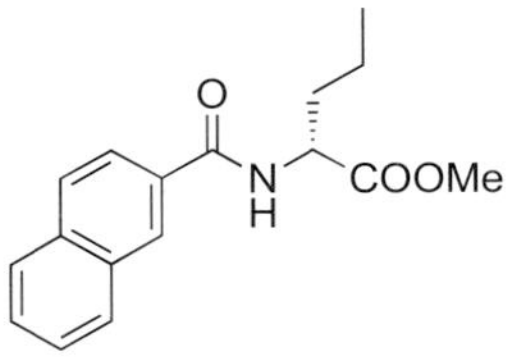

A suspension of dehydroalanine derivative **78** (38 mg, 0.15 mmol), magnesium perchlorate (45 mg, 0.2 mmol), and ligand **79** (71 mg, 0.2 mmol) in $CH_2Cl_2$ (3 ml) was stirred at room temperature for 1 h. During this time all the suspension dissolved to provide a homogeneous solution. To this solution was added ethyl iodide (0.06 ml, 0.75 mmol), tributyltin hydride (0.1 ml, 0.3 mmol), and triethylborane (1 ml, 1 mmol, 1 M in hexane) at −78 °C. The reaction vial was charged with oxygen. After stirring for 3 h at −78 °C, saturated aqueous $NH_4Cl$ (1 ml) was added. The mixture was extracted with ether (10 ml). The organic layers were washed with water (2 ml), saturated aqueous potassium fluoride (3 ml × 3), and brine (3 ml). The mixture was poured into silica gel and concentrated *in vacuo*. The silica gel was washed with hexane to remove tin byproducts. It was further eluted with ether and the resulting organics were concentrated *in vacuo* to give 126 mg of a colorless oil, which was chromatographed over silica gel (10–20%, EtOAc/hexanes) to give **80** (30.7 mg, 72%). $[\alpha]_D$ −26.2 (*c* 0.97, 85% e.e., $CH_2Cl_2$); melting point: 107–108 °C; $^1$H-NMR ($CDCl_3$, 500 MHz): δ = 8.33 (s, 1H), 7.82-7.98 (m, 5H), 7.52–7.62 (m, 2H), 6.84 (d, $J$ = 7.3 Hz, 1H), 4.91 (m, 1H), 3.81 (s, 3H), 1.99 (m, 1H), 1.83 (m, 1H), 1.38–1.56 (m, 2H), 0.98 (t, $J$ = 7.3 Hz, 3H); $^{13}$C-NMR (125 MHz, $CDCl_3$): δ = 173.6, 167.3, 135.1, 132.8, 131.4, 129.2, 128.7, 128.0, 127.8, 127.0, 123.8, 52.8, 52.7, 35.1, 18.9, 14.0; $R_f$ 0.30 (30% EtOAc/hexanes); analytically calculated for $C_{17}H_{19}NO_3$: C 71.56, H 6.71, N 4.91, found: C 71.94, H 6.58, N 5.09; HPLC: $t_R$ 35.1 min (*R*-isomer); $t_R$ 48.4 min (*S*-isomer) [Chiralpak AD (0.46 cm × 50 cm) hexane/*i*PrOH, 95: 5, 0.8 ml/min].

## References

1 Easton, C.J. (1997) *Chemical Reviews*, **97**, 53–82.

2 (a) Easton, C.J. and Hay, M.P. (1986) *Journal of the Chemical Society, Chemical Communications*, 55–57; (b) Burgess, V.A., Easton, C.J., and Hay, M.P. (1989) *Journal of the American Chemical Society*, **111**, 1047–1052.

3 Elad, D. and Sinnreich, J. (1965) *Journal of the Chemical Society, Chemical Communications*, 471–472.

4 (a) Elad, D. and Sperling, J. (1971) *Journal of the American Chemical Society*, **93**, 967–971; (b) Schwarzberg, M., Sperling, J., and Elad, D. (1973) *Journal of the American Chemical Society*, **95**, 6418–6426.

5 Elad, D. and Sperling, J. (1971) *Journal of the American Chemical Society*, **93**, 3839–3840.

6 Easton, C.J., Scharfbillig, I.M., and Tan, E.W. (1988) *Tetrahedron Letters*, **29**, 1565–1568.

7 Blakskjær, P., Pedersen, L., and Skrydstrup, T. (2001) *Journal of the Chemical Society, Perkin Transactions 1*, 910–915.

8 (a) Badran, T.W., Easton, C.J., Horn, E., Kociuba, K., May, B.L., Schliebs, D.M. and Tiekink, E.R.T. (1993) *Tetrahedron: Asymmetry*, **4**, 197–200; (b) Easton, C.J. (1997) *Pure and Applied Chemistry*, **69**, 489–494.

9 Ermert, P., Meyer, J., Stucki, C., Schneebeli, J., and Obrecht, J.-P. (1988) *Tetrahedron Letters*, **29**, 1265–1268.

10 (a) Hamon, D.P.G., Massy-Westropp, R.A., and Razzino, P. (1991) *Journal of the Chemical Society, Chemical Communications*, 722–724; (b) Hamon, D.P.G., Massy-Westropp, R.A., and Razzino, P. (1993) *Tetrahedron*, **49**, 6419–6428; (c) Hamon, D.P.G., Massy-Westropp, R.A., and Razzino, P. (1995) *Tetrahedron*, **51**, 4183–4194.

11 Yamamoto, Y., Onuki, S., Yumoto, M., and Asao, N. (1994) *Journal of the American Chemical Society*, **116**, 421–422.

12 (a) Wessig, P., Wettstein, P., Giese, B., Neuburger, M., and Zehnder, M. (1994) *Helvetica Chimica Acta*, **77**, 829–837; (b) Giese, B., Müller, S.N., Wyss, C., and Steiner, H. (1996) *Tetrahedron: Asymmetry*, **7**, 1261–1262.

13 Wyss, C., Batra, R., Lehmann, C., Sauer, S., and Giese, B. (1996) *Angewandte Chemie*, **108**, 2660–2662; (1996) *Angewandte Chemie (International Edition in English)*, **35**, 2529–2531.

14 Wood, M.E., Penny, M.J., Steere, J.S., Horton, P.N., Light, M.E., and Hursthouse, M.B. (2006) *Journal of the Chemical Society, Chemical Communications*, 2983–2985.

15 Seebach, D., Sting, A.R., and Hoffmann, M. (1996) *Angewandte Chemie*, **108**, 2881–2921; (1996) *Angewandte Chemie (International Edition in English)*, **35**, 2708–2748.

16 Barton, D.H.R., Hervé, Y., Potier, P., and Thierry, J. (1988) *Tetrahedron*, **44**, 5479–5486.

17 Barton, D.H.R., Bridon, D., Hervé, Y., Potier, P., Thierry, J., and Zard, S.Z. (1986) *Tetrahedron*, **42**, 4983–4990.

18 (a) Barton, D.H.R., Crich, D., Hervé, Y., Potier, P., and Thierry, J. (1985) *Tetrahedron*, **41**, 4347–4357; (b) Ciapetti, P., Mann, A., Shoenfelder, A., and Taddei, M. (1998) *Tetrahedron Letters*, **39**, 3843–3846.

19 Barton, D.H.R., Hervé, Y., Potier, P., and Thierry, J. (1987) *Tetrahedron*, **43**, 4297–4308.

20 Hiebl, J., Blanka, M., Guttman, A., Kollmann, H., Leitner, K., Mayrhofer, G., Rovenszky, F., and Winkler, K. (1998) *Tetrahedron*, **54**, 2059–2074.

21 Nutt, R.F., Strachan, R.G., Veber, D.F., and Holly, F.W. (1980) *The Journal of Organic Chemistry*, **45**, 3078–3080.

22 (a) Spantulescu, M.D., Jain, R.P., Derksen, D.J., and Vederas, J.C. (2003) *Organic Letters*, **5**, 2963–2965; (b) Jain, R.P. and Vederas, J.C. (2003) *Organic Letters*, **5**, 4669–4672.

23 (a) Kiyota, H., Takai, T., Saitoh, M., Nakayama, O., Oritani, T., and Kuwahara, S. (2004) *Tetrahedron Letters*, **45**, 8191–8194; (b) Kiyota, H., Takai, T., Shimasaki, Y., Saitoh, M., Nakayama, O., Takada, T., and Kuwahara, S. (2007) *Synthesis*, 2471–2480.

24 (a) Baldwin, J.E., Adlington, R.M., Birch, D.J., Crawford, J.A., and Sweeney, J.B. (1986) *Journal of the Chemical Society, Chemical Communications*, 1339–1340; (b) Baldwin, J.E., Fieldhouse, R., and Russell, A.T. (1993) *Tetrahedron Letters*, **34**, 5491–5494.

25 (a) Blaksjær, P., Gavrila, A., Andersen, L., and Skrydstrup, T. (2004) *Tetrahedron Letters*, **45**, 9091–9094; (b) Ebran, J.-P., Jensen, C.M., Johannesen, S.A., Karaffa, J., Lindsay, K.B., Taaning, R., and Skrydstrup, T. (2006) *Organic & Biomolecular Chemistry*, **4**, 3553–3564.

26 (a) Ricci, M., Madariaga, L., and Skrydstrup, T. (2000) *Angewandte Chemie*, **112**, 248–252; (2000) *Angewandte Chemie International Edition*, **39**, 242–246; (b) Ricci, M., Blakskjær, P., and Skrydstrup, T. (2000) *Journal of the American Chemical Society*, **122**, 12413–12421.

27 Kazmaier, U., Deska, J., and Watzke, A. (2006) *Angewandte Chemie*, **118**, 4973–4976; (2006) *Angewandte Chemie International Edition*, **45**, 4855–4858.

28 Friestad, G.K. (2001) *Tetrahedron*, **57**, 5461–5496.

29 Miyabe, H., Shibata, R., Ushiro, C., and Naito, T. (1998) *Tetrahedron Letters*, **39**, 631–634.
30 (a) Miyabe, H., Shibata, R., Sangawa, M., Ushiro, C., and Naito, T. (1998) *Tetrahedron*, **54**, 11431–11444; (b) Miyabe, H., Ueda, M., Yoshioka, N., and Naito, T. (1999) *Synlett*, 465–467.
31 (a) Friestad, G.K. and Qin, J. (2001) *Journal of the American Chemical Society*, **123**, 9922–9923; (b) Friestad, G.K. and Ji, A. (2008) *Organic Letters*, **10**, 2311–2313; (c) Friestad, G.K. and Banerjee, K. (2009) *Organic Letters*, **11**, 1095–1098.
32 (a) Ueda, M., Miyabe, H., Nishimura, A., Sugino, H., and Naito, T. (2003) *Tetrahedron: Asymmetry*, **14**, 2857–2859; (b) Ueda, M., Miyabe, H., Sugino, H., and Naito, T. (2005) *Organic & Biomolecular Chemistry*, **3**, 1124–1128.
33 Miyabe, H., Ueda, M., Nishimura, A., and Naito, T. (2002) *Organic Letters*, **4**, 131–134.
34 Nakakoshi, M., Ueda, M., Sakurai, S., Miyata, O., Sugiura, M., and Naito, T. (2006) *Magnetic Resonance in Chemistry*, **44**, 807–812.
35 Bertrand, M.P., Feray, L., Nouguier, R., and Stella, L. (1998) *Synlett*, 780–782.
36 (a) Bertrand, M.P., Feray, L., Nouguier, R., and Perfetti, P. (1999) *Synlett*, 1148–1150; (b) Bertrand, M.P., Feray, L., Nouguier, R., and Perfetti, P. (1999) *The Journal of Organic Chemistry*, **64**, 9189–9193; (c) Bertrand, M.P., Coantic, S., Feray, L., Nouguier, R., and Perfetti, P. (2000) *Tetrahedron*, **56**, 3951–3961.
37 Cannella, R., Clerici, A., Panzeri, W., Pastori, N., Punta, C., and Porta, O. (2006) *Journal of the American Chemical Society*, **128**, 5358–5359.
38 Miyabe, H., Ueda, M., and Naito, T. (2004) *Synlett*, 1140–1157.
39 Miyabe, H., Ushiro, C., Ueda, M., Yamakawa, K., and Naito, T. (2000) *The Journal of Organic Chemistry*, **65**, 176–185.
40 (a) Miyabe, H., Fujishima, Y., and Naito, T. (1999) *The Journal of Organic Chemistry*, **64**, 2174–2175; (b) Miyabe, H., Konishi, C., and Naito, T. (2000) *Organic Letters*, **2**, 1443–1445; (c) Miyabe, H., Konishi, C., and Naito, T. (2003) *Chemical & Pharmaceutical Bulletin*, **51**, 540–544.
41 Miyabe, H., Ueda, M., Yoshioka, N., Yamakawa, K., and Naito, T. (2000) *Tetrahedron*, **56**, 2413–2420.
42 Miyabe, H., Ushiro, C., and Naito, T. (1997) *Journal of the Chemical Society, Chemical Communications*, 1789–1790.
43 Bragnier, N., Guillot, R., and Scherrmann, M.-C. (2009) *Organic & Biomolecular Chemistry*, **7**, 3918–3921.
44 Fernandez, M. and Alonso, R. (2003) *Organic Letters*, **5**, 2461–2464.
45 Torrente, S. and Alonso, R. (2001) *Organic Letters*, **3**, 1985–1987.
46 Cho, D.H. and Jang, D.O. (2006) *Journal of the Chemical Society, Chemical Communications*, 5045–5047.
47 (a) Ueda, M., Miyabe, H., Teramachi, M., Miyata, O., and Naito, T. (2003) *Journal of the Chemical Society, Chemical Communications*, 426–427; (b) Ueda, M., Miyabe, H., Teramachi, M., Miyata, O., and Naito, T. (2005) *The Journal of Organic Chemistry*, **70**, 6653–6660.
48 Yoshimitsu, T., Matsuda, K., Nagaoka, H., Tsukamoto, K., and Tanaka, T. (2007) *Organic Letters*, **9**, 5115–5118.
49 Srikanth, G.S.C. and Castle, S.L. (2005) *Tetrahedron*, **61**, 10377–10441.
50 Giese, B. and Kretzschmar, G. (1982) *Chemische Berichte*, **115**, 2012–2014.
51 Crich, D., Davies, J.W., Negron, G., and Quintero, L. (1988) *Journal of Chemical Research (S)*, 140–141.
52 Crich, D. and Davies, J.W. (1989) *Tetrahedron*, **45**, 5641–5654.
53 (a) Colombo, L., Di Giacomo, M., Papeo, G., Carugo, O., Scolastico, C., and Manzoni, L. (1994) *Tetrahedron Letters*, **35**, 4031–4034; (b) Colombo, L., Di Giacomo, M., Scolastico, C., Manzoni, L., Belvisi, L., and Molteni, V. (1996) *Tetrahedron Letters*, **36**, 625–628.
54 Manzoni, L., Belvisi, L., and Scolastico, C. (2000) *Synlett*, 1287–1288.
55 Kessler, H., Wittmann, V., Köck, M., and Kottenhahn, M. (1992) *Angewandte Chemie*, **104**, 874–877; (1992) *Angewandte Chemie (International Edition in English)*, **31**, 902–904.
56 (a) Beckwith, A.L.J. and Chai, C.L.L. (1990) *Journal of the Chemical Society, Chemical Communications*, 1087–1088; (b) Axon, J.R. and Beckwith, A.L.J. (1995) *Journal of*

*the Chemical Society, Chemical Communications*, 549–550.

57 Adamson, G.A., Beckwith, A.L.J., and Chai, C.L.L. (2004) *Australian Journal of Chemistry*, **57**, 629–633.

58 Pyne, S.G. and Schafer, K. (1998) *Tetrahedron*, **54**, 5709–5720.

59 Herpin, T.F., Motherwell, W.B., and Weibel, J.-M. (1997) *Journal of the Chemical Society, Chemical Communications*, 923–924.

60 (a) Jones, R.C.F., Berthelot, D.J.C., and Iley, J.N. (2000) *Journal of the Chemical Society, Chemical Communications*, 2131–2132; (b) Jones, R.C.F., Berthelot, D.J.C., and Iley, J.N. (2001) *Tetrahedron*, **57**, 6539–6555.

61 (a) Chai, C.L.L. and King, A.R. (1995) *Tetrahedron Letters*, **36**, 4295–4298; (b) Chai, C.L.L. and King, A.R. (1999) *Journal of the Chemical Society, Perkin Transactions 1*, 1173–1182.

62 Gasanov, R.G., Il'inskaya, L.V., Misharin, M.A., Maleev, V.I., Raevski, N.I., Ikonnikov, N.S., Orlova, S.A., Kuzmina, N.A., and Belokon, Y.N. (1994) *Journal of the Chemical Society, Perkin Transactions 1*, 3343–3348.

63 Sibi, M.P., Asano, Y., and Sausker, J.B. (2001) *Angewandte Chemie*, **113**, 1333–1336; (2001) *Angewandte Chemie International Edition*, **40**, 1293–1296.

64 Renaud, P. and Stojanovic, A. (1996) *Tetrahedron Letters*, **37**, 2569–2572.

65 Srikanth, G.S.C. and Castle, S.L. (2004) *Organic Letters*, **6**, 449–452.

66 (a) He, L., Srikanth, G.S.C., and Castle, S.L. (2005) *The Journal of Organic Chemistry*, **70**, 8140–8147; (b) Banerjee, B., Capps, S.G., Kang, J., Robinson, J.W., and Castle, S.L. (2008) *The Journal of Organic Chemistry*, **73**, 8973–8978.

67 Ma, B., Litvinov, D.N., He, L., Banerjee, B., and Castle, S.L. (2009) *Angewandte Chemie*, **121**, 6220–6223; (2009) *Angewandte Chemie International Edition*, **48**, 6104–6107.

# 4
# Synthesis of β-Lactams (Cephalosporins) by Bioconversion

*José Luis Barredo, Marta Rodriguez- Sáiz, José Luis Adrio, and Arnold L. Demain*

## 4.1
## Introduction

Ring expansion of the five-membered thiazolidine ring of the intermediate penicillin N to the six-membered dihydrothiazine ring of deacetoxycephalosporin C (DAOC) (Figure 4.1) was discovered by Kohsaka and Demain [1] working with cell-free extracts of *Acremonium chrysogenum* (formerly *Cephalosporium acremonium*). DAOC synthase (deacetoxycephalosporin C synthaseDAOCS; also known as expandase) is a crucial enzyme responsible for converting penicillins to cephalosporins. In cephalosporin producers, *A. chrysogenum* and *Streptomyces clavuligerus*, expandase is an iron- and α-ketoglutarate-dependent enzyme. In the bacterium, *S. clavuligerus*, the subsequent hydroxylation of the methyl group of DAOC to give deacetylcephalosporin C (DAC) is catalyzed by a closely related enzyme, DAC synthase (deacetylcephalosporin C synthase DACS), whereas in the fungus *A. chrysogenum*, the activities of expandase and hydroxylase reside in a single functional protein (Figure 4.1).

The bifunctional expandase of *A. chrysogenum* is a monomer with a molecular weight of 41 kDa and an isoelectric point (p*I*) of 6.3 [2]. It is a bifunctional enzyme that catalyzes not only ring expansion but also the hydroxylation of the methyl group of DAOC to DAC. Expandase exhibited properties of α-ketoglutarate-linked dioxygenases [3]. Ring expansion activity was markedly increased by ascorbic acid, $Fe^{+2}$ [4, 5], α-ketoglutarate [4, 6], oxygen [6, 7], dithiothreitol (DTT), and ATP. The enzyme showed an absolute requirement for α-ketoglutarate [4, 8–10] and this cosubstrate could not be replaced by chemically related compounds such as α-ketoadipate, pyruvate, oxaloacetate, glutamate, succinate, α-ketovalerate, or α-ketobutyrate [11–13]. The properties of the expandase are similar to those of other α-ketoglutarate-dependent dioxygenases that incorporate one atom of oxygen into α-ketoglutarate to form succinate and the other atom into the product molecule. However, expandase differs in that no oxygen atom appears in its ring expansion product, DAOC. When isopenicillin N, penicillin G, penicillin V, ampicillin, and 6-aminopenicillanic acid (6-APA) were tested as substrate analogs of penicillin N, no ring expansion was observed [2, 11, 14].

*Amino Acids, Peptides and Proteins in Organic Chemistry.*
*Vol.3 – Building Blocks, Catalysis and Coupling Chemistry.* Edited by Andrew B. Hughes

ISBN: 978-3-527-32102-5

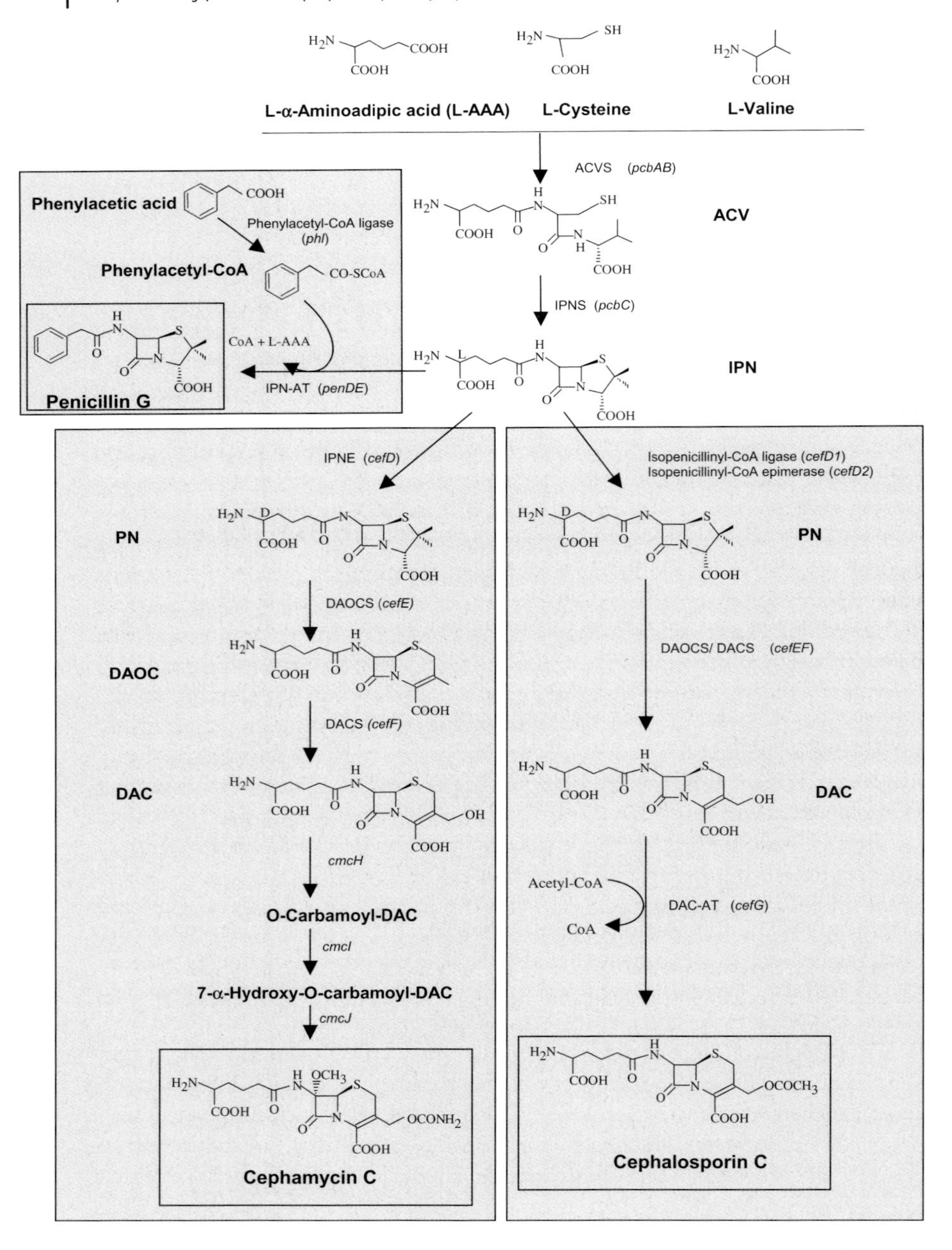


**Figure 4.1** Biosynthetic pathways of the main β-lactam families: penicillins (penicillin G by *P. chrysogenum*), cephalosporins (cephalosporin C by *A. chrysogenum*), and cephamycins (cephamycin C by *Streptomyces* spp.) (see text for main abbreviations). ACV biosynthesis is catalyzed by ACVS, which is encoded by the *pcbAB* gene. IPN is synthesized from ACV by IPNS, which is encoded by the *pcbC* gene. *pcbAB* and *pcbC* genes are common to the three

(*Continued*)

Both activities of the fungal enzyme were evident when $Fe^{2+}$ was replaced by $Fe^{3+}$ in the presence of a reducing agent; however, no activity was observed when $Fe^{2+}$ was replaced by $Mg^{2+}$, $Ca^{2+}$, $Cu^{2+}$, $Ni^{2+}$, $Co^{2+}$, $Zn^{2+}$, $Na^{+}$, or $K^{+}$. The $K_m$ value obtained for penicillin N was 29 μM, whereas for α-ketoglutarate, it was 22 μM [2, 11, 14]. Both activities were similarly sensitive to inhibition by metal chelators like EDTA or *o*-phenanthroline, as would be expected with chelators of $Fe^{2+}$. Inhibition by $Zn^{2+}$ and sulfhydryl reagents like *p*-hydroxymercuribenzoate (PHMB), 5,5′-dithio-bis(2-nitrobenzoic acid) (DTNB), or *N*-ethylmaleimide (NEM) suggests that at least one sulfhydryl group is essential for ring expansion activity.

In the fungus, the cephalosporin biosynthetic genes are located in two different clusters on separate chromosomes [15–17]. One of them is formed by the genes encoding the first two steps of the pathway, *pcbAB* and *pcbC* [18], found on chromosome VI, whereas the cluster containing the late genes (*cefEF* and *cefG*) is located on chromosome II [19]. The two *cef* genes are oppositely oriented and expressed from divergent promoters located within a 938-bp intergenic region [20]. Cloning of *cefEF* from *A. chrysogenum* in *Escherichia coli* was achieved using probes based on amino acid sequences of fragments of the purified protein [21]. The results obtained revealed an open reading frame of 996 bp encoding a protein of 332 amino acids with a predicted molecular weight of 36 462 Da. However, the protein migrated during sodium dodecylsulfate–polyacrylamide gel electrophoresis with an apparent molecular weight of 40–41 kDa. This could be due to posttranslational modification.

Owing to the effectiveness of the cephalosporins against penicillin-resistant pathogens, there has been great interest in producing large quantities of intermediates that can be used to prepare semisynthetic analogs of the natural cephalosporin C. One such intermediate is 7-aminodeacetoxycephalosporanic acid (7-ADCA); another is 7-aminocephalosporanic acid (7-ACA).

Industrial 7-ACA and 7-ADCA were traditionally produced by complex chemical processes, which were expensive and environmentally unfriendly. New microbiological procedures are now available, and are discussed in Sections 4.3 and 4.4, respectively.

The development of recombinant DNA techniques and their application to industrial microorganisms in the last few decades has led to the design of new

(*Continued*)

pathways. Epimerization of IPN to penicillin N (PN) is catalyzed by IPNE, encoded by a single gene (*cefD*) in *Streptomyces*, whereas two genes (*cefD1* and *cefD2*) encoding isopenicillinyl-CoA ligase and isopenicillinyl-CoA epimerase are involved in *A. chrysogenum*. Prokaryotic species possess separate *cefE* and *cefF* genes and enzymes (encoding DAOCS and DACS, respectively) that convert penicillin N into DAOC, which is then transformed into DAC. In the fungus *A. chrysogenum*, the *cefEF* gene encodes an enzyme with both DAOCS/DACS activities, that directly biosynthesizes DAC from penicillin N. DAC is finally converted into cephalosporin C by the DAC-AT encoded by the *cefG* gene. In the penicillin G pathway, the *penDE* gene codes for IPN-AT, encoded by the *penDE* gene, which exchanges the L-α-aminoadipic acid side-chain of IPN by phenylacetic acid, previously activated as phenylacetyl-CoA. *cmcH*, *cmcI*, and *cmcJ* genes code for enzymatic activities involved in the last three steps of cephamycin C biosynthesis: *O*-carbamoylation, hydroxylation, and *O*-methylation.

biosynthetic pathways and to the modification of existing pathways. This has resulted in yield improvement of new molecules of interest.

The product of the expandase acting on penicillin G, deacetoxycephalosporin G (DAOG), could be easily deacylated to 7-ADCA – the important compound for the manufacture of many semisynthetic cephalosporins of medical use. Thus, a bioconversion of the inexpensive and available penicillin G to DAOG is of great economic interest. Several research groups have reported on the narrow substrate specificity and lack of detectable activity of expandase from *S. clavuligerus* on inexpensive and available penicillins such as penicillin V and G produced by *Penicillium chrysogenum* [1, 11, 22, 23]. However, progress has been made and these developments are described in this chapter, which presents an overview on β-lactam synthesis by bioconversion to construct new and environmentally safer "green" routes for commercially valuable cephalosporins and penicillins.

## 4.2
## Biosynthetic Pathways of Cephalosporins and Penicillins

Cephalosporins are chemically characterized by a cephem nucleus composed of a β-lactam ring fused to a dihydrothiazine ring (Figure 4.1). The different cephalosporin derivatives are synthesized by side-chain substitutions from cephalosporin C – a cephalosporin with low antibacterial activity, industrially produced by submerged fermentation of *A. chrysogenum*.

The cephalosporin C biosynthetic pathway has been investigated in depth in *A. chrysogenum*. The route begins with three enzymatic reactions common to penicillins and cephalosporins (Figure 4.1). The first one is the condensation of L-cysteine and L-α-aminoadipic acid to form the dipeptide L-α-aminoadipyl-L-cysteine (AC). Then, L-valine is epimerized to D-valine, activated, and condensed with AC to form the tripeptide δ-(L-α-aminoadipyl)-L-cysteinyl-D-valine (ACV) [24, 25]. The first two reactions are catalyzed by the activity of ACV synthetase (δ-(L-α-aminoadipyl)-L-cysteinyl-D-valine synthetase; ACVS), encoded by the *pcbAB* gene [18, 26]. The purified ACVS has a molecular weight of 360 kDa and was suggested to be a dimeric structure. Its activity is dependent on L-α-aminoadipic acid, L-cysteine, ATP, and $Mg^{+2}$ or $Mn^{+2}$. The third reaction is the cyclization of ACV to isopenicillin N (IPN) catalyzed by the IPN synthase (isopenicillin N synthase IPNS or "cyclase"), encoded by the *pcbC* gene [27]. IPNS has a molecular weight of 41 kDa [28], and is stimulated by $Fe^{+2}$, reducing agents, and oxygen [10]. In *A. chrysogenum*, the IPN epimerase (isopenicillin N epimerase; IPNE) activity, responsible for the epimerization of the L-α-aminoadipic side-chain of IPN to the D-configuration of penicillin N, has been shown to be a two-component protein system catalyzed by the enzymes isopenicillinyl-CoA ligase and isopenicillinyl-CoA epimerase, encoded respectively by the *cefD1* and *cefD2* genes [29, 30], whereas the single *cefD* gene encodes IPNE in cephamycin or cephalosporin-producing bacteria, such as *S. clavuligerus*, *Nocardia lactamdurans*, and *Lysobacter lactamgenus* [31–33].

Although formerly, penicillin N and cephalosporin C were thought to be products of different biosynthetic pathways, the discovery of ring expansion of penicillin N to DAOC revealed the key step for a common pathway [1]. The ring expansion of penicillin N to form DAOC is catalyzed by an $\alpha$-ketoglutarate-dependent dioxygenase, DAOCS ("expandase") [1], requiring $Fe^{+2}$, ascorbate [10], oxygen, and $\alpha$-ketoglutarate [6, 11]. DAOCS and the following enzyme of the pathway, DACS, are encoded by separate genes in bacteria (*cefE* and *cefF*, respectively) [34, 35], but both activities are present as a single protein encoded by the *cefEF* gene located on chromosome II in *A. chrysogenum* [19, 21, 36, 37]. The DAOCS of *S. clavuligerus* is a monomer of 34.6 kDa [38] whose catalytic activity is stimulated by DTT and ascorbate, but not by ATP. On the other hand, the fungal bifunctional DAOCS–DACS enzyme is a monomer of 41 kDa with an isoelectric point of 6.3 [2]. It requires $Fe^{+2}$, $\alpha$-ketoglutarate, and oxygen, and is stimulated by ascorbate, DTT, and ATP. Finally, the hydroxymethyl group of DAC, located at C3, is acetylated by a DAC acetyltransferase (deacetylcephalosporin C acetyltransferase DAC-AT) encoded by the *cefG* gene [20, 39] to biosynthesize cephalosporin C.

The genes of *A. chrysogenum* encoding the first two enzymes of the cephalosporin biosynthetic pathway (*pcbAB* and *pcbC*) are clustered in chromosome VII, whereas the genes coding for the enzymes that catalyze the last two steps of the pathway (*cefEF* and *cefG*) are clustered in chromosome I [40]. Both clusters are present as a single copy per genome both in the wild-type and in the high cephalosporin-producing strains of *A. chrysogenum*. Altered electrophoretic karyotypes as a result of chromosomal rearrangements have been described in industrial strains of *A. chrysogenum* obtained by classical mutagenesis [41].

In the penicillin G biosynthetic pathway of *P. chrysogenum*, the activity of IPN acetyltransferase (isopenicillin N acyltransferase; IPN-AT) [42], encoded by the *penDE* gene of penicillin-producing fungi [43], exchanges the $\alpha$-aminoadipyl side-chain of IPN by aromatic acids such as phenylacetic or phenoxyacetic acids, resulting in the biosynthesis of penicillin G or V, respectively [44]. This activity is absent in *A. chrysogenum* and other cephalosporin producers because they lack a gene homologous to *penDE*.

## 4.3
## Production of 7-ACA by *A. chrysogenum*

The clinically important semisynthetic derivatives of cephalosporins are manufactured from 7-ACA and 7-ADCA. Important cephem antibiotics, such as cefazolin and ceftizoxime, are chemically synthesized from 7-ACA [45, 46], whereas production of oral cephem antibiotics, such as cefixime and cefdinir, utilizes as starting material 7-aminodeacetylcephalosporanic acid (7-ADACA), obtained by hydrolysis of 7-ACA [47, 48].

Industrial 7-ACA had been produced from cephalosporin C by a complex chemical process using harmful and toxic reagents to obtain a crystallized product [49]. This procedure was very expensive because important environmental protection measures were required. As an alternative to the chemical process, a two-step enzymatic procedure conducted in water was developed to convert cephalosporin C into 7-ACA. This process involves the conversion of cephalosporin C into 7-β-(5-carboxy-5-oxopentanamido)-cephalosporanic acid (ketoadipyl-7-ACA), catalyzed by the activity of D-amino acid oxidase (DAO) [50, 51], and then, the ketoadipyl-7-ACA reacts non-enzymatically with the hydrogen peroxide released in the previous reaction to give 7-β-(4-carboxy-butanamido)-cephalosporanic acid (glutaryl-7-ACA). Finally, the enzyme glutaryl acylase (GLA) hydrolyzes ketoadipyl-7-ACA and glutaryl-7-ACA to yield 7-ACA [52, 53] (Figure 4.2).

A fusion of the genes encoding DAO and GLA was expressed in *E. coli* obtaining a GLA–DAO protein, which includes both enzymatic activities and directly catalyzes the conversion of cephalosporin C into 7-ACA, thus simplifying the bioconversion process [54].

The industrial application of the two-step procedure has been implemented with high overall yield [53]. The filtered cephalosporin C solution obtained from the fermentation broth is sequentially treated with acrylic resins, including immobilized DAO and GLA, to produce 7-ACA in a very cost-competitive and environmentally friendly process. One of the major advantages of this enzymatic procedure over the chemical method is the high selectivity of the enzymes in recognizing a specific molecular group without interfering with any other reactive residues present in the substrate. This avoids the use of dangerous and toxic reagents as protective groups in

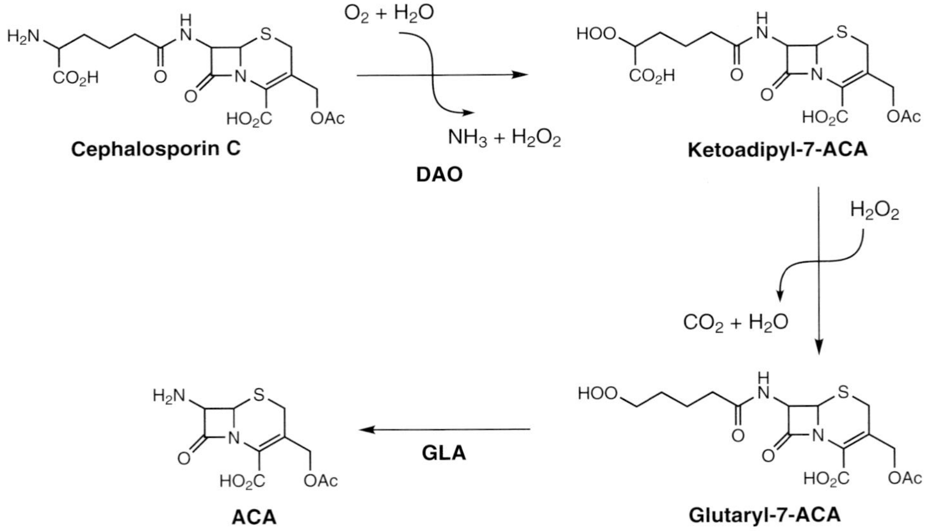

**Figure 4.2** Two-step enzymatic synthesis of 7-ACA from cephalosporin C using DAO and GLA. Cephalosporin C is transformed into 7-β-(5-carboxy-5-oxopentanamido)-cephalosporanic acid (ketoadipyl-7-ACA) by DAO and then the ketoadipyl-7-ACA reacts nonenzymatically with hydrogen peroxide to form glutaryl-7-ACA. Finally, glutaryl-7-ACA is hydrolyzed to 7-ACA by GLA.

the reaction. In addition, the product can be easily purified and concentrated by adsorption to resins, yielding highly pure 7-ACA. All these aspects make the enzymatic process more efficient, less expensive, and more environmentally suitable compared to the old chemical one.

This two-step enzymatic procedure for 7-ACA manufacturing involves an *in vitro* enzymatic conversion catalyzed by DAO and GLA. Different microorganisms have been reported as sources of DAO (e.g., *Trigonopsis variabilis, Rhodotorula gracilis, Fusarium solani*, etc.) [55–57] and of GLA (*Pseudomonas* spp., *Acinetobacter* spp., etc.) [52, 53]. On the basis of this procedure, a 7-ACA biosynthetic gene cluster was constructed in *A. chrysogenum* by the expression of the genes encoding DAO from *F. solani* and GLA from *Pseudomonas diminuta* in an attempt to directly produce 7-ACA by fermentation [57] (Figure 4.3a). The recombinant strains were able to produce detectable amounts of 7-ACA plus two side-products, 7-ADACA and 7-ADCA, probably synthesized from the biosynthetic intermediates DAC and DAOC, respectively. Although the low amount of 7-ACA produced by the recombinant strain of *A. chrysogenum* does not have commercial significance, this was the first report on direct microbial production of 7-ACA, and it proved the feasibility of introducing new biosynthetic capabilities into industrial microorganisms by combining fungal and bacterial genes.

## 4.4 Production of 7-ADCA by *A. chrysogenum*

7-ADCA is used as a substrate for the production of cephalexin and cephadroxyl – semisynthetic cephalosporins lacking the R2 side-chain at the C3 position. 7-ADCA has been traditionally synthesized from β-lactams obtained by fermentation via a complex chemical process that requires expensive reagents and prior purification of a highly impure starting material before chemical treatment, generating significant quantities of byproducts that could cause environmental problems [58]. The chemical process is based on a multistep ring expansion of penicillin G produced by fermentation of *P. chrysogenum* [59]. Chemical reactions are carried out in organic solvents to protect the free carboxyl groups of the penicillin G, oxidize the sulfur atom of the resulting intermediate, ring-expand the oxidized five-membered thiazolidine ring, and hydrolyze the protected carboxyl moiety of DAOG (phenylacetyl-7-ADCA). Finally, the phenylacetyl side-chain is enzymatically removed to give 7-ADCA using a penicillin acylase.

Fermentative processes to replace the chemical pathway to 7-ADCA have been developed in similar ways to those described above for 7-ACA to reduce costs and environmental problems. Three biological routes for 7-ADCA production have been described: the first one using *P. chrysogenum* [60], a second one based on *A. chrysogenum* [61], and a third method by enzymatic expansion of penicillins into cephalosporins [62, 63].

The development of a genetically modified *A. chrysogenum* strain for the biological production of 7-ADCA was reported by Velasco *et al.* [61]. In this case, the *cefEF* gene

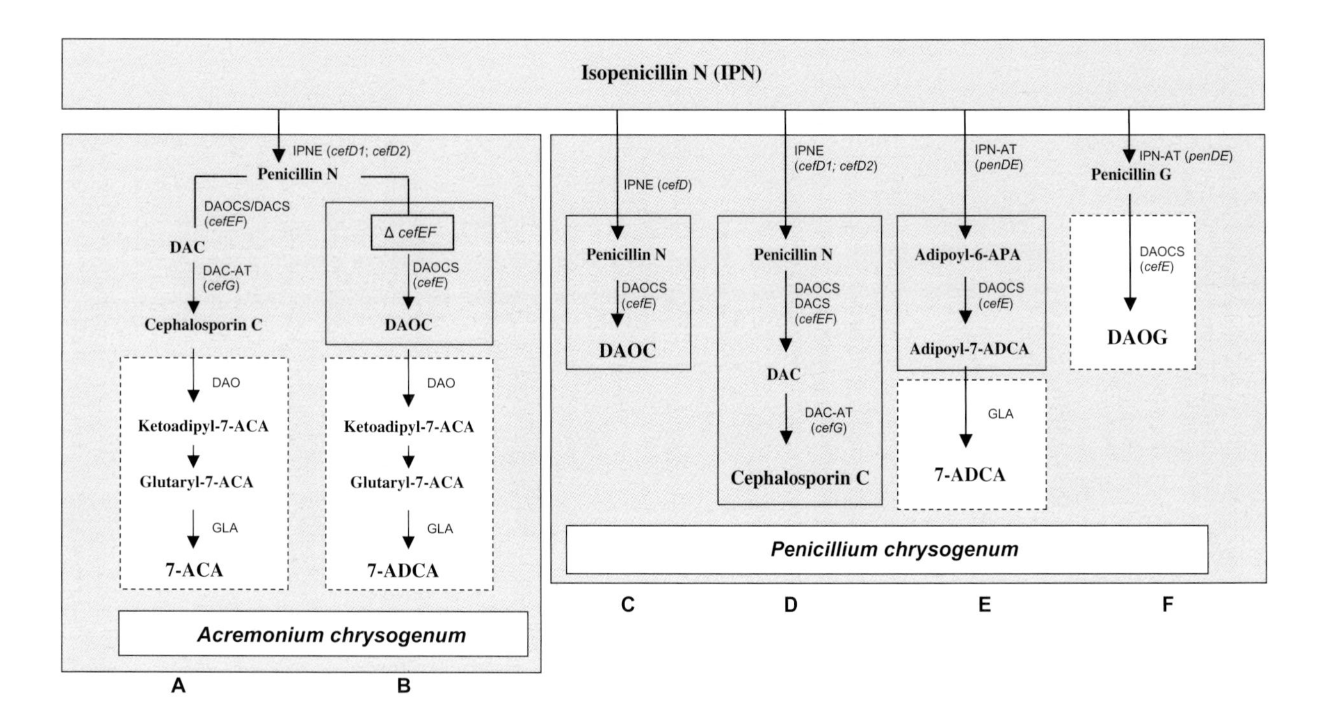


**Figure 4.3** Pathway engineering to production of 7-ACA, 7-ADCA, DAOC, cephalosporin C and DAOG. **(a)** The genes encoding DAO from *F. solani* and GLA from *P. diminuta* were expressed in *A. chrysogenum* to produce 7-ACA [55]. **(b)** The *cefEF* gene of *A. chrysogenum* was disrupted to biosynthesize penicillin N, which is expanded *in vivo* to DAOC by DAOCS encoded by the *cefE* gene of *S. clavuligerus*; DAOC is finally bioconverted into 7-ADCA by two enzymatic steps involving DAO and GLA [57]. **(c)** The genes encoding IPNE (*cefD* from *S. lipmanii*) and DAOCS (*cefE* from *S. clavuligerus*) were expressed in *P. chrysogenum* to produce DAOC [64]. **(d)** The *cefD1*, *cefD2*, *cefEF*, and *cefG* genes from *A. chrysogenum* were expressed in a strain of *P. chrysogenum* lacking IPNA to produce DAC (extracellular) and cephalosporin C (intracellular) [68]. **(e)** The *cefE* gene (encoding DAOCS) from *S. clavuligerus* was expressed in *P. chrysogenum* to produce adipoyl-7-ADCA in adipate-fed fermentations. Adipoyl-7-ADCA is converted into 7-ADCA by enzymatically removing the side-chain with a cephalosporin acylase [59]. **(f)** The penicillin G produced by *P. chrysogenum* is enzymatically bioconverted to DAOG by DAOCS [63]. Enzymatic bioconversions are boxed in white. *In vivo* biosyntheses are in gray.

disruption of an industrial strain of *A. chrysogenum* led to transformants accumulating large amounts of penicillin N. Further expression of the *cefE* gene from *S. clavuligerus* under the control of the *pcbC* promoter from *P. chrysogenum* caused *in vivo* expansion of penicillin N into DAOC (Figure 4.3b). DAOC constitutes the starting material for 7-ADCA production using the two-step enzymatic bioconversion involving the following: (i) DAO, which catalyzes the oxidative deamination of DAOC to 7-β-(5-carboxy-5-oxopentanamido)-deacetoxycephalosporanic acid (ketoadipyl-7-ADCA), (ii) the decarboxylation to glutaryl-7-ADCA by the side-product of the reaction (hydrogen peroxide) [51], and (iii) the hydrolysis of glutaryl-7-ADCA to 7-ADCA by GLA [52, 53]. In contrast to the data reported for recombinant strains of *P. chrysogenum* expressing ring-expansion activity, no detectable contamination with other cephalosporin intermediates occurs.

The process based on adipoyl-7-ADCA biosynthesis by a recombinant strain of *P. chrysogenum* expressing the *cefE* gene of *S. clavuligerus* and further removal of the adipoyl side-chain using a cephalosporin acylase [60] is described in Section 4.6. The method of expanding the ring of penicillins into cephalosporins involves ring expansion of the inexpensive penicillin G by *S. clavuligerus* cells or their expandase to yield DAOG (phenylacetyl-7-ADCA), followed by removal of the phenylacetyl side-chain yielding DAOC [62, 63]. As described above, DAOC can be enzymatically converted to 7-ADCA. This process is discussed in Section 4.7.

Using the above approaches, the entire synthesis of 7-ADCA is carried out avoiding chemical steps, and constitutes a very efficient bioprocess from the industrial and environmental points of view.

## 4.5
## Production of Penicillin G by *A. chrysogenum*

The *penDE* gene of *P. chrysogenum* [43] encodes the enzymatic activity IPN-AT [42], which exchanges the α-aminoadipyl side-chain of IPN by phenylacetic acid, resulting in the biosynthesis of penicillin G. This activity is absent in *A. chrysogenum* and other cephalosporin-producing microorganisms. The heterologous expression in *A. chrysogenum* of the *penDE* gene of *P. chrysogenum* caused simultaneous production of benzylpenicillin (penicillin G) and cephalosporin C in phenylacetate-fed fermentations of the transformant strains [64]. The branching of the biosynthetic pathway at the IPN step (Figure 4.1) lowered the cephalosporin C yield with an equivalent production of β-lactams (cephalosporin C plus penicillin G). The relative amount of each antibiotic was determined by the specific activities and substrate affinities of the two competing enzymes IPNE and IPN-AT, showing a decrease in the cephalosporin C level equivalent to the penicillin G synthesized. Northern analysis of the transformants revealed a transcript corresponding to the *penDE* gene identical in size to that of *P. chrysogenum*, showing that the three introns of the *penDE* gene were properly processed by *A. chrysogenum*. This suitable heterologous expression of the *penDE* gene under the control of its own promoter is in all probability due to the conservation of the transcription signals and expression motifs in both filamentous fungi. A potential application of these transformants could be adipoyl-cephalosporin production in adipate-fed fermentations [60] that could simplify the enzymatic production of 7-ACA.

## 4.6 Production of Cephalosporins by *P. chrysogenum*

The genes of *P. chrysogenum* encoding the penicillin biosynthetic enzymes have been characterized (Figure 4.1): *pcbAB* encoding ACVS [65, 66], *pcbC* encoding IPNS [67, 68], *penDE* coding for IPN-AT [47], and *phl* encoding a phenylacetyl-CoA ligase [69]. The first three genes are clustered together [65, 70, 71] in chromosome I [72] and tandem amplified in penicillin-overproducing strains [73–75]. It is also possible that some high-producing strains hold chromosome amplification and segregate to lower-producing strains probably by chromosome losses [76]. Nevertheless, the *phl* gene is not linked to the penicillin gene cluster. Although the phenylacetyl-CoA ligase encoded by the *phl* gene is involved in penicillin production, a second aryl-CoA ligase appears to contribute partially to phenylacetic acid activation.

*In silico* analysis of the *P. chrysogenum* genome revealed the presence of a gene called *ial* (IPN-AT-like, isopenicillin N acyltransferase-like IAL; Pc13g09140) initially described as a paralog of the *penDE* gene [77]. Nevertheless, no activity related to penicillin biosynthesis was detected for IAL. Sequence comparison between the *P. chrysogenum* IAL, the *Aspergillus nidulans* IAL homolog, and the IPN-AT, revealed that the lack of enzyme activity seems to be due to an alteration of the essential Ser309 in the thioesterase active site. It is possible that the *ial* and *penDE* genes might have been formed from a single ancestral gene that became duplicated during evolution.

The penicillin production level of *P. chrysogenum* is higher than that of cephalosporin production by *A. chrysogenum* [78]. This difference has led to the construction of genetically-modified strains of *P. chrysogenum* producing cephalosporin C derivatives.

Earlier studies reported the possibility of producing detectable amounts of DAOC in *P. chrysogenum* by the heterologous expression of the *cefD* gene of *S. lipmanii* and the *cefE* gene of *S. clavuligerus*, encoding IPNE and DAOCS, respectively [79]. The enzymatic ring expansion of penicillin G was not possible *in vivo*, although low levels of DAOC corresponding to expanded IPN were found in the fermentation broths (Figure 4.3c).

Old experiments showed that feeding *P. chrysogenum* fermentations with adipic acid (instead of phenylacetic or phenoxyacetic acids) as a side-chain precursor, forced the fungus to biosynthesize adipoyl-6-APA (carboxy-*n*-butyl-penicillin) [80]. Adipoyl-6-APA is a potential substrate of the DAOCS for the production of adipoyl-7-ADCA. On the basis of this idea, Crawford *et al.* [60] expressed the *cefE* gene of *S. clavuligerus* in an industrial strain of *P. chrysogenum*. As a result, significant amounts of adipoyl-7-ADCA were produced by selected transformants and the adipoyl side-chain was efficiently removed using a cephalosporin acylase from *Pseudomonas* (Figure 4.3e). Physiological studies of the *P. chrysogenum* strain expressing DAOCS showed slightly lower productivity when compared to the penicillin-producing host [81]. Finally, an industrial plant for the production of 7-ADCA using this technology was set up [82], exemplifying the power of metabolic engineering for developing new fermentative processes that are more competitive and environmentally friendly.

In a similar way, the *cefEF* and *cefG* genes (encoding DAOCS/DACS and DAC-AT, respectively) from *A. chrysogenum* were expressed in *P. chrysogenum* and the transformants, when grown in media containing adipic acid as the side-chain precursor, carried out a novel and efficient bioprocess for production of adipoyl-cephalosporins. Strains expressing DAOCS/DACS produced both adipoyl-7-ADCA and adipoyl-7-ADAC, whereas strains expressing DAOCS/DACS and DAC-AT produced adipoyl-7-ADCA, adipoyl-7-ADAC, and adipoyl-7-ACA. The adipoyl side-chain of these cephalosporins was easily removed with a cephalosporin acylase from *Pseudomonas* to yield the cephalosporin intermediates [60]. Although these adipoyl-cephalosporins were obtained in relatively low yields, the authors stated that application of classical strain improvement technology to the highest producing transformants resulted in significant further titer increases, thus forecasting a commercial process.

DAC was produced by a mutant strain of *P. chrysogenum* lacking IPN-AT activity (accumulating penicillin N) that was used as a host for the heterologous expression of the genes *cefD1*, *cefD2*, *cefEF*, and *cefG* from *A. chrysogenum* [83]. The transformant strains secreted significant amounts of DAC, but cephalosporin C was not detected in the culture broths. High-performance liquid chromatography (HPLC) analysis of cell extracts of three transformants showed that two of them accumulated DAC intracellularly, and the third accumulated both DAC and cephalosporin C. Even when accumulated intracellularly, cephalosporin C was not found in the culture broth (Figure 4.3d)

## 4.7 Conversion of Penicillin G and other Penicillins to DAOG by *Streptomyces clavuligerus*

### 4.7.1 Expandase Proteins and Genes

The *S. clavuligerus* expandase is a monomer of 34 600 Da with two isoelectric points of 6.1 and 5.3 [38]. Based on a 22-residue N-terminal sequence of the purified protein, the *cefE* gene coding for expandase was cloned and expressed in *E. coli* [34]. Comparison of this gene and the *cefEF* gene of *A. chrysogenum* showed 67% similarity, whereas at the protein level there was 57% similarity. The bacterial protein is 21 amino acids shorter than the fungal enzyme. Purified expandase was not able to expand isopenicillin N, penicillin G, penicillin V, ampicillin, or 6-APA [14, 38].

Based on the reported crystal structure of the *S. clavuligerus* expandase [84], a mechanism for the formation of the reactive ferryl species, a catalytic intermediate common to many nonheme oxygenases [85], was proposed. The apo-expandase, a crystallographic trimer, reacts first with the iron, which is ligated by three protein ligands (His183, Asp185, and His243) and three solvent molecules, causing dissociation of the apoenzyme into monomers – the catalytically active form. Then, the cosubstrate $\alpha$-ketoglutarate binds to the enzyme-$Fe^{2+}$, replacing two solvent molecules around the iron, and allowing dioxygen binding. The ferryl form of expandase, which reacts with penicillin N, is created by the splitting of dioxygen and oxidative decarboxylation of $\alpha$-ketoglutarate to succinate. Studies employing site-directed mutagenesis were carried out to replace the residues His183, Asp185, and His243 [86]. Substitution at these sites completely abolished ring expansion, confirming that these residues are essential for catalysis.

The utilization of alternative cosubstrates for the ring expansion reaction was investigated earlier [11, 13]. The expandase crystal structure revealed the arginine residues within the active site, with Arg258 being involved in cosubstrate binding [84, 87]. Site-directed mutagenesis of this residue resulted in mutant enzyme R258Q whose activity was reduced in the presence of $\alpha$-ketoglutarate, but could be fully restored using aliphatic 2-oxoacids as alternative cosubstrates [88]. Wild-type expandase has, at most, only traces of activity with the 2-oxoacids tested (other than $\alpha$-ketoglutarate and 2-oxoadipate). These results show that the side-chain of Arg258 is a major determinant of the 2-oxoacid cosubstrate selectivity. The decrease in activity of the R258Q mutant observed with $\alpha$-ketoglutarate seems to be due to the loss of a favorable ionic interaction between the 5-carboxyl group of $\alpha$-ketoglutarate and the guanidine group of Arg258.

The crystal structure of expandase has revealed the formation of a trimeric unit in which the C-terminus of one expandase molecule is inserted into its neighbor in a cyclical fashion [84, 87]. Further studies suggested that residues located at the C-terminus might be involved in orienting and/or binding of the penicillin substrates during the reaction. Construction of different mutants with truncated C-termini showed that deletion of up to 11 residues does not affect binding of $\alpha$-ketoglutarate but there were significant differences in the way in which the enzyme catalyzes penicillin N oxidation as compared to penicillin G oxidation [88]. With penicillin N, deletion of five to six residues did not significantly affect activity but when penicillin G was used as substrate, activity was very much reduced.

Expandase is inactivated by compounds such as PHMB, DTNB, or NEM [38], suggesting that at least one sulfhydryl group may be important for structural integrity and/or catalysis. Lee *et al.* [89] mutated three cysteine residues (Cys100, Cys155, and Cys197) that were possible candidates for involvement in disulfide bond formation. Mutation at these positions (singly or in combination) to alanine residues led to mutant enzymes with activity on penicillin N.

Superimposition analysis of *S. clavuligerus* expandase and the *A. nidulans* IPNS allowed Chin *et al.* [90] to identify several residues located in the substrate-binding pocket of *S. clavuligerus* expandase. Four of them (R74, R160, R266, and N304) were classified as the most hydrophilic. In order to modify the substrate specificity of expandase towards hydrophilic penicillins, these four positions were selected for single mutations to replace R (arginine) or N (asparagine) with L (leucine). Only mutant enzyme N304L was able to convert the three substrates tested (penicillin G, ampicillin, and amoxicillin) and improvements in ring expansion activity from 5- to

40-fold were observed relative to wild-type expandase. The improved enzymatic activity shown by this mutant could be due to the orientation of the L residue at the C-terminus guiding the entrance of the substrate into the catalytic center of expandase, confirming the proposal made by Lee *et al.* [88] that modifications at this part of the enzyme might be involved in orienting and/or binding of the penicillin substrates during the reaction.

Cloning and expression of the *Nocardia lactamdurans cefE* gene in *Streptomyces lividans* revealed an enzyme of 34 532 Da and an isoelectric point of 4.9 [32]. The gene has high similarity to the genes of *S. clavuligerus* (74.7%) and *A. chrysogenum* (69.2%). At the protein level, there is also high similarity to the expandase of *S. clavuligerus* (70.4%) and the expandase/hydroxylase of *A. chrysogenum* (59.5%). This expandase also showed a high specificity for the nature of the side-chain and only penicillin N was expanded.

## 4.7.2 Bioconversion of Penicillin G to DAOG

When IPN, penicillin G, penicillin V, ampicillin, carboxy-*n*-butyl penicillin, and 6-APA were tested as substrate analogs [2, 11, 14], no ring expansion by *A. chrysogenum* expandase was observed.

The catalytic activity of the *S. clavuligerus* expandase requires α-ketoglutarate, $Fe^{2+}$, and oxygen, and it is stimulated by DTT and ascorbate, but not by ATP [14, 38, 91]. The synthase was equally active with $Fe^{2+}$ or $Fe^{3+}$ in the presence of ascorbate and DTT; however, these metal ions could not be replaced by $Mg^{2+}$, $Ca^{2+}$, $Cu^{2+}$, $Ni^{2+}$, $Co^{2+}$, $Zn^{2+}$, $Na^{+}$, or $K^{+}$. As mentioned above for the fungal synthetase, the *S. clavuligerus* expandase activity is also very sensitive to metals chelators such as EDTA or *o*-phenanthroline, and to sulfhydryl reagents like PHMB, DTNB, and NEM. As was previously observed by other authors working with cell-free extracts [8, 23], purified expandase was unable to expand IPN, penicillin G, penicillin V, ampicillin, or 6-APA [14, 38].

## 4.7.3 Broadening the Substrate Specificity of Expandase

### 4.7.3.1 Resting Cells

*S. clavuligerus* NP1, a mutant producing only trace levels of cephalosporins [92], was used, since the absence of significant levels of cephalosporins in this strain facilitated detection of cephalosporins produced by ring expansion from added penicillins. Mycelia were developed in two shaken stages of medium MST [93] and were washed twice before being used for the reaction. The standard reaction mixture for expandase action was that described by Maeda *et al.* [23] except that penicillin G replaced penicillin N. Additions were made in the order established by Shen *et al.* [13]. At various times during the shaking incubation, samples were taken, centrifuged to remove cells, and supernatant fluids were assayed by the paper disk-agar diffusion bioassay, using *E. coli* Ess (a β-lactam supersensitive mutant) [94]. The formation of

DAOG was determined by inclusion of penicillinase in the assay agar (this narrow spectrum β-lactamase attacks penicillins, but not cephalosporins) and measuring zones of growth inhibition.

Although previous attempts by several research groups to expand the thiazolidine ring of penicillin G and other penicillins had failed [1, 11, 23, 38], success was achieved with resting cells of *S. clavuligerus* [95]. The cofactor requirements for penicillin G expansion to DAOG were examined with resting cells of *S. clavuligerus* NP1. When $Fe^{2+}$, α-ketoglutarate, or ascorbic acid was absent, the amount of product obtained was about 30% of that in the control. On the other hand, ATP, $Mg^{2+}$, $K^{+}$, and DTT did not play a significant role in the reaction with resting cells. The omission of DTT actually increased production by 50%. When the α-ketoglutarate concentration was raised from the previously used 0.64 [23] to 1.28 mM, activity of resting cells was doubled. When the concentration of $Fe^{2+}$ was raised 45-fold, the ring-expansion activity was markedly increased. The optimum ascorbate concentration for conversion was found to be 4–8 mM. Maeda *et al.* [23] had used 5 mM ascorbate and Lübbe *et al.* [96] had used 4 mM ascorbate for cell-free conversion of penicillin N to DAOC. ATP did not improve resting cell activity on penicillin G, whereas cell-free extracts acting on penicillin N had been stimulated by ATP [32]. Increasing cell mass enhanced the concentration of DAOG formed, the optimum concentration being 19 mg/ml. Higher cell concentrations inhibited the reaction, probably because oxygen supply became limiting. In the studies of Cho *et al.* [95], the buffer used for bioconversion had been 50 mM Tris–HCl at pH 7.4. It was later found that 50 mM MOPS buffer or 50 mM HEPES buffer at pH 6.5 improved activity [97]. Increasing the concentration of penicillin G (from 2 mg/ml) increased the concentration of DAOG produced but decreased the yield. Decreasing penicillin G concentration decreased the DAOG concentration but increased yield from below 1% to as high as 16.5%.

To identify the products of the biotransformation, a HPLC system for the separation of penicillin G, DAOG and DAG was used [98]. After 1 h of incubation, two new peaks (at 3.65 and 15.3 min) appeared on the chromatogram. The 15.3-min peak was DAOG but the 3.65-min peak was unidentified. During the subsequent 2 h of biotransformation, these two peaks increased in size. No peak corresponding to DAG was detected during the reaction. In *S. clavuligerus*, DAC is a precursor of three compounds: *O*-carbamoyldeacetylcephalosporin C, 7-α-hydroxy-*O*-carbamoyldeacetylcephalosporin C, and cephamycin C. The new unidentified peak might represent the phenylacetyl version of any one of these products.

#### 4.7.3.2 **Cell-Free Extracts**

Using the high concentrations of $FeSO_4$ and α-ketoglutarate established for resting cell conversion of penicillin G to DAOG, activity was observed with cell-free extracts. Furthermore, high protein (6 mg/ml) and high substrate (5 mg/ml) concentrations increased the concentration of product made. Using the conditions employed earlier with penicillin N [95], in which protein concentration in the cell-free extracts was 1–2 mg/ml, only a low level of activity on penicillin G was observed. The penicillin G concentrations previously used by Maeda *et al.* [23] were only 0.01–0.36 mg/ml. The importance of $Fe^{2+}$, α-ketoglutarate, and ascorbic acid, and the lack of importance of ATP, $MgSO_4$, and KCl, mentioned above with resting cells, were also observed with cell-free extracts. However, DTT did not decrease the activity of extracts on penicillin G and was included in later experiments to protect the enzyme [99]. All 15 penicillins tested were successfully bioconverted.

### 4.7.4 Inactivation of Expandase during the Ring-Expansion Reaction

The extent of expansion by *S. clavuligerus* extracts increased with increases in concentration of both protein and substrate [95]. However, in all cases, including purified recombinant expandase [100], the concentration of product increased during the first 1–3 h and then either remained stable or decreased. To determine if the limiting factor after 2 h was the exhaustion of one or more components of the reaction mixture, different concentrations and combinations of reaction constituents were added to the reaction at 2 h, but no additive was able to reactivate the system [101]. Lack of activity could not be attributed to enzyme instability during shaking. Preincubation of the cell-free extracts under different conditions of temperature, and agitation for 2 h did not affect subsequent ring expansion activity.

To determine whether one or more reaction mixture components might inactivate the enzyme, the cell-free extract was preincubated for 2 h in the presence of different reaction components. After this preincubation, the remaining reaction components as well as the substrate (penicillin G) were added. The amount of product obtained was markedly different depending on which component was present during preincubation. When buffer alone was present, the subsequent reaction yielded up to 95% of the amount of product formed in the control reaction. When certain individual components ($FeSO_4$, ascorbate, α-ketoglutarate, $MgSO_4$, or KCl) were present during incubation, production remained between 64 and 85% of control. However, when preincubation was with $Fe^{2+}$ plus ascorbic acid or $Fe^{2+}$ plus α-ketoglutarate, no detectable product was obtained. The same phenomenon was also observed with resting cells [97]. As with extracts, inactivation was not merely due to the presence of oxidized Fe because when the normal bioconversion reaction was carried out with $Fe^{3+}$ (as ferric sulfate) instead of $Fe^{2+}$, there was no inhibition.

Inactivation during aerobic incubation with $Fe^{2+}$ plus ascorbic acid and/or α-ketoglutarate might have been due to formation of hydrogen peroxide [102–104]. As for other α-ketoglutarate-dependent dioxygenases, expandase requires a reducing agent in addition to α-ketoglutarate and iron. This requirement could be fulfilled by several substances, but ascorbate is the most effective and is the reducing agent usually used with this type of enzyme [105]. However, ascorbate can have adverse effects, since incubation of some enzymes with ascorbate and oxygen leads to rapid loss of enzymatic activity due to production of hydrogen peroxide during autooxidation of this reducing agent [105, 106]. Catalase is known to stimulate the activity of such enzymes [99]; however, when catalase was added to the preincubation mixture with cell-free extract, no protection was observed [101]. Nevertheless, when 15 mM hydrogen peroxide was added to a reaction mixture, activity was markedly reduced, and when catalase was also added, the inactivation due to peroxide was reversed, indicating that inactivation normally observed with $Fe^{2+}$ plus ascorbate was not due, at least exclusively, to formation of hydrogen peroxide.

Oxygen-dependent inactivation of several key biosynthetic enzymes that involve mixed-function oxidation systems has been reported [102]. These systems catalyze synthesis of hydrogen peroxide and reduction of $Fe^{3+}$ to $Fe^{2+}$, followed by oxidation of enzyme-bound $Fe^{2+}$ to generate oxygen radicals that attack a histidine (or other oxidizable amino acids) at the metal-binding site of the enzyme. Golan-Goldhirst *et al.* [107] reported that preincubation of different proteins in the presence of ascorbate and copper (or iron) led to their inactivation due to the auto-oxidation of ascorbic acid. In this reaction, four strong oxidant species were formed and, after preincubation, changes in the amino acid composition of all proteins were observed. There was a major loss of histidine (His183 and His243) and methionine residues, which ligate ferrous ions (with Asp185) to give a catalytically active form [84].

In order to check if inactivation could be due to formation of reactive oxygen species such as superoxide, the effects of different radical scavengers such as mannitol, dimethylsulfoxide, as well as superoxide dismutase were examined. However, none of these was able to reactivate the enzyme after preincubation [101]. Although DTT is not necessary in the enzymatic conversion of penicillin G [95], this compound is known to stimulate the expandase activity of frozen crude extracts of *S. clavuligerus* and *A. chrysogenum* on penicillin N [99]. However, neither this reagent nor β-mercaptoethanol was able to stimulate the subsequent reaction with penicillin G when added during preincubation with $FeSO_4$ and ascorbate [101].

### 4.7.5 Further Improvements in the Bioconversion of Penicillin G to DAOG

#### 4.7.5.1 Stimulatory Effect of Growth in Ethanol

Biosynthesis of antibiotics rarely takes place during periods of rapid growth in rich media [108]. Rather, their production occurs best under conditions of nutrient imbalance brought about by limitation of carbon, nitrogen, or phosphorus and at low growth rates. The discovery of heat shock-like proteins (GroEL-like proteins), potentially important in antibiotic export and in the assembly of multienzyme complexes for polyketide antibiotic synthesis in a variety of streptomycetes [109], led to the first study on the relationship between non-nutritional stresses, such as heat shock or ethanol treatment, and antibiotic biosynthesis [110]. In that study, jadomycin production was induced by heat shock or by ethanol.

The effect of growth in the presence of alcohols of *S. clavuligerus* NP1 was studied [98] with regard to the ability of resting cells to biotransform penicillin G into cephalosporin-type antibiotics. Cultures upon growth showed the typical mycelial masses of tangled hyphae. However, when the growth medium was supplemented with ethanol, different morphologies were observed depending on the alcohol concentration. In 1% ethanol, the hyphae were somewhat more dispersed, whereas in 2% ethanol, the hyphae were extensively fragmented and dispersed. In the presence of 1% ethanol or 1–2% methanol, growth extent was slightly less than normal, but with 2% ethanol, growth was severely restricted; higher concentrations totally inhibited growth. Omission of starch from the growth medium led to a slight increase in specific production of DAOG. When the medium was supplemented with 1% ethanol, there was a marked increase in specific production and a 6- to 7-fold increase when 2% ethanol was added [98]. Addition of alcohols at later times during growth (at 2, 6, and 12 h) did not have any stimulatory effect.

The mechanism of alcohol stimulation on biotransformation of penicillin G remains a mystery although it could be related to the known ability of ethanol to trigger a heat-shock (stress) response [110–112]. Alternatively, growth in alcohol might yield cells with increased membrane permeability. The morphological effects observed could reflect this latter mechanism.

#### 4.7.5.2 **Use of Immobilized Cells**

For bioprocessing purposes, increases in the stability of biocatalysts are quite often achieved by immobilization of cells or enzymes [113, 114]. This technology is an attractive alternative to the use of expensive free enzymes and cofactors and can coordinate multistep enzymatic processes into a single operation. Furthermore, fermentative biosynthesis of cephamycin C using immobilized cells of *S. clavuligerus* NRRL 3585 had been accomplished [115]. Jensen *et al.* [116] reported on the immobilization of β-lactam-synthesizing enzymes from the same wild-type culture. However, neither of these early studies used penicillin substrates other than the normal intermediate (i.e., penicillin N).

The oxidative ring expansion of penicillin G by free and entrapped resting cells of *S. clavuligerus* NP1 was compared [117], and immobilized cells were found to perform the expandase reaction more slowly and less extensively than free cells, probably due to strong diffusional limitations. Both types of cells virtually ceased production after 2 h of reaction. Increasing resting cell concentration yielded increased product formation although, again, the reaction markedly decreased in rate after 2 h. To examine multiple cycles of ring expansion, free or resting cells immobilized by entrapment in polyethyleneimine-barium alginate were allowed to carry out oxidative ring expansion for 2 h, followed by centrifugation in the cold for 5 min. The cells were washed with 50 mM MOPS (pH 6.5) and recentrifuged. The expandase reaction was again initiated. Two-hour cycles were carried out up to 4 times with assays for product formation done at the end of each cycle. The activity of free cells was reduced by about 60% from the first to the second cycle and was completely lost after the second cycle. On the other hand, immobilized cells showed only a small reduction of activity at each cycle and still had activity through four cycles.

#### 4.7.5.3 **Elimination of Agitation and Addition of Water-Immiscible Solvents**

Since inactivation of expandase was thought to be an oxidative process [89], the effect of eliminating shaking during the ring expansion reaction was examined. It was found that the bioconversion rate was lower but, more importantly, the conversion yield increased [118]. The positive effect was observed with cells grown with or without ethanol.

A large number of different additives were tested for their effect on bioconversion by resting cells. Polyethylene glycols, alginate, Tween 80, DTT, and DL-methionine showed moderate or slight stimulatory effects [118]. Water-immiscible solvents were tested because they often increase enzyme activity and stability, and prevent the hydrolysis of the substrate or product [119, 120]. In order to check if they might decrease the inactivation of expandase by $Fe^{2+}$ plus ascorbate or $Fe^{2+}$ plus $\alpha$-ketoglutarate, some of these solvents were added to the reaction. When compared to other solvents used for enzyme catalyzed reactions, alkanes are often best because of their high log *P* values (octanol/water partition coefficient) which are favorable for activity and stability of cells and enzymes [121, 122]. Among them, *n*-decane has been reported to have minimal deleterious effects on microbial viability [123]. Based on these facts, the effect of adding *n*-decane to the reaction was examined [118]. Addition of 32% (v/v) *n*-decane led to an increase in DAOG formation that was especially significant in those reactions carried out without agitation.

A series of other solvents were compared to decane at the 32% concentration. *n*-Butanol, dimethyl sulfoxide, dioxane, isopropyl alcohol, and cyclohexanol totally inhibited the reaction [124]. Solvents that were as effective as *n*-decane were all alkanes (i.e., hexane, heptane, octane, dodecane, and hexadecane).

#### 4.7.5.4 **Addition of Catalase**

As mentioned above, oxidative inactivation of expandase during the ring-expansion reaction by cofactors $Fe^{2+}$, $\alpha$-ketoglutarate, and ascorbate might possibly be due to the formation of hydrogen peroxide and/or superoxide anion [87, 97]. The effect of enzymes capable of destroying these types of compounds was studied [124]. Although addition of catalase or scavengers did not prevent inactivation with cell-free extracts, a positive effect was surprisingly observed with resting cells. Addition of catalase to the bioconversion system, conducted in the absence of agitation and in the presence of 50% (v/v) hexane, was found to increase the bioconversion of penicillin G to DAOG. When other enzymes such as peroxidase or superoxide dismutase were added to the reaction, no enhancement was observed.

#### 4.7.5.5 **Recombinant *S. clavuligerus* Expandases**

Expression of the *S. clavuligerus* NRRL 3585 *cefE* gene in *E. coli* was achieved for the first time using the $\lambda P_L$ promoter [34] although the protein produced was predominantly insoluble. Cloning under control of the T7 promoter using the pET expression system resulted in production of large amounts of soluble expandase [86, 87, 100]. Although induction at 37 °C led to low levels of the glutathione–expandase fusion protein (below 5% of total soluble protein), when cultures were induced at 20 and 28 °C, high levels of recombinant protein were obtained (above 20% of total soluble protein) [100]. Expression and secretion of this synthase was also achieved in a eukaryotic background using the *Pichia pastoris* expression system [125].

The recombinant expandase obtained by Sim and Sim [100] was able to convert penicillin G not only under the conditions reported by Cho *et al.* [95], where the concentration of the two most important cofactors ($FeSO_4$ and $\alpha$-ketoglutarate) were 1.8 and 1.28 mM, respectively, but also using the conditions used by Maeda *et al.* [23]

with lower concentrations of both cofactors (0.04 and 0.64 mM, respectively). The rate of conversion was 2-fold higher using the conditions of Cho *et al.* [95]. Ring expansion activity was also observed when ampicillin, amoxicillin or penicillin V was used as substrate [100, 126]. Studies on the effect of cofactors on ring expansion with penicillin G or ampicillin as substrate revealed that $Fe^{2+}$ and $\alpha$-ketoglutarate omission results in more than 50% reduction of activity. Conversion of penicillin G by recombinant expandase was enhanced by 11-fold when DTT was omitted [100]. On the other hand, omission of ascorbate had a greater negative effect on expandase activity in resting cells of *S. clavuligerus* [95] than with the recombinant enzyme. When both DTT and ascorbate were omitted in the ring expansion reaction using the recombinant enzyme, the rate of conversion of penicillin G and ampicillin was enhanced by 21- and 35-fold, respectively [100].

Further improvements in the conversion of hydrophobic penicillins were obtained with a mutant enzyme in which the hydrophilic asparagine residue N304 was replaced by leucine (a noncharged, hydrophobic amino acid) [90]. Using penicillin G as substrate, an improvement of 83%, relative to that of wild-type *S. clavuligerus* expandase activity, was observed. Further modifications by Chin and Sim [127] involved the substitution of polar residues R306 and R307, also located at the C-terminus of the enzyme, each with a hydrophobic leucine residue. This resulted in improvements in the conversion of penicillin G of 102 and 24%, respectively. Further substitution of the asparagine residue N304 was carried out with 18 aliphatic amino acids in addition to leucine [128]. Substitution with alanine showed a 63% increase in activity, whereas replacement with K (lysine) and R (arginine) yielded increases of 142 and 163%, respectively. Modifications of residue C281 to C281Y increased activity by 151%, and to C281F by 164% [129] whereas modification of N304 to N304K increased activity by 137% and to N304R by 163%. Modification of I305 to I305M raised activity by 134%. Double mutations even had a greater effect as follows: C281Y N304A = 283% increase; V275I N304K = 194%; C281Y N304K = 328%; C281Y N304M = 177%; V275I N304R = 281%; C281Y N304R = 341%; V275I I305M = 320%; C281Y I305M = 395%; and C281Y R307L = 234% increase. These studies have been recently reviewed by Goo *et al.* [130].

Experiments by Wei *et al.* [131, 132] featured random mutagenesis followed by site-directed mutagenesis, which produced three mutants (N304K, I305L, and I305M) which possessed a 6- to 14-fold increase in $k_{cat}/K_m$ values. Mutants were made with all the possible combinations of these six sites, resulting in a double mutant (V275I, I305M) with a 32-fold increase in $k_{cat}/K_m$ and a 4-fold increase in penicillin G conversion activity. They further isolated the triple mutant (V275I, C281Y, I305M) with a 12-fold increase in activity on penicillin G. The same group [133], using DNA family shuffling of expandase genes from *S. clavuligerus* and newly isolated strains of *Streptomyces ambofaciens* and *Streptomyces chartreusis*, obtained an enzyme with 118-fold higher $k_{cat}/K_m$ on penicillin G than the wild-type *S. clavuligerus* enzyme.

Based on the high degree of similarity at the nucleotide (74%) as well as at the amino acid (70%) sequence levels between the expandase genes of *S. clavuligerus* and *N. lactamdurans* [32], it was of interest to determine whether homologous recombination (recombination between partially homologous sequences) could take place at a significant frequency between these two genes when cloned in the same orientation on a plasmid. If so, hybrid expandases with potentially altered activity/specificity could be generated *in vivo*. Studies were conducted by Adrio *et al.* [134] in a *recA*$^+$ *E. coli* strain. Seventeen colonies (27%) isolated after four rounds of propagation in such a strain revealed that two major types of DNA rearrangements had occurred in these plasmids. Sequencing of the cross-over junctions in representative recombinant plasmids showed that recombination had taken place in fully conserved sequence stretches of 2–17 bp at five different positions within the first 550 bp of the chimeric *cefE* genes. All clones had undergone in-frame recombination events.

Recombination was also attempted in *S. lividans* 1326. A rapid screen to detect the putative clones harboring a chimeric expandase gene was developed using as marker the *melC* gene from *Streptomyces glaucescens* which encodes the black pigment melanin [134]. Analysis of several white colonies ("melanin-negative") revealed the presence of plasmids of different sizes, ranging from 4 to 8 kb (the expected size was 6.3 kb). When the ring expansion abilities of several of these recombinants were determined, three white strains as well as controls (melanin-positive) showed activity producing growth inhibition zones in the agar bioassay and a peak corresponding to DAOG when analyzed by HPLC. Strain W25 produced approximately 20-fold more DAOG than *S. clavuligerus* NP1 and even more than B18, which contains the two intact cloned expandase genes [134].

When different growth conditions were tested for recombinant strain W25, it was found that the best conversion was obtained with cells grown in MT4E medium (MT medium plus 4% ethanol) for 1 day [135]. When the bioconversion was carried out without shaking and in the presence of 50% (v/v) hexane, production was markedly increased as previously observed for *S. clavuligerus* NP1 cells. The hybrid strain performed better at 16–25 °C rather than at 28 °C. Furthermore, lowering the iron concentration to 0.45 mM and raising α-ketoglutarate concentration to 1.92 mM improved the activity of W25 cells. In a statistical study, the $FeSO_4$ concentration was found to be the most important factor whereas ascorbate was the least. The 4-fold lower iron concentration preferred by the hybrid culture is probably the main reason for the improved bioconversion. This is based on the action of iron, not as cofactor in the biological oxidation reaction, but more importantly, as an inactivator of the expandase [101]. Catalase stimulated the bioconversion and could be replaced by the inexpensive additive yeast extract, which contains a component known to destroy hydrogen peroxide [136]. In a number of experiments, DAOG concentrations higher than 200 mg/liter were obtained.

## 4.8
## Conclusions

The directed evolution of DAOCS is providing new insights into the structure–function relationship of the protein that should lead to further rational engineering for new possibilities of manufacture of improved oral cephalosporins. In this sense, metabolic engineering and classical strain improvement are being jointly used by the β-lactam industry to develop new processes for production of cephalosporins. Additionally, the developments in functional genomics are improving the detailed phenotypic characterization of the microorganisms, providing excellent information to construct enhanced strains.

Since unwise recombinant DNA practices may have potential environmental impacts, the natural response to the uncertainty of manipulating genetically modified organisms (GMOs) was the development of regulations to prevent human health and environmental risks prior to confined manipulation and/or field release of GMOs. Knowledge and experience accumulated with GMOs have not revealed any particular safety and/or environmental problems. These encouraging results will promote deregulation programs to simplify the industrial application of many GMOs. The power and potential of metabolic engineering is great, and it must be exploited wisely and safely for the benefit of humans and the environment.

## References


1 Kohsaka, M. and Demain, A.L. (1976) *Biochemical and Biophysical Research Communications*, **70**, 465.

2 Dotzlaf, J.E. and Yeh, W.K. (1987) *Journal of Bacteriology*, **169**, 1611.

3 Abbott, M.T. and Udenfriend, S. (1974) In *Molecular Mechanisms of Oxygen Activation*, Academic Press, New York, p. 167.

4 Hook, D.J., Chang, L.T., Elander, R.P., and Morin, R.B. (1979) *Biochemical and Biophysical Research Communications*, **87**, 258.

5 Sawada, Y., Hunt, N.A., and Demain, A.L. (1979) *Journal of Antibiotics*, **32**, 1303.

6 Felix, H.R., Peter, H.H., and Treichler, H.J. (1981) *Journal of Antibiotics*, **34**, 567.

7 Rollins, M.J., Jensen, S.E., Wolfe, S., and Westlake, D.W.S. (1990) *Enzyme and Microbial Technology*, **12**, 40.

8 Jensen, S.E., Westlake, D.W.S., Bowers, R.J., and Wolfe, S. (1982) *Journal of Antibiotics*, **35**, 1351.

9 Demain, A.L. and Elander, R.P. (1999) *Antonie van Leeuwenhoek*, **75**, 5.

10 Sawada, Y., Baldwin, J.E., Singh, P.D., Solomon, N.A., and Demain, A.L. (1980) *Antimicrobial Agents and Chemotherapy*, **18**, 465.

11 Kupka, J., Shen, Y.-Q., Wolfe, S., and Demain, A.L. (1983) *FEMS Microbiology Letters*, **16**, 1.

12 Sawada, Y., Solomon, N.A., and Demain, A.L. (1980) *Biotechnology Letters*, **2**, 43.

13 Shen, Y.-Q., Wolfe, S., and Demain, A.L. (1984) *Enzyme and Microbial Technology*, **6**, 402.

14 Yeh, Y.H., Dotzlaf, J.E., and Huffman, G.W. (1994) In *50 Years of Penicillin Application: History and Trends*, Public, Prague, p. 208.

15 Aharonowitz, Y., Cohen, G., and Martín, J.F. (1992) *Annual Review of Microbiology*, **46**, 461.

16 Paradkar, A., Jensen, S.E., and Mosher, R.H. (1997) In *Biotechnology of Antibiotics*, 2nd edn, Dekker, New York, p. 241.

17 Martín, J.F., Gutierrez, S., and Demain, A.L. (1997) In *Fungal Biotechnology*, Chapman & Hall, Weinheim, p. 91.

18 Gutierrez, S., Díez, B., Montenegro, E., and Martín, J.F. (1991) *Journal of Bacteriology*, **173**, 2354.

19 Skatrud, P.L. and Queener, S.W. (1989) *Gene*, **78**, 331.

20 Gutierrez, S., Velasco, J., Fernández, F.J., and Martín, J.F. (1992) *Journal of Bacteriology*, **174**, 3056.

21 Samson, S.M., Dotzlaf, J.E., Slisz, M.L., Becker, G.W., Van Frank, R.M., Veal, L.E., Yeh, W.-K., Miller, J.R., Queener, S.W., and Ingolia, T.D. (1987) *Bio/Technology*, **5**, 1207.

22 Baldwin, J.E., Adlington, R.M., Coates, J.B., Crabbe, M.J.C., Keeping, J.W.,

Knight, G.C., Nomoto, T., Schofield, C.J., and Ting, H.H. (1987) *Journal of the Chemical Society, Chemical Communications*, 374.

23 Maeda, K., Luengo, J.M., Ferrero, O., Wolfe, S., Lebedev, M.Y., Fang, A., and Demain, A.L. (1995) *Enzyme and Microbial Technology*, **17**, 231.

24 Banko, G., Wolfe, S., and Demain, A.L. (1986) *Biochemical and Biophysical Research Communications*, **137**, 528.

25 Banko, G., Demain, A.L., and Wolfe, S. (1987) *Journal of the American Chemical Society*, **109**, 2858.

26 Rodriguez-Saiz, M., de la Fuente, J.-L., and Barredo, J.L. (2004) In *Methods in Biotechnology, Microbial Processes and Products*, vol. 18, Humana Press, Totowa, NJ, p. 41.

27 Zhang, J. and Demain, A.L. (1990) *Biochemical and Biophysical Research Communications*, **169**, 1145.

28 Hollander, I.J., Shen, Y.-Q., Heim, J., Demain, A.L., and Wolfe, S. (1984) *Science*, **224**, 616.

29 Ullán, R.V., Casqueiro, J., Bañuelos, O., Fernández, F.J., Gutiérrez, S., and Martín, J.F. (2002) *Journal of Biological Chemistry*, **277**, 46216.

30 Martín, J.F., Ullán, R.V., and Casqueiro, J. (2004) *Advances in Biochemical Engineering/Biotechnology*, **88**, 91.

31 Kovacevic, S., Tobin, M.B., and Miller, J.R. (1990) *Journal of Bacteriology*, **172**, 3952.

32 Coque, J.J.R., Martín, J.F., and Liras, P. (1993) *Molecular and General Genetics*, **236**, 453.

33 Kimura, H., Miyashita, H., and Sumino, Y. (1996) *Applied Microbiology and Biotechnology*, **45**, 490.

34 Kovacevic, S., Weigel, B.J., Tobin, M.B., Ingolia, T.D., and Miller, J.R. (1989) *Journal of Bacteriology*, **171**, 754.

35 Kovacevic, S. and Miller, J.R. (1991) *Journal of Bacteriology*, **173**, 398.

36 Scheidegger, A., Kuenzi, M.T., and Nüesch, J.J. (1984) *Journal of Antibiotics*, **37**, 522.

37 Wu, X.-B., Fan, K.-Q., Wang, Q.-H., and Yang, K.-Q. (2005) *FEMS Microbiology Letters*, **246**, 103.

38 Dotzlaf, J.E. and Yeh, W.K. (1989) *Journal of Biological Chemistry*, **264**, 10219.

39 Martin, J.F., Gutierrez, S., Fernandez, F.J., Velasco, J., Fierro, F., Marcos, A.T., and Kasalkova, K. (1994) *Antonie van Leeuwenhoek*, **65**, 227.

40 Gutiérrez, S., Fierro, F., Casqueiro, J., and Martín, J.F. (1999) *Antonie van Leeuwenhoek*, **75**, 81.

41 Walz, M. and Kück, U. (1991) *Current Genetics*, **19**, 73.

42 Alvarez, E., Meesschaert, B., Montenegro, E., Gutiérrez, S., Díez, B., Barredo, J.L., and Martín, J.F. (1993) *European Journal of Biochemistry*, **215**, 323.

43 Barredo, J.L., van Solingen, P., Díez, B., Alvarez, E., Cantoral, J.M., Kattevilder, A., Smaal, E.B., Groenen, M.A.M., Veenstra, A.E., and Martín, J.F. (1989) *Gene*, **83**, 291.

44 Luengo, J.M. (1995) *Journal of Antibiotics*, **48**, 1195.

45 Kariyone, K., Harada, H., Kurita, M., and Takano, T. (1970) *Journal of Antibiotics*, **23**, 131.

46 Kamimura, T., Matsumoto, Y., Okada, N., Mine, Y., Nishida, M., Goto, S., and Kuwahara, S. (1979) *Antimicrobial Agents and Chemotherapy*, **16**, 540.

47 Sakane, K., Inamoto, Y., and Takaya, T.J. (1992) *Japanese Journal of Antibiotics*, **45**, 909.

48 Yamanaka, H., Chiba, T., Kawabata, K., Takasugi, H., Masugi, T., and Takaya, T. (1985) *Journal of Antibiotics*, **38**, 1738.

49 Inamoto, Y., Chiba, T., Kamimura, T., and Takaya, T. (1988) *Journal of Antibiotics*, **41**, 828.

50 Morin, R.B., Jackson, B.G., Flynn, E.H., Roeske, R.W., and Andrews, S.L. (1969) *Journal of the American Chemical Society*, **91**, 1396.

51 Alonso, J., Barredo, J.L., Armisén, P., Díez, B., Salto, F., Guisán, J.M., García, J.L., and Cortés, E. (1999) *Enzyme and Microbial Technology*, **25**, 88.

52 Matsuda, A. and Komatsu, K.I. (1985) *Journal of Bacteriology*, **163**, 1222.

53 Croux, C., Costa, J., Barredo, J.L., and Salto, F. (1994) U. S. Pat. 5,354,667; Croux, C., Perez, J.C., Fuentes, J.L., and

Maldonado, F. (1995) *Biotechnology Advances*, **13**, 270;Croux, C., Perez, J.C., Fuentes, J.L.B., and Maldonado, F.S. (1991) *Chemical Abstracts*, **117**, 190273.

**54** Luo, H., Li, Q., Yu, H., and Shen, Z. (2004) *Biotechnology Letters*, **26**, 939.

**55** Pollegioni, L., Langkau, B., Tischer, W., Ghisla, S., and Pilone, M.S. (1993) *Journal of Biological Chemistry*, **268**, 13850.

**56** Alonso, J., Barredo, J.L., Díez, B., Mellado, E., Salto, F., García, J.L., and Cortés, E. (1998) *Microbiology*, **144**, 1095.

**57** Isogai, T., Fukagawa, M., Aramori, I., Iwami, M., Kojo, H., Ono, T., Ueda, Y., Kohsaka, M., and Imanaka, H. (1991) *Bio/Technology*, **9**, 188.

**58** Bunnell, C.A., Luke, W.D., and Perry, R.M. Jr (1986) *Beta-lactam Antibiotics for Clinical Use*, Dekker, New York, p. 255.

**59** Hersbach, G.J.M., van der Beek, C.P., and van Dijck, P.W.M. (1984) In *Biotechnology of Industrial Antibiotics*, Dekker, New York, p. 45.

**60** Crawford, L., Stepan, A.M., McAda, P.C., Rambosek, J.A., Condor, M.J., Vinci, V.A., and Reeves, C.D. (1995) *Bio/Technology*, **13**, 58.

**61** Velasco, J., Adrio, J.L., Moreno, M.A., Díez, B., Soler, G., and Barredo, J.L. (2000) *Nature Biotechnology*, **18**, 857.

**62** Demain, A.L., Adrio, J.L., and Piret, J.M. (2001) In *Enzyme Technology for Pharmaceutical and Biotechnological Applications*, Dekker, New York, p. 61.

**63** Adrio, J.L. and Demain, A.L. (2002) *Organic Process Research & Development*, **6**, 427.

**64** Gutiérrez, S., Díez, B., Alvarez, E., Barredo, J.L., and Martín, J.F. (1990) *Molecular and General Genetics*, **225**, 56.

**65** Díez, B., Gutiérrez, S., Barredo, J.L., van Solingen, P., van der Voort, L.H., and Martín, J.F. (1990) *Journal of Biological Chemistry*, **265**, 16358.

**66** Smith, D.J., Earl, A.J., and Turner, G. (1990) *The EMBO Journal*, **9**, 2743.

**67** Barredo, J.L., Cantoral, J.M., Alvarez, E., Díez, B., and Martín, J.F. (1989) *Molecular and General Genetics*, **216**, 91.

**68** Carr, L.G., Skatrud, P.L., Scheetz, M.E., Queener, S.W., and Ingolia, T.D. (1986) *Gene*, **48**, 257.

**69** Lamas-Maceiras, M., Vaca, I., Rodríguez, E., Casqueiro, J., and Martín, J.F. (2006) *Biochemical Journal*, **395**, 147.

**70** Díez, B., Barredo, J.L., Alvarez, E., Cantoral, J.M., Solingen, P., Groenen, M.A.M., Veenstra, A.E., and Martín, J.F. (1989) *Molecular and General Genetics*, **218**, 572.

**71** Smith, D.J., Burnham, M.K., Bull, J.H., Hodgson, J.E., Ward, J.M., Browne, P., Brown, J., Barton, B., Earl, A.J., and Turner, G. (1990) *The EMBO Journal*, **9**, 741.

**72** Fierro, F., Gutiérrez, S., Díez, B., and Martín, J.F. (1993) *Molecular and General Genetics*, **241**, 573.

**73** Barredo, J.L., Díez, B., Alvarez, E., and Martín, J.F. (1989) *Current Genetics*, **16**, 453.

**74** Fierro, F., Barredo, J.L., Díez, B., Gutiérrez, S., Fernández, F.J., and Martín, J.F. (1995) *Proceedings of the National Academy of Sciences of the United States of America*, **92**, 6200.

**75** Smith, D.J., Bull, J.H., Edwards, J., and Turner, G. (1989) *Molecular and General Genetics*, **216**, 492.

**76** Künkel, W., Berger, D., Risch, S., and Wittmann-Bresinsky, B. (1992) *Applied Microbiology and Biotechnology*, **36**, 499.

**77** García-Estrada, C., Vaca, I., Ullán, R.V., van den Berg, M.A., Bovenberg, R.A.L., and Martín, J.F. (2009) *BMC Microbiology*, **9**, 104.

**78** Brakhage, A.A. (1998) *Microbiology and Molecular Biology Reviews*, **62**, 547.

**79** Cantwell, C., Beckmann, R., Whiteman, P., Queener, S.W., and Abraham, E.P. (1992) *Proceedings of the Royal Society, London B*, **248**, 283.

**80** Ballio, A., Chain, E.B., di Accadia, F.D., Mastropietro-Cancellieri, M.F., Morpurgo, G., Serlupi-Crescenzi, G., and Sermonti, G. (1960) *Nature*, **185**, 97.

**81** Robin, J., Lettier, G., McIntyre, M., Noorman, H., and Nielsen, J. (2003) *Biotechnology and Bioengineering*, **83**, 361.

**82** van de Sandt, E.J.A.X. and de Vroom, E. (2000) *Chimica Oggi/Chemistry Today*, **18**, 72.

**83** Ullán, R.V., Campoy, S., Casqueiro, J., Fernández, F.J., and Martín, J.F. (2007) *Chemistry and Biology*, **14**, 329.

84 Valegard, K., van Scheltinga, A.C.T., Lloyd, M.D., Hara, T., Ramaswamy, S., Perrakis, A., Thompson, A., Lee, H.-J., Baldwin, J.E., Schofield, C.J., Hadju, J., and Andersson, I. (1998) *Nature*, **394**, 805.

85 Hegg, E.L. and Que, L. (1997) *European Journal of Biochemistry*, **250**, 625.

86 Sim, J. and Sim, T.-S. (2000) *Bioscience, Biotechnology and Biochemistry*, **64**, 828.

87 Lloyd, M.D., Lee, H.-J., Harlos, K., Zhang, Z.-H., Baldwin, J.E., Schofield, C.J., Charnock, J.M., Garner, C.D., Hara, T., van Scheltinga, A.C.T., Valegard, K., Viklund, J.A.C., Hadju, J., Andersson, I., Danielsson, A., and Bhikhabhai, R. (1999) *Journal of Molecular Biology*, **287**, 943.

88 Lee, H.-J., Lloyd, M.D., Harlos., K., Clifton, I.J., Baldwin, J.E., and Schofield, C.J. (2001) *Journal of Molecular Biology*, **308**, 937.

89 Lee, H.-J., Lloyd, M.D., Harlos, K., and Schofield, C.J. (2000) *Biochemical and Biophysical Research Communications*, **267**, 445.

90 Chin, H.S., Sim, J., and Sim, T.S. (2001) *Biochemical and Biophysical Research Communications*, **287**, 507.

91 Rollins, M.J., Westlake, D.W.S., Wolfe, S., and Jensen, S.E. (1988) *Canadian Journal of Microbiology*, **34**, 1196.

92 Mahro, B. and Demain, A.L. (1987) *Applied Microbiology and Biotechnology*, **27**, 272.

93 Jensen, S.E., Leskiw, B.K., Vining, L.C., Aharonowitz, Y., Westlake, D.W.S., and Wolfe, S. (1986) *Canadian Journal of Microbiology*, **32**, 953.

94 Yoshida, M., Konomi, T., Kohsaka, M., Baldwin, J.E., Herchen, S., Singh, P., Hunt, N.A., and Demain, A.L. (1978) *Proceedings of the National Academy of Sciences of the United States of America*, **75**, 6253.

95 Cho, H., Adrio, J.L., Luengo, J.M., Wolfe, S., Ocran, S., Hintermann, G., Piret, J.M., and Demain, A.L. (1998) *Proceedings of the National Academy of Sciences of the United States of America*, **95**, 11544.

96 Lübbe, C., Wolfe, S., and Demain, A.L. (1985) *Archives of Microbiology*, **140**, 317.

97 Baez-Vasquez, M.A., Adrio, J.L., Piret, J.M., and Demain, A.L. (1999) *Applied Biochemistry and Biotechnology*, **81**, 145.

98 Fernández, M.-J., Adrio, J.L., Piret, J.M., Wolfe, S., Ro, S., and Demain, A.L. (1999) *Applied Microbiology and Biotechnology*, **52**, 484.

99 Lübbe, C., Wolfe, S., and Demain, A.L. (1985) *Enzyme and Microbial Technology*, **7**, 353.

100 Sim, J. and Sim, T.-S. (2001) *Enzyme and Microbial Technology*, **29**, 240.

101 Adrio, J.L., Cho, H., Piret, J.M., and Demain, A.L. (1999) *Enzyme and Microbial Technology*, **25**, 497.

102 Fucci, L., Oliver, C.N., Coon, M.J., and Stadtman, E.R. (1983) *Proceedings of the National Academy of Sciences of the United States of America*, **80**, 1521.

103 Nakamura, S. (1970) *Biochemical and Biophysical Research Communications*, **41**, 177.

104 Levine, R.L., Oliver, C.N., Fulks, R.M., and Stadtman, E.R. (1981) *Proceedings of the National Academy of Sciences of the United States of America*, **78**, 2120.

105 Popenoe, E.A., Aronson, R.B., and Van Slyke, D.D. (1969) *Archives of Biochemistry and Biophysics*, **133**, 286.

106 Rhoads, R.E. and Udenfriend, S. (1970) *Archives of Biochemistry and Biophysics*, **139**, 329.

107 Golan-Goldhirsh, A., Osuga, D.T., Chen, A.O., and Whitaker, J.R. (1992) In *The Bioorganic Chemistry of Enzymatic Catalysis: A Homage to Myron L. Bender*, CRC Press, Boca Raton, FL, p. 61.

108 Demain, A.L. and Fang, A. (1995) *Actinomycetologica*, **9**, 98.

109 Guglielmi, G., Mazodier, P., Thompson, C.J., and Davies, J. (1991) *Journal of Bacteriology*, **173**, 7374.

110 Doull, J.L., Singh, A.K., Hoare, M., and Ayer, S.W. (1994) *Journal of Industrial Microbiology*, **13**, 120.

111 Lee, P.C., Bochner, B.R., and Ames, B.N. (1983) *Proceedings of the National Academy of Sciences of the United States of America*, **80**, 7496.

112 Arnosti, D.N., Singer, V.L., and Chamberlin, M.J. (1986) *Journal of Bacteriology*, **168**, 1243.

113 Fukui, S. and Tanaka, A. (1982) *Annual Review of Microbiology*, **36**, 145.

**114** Vandamme, E.J. (1983) *Enzyme and Microbial Technology*, **5**, 403.

**115** Freeman, A. and Aharonowitz, Y. (1981) *Biotechnology and Bioengineering*, **23** 2747.

**116** Jensen, S.E., Westlake, D.W.S., and Wolfe, S. (1984) *Applied Microbiology and Biotechnology*, **20**, 155.

**117** Demain, A.L. and Baez-Vásquez, M.A. (2000) *Applied Biochemistry and Biotechnology*, **87**, 135.

**118** Gao, Q. and Demain, A. (2001) *Applied Microbiology and Biotechnology*, **57**, 511.

**119** Nikolova, P. and Ward, O.P. (1993) *Journal of Industrial Microbiology*, **12**, 76.

**120** Salter, G.J. and Kelt, D.B. (1995) *Critical Reviews in Biotechnology*, **15**, 139.

**121** Zaks, A. and Klibanov, A.M. (1985) *Proceedings of the National Academy of Sciences of the United States of America*, **82**, 3192.

**122** Dordick, J.S. (1989) *Enzyme and Microbial Technology*, **11**, 194.

**123** Spinnler, H.E., Ginies, C., Khan, J.A., and Vulfson, E.N. (1996) *Proceedings of the National Academy of Sciences of the United States of America*, **93**, 3373.

**124** Gao, Q. and Demain, A.L. (2002) *Letters in Applied Microbiology*, **34**, 290.

**125** Adrio, J.L., Velasco, J., Soler, G., Rodriguez-Saiz, M., Barredo, J.L., and Moreno, M.A. (2001) *Biotechnology and Bioengineering*, **75**, 485.

**126** Dubus, A., Lloyd, M.D., Lee, H.-J., Schofield, C.J., Baldwin, J.E., and Frere, J.-M. (2001) *Cellular and Molecular Life Sciences*, **58**, 835.

**127** Chin, H.S. and Sim, T.S. (2002) *Biochemical and Biophysical Research Communications*, **295**, 55.

**128** Chin, H.S., Goo, K.S., and Sim, T.S. (2004) *Applied and Environmental Microbiology*, **70**, 607.

**129** Goo, K.S., Chua, C.S., and Sim, T.-S. (2008) *Applied and Environmental Microbiology*, **74**, 1167.

**130** Goo, K.-S., Chua, C.-S., and Sim, T.-S. (2009) *Journal of Industrial Microbiology and Biotechnology*, **36**, 619.

**131** Wei, C.-L., Yang, Y.-B., Wang, W.-C., Liu, W.-C., Hsu, J.-S., and Tsai, Y.-C. (2003) *Applied and Environmental Microbiology*, **69**, 2306.

**132** Wei, C.-L., Yang, Y.-B., Deng, C.-H., Liu, W.-C., Hsu, J.-S., Lin, Y.-C., Liaw, S.-H., and Tsai, Y.-C. (2005) *Applied and Environmental Microbiology*, **71**, 8873.

**133** Hsu, J.-S., Yang, Y.-B., Deng, C.-H., Wei, C.-L., Liaw, S.-H., and Tsai, Y.-C. (2004) *Applied and Environmental Microbiology*, **70**, 6257.

**134** Adrio, J.L., Hintermann, G.A., Demain, A.L., and Piret, J.M. (2002) *Enzyme and Microbial Technology*, **31**, 932.

**135** Gao, Q., Piret, J.M., Adrio, J.L., and Demain, A.L. (2003) *Journal of Industrial Microbiology and Biotechnology*, **30**, 190.

**136** Smith, J.S., Hillier, A.J., Lees, G.J., and Jago, G.R. (1975) *Journal of Dairy Research*, **42**, 123.

# 5
# Structure and Reactivity of β-Lactams

*Michael I. Page*

## 5.1
## Introduction

β-Lactams are the four-membered ring condensation products of β-amino acids. Although they occur relatively rarely in nature, the naturally occurring β-lactam antibiotics, such as the penicillins [1] and cephalosporins [2], are well known. Due to this unusual occurrence, it is not surprising that the biological activity of these compounds was initially attributed to the expected high chemical reactivity of the four-membered β-lactam ring. Shortly after the introduction of penicillin to the medical world it was suggested that the antibiotic's activity was due to the inherent strain of the four-membered ring [1] or to reduced amide resonance [2]. The latter may occur because the butterfly shape of the penicillin molecule [3] prevents the normal planar arrangement of the oxygen, carbon, and nitrogen atoms assumed to be necessary for the effective delocalization of the nitrogen lone pair. Both of these ideas are, of course, intuitively appealing and they remained unchallenged for several decades. Indeed, these two proposals dominated the thoughts of many synthetic chemists who were convinced that more effective antibiotics could be made by making the β-lactam system more strained or nonplanar. However, there is little evidence to show that the β-lactam in penicillin is unusually strained or that amide resonance is inhibited.

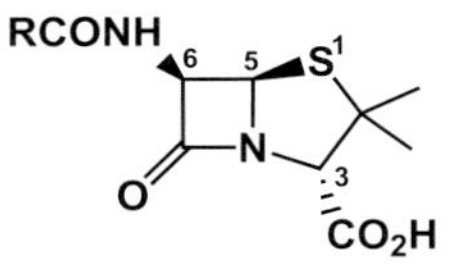

RCONH 6 5 S1 N 3 O CO2H

(1)

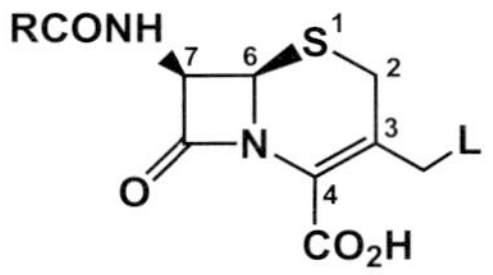

RCONH 7 6 S1 2 3 L N 4 O CO2H

(2)

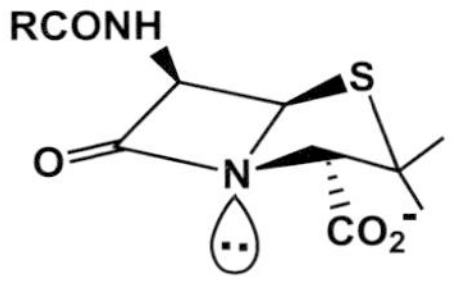

RCONH S O N CO2-

(3)

*Amino Acids, Peptides and Proteins in Organic Chemistry.*
Vol.3 – *Building Blocks, Catalysis and Coupling Chemistry.* Edited by Andrew B. Hughes

ISBN: 978-3-527-32102-5

## 5.2 Structure

Amide resonance is usually depicted by the canonical forms [4] and [5].

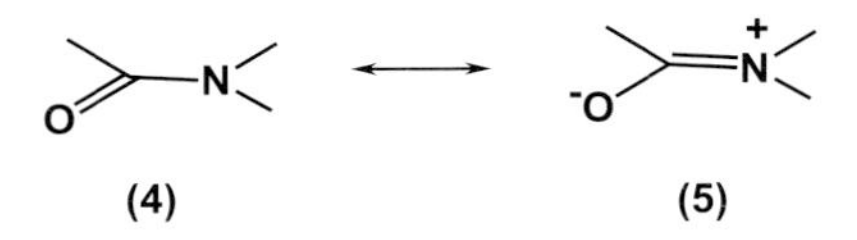


Any inhibition of amide resonance should make the amide resemble [4] at the expense of [5], so that this would:

i) Increase the C–N bond length and decrease the C–N bond strength.
ii) Decrease the C=O bond length and increase the C=O bond strength.
iii) Decrease the positive charge density on nitrogen.
iv) Decrease the negative charge density on oxygen.

X-ray crystallography has been invaluable in providing detailed three-dimensional structures of β-lactams and amides. Although the acyclic amide, acetamide, has $C_s$ symmetry in the gas phase and solution, the carbonyl carbon is actually pyramidalized in the crystalline state [3]. The degree of coplanarity of the β-lactam nitrogen with its three substituents can be expressed either by the perpendicular distance, *h*, of the nitrogen from the plane of its substituents or by the sum of the bond angles about nitrogen. The former is easier to visualize and the nitrogen ranges from being essentially in the plane of its three substituents in monocyclic β-lactams to being 0.5 Å out of the plane in bicyclic systems (Table 5.1). Until recently, it was assumed that a more pyramidal nitrogen would decrease amide resonance and lead to increased biological activity, consequently considerable effort was put into making

**Table 5.1** Structural parameters of some β-lactams [5].

| Compound | C=O stretch ($cm^{-1}$) | Distance on N from plane *h* (Å) | β-Lactam C=O bond length (Å) | β-Lactam C–N bond length (Å) |
|---|---|---|---|---|
| Penicillins | 1770–1790 | | | |
| Ampicillin | | 0.38 | 1.20 | 1.36 |
| Benzylpenicillin | | 0.40 | 1.17 | 1.34 |
| Phenoxymethylpenicillin | | 0.40 | 1.21 | 1.46 |
| $\Delta^3$-Cephalosporins | 1760–1790 | | | |
| Cephaloridine | | 0.24 | 1.21 | 1.38 |
| $\Delta^2$-Cephalosporins | 1750–1780 | | | |
| Phenoxymethyl $\Delta^2$-cephalosporin | | 0.06 | 1.22 | 1.34 |
| Anhydropenicillins | 1810 | | | |
| Phenoxymethylanhyropenicillin | | 0.41 | 1.18 | 1.42 |
| MBLs | 1730–1760 | 0 | 1.21 | 1.35 |
| Amides | 1600–1680 | 0 | 1.24 | 1.33 |

nonplanar β-lactams. 1-Carba-1-penem shows one of the highest $h$ values, 0.54 Å, and yet this compound is biologically inactive [4]. Furthermore, there is no direct correlation between $h$ values and chemical reactivity.

It has been claimed that the bond lengths seen in penicillins [1] and cephalosporins [2] indicate the inhibition of amide resonance [6]. The C−N bond length of planar monocyclic β-lactams (1.35 Å) is generally longer than that of amides (1.33 Å). The converse is true for C=O bond lengths, 1.24 Å for amides compared with 1.21 Å for monocyclic β-lactams. In nonplanar penicillins and cephalosporins there is a general trend for the C−N bond length to increase as the C=O bond length decreases [6, 7], but this trend is by no means linear. Bond lengths for C=O vary from 1.17 to 1.24 Å and for C−N from 1.33 to 1.46 Å, and there is a tendency for the C−N bond length to increase with $h$.

It is not easy to explain these bond length differences. Penicillin V shows [8] the longest C−N bond length of 1.46 Å and yet the C=O bond length is identical to that commonly found in planar monocyclic β-lactams (1.21 Å). In monocyclic β-lactams the nitrogen is coplanar with its three substituents and yet the bond length differences from those found in acyclic amides are also in the direction predicted by inhibition of amide resonance. The degrees of nonplanarity in penicillin V and ampicillin are similar ($h = 0.40$ and 0.38 Å, respectively), and yet the C−N bond length in the former is 0.10 Å longer than in the latter.

Structural data have also been used to support the suggestion that enamine resonance is important in cephalosporins [2] and that this also reduces amide resonance [6]. However, there is no significant difference in the C−O and C−N bond lengths of cephalosporins from that general trend exhibited by penicillins. The C(4)−N(5) of cephaloglycine is 1.51 Å [9] which is *longer* than the expected value of 1.47 Å for C−N. Furthermore, the C(4)−N(5) bond length in the $\Delta^3$-cephalosporin, cephaloglycin, is longer than that of 1.45 Å in $\Delta^2$-cephems [9] and that of 1.46 Å in cephams [10] where enamine resonance cannot occur.

It appears that variations in bond lengths in penicillins and cephalosporins are caused by the nature of substituents and the minimization of unfavorable strain energies caused by the geometry of the molecule. To attribute these differences to the inhibition of amide resonance seems speculative and is only supported by the selection of examples.

The conformation of the substituent on nitrogen relative to the carbonyl group has a significant effect on the carbonyl $^{13}C$ chemical shift in amides. For example, a difference of 4 ppm is observed in the carbonyl $^{13}C$ resonances of the *E*- and *Z*-isomers of *N*-methylformamide [11]. This reflects both steric and anisotropic differences in the environment. The β-lactam carbonyl carbon usually resonates between 160 and 167 ppm in a $^{13}C$-nuclear magnetic resonance (NMR) spectrum [12]. This is in the same region in which the carbonyl resonance of formamide and its *N*-methylated derivatives also appear [11]. The carbonyl resonances of five- or larger-membered lactams appear between 170 and 180 ppm. Replacement of an alkyl substituent on the β-lactam nitrogen by an aryl substituent causes shielding of the lactam carbonyl resonance by about 4 ppm [12]. Dipole moment [13] and UV spectral studies [14] indicate that the lone pair on the β-lactam nitrogen is conjugated with the aromatic ring. This resonance interaction presumably alters the electron density at the β-lactam carbonyl and could account for the shielding of the resonance due to that carbon.

There is little variation in the chemical shifts of the β-lactam carbonyl carbon of penicillins and cephalosporins [15]. The carbonyl carbon of the β-lactam in penicillins resonates about 10 ppm to lower field than that in cephalosporins. Surprisingly, the shifts in the biologically active $\Delta^3$- and the inactive $\Delta^2$-cephalosporins are similar [16]. Inhibition of amide resonance may be expected to make the carbonyl carbon more electron deficient [17] and although the difference in chemical shifts between penicillins and $\Delta^3$-cephalosporins support this proposal it is not apparent from the $\Delta^2$- to $\Delta^3$-cephalosporin comparison. In penicillins the nitrogen is 0.4 Å from the plane of its substituents compared with 0.2 Å in $\Delta^3$-cephalosporins, whereas the ceph-2-em systems are planar [18]. The similarity of the values of the $^{13}C$ shifts found for the β-lactam carbonyl carbons in ceph-3-ems and ceph-2-ems indicates that the charge density and bond order at the carbonyl carbons in both systems is approximately the same. $^{15}N$-Chemical shifts of the β-lactam nitrogen in ceph-3-ems are almost invariant (less than $\pm 1$ ppm) with the nature of the substituent at C(3) [19] and therefore also do not indicate significant enamine-type resonance in these systems. Not surprisingly there is a large difference of 15 ppm in the $^{15}N$ chemical shifts of the β-lactam nitrogen in ceph-2-ems and ceph-3-ems. There is an *upfield* shift of 30 ppm in the β-lactam nitrogen on going from nonplanar penicillins to planar ceph-2-ems [20] and yet increased amide conjugation in the planar system would be expected to induce a downfield shift.

The β-lactam IR carbonyl stretching frequency has been regarded as an important index for both inhibition of amide resonance and for investigating structure–activity relationships of the β-lactam antibiotics [9, 21–23]. In normal penams the β-lactam carbonyl stretching frequency occurs in the 1770–1790 $cm^{-1}$ range compared with 1730–1760 $cm^{-1}$ monocyclic unfused β-lactams and about 1600–1680 $cm^{-1}$ for amides. In general, the nonplanar 3-cephems show higher stretching frequencies, 1786–1790 $cm^{-1}$, than the planar 2-cephems, which absorb at 1750–1780 $cm^{-1}$. The frequency in cephalosporins increases by about 5 $cm^{-1}$ when the ring sulfur is replaced by oxygen, but decreases by a similar amount when the 7-α-hydrogen is substituted by the methoxy group [24]. Of course, the observed β-lactam frequency is dependent on the conditions of measurement (KBr, film, solution, etc.) which may cause variations comparable with those produced by structural changes. There is a tendency for a high carbonyl stretching frequency to be associated with a shorter β-lactam C=O bond length and a more pyramidal nitrogen. Furthermore, it has been tempting to associate a high carbonyl stretching frequency with increased strain, increased double bond character, and reduced amide resonance [7, 21, 25]. However, the evidence again is ambiguous. Although selected examples may show some of these inter-relationships there are many exceptions; for example, the carbonyl stretching frequency for some penems *decreases* 20 $cm^{-1}$ while the β-lactam nitrogen becomes *more* pyramidal by 0.12 Å [25].

Finally, the direct interpretation of carbonyl stretching frequencies in terms of bond order or electron density distributions is not straightforward. Many subtle effects can alter the frequency even if the force constant for C=O stretching, which is presumably the best indicator of bond strength, remains constant. For example, in the system X-C=O the carbonyl stretching frequency can be increased by decreasing the C–X bond length, by increasing the C–X stretching force constant, or by increasing the XCO bending force constant [26].

Theoretical geometry optimization of β-lactams at a semiempirical level and a limited *ab initio* study using minimal basis STO-3G calculations at fixed geometries [27] underestimates valence angles at heteroatoms and is expected to exaggerate the degree of nonplanarity in amides, while the split-valence 4–31G basis set characteristically overestimates these valence angles and consequently overestimates the tendency of amides to be planar. Given these reservations, the calculated STO-3G energy of formamide in a penicillin-like geometry is only 2.8 kcal/mol higher than the planar geometry [28]. Furthermore, in general, the geometrical parameters associated with the β-lactam ring vary only slightly with changes in the hybridization at nitrogen. An exception is the C−N bond length, which becomes longer as the nitrogen becomes pyramidal. Formamide lies in a potential well, which is very flat with respect to inversion at nitrogen. The inversion barrier is lower for molecules favoring a large angle at nitrogen (amides) and higher for systems adopting a small angle at nitrogen (e.g., aziridine) [29]. It appears that the nitrogen in amides can be made pyramidal without severe changes in energy.

A final aspect of structural effects is the basicity of β-lactam nitrogen. Inhibition of amide resonance in β-lactams is expected to increase localization of the lone pair on nitrogen and so increase the basicity of nitrogen. It is well known that torsional strain in amides can increase the basicity of nitrogen in amides. For example, 6,6-dimethyl-1-azabicyclo[2.2.2]octan-2-one has the nitrogen lone pair almost orthogonal to the carbonyl $\pi$ system and amide resonance is consequently inhibited. Amides are normally only very weakly basic and the p$K_a$ values of their conjugate acids are around zero. By contrast, 6,6-dimethyl-1-azabicyclo[2.2.2]octan-2-one is half protonated at pH 5.3, consistent with the increased basicity of the amide nitrogen [30].

Similarly, if amide resonance in penicillins is inhibited because of the pyramidal nature of the β-lactam nitrogen, penicillins should also show enhanced basicity compared with normal amides. There is no evidence to suggest that this is the case. In fact, penicillins appear to show reduced basicity and cannot be detectably protonated even in 12 M hydrochloric acid [31] and so *N*-protonated penicillins must have a p$K_a$ below −5. An indication of increased nitrogen basicity would be a large binding constant of penicillin to metal ions. However, the equilibrium constant for metal ion coordination between the carboxyl group and β-lactam nitrogen [6] is only about 100–200 $M^{-1}$ for Cu(II), Zn(II), Ni(II), and Co(II) [32, 33]. This is the order of magnitude expected for coordination between a normal amide and a carboxyl group.

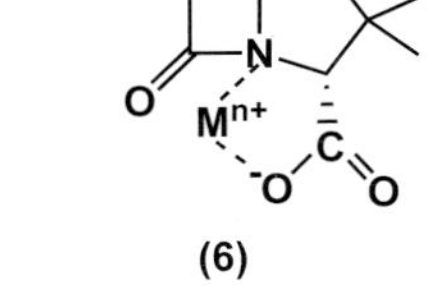

(6)

## 5.3 Reactivity

Nucleophilic substitution at the carbonyl group of an amide usually occurs in a stepwise manner by initial formation of a tetrahedral intermediate (Scheme 5.1). Conversion of the three-coordinate, $sp^2$-hybridized carbonyl carbon to a four-coordinate $sp^3$-hybridized carbon in the intermediate must be accompanied by the loss of amide resonance. This contribution to the activation energy will be reduced if amide resonance is inhibited and a rate enhancement is expected in such cases. Similarly, the release of strain energy will increase the rate if the four-membered ring is opened or has been opened in the transition state.

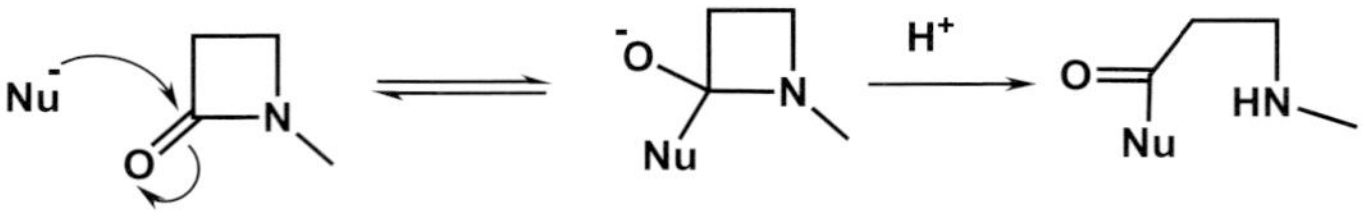

Nu⁻ ... N ⇌ ⁻O ... N ... Nu →(H⁺) O= ... Nu ... HN

Scheme 5.1

A simple reaction to see if either of these effects is apparent is the hydrolysis of the β-lactam antibiotics (Scheme 5.2). The rate of the alkaline hydrolysis of benzylpenicillin, which opens the β-lactam ring to give benzylpenicilloate is similar to that of ethyl acetate. The $pK_a$ of the protonated amine in the product is 5.2 and this weakly basic nitrogen is expected to improve the leaving group ability of the amine relative to more basic amines. Therefore, in order to assess any special reactivity of the β-lactam antibiotics, the dependence of the rate of hydrolysis of simple amides and β-lactams upon substituents must be known.

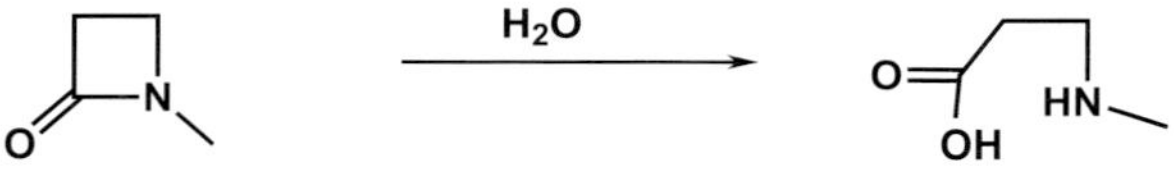

O ... N →($H_2O$) O= ... OH ... HN

Scheme 5.2

A Brønsted plot of the logarithm of the second-order rate constants for the hydroxide ion-catalyzed hydrolysis of acyclic amides, monocyclic β-lactams and bicyclic β-lactams against the $pK_a$ of the leaving group amine (in its protonated form) is a simple way to assess structure–reactivity relationships. The slope of these plots gives the Brønsted $\beta_{lg}$ value, which for *N*-substituted acyclic amides and anilides is only $-0.07$, as a result of the small dependence of reactivity on the basicity of the leaving group. The Brønsted $\beta_{lg}$ is compatible with rate-limiting *breakdown* of the tetrahedral intermediate, but is incompatible with rate-limiting expulsion of the amine anion or with a transition state in which the nitrogen has a unit positive charge (i.e., is fully protonated). The observations are consistent with water acting as a general acid catalyst in the breakdown of the tetrahedral intermediate [31, 34]. By contrast the rates of alkaline hydrolysis of β-lactams exhibit a first-order dependence on hydroxide ion concentration and show a Brønsted $\beta_{1g}$ value of −0.44, which is indicative of rate-limiting *formation* of the tetrahedral intermediate [35].

A consequence of the different dependence upon leaving group basicity for acyclic amides and monocyclic β-lactams and different rate-limiting steps is that the difference in reactivity of β-lactams compared with acyclic amides depends upon the basicity of the leaving group amine. β-Lactams of weakly basic amines are about 500-fold more reactive than an acyclic amide of the same amine but β-lactams of basic amines ($pK_a$ of $RNH_3^+$ >7) are only slightly more reactive than analogous acyclic amides. Crystallographic and spectroscopic evidence show that *N*-substituted β-lactams are planar and are resonance-stabilized similar to acyclic amides. The rate enhancement of 30–500-fold shown by β-lactams of amines of $pK_a < 6$ is simply due to relief of strain energy as the trigonal carbonyl carbon is converted to the tetrahedral intermediate [36]. The magnitude is similar to the 500-fold faster rate of reduction of cyclobutanone by borohydride compared with acetone [37]. The conversion of three- to four-coordinate carbon in four-membered rings is accompanied by the release of 11.4 kJ/mol of strain energy [36, 38]. A rate enhancement of 100-fold is therefore expected in the conversion of the β-lactam carbonyl carbon to a tetrahedral intermediate compared with the same process in an acyclic amide. The rate of alkaline hydrolysis of the simple *N*-methyl β-lactam is only 3-fold greater than that of *N*-dimethyl β-lactam.

There is nothing unusual about the chemical reactivity of the monocyclic β-lactam antibiotics nocardicin and the monobactams, and the second-order rate constants for their alkaline hydrolysis fit the Brønsted plot for other monocyclic β-lactams [31].

The release of strain energy accompanying the opening of the β-lactam ring (26–29 kcal/mol) could increase the rate by up to $10^{20}$ so, as β-lactams do not exhibit enhanced reactivity, this energy must still be present in the transition state for the hydrolysis of monocyclic β-lactams that is, there is little or no β-lactam C−N bond fission in the transition state. Fusing the β-lactam ring to a five-membered ring to make bicyclic 1-aza-bicyclo[3.2.0]heptan-2-ones increases the reactivity by about 100-fold but does not significantly change the Brønsted $\beta_{1g}$ value, which is −0.55 for the bicyclic system [31]. Although the rate enhancement is substantial, it is hardly of the magnitude expected from the release of strain energy in opening a four-membered ring or from a system in which amide resonance is significantly inhibited. Ring opening does not lower the activation energy because the rate-limiting step for the alkaline hydrolysis of penicillins is formation of the tetrahedral intermediate. The Brønsted $\beta_{1g}$ of −0.55 indicates that the nitrogen behaves as if it has no charge in the transition state and has lost all of the expected 0.6 positive charge present in the reactant resonance stabilized β-lactam. This is compatible with a transition state that very much resembles the tetrahedral intermediate.

In summary, both structure and reactivity do not indicate a significant degree of inhibition of amide resonance in penicillins and cephalosporins. The bicyclic β-lactam antibiotics do not exhibit exceptional chemical reactivity. Monocyclic β-lactams of weakly basic amines can be as chemically reactive as penicillins and cephalosporins, and it is not necessary to make the β-lactam part of a bicyclic system to have a reactive amide. A pyramidal geometry of the β-lactam nitrogen does not necessarily give a chemically more reactive β-lactam. Strained β-lactams are not necessarily better antibiotics and biological activity is not directly related to chemical reactivity.

## 5.4 Hydrolysis

### 5.4.1 Base Hydrolysis

It is well known that minor substituent changes in β-lactam antibiotics can have a dramatic effect upon antibacterial activity and susceptibility to β-lactamase-catalyzed hydrolysis. The mechanism of the alkaline hydrolysis of β-lactams was reviewed in the previous section where it was shown that decreasing the basicity of the leaving group amine increases the rate of alkaline hydrolysis of penicillins, which occurs with rate-limiting formation of a tetrahedral intermediate. Electron-withdrawing substituents at C(6) in penicillins [1] also increase the rate of hydroxide ion hydrolysis and give a Hammett $\rho$ value of $+2.0$ [31], which is only slightly less than the value of 2.7 for acyclic amides [39]. Substituents at C(6) affect the rate of nucleophilic substitution by their effect upon both the electrophilicity of the carbonyl carbon and the leaving group amine (Scheme 5.1). Although the acylamido side-chain at C(6) is important for biological activity and increases the rate of alkaline hydrolysis 20-fold relative to penicillanic acid, its effect on chemical reactivity is purely inductive. The carboxyl group at C(3) is very important for biological activity and it could, in principle, act as a general acid catalyst in the reaction of nucleophiles with penicillins. However [7], there is no evidence that the carboxyl group facilitates hydrolysis and, as expected on the basis of a purely inductive effect, the esterification of this group increases the rate of reaction 10- to 100-fold [31, 32].

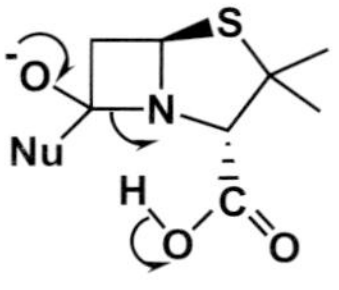

(7)

The replacement of the thiazolidine sulfur by $CH_2$ to give a carbapenam increases the rate by a factor of 3, whereas substitution by oxygen as in the oxapenams increases the rate about 5-fold as expected on the basis of an inductive effect. The incorporation of a double bond into the thiazolidine ring of a penam to give the corresponding penam system increases the rate of hydrolysis by about 25-fold. This is the order of magnitude expected from the decrease in basicity of the leaving group amine brought about by the introduction of a conjugated amine in the tetrahedral intermediate [31].

Conversion of a $\Delta^2$-carbapenem to a $\Delta^1$-carbapenem similarly decreases the reactivity 25-fold [25].

The rate of the alkaline hydrolysis of penicillins is not greatly affected by ionic strength, $I$, and the second-order rate constant increases by about 30% up to $I = 0.5$ M (KCl) but is then independent of $I$ up to $I = 4.0$ M. Kinetic solvent isotope effects, $k_{OH}/k_{OD} = 0.65$, for the alkaline hydrolysis of penicillins are consistent with rate-limiting formation of the tetrahedral intermediate [32, 40].

The major structural differences between cephalosporins [2] and penicillins [1] are that the five-membered thiazolidine ring of penicillins is replaced by a six-membered dihydrothiazine ring in cephalosporins and that the degree of pyramidalization of the β-lactam nitrogen is generally smaller in cephalosporins. In addition, many of the cephalosporins [2] have a leaving group for example, acetate, pyridine and thiol, at C(3′) and expulsion of these groups occurs during the hydrolysis of the β-lactam (Scheme 5.3) [41, 42].

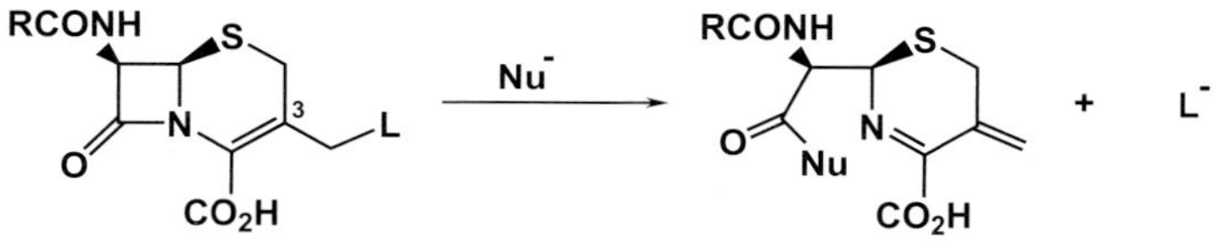

Scheme 5.3

There have been many suggestions [43], supported by theoretical calculations [44, 45], that a nucleophilic attack on the β-lactam carbonyl carbon is concerted with departure of the leaving group L at C(3′). It was thought that the presence of the leaving group at C(3′) enhanced chemical reactivity [46] and it was proposed that biological activity is related to the leaving group ability of the C(3′) substituent [47]. However, in general, the second-order rate constants for the hydroxide ion-catalyzed hydrolysis of cephalosporins are similar to those of penicillins [31, 48]. This similarity indicates that the nonplanarity of the β-lactam nitrogen does not significantly affect amide resonance since the nitrogen is 0.4 Å out of the plane defined by its substituents in penicillins [9] whereas in the cephalosporins it deviates by 0.2–0.3 Å [6]. The kinetic similarity also indicates that having a leaving group at C(3′) does not significantly affect the reactivity of cephalosporins. The rate-limiting step in the alkaline hydrolysis of cephalosporins appears to be formation of the tetrahedral intermediate. Electron-withdrawing substituents attached to the β-lactam nitrogen increase the rate of hydrolysis and give a Brønsted $\beta_{lg}$ of $-0.6$ [31]. In the stepwise mechanism, breakdown of the tetrahedral intermediate generates the enamine followed by expulsion of the leaving group at C(3′) to give the conjugated imine (Scheme 5.3) [49]. Several experimental observations indicate that the reaction is not concerted and that expulsion of the leaving group at C(3′) occurs *after* β-lactam ring opening.

The second-order rate constants for the hydroxide ion-catalyzed hydrolysis of cephalosporins are correlated with $\sigma_i$ for C3 substituents and give a Hammett $\rho_i$ of 2.5

for $CH_2L$ and of 1.35 for L. Several substituents at C3 (e.g., $CH_3$, H, and $CH_2CO_2Et$) are not expelled during hydrolysis or cannot be expelled directly by a concerted mechanism (e.g., Cl). Substituents that are and those that are not expelled are controlled by the same linear free energy relationship [22, 31, 43]. Leaving groups of different nucleofugalities [50] influence the rate of reaction only by their inductive effect [31, 43]. A series of cephalosporins with substituted pyridines and thiol leaving groups at C(3′) covering a range of 10 p$K_a$ units show a Brønsted $\beta_{1g}$ of 0.1, showing that there is little or no change in the effective charge on the leaving group on going from the ground to the transition state. Finally, the conversion of the enamine intermediate to the α,β-unsaturated imine product is reversible, and the addition of thiolate anions to the hydrolysis products of cephalosporins generated using β-lactamase as a catalyst indicates that the enamine and α,β-unsaturated imine are in equilibrium [51, 52].

Some structure–reactivity effects are replacement of the dihydrothiazine sulfur by oxygen, which increases the rate of alkaline hydrolysis about 6-fold [53], replacement by $CH_2$ increases the rate of hydrolysis about 3-fold. It is interesting to note that this increase in reactivity is accomplished by a *decrease* in the β-lactam carbonyl stretching frequency, which is contrary to the correlation described earlier [23]. Increased antibacterial activity of 1-oxacephalosporins may result from a higher rate of penetration through the bacterial cell membrane because of increased hydrophilicity [54]. The addition of a methyl, methoxy, or thiomethyl group at the 6-α-position of penicillin results in a reduction in antibacterial activity, whereas the addition of a 7-α-methoxy group to a cephalosporin results in compounds that are better transpeptidase enzyme inhibitors, although they do not necessarily show better antibacterial properties [55]. Substituent changes in the 7-β-acylamido side-chain have little effect upon chemical reactivity and yet can enormously change biological activity. For example, the incorporation of a *syn*-oxime function, as in cefuroxime, confers both high antibacterial activity and β-lactamase resistance [56, 57]. The oxime substituent, irrespective of its configuration, does not affect $k_{OH}$ for alkaline hydrolysis of the β-lactam, but is highly enzyme specific. For example, the *syn* isomer is 35-fold less reactive, as measured by $k_{cat}/K_m$, towards β-lactamase, whereas, the *anti* isomer is twice as reactive compared with an analogous cephalosporin lacking the oxime function [58].

### 5.4.2
### Acid Hydrolysis

There is no significant spontaneous hydrolysis of benzylpenicillin, but the β-lactam does undergo an acid-catalyzed degradation. By contrast, cephalosporins do exhibit a spontaneous pH-independent hydrolysis [31] and are less reactive towards acid than penicillins by a factor of about $10^4$.

In addition to the expected hydrolysis product, benzylpenicilloic acid, the acid-catalyzed degradation of benzylpenicillin gives benzylpenicillenic acid, benzylpenamaldic acid, benzylpenillic acid, and benzylpenilloic acid. The proportion of each product formed depends upon the pH [59, 60]. Although several kinetic studies have been reported on the degradation of penicillins in acidic media [61–64] there is still uncertainty about the details of the reaction pathway. It has been proposed [31, 63] that an oxazolone-thiazolidine intermediate, formed by nucleophilic attack of the acylamido side-chain on the β-lactam carbonyl carbon, is the precursor of the degradation products.

Unusually, the logarithms of the pseudo-first-order rate constants for the hydrolysis of β-lactam antibiotics and derivatives increase linearly with decreasing $H_o$ values up to $-5$ [31]. This behavior is not peculiar to bicyclic β-lactams since monocyclic β-lactams show similar behavior [31, 65]. This is quite unlike the behavior of other amides for which the rate of hydrolysis passes through a maximum, attributed to complete conversion of the amide into its *O*-conjugate acid and to decreasing water activity. Thermodynamically the most basic site for the protonation of normal amides is oxygen and the $pK_a$ of *O*-protonated amides is 0 to $-3$ [66]. The behavior of β-lactams towards acid hydrolysis indicates that neither the amide nitrogen nor the oxygen is sufficiently basic for substantial conversion to the conjugate acid and that the $pK_a$ for *O*- or *N*-protonation must be below $-5$. The β-lactams are far less basic than normal amides for *O*-protonation and a different mechanism of hydrolysis is operating. Similarly, cyclobutanones have a reduced basicity compared with other ketones [67] and the very weak basicity of β-lactams may have a similar origin [68]. The slopes of plots of the logarithms of the pseudo-first-order rate constants against $H_o$ are $-1$ to $-1.3$ and, since water activity decreases with increasing acidity, it appears that water is not involved in the transition state. All of these observations are compatible with a unimolecular A1-type mechanism with *N*-protonation of the β-lactam (Scheme 5.4). The introduction of the A1 mechanism must be an intrinsic property of β-lactams is probably due to the enhanced rate of C—N bond fission that occurs in β-lactams as a result of the relief of ring strain [31].

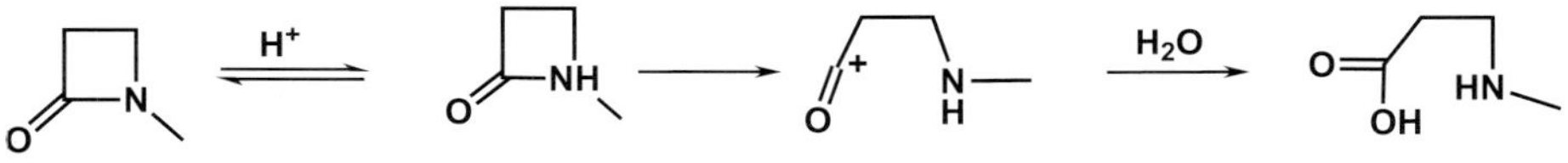

$H^+$     $H_2O$

**Scheme 5.4**

Electron-withdrawing substituents that cannot be involved in neighboring group participation greatly retard the rate with a Hammett $\varrho_i$ value of about $-4.0$ to $-5.0$ depending upon the acidity [31]. By contrast the effect of acyl substituents upon the rate of acid-catalyzed hydrolysis of acyclic amides is small, with electron-withdrawing substituents producing either a small increase or decrease in rate [39]. Electron-withdrawing substituents in the amine portion of the β-lactam decrease the rate of acid-catalyzed degradation of penicillins. The Brønsted β value is about 0.35 [31] compared with $-0.26$ for acyclic anilides and amides [69]. Although the effects of substituents are not large, they are significant and in the opposite direction for β-lactams compared with other amides, which again is indicative of a different mechanism.

Owing to the similarity of the rate constants for the degradation of benzylpenicillin and its methyl ester there is no evidence for neighboring group participation of the carboxy group in the fission of the β-lactam ring [31].

The acid hydrolysis of cephalosporins shows similar behavior to that of the penicillins, but they are about $10^4$-fold less reactive [31]. Electron-withdrawing substituents at C(7) in cephalosporins decrease the rate of acid hydrolysis and, as for penicillins, the Hammett $\rho_i$ value is about $-5$, but there is no evidence for neighboring group participation by the 7-acylamido group as seen for penicillins. There is no obvious explanation of the difference in behavior between the cephalosporins and penicillins.

Similar to alkaline hydrolysis there is no evidence for the group at C(3′) in cephalosporins (acetate or pyridine) affecting the rate of reaction. In fact, the 3-methyl derivative is more reactive than the cephalosporins with acetate or pyridine at C(3′), which again indicates that expulsion of these groups is not important in the rate-limiting step.

### 5.4.3 Spontaneous Hydrolysis

There is no significant pH-independent "uncatalyzed" hydrolysis of most penicillins. By contrast, most cephalosporins show a pH-independent reaction between pH 3 and 7 with $k_o$ in the range $5 \times 10^{-7}$ to $3 \times 10^{-6} s^{-1}$ [48, 70]. The solvent kinetic solvent isotope effect $k^{H_2O}/k^{D_2O}$ is 0.93 [48] is not typical of a water-catalyzed hydrolysis and the few penicillins that do show pH independent hydrolysis exhibit a significant solvent isotope effect, $k^{H_2O}/k^{D_2O}$ is 4.5 [71].

### 5.4.4 Buffer-Catalyzed Hydrolysis

The hydrolysis of both penicillins and cephalosporins are often catalyzed by buffers. The Brønsted plot for the hydrolysis of benzylpenicillin-catalyzed by oxygen bases shows a nonlinear dependence on the basicity of the buffer, which is indicative of a change in mechanism. The Brønsted β-value for weak bases is 0.39 and, together with a solvent isotope effect $k_B^{H_2O}/k_B^{D_2O}$ of 2.1, is indicative of general base-catalyzed hydrolysis. In contrast, the Brønsted β value for bases whose conjugate acids have a $pK_a$ above 7 is 0.95 and represents nucleophilic-catalyzed hydrolysis [5]. Evidence for an intermediate ester formed during the reaction has been obtained with alkoxide ions and phosphate dianion [72]. The mechanism of the reaction with alkoxide ions probably proceeds by rate-limiting breakdown of the tetrahedral intermediate.

### 5.4.5 Metal Ion-Catalyzed Hydrolysis

Transition metal ions cause an enormous increase in the rate of hydrolysis of penicillins and cephalosporins [32, 33]. For example, Cu(II) ions can enhance the rate of hydrolysis of benzylpenicillin $10^8$-fold – a change in the half-life from 11 weeks to 0.1 s at pH 7. In the presence of excess metal ions, the observed apparent first-order rate constants for the hydrolysis of the β-lactam derivatives are first order in hydroxide ion, but show a saturation phenomenon with respect to the concentration of metal ion that is indicative of the formation of an antibiotic/metal ion complex (Scheme 5.5)

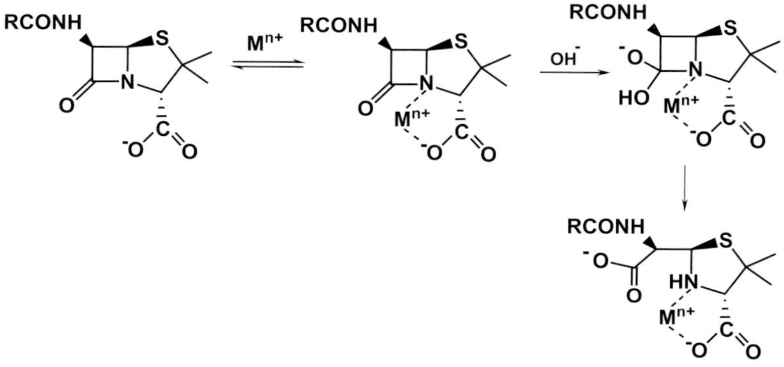

**Scheme 5.5**

The rate of hydroxide ion-catalyzed hydrolysis of benzylpenicillin bound to metal ion shows the following rate enhancements compared with the uncoordinated substrate: Cu(II), $8 \times 10^7$; Zn(II), $4 \times 10^5$; Ni(II), $4 \times 10^4$; Co(II), $3 \times 10^4$. The analogous data for cephaloridine [32] are: Cu(II), $3 \times 10^4$; Zn(II), $2 \times 10^3$. The Cu (II) ion coordinates to the carboxylate group and the β-lactam nitrogen of benzylpenicillin as shown in Scheme 5.5. Coordination occurring to the carboxylate group is indicated because esterification of this group decreases the rate enhancement by a factor of about $5 \times 10^3$. It has been suggested that Cu(II) ions coordinate to the 6-acylamino side-chain and the β-lactam carbonyl group [73]. However, replacement of the acylamino side-chain by the more basic amino group has little effect upon the binding constant and the rate enhancement for the hydroxide ion-catalyzed hydrolysis for 6-aminopenicillanic acid is very similar to that for benzylpenicillin. Furthermore, complete removal of the amido side-chain, as in penicillanic acid, also gives similar binding constants and rate enhancements. It appears that Cu(II) ions do not bind to the amido side-chain in penicillins, and that coordination probably occurs between the carboxylate oxygen and the β-lactam nitrogen [32, 74].

Cu(II) ions bind 10-fold more tightly to cephalosporins than to penicillins which would be surprising if the sites of coordination were similar. Molecular models indicate that one of the conformations of cephalosporins would be very suitable for metal ion coordination between the carboxylate group and the β-lactam carbonyl oxygen. Precipitation of the β-lactam/metal ion complex in the presence of excess ligand gives solids with different characteristics. For example, benzylpenicillin forms a 1: 1 complex with both Cu(II) and Zn(II) in which the asymmetric stretching frequencies of the β-lactam carbonyl and the carboxylate are decreased by about 30 $cm^{-1}$ compared with uncoordinated penicillin. The NMR spectrum of the Zn(II)/benzylpenicillin complex shows a downfield shift for the C(3) hydrogen consistent with the proposed mode of binding [75]. Cephalothin and 3-methyl-7β-phenylacetamidoceph-3-em-4-carboxylic acid form solid 2 : 1 complexes with transition metal ions in which both the asymmetric stretching frequency of the carboxylate and the β-lactam carbonyl stretching frequency are decreased by 10–30 $cm^{-1}$, depending upon the nature of the metal ion. The NMR spectrum of the Zn(II)/cephalothin complex shows a downfield shift of the C(7) hydrogen. The site of metal ion coordination could be different for cephalosporins and involve the β-lactam carbonyl oxygen, although the situation in solution may be different.

An important role of the metal ion in the hydroxide ion-catalyzed hydrolysis is to stabilize the tetrahedral intermediate (Scheme 5.5). The hydroxide ion-catalyzed hydrolysis of benzylpenicillin probably proceeds by the formation of the tetrahedral intermediate and there is an enormous change in the basicity (more than 12 $pK_a$ units) of the β-lactam nitrogen as it is converted from an amide to an amine. An estimate of the binding constant of Cu(II) ions to the tetrahedral intermediate of $10^{7.2}$ $M^{-1}$ can be made from a comparison with model compounds [33]. The rate of the hydroxide ion-catalyzed hydrolysis of Cu(II)-bound benzylpenicillin is $8 \times 10^7$ faster than that of uncoordinated benzylpenicillin [32] so Cu(II) ion stabilizes the transition state by 13.9 kcal/mol compared with an estimated value of 9.8 kcal/mol for the stabilization of the tetrahedral intermediate.

There is no correlation between the binding constant of the β-lactam antibiotic with the nature of the metal ion and the rate enhancement. The order of reactivity is that of the Irving–Williams series: Co(II) $<$ Ni(II) $<$ Cu(II) $<$ Zn(II).

Cu(II) ions bind about 10-fold more tightly to cephalosporins than to penicillins, which, at first, seems surprising in view of the greater nonplanarity of the penicillin molecule and so the assumed greater basicity of the β-lactam nitrogen. The rate of hydroxide ion-catalyzed hydrolysis of Cu(II)-bound cephaloridine is about $3 \times 10^4$-fold faster than that for the uncoordinated compound, compared with a rate enhancement of $8 \times 10^7$ seen for benzylpenicillin. The transition state for cephaloridine hydrolysis is stabilized by Cu(II) ions about 100-fold less than that for penicillin hydrolysis, but both transition states are greatly stabilized by the metal ion. Again, *ad hoc* explanations for this difference may be found in the lower basicity of the ring nitrogen in the tetrahedral intermediate formed from cephaloridine and/or a less favorable geometry.

It has been suggested that a ternary complex is formed between benzylpenicillin, Zn(II) and Tris buffers and that hydrolysis occurs by intramolecular nucleophilic attack of one of the coordinated buffer hydroxyl groups on the β-lactam [76, 77].

### 5.4.6 Micelle-Catalyzed Hydrolysis of Penicillins

The micelle-catalyzed hydrolysis of penicillins in alkaline solution is unusual because it involves the reaction between two anions – the hydroxide ion and the negatively charged benzylpenicillin [78]. The acid-catalyzed degradation of penicillins is inhibited in cationic micelles of cetyltrimethylammonium bromide (CTAB) [79] and, as expected, neither anionic micelles of sodium dodecylsulfate nor polyoxyethylene lauryl ether promote the hydroxide ion-catalyzed hydrolysis of benzylpenicillin [78]. There have been relatively few studies of the micellar-catalyzed hydrolysis of amides and the effects are small [80].

In the presence of CTAB the pseudo-first-order rate constants for the alkaline hydrolysis increase rapidly with surfactant concentration once above the critical micelle concentration of the surfactant [81]. Increasing the surfactant concentration eventually leads to a slow decrease in the observed rate. This general shape of surfactant-rate profile has been found for many bimolecular reactions catalyzed by cationic micelles. However, unusually, the observed pseudo-first-order rate constant is not independent of penicillin concentration. The binding constant between the micelle and substrate is unlikely to change significantly with concentration, and yet the lower the concentration of benzylpenicillin, the faster the rate increases and the greater the maximal rate obtained – the rate maximum shifting to a lower surfactant concentration. This observation could be explained if both hydroxide ion and benzylpenicillin compete for the same types of sites in the micelle, and if benzylpenicillin binds better than hydroxide ion. Increasing the hydroxide ion concentration inhibits the rate of the micellar-catalyzed reaction while the rate in the bulk aqueous phase increases. The observed pseudo-first-order rate constant for the micelle-catalyzed hydrolysis does not increase linearly with increasing hydroxide ion concentration at constant surfactant concentration, but reaches a maximum value [81]. The kinetics suggests that there must be binding of benzylpenicillin anion to the micelles of CTAB and this has been shown spectroscopically [82]. The maximum rate acceleration in the alkaline hydrolysis of benzylpenicillin by CTAB micelles is about 50. The rate increase observed for many reactions upon the addition of detergents above the critical micelle concentration has been explained on the basis of the substrate binding to the micelle to form a substrate–micelle complex and with an equilibrium constant $K_s$ ($300\,M^{-1}$ for benzylpenicillin). The substrate and substrate–micelle complex from the product P with rate constants $k_w$ and $k_m$ referring to bulk aqueous and micellar phases, respectively. The inhibitory effect of increasing benzylpenicillin concentration can be rationalized by the pseudo-phase ion-exchange model, but as the number of molecules of the antibiotic bound to the micelle increases (above 10 for concentrations above $2 \times 10^{-4}$ M) the behavior of a micelle covered with benzylpenicillin is probably different from a typical CTAB micelle [82].

Catalysis by micelles of the hydroxide ion-catalyzed hydrolysis of substrates appears to be qualitatively understood on the basis of a concentration effect of reactant on, or around, the micelle surface, and need not necessarily involve a difference in the free energies of activation in the micelle and bulk phase. This does not mean that the cationic micelles could not and do not cause electrostatic stabilization of the transition state. The cationic micelle surface can act as an electrostatic sink for the anionic intermediate leading to its stabilization, but a rate enhancement requires preferential stabilization of this intermediate compared with the reactant. The small rate enhancement of the micelle-catalyzed reaction, about 50-fold, is equally well explained by considering that the increased concentration of reactants at the micelle surface leads to a higher observed rate. Incorporation of the reactants into a limited volume decreases the entropy loss that is associated with bringing reactants together in the transition state and this leads to an increase in the pseudo-first-order rate constants in the presence of surfactant micelles. Cationic micelles of CTAB have also been shown to facilitate the alkaline hydrolysis of the cephalosporin cephalexin [83].

The hydrolysis reaction is inhibited by the addition of the hydrolysis product, the dianion benzylpenicilloate, which appears to bind no more tightly to the micelle than does benzylpenicillin itself. In benzylpenicilloate there are two carboxylate anions, yet the inhibition, which results from increasing its concentration is similar to that caused by increasing the benzylpenicillin concentration. The rate of the hydroxide ion-catalyzed hydrolysis of benzylpenicillin in the presence of micelles of CTAB is sensitive to electrolytes [78], supporting the idea of electrostatic interactions between substrates and the micelle surface. The importance of other effects has been demonstrated by modifying the 6-β-side-chain of penicillin to increase the substrate lipophilicity and hence the micelle–substrate hydrophobic interaction [81].

Increasing the hydrophobicity of the 6-β-side-chain increases the CTAB-catalyzed hydrolysis of penicillin derivatives, but once the 6-β-side has been extended to $CH_3(CH_2)_4CONH$-, further extension does not significantly increase the binding constant. The polar compound, 6-β-aminopenicillanic acid, is only weakly bound to the micelle, and electrostatic interactions may be all that exist between the substrate and micelle.

### 5.4.7 Cycloheptaamylose-Catalyzed Hydrolysis

Cycloamyloses (cyclic α-1,4-linked oligomers of D-glucose) have a toroidal ("doughnut")-shaped structure. The primary hydroxy groups are located on one side of the torus while the secondary ones lie on the other side. Relative to water, the interior of the cycloamylose torus is apolar. The catalytic properties of cycloamyloses depend on the formation of inclusion complexes with the substrate and subsequent catalysis by either the hydroxy, or other groups, located around the circumference of the cavity [84].

Under mildly alkaline conditions and in the presence of excess cycloheptaamylose the rate of degradation of penicillin is increased 20- to 90-fold compared with the rate of alkaline hydrolysis [85]. Michaelis–Menten kinetics are observed which are indicative of complex formation. The apparent binding constant of 6-substituted penicillins varies little with the length of the alkyl side-chain although it is increased about 10-fold for diphenylmethyl penicillin. The reaction is catalytic and hydrolysis proceeds by the formation of a penicilloyl-β-cyclodextrin covalent intermediate (i.e., ester formation) by nucleophilic attack of a carbohydrate hydroxyl on the β-lactam.

## 5.4.8 Enzyme-Catalyzed Hydrolysis

Most β-lactam antibiotics are susceptible to the β-lactamase hydrolytic enzymes, which are the most common, and growing, form of bacterial resistance. β-Lactamases catalyze the hydrolysis of the β-lactam to give the ring-opened and bacterially inert β-amino acid (Scheme 5.2). The main mechanistic division of β-lactamases is into serine enzymes, which have an active-site serine residue and the catalytic mechanism involves the formation of an acyl-enzyme intermediate, and zinc enzymes. On the basis of their amino acid sequences, the serine β-lactamases are subdivided into three classes (A, C and D), whereas the class B β-lactamases consist of the zinc enzymes [86].

### 5.4.8.1 Serine β-Lactamases

The class A and class C enzymes are monomeric medium-sized proteins with $M_r$ values of about 29 000 and 39 000, respectively [86], and show two major structural domains, all-α and α/β, with the active site situated in a groove between the two domains [87]. The class C β-lactamases have additional loops and secondary structure on the all-α-domain. The active site serine is situated at the N-terminus of the long, relatively hydrophobic, first α-helix of the all α-domain. There is very strong evidence for the formation of an acyl-enzyme intermediate (Scheme 5.6) – including electrospray mass spectrometry, IR measurements, trapping experiments, the determination of the rate constants for their formation and breakdown, and even an X-ray crystal structure [88].

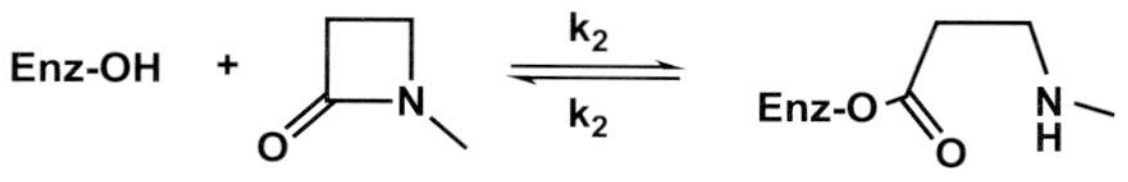

Enz-OH + [β-lactam] $\underset{k_2}{\overset{k_2}{\rightleftharpoons}}$ Enz-O–C(=O)–CH₂–CH₂–NH–

Scheme 5.6

Formation of the acyl-enzyme intermediate requires at least two proton transfers – proton removal from the attacking serine and proton donation to the departing β-lactam amine (Scheme 5.7). Despite the availability of a number of X-ray crystal structures of several class A and class C β-lactamases [87], and many site-directed mutagenesis studies [89], the identity of the catalytic groups involved in these proton transfer steps remains elusive.

The maximum second order rate constant for an enzyme-catalyzed reaction is that corresponding to diffusion control (about $10^8\ M^{-1}\ s^{-1}$), and some β-lactamases appear to be near this limit and are "perfect catalysts" [90].

In class A β-lactamases there are two serious contenders for the general base-acid –Glu166 and Lys73. The pH-dependence of $k_{cat}/K_m$ indicates two ionizing residues are important for catalysis – one of p$K_a$ about 5 and formally required in its basic form, and one of p$K_a$ about 9 and formally required in its acidic form. If the low p$K_a$ group corresponds to the general base then this is reasonable for the carboxylic

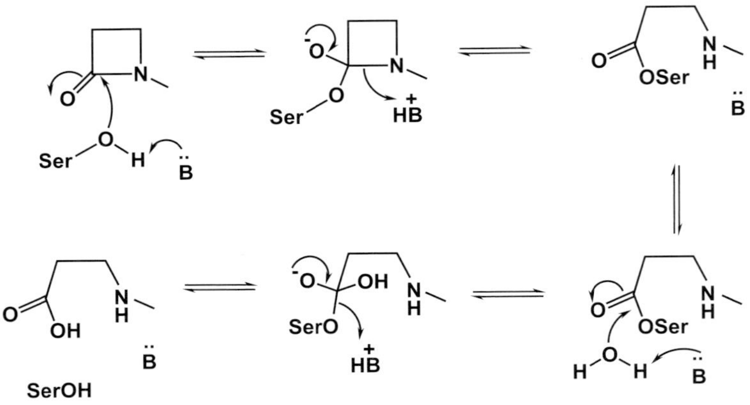

Scheme 5.7

acid of Glu166, but is rather low for an ammonium ion of Lys73. Although it has been suggested [88] that Lys73 may have a reduced p$K_a$ it is difficult to envisage the required reduction of 5.6 p$K_a$ units given the close proximity to Glu166 (2.8–3.4 Å). A normal p$K_a$ of Lys73 is supported by $^{13}$C-NMR studies [91] and theoretical calculations. Site-directed mutagenesis of Glu166 shows that the rates of both acylation and deacylation are affected, although the latter is more so [89]. On balance it appears that the evidence supports Glu166 (Scheme 5.7, B = $GluCO_2^-$) acting as the unique proton transfer agent – as a general base and acid in the formation and breakdown of the tetrahedral intermediate, respectively, and in the same role for the hydrolysis of the acyl-enzyme intermediate.

Historically, class C β-lactamases were often referred to as "cephalosporinases" because of the characteristic higher turnover numbers, $k_{cat}$, observed for cephalosporins compared with penicillins. However, the $k_{cat}/K_m$ values for the two classes of enzymes are generally similar and high for both penicillins and cephalosporins ($10^5$–$10^8$ $M^{-1}\,s^{-1}$) [92]. A major difference between the two classes is that for class C β-lactamases deacylation is often rate limiting so that the acyl-enzyme intermediate may accumulate giving rise to low values of $K_m$. In class C β-lactamase there is no equivalent glutamate residue but Tyr150 may take its role with Lys67 equivalent to Lys73 in the class A enzyme [87]. In addition, it has been suggested that the hydrolytic water involved in deacylation of the acyl-enzyme approaches from the β-face and that this hydrolysis may be substrate-assisted by the expelled amine, which was the β-lactam nitrogen, acting as a general base catalyst [93]. In class C β-lactamase it has been suggested that the phenol of Tyr150 has a severely reduced p$K_a$ and acts as a general base catalyst for proton removal from Ser64 [94] although this is not supported by site-directed mutagenesis of Tyr150 [95]. Owing to its relatively positive environment and strong hydrogen bonding of the phenoxide ion by lysine residues Tyr150 has a severely reduced p$K_a$. It appears that the Tyr150 residue is a very strong candidate for the role of a general base catalyst in class C β-lactamases, (Scheme 5.7, B = $TyrO^-$).

#### 5.4.8.2 Metallo β-Lactamases

Class B β-lactamases or metallo β-lactamases (MBLs) require one or two zinc ions to catalyze the hydrolysis of β-lactams. They have no sequence or structural homology to the serine β-lactamases and exhibit a broad spectrum substrate profile, hydrolyzing penicillins, cephalosporins, carbapenems, and even some mechanism-based inhibitors of class A β-lactamases [96]. The first MBL to be discovered was produced by an innocuous strain of *Bacillus cereus*, but in the last 20 years, MBL-mediated resistance has appeared in several pathogenic strains and is being rapidly spread by horizontal transfer, involving both plasmid and integron-borne genetic elements [97]. MBLs represent a huge potential clinical threat to β-lactam antibiotic therapy as presently there is no clinically useful inhibitor for this class of β-lactamases.

MBLs can be divided into three subclasses, B1, B2 and B3, according to their amino acid sequences, substrate profile, and metal ion requirement [98]. Subclass B1 is the largest and contains four well-studied β-lactamases: BcII from *B. cereus*, CcrA from *Bacteroides fragilis*, IMP-1 from *Pseudomonas aeruginosa*, and BlaB from *Cryseobacterium meningosepticum*. These enzymes efficiently catalyze the hydrolysis of a wide range of substrates, including penicillins, cephalosporins, and carbapenems. BcII hydrolyzes penicillins at significantly higher rates than cephalosporins and carbapenems, although CcrA does not show this preference, both enzymes exhibit lower $K_m$ values for cephalosporins [99].

The number of zinc ions required for MBL activity has been a matter of controversy and the different crystal structures reported has added to the confusion. The structures of several MBLs have been determined by X-ray diffraction and all show a similar αββα fold. The active site of MBLs is situated at the bottom of a wide shallow groove between two β-sheets and has two potential zinc ion binding sites at the active site often referred to as sites 1 and 2 [100–102]. The zinc ligands in the two sites are not the same and are not fully conserved between the different MBLs. In the subclass B1 such as the *B. cereus* enzyme BcII the zinc in site 1 (the histidine site or $His_3$ site) is tetra-coordinated by the imidazoles of three histidine residues (His116, His118, and His196) and a water molecule, $Wat_1$. In site 2 (or the Cys site) the metal is penta-coordinated by His263, Asp120, and Cys221, and one water molecule; the fifth ligand at site 2 is carbonate or water, often referred to as the apical water (or $Wat_2$) [101, 102]. The two metal ions are relatively close to each other, but the distance between them varies from 3.4 to 4.4 Å in different structures of the BcII and CcrA enzymes. Several structures of the CcrA enzyme show a bridging water ligand between the two metals, which is thought to exist as a hydroxide ion. In a structure of BcII containing two zinc ions determined at pH 7.5 there is also a similar bridging water molecule, but in structures of this enzyme at lower pH this solvent molecule is strongly associated with the zinc in site 1 [103].

A recent extensive study of BcII using circular dichroism, competitive chelation, mass spectrometry, and NMR concluded that the dizinc form is the only relevant species for catalysis. Similarly, isothermal titration calorimetry studies show only one binding event with two zinc ions bound to BcII, except below pH 5.6 where only one zinc binds to the enzyme. It now appears that there is no significant difference between BcII and the homologous class B1, CcrA from *B. fragilis* that binds both zinc ions very tightly [99].

Early kinetic studies of CcrA led to the proposal that both the mono- and dinuclear forms of the enzyme were catalytically active, with slightly different activities, at physiological pH [104]. However, later studies showed that only the dinuclear species was active, and that the previously observed "monozinc" CcrA was a mixture of the dizinc and the apo (metal-free) enzyme [105].

There are many potential mechanistic roles for the metal ion in metalloproteases and they may well vary from enzyme to enzyme [106, 107]. The role of zinc in catalysis is related to its ability to participate in tight but readily exchangeable ligand binding and its exceptional flexibility of its coordination number and geometry. In addition, zinc shows no redox properties and this facilitates its evolution in living systems without the risk of oxidative damage. Finally, its intermediate hard/soft behavior allows it to bind a variety of atoms, as seen, for example, in the second binding site of class B1 MBLs, which involve nitrogen, oxygen, and sulfur as ligands. The Lewis acidity, flexible geometry, and coordination number, and the lack of redox properties make zinc an ideal metal cofactor for many enzymes. The small energy difference between four, five, or six coordination geometries, and the rapid exchange of the kinetically labile zinc-bound water molecule is an important feature in all zinc hydrolases, including MBLs.

It is commonly suggested that the metal ion acts as a Lewis acid by coordination to the peptide carbonyl oxygen, giving a more electron-deficient carbonyl carbon, which facilitates nucleophilic attack. The metal ion thus stabilizes the negative charge developed on the carbonyl oxygen of the tetrahedral intermediate anion (Scheme 5.8). Many metalloproteases have a water molecule directly coordinated to the metal ion, which may act as the nucleophile to attack the carbonyl carbon. The role of the metal ion is to lower the p$K_a$ of the coordinated water so that the concentration of metal-bound hydroxide ion, albeit different, is increased relative to bulk solvent hydroxide ion at neutral pH and is a better nucleophile than water (Scheme 5.9). Although C–N bond fission is the most energetically difficult process in peptide hydrolysis, little attention is normally given to the mechanism of the breakdown of the tetrahedral intermediate. Breakdown of the tetrahedral intermediate could be facilitated by direct coordination of the departing amine nitrogen to the metal ion (Scheme 5.10). This is the mechanism adopted for the zinc-catalyzed hydrolysis of penicillin. Alternatively, a metal-bound water could act as a general acid catalyst protonating the amine nitrogen-leaving group to facilitate

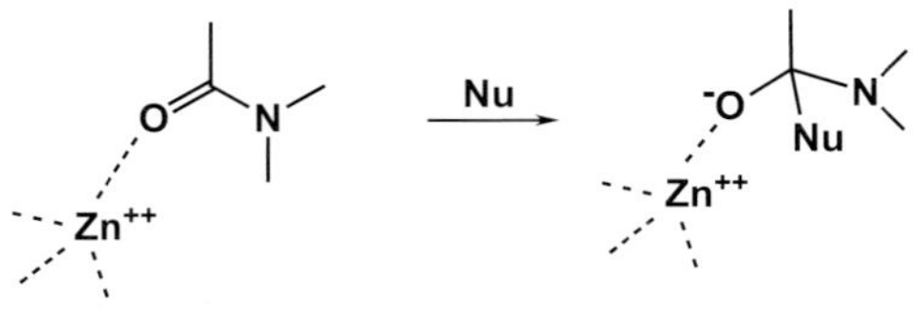

O N Zn$^{++}$ —Nu→ $^-$O N Nu Zn$^{++}$

Scheme 5.8

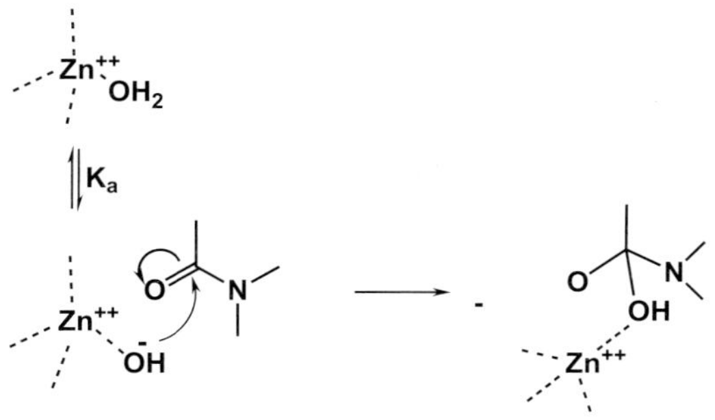

Scheme 5.9

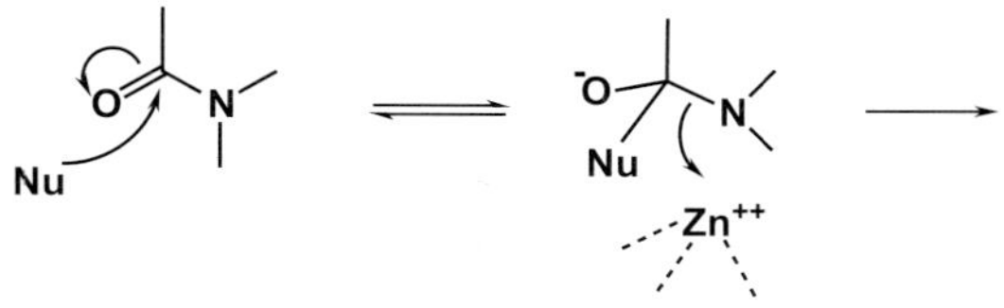

Scheme 5.10

C–N bond fission [108] (Scheme 5.11). Despite intense mechanistic studies, the detailed roles of the metal ion in metalloproteases remain controversial and distinguishing between the relative importance of the possible roles for zinc is complex.

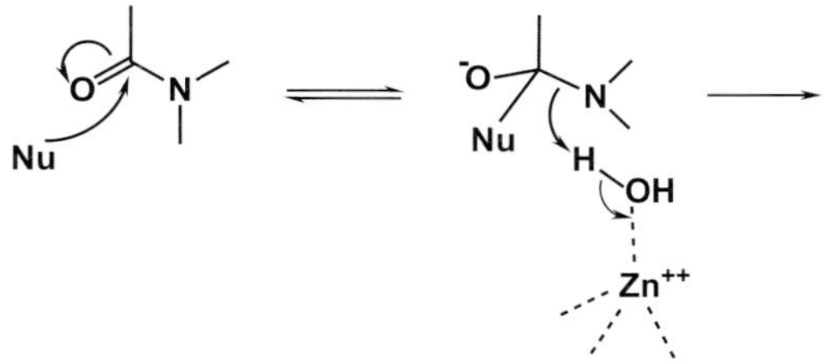

Scheme 5.11

The zinc of *B. cereus* β-lactamase is coordinated to three protein ligands in the zinc1 site (His86, His88, and His 149) and a water molecule [102]. In aqueous solution, zinc is coordinated to six water molecules and the p$K_a$ of the zinc bound water is 9.5. The reduced coordination number of 4 in the zinc1 site of β-lactamase reduces the p$K_a$ to less than 6. The pH-rate profile for the BcII-catalyzed hydrolysis of benzylpenicillin and cephaloridine [109] was taken to indicate that the zinc ion bound water has a low p$K_a$ of below 5 and is therefore fully ionized at neutral pH. Nucleophilic attack by the metal-bound hydroxide ion on the carbonyl followed by a proton abstraction from the Asp120 gives a dianionic tetrahedral intermediate (Scheme 5.12). It was suggested that the same aspartate residue functions as proton donor to facilitate C–N bond

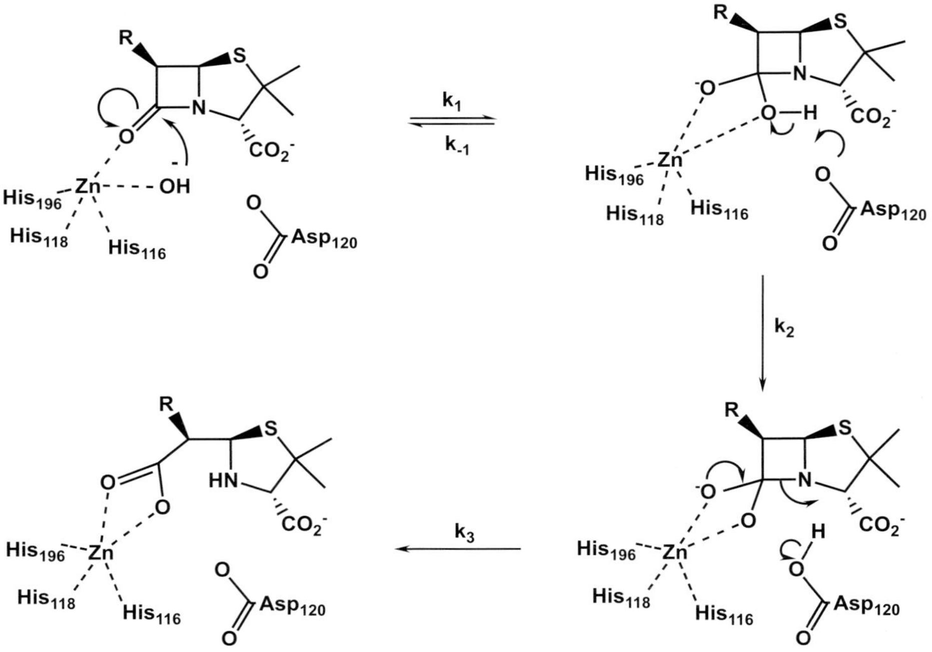

Scheme 5.12

fission, and either step $k_2$ or $k_3$ could be rate limiting. A dianionic intermediate assists β-lactam ring opening and generates a carboxylate anion rather than the undissociated acid.

A mechanism of hydrolysis for the dizinc enzyme (Scheme 5.13) uses the bridging hydroxide ion for the nucleophilic attack, which results in a negatively charged intermediate stabilized by the oxyanion hole. The apical water molecule bound to zinc is positioned to donate a proton to the leaving nitrogen.

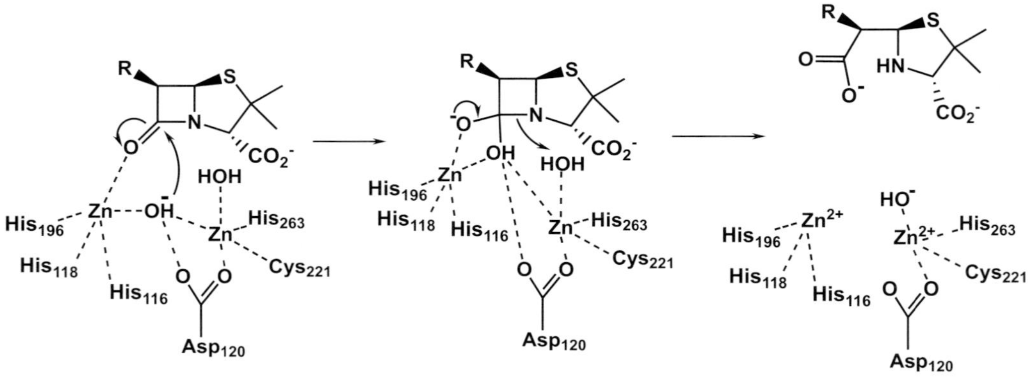

Scheme 5.13

The exchange of the spectroscopically silent zinc in zinc enzymes with probes, such as cobalt, copper, and cadmium, enables the study of the metal interactions in the enzyme active site with its ligands, substrates and inhibitors of the metalloenzyme, using techniques such as electronic spectroscopy, NMR, electron paramagnetic resonance, and perturbed angular correlation (PAC) spectroscopy. The zinc of MBLs can be exchanged with cadmium, cobalt, and manganese to give catalytically active enzymes. The use of a combination of NMR and PAC spectroscopy to study cadmium binding to *B. cereus* MBL has revealed a rapid intramolecular exchange of the metal between the two sites in the mono-cadmium enzyme and negative cooperativity in metal binding [110]. The metal-substituted enzymes have similar or higher catalytic activities compared with the native zinc enzyme, albeit at pHs above 7 and, for the cobalt enzyme, at all pHs. A higher p$K_a$ for the metal-bound water for cadmium and manganese BcII leads to more reactive enzymes than the native zinc BcII, suggesting that the role of the metal ion is predominantly to provide the nucleophilic hydroxide, rather than to act as a Lewis acid to polarize the carbonyl group and stabilize the oxyanion tetrahedral intermediate [111].

The bridging water is an important ligand binding the second metal ion to the protein, but it is *consumed* during the catalytic cycle of hydrolysis. Consequently, regeneration of the catalyst requires the coordination of a new water molecule to the active-site zinc and its deprotonation, and so it is possible that the second metal ion could be lost during turnover. In fact, the kinetics of the hydrolysis of benzylpenicillin catalyzed by the cobalt-substituted β-lactamase from BcII are biphasic with an initial burst of product formation followed by a steady-state rate of hydrolysis [112, 113]. This is due to a branched kinetic pathway with two enzyme intermediate species, $ES^1$ and $ES^2$, which have different metal: enzyme stoichiometries. $ES^1$ is a dimetal ion enzyme intermediate and is catalytically active, but it slowly loses one bound metal ion during turnover via the branching route, to give the mononuclear and inactive enzyme intermediate $ES^2$.

## 5.5 Aminolysis

The reaction of amines with penicillins to give penicilloyl amides (Scheme 5.7) is of interest because the major antigenic determinant of penicillin allergy is the penicilloyl group bound by an amide linkage to ε-amino groups of lysine residues in proteins [114–116]. The aminolysis of penicillin is an amide exchange – a normally difficult process, but one which occurs readily with β-lactams [117]. The C–N bond fission in amides usually requires protonation of the nitrogen to avoid expulsion of the unstable anion, but in β-lactams this process is accompanied by a large release of strain energy that modifies the requirements for catalysis compared with normal amides. Another important difference between C–N bond fission in β-lactams compared with that in amides is that the latter may be accompanied by a more favorable entropy change as the molecule fragments into two separate entities [36].

The aminolysis of β-lactams occurs by a stepwise mechanism, involving the reversible formation of a tetrahedral intermediate, the breakdown of which to products is catalyzed by bases (Scheme 5.14) [118–120].

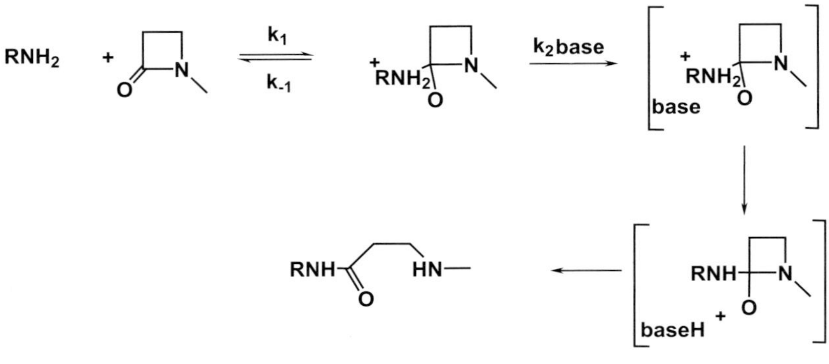

Scheme 5.14

The shape and rigidity of the penicillin molecule makes it a suitable substrate to study the effectiveness of intramolecular catalysis and, in particular, to elucidate any preferred direction of nucleophilic attack upon the β-lactam carbonyl group [121].

The rate of the reaction of penicillin with the monocation of 1,2-diaminoethane is about 100-fold greater than that predicted from the Brønsted plot for a monoamine of the same basicity. The rate enhancement is attributed to intramolecular general acid catalysis of aminolysis by the protonated amine [118, 122]. Breakdown of the tetrahedral intermediate is facilitated by proton donation from the terminal protonated amino group to the β-lactam nitrogen.

An interesting difference between nucleophilic substitution in penicillins and peptides/amides is the preferred direction of attack and the geometry of the initially formed tetrahedral intermediate. It is usually assumed, based on the theory of stereoelectric control [123], that nucleophilic attack on the carbonyl carbon of a planar peptide will generate a tetrahedral intermediate with the lone pair on nitrogen *anti* to the incoming nucleophile (**8** and **9**). Conversely, nucleophilic attack on β-lactams occurs from the least hindered α-face (*exo*) so that the β-lactam nitrogen lone pair is *syn* to the incoming nucleophile in the tetrahedral intermediate (**10**) [119, 122]. This has obvious consequences for the placement of catalytic groups, particularly the general acid donating a proton to the departing amine of the β-lactam.

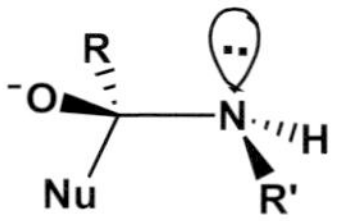

**(8)** 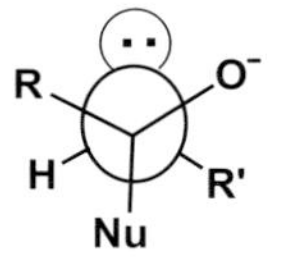 **(9)** 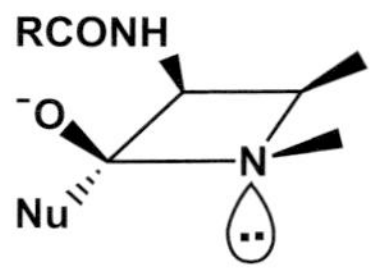 **(10)**

The observation of intramolecular general acid catalysis in the reaction with the monocation of 1,2-diaminoethane gives an indication of the direction of nucleophilic attack upon penicillin. In order that ready proton transfer takes place from the protonated amine to the β-lactam nitrogen, it is essential that the tetrahedral intermediate has the geometry (**10**) shown. Further evidence for nucleophilic attack taking place from the α-face comes from the *absence* of intramolecular general base catalysis in the aminolysis of 6-β-aminopenicillanic acid. That the lone pair of β-lactam nitrogen takes up the geometry shown in (**10**) is supported by the observation that Cu(II) ions catalyze the aminolysis of penicillin by coordination to the β-lactam nitrogen and the carboxy group, thus stabilizing the tetrahedral intermediate [32].

The rate of aminolysis of benzylpenicillin and cephaloridine by hydroxylamine, unlike other amines, shows only a first-order dependence on amine concentration [124]. The rate enhancement compared with that predicted from a Brønsted plot for other primary amines with benzylpenicillin is greater than $10^6$ and is attributed to rate-limiting formation of the tetrahedral intermediate due to a rapid intramolecular general acid-catalyzed breakdown of the intermediate. For cephaloridine, the rate enhancement is greater than $10^4$, which demonstrates that β-lactam C–N bond fission and expulsion of the leaving group at C3′ are not concerted.

*N*-Aroyl β-lactams (**11**) are imides with exo- and endo-cyclic acyl centers that react with amines in aqueous solution to give the ring-opened β-lactam aminolysis product. Unlike the strongly base-catalyzed aminolysis of β-lactam antibiotics, the rate law for the aminolysis of *N*-aroyl β-lactams is dominated by a term with a first-order dependence on amine concentration in its free base form, indicative of an uncatalyzed aminolysis reaction. The rate constants for this uncatalyzed aminolysis of *N*-*p*-methoxybenzoyl β-lactam with a series of substituted amines generate a Brønsted $\beta_{nuc}$ value of $+0.90$, indicative of a large development of positive effective charge on the amine nucleophile in the transition state. Similarly, the rate constants for the reaction of 2-cyanoethylamine with substituted *N*-aroyl β-lactams give a Brønsted $\beta_{lg}$ value of $-1.03$ for different amide leaving groups and is indicative of considerable change in effective charge on the leaving group in the transition state. These observations indicate a concerted mechanism with simultaneous bond formation and fission (**12**) in which the amide leaving group is expelled as an anion (**13**). The sensitivity of the Brønsted $\beta_{nuc}$ and $\beta_{lg}$ values to the nucleofugality of the amide leaving group and the nucleophilicity of the amine nucleophiles, respectively, indicate a coupled bond formation and bond fission processes [125].

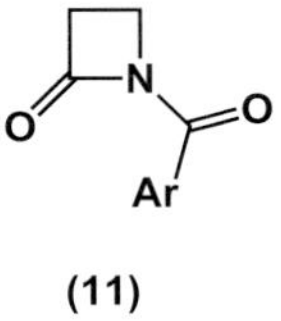

N O O Ar

**(11)**

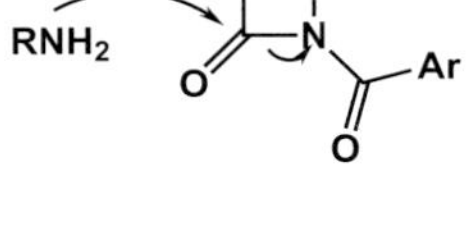

RNH2 N O Ar O

**(12)**

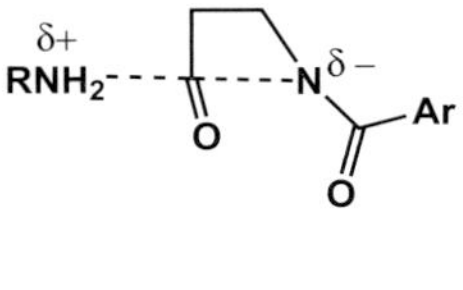

δ+ RNH2 N δ− O Ar O

**(13)**

Amide resonance is usually described in terms of charge transfer from nitrogen to oxygen as shown in the traditional resonance formulation (**4**) and (**5**), which is used to rationalize both the large barrier to rotation in amides and their low chemical reactivity towards nucleophilic substitution at the acyl center. In thioamides it is expected that there would be greater π charge transfer from nitrogen to sulfur due to the larger size of sulfur and weaker π bond of C=S, despite the smaller electronegativity of sulfur compared with oxygen. The effect of replacing the β-lactam carbonyl oxygen in cephalosporins by sulfur has a minimal effect upon the rate of alkaline hydrolysis – the sulfur analog is only 2-fold less reactive than the natural cephalosporin. However, the thioxo-derivative of cephalexin (**14**), with an amino group in the C7 side-chain, undergoes β-lactam ring opening with intramolecular aminolysis to give (**15**) – a reaction seen with cephalexin itself [126]. However, the rate of intramolecular aminolysis for the sulfur analog is three orders of magnitude greater than that for cephalexin, although their rates of hydrolysis are similar. Furthermore, unlike cephalexin, intramolecular aminolysis in the sulfur analog occurs up to pH 14 with no competitive hydrolysis.

The rate of intermolecular aminolysis of cephalosporins is normally dominated by a second-order dependence on amine concentration, due to the second amine acting as a general base catalyst. In contrast, the aminolysis of thioxo-cephalosporins (**16**) shows only a first-order term in amine. The Brønsted $\beta_{nuc}$ for the aminolysis of thioxo-cephalosporin is $+0.39$, indicative of rate-limiting formation of the tetrahedral intermediate with an early transition state with relatively little C−N bond formation. The change in mechanism is due to a slower rate of breakdown of the tetrahedral intermediate to regenerate the reactants for thioxo-β-lactam (Scheme 5.14). It is likely that a dominant feature controlling the relative stabilities of the zwitterionic tetrahedral intermediates is the more favorable sulfur anion compared with oxygen [126].

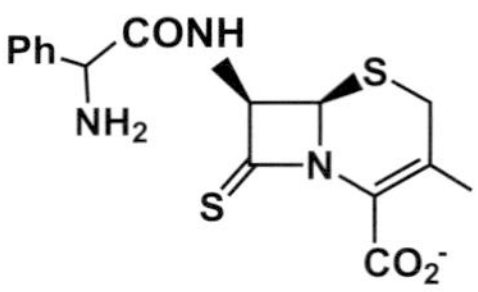

(**14**) 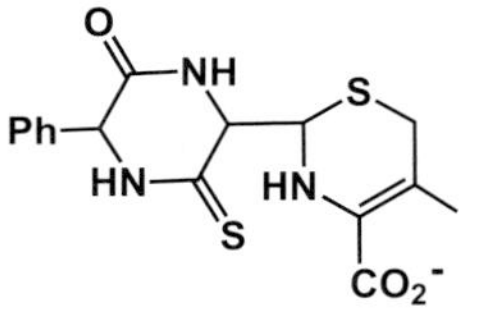 (**15**) 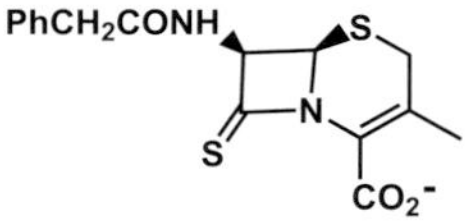 (**16**)

The alcoholysis and thiolysis of β-lactam antibiotics have also been studied [127–129]. Both alcohols and thiols catalyze the hydrolysis of benzylpenicillin through the formation of a ester intermediates. The catalytically reactive form of the nucleophile is the anion and the rate-limiting step is the breakdown of the tetrahedral intermediate, as occurs in aminolysis. Solvent kinetic isotope effects of 2.2–2.4 indicate that the solvent water probably acts as a general acid catalysis in the breakdown of the tetrahedral intermediate.

## 5.6 Epimerization

The initial product of alkaline hydrolysis of benzylpenicillin is (5*R*,6*R*)-benzylpenicilloic acid. However, epimerization then occurs at C(5) to give a mixture of the (5*R*,6*R*)- and (5*S*,6*R*)-penicilloic acids (**17**) [64, 130–133]. The equilibrium constant for the ratio of the (5*S*,6*R*)- to that of the (5*R*,6*R*)-benzylpenicilloate is 4. The rate constants for epimerization are pH- and buffer-independent from pH 6 to 12.5, but become first order in hydroxide ion at higher pH. The pH-independent epimerization in $D_2O$ occurs without deuterium incorporation at C(5) or C(6) [130, 134, 135] indicative of a mechanism involving unimolecular ring opening and closing of the thiazolidine to form the iminium ion (**18**). The base-catalyzed epimerization probably occurs by elimination across C(6)–C(5) and thiazolidine ring opening to give the enamine (**19**) as shown to occur for α-penicilloyl esters and supported by the appearance of a chromophore at 280 nm and deuterium exchange at C(6) [134, 135].

RCONH, $^-O_2C$, HN, S, $CO_2H$ (**17**)

RCONH, $^-O_2C$, $HN^+$, $^-S$, $CO_2H$ (**18**)

RCONH, $^-O_2C$, NH, $^-S$, $CO_2H$ (**19**)

Proton abstraction at C(6) in penicillins appears to give unusual carbanion intermediates. Although the treatment of a penicillin derivative with an imine at C(6) with phenyllithium in tetrahydrofuran appears to lead to abstraction of the C(6) proton, reprotonation regenerates the 6-β epimer despite being the less stable epimer [136, 137]. Even more strange is the observation that reprotonation with $D_2O/CD_3CO_2D$ does not incorporate deuterium. However, epimerization by triethylamine in acetonitrile containing $D_2O$ is accompanied by deuteriation of the presumed carbanion intermediate and occurs preferentially from the least hindered α-face to give the less stable β-epimer. Deuterium exchange at C(6) of the 6-β epimer occurs faster than epimerization at C(6) [138]. The calculated isotope effect $k_H/k_D$ for protonation of the carbanion is 12.7, which is ascribed to proton tunneling. However, 6-α-chloropenicillanic acid undergoes deuterium exchange at C(6) in $NaOD/D_2O$ without epimerization and at a rate faster than β-lactam ring opening, whereas the 6-β-epimer undergoes epimerization and deuteriation at C(6) at the same rate [139].

## References


1 Strominger, J.L. (1967) *Antibiotics*, **1**, 705–713.
2 Woodward, R.B. (1949) In *The Chemistry of Penicillin* (eds H.T. Clarke, J.R., Johnson, and R. Robinson), Princeton University Press, Princeton, NJ, p. 443.
3 Jeffrey, G.A., Ruble, J.R., McMullan, R.K., DeFrees, D.J., Binkley, J.S., and

Pople, J.A. (1980) *Acta Crystallographica B*, **36**, 2242–2299.
4 Woodward, R.B. (1980) *Philosophical Transactions of the Royal Society, London B*, **289**, 239–250.
5 Page, M.I. (1987) *Advances in Physical Organic Chemistry*, **23**, 165–270.
6 Sweet, R.M. (1973) In *Cephalosporins and Penicillins: Chemistry and Biology* (ed E.H. Flynn), Academic Press, New York, p. 280.
7 Simon, G.L., Morin, R.B., and Dahl, L.F. (1972) *Journal of the American Chemical Society*, **94**, 8557–8563.
8 Abrahamsson, S., Hodkgin, D.C., and Maslen, E.N. (1963) *Biochemical Journal*, **86**, 514–535.
9 Sweet, R.M. and Dahl., L.F. (1970) *Journal of the American Chemical Society*, **92**, 5489–5507.
10 Vijayan, K., Anderson, B.F., and Hodgkin, D.C. (1973) *Journal of the Chemical Society, Perkin Transactions 1*, 484–488.
11 Levy, G.C. and Nelson, G.L. (1972) *Carbon-13 Nuclear Magnetic Resonance for Organic Chemists*, Wiley Interscience, New York.
12 Bose, A.K. and Srinivasan, P.R. (1979) *Organic Magnetic Resonance*, **12**, 34–38.
13 Malinowski, E.R., Manhas, M.S., Goldberg, M., and Fanelli, V. (1974) *Journal of Molecular Structure*, **23**, 321.
14 Manhas, M.S., Jeng, S.J., and Bose, A.K. (1968) *Tetrahedron*, **24**, 1237–1245.
15 Schanck, A., Coene, B., Van Meerssche, M., and Dereppe, J.M. (1979) *Organic Magnetic Resonance*, **12**, 337–338.
16 Mondelli, R. and Ventura, P. (1977) *Journal of the Chemical Society, Perkin Transactions 2*, 1749–1752.
17 Dhami, K.S. and Stothers, J.B. (1964) *Tetrahedron Letters*, 631–639.
18 Flynn, E.H. (1972) *Cephalosporins and Penicillins: Chemistry and Biology*, Academic Press, New York.
19 Paschal, J.W., Dorman, D.E., Srinivasan, P.R., and Lichter, R.L. (1978) *Journal of Organic Chemistry*, **43**, 2013–2016.
20 Lichter, R.L. and Dorman, D.E. (1976) *Journal of Organic Chemistry*, **41**, 582–583.
21 Morin, R.B., Jackson, B.G., Mueller, R.A., Lavagnino, E.R., Scanlon, W.B., and Andrews, S.L. (1969) *Journal of the American Chemical Society*, **91**, 1401–1407.
22 Indelicato, J.M., Norvilas, T.T., Pfeiffer, R.R., Wheeler, W.J., and Wilham, W.L. (1974) *Journal of Medicinal Chemistry*, **17**, 523–527.
23 Nishikawa, J., Tori, K., Takasuka, M., Onoue, H., and Narisada, M. (1982) *Journal of Antibiotics*, **35**, 1724–1728.
24 Takasuka, M., Nishikawa, J., and Tori, K. (1982) *Journal of Antibiotics*, **35**, 1729–1733.
25 Pfaendler, H.R., Gosteli, J., Woodward, R.B., and Rihs, G. (1981) *Journal of the American Chemical Society*, **103**, 4526–4531.
26 Collings, A.J., Jackson, P.F., and Morgan, K.J. (1970) *Journal of the Chemical Society B*, 581–584.
27 Glidewell, C. and Mollison, G.S.M. (1981) *Journal of Molecular Structure*, **72**, 203–208.
28 Vishveshwara, S. and Rao, V.S.R. (1983) *Journal of Molecular Structure*, **92**, 19–29.
29 Stackhouse, J., Baechler, R.D., and Mislow, K. (1971) *Tetrahedron Letters*, 3437–3440.
30 Pracejus, H., Kehlen, M., Kehlen, H., and Matschiner, H. (1965) *Tetrahedron*, **21**, 2257–2270.
31 Proctor, P., Gensmantel, N.P., and Page, M.I. (1982) *Journal of the Chemical Society, Perkin Transactions 2*, 1185–1192.
32 Gensmantel, N.P., Gowling, E.W., and Page, M.I. (1978) *Journal of the Chemical Society, Perkin Transactions 2*, 335–342.
33 Gensmantel, N.P., Proctor, P., and Page, M.I. (1980) *Journal of the Chemical Society, Perkin Transactions 2*, 1725–1732.
34 Morris, J.J. and Page, M.I. (1980) *Journal of the Chemical Society, Perkin Transactions 2*, 679–684. and 685-L 692.
35 Blackburn, G.M. and Plackett, J.D. (1972) *Journal of the Chemical Society, Perkin Transactions 2*, 1366–1370.
36 Page, M.I. (1973) *Chemical Society Reviews*, 295–323.
37 Brown, H.C. and Ichikawa, K. (1957) *Tetrahedron*, **1**, 221–230.
38 Allinger, N.L., Tribble, M.T., and Miller, M.A. (1972) *Tetrahedron*, **28**, 1173–1190.

**39** Bruylants, A. and Kezdy, F. (1960) *Record of Chemical Progress*, **21**, 213–240.

**40** Yamana, T., Tsuji, A., Kiya, E., and Miyamoto, E. (1977) *Journal of Pharmaceutical Science*, **66**, 861–866.

**41** Hamilton-Miller, J.M.T., Richards, E., and Abraham, E.P. (1970) *Biochemical Journal*, **116**, 385–395.

**42** O'Callaghan, C.H.O., Kirby, S.M., Morris, A., Waller, R.E., and Duncombe, R.E. (1972) *Journal of Bacteriology*, **110**, 988–991.

**43** Bundgaard, H. (1975) *Archiv fur Pharmacy og Chemi, Scientific Edition*, **3**, 94–123; *Chemical Abstracts*, **84**, 89153.

**44** Boyd, D.B., Hermann, R.B., Presti, D.E., and Marsh, M.M. (1975) *Journal of Medicinal Chemistry*, **18**, 408–417.

**45** Boyd, D.B. and Lunn, W.H.W. (1979) *Journal of Medicinal Chemistry*, **22**, 778–784.

**46** Wei, C.-C., Borgese, J., and Weigele, M. (1983) *Tetrahedron Letters*, **24**, 1875–1878.

**47** Boyd, D.B., Herron, D.K., Lunn, W.H.W., and Spitzer, W.A. (1980) *Journal of the American Chemical Society*, **102**, 1812–1814.

**48** Yamana, T. and Tsuji, A. (1976) *Journal of Pharmaceutical Science*, **65**, 1563–1574.

**49** Agathocleous, D., Buckwell, S., Proctor, P., and Page, M.I. (1985) In *Recent Advances in the Chemistry of β-lactam Antibiotics* (eds A.G. Brown and S.M. Roberts), Royal Society of Chemistry, London, p. 18.

**50** Stirling, C.J.M. (1979) *Accounts of Chemical Research*, **12**, 198–203.

**51** Buckwell, S., Page, M.I., and Longridge, J.L. (1986) *Journal of the Chemical Society, Chemical Communications*, 1039–1040.

**52** Buckwell, S.C., Page, M.I., Longridge, J.L., and Waley, S.G. (1988) *Journal of the Chemical Society, Perkin Transactions 2*, 1823–1828.

**53** Narisada, M., Yoshida, T., Ohtani, M., Ezumi, K., and Takasuka, M. (1983) *Journal of Medicinal Chemistry*, **26**, 1577–1582.

**54** Murakami, K. and Yoshida, T. (1982) *Antimicrobial Agents and Chemotherapy*, **21**, 254–258.

**55** Indelicato, J.M. and Wilham, W.L. (1974) *Journal of Medicinal Chemistry*, **17**, 528–529.

**56** Bucourt, R., Heymès, R., Lutz, A., Pénasse, L., and Perronnet, J. (1978) *Tetrahedron*, **34**, 2233–2243.

**57** Schrinner, E., Limbert, M., Pénasse, L., and Lutz, A. (1980) *Journal of Antimicrobial Chemotherapy*, **6** (Suppl. A), 25–30.

**58** Laurent, G., Durant, F., Frère, J.M., Klein, D., and Ghuysen, J.M. (1984) *Biochemical Journal*, **218**, 933–937.

**59** Schwartz, M.A. (1965) *Journal of Pharmaceutical Science*, **54**, 472–473.

**60** Dennen, D.W. and Davis, W.W. (1962) *Antimicrobial Agents and Chemotherapy*, 531–536.

**61** Longridge, J.L. and Timms, D. (1971) *Journal of the Chemical Society B*, 852–857.

**62** Degelaen, J.P., Loukas, S.L., Feeney, J., Roberts, G.C.K., and Burgen, A.S.V. (1979) *Journal of the Chemical Society, Perkin Transactions 2*, 86–90.

**63** Bundgaard, H. (1980) *Archiv fur Pharmacy og Chemi, Scientific Edition*, **8**, 161–180.

**64** Kessler, D.P., Cushman, M., Ghebre-Sellassie, I., Knevel, A.M., and Hem, S.L. (1983) *Journal of the Chemical Society, Perkin Transactions 2*, 1699–1704.

**65** Wan, P., Modro, T.A., and Yates, K. (1980) *Canadian Journal of Chemistry*, **58**, 2423–2432.

**66** Liler, M. (1969) *Journal of the Chemical Society B*, 385–389.

**67** McLelland, R.A. and Reynolds, W.F. (1976) *Canadian Journal of Chemistry*, **54**, 718–725.

**68** Bouchoux, G. and Houriet, R. (1984) *Tetrahedron Letters*, **25**, 5755–5758.

**69** Giffney, C.J. and O'Connor, C.J. (1975) *Journal of the Chemical Society, Perkin Transactions 2*, 1357–1360.

**70** Fujita, T. and Koshiro, A. (1984) *Chemical and Pharmaceutical Bulletin*, **32**, 3651–3661.

**71** Yamana, T., Tsuji, A., and Mizukami, Y. (1974) *Chemical and Pharmaceutical Bulletin*, **22**, 1186–1197.

**72** Bundgaard, H. and Hansen, J. (1981) *International Journal of Pharmaceutics*, **9**, 273–283.

**73** Cressman, W.A., Sugita, E.T., Doluisio, J.T., and Niebergall, P.J. (1969) *Journal of Pharmaceutical Science*, **58**, 1471–1476.

**74** Fazakerley, G.V. and Jackson, G.E. (1977) *Journal of Pharmaceutical Science*, **66**, 533–535.

**75** Asso, M., Panossian, R., and Guiliano, M. (1984) *Spectroscopy Letters*, **17**, 271–283.

**76** Schwartz, M.A. (1982) *Bioorganic Chemistry*, **11**, 4–18.

**77** Tomida, H. and Schwartz, M.A. (1983) *Journal of Pharmaceutical Science*, **72**, 331–335.

**78** Gensmantel, N.P. and Page, M.I. (1982) *International Journal of Pharmaceutics*, 147–153.

**79** Tsuji, A., Miyamoto, E., Matsuda, M., Nishimura, K., and Yamana, T. (1982) *Journal of Pharmaceutical Science*, **71**, 1313–1318.

**80** O'Connor, C.J. and Tan, A. (1980) *Australian Journal of Chemistry*, **33**, 747–755.

**81** Gensmantel, N.P. and Page, M.I. (1982) *Journal of the Chemical Society, Perkin Transactions 2*, 155–159.

**82** Chaimovich, H., Correia, V.R., Araujo, P.S., Aleixo, R.M.V., and Cuccovia, I.M. (1985) *Journal of the Chemical Society, Perkin Transactions 2*, 925–928.

**83** Yasuhara, M., Sato, F., Kimura, T., Muranishi, S., and Sezaki, H. (1977) *Journal of Pharmacy and Pharmacology*, **29**, 638–640.

**84** Komiyama, M. and Bender, M.L. (1984) In *The Chemistry of Enzyme Action* (ed. M.I. Page), Elsevier, Amsterdam. p. 505

**85** Tutt, D.E. and Schwartz, M.A. (1971) *Journal of the American Chemical Society*, **93**, 767–772.

**86** Waley, S.G. (1992) In *The Chemistry of β-Lactams* (ed. M.I. Page), Blackie, Glasgow, pp. 198–226;Page, M.I. and Laws, A.P. (1998) *Journal of the Chemical Society, Chemical Communications*, 1609–1617.

**87** Kelly, J.A., Dideberg, O., Charlier, P., Wery, J.P., Libert, M., Moews, P.C., Knox, J.R., Duez, C., Fraipont, C., Joris, B., Dusart, J., Frère, J.-M., and Ghuysen, J.M. (1986) *Science*, **231**, 1429–1431; Samraoui, B., Sutton, B., Todd, R.J., Artymiuk, P.J., Waley, S.G., and Phillips, D.C. (1986) *Nature*, **320**, 378–380; Knox, J.R. and Moews, P.C. (1991) *Journal of Molecular Biology*, **220**, 435–455; Moews, P.C., Knox, J.R., Dideberg, O., Charlier, P., and Frère, J.-M. (1990) *Proteins: Structure, Function and Genetics*, **7**, 156–171; Herzberg, O. and Moult, J. (1987) *Science*, **236**, 694–701; Herzberg, O. (1991) *Journal of Molecular Biology*, **217**, 701–719; Dideberg, O., Charlier, P., Wéry, J.-P., Dehottay, P., Dusart, J., Erpicum, T., Frère, J.-M., and Ghuysen, J.-M. (1987) *Biochemical Journal*, **245**, 911–913; Lamotte-Brasseur, J., Dive, G., Dideberg, O., Charlier, P., Frère, J.-M., and Ghuysen, J.-M. (1991) *Biochemical Journal*, **279**, 213–221; Jelsch, C., Mourey, L., Masson, J.-M., and Samama, J.-P. (1993) *Proteins: Structure, Function and Genetics*, **16**, 364–383; Oefner, C., D'Arcy, A., Daly, J.J., Gubernator, K., Charnas, R.L., Heinze, I., Hubschwerlen, C., and Winkler, F.K. (1990) *Nature*, **343**, 284–288; Lobkovsky, E., Moews, P.C., Liu, H., Zhao, H., Frère, J.-M., and Knox, J.R. (1993) *Proceedings of the National Academy of Sciences of the United States of America*, **90** 11257–11261.

**88** Strynadka, N.C.J., Adachi, H., Jensen, S.E., Johns, K., Sielecki, A., Betzel, C., Sutoh, K., and James, M.N.G. (1992) *Nature*, **359**, 700–705.

**89** Matagne, A., and Frère, J.-M. (1995) *Biochimica et Biophysica Acta: Protein Structure*, **1246**, 109–127.

**90** Christensen, H., Martin, M.T., and Waley, S.G. (1990) *Biochemical Journal*, **266**, 853–861.

**91** Damblon, C., Raquet, X., Lian, L.-Y., Lamotte-Brasseur, J., Fonzé, E., Charlier, P., Roberts, G.C., and Frère, J.-M. (1996) *Proceedings of the National Academy of Sciences of the United States of America*, **93**, 1747–1752.

**92** Galleni, M., Amicosante, G., and Frère, J.-M. (1988) *Biochemical Journal*, **255**, 1233.

**93** Bulychev, A., Massova, I., Miyashita, K., and Mobashery, S. (1997) *Journal of the American Chemical Society*, **119**, 7619–7625.

**94** Damblon, C., Zao, G.H., Jamin, M., Ledent, P., Dubus, A., Vanhove, M., Raquet, X., Christiaens, L., and Frère, J.-M. (1995) *Biochemical Journal*, **309**, 431–436; Xu, Y., Soto, G., Hirsch, K.R., and Pratt, R.F. (1996) *Biochemistry*, **35**, 3595–3603; Murphy, B.P. and Pratt, R.F., (1991) *Biochemistry*, **30**, 3640–3649; Page, M.I., Vilanova, B., and Layland, N.J. (1995) *Journal of the American Chemical Society*, **117**, 12092–12095.

95 Dubus, A., Normark, S., Kania, M., and Page, M.G.P. (1994) *Biochemistry*, **33**, 8577–8586.

96 Frere, J.-M. (1995) *Molecular Microbiology*, **16**, 385–395.

97 Payne, D.J. (1993) *Journal of Medical Microbiology*, **39**, 93–99.

98 Galleni, M., Lamotte-Brasseur, J., Rossolini, G.M., Spencer, J., Dideberg, O., and Frère, J.-M. (2001) *Antimicrobial Agents and Chemotherapy*, **45**, 660–663.

99 Crowder, M.W., Wang, Z., Franklin, S.L., Zovinka, E.P., and Benkovic, S.J. (1996) *Biochemistry*, **35**, 12126–12132.

100 Fitzgerald, P.M.D., Wu, J.K., and Toney, J.H. (1998) *Biochemistry*, **37**, 6791–6800.

101 Concha, N.O., Rasmussen, B.A., Bush, K., and Herzberg, O. (1997) *Protein Science*, **6**, 2671–2676.

102 Carfi, A., Duee, E., Galleni, M., Frère, J.-M., and Dideberg, O. (1998) *Acta Crystallographica D*, **D54**, 313–323.

103 Fabiane, S.M., Sohi, M.K., Wan, T., Payne, D.J., Bateson, J.H., Mitchell, T., and Sutton, B.J. (1998) *Biochemistry*, **37**, 12404–12411.

104 Paul-Soto, R., Hernandez-Valladares, M., Galleni, M., Bauer, R., Zeppezauer, M., Frère, J.-M., and Adolph, H.-W. (1998) *FEBS Letters*, **438**, 137–140.

105 Fast, W., Wang, Z., and Benkovic, S.J. (2001) *Biochemistry*, **40**, 1640–1650.

106 Weston, J. (2005) *Chemical Reviews*, **105**, 2151–2174.

107 Parkin, G. (2004) *Chemical Reviews*, **104**, 699–767.

108 Gensmantel, N.P., Proctor, P., and Page, M.I. (1980) *Journal of the Chemical Society, Perkin Transactions 2*, 1725–1732.

109 Bounaga, S., Laws, A.P., Galleni, M., and Page, M.I. (1998) *Biochemical Journal*, **331**, 703–711.

110 Hemmingsen, L., Damblon, C., Antony, J., Jensen, M., Adolph, H.W., Wommer, S., Roberts, G.C.K., and Bauer, R. (2001) *Journal of the American Chemical Society*, **123**, 10329–10335.

111 Badarau, A. and Page, M.I. (2006) *Biochemistry*, **45**, 10654–10666.

112 Badarau, A., Damblon, C., and Page, M.I. (2007) *Biochemical Journal*, **401**, 197–203.

113 Badarau, A. and Page, M.I. (2008) *Journal of Biological Inorganic Chemistry*, **13**, 919–928; Badarau, A. and Page, M.I. (2006) *Biochemistry*, **45**, 11012.

114 Levine, B.B. and Ovary, Z. (1961) *Journal of Experimental Medicine*, **114**, 875–904.

115 DeWeck, A.L. and Blum, G. (1965) *International Archives of Allergy and Immunology*, **27**, 221–256.

116 Parker, C.W., Shapiro, J., Kern, M., and Eisen, H.N. (1962) *Journal of Experimental Medicine*, **115**, 821–838.

117 Blackburn, G.M. and Plackett, J.D. (1973) *Journal of the Chemical Society, Perkin Transactions 2*, 981–985.

118 Morris, J.J. and Page, M.I. (1980) *Journal of the Chemical Society, Perkin Transactions 2*, 212–219.

119 Gensmantel, N.P. and Page, M.I. (1979) *Journal of the Chemical Society, Perkin Transactions 2*, 137–142.

120 Proctor, P. and Page, M.I. (1984) *Journal of the American Chemical Society*, **106**, 3820–3825.

121 Martin, A.F., Moriss, J.J., and Page, M.I. (1976) *Journal of the Chemical Society, Chemical Communications*, 495–496.

122 Martin, A.F., Morris, J.J., and Page, M.I. (1979) *Journal of the Chemical Society, Chemical Communications*, 298–299.

123 Deslongchamps, P. (1975) *Tetrahedron*, **31**, 2463–2490.

124 Llinas, A. and Page, M.I. (2004) *Organic & Biomolecular Chemistry*, **2**, 651–654.

125 Tsang, W.Y., Ahmed, N., and Page, M.I. (2007) *Organic & Biomolecular Chemistry*, **5**, 485–493.

126 Tsang, W., Dhanda, A., Schofield, C.J., and Page, M.I. (2004) *The Journal of Organic Chemistry*, **69**, 339–344.

127 Davis, A.M., Proctor, P., and Page, M.I. (1991) *Journal of the Chemical Society, Perkin Transactions 2*, 1213–1217.

128 Llinas, A., Donoso, J., Vilanova, B., Frau, J., Munoz, F., and Page, M.I. (2000) *Journal of the Chemical Society, Perkin Transactions 2*, 1521–1525.

**129** Llinas, A., Vilanova, B., and Page, M.I. (2004) *Journal of Physical Organic Chemistry*, **17**, 521–528.

**130** Davis, A.M. and Page, M.I. (1985) *Journal of the Chemical Society, Chemical Communications*, 1702–1704.

**131** Carroll, R.D., Jung, S., and Sklavounos, C.G. (1977) *Journal of Heterocyclic Chemistry*, **14**, 503–505.

**132** Busson, R., Vanderhaeghe, H., and Toppet, S. (1976) *Journal of Organic Chemistry*, **41**, 3054–3056.

**133** Bird, A.E., Cutmore, E.A., Jennings, K.R., and Marshall, A.C. (1983) *Journal of Pharmacy and Pharmacology*, **35**, 138–143.

**134** Davis, A.M., Jones, M., and Page, M.I. (1991) *Journal of the Chemical Society, Perkin Transactions 2*, 1219–1223.

**135** Davis, A.M., Layland, N.J., Page, M.I., Martin, F., and More, R.M. (1991) *Journal of the Chemical Society, Perkin Transactions 2*, 1225–1229.

**136** Firestone, R.A., Schelechow, N., Johnston, D.B.R., and Christensen, B.G. (1972) *Tetrahedron Letters*, 375–378.

**137** Firestone, R.A., Maciejewicz, N.S., Ratcliffe, R.W., and Christensen, B.G. (1974) *Journal of Organic Chemistry*, **39**, 437–440.

**138** Firestone, R.A. and Christensen, B.G. (1977) *Journal of the Chemical Society, Chemical Communications*, 288–289.

**139** Clayton, J.P., Nayler, J.H.C., Southgate, R., and Stove, E.R. (1969) *Journal of the Chemical Society, Chemical Communications*, 129–130.

# Part Two
# Amino Acid Coupling Chemistry

*Amino Acids, Peptides and Proteins in Organic Chemistry.*
*Vol.3 – Building Blocks, Catalysis and Coupling Chemistry.* Edited by Andrew B. Hughes

ISBN: 978-3-527-32102-5

# 6
# Solution-Phase Peptide Synthesis

*Yuko Tsuda and Yoshio Okada*

## 6.1
## Principle of Peptide Synthesis

Early in the twentieth century, Curtius [1] and Fischer [2] independently began the chemical synthesis of simple peptides, with the dream to eventually synthesize biologically active peptides and proteins. Since then, the chemistry of peptide synthesis has been a challenge for over a century. The chemistry has been developed using the following chemical approaches: (i) selection of protecting groups for amino acids and their deprotection, and (ii) peptide bond formation. In recent peptide synthesis methodologies, two procedures are primarily applied – one is based on synthesis in solution and the other involves synthesis on a solid support (solid phase), although they are based on fundamentally the same principles as shown in Figure 6.1.

In this chapter, we focus on the chemistry for protecting functional groups and formatting the amide bond between the amino group of an amino acid and the carboxyl group of a second amino acid.

Furthermore, it is required to prepare optically pure products; however, racemization can occur during protection of amino acids and also during the procedures for the removal of protecting groups as well as at the activation step. The following two racemization mechanisms were considered:

i) **Direct proton abstraction** [3]. This mechanism is the pathway only in very special cases such as the rapid racemization of derivatives of phenylglycine (Scheme 6.1). As shown in Scheme 6.1, the racemization occurs by the direct abstraction of the α-proton by strong bases. This can be prevented by application of suitable reaction conditions, especially by use of a tertiary amine [4, 5].
ii) **Racemization through oxazol-5(4*H*)-ones** [6, 7]. The formation of oxazol-5(4*H*)-ones [6] was proposed to explain the mechanism of racemization. In this mechanism, it was shown that the properties of the amino protecting group might be associated with the tendency to racemization. The activation of the carboxyl moiety of *N*-acyl amino acids (acyl = acetyl, benzoyl, peptidyl, etc.) readily results in the formation of oxazol-5(4*H*)-ones, which give rise to

*Amino Acids, Peptides and Proteins in Organic Chemistry.*
*Vol.3 – Building Blocks, Catalysis and Coupling Chemistry.* Edited by Andrew B. Hughes

ISBN: 978-3-527-32102-5

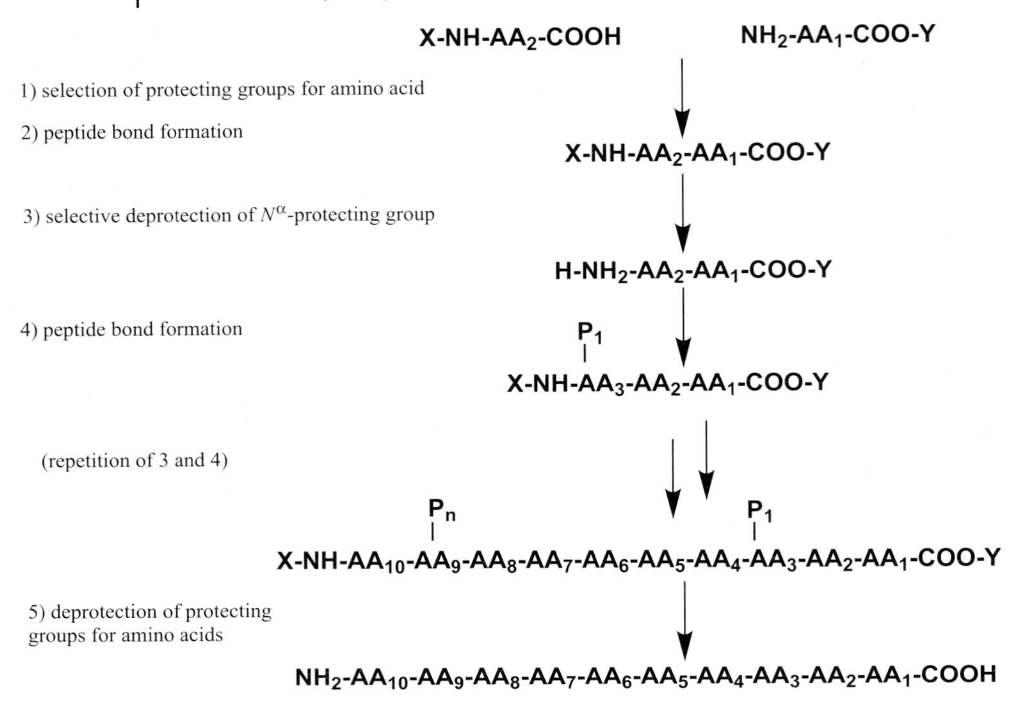


**Figure 6.1** Principle of peptide synthesis.

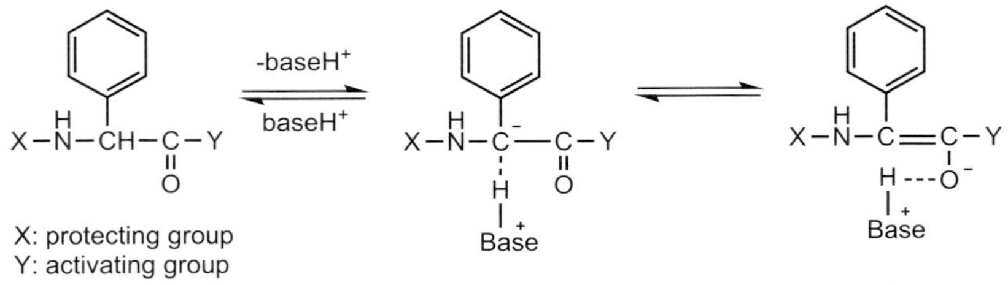

**Scheme 6.1** Proton abstraction mechanism [3].

chirally unstable intermediates via tautomerization of the oxazol-5(4*H*)-ones (Scheme 6.2). In the case of urethane-protected amino acids, such as benzyloxycarbonyl (Z or Cbz)- or *tert*-butyloxycarbonyl (Boc)-amino acids, it has been the accepted view that oxazol-5(4*H*)-ones are not produced from such amino acid derivatives, resulting in resistance to racemization during activation and coupling. Nonetheless, optically pure oxazol-5(4*H*)-ones were

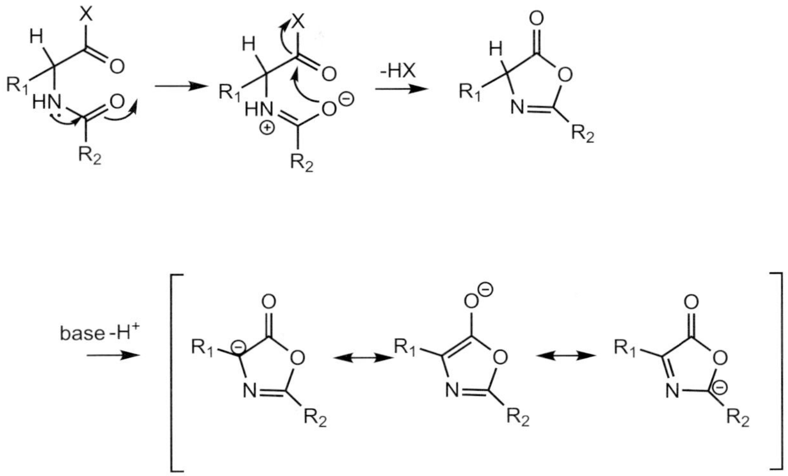

Scheme 6.2 Chirally unstable oxazol-5(4*H*)-one [3].

isolated when *Z*- or Boc-amino acids were activated by carbodiimide, mixed anhydride, or thionylchloride [8, 9].

## 6.2 Protection Procedures

As shown in Figure 6.1, the controlled formation of a peptide bond in solution requires the coupling between one protected at its N-terminus and the other at its C-terminus. After amide bond formation, the protected dipeptide is isolated, purified, and characterized. After deprotection of the $N^{\alpha}$-protecting group, the resulting dipeptide amine is then coupled with another amino acid protected at its $N^{\alpha}$-amino function. Thus, in order to obtain the desired peptide, $N^{\alpha}$-amino or carboxyl function and side-chain functional groups should be protected with an appropriate protecting group.

### 6.2.1 Amino Group Protection

#### 6.2.1.1 Z Group

The Z group **1** was introduced by Bergmann and Zervas [10] for the protection of amino groups. For the preparation of Z-amino acids, benzyloxycarbonyl chloride is allowed to react with amino acids under conditions of the Schotten–Bauman reaction [10] as shown in Scheme 6.3.

This protecting group can be removed by a variety of methods, including hydrogenolysis, reduction with sodium in liquid ammonia, liquid HF, and so on. However,

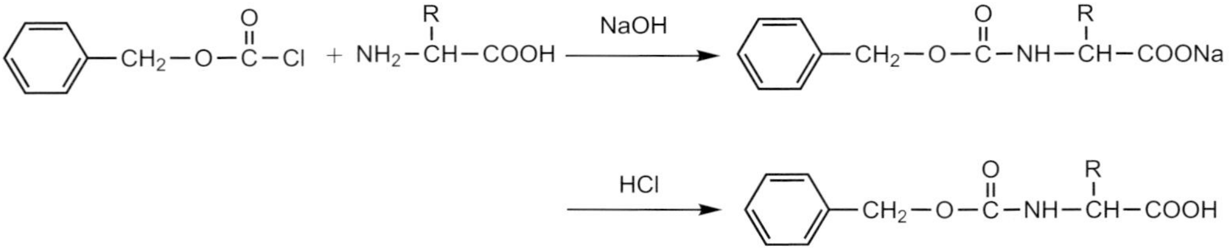

$$C_6H_5\text{-}CH_2\text{-}O\text{-}C(=O)\text{-}Cl + NH_2\text{-}CH(R)\text{-}COOH \xrightarrow{NaOH} C_6H_5\text{-}CH_2\text{-}O\text{-}C(=O)\text{-}NH\text{-}CH(R)\text{-}COONa$$

$$\xrightarrow{HCl} C_6H_5\text{-}CH_2\text{-}O\text{-}C(=O)\text{-}NH\text{-}CH(R)\text{-}COOH$$

**Scheme 6.3** Synthetic route to the Z-amino acids.

hydrogenolysis [10] and acidolysis with HBr [11] are the most widely used procedures. The Z group can be removed by catalytic hydrogenolysis at room temperature and atmospheric pressure – a process that leaves the peptide bond and the various side-chain functions unaffected, and generates only relatively harmless byproducts, toluene and carbon dioxide, as shown in Scheme 6.4.

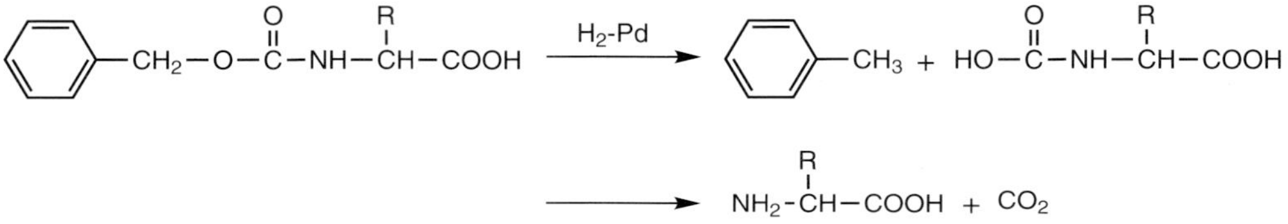

$$C_6H_5\text{-}CH_2\text{-}O\text{-}C(=O)\text{-}NH\text{-}CH(R)\text{-}COOH \xrightarrow{H_2\text{-}Pd} C_6H_5\text{-}CH_3 + HO\text{-}C(=O)\text{-}NH\text{-}CH(R)\text{-}COOH$$

$$\longrightarrow NH_2\text{-}CH(R)\text{-}COOH + CO_2$$

**Scheme 6.4** Removal of the Z group by catalytic hydrogenolysis.

The Z group can be also removed by acidolysis with HBr. As the initial step, the carbonyl oxygen is protonated, followed by fission of the benzyl-oxygen bond. Decarboxylation of the carbamic acid intermediate leads to the liberation of the amino group in a protonated form (Scheme 6.5).

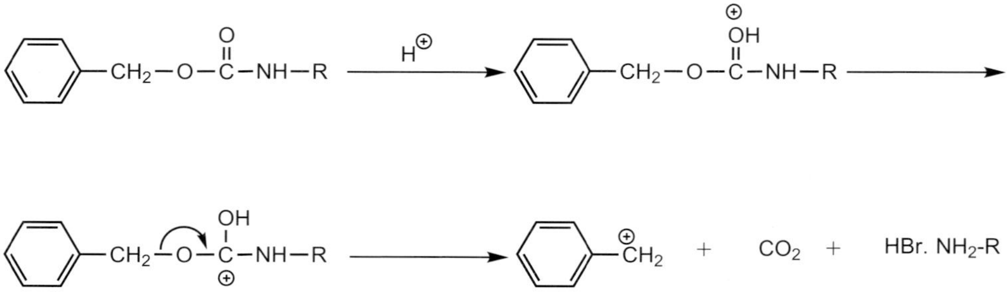

$$C_6H_5\text{-}CH_2\text{-}O\text{-}C(=O)\text{-}NH\text{-}R \xrightarrow{H^{\oplus}} C_6H_5\text{-}CH_2\text{-}O\text{-}C(=\overset{\oplus}{O}H)\text{-}NH\text{-}R \longrightarrow$$

$$C_6H_5\text{-}CH_2\text{-}O\text{-}\overset{\oplus}{C}(OH)\text{-}NH\text{-}R \longrightarrow C_6H_5\text{-}\overset{\oplus}{C}H_2 + CO_2 + HBr.\ NH_2\text{-}R$$

**Scheme 6.5** Removal of the Z group by HBr.

Catalytic hydrogenolysis of the Z groups of sulfur-containing amino acids such as Cys and Met fails because of catalyst poisoning. Several attempts were made in order to overcome this restriction; for example, an additive, such as $BaSO_4$ [12] or boron trifluoride etherate [13] was reported to improve hydrogenolysis only in the case of Met-containing peptides, but not Cys-containing peptides. Likewise, catalytic transfer hydrogenation with a proton donor, such as 1,4-cyclohexadiene [14], ammonium formate [15], or cyclohexene [16], worked well with Met-containing peptides. Kuromizu and Meienhofer [17] reported that the Z group from Cys-containing peptides could be removed by catalytic hydrogenolysis in liquid ammonia. This procedure was applied to the synthesis of somatostatin [18], but further applications are not available at the present time.

#### 6.2.1.2 **Substituted Z and other Urethane-Type Protecting Groups**

The isonicotinyloxycarbonyl (*i*Noc) group **2** [19] can be also removed by hydrogenolysis over a palladium catalyst due to the presence of the benzylic C–O bond, while **2** is more stable than **1** under acidic conditions (Figure 6.2). The incorporation of an electron-donating group on the aromatic ring, such as the 4-methoxybenzyloxycarbonyl Z(OMe) group **3**, increased the acid lability of the Z group. It can be removed by trifluoroacetic acid (TFA) treatment [20–23] as well as by hydrogenolysis. Even more acid labile are the α,α-dimethyl-3,5-dimethoxybenzyloxycarbonyl (Ddz) **4** [24] and 2-(4-biphenyl)isopropyloxycarbonyl (Bpoc) **5** [25] groups. These are removed under much milder conditions than those necessary for the Z group, or even for the Boc group; a solution of chloroacetic acid in dichloromethane is normally sufficient.

#### 6.2.1.3 **Boc Group**

The Boc group **6** was introduced by McKay and Albertson [26]. Unlike benzyl chloroformate, *tert*-butylchloroformate is unstable except at low temperature. Thus, various alternative methods for the preparation of Boc-amino acids were investigated. Of these, di-*tert*-butyl dicarbonate [27, 28] is popularly employed as a reagent for preparation of Boc-amino acids as shown in Scheme 6.6.

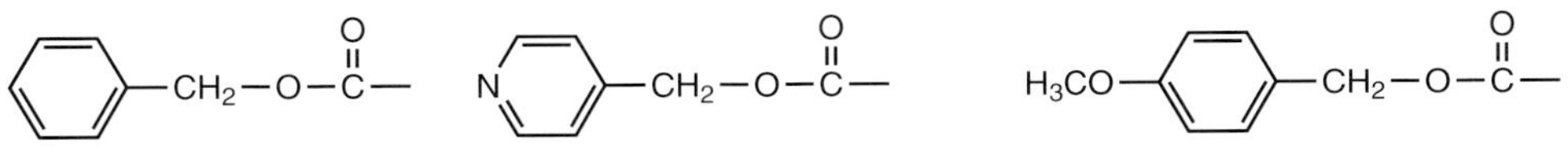

**1**: benzyloxycarbonyl(Z) **2**: isonicotinyloxycarbonyl (*i*Noc) [19] **3**: 4-methoxybenzyloxycarbonyl[Z(OMe)] [20–23]

**4**: α,α-dimethyl-3,5-dimethoxybenzyloxycarbonyl(Ddz) [24] **5**: 2-(4-biphenyl)isopropyloxycarbonyl(Bpoc) [25]

**Figure 6.2** Structures of the Z and substituted Z groups.

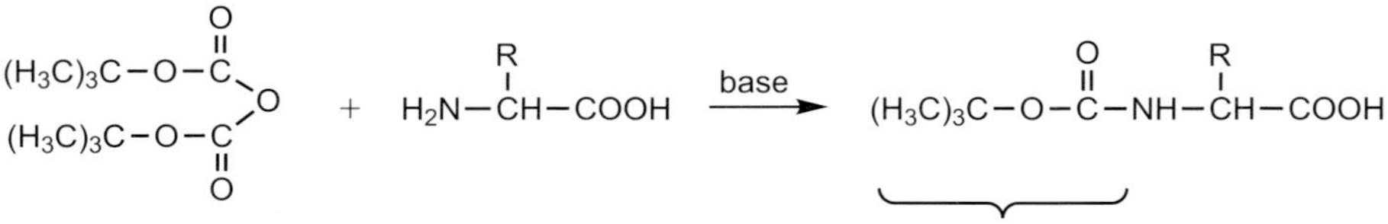

$$(H_3C)_3C{-}O{-}C({=}O){-}O{-}C({=}O){-}O{-}C(CH_3)_3 + H_2N{-}CH(R){-}COOH \xrightarrow{\text{base}} (H_3C)_3C{-}O{-}C({=}O){-}NH{-}CH(R){-}COOH$$

di-*tert*-butyl dicarbonate; **6**: *tert*-butyloxycarbonyl(Boc)

**Scheme 6.6** Synthetic route to the Boc-amino acids.

The Boc group is stable to catalytic hydrogenation, sodium in liquid ammonia and alkali, but can be removed under mild acidic conditions such as TFA. The Boc group is cleaved by treatment with TFA at room temperature for 30–60 min [29]. It is also cleaved by HBr within 1 min. Treatment of Boc derivatives with HCl (about 10 equiv.) in acetic acid, AcOEt, or dioxane can remove the Boc group within 30 min [26, 30]. Under these conditions, the Z group remains intact or cleavage is negligible. Since the Z group is removed by catalytic hydrogenolysis and is stable to mild acidic conditions, those Z and Boc groups are used as orthogonal protecting groups in the synthesis of complex peptides. Additionally, Carpino [29] and Sakakibara *et al.* [31] reported that the Boc group can be removed by 48% HF and anhydrous HF at 0 °C within a few minutes.

The cleavage by acidic treatment proceeds through the formation of the *tert*-butyl (*t*Bu) cation, which undergoes loss of $H^+$ to form isobutylene. The other products are carbon dioxide and a salt of the free amino acid or peptide (Scheme 6.7).

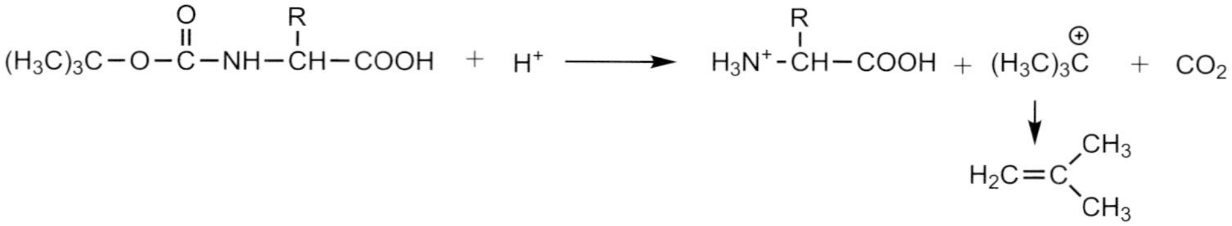

$$(H_3C)_3C{-}O{-}C({=}O){-}NH{-}CH(R){-}COOH + H^+ \longrightarrow H_3N^+{-}CH(R){-}COOH + (H_3C)_3C^{\oplus} + CO_2$$

$$(H_3C)_3C^{\oplus} \downarrow H_2C{=}C(CH_3)_2$$

**Scheme 6.7** Cleavage of the Boc group by acidolysis.

Not only is the *t*Bu cation active in alkylating functional amino acids, such as Tyr and Trp [32, 33], but also isobutylene can react with TFA to produce *t*Bu trifluoroacetate, which also has an ability to alkylate functional amino acids. Thus, in order to suppress such alkylation, anisole is added as a cation scavenger. Boron trifluoride etherate [34], formic acid [35], and mercaptoethane sulfuric acid [36] have also been reported to be able to remove the Boc group. $(CH_3)_3SiClO_4$ [37], 1 M *p*-toluenesulfonic acid [38], and 80% formic acid [39] can also remove Boc groups quantitatively, leaving other protecting groups, such as Z, benzyl, and even *t*Bu esters unaffected.

#### 6.2.1.4 **Tri Group**

When Lys(Boc), Glu(O*t*Bu), and Asp(O*t*Bu) are employed, the peptide chain can be elongated by removing the $N^{\alpha}$-Z group by catalytic hydrogenolysis. However, the

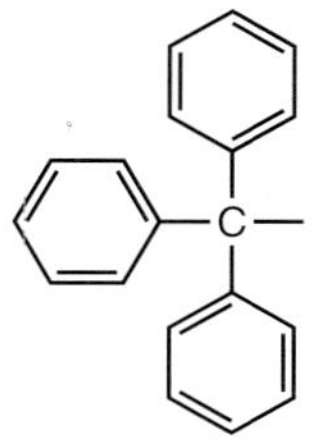

7: triphenylmethyl(Tri) [40–42]

**Figure 6.3** Structure of the Tri group.

procedure cannot be applied for the introduction of sulfur-containing amino acids, Met and Cys, due to catalyst poisoning of the sulfur atom. Thus efforts were concentrated to explore novel type α-amino protecting groups that can be removed selectively leaving the side-chain protecting groups intact, preferably by milder acids than TFA. The triphenylmethyl (trityl: Tri) group **7** [40–42] can be introduced into amino acids by the action of trityl chloride with amino acid esters in the presence of pyridine, followed by saponification of the ester group (Figure 6.3).

Since every carboxyl group of the Tri-protected amino acid is sterically hindered, peptide bond formation is achieved only by the *N,N'*-dicyclohexylcarbodiimide (DCC) method, but in low yield. Thus, it seems advantageous to use this protecting group for α-amino protection of peptides, rather than that of amino acids [43]. The Tri group can be removed by catalytic hydrogenolysis, dilute acetic acid or 1 equiv. of HCl.

Sieber *et al.* [44] synthesized human insulin by using the Tri group as an α-amino protecting group of necessary peptide fragments. The Tri group was selectively removed by HCl in trifluoroethanol (pH 3.5–4.0), in the presence of the Bpoc, Boc, and *t*Bu groups.

#### 6.2.1.5 **Fmoc Group**

In principle, protecting groups removable by strong bases are not attractive for peptide synthesis, since racemization may take place under the basic conditions. However, protecting groups removable under mild basic treatment have great values for practical peptide synthesis, since selective removal of the Bpoc or Tri group in the presence of Lys(Boc) seems to be particularly difficult. The principle for removal of protecting groups by base is based on β-elimination (Scheme 6.8). As shown in Figure 6.4, an electron-withdrawing substituent at the β-position of the ethyl group of the 2-*p*-tosylethoxycarbonyl group **8** [45] and the 2-nitroethoxycarbonyl group **9** [46] activates the adjacent C–H bond. The activated hydrogen atom is then easily subtracted with base for the β-elimination to proceed. Of various base-labile protecting groups, the 9-fluorenylmethyloxycarbonyl (Fmoc) group **10** [47] has been widely used in solid phase synthesis.

Preparation of Fmoc-amino acids by the reaction of free amino acids with Fmoc-Cl proceeds with the formation of a small amount of Fmoc-dipeptides as a side-product [48]. In order to avoid this side-reaction, several additional Fmoc reagents were introduced: Fmoc-*O*Su **11** [49, 50], Fmoc-*O*PCP **12** [49] and Fmoc-*O*Bt **13** [49]

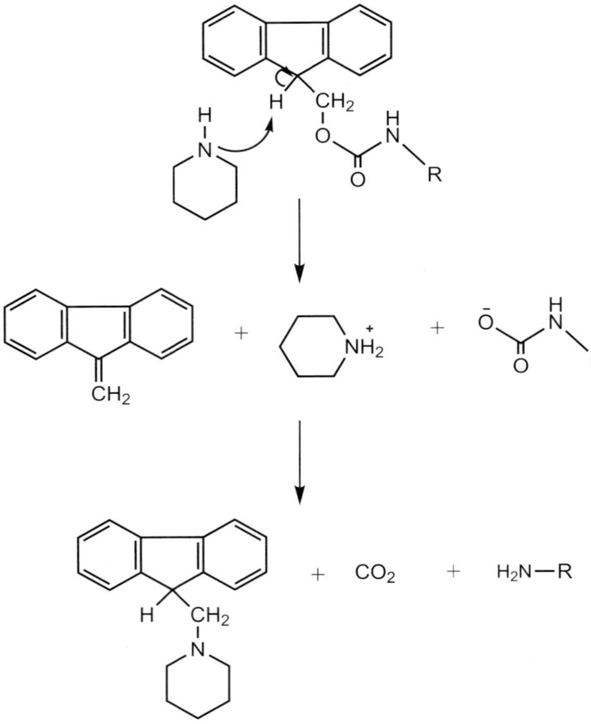

Scheme 6.8 Removal of the Fmoc group with piperidine [54].

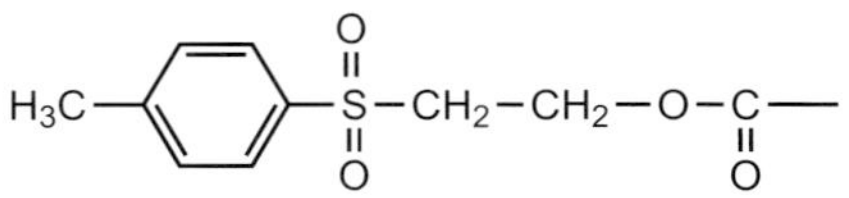

8: 2-*p*-tosylethoxycarbonyl(Tos)

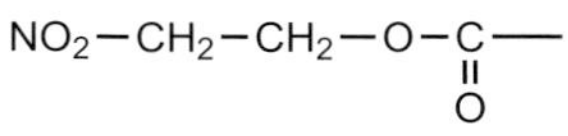

9: 2-nitroethoxycarbonyl

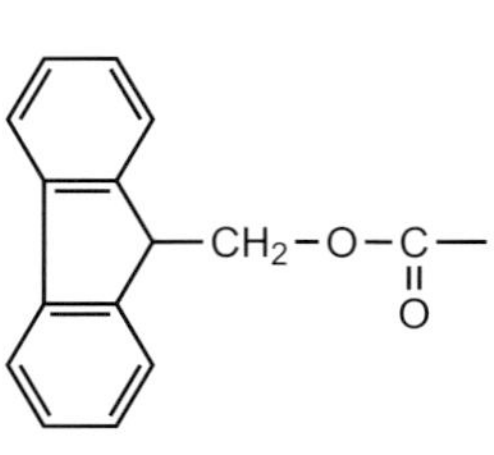

10: 9-fluorenylmethyloxycarbonyl(Fmoc)

Figure 6.4 Structures of the Fmoc and related groups.

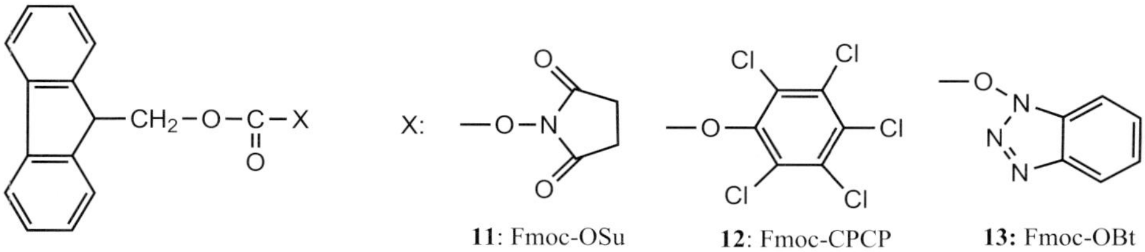

11: Fmoc-OSu 12: Fmoc-CPCP 13: Fmoc-OBt

**Figure 6.5** Structures of Fmoc reagents.

(Figure 6.5). Alternatively, amino acids are converted *in situ* to the corresponding trimethylsilyl ester, which can be acylated by Fmoc-Cl without the side-reaction described above [51].

Deblocking of the Fmoc group proceeds readily by various amines, most rapidly by unhindered cyclic secondary amines, such as piperidine [52, 53]. As shown in Scheme 6.8, Fmoc deprotection by piperidine eventually leads to the formation of a fulvene–piperidine adduct, which presents difficulty in removal from the reaction mixture in solution-phase synthesis. To alleviate this problem, polymer-bonded amines [55–58] have been shown to efficiently remove the Fmoc group in solution. Tetrabutyl ammonium fluoride in dimethylformamide (DMF) is also an effective alternative to piperidine for removal of Fmoc groups [59]. This protecting group is stable in acidic reagents, but unexpectedly cleaved by catalytic hydrogenolysis [60, 61], particularly in combination with ammonium formate over a palladium/charcoal catalyst [15].

#### 6.2.1.6 Other Representative Protecting Groups

Effective $N^{\alpha}$-protection is also provided by the allyloxycarbonyl (Alloc) group **14** [61, 62]. The Alloc group can be readily removed through palladium catalyzed transfer of the allyl entry to various nucleophilic species under neutral conditions as shown in Scheme 6.9. The *t*Bu group and fluorene-9-ylmethyl (Fm) groups are not affected by these procedures.

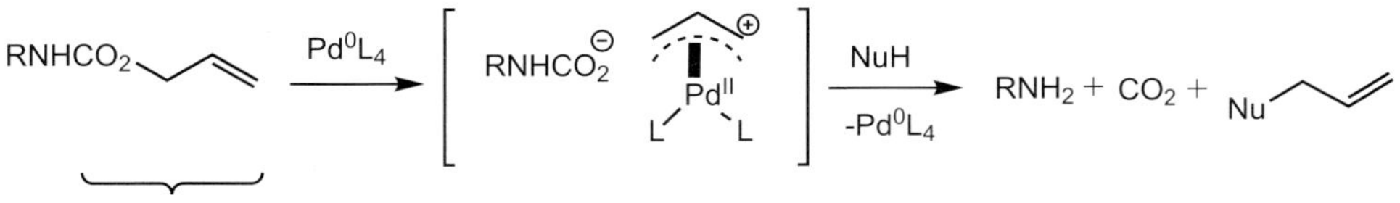

**14**: allyloxycarbonyl(Alloc)

**Scheme 6.9** Palladium-catalyzed deprotection of the Alloc group [62b].

The 2,2,2-trichloroethoxycarbonyl (Troc) group **15** [63–66] presents a useful property of being stable in both acid and base, and removed under mild conditions on treatment with zinc dust in acetic acid, following the mechanism outlined in Scheme 6.10. The stability of the Troc group to catalytic hydrogenolysis is somewhat questionable.

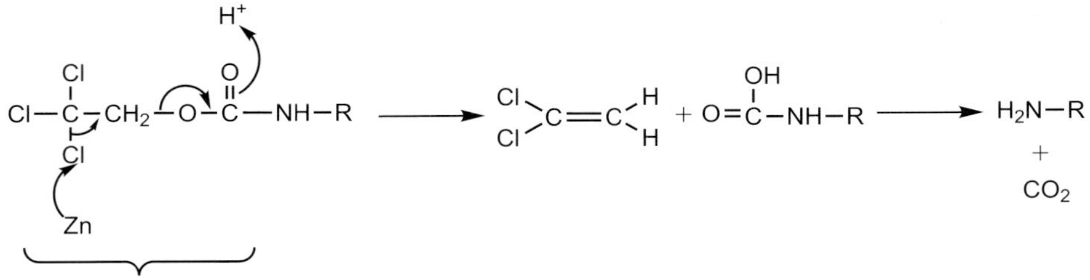

15: 2,2,2-trichloroethoxycarbonyl(Troc)

**Scheme 6.10** Removal of the Troc group with zinc dust in acetic acid [67].

Many other urethane type $N^{\alpha}$-protecting groups have been used in solution synthesis, such as the *tert*-amyloxycarbonyl (Aoc) group **16**, which is removed by TFA [68–71], and the 1-adamantyloxycarbonyl (1-Adoc) group **17** [72] as well, although neither have been extensively applied in peptide synthesis.

Nonurethane $N^{\alpha}$-protecting groups have been applied far less frequently in peptide synthesis. This is mainly because they do not provide adequate protection against amino acid racemization when the carboxyl group is activated. Nevertheless, a number of nonurethane groups provide sufficient protection against racemization and perhaps the most important is the *o*-nitrophenylsulfenyl (Nps) group **18**, which has been used in complex peptide synthesis [73, 74] (Figure 6.6). It can be removed under mild conditions, on treatment with 2-mercaptopyridine in acetic acid [75], which does not affect *t*Bu- or benzyl-based protecting groups.

### 6.2.2
### Carboxyl Group Protection

As described, protection of the C-terminus is the major difference between solid-phase peptide synthesis and peptide synthesis in solution. In the former, the insoluble polymeric support may be considered to be the C-terminus protecting group, whereas in peptide synthesis in solution, more conventional protecting groups are used. The C-terminus is most commonly protected as an alkyl or aryl ester. Hydrazides or protected hydrazides are also C-terminal protecting groups. These hydrazides were important in classical peptide synthesis in solution because they are converted into azides for segment coupling reactions.

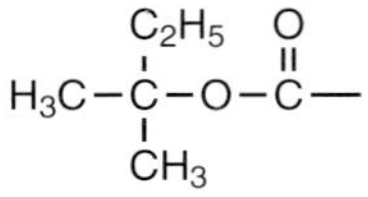

**16**: *tert*-amyloxycarbonyl (Aoc) [68–71]

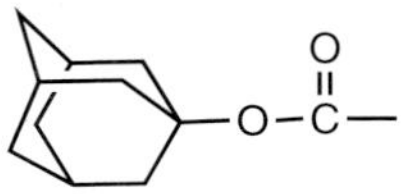

**17**: 1-Adamantyloxycarbonyl (1-Adoc) [72]

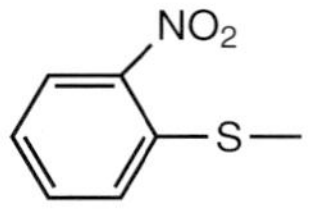

**18**: *O*-nitrophenylsulfenyl (Nps) [73, 74]

**Figure 6.6** Other protecting groups.

#### 6.2.2.1 Methyl Ester (-*O*Me) and Ethyl Ester (-*O*Et)

The preparation of methyl ester (-*O*Me) **19** and ethyl ester (-*O*Et) **20** [1] can be achieved by the route shown in Scheme 6.11 [76].

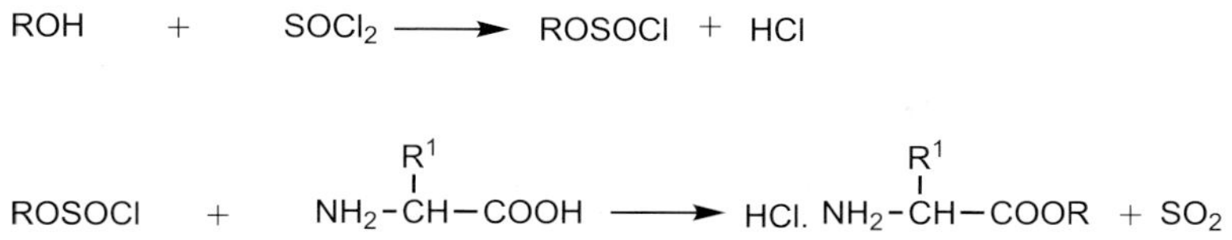

$$\mathrm{ROH} + \mathrm{SOCl_2} \longrightarrow \mathrm{ROSOCl} + \mathrm{HCl}$$

$$\mathrm{ROSOCl} + \mathrm{NH_2{-}CH(R^1){-}COOH} \longrightarrow \mathrm{HCl.\ NH_2{-}CH(R^1){-}COOR} + \mathrm{SO_2}$$

**19**: $R=CH_3$
**20**: $R=C_2H_5$

**Scheme 6.11** Preparation of methyl or ethyl esters by thionyl chloride.

Methyl and ethyl esters can be removed by saponification and are also converted to the corresponding peptide C-terminal amide by treatment with ammonia [77] and to the corresponding peptide hydrazide by hydrazinolysis [78]. Peptide hydrazides can be converted into the corresponding acyl azides for segment coupling.

#### 6.2.2.2 Benzyl Ester (-*O*Bzl)

The benzyl ester (-*O*Bzl) **21** of amino acids is easily prepared under mild conditions by acid (*p*-toluenesulfonic acid)-catalyzed esterification with benzyl alcohol [79, 80] or by reaction of the amino acid cesium carboxylate with benzyl bromide [81]. Cleavage of the benzyl ester can be brought about by acidolysis with strong acid or a more frequently used milder alternative is catalytic hydrogenolysis [82]. The benzyl ester may also be removed by saponification or it can be converted into the corresponding C-terminal amide by treatment with ammonia [83] or the corresponding hydrazide by treatment with hydrazine hydrate [84, 85].

#### 6.2.2.3 *t*Bu Ester (-*Ot*Bu)

Amino acid *t*Bu esters **22** can be prepared by the reaction of amino acids with isobutylene in the presence of sulfuric acid as a catalyst [86], by transesterification of the amino acid with *tert*-butylacetate in the presence of perchloric acid [87, 88] or by using *t*Bu 2,2,2-trichloroacetimidate [89]. *N*-Protected amino acids are soluble in organic solvents. Therefore, *N*-protected amino acids can be easily esterified by isobutylene methods [90, 91].

*t*Bu esters are stable to base-catalyzed hydrolysis, to hydrogenolysis over a palladium catalyst and to nucleophiles in general; the formation of diketopiperazines from dipeptide *t*Bu esters does not usually occur. *t*Bu esters are useful C-terminus protecting groups and may be removed by acidolysis with moderately strong acid, such as TFA or solutions of HCl in organic solvents. They are, however, sufficiently stable to weak acids to allow the washing of organic solutions of the peptides with dilute aqueous acid.

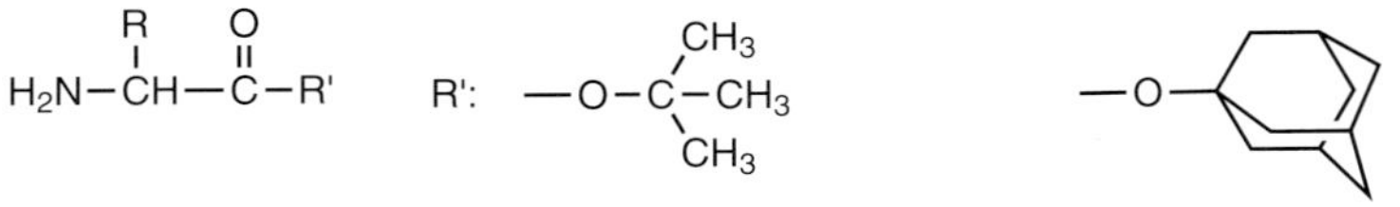

$$H_2N-CH(R)-C(=O)-R' \qquad R': -O-C(CH_3)_3$$

**22**: *tert*-butyl ester(-*O*-*t*Bu) [86–88] **23**: 1-adamantyl ester(-*O*-1-Ada) [92]

**Figure 6.7** Structures of *t*Bu and 1-adamantyl esters.

The property of 1-adamantyl esters of amino acids **23** [92] is similar to that of *t*Bu esters and the 1-adamantyl esters can be removed by the action of TFA (Figure 6.7).

#### 6.2.2.4 Phenacyl Ester (-*O*Pac)

Amino acid phenacyl esters **24** are produced in high yield upon treatment of *N*-protected amino acid carboxylates with bromoacetophenone [93]. The phenacyl ester may be regarded as a methyl ester substituted with a benzoyl group. The phenacyl group can be removed by treatment with zinc in acetic acid [93] or with sodium thiophenoxide in an inert solvent at or below room temperature [94]. It can be only partially removed by catalytic hydrogenolysis [95]. The phenacyl group is stable to acid, even to liquid HF. This property makes the phenacyl group suitable for use with the Boc as *N*-protection. Its use has been advocated by Sakakibara [96] as part of a general approach to the synthesis of large peptides and proteins (Figure 6.8).

#### 6.2.2.5 Hydrazides

Peptidyl hydrazides **25** are prepared by treating the peptide methyl, ethyl, or benzyl esters with hydrazine hydrate [84, 85]. The hydrazide can serve as a protecting group for the carboxyl function and may be converted into the acyl azide by treatment with nitrous acid [97] or by treatment with organic nitrites [98]. In order for the protection of the C-terminus to be useful as the hydrazide, only moderately active acylating derivatives, such as active esters, can be used for chain elongation [99].

In the case of the protected peptide alkyl ester containing Arg($NO_2$), Asp(*O*Bzl), Asp(*Ot*Bu), or Glu(*O*Bzl), treatment with hydrazine hydrate gives side-reactions, such as hydrazinolysis of side-chain functional groups. In order to avoid these side-reactions, it is preferable to synthesize protected peptide hydrazides from the corresponding substituted hydrazide. The most useful in this class of compounds are the Z **26** [100], Boc **27** [101], Tri **28** [102], Troc **29** [103], and *i*Noc **30** [104] protected derivatives.

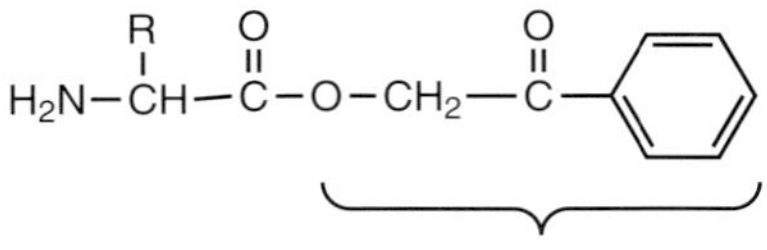

$$H_2N-CH(R)-C(=O)-O-CH_2-C(=O)-C_6H_5$$

**24**: phenacyl ester(-*O*Pac) [93, 94]

**Figure 6.8** Structure of phenacyl ester.

These C-terminus protecting groups can be prepared by reaction of the appropriately protected hydrazide derivative and *N*-protected amino acid, using standard coupling procedures [105]. With the C-terminus protected, the peptide is synthesized by chain elongation at the N-terminus until, at the appropriate time, the hydrazide protecting group is removed, releasing the hydrazide that is then converted into the azide for coupling at the C-terminus.

### 6.2.3
### Side-Chain Protection

Some naturally occurring amino acids possess a third functional group. In the construction of peptides, the ε-amino group of Lys (or δ-amino group of Orn) in either the carboxyl or amino component and the β-mercapto group of Cys in either component require protecting groups. On the other hand, there is no general rule as to the requirement of protection for other side-chain functional groups; such as the carboxyl groups of Asp and Glu, the guanidine group of Arg, the phenolic hydroxyl group of Tyr, the aliphatic hydroxyl group of Ser and Thr, the imidazole of His, the primary carboxamide of Asn and Gln, the aliphatic thioether of Met, and the indole ring of Trp. Okada *et al.* synthesized eglin c, which consists of 70 amino acid residues, by a minimum protection method using an excess amount of small segment azides [106, 107].

In contrast, in the "maximum protection" strategy, in which the maximum numbers of functional groups are masked in order to avoid side-reaction. Sakakibara recommended "solution synthesis of peptides by the maximum protection procedure" using the DCC coupling method and a HF final deprotection method [108]. As described above, peptide segments with full side-chain protection have less byproduct formation. Nonetheless, it is difficult to construct highly soluble segments with full protection; however, if the principle of maximum protection is followed, the carboxyl component can be activated by various methods to almost any extent. In fact, Nakao *et al.* synthesized osteocalcin, which consists of 49 amino acid residues, by the maximum protection procedure [109].

#### 6.2.3.1 ε-Amino Function of Lys (δ-Amino Function of Orn)

As described above, protection of the ε-amino group of Lys is indispensable during peptide synthesis, otherwise its acylation will lead to branching of the peptide chain. Similarly, protecting groups for the α-amino function can be used to also protect the ε-amino function of Lys. Moreover, the protecting group for the ε-amino function should be stable to the conditions employed to remove the α-amino protecting group. The former protecting group, different from the latter, does not lead to racemization. Therefore, an acyl-type protecting group is quite acceptable.

**Z Group** The combination of $N^{\varepsilon}$-Z group and $N^{\alpha}$-Boc group has been widely employed for peptide synthesis. Lys(Z) is prepared by the reaction of Z-Cl with $Lys_2/Cu^{2+}$, followed by removal of $Cu^{2+}$ with $H_2S$ or EDTA [110–112]. The Z group is removed by catalytic hydrogenolysis or strong acids such as HF. In solid-phase peptide synthesis, the $N^{\varepsilon}$-Z group was used at an early time in the history of the technique. However, the $N^{\varepsilon}$-Z group was partially cleaved under the condition to remove the $N^{\alpha}$-Boc group during the peptide synthesis. Therefore, protecting groups more stable to TFA were required and the following Z derivatives are now available. Use of the 2-chlorobenzyloxycarbonyl (2-ClZ) group **31** [113], the 3-chlorobenzyloxycarbonyl (3-ClZ) group **32** [114], the 4-chlorobenzyloxycarbonyl (4-ClZ) group **33** [113], the 2,4-dichlorobenzyloxycarbonyl (2,4-$Cl_2$Z) group **34** [113], the 2,6-dichlorobenzyloxycarbonyl (2,6-$Cl_2$Z) group **35** [113], and the 3,4-dichlorobenzyloxycarbonyl (3,4-$Cl_2$Z) group **36** [113] has been reported.

**Boc Group** This derivative was introduced by Schweyzer *et al.* [115] for preparation of α-melanocyte stimulating hormone. Recently, it was found to be important for the Fmoc strategy in solid-phase peptide synthesis. Lys(Boc) is prepared by the reaction of the $Lys_2/Cu^{2+}$ complex with $(Boc)_2O$ [27].

**2-Adoc Group** Lys(2-Adoc) **37** (Figure 6.9) was developed by Nishiyama and Okada [116], and it was prepared using $Lys_2/Cu^{2+}$ complex and 2-adamantyl chloroformate in the usual manner [117]. The 2-Adoc group was stable to 7.6 M HCl in dioxane, TFA, 25% HBr in acetic acid and 1 M trimethylsilylbromide (TMSBr)–thioanisole–TFA for up to 24 h. It could be removed by 1 M trifluoromethanesulfonic acid (TFMSA)–thioanisole–TFA or anhydrous HF in a few minutes at 0 °C, but was cleaved very slowly by methanesulfonic acid (MSA). Further, the 2-Adoc group was stable to 20% piperidine in DMF, 10% triethylamine in DMF, 10% $NaHCO_3$, and 2 M NaOH for up to 24 h.

#### 6.2.3.2 β-Mercapto Function of Cys

**S-Benzyl (S-Bzl)cysteine and Related Derivatives** Cys(*S*-benzyl, *S*-Bzl) was prepared by the reaction of Cys hydrochloride with benzyl chloride [118]. The *S*-Bzl **38** group can be removed by Na/liquid $NH_3$ [119] or HF (20 °C for 30 min) [120]. However, such

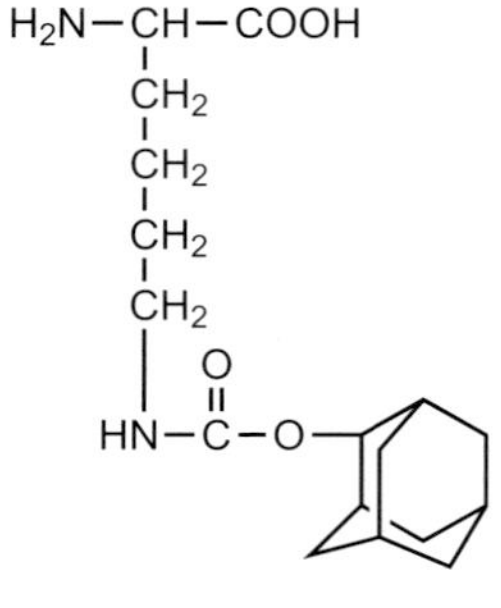

**37**: H-Lys(2-Adoc)-OH

**Figure 6.9** Structure of H-Lys(2-Adoc)-OH.

conditions are incompatible with sensitive functionalities such that current protection schemes with benzyl are seldom conducted. This protecting group was introduced by du Vigneaud [121]. In order to increase acid lability such that its removal might be effected under less drastic conditions, substituted benzyl thioethers, like the compounds *p*-methylbenzyl (*S*-MeBzl) **39** [122], *p*-methoxybenzyl (*S*-MBzl) **40** [123], *S*-trimethylbenzyl (*S*-Tmb) **41** [124], *S*-benzhydryl (*S*-Bzh) **42** [125], and *S*-Tri **43** [126, 127], were investigated and evaluated for their efficacy in complex peptide synthesis. The cleavage of *S*-MeBzl and *S*-MBzl still requires treatment with liquid HF, albeit at a lower temperature than that required for the benzyl group. The *S*-Tmb is stable to TFA at 23 °C for 2 h and cleaved by HF–anisole at 0 °C for 30 min. *S*-Bzh and *S*-Tri are labile in TFA. The *S*-MeBzl and *S*-MBz groups are compatible with the Boc/Bzl strategy, while the *S*-Bzh and *S*-Tri groups can be used in the Fmoc/*t*Bu approach.

**S-Acetamidomethyl (*S*-Acm)cysteine and Related Derivatives** Cys(*S*-acetamidomethyl Acm) **44** can be prepared from acetoamide, formaline, and Cys [128]. The Cys(*S*-Acm) is stable to acids, such as HF, TFA and HBr/AcOH and bases, such as $NH_2NH_2$ and 1 M NaOH. The *S*-Acm group is removed by $Hg(OAc)_2$ or AgOTf [129] or $I_2$ [130]. Cys (*S*-Tacm: *S*-trimethylacetoamidomethyl) **45** was developed by Kiso *et al.* [131, 132]. It is stable to acids and bases as is *S*-Acm and, it can be removed by a similar procedure to *S*-Acm. *S*-Tacm is also more easily prepared than *S*-Acm (Figure 6.10).

**S-3-Nitro-2-pyridylsulfenyl (*S*-Npys)cysteine and Other Derivatives** 3-nitro-2-pyridylsulfenyl (Npys) was introduced by Matsueda *et al.* [133]. Cys(*S*-Npys) **46** can be prepared by the reaction of *N*-protected Cys (Boc-Cys) with Npys-Cl. This protecting group is stable in acids, such as HF or TFA and can be removed by tri-*n*-butylphosphine or β-mercaptoethanol. A particularly useful aspect of the Npys group is that it can be displaced by the free sulfhydryl group of another Cys residue, allowing the directed formation of disulfide bridge. Other mixed disulfides, such as the *S*-*tert*-butylmercapto (*S*-*S*-*t*Bu) **47** [134] and *S*-*S*-Et **48** [135], can be removed by thiolysis.

### 6.2.3.3 β- and γ-Carboxyl Functions of Asp and Glu

In the case of the minimum protection procedure, it is not necessary to protect side-chain carboxyl functions after introduction into a peptide segment. However, in the case of the maximum protection procedure, both carboxyl functions of Asp and Glu should be protected during peptide synthesis.

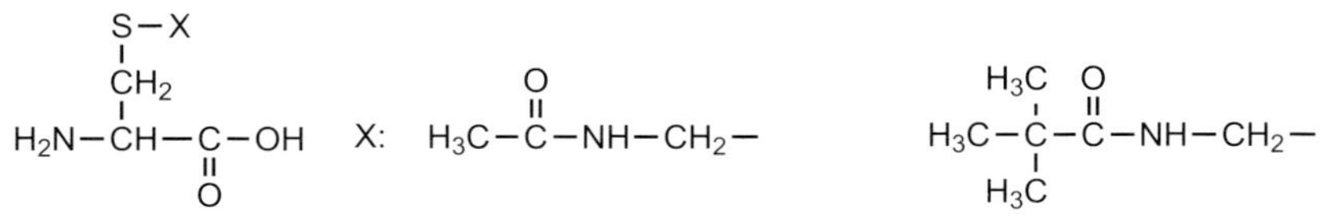

$H_2N-CH(CH_2-S-X)-C(=O)-OH$  X: $H_3C-C(=O)-NH-CH_2-$  $(H_3C)_3C-C(=O)-NH-CH_2-$

**44**: acetoamidomethyl(Acm) [128] **45**: trimethylacetoamidomethyl(Tacm) [131, 132]

**Figure 6.10** Structures of Acm and Tacm groups.

**Aspartic Acid β-Benzyl Ester, Asp(*O*Bzl)** Asp(*O*Bzl) **49** can be prepared from Asp and benzylalcohol in the presence of sulfuric acid [136]. The benzyl group can be easily removed by catalytic hydrogenolysis or HF [17], TFMSA [137], trimethylsilyltrifluoromethanesulfonate (TMSOTf) [138], and TMSBr [139]. When Asp(*O*Bzl) is used for peptide synthesis, formation of succinimide is often observed (Scheme 6.12) [140–142]. This succinimide is very sensitive to base and rearrangement of the peptide bond (from α to β) is also observed [143–148].

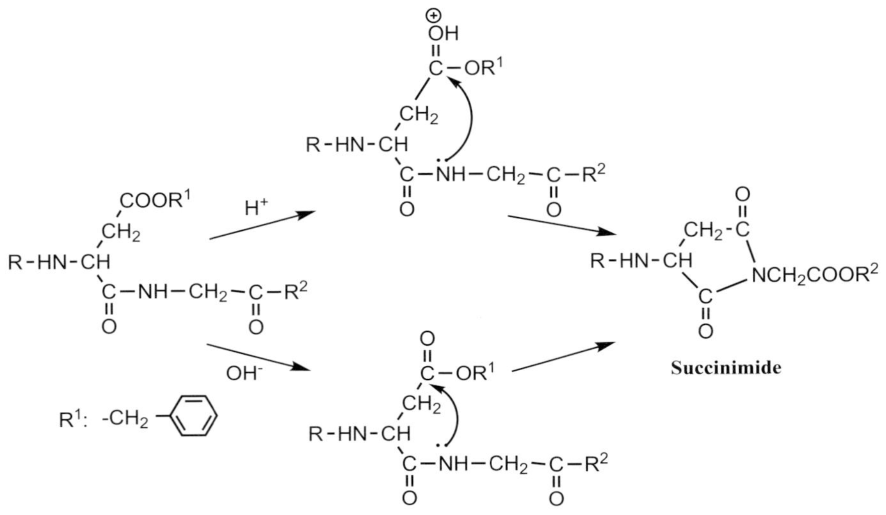

**Scheme 6.12** Acid- and base-catalyzed ring closure in β-benzylaspartyl peptides [140].

**Aspartic Acid β-*tert*-Butyl Ester, Asp(*Ot*Bu)** Z-Asp(*Ot*Bu)-OH can be prepared from Z-Asp-OEt [149, 150]. Asp(*Ot*Bu) **50** is stable to catalytic hydrogenation and is easily removed by acids (TFA, HCl in organic solvent, HBr/AcOH, etc.) [39, 151]. It should be kept in mind that Asp(*Ot*Bu) is partially saponified under alkaline conditions [152–154] and is partially converted to the corresponding hydrazide by hydrazine hydrate [103, 154].

**Aspartic Acid β-Cyclohexyl Ester, Asp(*O*cHx)** As described above, on the introduction of Asp(*O*Bzl) to the peptide, or upon removal of the β-*O*Bzl [155], especially when a Thr, Gly, His, or Asn residue is located at the C-terminus of Asp(*O*Bzl) [140–142], formation of succinimide from Asp residue readily occurs (Scheme 6.12), followed by conversion of an α-peptide bond to an α,β-peptide bond. In order to suppress the above side-reaction, various protecting groups for the β-carboxyl function were developed. The cyclopentyl ester, Asp(OcPe) **51**, removable by HF is more stable to TFA than Asp(*O*Bzl) [156]. However, the tendency to form succinimide is similar to Asp(*O*Bzl). The cyclohexyl ester, Asp(OcHx) **52** [157], is stable to TFA and removed by

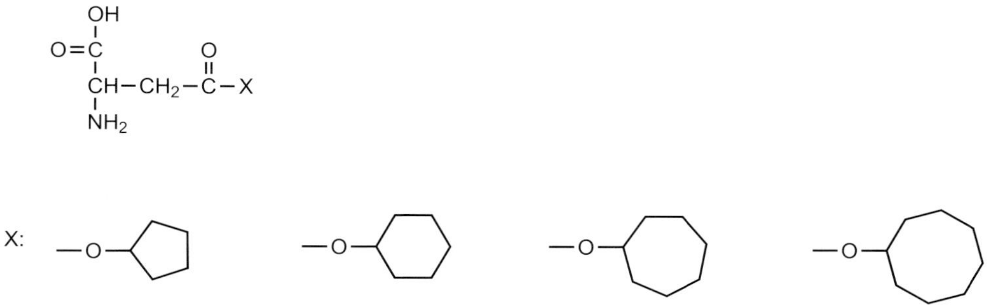

$HO-C(=O)-CH(NH_2)-CH_2-C(=O)-X$

X: —O—cyclopentyl, —O—cyclohexyl, —O—cycloheptyl, —O—cyclooctyl

**51**: cyclopentyl(-*O*cPe) [151] **52**: cyclohexyl(-*O*cHx) [157] **53**: cycloheptyl(-*O*cHp) [158] **54**: cyclooctyl(-*O*cOc) [158]

**Figure 6.11** Structures of β-aspartates.

HF. The above side-reaction of the peptide with this protecting group was reduced dramatically compared with the Asp(*O*Bzl) containing peptide. Cycloheptyl ester, Asp(*O*cHp) **53** and cyclooctyl ester, Asp(*O*cOc) **54** [158] were also developed by Yajima *et al.* (Figure 6.11).These protecting groups are stable to TFA and removed by HF or 1 M TFMSA–thioanisole–TFA and can reduce the formation of succinimide [159].

**Aspartic Acid β-1- or 2-Adamantyl Ester, Asp(*O*-1- or 2-Ada)** β-1- or 2-Adamantylaspartate, Asp(*O*-1-Ada) **55** and Asp(*O*-2-Ada) **56** were synthesized [160, 161]. The 1-Ada group is susceptible to TFA, whereas the 2-Ada group is stable during TFA treatment and easily removable by MSA or HF. Both groups are stable to 50% piperidine in dichloromethane (DCM). These groups suppress aspartimide formation as a side-reaction under acidic and basic conditions during the synthesis of an aspartyl peptide.

#### 6.2.3.4 Protecting Groups for the γ-Carboxyl Function of Glu

These protecting groups are similar to those of Asp. The following protecting groups were reported: γ-Benzylglutamate, Glu(*O*Bzl) **57** [162], γ-*tert*-butylglutamate, Glu(*Ot*Bu) **58** [149], γ-cyclopentylglutamate, Glu(*O*cPe) **59** [163], γ-cyclohexylglutamate, Glu(*O*cHx) **60** [160], γ-cycloheptylglutamate, Glu(*O*cHp) **61** [164], and γ-2-adamantylglutamate Glu(*O*-2-Ada) **62** [165].

#### 6.2.3.5 δ-Guanidino Function of Arg

The δ-guanidino group of Arg is strongly basic (p$K_a$ = 12.5) and protonation provides a simple method of protection that was often used in peptide synthesis by the solution method [166, 167]. However, the poor solubility associated with the intermediates and problems arising as a result of the protonated peptides behaving like an ion-exchange resin, causing them to change anions during the course of the peptide synthesis, prompted the search for a more satisfactory method for protecting the side-chain of Arg.

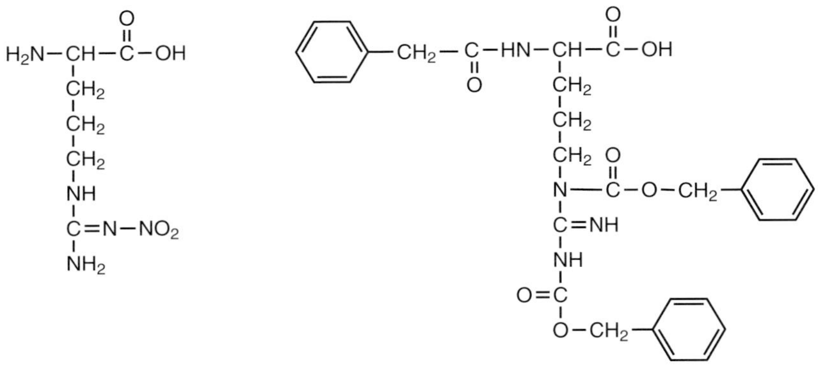

**63**: Arg($NO_2$) [168, 169] **64**: Z-Arg(Z)$_2$ [190]

**Figure 6.12** Structures of H-Arg($NO_2$)-OH and Z-Arg(Z)$_2$-OH.

***N*$^G$-Nitro($NO_2$)arginine** Arg($NO_2$) **63** is prepared by addition of ammonium nitrate to a solution of arginine hydrochloride in concentrated sulfuric acid [168, 169]. This protecting group is stable to acids, such as HBr/AcOH, TFA, and removed by HF (0 °C, 30 min) [31]. It takes a relatively long reaction time to remove the protecting group by catalytic hydrogenation and sometimes, stable, partially reduced products are formed.

***N*$^G$-Benzyloxycarbonyl(Z)$_2$arginine** The reaction of arginine with Z-Cl (4 equiv.) under strong basic conditions yields Z-Arg(Z)$_2$**64** [170] (Figure 6.12). In this case, basicity of the guanidino function is completely protected and can be used as an acid component. This group can be removed by catalytic hydrogenolysis and is stable to HBr/AcOH [170].

***N*$^G$-(1-Adoc)$_2$arginine** Two 1-Adoc residues are introduced into the guanidino function of Arg by reaction of $N^{\alpha}$-protected arginine with 1-adamantyl chloroformate to give Boc-Arg(1-Adoc)$_2$-OH or Z-Arg(1-Adoc)$_2$-OH [171]. The solubility of Arg(1-Adoc)$_2$-containing peptides in organic solvents is considerably increased. Cleavage of the 1-Adoc group is achieved with TFA or HCl treatment [172].

***N*$^G$-*p*-Toluenesulfonyl(Ts) arginine** The *p*-toluenesulfonyl Tos group **65** is currently the most commonly applied to peptide synthesis. This protecting group is stable to acids such as HBr/AcOH and TFA and bases such as $NH_2NH_2$ and NaOH [173, 174] and it is cleaved by HF (0 °C, 30 min) [175] or TFMSA (25 °C, 90 min) [176, 177].

The 4-methoxy-2,3,6-trimethylbenzenesulfonyl group (Mtr) **66** [173, 178] is removed with TFA-thioanisole (9 : 1) [177]. This protecting group is also removed by 1 M TMSBr–thioanisole–TFA [179] and 1 M tetrafluoroboric acid–thioanisole–TFA [180]. The 2,2,5,7,8-pentamethylchroman-5-6-sulfonyl group (Pmc) **67** [181]

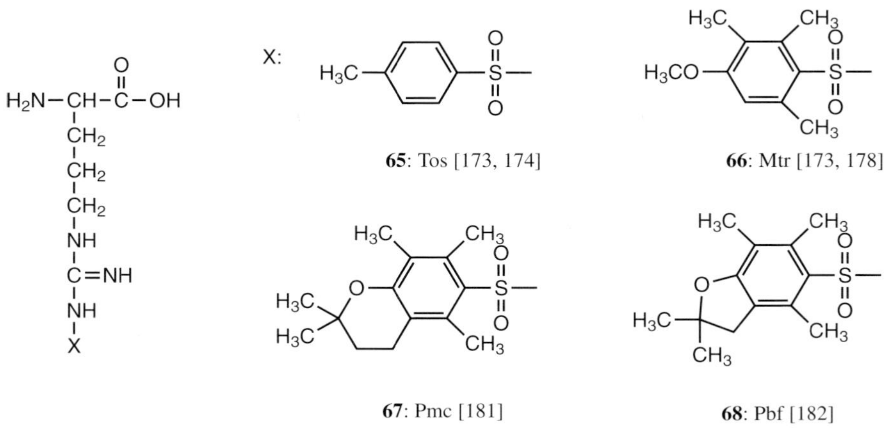

**65**: Tos [173, 174]

**66**: Mtr [173, 178]

**67**: Pmc [181]

**68**: Pbf [182]

**Figure 6.13** Protecting groups for δ-guanidino function of Arg.

and the closely related 2,2,4,6,7-pentamethyldihydrobenzofuran-5-sulfonyl group (Pbf) **68** [182] are more easily removed by TFA (Figure 6.13).

#### 6.2.3.6 **Phenolic Hydroxy Function of Tyr**

The phenolic hydroxyl group of Tyr can react with acylating agents especially under basic conditions where the formation of the phenolate anion generates a potent nucleophile [181, 182]. Therefore, it is preferable to protect the phenolic hydroxyl function. However, it should be kept in mind that the phenol ring of Tyr is susceptible to electrophilic aromatic substitution in the phenol ring, particularly at the position *ortho* to the hydroxyl group. In peptide synthesis, such alkylation usually occurs on acidolytic removal of protecting groups.

***O*-Benzyl(*O*Bzl)tyrosine** Originally, the hydroxyl function of Tyr was protected with a benzyl ether [183]. Tyr(*O*Bzl) **69** is stable in AcOH and TFA and removed preferably by Na/liquid $NH_3$ or catalytic hydrogenolysis [184]. Significant amounts of 3-benzyltyrosine were observed on acidolytic removal by HBr/AcOH or HF [185]. In order to reduce this rearrangement reaction, the 2,6-dichlorobenzyl ether was developed [186].

**Other Protecting Groups** *O*-2-Bromobenzyloxycarbonyltyrosine, Tyr(*O*-2BrZ) **70** [187], is more stable to TFA than **69** by 50-fold and it is removed by HF or catalytic hydrogenolysis. The *O-tert*-butyl ether, Tyr(*Ot*Bu) **71**, is removed by TFA or HF. However, 3-*tert*-butyltyrosine was observed upon acidolysis [188, 189]. *O*-Cyclohexyl ether, Tyr(*O*cHx) **72** [25], is removed by HF (0 °C, 30 min) and reduces the alkylation that occurs on the phenol ring. *O*-2-Adamantyloxycarbonyltyrosine, Tyr(*O*-2-Adoc) **73**, was prepared [82]. The 2-Adoc group is stable to TFA, 5 M HCl in dioxane, and catalytic hydrogenolysis, but removed by treatment of 1 M TFMSA–thioanisole–TFA and HF.

#### 6.2.3.7 **Aliphatic Hydroxyl Function of Ser and Thr**

*O*-Benzylserine, Ser(*O*Bzl) **74** [190], and *O*-benzylthreonine, Thr(*O*Bzl) **75** [190], were developed. This protecting group is stable to alkaline condition and removed by catalytic hydrogenolysis [191], HBr/dioxane, HBr/TFA [192], or HF [31]. The *O*-Bzl is stable to TFA; however, it is partially cleaved during repetitive treatments of the peptide with TFA. Bromobenzyl ethers [193] or chlorobenzyl ethers [193] are more stable to TFA than the benzyl ether.

*O-tert*-Butylserine, Ser(*Ot*Bu) **76**, and *O-tert*-butylthreonine, Thr(*Ot*Bu) **77**, were developed [194]. This protecting group is removed by TFA (at room temperature, 30–60 min) [195], HCl/TFA (at room temperature, 30 min) [196], and HF [31].

#### 6.2.3.8 **Imidazole Nitrogen of His**

The protection of the side-chain of His was problematic for many years and even today it has still not been resolved in a completely satisfactory manner. Difficulties arise because the imidazole ring has two nonequivalent, but similarly reactive nitrogen atoms, designated $\pi$ and $\tau$ as shown in Figure 6.14.

**$N^{im}$-Tosyl($N^{im}$-Tos)histidine** His($N^{im}$-tosyl, Tos) **78** was prepared by the reaction of acylhistidine (acyl = *tert*-amyloxycarbonyl, Nps, etc.) with tosyl chloride [197, 198]. This protecting group is stable to TFA and catalytic hydrogenolysis and removed by HF [197] and TFMSA [178]. It is also removed by 1 M NaOH and partially removed by triethylamine in organic solvent or *N*-hydroxybenzotriazole [197, 198].

**$N^{im}$-Dinitrophenyl($N^{im}$-Dnp)histidine** His($N^{im}$-dinitrophenyl Dnp) **79** was prepared by the reaction of Z-histidine with 2,4-dinitrofluorobenzene [199]. The Dnp group is stable to acid, such as TFA or HF [200], and removed by thiolysis [201].

**$N^{\pi}$-Benzyloxymethyl($N^{\pi}$-Bom)histidine and the Other Derivatives** Jones *et al.* pointed out that protection of the $\pi$-nitrogen of imidazole function was required to prevent racemization of the His residue [202, 203]. They introduced the benzyloxymethyl (Bom) group **80** to the $\pi$-nitrogen, which prevents racemization [204]. This group is removed by catalytic hydrogenolysis, HBr/TFA, and HF, and is stable to TFA. The problem associated with this derivative, however, is that the Bom group generates formaldehyde upon strong acid-mediated cleavage, which can lead to the irreversible blocking of the amino group or to the modification of Cys residues [205–208].

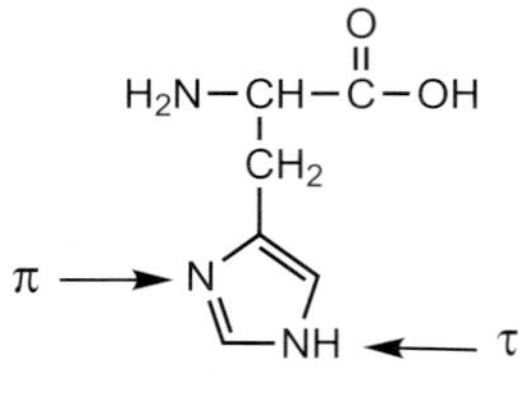

**Figure 6.14** Imidazole nitrogens of His.

The $N^{\pi}$-*tert*-butyloxymethyl (Bum) group **81** [209] is stable to catalytic hydrogenolysis and alkaline conditions and it is removed by TFA.

The $N^{\pi}$-1-adamantyloxymethylhistidine, His($N^{\pi}$-1-Adom) **82**, is stable to alkaline conditions and it is cleaved by TFA [210, 211]. The $N^{\pi}$-2-adamantyloxymethylhistidine, His($N^{\pi}$-2-Adom) **83** is stable to both TFA and alkaline conditions and it is cleaved by 1 M TFMSA–thioanisole–TFA and HF [212, 213].

$N^{\tau}$-Tritylhistidine, His($N^{\tau}$-Tri) **84**, is currently the method of choice. It is completely stable to basic conditions and is readily removed by mild acidolysis.

#### 6.2.3.9 Indole Nitrogen of Trp

The most popular choice for the protection of the Trp side-chain is the $N^{ind}$-formyl (For) **85** group, which is stable to HF [214]. Its removal can be accomplished by treatment with a solution of piperidine in water, or with hydrazine or hydroxylamine. Trp($N^{ind}$-Mtr) **86**, Trp($N^{ind}$-Mtb: 2,4,6-trimethoxybenzenesulfonyl) **87** [215], and Trp ($N^{ind}$-Mts: mesitylsulfonyl) **88** [216] were also developed. Those protecting groups can be removed by HF or 1 M TFMSA–thioanisole–TFA.

## 6.3 Chain Elongation Procedures

A combination of stepwise elongation and segment condensation strategies may be used when peptide molecules are synthesized in solution. Chemical methods used on both of those types of couplings are principally similar.

### 6.3.1 Methods of Activation in Stepwise Elongation

It is very important to avoid epimerization during the peptide synthesis. Stepwise elongation from the C-terminal by one amino acid at a time using urethane-protected amino acids such as Z-amino acids and Boc-amino acids, is advantageous in avoiding epimerization during the peptide bond-forming reaction [77, 217, 218]. The stepwise elongation procedure is effective for the synthesis of small peptides, and for preparing peptide segments to construct large peptides and proteins.

#### 6.3.1.1 Carbodiimides

In 1952, Khorana employed DCC **89** for nucleotide synthesis [219] and, in 1955, Sheehan and Hess employed DCC for peptide synthesis [220]. For many years, DCC has been the most useful and popular coupling reagent in peptide synthesis for both solid-phase and solution methods. It is highly reactive and gives high yields within a short time. The disadvantage to the use of DCC is the formation of the insoluble dicyclohexylurea during activation/coupling. To overcome the problems of separation of the insoluble urea byproduct, the use of diisopropylcarbodiimide (DIPCDI) **90**, which is equally effective and forms a more soluble urea byproduct [220], has been advocated. A reaction mechanism with DCC is summarized in Scheme 6.13.

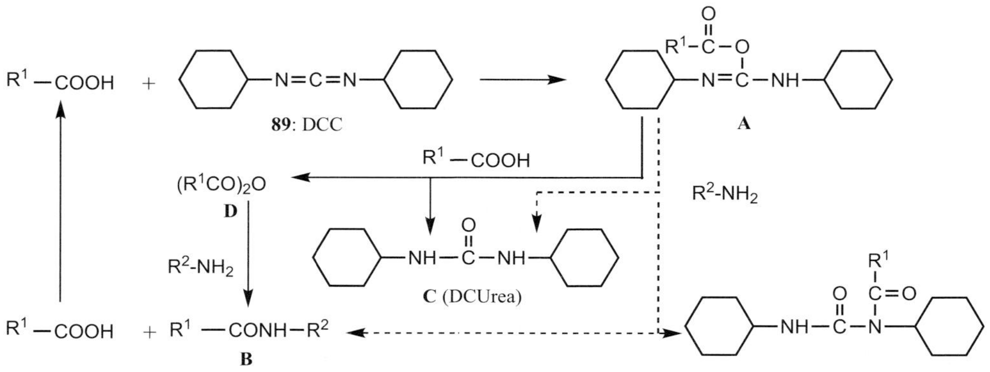

Scheme 6.13 DCC-mediated coupling pathway [67].

Addition of the *N*-protected amino acid to one of the N=C bonds of DCC gives the *O*-acylisourea, intermediate **A**, the first active species formed in the coupling reaction [221–223]. The highly reactive **A** can then undergo aminolysis by the amino component, leading to formation of amide **B** and the dicyclohexylurea byproduct **C**. Alternatively, another molecule of carboxylic acid can react with *O*-acylisourea, forming the amino acid symmetrical anhydride **D**. This is another potent acylating species that also reacts with the amino component, again leading to amide bond formation [224–226]. Water-soluble carbodiimide (WSCI), 1-ethyl-3-(3′-dimethylaminopropyl)carbodiimide (EDC) **91**, was also developed [227] (Figure 6.15).

### 6.3.1.2 Mixed Anhydride Method

This principle was already reported more than 100 years ago by Curtius [1]; however, in fact, Wieland was first to employ this method for peptide synthesis [228]. There is an inherent problem in the reaction using a mixed anhydride formed from protected amino acids or peptides and carboxylic acids. Since the anhydride has two carbonyl groups, there is a regiochemical problem. Attack of the amino component must take place at the desired place if the correct product is to be obtained (Scheme 6.14).

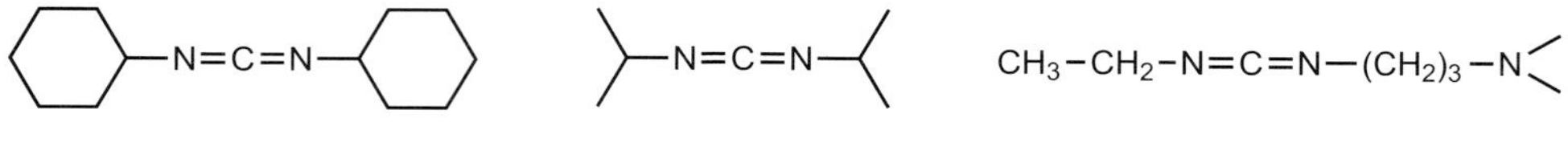

Figure 6.15 Structures of carbodiimides.

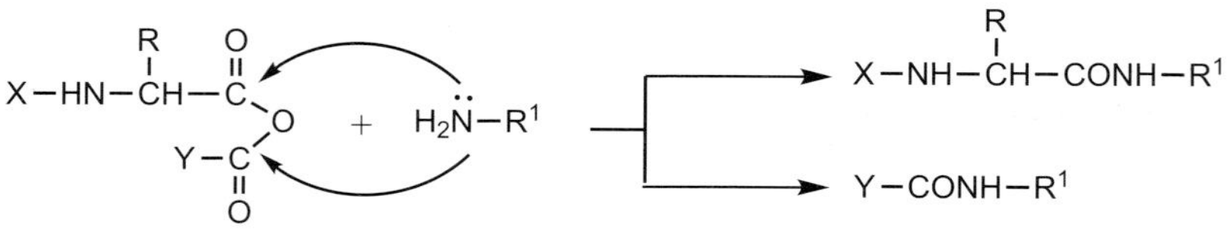

Scheme 6.14 Two possible products via a mixed anhydride.

92

93

Figure 6.16 Structures of the mixed anhydrides with carboxylic acids.

The success of the method depends on the ability to completely reduce the attack at the wrong carbonyl group, which would lead to the formation of *N*-acylated amino derivatives that can no longer take part in peptide synthesis. This can be achieved by using carboxylic acids, or their derivatives, which present severe steric hindrances. Mixed anhydrides containing isovaleryl **92** [229] or pivaloyl residues **93** [230] are the most successful of this class of coupling reagents (Figure 6.16).

The most popular type of mixed anhydride is that formed with carbonic acids instead of carboxylic acids. Typically, ethyl chloroformate **94** [231, 232] and isobutyl chloroformate **95** [233] are used. The pathway for the amide bond formation is summarized in Scheme 6.15.

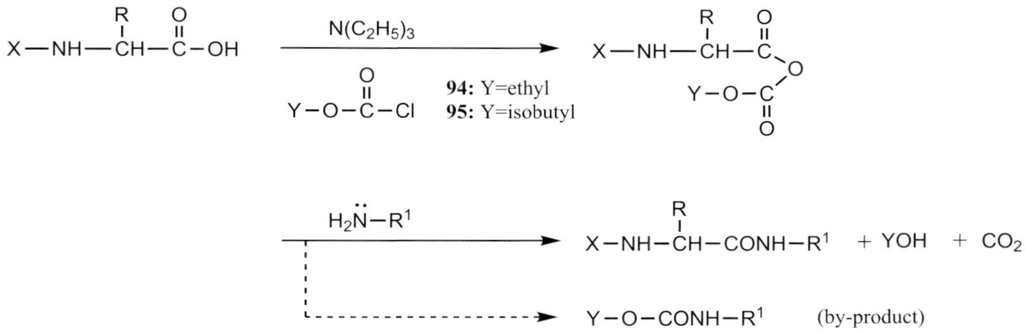

$N(C_2H_5)_3$

**94:** Y=ethyl
**95:** Y=isobutyl

(by-product)

Scheme 6.15 Coupling via a mixed anhydride with carbonic acids.

The advantage of this method is that the byproducts generated, carbon dioxide and the corresponding alcohol, are easy to remove from the reaction mixture.

#### 6.3.1.3 **Active Esters**

Methyl esters or ethyl esters are converted to the corresponding amide or hydrazide by ammonia or hydrazine hydrate, respectively. Therefore, these esters have a tendency to suffer ammonolysis. In order to form the peptide bond from an amino ester under mild conditions, the ester moiety should be activated [234]. Based on the principle of the mixed anhydride procedure, Wieland *et al.* developed an amino acid thiophenyl ester [235]. It can be considered as a type of mixed anhydride from the carboxylic acid and thiophenol. Electrophilicity of the carbonyl carbon of carboxyl function is raised by the electron-withdrawing property of the alcohol moiety of the ester, thereby the carbonyl carbon becomes more vulnerable to attack by the amino component. Generally, phenols with electron withdrawing substituent at *ortho* or *para* positions can be converted to active esters by the DCC procedure or a mixed anhydride procedure. Of these, *p*-nitrophenyl ester **96** [77], 2,4,5-trichlorophenyl ester **97** [236], pentachlorophenyl ester **98** [237], and pentafluorophenyl ester **99** [238] are widely used.

Anderson *et al.* developed *N*-hydroxysuccinimide (HOSu) ester **100** as one type of *N*-hydroxyamine type active ester [239]. These types of active ester can suppress epimerization. The advantage over the phenol type of active esters is that the *N*-hydroxyamine derivative can be removed simply by washing with water. The HOSu ester is easily prepared from the *N*-protected amino acid and HOSu with DCC. However, it was revealed that HOSu reacted with DCC (in a molar proportion of 3: 1) forming dicyclohexylurea and succinimideoxycarbonyl-β-alanine-hydroxysuccinimide ester through a Lossen rearrangement [240]. *N*-Hydroxybenzotriazole (HOBt) ester **101** can be easily prepared by the DCC method, it suppresses epimerization and accelerated aminolysis, and therefore it is widely used in peptide synthesis [241, 242]. The typical structures of the active esters are shown in Figure 6.17.

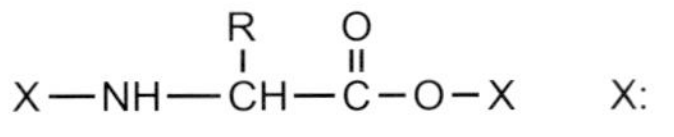


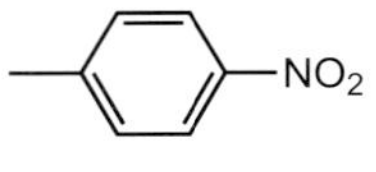


**96**: *p*-nitophenyl ester

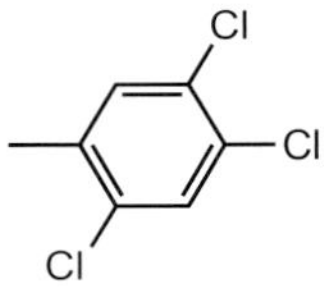


**97**: 2,4,5-trichlorophenyl ester

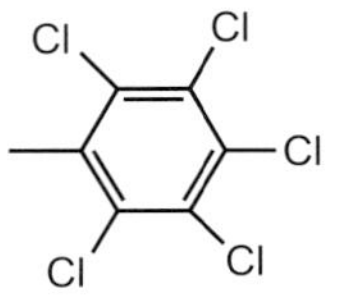


**98**: pentachlorophenyl ester

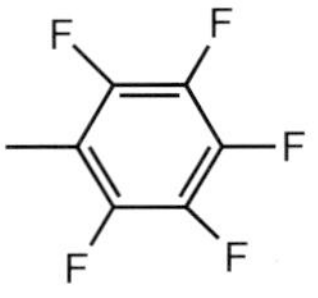


**99**: pentafluorophenyl ester

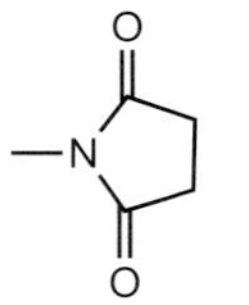


**100**: *N*-hydroxysuccinimide ester (HOSu)

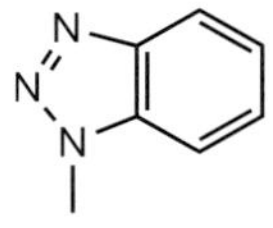


**101**: 1-hydroxybenzotriazole ester (HOBt)

**Figure 6.17** Structures of the active esters.

#### 6.3.1.4 **Phosphonium and Uronium Reagents**

The use of acylphosphonium salts as reactive intermediates for the formation of amide bonds was first postulated by Kenner in 1969 [243]. However, compounds that promote peptide coupling by the formation of acylphosphonium salts did not become widely used until Castro developed the (benzotriazol-1-yloxy)tris(dimethylamino) phosphonium hexafluorophosphate (BOP) reagent **102** [244]. Since then, the use of phosphonium salts in peptide chemistry has increased dramatically. BOP is easy to handle and promotes rapid coupling. As the reagent does not react with $N^{\alpha}$-amino group, it can be added directly to the amino and carboxyl components that are to be coupled, meaning that the BOP reagent is able to produce the active ester *in situ*. A drawback of its use is that BOP reagent produces the toxic hexamethylphosphorotriamide (HMPA) **A** as a byproduct. The proposed reaction mechanism of BOP reagent is shown in Scheme 6.16 [245a,b]. The initial attack of carboxylate on the phosphonium salts leads to the acyloxyphosphonium salts **B**. This is extremely reactive and is attacked by the amino component or by the oxyanion of HOBt **C** to form benzotriazolyl ester **D**, which is thought to be the predominant species suffering aminolysis. A tertiary amine is normally used as base to form the carboxylate ion of carboxyl component.

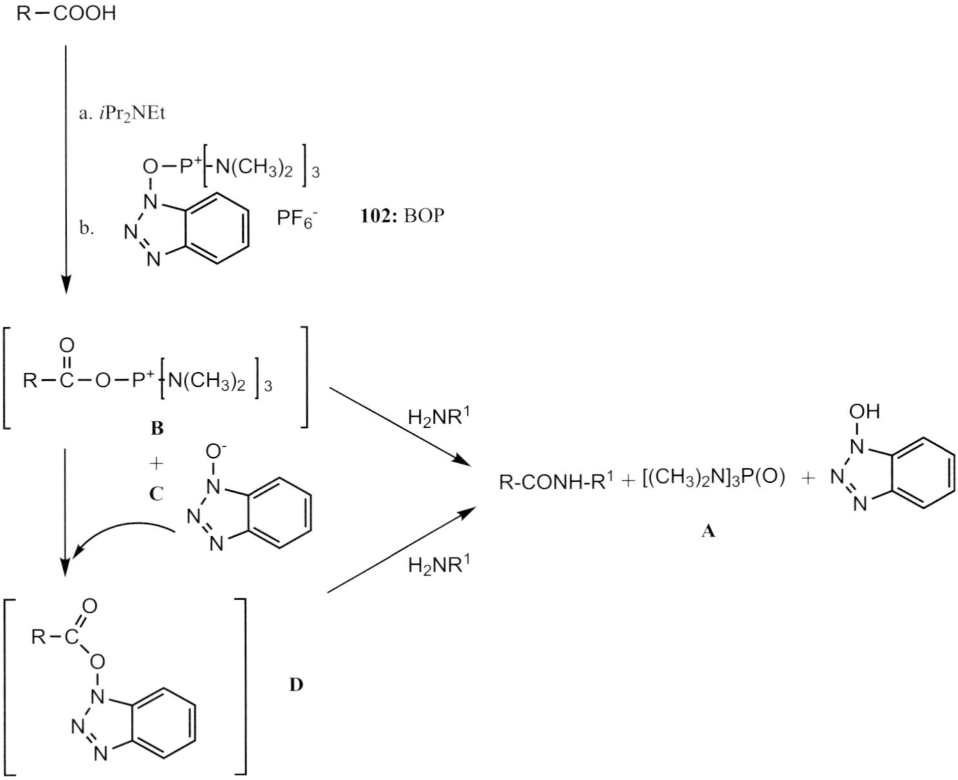

R−COOH

a. $iPr_2NEt$

b. **102:** BOP

**B** + **C**

**D**

$H_2NR^1$

R-CONH-$R^1$ + $[(CH_3)_2N]_3P(O)$ + HOBt

**A**

**Scheme 6.16** BOP-coupling pathway [245b].

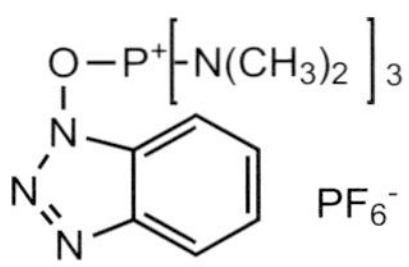

**102:** BOP

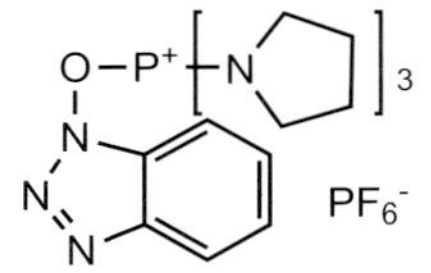

**103:** PyBOP

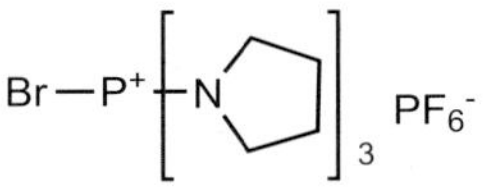

**104:** PyBroP

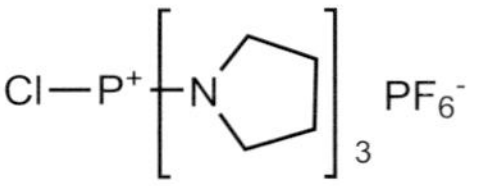

**105:** PyCloP

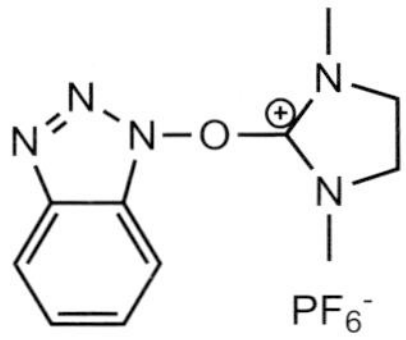

**106:** BOI

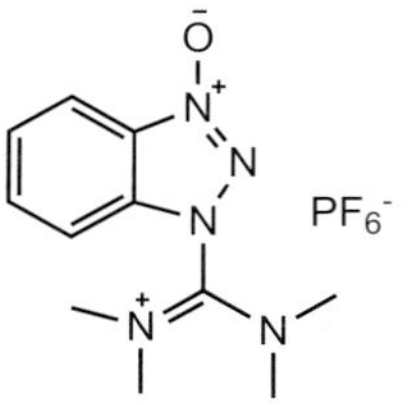

**107:** HBTU

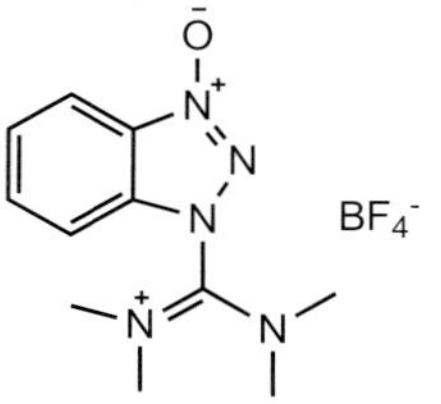

**108:** TBTU

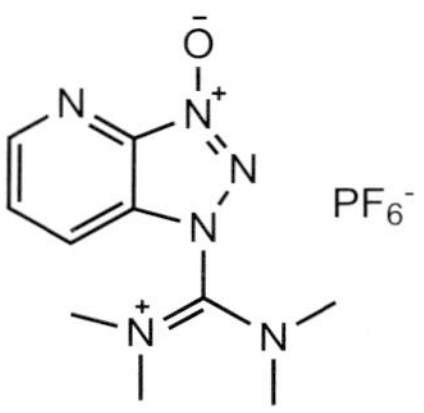

**110:** HATU

**Figure 6.18** Phosphonium, uronium, and guanidium salts.

Based on the structure of BOP reagent, Castro developed (benzotriazol-1-yloxy) tripyrrolidinophosphonium hexafluorophosphate (PyBOP) **103** [251] (Figure 6.18). PyBOP contains pyrrolidine instead of dimethylamine, therefore, it does not form a noxious byproduct [246]. Halo (bromo or chloro)tripyrrolidinophosphonium hexafluorophosphate, bromotripyrrolidinophosphonium hexafluorophosphate PyBroP **104** or chlorotripyrrolidinophosphonium hexafluorophosphate PyCloP **105** were also developed [247, 248], which are effective on the coupling reaction of *N*-methylamino acids or $\alpha,\alpha$-disubstituted amino acids. Kiso *et al.* introduced 2-(benzotriazol-1-yloxy)-1,3-dimethylimidazolidinium hexafluorophosphate (BOI) **106** [169].

The related uronium salts have also been developed for peptide synthesis in solid-phase methods. Since the uronium salts react with the $N^{\alpha}$-amino group, the pre-activation of the carboxyl component is required. In uronium activation, the active species are formed quickly, even in the polar solvents required for optimal peptide resin activation. Side-reactions such as epimerization and dehydration of amide side-chains can be minimized [249]. The reagent, 2-(1*H*-benzotriazol-1-yl)-1,1,3,3-tetramethyluronium hexafluorophosphate (HBTU) **107**, was introduced by Dourtoglou *et al.* [250, 251]. Several other uronium salts were introduced subsequently. Knorr *et al.* reported the usefulness of 2-(1*H*-benzotriazol-1-yl)-1,1,3,3,-tetramethyluronium tetrafluoroborate (TBTU) **108** [249]. More recently, a new generation of phiosphonium and uronium coupling reagents, based on 1-hydroxy-7-azabenzotriazole (HOAt) **109**, have been developed [252]. The aza analog of HBTU, 2-(1*H*-7-azabenzotriazol-1-yl)-1,1,3,3,-tetramethyluronium hexafluorophosphate (HATU) **110**, was introduced [252]. For years, HBTU and HATU were thought to exist as uronium salts; however, X-ray crystallography analysis revealed their true structure as guanidinium salts [253], but they are still commonly called uronium salts.

### 6.3.2
### Methods of Activation in Segment Condensation

For the synthesis of larger peptides or proteins by solution methods, the condensation of peptide segments is an attractive strategy. The separation of the product from truncated or deleted sequences is considerably facilitated. The main and serious problems of this procedure lies in the easy epimerization of the C-terminal residue of the carboxylic segment when the residue is neither glycine nor proline. This epimerization proceeds primarily through an oxazolone intermediate by the reaction of the penultimate residue on the terminal activated carboxylate (Scheme 6.2). Therefore, the most important requirement for coupling applied to the formation of an amide bond between two protected peptide segments is to reduce epimerization of the C-terminal amino acid of the carboxyl component to a minimum extent. In this chapter, suitable coupling procedures to suppress epimerization in segment condensation will be described: (i) the azide procedure, (ii) DCC methodology using an acidic additive, and (iii) native chemical ligation.

#### 6.3.2.1 **Azide Procedure**

The application of the azide procedure (Scheme 6.17) is preferable when the segment to be activated does not contain a Gly or Pro residue at the C-terminus, although epimerization using the azide procedure was observed by Kemp *et al.* [4]. However, it is still true that the amount of epimerization is usually relatively low and, if certain precautions are taken, it can be maintained at very low levels [254a]. It is possible to explain the reason why so little epimerization occurs with this method compared with others based on the fact that aminolysis of the peptidyl azide is governed by

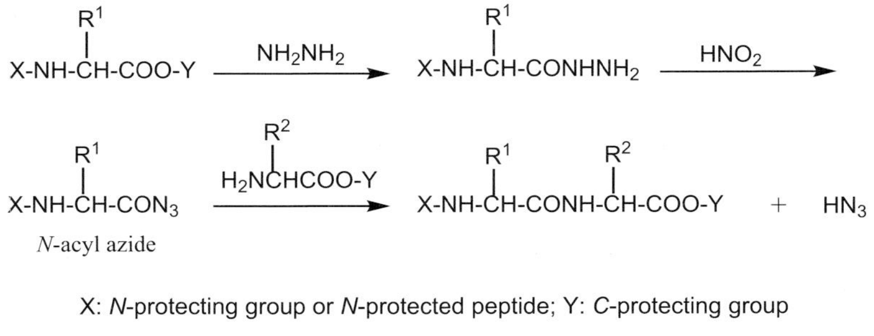

$$\text{X-NH-CH(R}^1\text{)-COO-Y} \xrightarrow{NH_2NH_2} \text{X-NH-CH(R}^1\text{)-CONHNH}_2 \xrightarrow{HNO_2}$$

$$\underset{N\text{-acyl azide}}{\text{X-NH-CH(R}^1\text{)-CON}_3} \xrightarrow{H_2NCH(R^2)COO\text{-}Y} \text{X-NH-CH(R}^1\text{)-CONH-CH(R}^2\text{)-COO-Y} + HN_3$$

X: *N*-protecting group or *N*-protected peptide; Y: *C*-protecting group

**Scheme 6.17** Peptide bond formation via an *N*-acylamino acid azide.

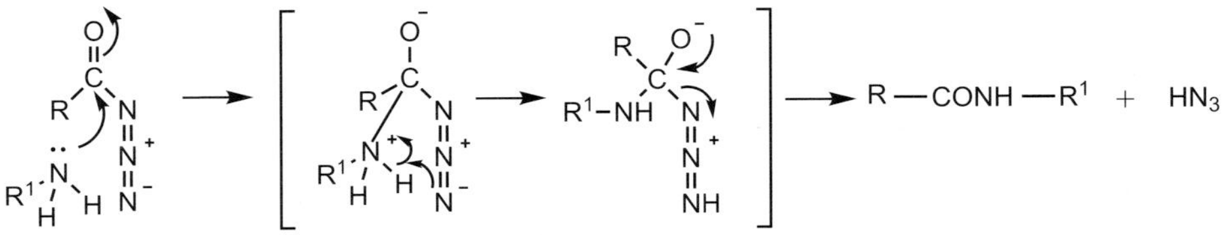

Scheme 6.18 Intramolecular base catalysis mechanism [261b].

intramolecular general base catalysis as shown in Scheme 6.18 and this type of catalysis does not lead to oxazolone formation.

A further advantage is that the azides do not acylate hydroxyl functions, thus the principle of minimum protection can be applied. However, its main disadvantage is the considerably decreased rate of coupling which favors the rearrangement-reaction leading to carbamide-peptides which are difficult to separate [97].

Diphenylphosphoryl azide (DPPA) **111**, which was developed as a reagent forming acyl azides by Shioiri *et al.* [255], gives a peptide bond without protection of the side-chain functional groups in a low yield of the epimeric peptide [255–257]. DPPA also proves to be effective in the synthesis of cyclic peptides [258]. The reaction mode postulated is shown in Scheme 6.19 [257].

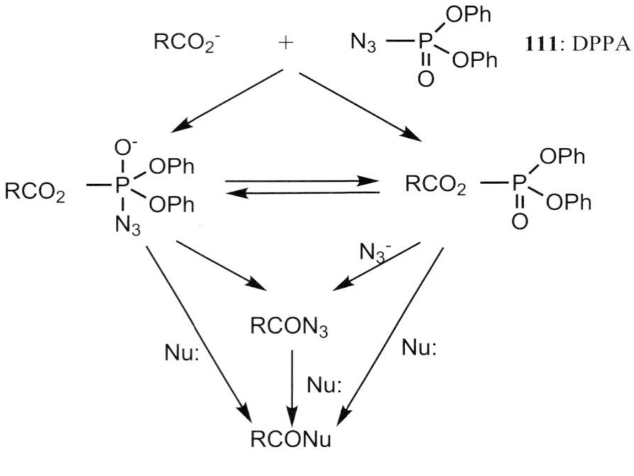

Scheme 6.19 Reaction mode of DPPA [257].

### 6.3.2.2 Carbodiimides in the Presence of Additives

It was revealed that the coupling of protected peptide segments using carbodiimides alone proceeds with extensive epimerization of the C-terminal amino acid of the segment [259].

Weygand *et al.* found that segment coupling with DCC in the presence of HOSu drastically decreased epimerization (less than 1%) [260, 261] and Wuensch *et al.* successfully applied this procedure to the synthesis of glucagon [262–264], although Schroeder had previously given up on its synthesis by the azide procedure.

Later, Koenig and Geiger introduced a more effective reagent than HOSu, namely HOBt [242] and 3,4-dihydro-3-hydroxy-4-oxo-1,2,3-benzotriazine (HOOBt), in

1970 [265, 266]. The DCC/HOBt method was successfully employed for the synthesis of human adrenocorticotropic hormone (ACTH) [267]. Sakakibara *et al.* synthesized human parathyroid hormone [268] by using the WSCI/HOOBt segment coupling method in addition to calciseptine – an L-type specific calcium channel blocker [269], $Na^+/K^+$-ATPase inhibitor-1 (SPAI-1) [270] and *O*-palmitoylated 44-residue peptide amide (PLTXII) blocking presynaptic calcium channels in *Drosophila* [271] by using the WSCI/HOOBt segment coupling method. In segment-coupling model studies, even lower levels of epimerization have been reported [272] when WSCI is used in conjunction with additional HOAt.

The segment coupling method accompanies solubility problems, even though powerful solvents such as DMF, HMPA, and *N*-methylpyrrolidone (NMP) [273] are used. In peptide synthesis by solution methods, the difficulty of obtaining homogeneous products increases drastically as the size of the target peptide increases. The low solubility of large segments brings about the low concentrations in the reaction mixtures, resulting in slow reaction rates. In order to maintain the solubility of peptide segments suitable for the coupling reactions, two main approaches have been attempted. One is the development of protecting groups, which can lead to peptide intermediates that are soluble in organic solvent systems that have high solubilizing potential. The other is the choice of proper reaction solvent systems. We focused on the latter, namely effective solvent systems. The insolubility of the protected peptide intermediates is caused by β-sheet aggregation of the peptide chain resulting from intra- or inter-chain hydrogen bonds [274, 275]. In order to dissolve such peptides, the β-sheet structure must be disrupted and transformed into a helical or random structure. Trifluoroethanol (TFE) and hexafluoroisopropanol (HFIP) have been recommended for disrupting β-sheet aggregation [276, 277]. Narita *et al.* reported that a mixture of chloroform (CHL) or DCM with HFIP, phenol, TFE, or some other polar solvent is an extremely good solvent for dissolving fully protected peptides [278]. Sakakibara *et al.* tried to couple Boc-(1–10)-OH with H-(11–28)-OBzl in order to synthesize protected rat atrial natriuretic peptide (1–28). In CHL/TFE (3: 1), the coupling reaction proceeded smoothly by the EDC/HOOBt method without any detectable epimerization. The reaction was completed within 5 h and the reaction mixture remained a clear solution [279]. Similar results were obtained in the synthesis of human parathyroid hormone [268, 279], calcicludine [280], muscarinic toxin I [281], and β-amyloid peptides [282].

#### 6.3.2.3 Native Chemical Ligation

Native chemical ligation, the highly chemoselective reaction between unprotected peptides bearing a C-terminal thioester and N-terminal Cys results in polypeptides under mild conditions in aqueous solution, is the most practical and most widely used for creating long peptide chains [283] (Scheme 6.20). Dawson *et al.* extended the utility of this approach to non-Cys containing peptides [284]. After ligation, selective desulfurization of the resulting unprotected polypeptide with $H_2$/metal reagents converts the Cys residue to Ala [284]. Since Ala is a common amino acid in proteins as compared with Cys, ligation at Ala residues enables universal application. The convergent chemical synthesis of a polypeptide chain with 203 amino acids by sequential ligation of four peptide segment has been achieved [285].

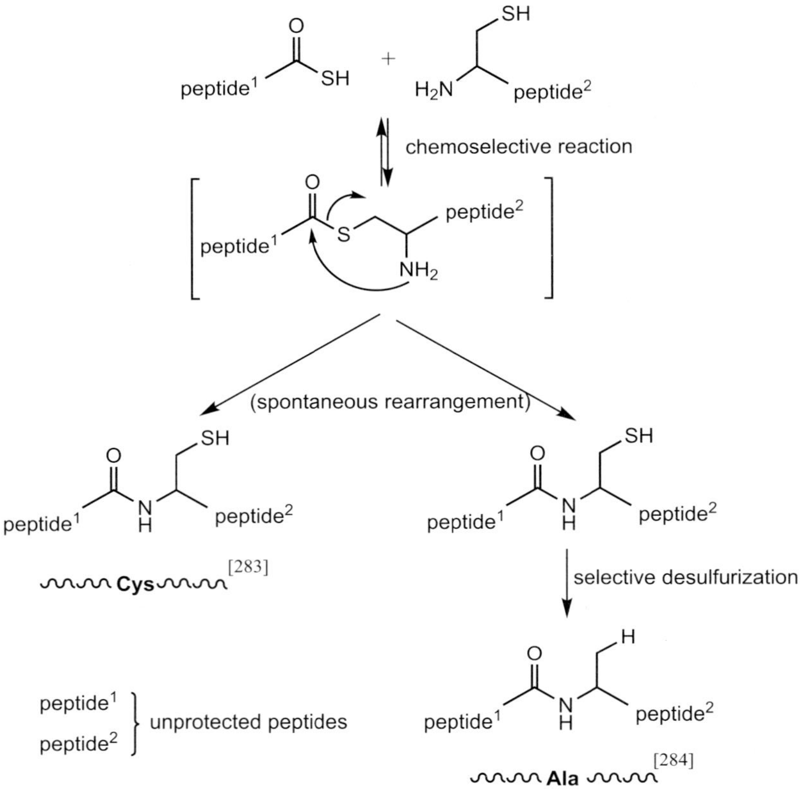

Scheme 6.20 Synthesis of peptides by native chemical ligation.

## 6.4 Final Deprotection Methods

The final stage of peptide synthesis consists of deprotection reactions to remove all protecting groups. As described above, each protecting group on an α-amino or carboxyl function or side-chain of an amino acid has properties suitable for peptide synthesis. By a combination of protecting groups and the selection of an appropriate activation method on carbonyl groups, protected peptides can be synthesized by both solution and solid-phase methods through stepwise condensation and segment condensation procedures. The solution method enables the purification of a reaction product after each coupling step. Therefore, protected peptides can be obtained by this method with a high degree of purity. At the final deprotection step, all protecting groups should be completely removed. Obviously, the choice of the reagents used in the final deprotection step determines the main strategy of peptide synthesis. These final deprotection procedures are mainly: (i) catalytic hydrogenolysis in the presence of palladium [7, 286], (ii) sodium treatment in liquid ammonia for 10–15 s repetitively over 30 min [287–289], (iii) TFA treatment at room temperature for 1 h [115, 290], (iv) HF treatment at 0 °C for 1 h [34, 291], or (v) the "hard–soft acid–base (HSAB)" procedure [166, 292–298].

### 6.4.1 Final Deprotection by Catalytic Hydrogenolysis

Generally, peptides which do not contain Lys and sulfur-containing amino acids (Met and Cys) can be prepared by a combination of the Z group with amino acid derivatives containing protecting groups which are removable by catalytic hydrogenolysis, such as Arg($NO_2$), Asp(*O*Bzl), and Glu(*O*Bzl). Boissonnas *et al.* synthesized bradykinin by this procedure [7].

### 6.4.2 Final Deprotection by Sodium in Liquid Ammonia

This procedure was first employed by du Vigneud for the synthesis of oxytocin [83] and later for the synthesis of insulin [287–289]. As a side-reaction in this procedure, Hofmann and Yajima reported that the peptidylproline bond was cleaved during the synthesis of ACTH [299]. In the synthesis of insulin, the Thr–Pro bond in the B-chain was partially cleaved. The procedure is no longer employed for the synthesis of large peptides or proteins due to the occurrence of these detrimental side-reactions.

### 6.4.3 Final Deprotection by TFA

For the final deprotection by acid, novel amino protection groups were required beside the Z group. Deprotection of the Boc group by McKay and Albertson resolved this problem [26]. Schwyzer *et al.* established this novel strategy by applying Lys(Boc) to the synthesis of porcine ACTH (1–19) [115]. The synthesis of calcitonin [290] or porcine glucagon [262–264] also employed this procedure.

TFA cleavage of *t*Bu and Boc groups results in *t*Bu cation and *t*Bu trifluoroacetate formation (Scheme 6.7). These species are responsible for the alkylation of Trp [300], Tyr [301], and Met [302]. 1,2-Ethanedithiol has been shown to be most effective scavenger for *t*Bu trifluoroacetate, compared to anisole, phenol, and so on [300]. However, 1,2-ethanedithiol alone is unable to fully prevent *tert*-butylation such that a second scavenger, usually anisole is used [303].

### 6.4.4 Final Deprotection by HF

Anhydrous HF can easily cleave benzyl-based protecting groups as well as *t*Bu-based protecting groups [31]. It should be noted that HF is a typical reagent in the Friedel–Crafts reaction. In the final deprotection procedure using HF, scavengers should be added to the reaction mixture to avoid the alkylation by the cation produced in the acidolytic cleavage of protecting groups, such as Tyr(*O*Bzl) and Tyr(*O*$Cl_2$Bzl). Anisole remains one of most widely used scavengers for HF cleavage. The Boc/Bzl strategy is widely employed for both solution- and solid-phase peptide synthetic methods. The solid-phase peptide synthetic methods could not have been developed without the discovery of the HF deprotection procedure.

### 6.4.5
### Final Deprotection by HSAB Procedure

The HF deprotection procedure is more suitable for the preparation of large peptides and proteins than using TFA and is employed more widely. However, special equipment is required in the HF procedure and TFMSA is an alternative to HF cleavage. Organic sulfonic acids, such as TFMSA and MSA, are 47- and 2-fold as acidic as HCl, respectively. Yajima *et al.* understood these phenomena and demonstrated the possibility of using organic sulfonic acids as the final deprotection reagent [176]. They successfully employed a 1 M TFMSA–thioanisole–TFA system for the synthesis of large peptides and succeeded in synthesizing a catalytically active bovine RNase, which consists of 124 amino acids [136]. This reaction mechanism can be explained by the HSAB theory reported by Pearson [304] and summarized as shown in Scheme 6.21 [304–306].

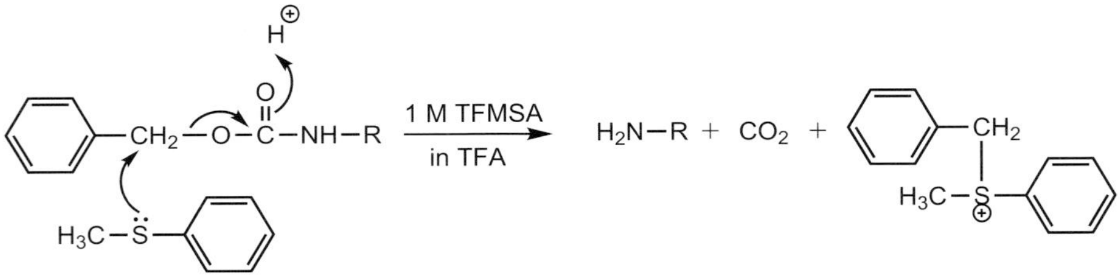

**Scheme 6.21** Deprotection by the HSAB theory [306].

As the first step, the carbonyl oxygen is protonated, followed by fission of the benzyl oxygen bond. The fission is accelerated by the attack of thioanisole, the so-called "soft" nucleophile. Later, TMSOTf was also employed as a super hard acid [307] with thioanisole as a soft base for the final deprotection of neuromedin U-25 [308], sauvagin [309], and human pancreatic polypeptide [310] by a solution method, and later magainin [311] by a solid-phase method.

## References

1 Curtius, T. (1881) Über die einwirkung von chlorbenzoyl auf glycocollsilber. *Journal fur Praktische Chemie*, **24**, 239–240.

2 Fischer, E. (1902) On some derivatives of glycine, alanine and leucine. *Berichte der Deutschen Chemischen Gesellschaft*, **35**, 1095–1106.

3 Bodanszky, M., Klausner, Y.S., and Ondetti, M.A. (1976) *Peptide Synthesis*, John Wiley & Sons, Ltd, New York, pp. 137–157.

4 Kemp, D.S., Wang, S.W., Busby, G., and Hugel, G. (1970) Microanalysis by successive isotopic dilution. A new assay for racemic content. *Journal of the American Chemical Society*, **92**, 1043–1055.

5 Bodanszky, M., Bodanszky, A., Casaretto, M., and Zahn, H. (1985) Coupling in the absence of tertiary amines IV. Tetrazole as acidolytic reagent. *International Journal of Peptide and Protein Research*, **26**, 550–556.

6 Bergmann, M. and Zervas, L. (1928) Catalytic racemization of amino acids and peptides. *Biochemische Zeitschrift*, **203**, 280–292.

7 Goodman, M. and Levine, L. (1964) Peptide synthesis via active esters. IV. Racemization and ring-opening reactions of optically active oxazolones. *Journal of the American Chemical Society*, **86**, 2918–2922.

8 Benoiton, N.L. and Chen, F.M.F. (1981) 2-Alkoxy-5(4*H*)-oxazolones from *N*-alkoxycarbonylamino acids and their implication in carbodiimide-mediated reactions in peptide synthesis. *Canadian Journal of Chemistry*, **59**, 384–389.

9 Jones, J.H. and Witty, M.J. (1979) The formation of 2-benzyloxyoxazol-5(4*H*)-ones from benzyloxycarbonylamino-acids. *Journal of the Chemical Society, Perkin Transactions 1*, 3203–3206.

10 Bergmann, M. and Zervas, L. (1932) A general process for the synthesis of peptides. *Berichte der Deutschen Chemischen Gesellschaft*, **65**, 1192–1201.

11 Ben-Ishai, D. and Berger, A. (1952) Cleavage of *N*-carbobenzoxy groups by hydrogen bromide and hydrogen chloride. *The Journal of Organic Chemistry*, **17**, 1564–1570.

12 Guttmann, S. and Boissonnas, R.A. (1961) Synthesis of structural analogs of bradykinin. *Helvetica Chimica Acta*, **44**, 1713–1723.

13 Okamoto, M., Kimoto, S., Oshima, T., Kinomura, Y., Kawasaki, K., and Yajima, H. (1967) Use of boron trifluoride etherate for debenzyloxycarbonylation of methionine-containing peptides by catalytic hydrogenolysis. *Chemical & Pharmaceutical Bulletin*, **15**, 1618–1620.

14 Felix, A.M., Heimer, E.P., Edger, P., Lambros, T.J., Tzougraki, C., and Meienhofer, J. (1978) Rapid removal of protecting groups from peptides by catalytic transfer hydrogenation with 1,4-cyclohexadiene. *The Journal of Organic Chemistry*, **43**, 4194–4196.

15 Anwer, M.K. and Spatola, A.F. (1980) An advantageous method for the rapid removal of hydrogenolyzable protecting groups under ambient conditions; synthesis of leucine-enkephalin. *Synthesis*, 929–932.

16 Okada, Y. and Ohta, N. (1982) Amino acid and peptides. VII. Synthesis of methionine-enkephalin using transfer hydrogenation. *Chemical & Pharmaceutical Bulletin*, **30**, 581–585.

17 Kuromizu, K. and Meienhofer, J. (1974) Reactions in liquid ammonia. V. Removal of the $N^{\alpha}$-benzyloxycarbonyl group from cysteine-containing peptides by catalytic hydrogenolysis in liquid ammonia, exemplified by a synthesis of oxytocin. *Journal of the American Chemical Society*, **96**, 4978–4981.

18 Felix, A.M., Jimenez, M.H., Mowles, T., and Meienhofer, J. (1978) Catalytic hydrogenolysis in liquid ammonia. Cleavage of $N^{\alpha}$benzyloxycarbonyl groups from cysteine-containing peptides with *tert*-butyl side chain protection. *International Journal of Peptide and Protein Research*, **11**, 329–339.

19 Veber, D.F., Paleveda, W.J. Jr, Lee, Y.C., and Hirschmann, R. (1977) Isonicotinyloxycarbonyl, a novel amino protecting group for peptide synthesis. *The Journal of Organic Chemistry*, **42**, 3286–3288.

20 Weygand, F. and Hunger, K. (1962) Acylation of amino acids with *p*-methoxybenzyloxy-carbonyl-azide. *Chemische Berichte*, **95**, 1–6.

21 Weygand, F. and Hunger, K. (1962) Synthesis of *p*-methoxybenzyloxy-carbonyl-γ-L-hexaglutamyl L-glutamic acid octabenzyl ester. *Chemische Berichte*, **95**, 7–16.

22 Ohno, M., Kuromizu, K., Ogawa, H., and Izumiya, N. (1971) Improved synthesis of

gramicidin S via solid phase synthesis and cyclization by the azide method. *Journal of the American Chemical Society*, **93**, 5251–5254.

23 Wang, S.S., Chen, S.T., Wang, K.T., and Merrifield, R.B. (1987) 4-Methoxybenzyloxycarbonyl amino acids in solid phase peptide synthesis. *International Journal of Peptide and Protein Research*, **30**, 662–667.

24 Birr, C., Lochinger, W., Stahnke, G., and Lang, P. (1972) The α,α,-Dimethyl-3,5-dimethoxybenzyloxycarbonyl (Ddz) residue, an *N*-protecting group labile toward weak acids and irradiation. *Liebigs Annalen der Chemie*, **763**, 162–172.

25 Sieber, P. and Iselin, B.M. (1968) Peptide synthesis using the 2-(*p*-biphenyl) isopropyloxycarbonyl (Dpoc) radical for amino group protection. *Helvetica Chimica Acta*, **51**, 622–632.

26 McKay, F.C. and Albertson, N.F. (1957) New amine-masking groups for peptide synthesis. *Journal of the American Chemical Society*, **79**, 4686–4690.

27 Tarbell, D.S., Yamamoto, Y., and Pope, B.M. (1972) New method to prepare *N-tert*-butoxycarbonyl derivatives and the corresponding sulfur analogs from di-*tert*-butyl dicarbonate or di-*tert*-butyldithiolodicarbonates and amino acids. *Proceedings of the National Academy of Sciences of the United States of America*, **69**, 730–732.

28 Moroder, L., Hallett, A., Wuensch, E., Keller, O., and Wersin, G. (1976) Di-*tert*-butyldicarbonate, a useful reagent for the introduction of the *tert*-butyloxycarbonyl protecting group. *Hoppe-Seyler's Zeitschrift fur Physiologische Chemie*, **357**, 1651–1653.

29 Carpino, L.A. (1957) Oxidative reactions of hydrazines. II. Isophthalimides. New protective groups on nitrogen. *Journal of the American Chemical Society*, **79**, 98–101.

30 Klee, W. and Breener, M. (1950) tert-Butyloxycarbonylimidazole and *tert*-butyloxycarbonylhydrazine. *Helvetica Chimica Acta*, **44**, 2151–2153.

31 Sakakibara, S., Shimonishi, Y., Kishida, Y., Okada, M., and Sugihara, H. (1967) Use of anhydrous hydrogen fluoride in peptide synthesis. I. Behavior of various protecting groups in anhydrous hydrogen fluoride. *Bulletin of the Chemical Society of Japan*, **40**, 2164–2167.

32 Ogawa, H., Sugiura, M., Yajima, H., Sakurai, H., and Tsuda, K. (1978) Studies on peptides. LXXVIII. Synthesis of the nonacosapeptide corresponding to the entire amono acid sequence of avian glucagon (duck). *Chemical & Pharmaceutical Bulletin*, **26**, 1549–1557.

33 Wüensch, E., Jaeger, E., Kisfaludy, L., and Löew, M. (1977) Side reactions in peptide synthesis: *tert*-butylation of tryptophan. *Angewandte Chemie*, **89**, 330–331.

34 Schnabel, E., Klostermeyer, H., and Berndt, H. (1971) Selective acidolytic cleavage of the *tert*-butoxycarbonyl group. *Annals of Chemistry*, **749**, 90–108.

35 Halpern, B. and Nitecki, D.E. (1967) The deblocking of *tert*-butoxycarbonyl-peptides with formic acid. *Tetrahedron Letters*, **8**, 3031–3033.

36 Loffet, A. and Dremier, C. (1971) New reagent for the cleavage of the tertiary butyloxycarbonyl protecting group. *Experentia*, **27**, 1103–1104.

37 Vorbrüggen, H. and Krolikiewicz, K. (1975) Selective acidolytic cleavage of the *tert*-butoxycarbonyl group. *Angewandte Chemie*, **87**, 877.

38 Goodacre, J., Ponsford, H.J., and Stirling, I. (1975) Selective removal of the *tert*-butyloxycarbonyl protecting group in the presence of *tert*-butyl and *p*-methoxybenzyl esters. *Tetrahedron Letters*, **16**, 3609–3612.

39 Kinoshita, H. and Kotake, H. (1974) Selective cleavage of *N-t*-butoxycarbonyl protecting group. *Chemistry Letters*, **3**, 631–634.

40 Helferich, B., Moog, L., and Junger, A. (1925) Replacement of reactive hydrogen atoms in sugars, hydroxyl and amino acids by the triphenylmethyl residue. *Berichte der Deutschen Chemischen Gesellschaft*, **58**, 872–886.

41 Zervas, L. and Theodoropoulos, D.M. (1956) *N*-Tritylamino acids and peptides. A new method of peptide synthesis. *Journal of the American Chemical Society*, **78**, 1359–1363.

42 Barlos, K., Papaioannou, D., Patrianakou, S., and Tsegenidis, T. (1986) Peptide synthesis at elevated temperature free of racemization. *Liebigs Annalen der Chemie*, 1950–1955.

43 Hillmann-Elies, A., Hillmann, G., and Jatzkewitz, H. (1953) *N*-(Triphenylmethyl) amino acids and peptides. *Zeitschrift für Naturforschung*, **8b**, 445–446.

44 Sieber, P., Kamber, B., Hartmann., A., Jöhl, A., Riniker, B., and Rittel, W. (1974) Total synthesis of human insulin involving directed formation of disulfide bonds. *Helvetica Chimica Acta*, **57**, 2617–2621.

45 Kader, A.T. and Stirling, C.J.M. (1964) Elimination-addition. III. New procedures for the protection of amino groups. *Journal of the Chemical Society*, 258–266.

46 Wieland, T., Schmitt, G.J., and Pfaender, P. (1966) Synthesis and properties of various acid nitroethyl esters. *Liebigs Annalen der Chemie*, **694**, 38–43.

47 Carpino, L.A. and Han, G.Y. (1970) 9-Fluorenylmethoxycarbonyl function, a new base-sensitive amino-protecting group. *Journal of the American Chemical Society*, **92**, 5748–5749.

48 Tessier, M., Albericio, F., Pedroso, E., Grandas, A., Eritja, R., Giralt, E., Granier, C., and Pietschoten, J. (1983) Amino-acids condensations in the preparation of $N^{\alpha}$-9-fluorenylmethyloxycarbonylamino-acids with 9-fluorenylmethylchloro-formate. *International Journal of Peptide and Protein Research*, **22**, 125–128.

49 Paquet, A. (1982) Introduction of 9-fluorenylmethyloxycarbonyl, trichloroethoxycarbonyl, and benzyloxycarbonyl amine protecting groups into *O*-unprotected hydroxyl amino acids using succinimidyl carbonates. *Canadian Journal of Chemistry*, **60**, 976–980.

50 Ten Kortenaar, P.B.W., Van Dijk, B.G., Peeters, J.M., Raaben, B.J., Adams, P.J.H.M., and Tesser, G.I. (1986) Rapid and efficient method for the preparation of Fmoc-amino acids starting from 9-fluorenylmethanol. *International Journal of Peptide and Protein Research*, **27**, 398–400.

51 Artherton, E., Logan, C.J., and Sheppard, R.C. (1981) Peptide synthesis. Part 2. Procedures for solid-phase synthesis using $N^{\alpha}$-fluorenylmethoxycarbonyl amino-acids on polyamide supports. Synthesis of substance P and of acyl carrier protein 65–74 decapeptide. *Journal of the Chemical Society, Perkin Transactions 1*, 538–546.

52 Carpino, L.A., and Han, G.Y. (1972) 9-Fluorenylmethoxycarbonyl amino-protecting group. *The Journal of Organic Chemistry*, **37**, 3404–3409.

53 Carpino, L.A. (1987) The 9-fluorenylmethoxycarbonyl family of base-sensitive amino-protecting group. *Accounts of Chemical Research*, **20**, 401–407.

54 Fields, G.B. and Noble, R.L. (1990) Solid phase peptide synthesis utilizing 9-fluorenylmethoxycarbonyl amino acids. *International Journal of Peptide and Protein Research*, **35**, 161–214.

55 Carpino, L.A., Willams, J.R., and Lopusinksi, A. (1978) Polymeric deblocking agents for the fluorenyl-9-ylmethoxycarbonyl (FMOC) amino-protecting group. *Journal of the Chemical Society, Chemical Communications*, 450–451.

56 Arshady, R., Artherton, E., and Sheppard, R.C. (1979) Basic polymers for the cleavage of fluorenylmethoxycarbonyl amino-protecting groups in peptide synthesis. *Tetrahedron Letters*, **20**, 1521–1524.

57 Carpino, L.A., Mansour, E.M.E., Cheng, C.-H., Williams, J.R., MacDonald, R., Knapczyk, J., Carman, M., and Lopusinski, A. (1983) Polystyrene-based deblocking-scavenging agents for the 9-fluorenylmethoxycarbonyl amino-protecting group. *The Journal of Organic Chemistry*, **48**, 661–665.

58 Carpino, L.A., Mansour, E.M.E., and Knapczyk, J. (1983) Piperazino-functionalized silica gel as a deblocking-scavenging agent for the 9-fluorenylmethoxycarbonyl amino-protecting group. *The Journal of Organic Chemistry*, **48**, 666–669.

59 Ueki, M. and Amemiya, M. (1987) Removal of the 9-fluorenylmethoxy-

carbonyl (Fmoc) group with tetrabutylammonium fluoride. *Tetrahedron Letters*, **28**, 6617–6620.

60 Atherton, E., Bury, C., Sheppard, R.C., and Williams, B.J. (1979) Stability of fluorenylmethoxycarbonylamino group in peptide synthesis. Cleavage by hydrogenolysis and by dipolar aprotic solvents. *Tetrahedron Letters*, **20**, 3041–3042.

61 Martinez, J., Tolle, J.C., and Bodanszky, M. (1979) Side reactions in peptide synthesis. 12. Hydrogenolysis of the 9-fluorenylmethoxycarbonyl group. *The Journal of Organic Chemistry*, **44**, 3596–3598.

62 (a) Kunz, H. and Unverzagt, C. (1984) The allyloxycarbonyl (Alloc) moiety-conversion of an unsuitable into a valuable amino protecting group for pepetide synthesis. *Angewandte Chemie (International Edition in English)*, **23**, 436–437; (b) Gomez-Martinez, P., Dessolin, M., Guibe, F., and Albericio, F. (1999) $N^{\alpha}$-Alloc temporary protection in solid-phase peptide synthesis. The use of amino-borane complexes as allyl group scavengers. *Journal of the Chemical Society, Perkin Transactions 1*, 2871–2874.

63 Woodward, R.B., Heusler, K., Gosteli, J., Naegeli, P., Oppolzer, W., Ramage, R., Ranganathan, S., and Vorbrüggen, H. (1966) Total synthesis of cephalosporin C. *Journal of the American Chemical Society*, **88**, 852–853.

64 Windholz, T.B. and Johnston, D.B.R. (1967) Trichloroethoxycarbonyl: a generally applicable protecting group. *Tetrahedron Letters*, **8**, 2555–2557.

65 Lapatsanis, L., Milias, G., Froussios, K., and Kolovos, M. (1983) Synthesis of *N*-2,2,2-(Trichloroethoxycarbonyl)-L-amino acids and *N*-(9-fluorenylmethoxy-carbonyl)-L-amino acids involving succinimidoxy anion as a leaving group in amino acid protection. *Synthesis*, 671–673.

66 Dong, Q., Anderson, E.C., and Ciufolini, M.A. (1995) Reductive cleavage of Troc groups under neutral conditions with cadmium-lead couple. *Tetrahedron Letters*, **36**, 5681–5682.

67 Okada, Y. (2001) Synthesis of peptides by solution methods. *Current Organic Chemistry*, **5**, 1–43.

68 Sakakibara, S. and Fujino, M. (1966) *tert*-Amyloxycarbonyl as a new protecting group in peptide synthesis. II. Synthesis of a hexapeptide amide related to cledoisin by the *tert*-amyloxycarbonyl method. *Bulletin of the Chemical Society of Japan*, **39**, 947–951.

69 Sakakibara, S. and Itoh, M. (1967) *tert*-Amyloxycarbonyl as a new protecting group in peptide synthesis. III. Unexpected side reaction during the synthesis of tert-amyloxycarbonylamino acids. *Bulletin of the Chemical Society of Japan*, **40**, 646–649.

70 Honda, I., Shimonishi, Y., and Sakakibara, S. (1967) *tert*-Amyloxycarbonyl as a new protecting group in peptide synthesis. IV. Synthesis and use of *tert*-amyl azidoformate. *Bulletin of the Chemical Society of Japan*, **40**, 2415–2418.

71 Sakakibara, S., Honda, I., Takada, K., Miyoshi, M., Ohnishi, T., and Okumura, K. (1969) *tert*-Amyloxycarbonyl as a new protecting group in peptide synthesis. V. Direct synthesis of *tert*-amyloxycarbonyl and *tert*-butyloxycarbonyl amino acids using the respective *tert*-alkyl chloroformates. *Bulletin of the Chemical Society of Japan*, **42**, 809–811.

72 Haas, W.L., Krumkalns, E.V., and Gerzon, K. (1966) Adamantyloxycarbonyl, a new blocking group. Preparation of 1-adamantyl chloroformate. *Journal of the American Chemical Society*, **88**, 1988–1992.

73 Wünsch, E. and Spangenberg., R. (1972) Deacylation of *N*-sulfenylpeptides by thiocyanate ions. *Chemische Berichte*, **105**, 740–742.

74 Zervas, L., Borovas, D., and Gazis, E. (1963) New methods in peptide synthesis. 1. Tritylsulfenyl and *o*-nitrophenylsulfenyl groups as *N*-protecting groups. *Journal of the American Chemical Society*, **85**, 3660–3666.

75 Tun-Kyi, A. (1978) Selective removal of the *o*-nitrophenylsulfenyl protecting group in peptide synthesis. *Helvetica Chimica Acta*, **61**, 1086–1090.

76 Brenner, M. and Huber, W. (1953) Preparation of α-amino acid esters by alcoholysis of the methyl esters. *Helvetica Chimica Acta*, **36**, 1109–1115.

77 Bodanszky, M. and du Vigneaud, V. (1959) A method of synthesis of long peptide chains using a synthesis of oxytocin as an example. *Journal of the American Chemical Society*, **81**, 5688–5691.

78 Fujii, N., Shimokura, M., Nomizu, M., Yajima, H., Shono, F., Tsuda, M., and Yoshitake, A. (1984) Studies on peptides. CXVII. Solution synthesis of the tetratetracontapeptide amide corresponding to the entire amino acid sequence of growth hormone releasing factor, somatocrinin. *Chemical & Pharmaceutical Bulletin*, **32**, 520–529.

79 Miller, H.K. and Waelsch, H. (1952) Benzyl esters of amino acids. *Journal of the American Chemical Society*, **74**, 1092–1093.

80 Cipera, J.D. and Nicholls, R.V.V. (1955) Preparation of benzyl esters of amino acids. *Chemistry & Industry*, 16–17.

81 Wang, S.-S., Gisin, B.F., Winter, D.P., Makofske, R., Kulesha, I.D., Tzougraki, C., and Meienhofer, J. (1977) Facile synthesis of amino acid and peptide esters under mild conditions via cesium salts. *The Journal of Organic Chemistry*, **42**, 1286–1290.

82 Okada, Y., Shintomi, N., Kondo, Y., Yokoi, T., Joshi, S., and Li, W. (1997) Amino acids and peptides. LI. Application of the 2-adamantyloxycarbonyl (2-Adoc) group to the protection of the hydroxyl function of tyrosine in peptide synthesis. *Chemical & Pharmaceutical Bulletin*, **45**, 1860–1864.

83 du Vigneaud, V., Ressler, C., Swan, J.M., Roberts, C.W., Katsoyannis, P.G., and Gordon, S. (1953) The synthesis of an octapeptide amide with the hormonal activity of oxytocin. *Journal of the American Chemical Society*, **75**, 4879–4880.

84 Gillessen, D., Schnabel, E., and Meienhofer, J. (1963) Synthesis of the insulin sequence B 13–20. *Liebigs Annalen der Chemie*, **667**, 164–171.

85 Strachan, R.G., Paleveda, W.J., Nutt, R.F., Vitali, R.A., Veber, D.F., Dickinson, M.J., Garsky, V., Deak, J.E., Walton, E., Jenkins, S.R., Holly, F.W., and Hirschmann, R. (1969) Studies on the total synthesis of an enzyme. II. Synthesis of a protected tetratetracontapeptide corresponding to the 21–64 sequence of ribonuclease A. *Journal of the American Chemical Society*, **91**, 503–505.

86 Roeske, R.W. (1959) Amino acid *tert*-butyl esters. *Chemistry & Industry*, 1121–1222.

87 Taschner, E., Wasielewski, C., Sokolowska, T., and Biernat, J.F. (1961) New esterification procedures in peptide chemistry. VII. Synthesis of tosyl- and carbobenzoxyglutamic acid-α tert-butyl esters and use in α-glutamyl peptide synthesis. *Liebigs Annalen der Chemie*, **646**, 127–133.

88 Taschner, E., Chimiak, A., Bator, B., and Sokolowska, T. (1961) New esterification procedures in peptide chemistry. VIII. Synthesis of *tert*-butyl esters of free amino acids. *Liebigs Annalen der Chemie*, **646**, 134–136.

89 Armstrong, A., Brackenridge, I., Jackson, R.F.W., and Kirk, J.M. (1988) A new method for the preparation of tertiary butyl ethers and esters. *Tetrahedron Letters*, **29**, 2483–2486.

90 Anderson, G.W. and Callahan, F.M. (1960) *tert*-Butyl esters of amino acids and peptides and their use in peptide synthesis. *Journal of the American Chemical Society*, **82**, 3359–3363.

91 Schröder, E. and Gibian, H. (1962) Peptide syntheses XII. Synthesis of partial sequences of glucagon. *Liebigs Annalen der Chemie*, **656**, 190–204.

92 Iossifidou, S.M. and Froussios, C.C. (1996) Facile synthesis of 1-adamantyl esters of L-α-amino acids, a new class of carboxy protected derivatives. *Synthesis*, 1355–1358.

93 Hendrickson, J.B. and Kandall, C. (1979) Phenacyl protecting group for acids and phenols. *Tetrahedron Letters*, 343–344.

94 Stelakatos, G.C., Paganou, A., and Zervas, L. (1966) New methods in peptide synthesis. III. Protection of carboxyl group. *Journal of the Chemical Society (C)*, 1191–1199.

95 Taylor-Papadimitriou, J., Yovanidis, C., Paganou, A., and Zervas, L. (1967) New

methods in peptide synthesis. Part V. On α- and γ-diphenylmethyl and phenyl esters of L-glutamic acid. *Journal of the Chemical Society (C)*, 1830–1836.

96 Sakakibara, S. (1999) Chemical synthesis of proteins in solution. *Biopolymers (Peptide Science)*, **51**, 279–296.

97 Curtius, T. (1902) Synthetic experiments with hippur azide. *Berichte der Deutschen Chemischen Gesellschaft*, **35**, 3226–3229.

98 Honzl, J. and Rudinger, J. (1961) Amino acids and peptides. XXXIII. Nitrosyl chloride and butyl nitrite as reagents in peptide synthesis by the azide method; suppression of amide formation. *Collection of Czechoslovak Chemical Communications*, **26**, 2333–2344.

99 Cheung, H.T. and Blout, E.R. (1965) The hydrazides as a carboxylic-protecting group in peptide synthesis. *The Journal of Organic Chemistry*, **30**, 315–316.

100 Hofmann, K., Magee, M.Z., and Lindenmann, A. (1950) Studies on polypeptides. II. The preparation of α-amino acid carbobenzoxyhydrazides. *Journal of the American Chemical Society*, **72**, 2814–2815.

101 Boissonas, R.A., Guttmann, S., and Jaquenoud, P.-A. (1960) Synthesis of a nonapeptide with properties of bradykinin. *Helvetica Chimica Acta*, **43**, 1349–1358.

102 Weygand, F. and Steglich, W. (1959) *N*-Trifluoroacetylamino acids. XII. Peptide syntheses with *N*-trifluoroacetylamino acids and *N*-trifluoroacetylpeptide tritylhydrazides. *Chemische Berichte*, **92**, 313–319.

103 Yajima, H. and Kiso, Y. (1971) Peptides. XXX. Trichloroethoxycarbonylhydrazine. *Chemical & Pharmaceutical Bulletin*, **19**, 420–423.

104 Macrae, R. and Young, G.T. (1974) Extension of the "handle" method of peptide synthesis. Use of 4-picolyloxycarbonylhydrazides. *Journal of the Chemical Society, Chemical Communications*, 446–447.

105 Wang, S.S., Kulesha, I.D., Winter, D.P., Makofske, R., Kutney, R., and Meienhoffer, J. (1978) Preparation of protected peptide hydrazides from the acids and hydrazine by dicyclohexylcarbodiimide–hydroxybenzotriazole coupling. *International Journal of Peptide and Protein Research*, **11**, 297–300.

106 Okada, Y. and Tsuboi, S. (1991) Amino Acids and peptides. Part 31. Total synthesis of eglin c. Part 1. Synthesis of a triacontapeptide corresponding to the C-terminal sequence 41–70 of eglin c and related peptides and studies on the relationship between the structure and inhibitory activity against human leukocyte elastase, cathepsin G and α-chymotrypsin. *Journal of the Chemical Society, Perkin Transactions 1*, 3315–3319.

107 Okada, Y. and Tsuboi, S. (1991) Amino Acids and peptides. Part 32. Total synthesis of eglin c. Part 2. Synthesis of a heptacontapeptide corresponding to the entire amino acid sequence of eglin c and of related peptides, and studies on the relationship between the structure and inhibitory activity against human leukocyte elastase, cathepsin G and α-chymotrypsin. *Journal of the Chemical Society, Perkin Transactions 1*, 3321–3328.

108 Goodman, M. and Meienhofer, J. (eds) (1977) *Peptides: Proceedings of the Fifth American Peptide Symposium, San Diego*, John Wiley & Sons, Inc., New York.

109 Nakao, M., Nishiuchi, Y., Nakata, M., Kimura, T., and Sakakibara, S. (1994) Synthesis of human osteocalcins: γ-carboxyglutamic acid at position 17 is essential for a calcium-dependent conformational transition. *Peptide Research*, **7**, 171–174.

110 Neuberger, A. and Sanger, F. (1943) The availability of the acetyl derivatives of lysine for growth. *The Biochemical Journal*, **37**, 515–518.

111 Synge, R.L.M. (1948) Synthesis of some dipeptides related to gramicidin S. *The Biochemical Journal*, **42**, 99–104.

112 Poduska, K. and Rudinger, J. (1959) Amino acids and peptides. XXVII. Derivatives of 1-tosyl-L-3-amino-2-pyrrolidone. Preparation and synthetic potentialities. *Collection of Czechoslovak Chemical Communications*, **24**, 3449–3467.

113 Erickson, B.W. and Merrifield, R.B. (1973) Use of chlorinated benzyloxycarbonyl protecting groups to

eliminate $N^{\varepsilon}$-branching at lysine during solid-phase peptide synthesis. *Journal of the American Chemical Society*, **95**, 3757–3763.

**114** Noda, K., Terada, S., and Izumiya, N. (1970) Modified benzyloxycarbonyl groups for protection of ε-amino group of lysine. *Bulletin of the Chemical Society of Japan*, **43**, 1883–1885.

**115** Schweyzer, R. and Rittel, W. (1961) Synthesis of intermediates for a corticotropic nonadecapeptide. I. $N^{\varepsilon}$-*tert*-Butoxycarbonyl-L-lysine, $N^{\alpha}$-($N^{\varepsilon}$-*tert*-butoxycarbonyl-L-lysyl)-$N^{\varepsilon}$-*tert*-butoxycarbonyl-L-lysine, $N^{\varepsilon}$-(*tert*-butoxycarbonyl)-L-lysyl-L-prolyl-L-valylglycine and derivatives. *Helvetica Chimica Acta*, **44**, 159–169.

**116** Nishiyama, Y., and Okada, Y. (1993) Development of a new amino-protecting group, 2-adamantyloxycarbonyl (2-Adoc), and its application to the solid-phase synthesis of protected peptides. *Journal of the Chemical Society, Chemical Communications*, 1083–1084.

**117** Nishiyama, Y., Shintomi, N., Kondo, Y., and Okada, Y. (1994) Amino acids and peptides. Part 38. Development of a new amino-protecting group, 2-adamantyloxycarbonyl, and its application to peptide synthesis. *Journal of the Chemical Society, Perkin Transactions 1*, 3201–3207.

**118** Lutz, W.R., Ressler, C., Nettleton, D.E. Jr, and du Vigneaud, V. (1959) Isoasparagine-oxytocine: the isoasparagine isomer of oxytocin. *Journal of the American Chemical Society*, **81**, 167–173.

**119** Sifferd, R.H. and du Vigneaud, V. (1935) A new synthesis of carnosine, with some observations of the benzyl group from carbobenzoxy derivatives and from benzylthio ethers. *The Journal of Biological Chemistry*, **108**, 753–761.

**120** Sakakibara, S., Shimonishi, Y., Kishida, Y., Okada, M., and Sugihara, H. (1967) Use of anhydrous hydrogen fluoride in peptide synthesis. I. Behavior of various protective groups in anhydrous hydrogen fluoride. *Bulletin of the Chemical Society of Japan*, **40**, 2164–2167.

**121** Wood, J.L. and du Vigneaud, V. (1939) Racemization of benzyl-L-cysteine with a new method of preparing D-cystine. *The Journal of Biological Chemistry*, **130**, 109–114.

**122** Erickson, B.W. and Merrifield, R.B. (1973) Acid stability of several benzylic protecting groups used in solid-phase peptide synthesis. Rearrangement of *O*-benzyltyrosine to 3-benzyltyrosine. *Journal of the American Chemical Society*, **95**, 3750–3751.

**123** Akabori, S., Sakakibara, S., Shimonishi, Y., and Nobuhara, Y. (1964) A new method for the protection of the sulfhydryl group during peptide synthesis. *Bulletin of the Chemical Society of Japan*, **37**, 433–434.

**124** Brtnik, F., Krojidlo, M., Barth, T., and Jost, K. (1981) Amino acids and peptides. CLXIX. Synthesis of oxytocin and arginine-vasopressin and its deamino analog using the 2,4,6-trimethylbenzyl group for protection of the cysteine sulfur. *Collection of Czechoslovak Chemical Communications*, **46**, 286–299.

**125** Kamber, B. (1981) Synthesis of cystine peptides illustrated by the total synthesis of human insulin. *Zeitschrift fur Naturforschung B*, **36**, 508–514.

**126** Photaki, I., Taylor-Papadimitriou, J., Sakarellos, C., Mazarakis, P., and Zervas, L. (1970) On cysteine and cystine peptides. V. *S*-trityl- and *S*-diphenyl-methylcysteine and -cysteine peptides. *Journal of the Chemical Society (C)*, 2683–2687.

**127** Hiskey, R.G., Li, C.-D., and Vunnam, R.R. (1975) Sulfur-containing polypeptides. XVIII. Unambiguous synthesis of the parallel and antiparallel isomers of some bis-cystine peptides. *The Journal of Organic Chemistry*, **40**, 3697–3703.

**128** Veber, D.F., Milkowski, J.D., Varga, S.L., Denkewalter, R.G., and Hirschmann, R. (1972) Acetamidomethyl. A novel thiol protecting group for cysteine. *Journal of the American Chemical Society*, **94**, 5456–5461.

**129** Fujii, N., Otaka, A., Watanabe, T., Okamachi, A., Tamamura, H., Yajima, H., Inagaki, Y., Nomizu, M., and Asano, K. (1989) Silver trifluoromethanesulfonate as an *S*-deprotecting reagent for the synthesis of cystine peptides. *Journal of* *the Chemical Society, Chemical Communications*, 283–284.

130 Kamber, B. and Rittel, W. (1968) Cystine peptide synthesis. *Helvetica Chimica Acta*, **51**, 2061–2064.

131 Kiso, Y., Yoshida, Y., Kimura, T., Fujiwara, Y., and Shimokura, M. (1989) A new thiol protecting trimethylacetamidomethyl group. Synthesis of a new porcine brain natriuretic peptide using the *S*-trimethylacetamidomethyl-cysteine. *Tetrahedron Letters*, **30**, 1979–1982.

132 Kiso, Y., Yoshida, M., Fujiwara, Y., Kimura, T., Shimokura, M., and Akaji, K. (1990) Trimethylacetamidomethyl (Tacm) group, a new protecting group for the thiol function of cysteine. *Chemical & Pharmaceutical Bulletin*, **38**, 673–675.

133 Matsueda, R., Kimura, T., Kaiser, E.T., and Matsueda, G.R. (1981) 3-Nitro-2-pyridinesulfenyl group for protection and activation of the thiol function of cysteine. *Chemistry Letters*, 737–740.

134 Wieland, T., Abel, K.-J., and Birr, C. (1977) Antamanide, XXI. Synthesis of disulfide bridged analogs of antamanide. *Liebigs Annalen der Chemie*, 371–380.

135 Inukai, N., Nakano, K., and Murakami, M. (1967) Peptide synthesis. I. Use of the *S*-ethylmercapto group for the protection of the thiol function of cysteine. *Bulletin of the Chemical Society of Japan*, **40**, 2913–2918.

136 Benoiton, L., (1962) A synthesis of isoasparagine from β-benzyl aspartate. *Canadian Journal of Chemistry*, **40**, 570–572.

137 Yajima, H. and Fujii, N. (1981) Studies on peptides. 103. Chemical synthesis of a crystalline protein with the full enzymic activity of ribonuclease A. *Journal of the American Chemical Society*, **103**, 5867–5871.

138 Fujii, N., Otaka, A., Ikemura, O., Akaji, K., Funakoshi, S., Hayashi, Y., Kuroda, Y., and Yajima, H. (1987) Trimethylsilyl trifluoromethanesulfonate as a useful deprotecting reagent in both solution and solid phase peptide syntheses. *Journal of the Chemical Society, Chemical Communications*, 274–275.

139 Fujii, N., Otaka, A., Sugiyama, N., Hatano, M., and Yajima, H. (1987) Studies on peptides. CLV. Evaluation of trimethylsilyl bromide as a hard-acid deprotecting reagent in peptide synthesis. *Chemical & Pharmaceutical Bulletin*, **35**, 3880–3883.

140 Bodanszky, M. and Kwei, J.Z. (1978) Side reactions in peptide synthesis. VII. Sequence dependence in the formation of aminosuccinyl derivatives from β-benzyl-aspartyl peptides. *International Journal of Peptide and Protein Research*, **12**, 69–74.

141 Bodanszky, M., Tolle, J.C., Deshmane, S.S., and Bodanszky, A. (1978) Side reactions in peptide synthesis. VI. A reexamination of the benzyl group in the protection of the side chains of tyrosine and aspartic acid. *International Journal of Peptide and Protein Research*, **12**, 57–68.

142 Teno, N., Tsuboi, S., Shimamura, T., Okada, Y., Yanagida, Y., Yoshinaga, M., Ohgi, K., and Irie, M. (1987) Amino Acids and peptides. XIV. Synthesis and biological activity of three S-peptide analogs of bovine pancreatic ribonuclease A (RNase A). *Chemical & Pharmaceutical Bulletin*, **35**, 468–478.

143 Schneider, F. (1963) Side-chain interaction of serine- or threonine-containing compounds. *Hoppe-Seyler's Zeitschrift fur Physiologische Chemie*, **332**, 38–53.

144 Battersby, A.R. and Robinson, J.C. (1955) Studies on specific chemical fission of peptide links. I. Rearrangements of aspartyl and glutamyl peptides. *Journal of the Chemical Society*, 259–269.

145 Hanson, R.W. and Rydon, H.N. (1964) Polypeptides. IX. Derivatives of aspartylserine. *Journal of the Chemical Society*, 836–842.

146 Schellenberg, P. and Ullrich, J. (1959) Synthesis of further oligopeptides from L-glutamic acid and glycine as well as L-tyrosine. *Chemische Berichte*, **92**, 1276–1287.

147 Liefländer, M. (1960) Preparation and zinc-binding capacity of some glutamyl peptides. *Hoppe-Seyler's Zeitschrift fur Physiologische Chemie*, **320**, 35–57.

148 Iselin, B. and Schwyzer, R. (1962) Synthesis of peptide intermediates for the preparation of bovine melanophore stimulating hormone (β-MSH). I.

Protected peptide sequences 1 to 6 and 1 to 7. *Helvetica Chimica Acta*, **45**, 1499–1509.

**149** Kovacs, J., Kovacs, H.N., and Ballina, R. (1963) Glutamic and aspartic anhydrides. Rearrangement of *N*-carboxyglutamic 1, 5-anhydride to the Leuchs' anhydride and conversion of the latter to pyroglutamic acid. *Journal of the American Chemical Society*, **85**, 1839–1844.

**150** Schröder, E. and Klieger, E. (1964) Peptide syntheses. XVIII. Preparation and reactions of *N*-substituted L-glutamic acid derivatives. *Liebigs Annalen der Chemie*, **673**, 196–207.

**151** Wünsch, E. and Zwick, A. (1966) Synthesis of glucagon. IX. Preparation of sequence 9–15. *Chemische Berichte*, **99**, 105–109.

**152** Schwyzer, R., Iselin, B., Kappeler, H., Riniker, B., Rittel, W., and Zuber, H. (1963) Synthesis of β-melanotropin (β-MSH) with the amino acid sequence of the bovine hormone. *Helvetica Chimica Acta*, **46**, 1975–1996.

**153** Bajusz, S., Lazar, T., and Paulay, Z. (1964) An anomalous reaction of β-*tert*-butyl aspartate. *Acta Chimica Academiae Scientiarum Hungaricae*, **41**, 329–330.

**154** Chillemi, F. (1966) Synthesis of peptide-hydrazides containing a β-*tert*-butyl-L-aspartic acid residue. *Gazzetta Chimica Italiana*, **96**, 359–374.

**155** Yang, C.C. and Merrifield, R.B. (1976) β-Phenacyl ester as a temporary protecting group to minimize cyclic imide formation during subsequent treatment of aspartyl peptides with hydrofluoric acid. *The Journal of Organic Chemistry*, **41**, 1032–1041.

**156** Blake, J. (1979) Use of cyclopentyl ester protection for aspartic acid to reduce base catalyzed succinimide formation in solid-phase peptide synthesis. *International Journal of Peptide and Protein Research*, **13**, 418–425.

**157** Tam, J.P., Wong, T.W., Reimen, M.W., Tjoeng, F.S., and Merrifield, R.B. (1979) Cyclohexyl ester as a new protecting group for aspartyl peptides to minimize aspartimide formation in acidic and basic treatments. *Tetrahedron Letters*, **20**, 4033–4036.

**158** Yajima, H., Futaki, S., Fujii, N., Akaji, K., Funakoshi, S., Sakurai, M., Katakura, S., Inoue, K., Hosotani, R., Tobe, T., Segawa, T., Inoue, A., Tatemoto, K., and Mutt, V. (1985) Synthesis of galanin, a new gastrointestinal polypeptide. *Journal of the Chemical Society, Chemical Communications*, 877–878.

**159** Yajima, H., Futaki, S., Fujii, N., Akaji, K., Funakoshi, S., Sakurai, M., Katakura, S., Hosotani, R., Tobe, T., Inoue, K., Hosotani, R., Tatemoto, K., and Mutt, V. (1986) Studies on peptides. CXXXIII. Synthesis and biological activity of galanin, a novel porcine intestinal polypeptide. *Chemical & Pharmaceutical Bulletin*, **34**, 528–539.

**160** Okada, Y., Iguchi, S., and Kawasaki, K. (1987) Synthesis of β-1- and β-2-adamantyl aspartates and their evaluation for peptide synthesis. *Journal of the Chemical Society, Chemical Communications*, 1532–1534.

**161** Okada, Y. and Iguchi, S. (1988) Amino acids and peptides. Part 19. Synthesis of β-1- and β-2-adamantyl aspartates and their evaluation for peptide synthesis. *Journal of the Chemical Society, Perkin Transactions 1*, 2129–2136.

**162** Klieger, E. and Gibian, H. (1962) Peptide synthesis. X. Simplified preparation and reactions of carbobenzoxy-L-glutamic acid a-(half-esters). *Annals of Chemistry*, **655**, 195–210.

**163** Yamashiro, D., Garzia, R., Hammonds, R.G. Jr, and Li, C.H. (1982) Synthesis and properties of human β-endorphin-(1–9) and its analogs. *International Journal of Peptide and Protein Research*, **19**, 284–289.

**164** Fujii, N., Sakurai, M., Akaji, K., Nomizu, M., Yajima, H., Mizuta, K., Aono, M., Moriga, M., Inoue, K., Hosotani, R., and Tobe, T. (1986) Studies on peptides. CXXXIX. Solution synthesis of a 42-residue peptide corresponding to the entire amino acid sequence of human glucose-dependent insulinotropic polypeptide (GIP). *Chemical & Pharmaceutical Bulletin*, **34**, 2397–2410.

**165** Okada, Y. and Mu, Y. (1997) Amino acids and peptides. XLIX. Synthesis of γ-2-adamantylglutamate and its evaluation for peptide synthesis. *Chemical & Pharmaceutical Bulletin*, **45**, 88–92.

**166** Allen, M.C., Brundish, D.E., Wade, R., Sandberg, B.E.B., Hanley, M.R., and Iversen, L.L. (1982) Tritiated peptides. 12. Synthesis and biological activity of [4-$^3$H-Phe$^8$] substance P. *Journal of Medicinal Chemistry*, **25**, 1209–1213.

**167** Kiso, Y., Kimura, T., Fujiwara, Y., Sakikawa, H., and Akaji, K. (1990) Efficient solid phase peptide synthesis on a phenacyl-resin by a methanesulfonic acid α-amino deprotecting procedure. *Chemical & Pharmaceutical Bulletin*, **38**, 270–272.

**168** Hayakawa, T., Fujikawa, Y., and Noguchi, J. (1967) A new method of reducing nitroarginine-peptide into arginine-peptide, with reference to the synthesis of poly-L-arginine hydrochloride. *Bulletin of the Chemical Society of Japan*, **40**, 1205–1208.

**169** Hofmann, K., Peckham, W.D., and Rheiner, A. (1956) Studies on polypeptides. VII. The synthesis of peptides containing arginine. *Journal of the American Chemical Society*, **78**, 238–242.

**170** Zervas, L., Winitz, M., and Greenstein, J.P. (1956) The percarbobenzoxylation of L-arginine. *Archives of Biochemistry and Biophysics*, **65**, 573–574.

**171** Jäger, G. and Geiger, R. (1970) Adamantyl-1-oxycarbonyl residues as blocking group for the guanidine function of arginine. *Chemische Berichte*, **103**, 1727–1747.

**172** Paulay, Z. and Bajusz, S. (1965) A novel protection for the guanidine group of arginine. *Acta Chimica Academiae Scientiarum Hungaricae*, **43**, 147–148.

**173** Ramachandran, J. and Li, C.H. (1962) Preparation of crystalline $N^G$-tosylarginine derivatives. *The Journal of Organic Chemistry*, **27**, 4006–4009.

**174** Schnabel, E. and Li, C.H. (1960) The synthesis of L-histidyl-D-phenylalanyl-L-arginyl-L-tryptophylglycine and its melanocyte-stimulating activity. *Journal of the American Chemical Society*, **82**, 4576–4579.

**175** Mazur, R.H. and Plume, G. (1968) Synthesis of bradykinin. *Experientia*, **24**, 661.

**176** Yajima, H., Fujii, N., Ogawa, H., and Kawatani, H. (1974) Trifluoromethanesulphonic acid as a deprotecting reagent in peptide chemistry. *Journal of the Chemical Society, Chemical Communications*, 107–108.

**177** Kiso, Y., Satomi, M., Ukawa, K., and Akita, T. (1980) Efficient deprotection of $N^G$-tosylarginine with a thioanisole-trifluoromethanesulphonic acid system. *Journal of the Chemical Society, Chemical Communications*, 1063–1064.

**178** Fujino, M., Wakimasu, M., and Kitada, C. (1981) Further studies on the use of multi-substituted benzenesulfonyl groups for protection of the guanidine function of arginino. *Chemical & Pharmaceutical Bulletin*, **29**, 2825–2931.

**179** Fujii, N., Otaka, A., Sugiyama, N., Hatano, M., and Yajima, H. (1987) Studies on peptides. CLV. Evaluation of trimethylsilyl bromide as a hard-acid deprotecting reagent in peptide synthesis. *Chemical & Pharmaceutical Bulletin*, **35**, 3880–3883.

**180** Akaji, K., Yoshida, M., Tatsumi, T., Kimura, T., Fujiwara, Y., and Kiso, Y. (1990) Tetrafluoroboric acid as a useful deprotecting reagent in Fmoc-based solid-phase peptide synthesis (Fmoc = 9-fluorenylmethoxycarbonyl). *Journal of the Chemical Society, Chemical Communications*, 288–290.

**181** Ramachandran, J., and Li, C.H. (1963) The synthesis of L-valyl-L-lysyl-L-valyl-L-tyrosyl-L-proline. *The Journal of Organic Chemistry*, **28**, 173–177.

**182** Paul, R. (1963) O-Acylation of tyrosine during peptide synthesis. *The Journal of Organic Chemistry*, **28**, 236–237.

**183** Wünsch, E., Fries, G., and Zwick, A. (1958) Peptide synthesis. V. Preparation and use of O-benzyl-L-tyrosine in the synthesis of peptides. *Chemische Berichte*, **91**, 542–547.

**184** Katsoyannis, P.G. and Suzuki, K. (1962) Insulin peptides. III. Synthesis of a protected nanopeptide containing the C-terminal sequence of the B-chain of insulin. *Journal of the American Chemical Society*, **84**, 1420–1423.

**185** Erickson, B.W. and Merrifield, R.B. (1973) Acid stability of several benzylic protecting groups used in solid-phase peptide synthesis. Rearrangement of *O*-benzyltyrosine to 3-benzyltyrosine. *Journal of the American Chemical Society*, **95**, 3750–3756.

**186** Erickson, B.W., and Merrifield, R.B. (1973) Use of chlorinated benzyloxycarbonyl protecting groups to eliminate $N^{\varepsilon}$-branching at lysine during solid-phase peptide synthesis. *Journal of the American Chemical Society*, **95**, 3757–3763.

**187** Yamashiro, D. and Li, C.H. (1973) Protection of tyrosine in solid-phase peptide synthesis. *The Journal of Organic Chemistry*, **38**, 591–592.

**188** Engelhard, M. and Merrifield, R.B. (1978) Tyrosine protecting groups: minimization of rearrangement to 3-alkyltyrosine during acidolysis. *Journal of the American Chemical Society*, **100**, 3559–3563.

**189** Lundt, B.F., Johansen, N.L., and Markussen, J. (1979) Formation and synthesis of 3′-*t*-butyltyrosine. *International Journal of Peptide and Protein Research*, **14**, 344–346.

**190** Sugano, H. and Miyoshi, M. (1976) A convenient synthesis of *N-tert*-butyloxycarbonyl-*O*-benzyl-L-serine. *The Journal of Organic Chemistry*, **41**, 2352–2353.

**191** Grassmann, W., Wünsch, E., Deufel, P., and Zwick, A. (1958) Peptide synthesis. IV. Preparation and use of *O*-benzylserine in the synthesis of peptides. *Chemische Berichte*, **91**, 538–541.

**192** Bodanszky, M., Ondetti, M.A., Rubin, B., Piala, J.J., Fried, J., Sheehan, J.T., and Birkhimer, C.A. (1962) Biologically active citrulline peptides. *Nature*, **194**, 485–486.

**193** Yamashiro, D. (1977) Protection of aspartic acid, serine, and threonine in solid-phase peptide synthesis. *The Journal of Organic Chemistry*, **42**, 523–525.

**194** Beyerman, H.C. and Bontekoe, J.S. (1962) The *tert*-butoxy group, a novel hydroxy-protecting group for use in peptide synthesis with hydroxyl amino acids. *Recueil des Travaux Chimiques des Pays-Bas*, **81**, 691–698.

**195** König, W. (1973) Peptides with the properties of human-proinsulin-C-peptide (hC-peptide). II. Sequences 18–23 and 24–27 of human-proinsulin-C-peptide. *Chemische Berichte*, **106**, 193–198.

**196** Schröder, E. (1963) Peptide synthesis. XV. New *O-tert*-butylhydroxyamino acid derivatives and their use in the syntheses of a glucagon fragment. *Annals of Chemistry*, **670**, 127–136.

**197** Sakakibara, S. and Fujii, T. (1969) Synthesis and use of $N^{im}$-tosyl-L-histidine. *Bulletin of the Chemical Society of Japan*, **42**, 1466.

**198** Beyerman, H.C., Hirt, J., Kranenbuerg, P., Syrier, J.L.M., and Van Zon, A. (1974) Excess mixed anhydride peptide synthesis with histidine derivatives. *Recueil des Travaux Chimiques des Pays-Bas*, **93**, 256–257.

**199** Siepmann, E. and Zahn, H. (1964) $N^{im}$-2,4-Dinitrophenyl-L-histidine. *Biochimica et Biophysica Acta*, **82**, 412–415.

**200** Losse, G. and Krychowshi, U. (1971) New histidine derivatives for Merrifield peptide synthesis. *Tetrahedron Letters*, **12**, 4121–4124.

**201** Shalteil, S. (1967) Thiolysis of some dinitrophenyl derivatives of amino acids. *Biochemical and Biophysical Research Communications*, **29**, 178–183.

**202** Jones, J.H. and Ramage, W.I. (1978) An approach to the prevention of racemization in the synthesis of histidine-containing peptides. *Journal of the Chemical Society, Chemical Communications*, 472–473.

**203** Fletcher, A.R., Jones, J.H., Ramage, W.I., and Stachulski, A.V. (1979) The use of the $N^{\pi}$-phenacyl group for the protection of the histidine side chain in peptide synthesis. *Journal of the Chemical Society, Perkin Transactions 1*, 2261–2267.

**204** Brown, T. and Jones, J.H. (1981) Protection of histidine side-chains with π-benzyloxymethyl or π-bromobenzyloxymethyl-groups. *Journal of the Chemical Society, Chemical Communications*, 648–649.

**205** Brown, T., Jones, J.H., and Richards, J.D. (1982) Further studies on the protection of histidine side chains in peptide synthesis: the use of the π-benzyloxymethyl group. *Journal of the Chemical Society, Perkin Transactions 1*, 1553–1561.

206 Gesquiee, J.-C., Diesis, E., and Tartar, A. (1990) Conversion of *N*-terminal cysteine to thiazolidine carboxylic acid during hydrogen fluoride deprotection of peptides containing $N^{\pi}$-Bom protected histidine. *Journal of the Chemical Society, Chemical Communications*, 1402–1403.

207 Mitchell, M.A., Runge, T.A., Mathews, W.R., Ichhpurani, A.K., Harn, N.K., Dobrowolski, P.J., and Eckenrode, F.M. (1990) Problems associated with use of the benzyloxymethyl protecting group for histidines. Formaldehyde adducts formed during cleavage by hydrogen fluoride. *International Journal of Peptide and Protein Research*, **36**, 350–355.

208 Kumagaya, K.Y., Inui, T., Nakajima, K., Kimura, T., and Sakakibara, S. (1991) Suppression of a side reaction associated with $N^{im}$-benzyloxymethyl group during synthesis of peptides containing cysteinyl residue at the *N*-terminus. *Peptide Research*, **4**, 84–87.

209 Colombo, R., Colombo, F., and Jones, J.H. (1984) Acid-labile histidine side-chain protection: the $N^{(\pi)}$-*tert*-butoxymethyl group. *Journal of the Chemical Society, Chemical Communications*, 292–293.

210 Okada, Y., Wang, J., Yamamoto, T., and Mu, Y. (1996) Development of a new $N^{\pi}$-protecting group for histidine, $N^{\pi}$-1-adamantyloxymethylhistidine. *Chemical & Pharmaceutical Bulletin*, **44**, 871–873.

211 Okada, Y., Wang, J., Yamamoto, T., Mu, Y., and Yokoi, T. (1996) Amino acids and peptides. Part 45. Development of a new $N^{\pi}$-protecting group of histidine, $N^{\pi}$-(1-adamantyloxymethyl)histidine, and its evaluation for peptide synthesis. *Journal of the Chemical Society, Perkin Transactions 1*, 2139–2143.

212 Okada, Y., Wang, J., Yamamoto, T., Yokoi, T., and Mu, Y. (1996) Synthesis of $N^{\pi}$-2-adamantyloxymethylhistidine, His($N^{\pi}$-2-Adom), and its evaluation for peptide synthesis. *Journal of the Chemical Society, Perkin Transactions 1*, 753–754.

213 Okada, Y., Wang, J., Yamamoto, T., Yokoi, T., and Mu, Y. (1997) Amino acids and peptides. L. Development of a novel $N^{\pi}$-protecting group for histidine, $N^{\pi}$-2-adamantyloxymethylhistidine, and its application to peptide synthesis. *Chemical & Pharmaceutical Bulletin*, **45**, 452–456.

214 Previero, A., Coletti-Previero, M.A., and Cavadore, J.-C. (1967) A reversible chemical modification of the tryptophan residue. *Biochimica et Biophysica Acta*, **147**, 453–461.

215 Fukuda, T., Wakimasu, M., Kobayashi, S., and Fujino, M., (1982) New protecting groups for the indole ring of tryptophan in peptide synthesis; 2,4,6-trimethoxybenzenesulfonyl groups. *Chemical & Pharmaceutical Bulletin*, **30**, 2825–2835.

216 Fujii, N., Futaki, S., Yasumura, K., and Yajima, H. (1984) Studies on peptides CXXI. $N^{in}$-mesitylensulfonyltryptophan, a new derivative for peptide synthesis. *Chemical & Pharmaceutical Bulletin*, **32**, 2660–2665.

217 Bodanszky, M. and Williams, N.J. (1967) Synthesis of secretin. I. The protected tetradecapeptide corresponding to sequence 14–27. *Journal of the American Chemical Society*, **89**, 685–689.

218 Bodanszky, M., Ondetti, M.A., Levine, S.D., and Williams, N.J. (1967) Synthesis of secretin. II. The stepwise approach. *Journal of the American Chemical Society*, **89**, 6753–6757.

219 Khorana, H.G. (1952) Peptides. III. Selective degradation from the carboxyl end. The use of carbodiimides. *Journal of the Chemical Society*, 2081–2088.

220 Sheehan, J.C. and Hess, G.P. (1955) A new method of forming peptide bonds. *Journal of the American Chemical Society*, **77**, 1067–1068.

221 Sarantakis, D., Teichman, J., Lien, E.L., and Fenichel, R.L. (1976) A novel cyclic undecapeptide, WY-40,770, with prolonged growth hormone release inhibiting activity. *Biochemical and Biophysical Research Communications*, **73**, 336–342.

222 Khorana, H.G. (1955) The use of dicyclohexylcarbodiimide in the synthesis of peptides. *Chemistry & Industry*, 1087–1088.

223 Khorana, H.G. and Todd, A.R. (1953) Phosphorylation. XI. The reaction between carbodiimides and acid esters of phosphoric acid. A new method for the

preparation of pyrophosphates. *Journal of the Chemical Society*, 2257–2260.

**224** Smith, M., Moffatt, J.G., and Khorana, H.G. (1958) Carbodiimides. VIII. Observations on the reactions of carbodiimides with acids and some new applications in the synthesis of phosphoric acid esters. *Journal of the American Chemical Society*, **80**, 6204–6212.

**225** DeTar, D.F. and Silverstein, R. (1966) Reaction of carbodiimides. I. The mechanisms of the reactions of acetic acid with dicyclohexylcarbodiimide. *Journal of the American Chemical Society*, **88**, 1013–1019.

**226** Merrifield, R.B., Gisin, B.F., and Bach, A.N. (1977) The limits of reaction of radioactive dicyclohexylcarbodiimide with amino groups during solid-phase peptide synthesis. *The Journal of Organic Chemistry*, **42**, 1291–1295.

**227** Sheehan, J.C., Cruickshank, P.A., and Boshart, G.L. (1961) A convenient synthesis of water-soluble carbodiimides. *The Journal of Organic Chemistry*, **26**, 2525–2528.

**228** Wieland, T., Kern, W., and Sehring, R. (1950) Anhydrides of acylated amino acids. *Annals of Chemistry*, **569**, 117–121.

**229** Vaughan, J.R. Jr and Osato, R.L. (1951) Preparation of peptides using mixed carboxylic acid anhydrides. *Journal of the American Chemical Society*, **73**, 5553–5555.

**230** Zaoral, M. (1962) Amino acids and peptides. XXXVI. Pivaloyl chloride as a reagent in the mixed anhydride synthesis of peptides. *Collection of Czechoslovak Chemical Communications*, **27**, 1273–1277.

**231** Wieland, T. and Bernhard, H. (1951) Peptide syntheses. III. Use of anhydrides of *N*-acylated amino acids and derivatives of inorganic acids. *Annals of Chemistry*, **572**, 190–194.

**232** Boissonas, R.A. (1951) A new method of peptide synthesis. *Helvetica Chimica Acta*, **34**, 874–879.

**233** Vaughan, J.R. and Osato, R.L. (1952) The preparation of peptides using mixed carbonic-carboxylic acid anhydrides. *Journal of the American Chemical Society*, **74**, 676–678.

**234** Gordon, M., Miller, J.G., and Day, A.R. (1948) Effect of structure on reactivity. I. Ammonolysis of esters with special reference to the electron-release effects of alkyl and aryl groups. *Journal of the American Chemical Society*, **70**, 1946–1953.

**235** Wieland, T., Schäfer, W., and Bokelmann, E. (1951) Peptide syntheses. V. A convenient method for the preparation of acylthiophenols and their application in the syntheses of amides and peptides. *Annals of Chemistry*, **573**, 99–104.

**236** Sandrin, E. and Boissonnas, R.A. (1963) Synthesis of structural analogs of eledoisin. I. Preparation of intermediates. *Helvetica Chimica Acta*, **46**, 1637–1669.

**237** Kovacs, J., Ceprini, M.Q., Dupraz, C.A., and Schmit, G.N. (1967) Pentachlorophenyl esters of *N*-carbobenzoxy-L-amino acids. *The Journal of Organic Chemistry*, **32**, 3696–3698.

**238** Kovacs, J., Kisfaludy, L., and Ceprini, M.Q. (1967) On the optical purity of peptide active esters prepared by *N,N′*-dicyclohexylcarbodiimide and "complexes" of *N,N′*-dicyclohexylcarbodiimide-pentachloorphenol and *N,N′*-dicyclohexylcarbodiimde-pentafluorophenol. *Journal of the American Chemical Society*, **89**, 183–184.

**239** Anderson, G.W., Zimmerman, J.E., and Callahan, F.J. (1963) *N*-Hydroxysuccinimide esters in peptide synthesis. *Journal of the American Chemical Society*, **85**, 3039.

**240** Gross, H. and Bilk, L. (1968) Reaction of *N*-hydroxysuccinimide with dicyclohexylcarbodiimide. *Tetrahedron*, **24**, 6935–6939.

**241** Horiki, K. (1977) Behavior of acylated 1-hydroxybenzotriazole. *Tetrahedron Letters*, **18**, 1897–1900.

**242** König, W. and Geiger, R. (1970) New method for the synthesis of peptides: Activation of the carboxyl group with dicyclohexylcarbodiimide by using 1-hydroxybenzotriazoles as additives. *Chemische Berichte*, **103**, 788–798.

**243** Gawne, G., Kenner, G.W., and Sheppard, R.C. (1969) Acyloxyphosphonium salts as acylating agents. Synthesis of peptides.

*Journal of the American Chemical Society*, **91**, 5669–5671.

**244** Castro, B., Dormoy, J.R., Evin, G., and Selve, C. (1975) Peptide coupling reagents IV. *N*-[Oxytris (dimethylamino) phosphonium]benzotriazole hexafluorophosphate. *Tetrahedron Letters*, 1219–1222.

**245** (a) Kim, M.H. and Patel, D.V. (1994) "BOP" as a reagent for mild and efficient preparation of esters. *Tetrahedron Letters*, **35**, 5603–5606; (b) Le-Nguyen, D., Heitz, A., and Castro, B. (1987) Renin substrates. Part 2. Rapid solid phase synthesis of the ratine sequence tetradecapeptide using BOP reagent. *Journal of the Chemical Society, Perkin Transactions 1*, 1915–1919.

**246** Coste, J., Le-Nguyen, D., and Castro, B. (1990) PyBOP®: a new peptide coupling reagent devoid of toxic byproduct. *Tetrahedron Letters*, **31**, 205–208.

**247** Coste, J., Dufour, M.N., Pantaloni, A., and Castro, B. (1990) BroP: a new reagent for coupling *N*-methylated amino acids. *Tetrahedron Letters*, **31**, 669–672.

**248** Coste, J., Frerot, E., and Jouin, P. (1994) Coupling *N*-methylated amino acids using PyBroP and PyCloP halogenophosphonium salts: mechanism and fields of application. *The Journal of Organic Chemistry*, **59**, 2437–2446.

**249** Knorr, R., Trzeciak, A., Bannwarth, W., and Gillessen, D. (1989) New coupling reagents in peptide chemistry. *Tetrahedron Letters*, **30**, 1927–1930.

**250** Dourtoglou, V., Gross, B., Lambropoulou, V., and Zioudrou, C. (1984) *O*-Benzotriazolyl-*N,N,N′,N′*-tetramethyluronium hexafluorophosphate as coupling reagent for the synthesis of peptides of biological interest. *Synthesis*, 572–574.

**251** Fields, C.G., Lloyd, D.H., MacDonald, R.L., Otteson, K.M., and Noble, R.L. (1991) HBTU activation for automated Fmoc solid-phase peptide synthesis. *Peptide Research*, **4**, 95–101.

**252** Carpino, L.A. (1993) 1-Hydroxy-7-azabenzotriazole. An efficient peptide coupling additive. *Journal of the American Chemical Society*, **115**, 4397–4398.

**253** Carpino, L.A., Imazumi, H., El-Faham, A., Ferrer, F.J., Zhang, C., Lee, Y., Foxman, B.M., Henklein, P., Hanay, C., Mügge, C., Wenschuh, H., Klose, J., Beyermann, M., and Bienert, M. (2002) The uronium/guanidium peptide coupling reagents: finally the true uronium salts. *Angewandte Chemie International Edition*, **41**, 441–445.

**254** (a) Benoiton, N.L., Kuroda, K., and Chen, F.M.F. (1982) Examination of chiral stability during the preparation of hydrazides and coupling by the azide procedure using a series of model peptides. *International Journal of Peptide and Protein Research*, **20**, 81–86; (b) Jones, J.H. (1979) The formation of peptide bonds: a general survey, in *The Peptides: Analysis, Synthesis, Biology* (eds E. Gross and J. Meienhofer), Academic Press, New York, pp. 65–104.

**255** Shioiri, T., Ninomiya, K., and Yamada, S. (1972) Diphenylphosphoryl azide. New convenient reagent for a modified Curtius reaction and for peptide synthesis. *Journal of the American Chemical Society*, **94**, 6203–6205.

**256** Shioiri, T. and Yamada, S. (1974) Amino acids and peptides. XI. Phosphorus in organic synthesis. VI. Application of diphenyl phosphorazidate to the synthesis of peptides containing various functions. *Chemical & Pharmaceutical Bulletin*, **22**, 859–863.

**257** Shioiri, T. and Yamada, S. (1974) Amino acids and peptides. X. Phosphorus in organic synthesis. V. Mechanism for the peptide synthesis by diphenyl phosphorazidate. *Chemical & Pharmaceutical Bulletin*, **22**, 855–858.

**258** Brady, S.F., Varga, S.L., Freidinger, R.M., Schwenk, D.A., Mendlowski, M., Holly, F.W., and Veber, D.F. (1979) Practical synthesis of cyclic peptides, with an example of dependence of cyclization yield upon linea sequence. *The Journal of Organic Chemistry*, **44**, 3101–3105.

**259** Weygand, F., Prox, A., and König, W. (1966) Racemization during peptide synthesis. *Chemische Berichte*, **99**, 1451–1460.

**260** Weygand, F., Hoffmann, D., and Wünsch, E. (1966) Synthesis of peptides with dicyclohexylcarbodiimide by addition of N-hydroxysuccinimide. *Zeitschrift fur Naturforschung B*, **21**, 426–428.

**261** Wünsch, E. and Drees, F. (1966) Synthesis of glucagon. X. Synthesis of sequence 22–29. *Chemische Berichte*, **99**, 110–120.

**262** Wünsch, E. (1967) Total synthesis of the pancreatic hormone glucagon. *Zeitschrift fur Naturforschung B*, **22**, 1267–1276.

**263** Wünsch, E. and Wendlberger, G. (1968) Glucagon synthesis. XVIII. Preparation of the total sequence. *Chemische Berichte*, **101**, 3659–3663.

**264** Wünsch, E., Jaeger, E., and Scharf, R. (1968) Glucagon synthesis. XIX. Purification of synthetic glucagon. *Chemische Berichte*, **101**, 3664–3670.

**265** König, W. and Geiger, R. (1979) Racemization in peptide synthesis. *Chemische Berichte*, **103**, 2024–2033.

**266** König, W. and Geiger, R. (1979) Synthesis of peptides: activation of the carboxyl group with dicyclohexylcarbodiimide and 3-hydroxy-4-oxo-2,4-dihydro-1,2,3-benzotriazole. *Chemische Berichte*, **103**, 2034–2040.

**267** Koyama, K., Watanabe, H., Kawatani, H., Iwai, J., and Yajima, H. (1976) Studies on peptides. LXI. Alternate synthesis of human corticotropin (ACTH). *Chemical & Pharmaceutical Bulletin*, **24**, 2558–2563.

**268** Kimura, T., Takai, M., Masui, Y., Morikawa, T., and Sakakibara, S. (1981) Strategy for the synthesis of large peptides: an application to the total synthesis of human parathyroid hormone [hPTH (1–84)]. *Biopolymers*, **20**, 1823–1832.

**269** Kuroda, H., Chen, Y.N., Watanabe, T.X., Kimura, T., and Sakakibara, S. (1992) Solution synthesis of calciseptine, an L-type specific calcium channel blocker. *Peptide Research*, **5**, 265–268.

**270** Nishio, N., Kumagaye, S., Kuroda, H., Chino, N., Emura, J., Kimura, T., and Sakakibara, S. (1992) Solution synthesis of sodium-potassium ATPase inhibitor-1 (SPAI-1). *Peptide Research*, **5**, 227–232.

**271** Bodi, J., Nishio, H., Zhou, Y., Branton, W.D., Kimura, T., and Sakakibara, S. (1995) Synthesis of O-palmitoylated 44-residue peptide amide (PLTX II) blocking presynaptic calcium channels in *Drosophila*. *Peptide Research*, **8**, 228–235.

**272** Carpino, L.A., El-Faham, A., and Albericio, F. (1995) Efficiency in peptide coupling: 1-hydroxy-7-azabenzotriazole vs 3,4-dihydro-3-hydroxy-4-oxo-1,2,3-benzotriazine. *The Journal of Organic Chemistry*, **60**, 3561–3564.

**273** Lloyd-Williams, P., Albercio, F., and Giralt, E. (1993) Convergent solid-phase peptide synthesis. *Tetrahedron*, **49**, 11065–11133.

**274** Narita, M., Fukunaga, T., Wakabayashi, A., Ishikawa, K., and Nakano, H. (1984) Synthesis and properties of tertiary peptide bond-containing polypeptides. I. Syntheses and properties of oligo (L-leucine)s containing proline or glycyl-N-(2,4-dimethoxybenzyl)-L-leucine residues. *International Journal of Peptide and Protein Research*, **23**, 306–314.

**275** Narita, M., Ishikawa, K., Chen, J.-Y., and Kim, Y. (1984) Prediction and improvement of protected peptide solubility in organic solvents. *International Journal of Peptide and Protein Research*, **24**, 580–587.

**276** Narita, M., Honda, S., Umeyama, H., and Obana, S. (1988) The solubility of peptide intermediates in organic solvents. Solubilizing potential of hexafluoro-2-propanol. *Bulletin of the Chemical Society of Japan*, **61**, 281–284.

**277** Yamashiro, D., Brake, J., and Li, C.H. (1976) The use of trifluoroethanol for improved coupling in solid phase peptide synthesis. *Tetrahedron Letters*, **17**, 1469–1472.

**278** Narita, M., Kojima, Y., and Isokawa, S. (1989) Design of the synthetic route for helical peptides. Synthesis and solubility of model peptides having a helical structure. *Bulletin of the Chemical Society of Japan*, **62**, 1976–1981.

**279** Kuroda, H., Chen, Y.N., Kimura, T., and Sakakibara, S. (1992) Powerful solvent systems useful for synthesis of sparingly-soluble peptides in solution. *International Journal of Peptide and Protein Research*, **40**, 294–299.

**280** Nishi, N. (ed.) (1996) *Peptide Chemistry: Proceedings of the 33rd Symposium on*

*Peptide Chemistry*, Protein Research Foundation, Minoh.

281 Shimonishi, Y. (ed.) (1999) *Peptide Science – Present and Future: Proceedings of the 1st International Peptide Symposium*, Kluwer, Dordrecht.

282 Fields, G.B., Tam, J.P., and Barany, G. (eds) (2000) *Peptides for the New Millennium: Proceedings of the 16th American Peptide Symposium*, Kluwer, Dordrecht.

283 Dawson, P.E., Muir, T.W., Clark-Lewis, I., and Kent, S.B. (1994) Synthesis of proteins by native chemical ligation. *Science*, **266**, 776–779.

284 Yan, L.Z. and Dawson, P.E. (2001) Synthesis of peptides and proteins without cysteine residues by native chemical ligation combined with desulfurization. *Journal of the American Chemical Society*, **123**, 526–533.

285 Torbeev, V.Y. and Kent, S.B.H. (2007) Convergent chemical synthesis and crystal structure of a 203 amino acid "covalent dimer" HIV-1 protease enzyme molecule. *Angewandte Chemie International Edition*, **46**, 1667–1670.

286 Arakawa, K. and Bumpus, F.M. (1961) An improved synthesis of isoleucine[5] angiotensin octapeptide. *Journal of the American Chemical Society*, **83**, 728–732.

287 Kung, Y.T., Du, Y.C., Huang, W.T., Chen, C.C., and Ke, L.T. (1965) Total synthesis of crystalline bovine insulin. *Scientia Sinica*, **14**, 1710–1716.

288 Kung, Y.T., Du, Y.C., Huang, W.T., Chen, C.C., and Ke, L.T. (1966) Total synthesis of crystalline insulin. *Scientia Sinica*, **15**, 544–561.

289 Zahn, H., Brinkhoff, O., Meienhofer, J., Pfeiffer, E.F., Ditschuneit, H., and Gloxhuber, C. (1965) Peptides. XLIX. Combination of synthetic insulin chains into biologically active preparations. *Zeitschrift fur Naturforschung B*, **20**, 666–670.

290 Rittel, W., Brugger, M., Kamber, B., Riniker, B., and Sieber, P. (1968) Thyrocalcitonin. III. Synthesis of α-thyrocalcitonin. *Helvetica Chimica Acta*, **51**, 924–928.

291 Tam, J.P., Heath, W.F., and Merrifield, R.B. (1982) Improved deprotection in solid phase peptide synthesis: quantitative reduction of methionine sulfoxide to methionine during hydrogen fluoride cleavage. *Tetrahedron Letters*, **23**, 2939–2942.

292 Fujii, N. and Yajima, H. (1981) Total synthesis of bovine pancreatic ribonuclease A. Part 1. Synthesis of the protected pentadecapeptide ester (positions 110–124). *Journal of the Chemical Society, Perkin Transactions 1*, 789–796.

293 Fujii, N. and Yajima, H. (1981) Total synthesis of bovine pancreatic ribonuclease A. Part 2. Synthesis of the protected hexatriacontapeptide ester (positions 89–124). *Journal of the Chemical Society, Perkin Transactions 1*, 797–803.

294 Fujii, N. and Yajima, H. (1981) Total synthesis of bovine pancreatic ribonuclease A. Part 3. Synthesis of the protected hexapentacontapeptide ester (positions 69–124). *Journal of the Chemical Society, Perkin Transactions 1*, 804–810.

295 Fujii, N. and Yajima, H. (1981) Total synthesis of bovine pancreatic ribonuclease A. Part 4. Synthesis of the protected tetraoctacontapeptide ester (positions 41–124). *Journal of the Chemical Society, Perkin Transactions 1*, 811–818.

296 Fujii, N. and Yajima, H. (1981) Total synthesis of bovine pancreatic ribonuclease A. Part 5. Synthesis of the protected S-protein (positions 21–124) and the protected S-peptide (positions 1–20). *Journal of the Chemical Society, Perkin Transactions 1*, 819–830.

297 Fujii, N. and Yajima, H. (1981) Total synthesis of bovine pancreatic ribonuclease A. Part 6. Synthesis of the RNase A with full enzymic activity. *Journal of the Chemical Society, Perkin Transactions 1*, 831–841.

298 Yajima, H., Fujii, N., Akaji, K., Sakurai, M., Nomizu, M., Mizuta, K., Aono, M., Moriga, M., Inoue, K., Hosotani, R., and Tobe, T. (1985) Synthesis of a 42 residue peptide corresponding to the entire amino acid sequence of human GIP. *Chemical & Pharmaceutical Bulletin*, **33**, 3578–3581.

299 Hofmann, K., Yajima, H., Yanaihara, N., Liu, Y.-y., and Lande, S. (1961) Studies on polypeptides. XIII. The synthesis of a

tricosapeptide possessing essentially the full biological activity of natural ACTH[1-13]. *Journal of the American Chemical Society*, **83**, 487–489.

300 Lundt, B.F., Johansen, N.L., Volund, A., and Markussen, J. (1978) Removal of *t*-butyl and *t*-butoxycarbonyl protecting groups with trifluoroacetic acid. Mechanisms, byproduct formation and evaluation of scavengers. *International Journal of Peptide and Protein Research*, **12**, 258–268.

301 Lundt, B.F., Johansen, N.L., and Markussen, J. (1979) Formation and synthesis of 3′-*t*-butyltyrosine. *International Journal of Peptide and Protein Research*, **14**, 344–346.

302 Noble, R.L., Yamashiro, D., and Li, C.H. (1976) Synthesis of a nonadecapeptide corresponding to residues 37–55 of ovine prolactin. Detection and isolation of the sulfonium form of methionine-containing peptides. *Journal of the American Chemical Society*, **98**, 2324–2328.

303 Dryland, A. and Sheppard, R.C., (1986) Peptide synthesis. Part 8. A system for solid-phase synthesis under low pressure continuous-flow conditions. *Journal of the Chemical Society, Perkin Transactions 1*, 125–137.

304 Pearson, R.G. and Songstad, J. (1967) Application of the principle of hard and soft bases to organic chemistry. *Journal of the American Chemical Society*, **89**, 1827–1836.

305 Pearson, R.G. (1966) Acids and bases. *Science*, **151**, 172–177.

306 Yajima, H., Funakoshi, S., and Akaji, K. (1986) Several methodological improvements for the synthesis of biologically active polypeptides. *Biopolymers*, **25** (Suppl.), S39–S46.

307 Noyori, R., Murata, S., and Suzuki, M. (1981) Trimethylsilyl triflate in organic synthesis. *Tetrahedron*, **37**, 3899–3910.

308 Fujii, N., Ikemura, O., Funakoshi, O., Matsuo, H., Segawa, T., Nakata, Y., Inoue, A., and Yajima, H. (1987) Studies on peptides. CXLVII. Application of a new deprotecting procedure with trimethylsilyl trifluoromethanesulfonate for the syntheses of two porcine spinal cord peptides, neuromedin U-8 and neuromedin U-25. *Chemical & Pharmaceutical Bulletin*, **35**, 1076–1084.

309 Nomizu, M., Akaji, K., Fukata, J., Imura, H., Inoue, A., Nakata, Y., Segawa, T., Fujii, N., and Yajima, H. (1988) Studies on peptides. CLVII. Synthesis of a frog-skin peptide, sauvagine. *Chemical & Pharmaceutical Bulletin*, **36**, 122–133.

310 Sugiyama, N., Fujii, N., Funakoshi, S., Funakoshi, A., Miyasaka, K., Aono, M., Moriga, M., Inoue, K., Kogire, M., Sumi, S., Doi, R., Tobe, T., and Yajima, H. (1987) Studies on peptides. CLIV. Synthesis of a 36-residue peptide corresponding to the entire amino acid sequence of human pancreatic polypeptide. *Chemical & Pharmaceutical Bulletin*, **35**, 3585–3596.

311 Matsuzaki, K., Harada, M., Handa, T., Funakoshi, S., Fujii, N., Yajima, H., and Miyajima, K. (1989) Magainin 1-induced leakage of entrapped calcein out of negatively-charged lipid vesicles. *Biochimica et Biophysica Acta*, **981**, 130–134.

# 7
# Solid-Phase Peptide Synthesis: Historical Aspects

*Garland R. Marshall*

## 7.1
## Introduction

It is rare to be an observer/participant in the development of a technology that has had as much impact on scientific capabilities as has the solid-phase method of using a polymeric protecting group as a convenient handle for purification. While the potential impact was certainly apparent in 1959 to its inventor, Professor R. Bruce Merrifield (Figure 7.1) of Rockefeller University, and many others as they became exposed to the idea, no one could have predicted its dominance in combinatorial chemistry, a field not yet conceived. It is also interesting to reflect on the severe resistance that the concept generated by the synthetic organic chemistry community whom it would best serve. One might expect such reluctance as a common phenomenon for any paradigm shift that threatens the status quo [1]. What follows is a personal view of the historical development of solid-phase peptide synthesis as well as its acceptance by the scientific community at a time when the basic concept was being developed from a prototype into a reliable methodology.

## 7.2
## Selection of Compatible Synthetic Components

Merrifield's notebook entry of 5/26/59 – "A New Approach to the Continuous, Stepwise Synthesis of Peptides" records the first formal expression of the concept of solid-phase peptide synthesis (SPPS). In his autobiography written at the request of the American Chemical Society, Bruce outlined his scientific background and experiences at the bench with the solution synthesis of peptides that led him to this revolutionary concept, so trivial once expressed and so powerful once reduced to practice. With Professor D. Wayne Woolley's blessing as head of the laboratory, Bruce started the search for the right support, linker, amino-protecting group and procedure for cleaving the product from the polymeric support. Except for his autobiography, *Life During a Golden Age of Peptide Chemistry* [2], published by the ACS in 1993,

*Amino Acids, Peptides and Proteins in Organic Chemistry.*
*Vol.3 – Building Blocks, Catalysis and Coupling Chemistry.* Edited by Andrew B. Hughes

ISBN: 978-3-527-32102-5

**Figure 7.1** Bruce Merrifield reviewing his notebook entry of 5/26/59 "A New Approach to the Continuous, Stepwise Synthesis of Peptides" (Courtesy of Professor John M. Stewart).

very little has been revealed about the multiple combinations of the different components explored by Merrifield during this period to realize his concept of SPPS. A number of polymeric supports, linkers, protecting groups, cleavage reagents, and so on, were examined and found wanting, at least with the technology available at the time. The problem was solved when Bruce obtained a sample of the polymer, polystyrene cross-linked with divinylbenzene (Figure 7.2), used by Dow to make ion-exchange resins for column chromatography. The polymer beads had to be functionalized (Figure 7.3) to generate linkage sites for building the growing peptide

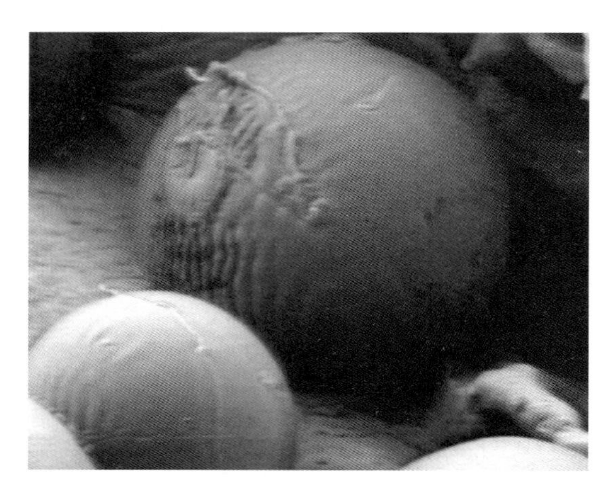

**Figure 7.2** Polystyrene beads cross-linked with divinylbenzene – the first successful polymeric support.

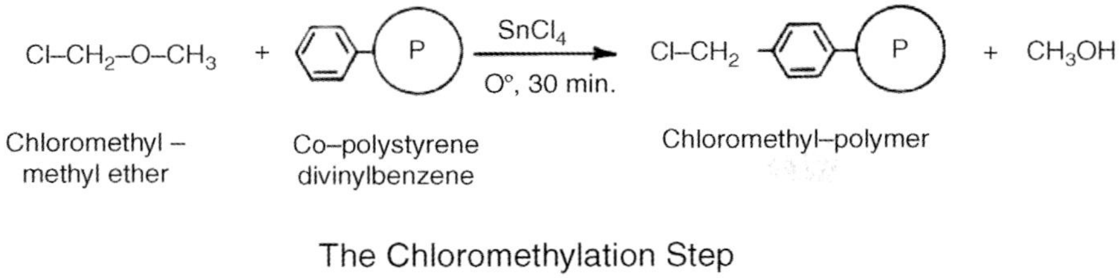

$Cl{-}CH_2{-}O{-}CH_3$ + (P) $\xrightarrow[\text{O°, 30 min.}]{SnCl_4}$ $Cl{-}CH_2$–(P) + $CH_3OH$

Chloromethyl – methyl ether | Co–polystyrene divinylbenzene | Chloromethyl–polymer

The Chloromethylation Step

**Figure 7.3** Polystyrene polymer beads cross-linked with divinylbenzene were functionalized by chloromethylation.

chain. His efforts with other supports including cellulose and multiple combinations of protecting groups, linkers and cleavage conditions during this development period were only revealed in any detail recently in this autobiography with direct quotations from his laboratory books.

The combination of chemically modified (chloromethylated) cross-linked polystyrene beads with stepwise addition of N-terminal carbobenzoxy (Z)-protected amino acids first led to a successful synthesis of small peptides. By the end of 1962, Bruce had demonstrated feasibility for SPPS by preparing a simple tetrapeptide, Leu–Ala–Gly–Val, utilizing the Z group for amino protection and HBr/HOAc for deprotecting the amino group of the growing peptide chain [3]. The strongly acidic conditions required for Z removal mandated a more resistant linkage to the polymeric support than the benzyl ester linkage originally tried. Bruce nitrated the polystyrene resin to generate, in effect, a nitrobenzyl carboxy-protecting group stable to HBr/HOAc that could be cleaved by saponification as shown in Figure 7.4. The ion-exchange chromatogram (Figure 7.5) of the cleaved product of the synthesis of Leu–Ala–Gly–Val already foreshadowed many of the problems that would have to be overcome to make SPPS a robust and reliable procedure. For example, the presence of Gly–Val and Ala–Gly–Val in the chromatogram implied that chain elongation was incomplete, leading to these truncated sequences due either to incomplete deprotection or coupling. The presence of the deletion sequences Leu–Ala–Val and Leu–Val that have omitted Gly and Ala–Gly, respectively, imply that incomplete deprotection/coupling was only temporary, and growing chains could resume participation in chain elongation (Figure 7.6). The tetrapeptide with D-Val at the C-terminus implies racemization, either during coupling to the resin or upon saponification. The presence of acetylated peptides implies incomplete washing after HBr/HOAc treatment with retention of acetic acid into the activation and coupling procedure. Nevertheless, despite these deficiencies, the major peak corresponded to the desired product Leu–Ala–Gly–Val; the ease of synthesis compared with the labor and time to produce a tetrapeptide by conventional solution approaches was sufficient to provoke both intense interest in SPPS as well as a backlash by the more classical solution peptide chemistry community. One of the referee's critiques of the first paper suggested that this SPPS should be shunned as it violated the basic principles of synthetic organic chemistry (i.e., isolation and characterization of intermediates). In view of the mixture of compounds in the cleaved product, such a reaction was not

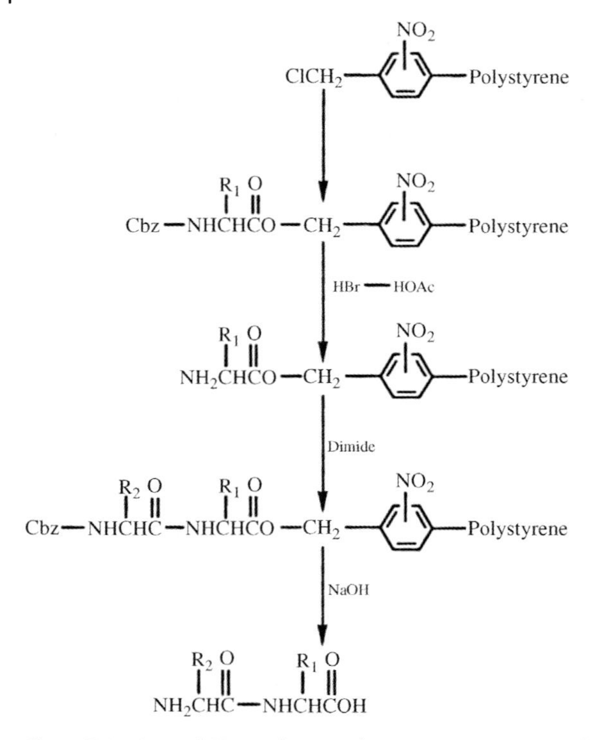


**Figure 7.4** Original SPPS scheme with Z-amino protection and a nitrobenzyl linkage to the polymeric support.

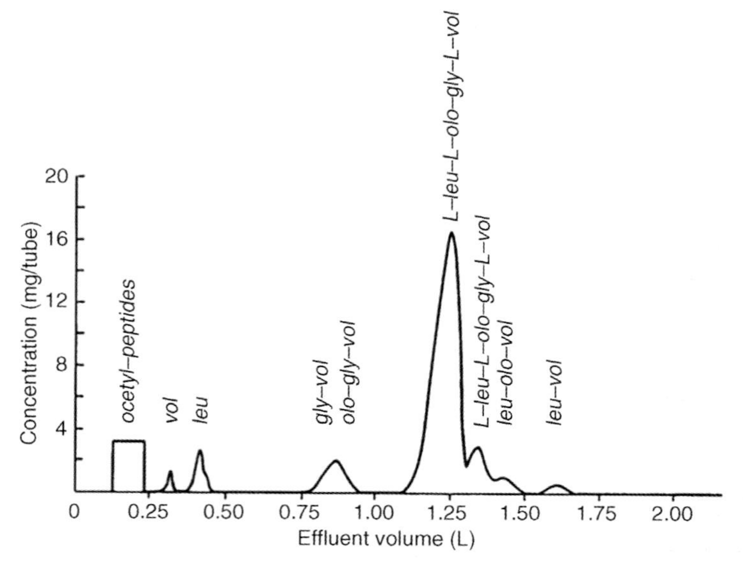


**Figure 7.5** Ion-exchange chromatogram of tetrapeptide Leu–Ala–Gly–Val obtained from polymer by saponification.

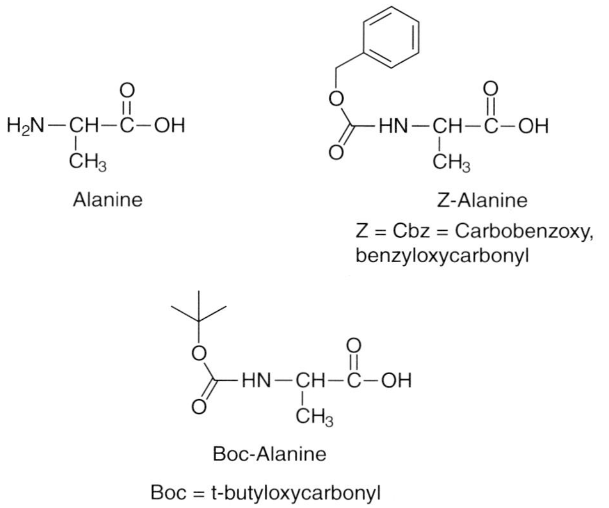

Alanine

Z-Alanine

Z = Cbz = Carbobenzoxy, benzyloxycarbonyl

Boc-Alanine

Boc = t-butyloxycarbonyl

**Figure 7.6** Use of the urethane-protecting groups, Z and Boc, for the amino group of amino acids to minimize racemization during stepwise activation and coupling.

unreasonable. This was not the last time that this criticism among others would be heard from synthetic organic chemists.

## 7.3 Racemization and Stepwise Peptide Assembly

The choice of urethane (benzyloxycarbonyl or carbobenzoxy (Z), *t*-butyloxycarbonyl, (Boc), fluorenylmethoxycarbonylamino (Fmoc), biphenylisopropoxycarbonyl (Bpoc), etc.)-protected amino acids was dictated by their resistance to racemization of the α-carbon hydrogen when activated. Other amino acid derivatives, such as *N*-trityl and *N*-*o*-nitrophenylsulfonyl, were also tried in alternative strategies because they too resisted racemization upon carbonyl activation. This reasoning also dominated the choice for stepwise addition of protected amino acids to elongate the growing peptide chain from the C-terminal residue; otherwise, peptide carboxyl groups would require activation if the growing peptide chain were attached to the polymer at the N-terminus, with a high probability of racemization. Prior to native chemical ligation, fragment-condensation strategies were chosen to have either proline or glycine residues at the C-terminus to minimize racemization. If this were not feasible, special activation procedures, such as the azide, of the C-terminal carboxylate of the peptide were often utilized to minimize racemization during fragment condensation. By choosing urethane protecting groups and stepwise addition of protected amino acids, the Merrifield strategy attempted to avoid racemization as a side-reaction during chain elongation.

## 7.4
## Optimization of Synthetic Components

I was essentially unaware of the many trials of alternative approaches that Bruce had endured before I showed up in his laboratory in the spring of 1963 at the suggestion of my research advisor, the prominent immunologist Henry Kunkel. Professor Kunkel had suggested that I try and develop an immunoassay for angiotensin II – an important octapeptide hormone involved with blood pressure regulation. He suggested that I work with Bruce on coupling the peptide to a carrier protein in order to use it as a hapten and develop antibodies for the immunoassay. The protein chemistry tradition (Max Bergman and Leonid Zervas came from Germany in 1933 just after their publication of the Z group in 1932) at the Rockefeller Institute for Medical Research (later Rockefeller University) was especially strong as was that of immunology (Landsteiner had thoroughly developed the use of haptens conjugated to proteins to generate antibodies that recognized the small-molecule hapten). At that time, the Merrifield section of the Woolley group consisted of only Bruce and his technician, Angela Corrigliano. The first paper on SPPS with Z-amino protection was in press in *Journal of the American Chemical Society*, and Bruce was already working on its improved successor using Boc-amino protection (Figure 7.7) and a benzyl ester linkage to the polymeric polystyrene support. Since each protected amino acid had to be generated in the laboratory, each new peptide presented its own set of solution synthetic problems. Bruce and Angela generated Boc-Arg($NO_2$), Boc-Pro, Boc-Phe, Boc-Gly, and Boc-Ser(*O*Bzl) for the synthesis of bradykinin, Arg–Pro–Pro–Gly–Ser–Pro–Phe–Arg. My initial job was to generate the Boc derivatives of Val and side-chain protected Asp, His, and Tyr for the synthesis of angiotensin II, Asp–Arg–Val–Tyr–Val–His–Pro–Phe.

Certainly, the SPPS of bradykinin reported in 1964 was revolutionary compared with traditional solution methods for preparing peptides. The nonapeptide was assembled in 4 days, purified and fully characterized in 68% overall yield in another 5 days (a large part of the latter being the purification by counter-current distribution and characterization by manual amino acid analysis – both techniques being tedious and time consuming). This synthesis was comparable to the work of the large synthetic group of Vincent du Vigneaud at Cornell on the synthesis of the two nonapeptide pituitary peptides, oxytocin and vasopressin, for which Vincent du Vigneaud had received the Nobel Prize only 10 years before. It was readily apparent to perceptive chemists around the world that a paradigm shift in peptide chemistry was underway. To some, the concept of a filterable protecting group was readily adaptable to synthetic chemistry in general.

## 7.5
## Foreshadowing of the Nobel Prize

Despite my background in biology at Caltech, or perhaps because of it, I instantly sought tutoring in synthetic organic chemistry by Professor John M. Stewart, then

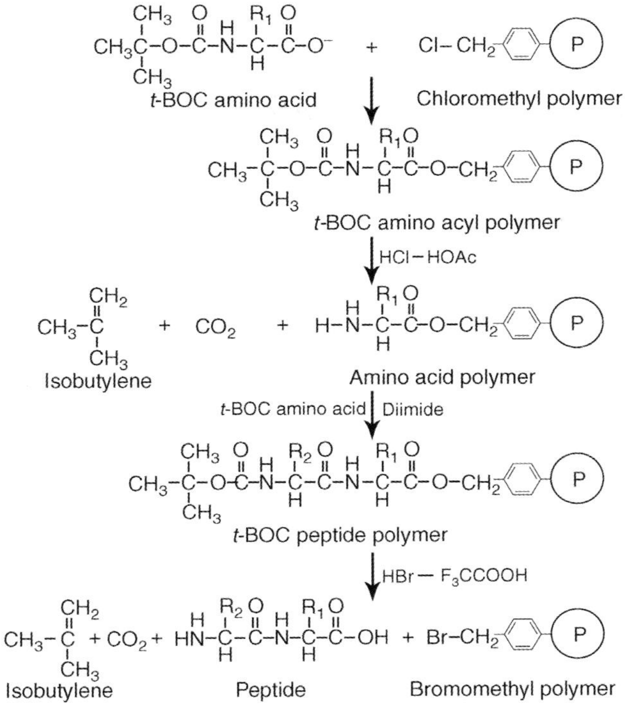

**Figure 7.7** Improved SPPS approach using Boc-protected amino acids and acidic cleavage of benzyl ester linkage to polymeric support. Both bradykinin and angiotensin II were successfully prepared by this protocol, often referred to as the Merrifield approach. Later HBr/TFA was often replaced by HF cleavage.

an assistant professor associated with the Woolley group. I tried not to expose my chemical ignorance during laboratory discussions. I was essentially an apprentice across the laboratory bench from Bruce and his technician, Angela, on a daily basis. Despite my naivety, Bruce and other members of the Woolley laboratories never treated me as anything but a colleague, which certainly put pressure on me to learn as quickly as possible. It was an ideal environment in which to mature as a scientist and my focus on SPPS intensified. I recognized the potential impact that SPPS would have on science once developed. The method brought numerous prominent scientists to visit, many of whom claimed to have conceived of SPPS, but none other than Bruce to my knowledge had seriously attempted to reduce the concept to practice. I particularly remember the visit of Sir Robert Robinson, Nobel Laureate, to the laboratory in 1964. Bruce was especially excited that such a prominent figure of synthetic chemistry would visit him. The Nobel Laureate pulled me aside to tell me how lucky I was to be working with Professor Merrifield; I agreed and then he said that he intended to nominate Bruce for the Nobel Prize. Every October for 20 years, I awaited the inevitable (at least to me) announcement.

## 7.6
## Automation of SPPS

It was obvious to Bruce that the simplicity of the steps involved in SPPS could be automated as only introduction of solvent/reagent, shaking, and filtering were involved (Figure 7.8) in an iterative synthetic cycle. He had witnessed the automation of amino acid analysis by Professors Moore and Stein at Rockefeller, and saw the dramatic improvement in both time savings and reproducibility (amino acid analyses were done manually in the Merrifield laboratory). Bruce got John Stewart to help with the electronics and drum programmer, and Nils Jernberg to design a novel rotary valve using Teflon for the liquid interface to select solvents/reagents while he focused on the plumbing, shaker, and so on. In a fairly short time, the prototype [4] (Figure 7.9) was working 24 h a day, much more reproducibly than a technician or graduate student. The prototype was eventually retired to the Smithsonian Museum in Washington, DC. I constructed the second automated synthesizer when I moved to Washington University School of Medicine in 1966. I tried to improve the design with a subroutine for washing built of relays to save space on the drum programmer, but quickly learned that computers built with relays were inherently unreliable, much to the amusement of my computer-experienced colleague, Professor Charles Molnar. Interaction with Professor Molnar eventually led to my involvement with computer-aided molecular design, but that is another story [5].

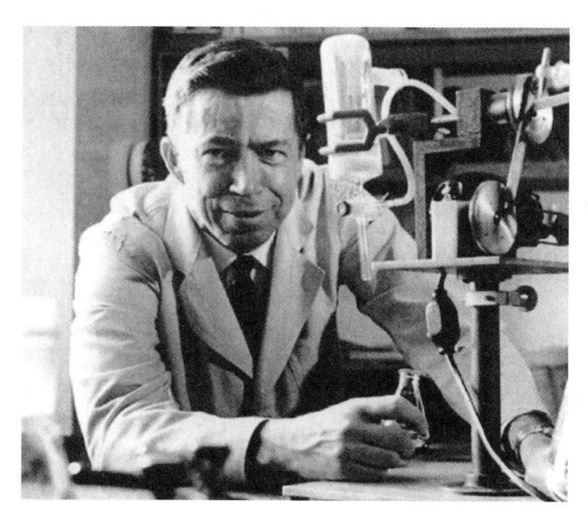

**Figure 7.8** Bruce Merrifield with new manual solid-phase shaker as seen from across the lab bench at his lab at Rockefeller Institute for Medical Research in 1963.

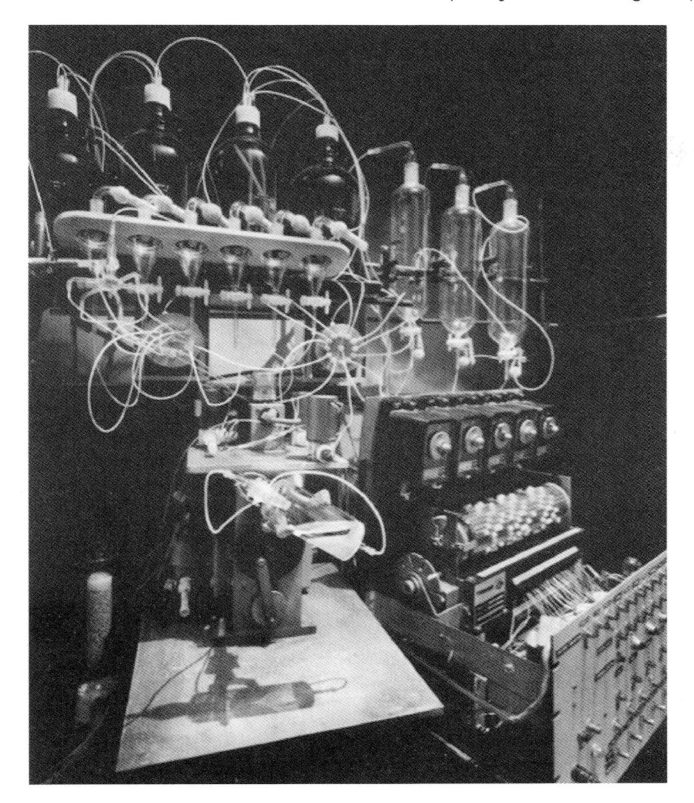

**Figure 7.9** Automated solid-phase peptide synthesizer (now in the Smithsonian Museum). Note drum programmer and timers within control module on right, two rotary valves above shaker with reagent vessels on top.

## 7.7 Impact of New Protecting Groups and Resin Linkages

There were still many opportunities for significant improvement to the Merrifield Boc/benzyl ester strategy. Boc amino protection and the benzyl ester side-chain protection and linkage to the polymeric support were both acid-labile. While the benzyl ester requires more rigorous acidic conditions, such as HBr/trifluoroacetic acid (TFA), for removal, multiple exposures to the milder acidic conditions used repetitively to remove Boc groups during peptide chain elongation was potentially problematic in cleaving the growing peptide chain from the polymeric support. In addition, cleavage of the product from the benzyl ester support and removal of side-chain protecting groups required rather harsh conditions, incompatible with the preparation of glycopeptides, for example. The introduction of the base-labile Fmoc group by Carpino in 1972 had a major impact on multiple applications of solid-phase peptide synthesis. When combined with a number of much more acid-labile linkages to the polymeric support, it became feasible to avoid strong-acid cleavage conditions such HBr/TFA or HF and prepare side-chain protected peptides for subsequent fragment condensations. While the benzyl ester linkage was amenable to ammonolysis to generate the peptide C-terminal amide, this precluded use of esters for side-chain protection of acidic side-chains such as Asp and Glu. As many peptide hormones such as oxytocin, vasopressin, luteinizing-hormone-releasing hormone, thyrotropin-releasing hormone, and so on, had C-terminal amide groups, Pietta and Marshall developed a modified benzyhydrylamine linkage to the resin to generate C-terminal peptide amides selectively. Methods for preparation of C-terminal amides, aldehydes, and so on, compatible with other protocols were rapidly adapted from solution chemistry.

## 7.8 Solid-Phase Organic Chemistry

It was also clear to Merrifield that filterable, polymeric protecting groups were generally applicable to synthetic organic chemistry. He stated in a review in 1969, "A gold mine awaits discovery by organic chemists," but it was not his objective to generalize the concept beyond biopolymers. Pioneers in this effort were Professor Robert L. Letsinger of Northwestern University and Professor Clifford C. Leznoff of York University in Canada. Letsinger [15] functionalized several different polymers as potential filterable protecting groups. Leznoff used insoluble polymeric supports to overcome a number of synthetic organic problems. For example, monoreactions of symmetrical bifunctional compounds were demonstrated using a functionalized diol to react with a symmetrical dialdehyde. Leznoff also used this approach to synthesize insect sex attractants and carotenoids [16, 17]. In his excellent review [17] on "The use of insoluble polymeric supports in general organic synthesis" published in *Accounts of Chemical Research* in 1978, Leznoff quotes Merrifield's comment on the gold mine awaiting organic chemists with the comment, "Many gold nuggets have now been mined... and some iron pyrites." Several other synthetic organic chemists also realized the potential advantages of SPPS, and their pioneering efforts need to be recognized considering the cold reception that most of the synthetic chemistry community gave SPPS. The Patchornik group in Israel developed a number of insoluble polymeric reagents [18, 19] including coupling reagents and demonstrated their utility in the synthesis of peptides. Crowley and Rapoport evaluated the hyperentropic utility of polymeric supports as an alternative to high dilution and concluded, based on experimental work, that co(polystyrene–2% divinylbenzene) did not provide adequate site isolation [20]. Depending on the cross-linking and loading of reactive groups on the polymeric support, intermolecular side-reactions including aggregation were readily demonstrated.

While Bruce recognized the obvious extension of solid-phase synthesis to other biologically important heteropolymers (Figure 7.10) such as nucleic acids and oligosaccharides, the underlying chemistry was less developed and Bruce was loath to claim any obvious potential for solid-phase synthesis that was not yet realized. As a

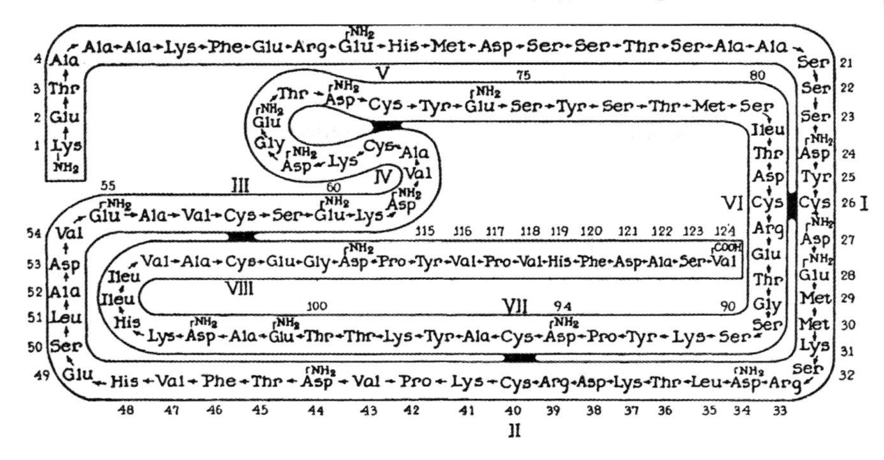


**Figure 7.10** Primary structure of pancreatic ribonuclease A – the first enzyme to have its primary sequence determined (Moore and Stein) and the first enzyme to be synthesized by SPPS (Gutte and Merrifield) at Rockefeller University.

feasibility experiment, I prepared the first nucleotide, dTT, using solid-phase synthesis in his laboratory in 1965 (Marshall and Merrifield, unpublished), but it was clear that coupling yields needed to be enhanced for a serious effort in nucleic acid synthesis, and further developments by Letsinger and Caruthers [6, 7] were necessary. Certainly, the ability to generate oligonucleotide sequences at will for probes enabled much of modern molecular biology. Bruce finally yielded to laboratory pressure to discuss the generalization of solid phase peptide chemistry to other heteropolymers in his review in *Science* in 1965 [8]. Solid-phase synthesis of oligosaccharides has finally reached an equivalent state of development [9–14]; the delay simply being due to the enhanced complexity of carbohydrates and the necessary developments of appropriate orthogonal protecting groups.

Only later did the majority of the synthetic organic community recognize that the use of polymeric protecting groups as an obvious improvement for a multiplicity of synthetic applications.

## 7.9
## Early Applications of SPPS to Small Proteins

The development of the automated synthesizer led to the synthesis of both chains of insulin (51 residues) and their recombination by Marglin and Merrifield in 1966 [21], ribonuclease (124 residues) by Gutte and Merrifield in 1969 [22, 23], and to our own synthesis of the 74-residue acyl carrier protein (ACP) [24] involved in fatty acid biosynthesis in collaboration with its discoverer, Professor P. Roy Vagelos. It was exciting to receive an invitation from the program committee of the 11th European Peptide Society to present our ACP work [24–26] in Vienna in 1971 [27], but I was not prepared for the warmth of the reception that I received. The established heads of many peptide laboratories, whose pioneering work I still venerate, were more than generous with their criticisms; the only real defense was that the Vagelos group could not distinguish our synthetic product from the protein isolated from liver using the biochemical and biophysical techniques available at the time. Fortunately, two giants of peptide chemistry, Ralph Hirschmann of Merck and Joseph Rudinger of the Czech Academy of Sciences, both befriended me, and reassured me that my treatment was a reaction to the threat of SPPS to the status quo of solution peptide chemistry and not to the experimental science that I had presented. Thus, it was an exceptional pleasure to have attended the 27th European Peptide Society Symposium in Sorrento some 30 odd years later to confirm that SPPS has become the dominant approach in peptide synthesis and combinatorial chemistry, even at the most reactionary laboratories.

## 7.10
## Side-Reactions and Sequence-Dependent Problems

Merrifield did an interesting experiment that dramatized the unanticipated sequence-dependent problems that occasionally plagued SPPS. Since bradykinin and angiotensin had each been successfully synthesized, he decided to synthesize a hybrid continuing the sequence of angiotensin appended to the grown bradykinin sequence on the polymeric support. Figure 7.11 shows the coupling yields for each

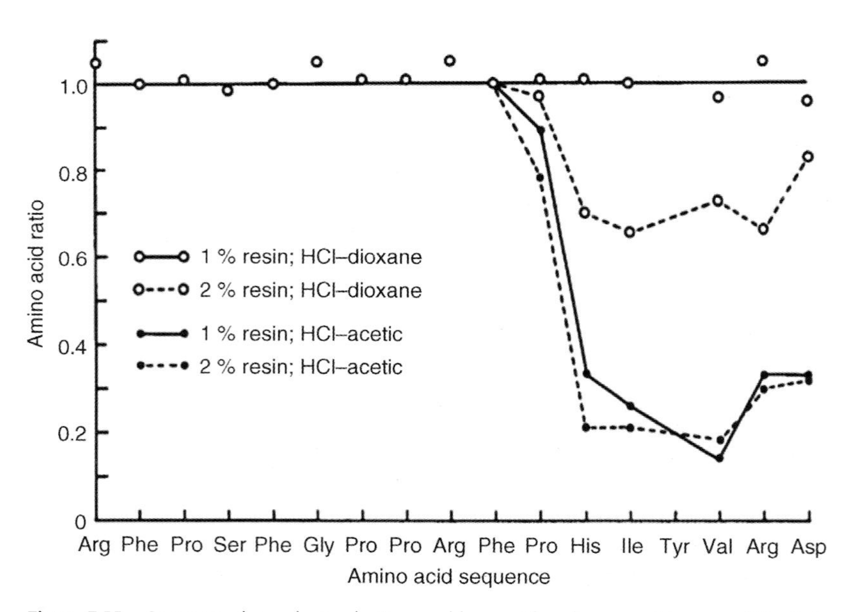


**Figure 7.11** Sequence-dependent solvation problems with polymer matrix as peptide chain was extended in synthesis of angiotensin–bradykinin hybrid. Both peptides had previously been prepared by SPPS without observing this problem.

residue. In this case truncation of the growing chain could be overcome by modifying the divinylbenzene cross-linking and deprotection reagent. From our work on acyl carrier protein came additional insight into one of the underlying problems of SPPS – changes in the physical properties of the polymeric support with chain elongation [28]. We utilized titration of the amine on the polymer with a radioactive isotope of chlorine $^{36}Cl$ following the method of Dorman [29]. The C-terminus of ACP showed a dramatic decrease in the growing chain that reappeared leading to a deletion sequence (deletion and truncation sequences were clearly defined and their origins discussed by Hancock *et al.* [28]). The titration showed that all the bound chloride could not be removed by simple washing with triethylamine in organic solvent, but additional chloride was rendered accessible only by shrinking the polymer with *tert*-butanol and then reswelling in dichloromethane, followed by washing with additional triethylamine in organic solvent (Figure 7.12). This implied that part of the polymer had become inaccessible to solvent due to the heterogeneity of sites and changes in the ratio of peptide to polystyrene; Steve Kent and Bruce Merrifield later attributed this, based on further experiments, to formation of β-sheet aggregation of the peptide consistent with the amino acid sequence of the ACP peptide. What was of some surprise was the ability of different laboratories to reproduce the difficulties with this ACP sequence despite the variations in support, linker, and loadings that were used, suggesting that aggregation of the growing peptide chain was the most likely explanation.

The disappearance and reappearance of the growing peptide chain leading to deletion sequences inspired the Merrifield group and others to explore alternative polymeric supports. Atherton and Sheppard realized that the growing peptide chain modified the hydrophobic nature of cross-linked polystyrene, and postulated that more hydrophobic polymeric supports might minimize deletion sequences. They and others have generated a plethora of alternative supports, each of which have advantages depending on the synthetic problem being addressed. Obviously, solid-phase organic chemistry (SPOC) is very different from assembly of large heteropolymers, such as proteins or nucleic acids.

## 7.11 Rapid Expansion of Usage Leading to the Nobel Prize

This period of the 1970s and 1980s saw a rapid acceptance of solid-phase synthesis by the molecular biology community for synthesis of both peptides and nucleic acids as biological probes. The availability of a number of reliable automated synthesizers with great effort paid to optimization of yields and elimination of side-reactions made reliability an achievable goal. There were the occasional "difficult sequences," but overall chemists were making products in good yield and acceptable purities. The increasing availability of high-performance liquid chromatography (HPLC) and mass spectrometers made purification and characterization more routine. The many chemists who helped optimize solid-phase peptide chemistry are too numerous to name, but there was a heavy concentration of very talented young scientists

TABLE IV

REEXPOSURE OF CHLORIDE BINDING SITES AT DIFFERENT STAGES OF THE SYNTHESIS OF THE PEPTIDE 74–67[a]

| Number of wash | Reagent | BOC-Gly$^{74}$ $^{24}$CL dpm in washes | BOC-Ala$_{57}$ $^{24}$CL dpm in washes | Deprotected Ala$_{67}$ $^{36}$Cl in washes |
|---|---|---|---|---|
| 12 | $Et_3N$ | | | |
| 13 | $Et_3N$ | | | |
| 14 | $Et_3N$ | | | |
| 15 | $CH_2Cl_2$ | | | |
| 16 | $CH_2Cl_2$ | | | |
| 17 | $CH_2Cl_2$ | | | |
| 18 | t-BuOH | | | |
| 19 | t-BuOH | | | |
| 20 | t-BuOH | | | |
| 21 | $CH_2Cl_2$ | | | |
| 22 | $CH_2Cl_2$ | | | |
| 23 | $CH_2Cl_2$ | | | |
| 24 | $Et_3N$ | 73 | 213 | 630 |
| 25 | $Et_3N$ | 16 | 75 | 112 |
| 26 | $CH_2Cl_2$ | | 20 | 19 |
| Total dpm in wash 11–15 | | 286 | 778 | 2,050 |
| Total dpm in wash 23–25 | | 89 | 308 | 761 |
| % of dpm in wash 23–25 relative to wash 11–15 | | 31 | 40 | 37 |

[a] The resin peptide was treated with pyridine hydrochloride (7.5 mmol, 1.4 × 10$^4$ dpm) and the excess pyridine hydrochloride was removed by the same washes (1a–11a) as described in Table II. All available chloride was then displaced by the washes 12-17 as shown in this table. The resin was then subjected to washes 18–23 in an attempt to expose buried regions of the resin that had bound chloride, and in fact the second triethylamine treatment washes 24–26 did liberate more chloride. Samples of the washes were counted for chloride-36 in the manner described for Table II.

**Figure 7.12** Sequestering of radioactive chloride ions by polymeric support to the extent that they are inaccessible to triethylamine. Shrinking and swelling of polymeric support allows access to chloride ions by base.

(DiMarchi, Erickson, Gisin, Hodges, Kent, Mitchell, Tam, Wang, etc.) in Bruce's laboratory at Rockefeller University that explored many variations of support, monitoring methods, deprotection schemes, side-reactions, and so on, to overcome technical difficulties with SPPS during this period. Finally, on 17 October 1984, the announcement that R. Bruce Merrifield had won the Nobel Prize in Chemistry came from Stockholm. The annual October vigil for me for 20 years that began with Sir Robert Robinson's comment in 1964 had finally come to fruition.

## 7.12 From the Nobel Prize Forward to Combinatorial Chemistry

All the effort focused on transforming the idea of SPPS into a prototype and finally, into the method of choice, was finally vindicated. The recognition signified by the Nobel Prize was certainly gratifying to all those who had worked on SPPS. However, there were always new horizons and opportunities for expanding the scope of application areas and the advent of combinatorial chemistry was just such an opportunity. The impetus for combinatorial chemistry was the development of high-throughput *in vitro* binding assays and tissue-culture screens that allowed testing of large numbers of compounds in the pharmaceutical industry. The traditional approach of medicinal chemistry to sequentially synthesize a logically designed set of analogs based on a pharmacological lead was surpassed by the ability to screen whole compound libraries accumulated over years by large pharmaceutical companies.

The groundwork for combinatorial chemistry had already been set in peptide chemistry by Geysen with his pin approach [30–32] and Houghten with his "teabag" approach to epitope mapping [33] of antigens. Both approaches used physical separation by polymers to control reaction sequences and thus peptide products. Lam [34] and Furka [35] conceived independently of the "one bead, one product" split-and-mix approach that has been so powerful. Houghten has shown that synthesis of large mixtures followed by screening and deconvolution to identify the active components is a viable and efficient technique [36–40]. In many ways, it is analogous to isolation of active natural products from fermentation broths. Nevertheless, the pressure from the medicinal chemistry community in the pharmaceutical industry has focused on combinatorial synthesis of single compounds, partially due to perceived problems with false positives in the deconvolution process.

Enhanced recognition by the medicinal chemistry community of the advantages of filterable, polymeric protecting groups (advocated in 1971 in a review by Marshall and Merrifield [41] and demonstrated so convincingly by Leznoff [16, 17]) was catalyzed by a paper in 1992 by Bunin and Ellman [42] on synthesis of a combinatorial library of benzodiazepines – a privileged class of structure thought to mimic β-turns. The generation of a compound library of direct interest to the pharmaceutical industry because of the many biological activities found with benzodiazepines was a turning point in acceptance of the overall approach. It has become difficult to find a chemical reaction, or class of compound, that has not been adapted to solid-phase chemistry. As examples of the pervasiveness of the approach in synthetic organic chemistry, two reviews, one on multiple approaches to traceless supports [43] for solid-phase synthesis and the other on heterocyclic chemistry [44], have appeared. Synthesis of complex natural products as diverse as sarcodictylins, chalcones, and epothilones [45] utilizing SPOC are becoming more commonplace as the advantages of a filterable, polymeric protecting group become more widely recognized. The paradigm shift has even extended to the search for metal-binding ligands, catalysts, and new materials [46, 47]. See Figure 7.13.

**Figure 7.13** Rapid adoption of combinatorial chemistry, often based on SPOC, by the pharmaceutical industry led to generation of multiple organic scaffolds for structure–activity studies and optimization. Shown are Marshall and Merrifield at the inaugural symposium celebrating new automated SPOC facility at SmithKlineBeckman.

## 7.13
## Protein Synthesis and Peptide Ligation

While a number of small proteins have been successfully assembled by SPPS, practical limitations regarding the ability to purify and characterize the mixtures that inevitably result from less than complete reactions and side-reactions during deprotection limit most efforts to below 100 residues. The advent of chemical ligation where purified fragments without side-chain protection can be stitched together has provided a viable hybrid strategy for synthesis of larger proteins. This approach evolved from the thiol-capture approach of Dan Kemp [48], and has been actively developed in the laboratories of Steve Kent [49, 50] and James Tam [51]. This is another example of the power of convergent synthesis rather than a brute force approach. A paradigm shift has come from the laboratory of Tom Muir where expressed protein fragments are ligated to synthetic peptides to generate hybrids [52, 53]. A student in my group, Lori Anderson, used expressed protein ligation to specifically label the C-terminal segment of the α-subunit of a G-protein for both magic angle spinning nuclear magnetic resonance (NMR) and electron spin resonance studies aimed at mapping the interface between the α-subunit and the activated G-protein-coupled receptor rhodopsin. The blend of synthetic chemistry with molecular biology has enormous potential as we attempt to understand the dynamics of the large complex multiprotein systems found in biology.

## 7.14 Conclusions

Often, it is the outsider who brings a fresh perspective to a problem that generates the insight necessary for a paradigm shift. Certainly, the invention by Merrifield of SPPS with its following automation is a classic example of scientific revolution as discussed by Thomas Kuhn [1]. Traditional synthetic organic chemistry required isolation and characterization of intermediates as concrete evidence supporting the chemical structure of the product. By substituting the use of excess reagents to force chemical reactions to completion (or as close as possible), solid-phase chemistry was anathema to traditional synthetic practice of the time. Resistance to change by synthetic chemists in general, and peptide chemists in particular, was both vehement and vitriolic. In addition, solid-phase chemistry required careful purification and characterization of its product that did not depend on the history of the synthetic process. In reality, this was only possible with a concomitant improvement in both purification techniques and analytical methods of structural characterization. Without modern HPLC, capillary electrophoresis, NMR, mass spectrometry, and so on, solid-phase chemistry would not have been so feasible. The practical advantages in handling and automation offered by a filterable, polymeric protecting group in automation of chemical synthesis far outweigh the increased needs for more effort in purification and characterization.

I presume that Emil Fischer would be both amused and impressed with the progress made in the last century since his synthesis of the first amide bond. The ability to synthesize peptides, small proteins and nucleic acids by solid-phase synthesis has enabled much of modern molecular biology. In turn modern molecular biology has provided us with the plethora of therapeutic targets through cloning and expression that has driven both combinatorial chemistry and structural biology. Technology enables us to ask relevant questions; there is little doubt that expression ligation combined with SPPS of labeled peptides will enable dissection of the molecular mechanisms of many significant biological systems.

### Acknowledgments

From the contents of this personal and historical review of SPPS, my deep debt of gratitude to my former mentor, colleague and friend, Professor R. Bruce Merrifield, should be obvious. In addition, Professor John M. Stewart of the University of Colorado Medical School exerted a strong positive influence on my scientific development. I would be remiss if I did not acknowledge the support over these many years from the National Institutes of Health that have allowed me to indulge my scientific interests and that supported the development of SPPS in so many different laboratories. To the many who have labored to make SPPS reliable and who were not specifically acknowledged in this overview, both my thanks for your efforts and my apologies for the omission. Be consoled that your efforts to make SPPS an accepted tool have been successful.

## References


1 Kuhn, T.S. (1962) *Structure of Scientific Revolutions*, University of Chicago Press, Chicago, IL.

2 Merrifield, R.B. (1993) *Life During a Golden Age of Peptide Chemistry: The Concept and Development of Solid-Phase Peptide Synthesis (Profiles, Pathways, and Dreams)*, American Chemical Society, Washington, DC.

3 Merrifield, R.B., Stewart, J.M., and Jernberg, N. (1966) Instrument for automated synthesis of peptides. *Analytical Chemistry*, **38**, 1905–1914.

4 Marshall, G.R. (1970) Conformation, computers and biological activity, in *Peptides: Chemistry and Biochemistry* (eds B. Weinstein and S. Lande), Dekker, New York.

5 Marshall, G.R., Barry, C.D., Bosshard, H.E., Dammkoehler, R.A., and Dunn, D.A. (1979) The conformational parameter in drug design: the active analog approach, in *Computer-Assisted Drug Design* (eds E.C. Olson and R.E. Christoffersen), American Chemical Society, Washington, DC.

6 Caruthers, M.H. (1982) The application of nucleic acid chemistry to studies on the functional organization of gene control regions. *Princess Takamatsu Symposium*, **12**, 295–306.

7 Caruthers, M.H. (1985) Gene synthesis machines: DNA chemistry and its uses. *Science*, **230**, 281–285.

8 Merrifield, R.B. (1965) Automated synthesis of peptides. *Science*, **150**, 178–185.

9 Plante, O.J., Palmacci, E.R., and Seeberger, P.H. (2001) Automated solid-phase synthesis of oligosaccharides. *Science*, **291**, 1523–1527.

10 Sears, P. and Wong, C.-H. (2001) Toward automated synthesis of oligosaccharides and glycoproteins. *Science*, **291**, 2344–2350.

11 Barkley, A. and Arya, P. (2001) Combinatorial chemistry toward understanding the function(s) of carbohydrates and carbohydrate conjugates. *Chemistry – A European Journal*, **7**, 555–563.

12 Hewitt, M.C. and Seeberger, P.H. (2001) Automated solid-phase synthesis of a branched Leishmania cap tetrasaccharide. *Organic Letters*, **3**, 3699–3702.

13 Plante, O.J., Palmacci, E.R., Andrade, R.B., and Seeberger, P.H. (2001) Oligosaccharide synthesis with glycosyl phosphate and dithiophosphate triesters as glycosylating agents. *Journal of the American Chemical Society*, **123**, 9545–9554.

14 Bartolozzi, A. and Seeberger, P.H. (2001) New approaches to the chemical synthesis of bioactive oligosaccharides. *Current Opinion in Structural Biology*, **11**, 587–592.

15 Letsinger, R.L., Kornet, M.J., Mahadevan, V., and Jerina, D.M. (1964) Reactions on polymer supports. *Journal of the American Chemical Society*, **86**, 5163–5165.

16 Leznoff, C.C. (1974) The use of insoluble polymer supports in organic chemical synthesis. *Chemical Society Reviews*, **3**, 65–85.

17 Leznoff, C.C. (1978) The use of insoluble polymer supports in general organic synthesis. *Accounts of Chemical Research*, **11**, 327–333.

18 Kalir, R., Warshawsky, A., Fridkin, M., and Patchornik, A. (1975) New useful reagents for peptide synthesis. Insoluble active esters of polystyrene-bound 1-hydroxybenzotriazole. *European Journal of Biochemistry*, **59**, 55–61.

19 Patchornik, A. and Kraus, M.A. (1975) The use of polymeric reagents in organic synthesis. *Pure and Applied Chemistry*, **43**, 503–526.

20 Crowley, J.I. and Rapoport, H. (1976) Solid-phase organic synthesis: novelty or fundamental concept? *Accounts of Chemical Research*, **9**, 135–144.

21 Marglin, B. and Merrifield, R.B. (1966) The synthesis of bovine insulin by the solid phase method. *Journal of the American Chemical Society*, **88**, 5051–5052.

22 Gutte, B. and Merrifield, R.B. (1969) The total synthesis of an enzyme with ribonuclease A activity. *Journal of the American Chemical Society*, **91**, 501–502.

23 Gutte, B. and Merrifield, R.B. (1971) The synthesis of ribonuclease A. *Journal of Biological Chemistry*, **246**, 1922–1941.

24 Hancock, W.S., Marshall, G.R., and Vagelos, P.R. (1973) Acyl carrier protein. XX. Chemical synthesis and characterization of analogues of acyl carrier protein. *Journal of Biological Chemistry*, **248**, 2424–2434.

25 Hancock, W.S., Prescott, D.J., Marshall, G.R., and Vagelos, P.R. (1972) Acyl carrier protein. XVIII. Chemical synthesis and characterization of a protein with acyl carrier protein activity. *Journal of Biological Chemistry*, **247**, 6224–6233.

26 Vagelos, P.R., Hancock, W.S., Prescott, D.J., Nulty, W.L., Weintraub, J., and Marshall, G.R. (1971) Synthesis of a protein with acyl carrier protein activity. *Journal of the American Chemical Society*, **93**, 1799–1800.

27 Marshall, G.R., Hancock, W.S., Prescott, D.J., Nulty, W.L., Weintraub, J., and Vagelos, P.R. (1973) Solid phase synthesis of acyl carrier protein, in *Peptides 1971: Proc. 11th European Peptide Symposium* (ed. H. Nesvadba), North-Holland, Amsterdam.

28 Hancock, W.S., Prescott, D.J., Vagelos, P.R., and Marshall, G.R. (1973) Solvation of the polymer matrix: source of truncated and deletion sequences in solid phase synthesis. *Journal of Organic Chemistry*, **38**, 774–78l.

29 Dorman, L.C. (1969) A non-destructive method for the determination of completeness of coupling reactions in solid phase peptide synthesis. *Tetrahedron Letters*, **10**, 2319–2321.

30 Geysen, H.M., Meloen, R.H., and Barteling, S.J. (1984) Use of peptide synthesis to probe viral antigens for epitopes to a resolution of a single amino acid. *Proceedings of the National Academy of Sciences of the United States of America*, **81**, 3998–4002.

31 Maeji, N.J., Bray, A.M., and Geysen, H.M. (1990) Multi-pin peptide synthesis strategy for T cell determinant analysis. *Journal of Immunological Methods*, **134**, 23–33.

32 Valerio, R.M., Bray, A.M., Campbell, R.A., Dipasquale, A., Margellis, C., Rodda, S.J., Geysen, H.M., and Maeji, N.J. (1993) Multipin peptide synthesis at the micromole scale using 2-hydroxyethyl methacrylate grafted polyethylene supports. *International Journal of Peptide and Protein Research*, **42**, 1–9.

33 Houghten, R.A. (1985) General method for the rapid solid-phase synthesis of large numbers of peptides: specificity of antigen–antibody interaction at the level of individual amino acids. *Proceedings of the National Academy of Sciences of the United States of America*, **82**, 5131–5135.

34 Lam, K.S., Sroka, T., Chen, M.-L., Zhao, Y., Lou, Q., Wu, J., and Zhao, Z.G. (1998) Application of "one-bead one-compound" combinatorial library methods in signal transduction research. *Life Sciences*, **62**, 1577–1583.

35 Furka, A., Sebestyen, F., Asgedom, M., and Dibo, G. (1991) General method for rapid synthesis of multicomponent peptide mixtures. *International Journal of Peptide and Protein Research*, **37**, 487–493.

36 Ostresh, J.M., Winkle, J.H., Hamashin, V.T., and Houghten, R.A. (1994) Peptide libraries: determination of relative reaction rates of protected amino acids in competitive couplings. *Biopolymers*, **34**, 1681–1689.

37 Eichler, J., Lucka, A.W., and Houghten, R.A. (1994) Cyclic peptide template combinatorial libraries: synthesis and identification of chymotrypsin inhibitors. *Peptide Research*, **7**, 300–307.

38 Houghten, R.A. (1994) Tools of the trade. Combinatorial libraries: Finding the needle in the haystack. *Current Biology*, **4**, 564–567.

39 Houghten, R.A. (2000) Parallel array and mixture-based synthetic combinatorial chemistry: tools for the next millennium. *Annual Review of Pharmacology and Toxicology*, **40**, 273–282.

40 Houghten, R.A., Pinilla, C., Appel, J.R., Blondelle, S.E., Dooley, C.T., Eichler, J., Nefzi, A., and Ostresh, J.M. (1999) Mixture-based synthetic combinatorial libraries. *Journal of Medicinal Chemistry*, **42**, 3743–3778.

41 Marshall, G.R. and Merrifield, R.B. (1971) Solid phase synthesis: the use of solid

supports and insoluble reagents in peptide synthesis, in *Biochemical Aspects of Reactions on Solid Supports* (ed. G. Stark), Academic Press, New York.

**42** Bunin, B.A. and Ellman, J.A. (1992) A general and expedient method for the solid-phase synthesis of 1,4-Benzodiazepine derivatives. *Journal of the American Chemical Society*, **114**, 10997–10998.

**43** Blaney, P., Grigg, R., and Sridharan, V. (2002) Traceless solid-phase organic synthesis. *Chemical Reviews*, **102**, 2607–2624.

**44** Krchnak, V. and Holladay, M.W. (2002) Solid phase heterocyclic chemistry. *Chemical Reviews*, **102**, 61–92.

**45** Nicolaou, K.C. and Pfefferkorn, J.A. (2001) Solid phase synthesis of complex natural products and libraries thereof. *Biopolymers*, **60**, 171–193.

**46** Francis, M.B., Jamison, T.F., and Jacobsen, E.N. (1998) Combinatorial libraries of transition-metal complexes, catalysts and materials. *Current Opinion in Chemical Biology*, **2**, 422–428.

**47** Reetz, M.T. (2001) Combinatorial and evolution-based methods in the creation of enantioselective catalysts. *Angewandte Chemie International Edition*, **40**, 284–310.

**48** Fotouhi, N., Bowen, B.R., and Kemp, D.S. (1992) Resolution of proline acylation problem for thiol capture strategy by use of a chloro-dibenzofuran template. *International Journal of Peptide and Protein Research*, **40**, 141–147.

**49** Dawson, P.E., Muir, T.W., Clark-Lewis, I., and Kent, S.B. (1994) Synthesis of proteins by native chemical ligation. *Science*, **266**, 776–779.

**50** Dawson, P.E. and Kent, S.B.H. (2000) Synthesis of native proteins by chemical ligation. *Annual Review of Biochemistry*, **69**, 923–960.

**51** Tam, J.P., Xu, J., and Eom, K.D. (2001) Methods and strategies of peptide ligation. *Biopolymers*, **60**, 194–205.

**52** Cowburn, D. and Muir, T.W. (2001) Segmental isotopic labeling using expressed protein ligation. *Methods in Enzymology*, **339**, 41–54.

**53** Muir, T.W., Sondhi, D., and Cole, P.A. (1998) Expressed protein ligation: a general method for protein engineering. *Proceedings of the National Academy of Sciences of the United States of America*, **95**, 6705–6710.

# 8
# Linkers for Solid-Phase Peptide Synthesis

*Miroslav Soural, Jan Hlaváč, and Viktor Krchňák*

## 8.1
## Introduction

Linkers, infrequently also referred to as handles, facilitate temporary immobilization of the first synthetic component – an amino acid in the case of peptide synthesis – to a solid support (Figure 8.1). From the onset of solid-phase peptide synthesis (SPPS), protecting groups have served as linkers for immobilization of amino acids.

A vast majority of solid-phase peptide syntheses use the original Merrifield support – polystyrene cross-linked with 1–2% divinylbenzene [1]. Figure 8.2 shows the structure of a section of aminomethyl-derivatized copolymer and its abbreviated representation in schemes and figures in this manuscript. Use of the TentaGel resin is indicated by a "TG" symbol typed inside the schematic ball. Figure 8.2 also depicts a representation of a generic peptide. In numerous schemes, we always represent a generic peptide with the N-terminus on the left-hand side.

Traditional synthesis of a dipeptide requires protection of the carboxyl group of the C-terminal amino acid and the amino group of the N-terminal amino acid. For this first solid-phase synthesis of peptides, Merrifield protected the C-terminal amino acid by a resin-bound benzyl ester. The ester served two simultaneous functions: (i) immobilization of the amino acid and (ii) protection of the carboxylate function from undesirable participation in subsequent reactions. Resin-bound protecting groups soon became useful intermediates for protection of different functional groups, such as alcohols [2–7], aldehydes [8–10], carboxylic acids [11, 12], and amines [13]. Contemporary solid-phase organic chemistry offers a variety of linkers for different functional groups [14–16].

Generally, a linker needs to fulfill several requirements. The most relevant features include the following:

i) Straightforward and high yielding attachment of the first synthetic component (an amino acid).
ii) Stability to all chemical transformations during solid-phase synthesis.
iii) High yielding cleavage of the target compound without significant side-reactions.

*Amino Acids, Peptides and Proteins in Organic Chemistry.*
*Vol.3 – Building Blocks, Catalysis and Coupling Chemistry.* Edited by Andrew B. Hughes

ISBN: 978-3-527-32102-5

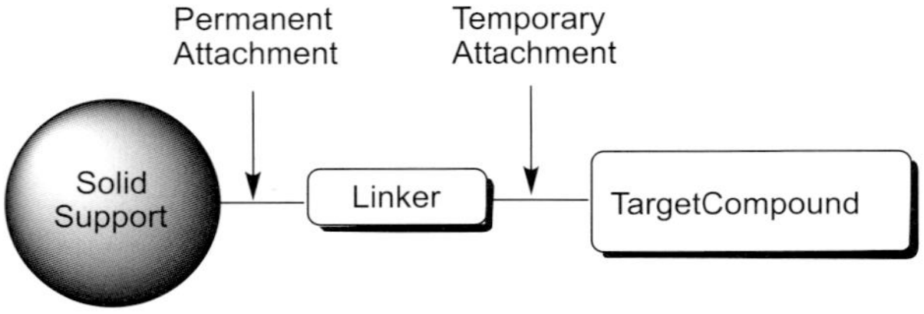

Figure 8.1 Attachment of the target compound to a solid support via a linker.

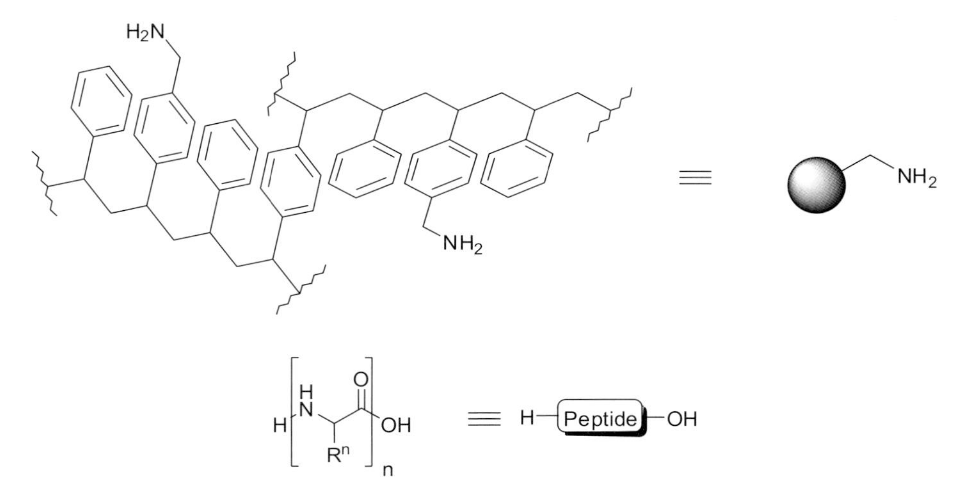


Figure 8.2 Representation of polystyrene cross-linked with divinylbenzene.

Furthermore, linkers have been incorporated onto solid support by three different routes:

i) Introduction of a linker onto existing solid support.
ii) Synthesis of solid support by copolymerization using one monomer suitably derivatized to serve as a future linker.
iii) The first synthetic component is coupled to a linker in solution and then the linker-compound moiety is attached to the resin.

The last two methods required time-consuming work prior to peptide synthesis and potentially included the preparation of numerous intermediates in solution. Irrespective of the preparation method, it is essential to carry out quantitative analysis of resin loading before the synthesis. Therefore, commercially available solid supports with linkers already attached are typically used for SPPS. It is important to point out that the quality of the same type of resin with linker can vary substantially even among different resin lots from the same source [17]. In the case of the most commonly used linkers, an amino acid loaded to the resin can be purchased as well.

### 8.1.1
### Immobilization Strategies

Linkers and the type of linkage used to attach the first synthetic component to the support can be classified in a variety of ways, such as the type of functionality produced when the compound is detached from the resin, the conditions required (acidic, basic, photolytic, etc.) to remove the compound from the support, or by the type of moiety used to attach the molecule of interest to the support. In this chapter, linkers are classified according to the type of group that is used to anchor the first building block, the amino acid, to the support. This classification allows various cleavage techniques to be used for a particular type of linkage.

i) Immobilization via carboxyl group.
ii) Immobilization via amino group.
iii) Backbone immobilization.
iv) Immobilization via amino acid side-chain:

a) A typical SPPS is carried out stepwise from the C- to N-termini by immobilization of the *N*-protected C-terminal amino acid. The amino group is usually protected as an ester or amide, but also less frequently as a hydrazide, *O*-substituted oxime, or thioester. This route is the most common linking strategy and it was used in the first Merrifield peptide synthesis [1]. It provides the widest selection of linkers, $\mathbf{L_C}$, and has been used for the synthesis of peptide acids (X = O), amides (X = NH), or *N*-alkylamides (X = NR). The N-terminus of the target peptide can be derivatized (alkylation, acylation, sulfonylation, etc.) while still attached to the resin. Additionally, numerous linkers are commercially available. See Scheme 8.1.

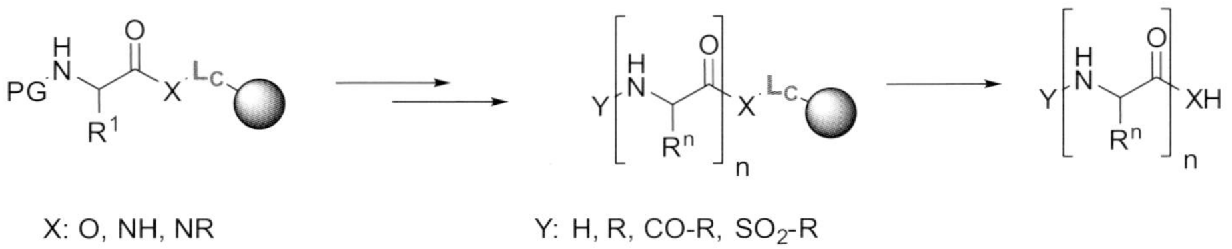

X: O, NH, NR

Y: H, R, CO-R, $SO_2$-R

**Scheme 8.1** Immobilization via carboxyl group.

b) Immobilization of the *O*-protected N-terminal amino acid and carrying out the synthesis from N- to C-terminus (referred to as inverse or reverse peptide synthesis), reported for the first time by Letsinger [18, 19], is considerably less common. Amino acid esters are immobilized via a carbamate-type amino-protecting group ($\mathbf{L_N}$). The carboxylate of the C-terminal amino acid is amenable to subsequent on-resin chemical transformations. See Scheme 8.2.

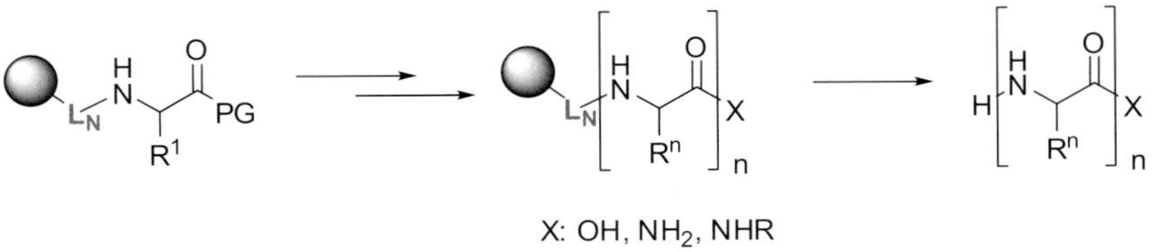

H N O PG $L_N$ $R^1$ → H N O X $L_N$ $R^n$ n → H H N O X $R^n$ n

X: OH, $NH_2$, NHR

Scheme 8.2 Immobilization via amino group.

c) Backbone immobilization enables extension of the peptide chain in both directions and it is the strategy of choice for synthesis of peptides with modified termini, cyclic peptides, and peptides that include a non-peptidic component at either terminus of the peptide. A typical linker, $\mathbf{L_B}$, for the backbone immobilization is an electron-rich benzyl group. See Scheme 8.3.

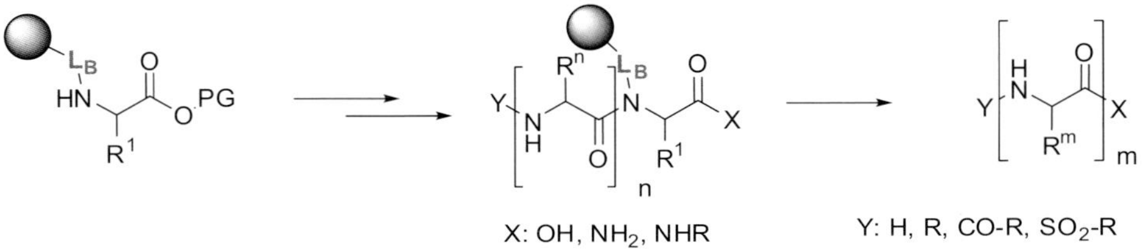

$L_B$ HN O O PG $R^1$ → Y N H $R^n$ O $L_B$ N O X $R^1$ n → Y H N O X $R^m$ m

X: OH, $NH_2$, NHR    Y: H, R, CO-R, $SO_2$-R

Scheme 8.3 Backbone immobilization of amino acid.

d) Immobilization via a side-chain of an amino acid represents a special category. This immobilization has been used for the synthesis of classes of compounds similar to those prepared using the backbone linker; however, it requires an amino acid with a side-chain suitable for immobilization. The selection of a linker, $\mathbf{L_S}$, depends on the side-chain character. See Scheme 8.4.

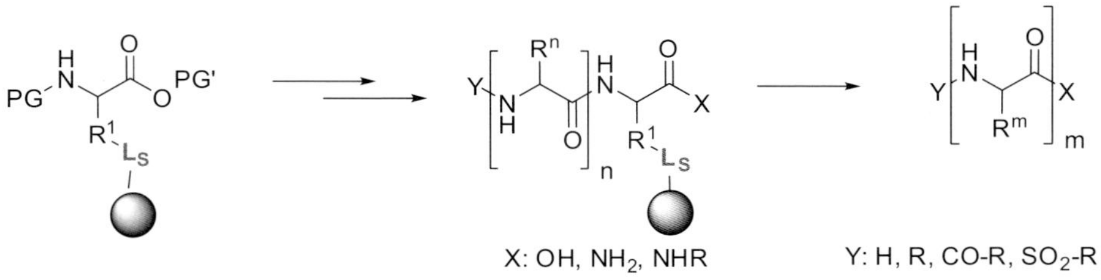

PG H N O O PG' $R^1$ $L_S$ → Y N H $R^n$ O H N O X $R^1$ $L_S$ n → Y H N O X $R^m$ m

X: OH, $NH_2$, NHR    Y: H, R, CO-R, $SO_2$-R

Scheme 8.4 Immobilization of amino acid via side-chain.

### 8.1.2
### Overview of Linker Types

The most common linkers used in contemporary solid-phase peptide and also organic syntheses are acid labile. Guillier *et al.* [15] reported that more than 60%

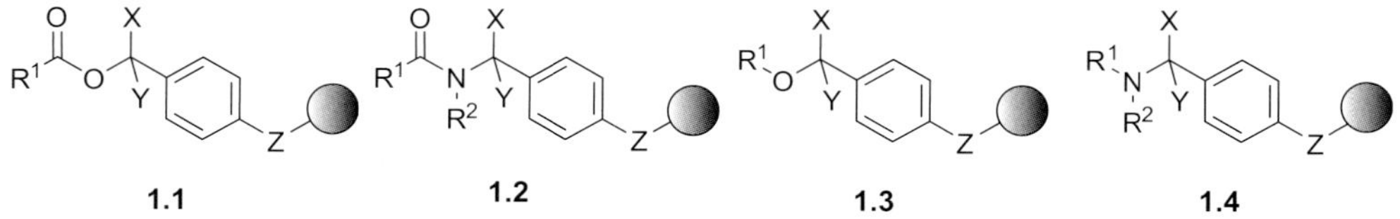

X = Y = H (benzyl)
X = H, Y = Ph (benzhydryl)
X = Y = Ph (trityl)

Z: anchoring group
$R^1$ = alkyl or aryl
$R^2$ = H or alkyl

**Figure 8.3** Acid-cleavable benzyl group-derived linkers.

of recently used linkers are acid cleavable. The most common reagents for acid cleavage are trifluoroacetic acid (TFA) and HF; both liquid reagents have the advantage of easy removal. However, HF attacks glass and requires a dedicated apparatus. HF has a boiling point 19 °C and is used for cleavage both as a liquid, usually at 0 °C [20, 21], or as a gas at 25 °C [22].

The most widespread core structure used in acid labile linkers is the benzyl-type linkage, which includes its numerous variations, such as the α-phenylbenzyl (benzhydryl) and trityl linkers (Figure 8.3). Esters **1.1** provide carboxylic acids upon cleavage from the resin, secondary and tertiary amides **1.2** are cleaved to primary and secondary carboxamides, ethers **1.3** provide alcohols, and secondary and tertiary amines **1.4** yield primary and secondary amines. Among those four types of substrates the acid stability increases in the order ester < amide < ether < amine. The acid lability of the carbon–oxygen/nitrogen bond also largely depends on the stability of the carbocation formed upon cleavage. Electron-donating groups contribute to the stability of the carbocation formed on the linker upon cleavage and thus contribute to the acid lability of the linker–oxygen/nitrogen bond. The ease of cleavage of the linkage can be fine-tuned by appropriate substitution on the methylene carbon of the benzyl group and the aromatic ring.

Polymer-supported benzyl groups are also very frequently attached to a polystyrene-based support via a heteroatom, in most cases oxygen (e.g., Wang resin, $Z = O\text{-}CH_2$). Owing to limited stability of this *O*-benzyl linkage towards acids, homobenzyl (phenethyl) derivatives have also been developed [23–26].

### 8.1.3
### Selection of a Linker

The primary criteria for linker selection are the structure of the target peptide and the protection strategy. The structure of the peptide dictates the type of the linker, while the protection strategy determines the necessary stability of the linker. The temporary protecting groups need to be removed after each iterative step and the linker needs to be stable towards the deprotection conditions. The side-chain protecting groups need to survive all steps of the solid-phase synthesis, but they need to be removed after the synthesis is finished (either concurrently with releasing the peptide or in an additional step, typically prior to the release from the resin).

Historically, the most frequently used protecting moiety of the $\alpha$-amino group for SPPS performed from the C- to N-terminus was the Boc (*tert*-butyloxycarbonyl) group, cleavable under acidic conditions. Most contemporary SPPS syntheses are carried out using the Fmoc (9*H*-fluoren-9-ylmethoxycarbonyl) protecting group, which is cleavable under basic conditions.

The removal of the Boc protecting group is performed in strong acidic conditions (e.g., 50% TFA/dichloromethane (DCM); 15 min; room temperature), and thus linkers providing substantial acid-stability have to be used. Sufficiently stable linkers include 4-hydroxymethyl-phenylacetamidomethyl (PAM), methylbenzhydrylamine (MeBHA), and benzhydrylamine (BHA), which are stable in TFA and cleavable by HF [27], trifluoromethanesulfonic acid (TFMSA) [28], trimethylsilyl trifluoromethanesulfonate (TMSOTf) [29] or HBr/TFA [30]. The need to use strong acids for cleavage together with minor cleavage of the peptide from the linker in individual Boc-deprotection steps has limited the use of this protecting group strategy.

The Fmoc-based strategy is employed significantly more frequently – it employs mild reagents for temporary protecting group removal and less vigorous cleavage conditions for side-chain protecting group removal with concurrent release of the target peptide from the resin. Almost any acid-cleavable linker can be used, with the Wang linker being the most popular. Deprotection of the Fmoc group from the amino acids is carried out under basic conditions, typically with 20% piperidine solution in *N,N*-dimethylformamide (DMF). Alternatively, 1,8-Diazabicyclo[5.4.0]undec-7-ene (DBU), morpholine, or tetraalkylammonium fluorides have been advocated to achieve more effective cleavage or to avoid some side-reactions [31–36].

The substantially less frequent N $\rightarrow$ C synthesis mode is used particularly for C-terminal carboxyl group-modified peptides such as peptide aldehydes, halomethyl ketones, boronic acids, and peptide mimetics. For example, C-terminally modified peptides and peptide libraries have been synthesized in the inverse direction mode using organic syntheses [37, 38] as well as enzymatic catalysis [39].

For this less-frequent N $\rightarrow$ C mode, the carboxyl is protected as an ester such as methyl ester [19], *t*Bu ester [40], or benzyl ester [18, 41]. The conditions required for the removal of the ester protecting group before commencing with the next amino acid addition include treatment with strong acid or hydroxide (e.g., 8% NaOH [19], 4 M HCl/dioxane [40]). Easily removable protecting groups such as tributoxysilyl (t-Bos) [42] have been suggested. The derivatives with this protecting group can be readily prepared, are inexpensive, are stable throughout the coupling reaction, and the protecting group can be selectively removed in high yield by treatment with 25% TFA/DCM before beginning the next cycle [42].

With respect to side-chain protection, the amino acids can be divided into three groups: no protection required (Gly, Ala, etc.), protection is optional (His, Trp), and protecting groups are needed (Asp, Lys, Orn). Typical side-chain functional groups that require protection are carboxyl and amino groups. Side-chain protecting groups are selected to be compatible with the chosen temporary protecting group. Thus,

**Table 8.1** Recommended combination of protecting groups.

| Amino acid | Boc strategy | Fmoc strategy |
|---|---|---|
| Arg | Ts, Mts | Mtr, Pmc, Pbf |
| Asn | — | Tr |
| Asp | Bn, Cy | *t*Bu |
| Cys | 4-MeBn, Acm | Tr, Acm |
| Gln | — | Tr |
| Glu | Bn | *t*Bu |
| His | Bom, Dnp | Tr |
| Lys | 2-Cl-Z | Boc |
| Ser | Bn | *t*Bu |
| Thr | Bn | *t*Bu |
| Trp | Formyl | Boc |
| Tyr | 2-Br-Z | *t*Bu |

2-Br-Z, 2-bromobenzyloxycarbonyl; 2-Cl-Z, 2-chlorobenzyloxycarbonyl; 4-MeBn, 4-methylbenzyl; Acm, acetamidomethyl; Dnp, 2,4-dinitrophenol; Mtr, 4-methoxy-2,3,6-trimethyl-benzenesulfonyl; Mts, 2-mesitylenesulfonyl; Pbf, 2,2,4,6,7-pentamethyldihydrobenzofuran-5-sulfonyl; Pmc, *N*-(2,2,5,7,8-pentamethylchroman-6-sulfonyl-).

carboxyl groups are typically protected as esters. For the Boc protection strategy, benzyl esters or sterically more extensive cyclohexyl esters (to avoid some side-reactions such as cyclization of Asp to aspartiimide [43]) are used. The *t*Bu esters, cleavable in TFA-containing cocktails, are used for Asp and Glu protection during the Fmoc protection strategy. Orthogonally cleavable protecting groups include allyl ester, removable by palladium-catalyzed hydrostannolysis [44], DCM/$PhSiH_3$/$Pd(Ph_3P)_4$ [45, 46] or by secondary amines, and also recently reported trialkoxylsilylesters (t-Bos) [42]. As the deprotection conditions are very mild, almost any linker can be used.

Typically protection of amino groups in the side-chain (Lys, Orn) is achieved using 2-chlorobenzyloxycarbonyl with the Boc protection strategy and Boc for the Fmoc protection strategy. Recommended side-chain protecting groups for Boc and Fmoc protection strategies are summarized in the Table 8.1 [47].

## 8.2 Immobilization via Carboxyl Group

Immobilization of the first amino acid via an *N*-protected amino acid carboxylate to a solid support is the most frequently used method of peptide synthesis. The selection of a linker primarily depends on the synthetic strategy, briefly discussed in the Introduction, and on the required target C-terminus, peptide acid, or amide. The carboxylic group of the amino acid is immobilized to the solid support as an ester, amide, hydrazide, or *O*-substituted oxime. Reaction conditions for the release of target peptides are then dictated by the linker type.

### 8.2.1
### Esters

There is a plethora of ester linkers, the majority of them are of the benzyl type (Figure 8.4), with fine-tuned electronic properties for acid- or base-mediated release

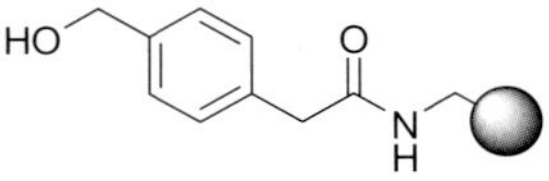

**2.1**
**PAM resin**
4-Hydroxymethyl-**p**hen**y**l**a**cetamido**m**ethyl Resin

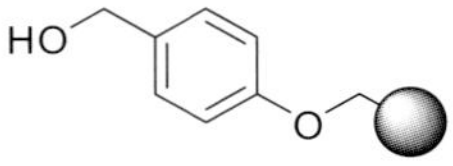

**2.2**
**Wang resin**
(4 - Hydroxymethyl-phenoxy)-methyl Resin

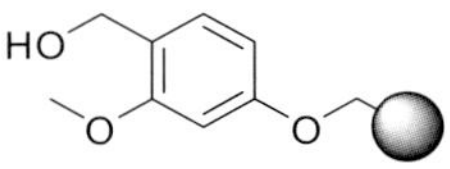

**2.3**
**SASRIN resin** (**s**uper **a**cid **s**ensitive **r**es**in**)
(4 - Hydroxymethyl-3-methoxy-phenoxy)-methyl Resin

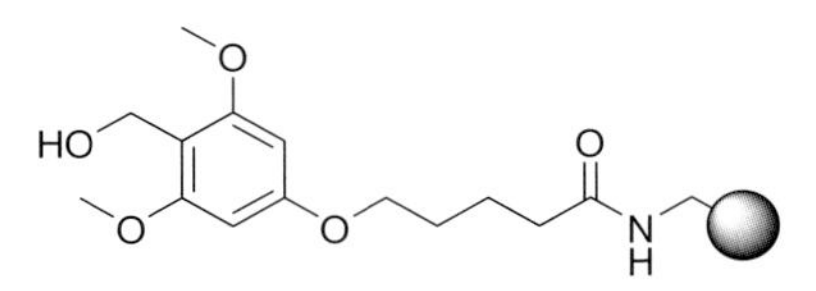


**2.4**
**HAL resin** (**h**ypersensitive **a**cid-**l**abile resin)
[4-(4-Hydroxymethyl-3,5-dimethoxy-phenoxy)-pentanoylamino]-methyl Resin

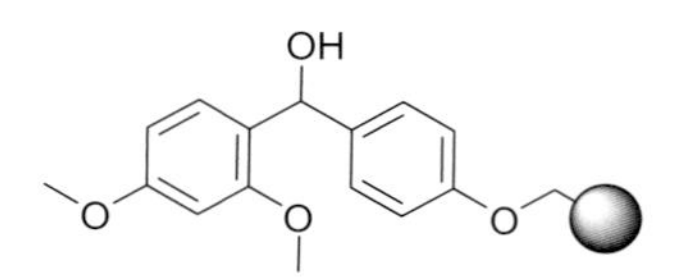


**2.5**
**Rink ester resin**
{4-[(2,4-Dimethoxy-phenyl)-hydroxy-methyl]-phenoxy}-methyl Resin

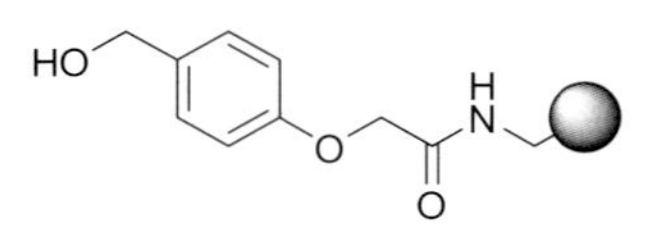


**2.6**
**HMPA resin**
[2-(4-**H**ydroxy**m**ethyl-**p**henoxy)-**a**cetylamino] Resin

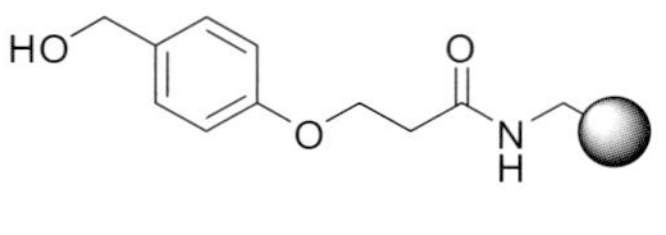

**2.7**
**HMPP resin**
[3-(4-**H**ydroxy**m**ethyl-**p**henoxy)-**p**ropionylamino] Resin

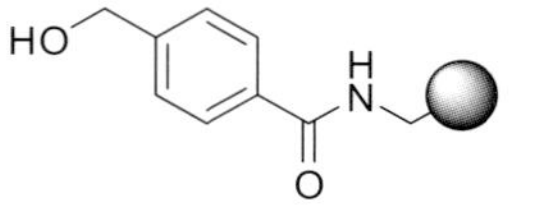

**2.8**
**HMBA resin**
(4-**H**ydroxy**m**ethyl-**b**enzoylaminomethyl) Resin

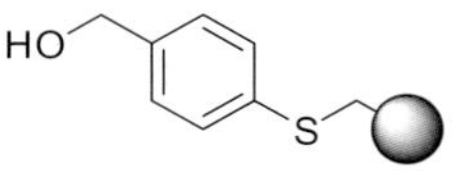

**2.9**
(4-Hydroxymethyl-phenylsulfanyl methyl)

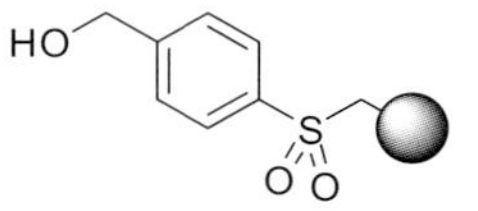


**2.10**
(4-Hydroxymethyl-benzenesulfonylmethyl)

**Figure 8.4** Linkers with a terminal hydroxy group.

of target compounds under a diverse range of reaction conditions. The effect of substitution on the aromatic ring(s) of a linker was discussed in the introductory section. The Wang linker and Fmoc/*t*Bu protection strategy are the most frequently used for SPPS. Owing to the extensive number of linkers and their high frequency of usage in peptide syntheses, we describe the preparation of polymer-supported esters first, followed by different ways of releasing the target peptide from the support.

The formation of esters from amino acids and linkers can be achieved through various derivatives. The most frequently used method for ester formation is esterification of resin-bound hydroxy linkers with *N*-protected amino acids. Alternatively, nucleophillic substitution of benzyl derivatives substituted by chloro [48–51], bromo [51, 52], mesyl [50, 52], tosyl [52], diazo [53], and trichloroacetimidate [54] groups have been used for ester formation.

#### 8.2.1.1 **Hydroxy Linkers for Preparation of Resin-Bound Esters**

The resin-bound hydroxy group for ester formation is typically found on a benzyl group. Since the first solid-phase synthesis in 1963 [1], a multiplicity of substituted benzyl alcohols have been suggested for this purpose. The substitution pattern influences the stability towards electrophiles and nucleophiles, and the properties of individual linkers can be fine-tuned for a particular synthetic strategy and target C-terminal functional group (carboxylic acid, ester, amide, etc.). Structures of hydroxy linkers are summarized in Figure 8.4.

Reaction conditions used for esterification of these linkers involve activation of *N*-protected amino acids with carbodiimides (Scheme 8.5). Typical coupling conditions for several frequently used linkers are summarized in Table 8.2.

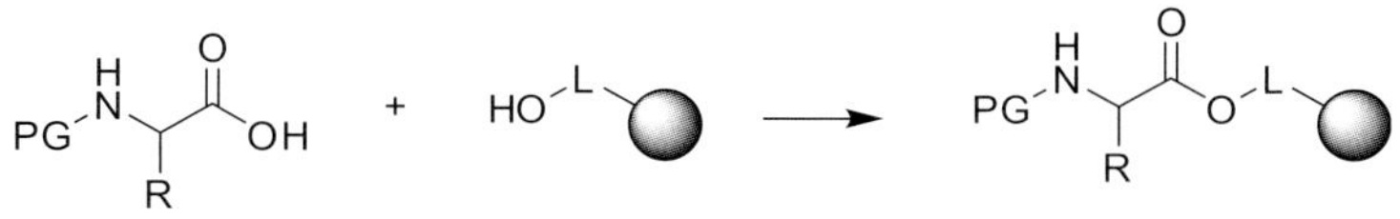

PG–NH–CH(R)–C(=O)OH + HO–L–(resin) → PG–NH–CH(R)–C(=O)–O–L–(resin)

**Scheme 8.5** Ester formation from resin-bound alcohols.

Suitable reaction conditions must also be found with respect to potential racemization of amino acids. The racemization is dependent on the type of activating agent and type and amount of base. Additionally, solvent and temperature have been shown to be crucial for the amount of formed epimeric amino acid bound to the linker. For example, when activation is achieved with 1-(2-mesitylone-sulfonyl)-3-nitro-1,2,4-triazole (MSNT) [64] or 1,4-(dicyanomethylene)benzoyl cyanide (DCBC) [65] as activating agents instead of widely used *N,N*′-dicyclohexylcarbodiimide (DCC), racemization is significantly suppressed.

The role of base can be exemplified by use of 4-(*N,N*-dimethylamino)pyridine (DMAP). When 1 equiv. of this base in DMF was used for immobilization of amino acids, racemization reached 1.5–20% [66]. If the amount of DMAP was reduced to 0.06 equiv., racemization was suppressed below 1.2% [67]. No racemization was

**Table 8.2** Reaction conditions for ester formation between linker and the first amino acid.

| Resin | Structure | Reagents | Solvent | Temperature | Time | Reference |
|---|---|---|---|---|---|---|
| PAM | **2.1** | CDI | DCM | room temperature | 8 h | [55] |
| Wang | **2.2** | DCC/ pyridine | DCM | room temperature | 150 min | [56] |
| SASRIN | **2.3** | DCC/ DMAP | DMF/ DCM | 0 °C | 20 h | [57] |
| HAL | **2.4** | HOBt | DMF | room temperature | 2 h | [58] |
| Rink | **2.5** | DCCI | DCE/ DMAP | room temperature | 4 h | [59] |
| HMPA | **2.6** | DIC/DMAP | DCM | room temperature | 12 h | [60] |
| HMPP | **2.7** | DSC/DMAP | DMF | room temperature | 2 h | [61] |
| HMBA | **2.8** | MSNT/ MeIm | DMF | room temperature | 45 min | [62] |
| Phenylsulfanyl | **2.9** | DCC | DCM | room temperature | overnight | [63] |
| Benzenesulfonyl | **2.10** | oxidation of **2.9** | AcOH | room temperature | 12 h | [63] |

CDI, 1,1′-carbonyldiimidazole; DCE, 1,2-dichloroethane; DIC, *N,N′*-diisopropylcarbodiimide; DSC, *N,N′*-disuccinimidyl carbonate; MeIm, 2-methylimidazole.

observed when coupling of the first amino acid was performed without DMAP or when DMAP was replaced by *N*-hydroxybenzotriazole (HOBt) [68]. Also, temperature played a significant role: coupling of amino acids in the presence of 1 equiv. of DMAP at 0–3 °C caused racemization in only 0.1–0.3% with acceptable loading during 16 h [69]. The racemization was diminished also by application of different methods of amino acid coupling, including Mitsunobu reaction [70, 71] or esterification with amino acid fluorides [70].

#### 8.2.1.2 Electrophilic Linkers for Preparation of Resin-Bound Esters

This type of resin-bound linker was used for the first peptide synthesis: the Merrifield resin was chloromethylated cross-linked polystyrene resin [1]. This original resin has been modified over time to achieve better stability of ester linkages during the peptide synthesis as well as easier cleavage from the beads in the end. The most frequent linkers of this type used for peptide synthesis are summarized in Figure 8.5.

Reaction conditions for preparation of polymer-supported esters from electrophilic resin-bound linkers are summarized in Scheme 8.6 and Table 8.3.

#### 8.2.1.3 Cleavage from the Resin

**Acid-Mediated Cleavage** **Benzyl esters** are the most commonly used acid-cleavable linkers for the synthesis of peptides. The original Merrifield protocol [1] used benzyloxycarbonyl groups for the protection of the *N*-α-amino moiety and its cleavage was accomplished using 10% HBr in acetic acid. The benzyl ester linkage to the resin was partially cleaved and therefore its acid stability was increased by nitration.

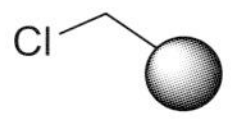

**2.11**
**Merrifield resin**
Chloromethylresin

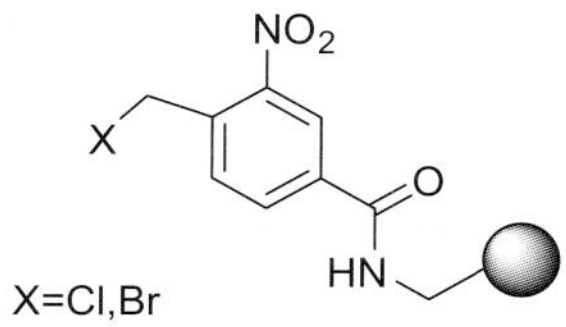

X=Cl,Br

**2.12**
4-Halogenmethyl-3-nitro-benzoylaminoresin

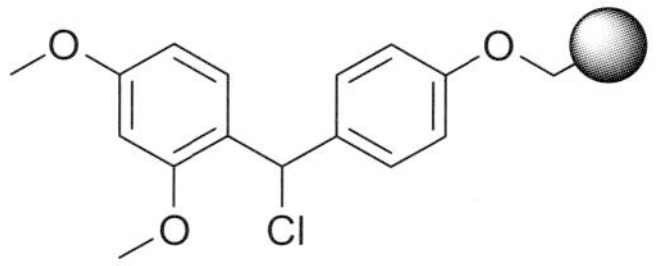

**2.13**
**Rink chloride resin**
{4-[Chloro-(2,4-dimethoxy-phenyl)-methyl]-phenoxy}-methylresin

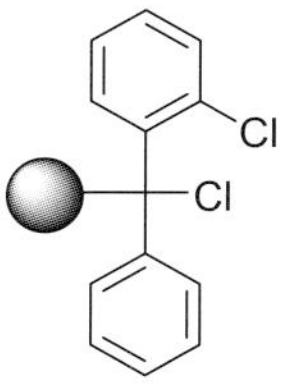

**2.14**
**2-Chlorotrityl resin**
1-Chloro-2-(chloro-diphenyl-methyl)resin

**Figure 8.5** Electrophilic linkers for preparation of resin-bound esters.

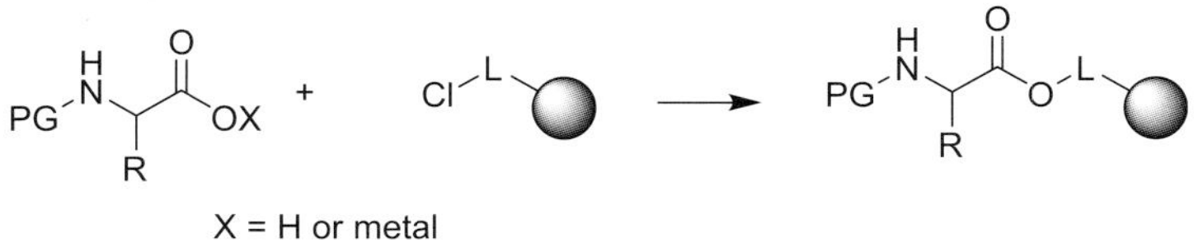

X = H or metal

**Scheme 8.6** Ester formation in SPPS.

**Table 8.3** Ester formation using chlorinated linkers.

| Resin | Structure | Amino acid | Base | Solvent | Temperature | Time | Reference |
|---|---|---|---|---|---|---|---|
| Merrifield | **2.11** | Cs salt | none | DMF | 50 °C | overnight | [1] |
| Nitro-halomethyl | **2.12** | Cs salt | none | DMF | 45 °C | overnight | [72] |
| Rink chloride | **2.13** | free acid | DIPEA | DCE | 20 °C | 18–26 h | [73] |
| 2-Chlorotrityl | **2.14** | free acid | DIPEA | DCE | 20 °C | 20 min | [74, 75] |

DIPEA, *N*,*N*-diisopropylethylamine.

Specifically, the electron-withdrawing nitro group destabilizes the carbocation and increases the acid stability. Later, the Boc group was used for peptide synthesis [76] and its removal was affected using 1 M HCl in AcOH. At the end of the synthesis the peptide was cleaved from the support by HBr in TFA. In subsequent publications, peptides were cleaved from the support and all side-chain protecting groups removed by treatment of the peptidyl resin with a strong acid, liquid HF at 0 °C [20]. From a practical point of view, liquid HF offers an advantage in that it has a boiling point of only 19 °C and is therefore evaporated very easily. However, it was found that cleavage of protecting groups generated carbocations that eventually alkylated aromatic rings of Tyr and Trp. To overcome this problem, scavengers that trap the reactive species were added to the cleavage mixture. The need to use scavengers required that after the evaporation of HF, the crude reaction mixture be extracted to remove the excess scavengers and scavenger byproducts, or that the peptide be precipitated and isolated by filtration or centrifugation. Later, a "low–high" cleavage protocol was developed [21] whereby the resin-bound peptides were treated with a mixture of dimethylsulfide (DMS)/HF (HF concentration of 30% by volume) and scavengers. With this "low" concentration of HF, the side-chain protecting groups were removed via an $S_N2$ mechanism, thus preventing undesired alkylation of amino acids with sensitive side-chains (Tyr, Trp). The HF concentration was then increased and the peptides were cleaved from the resin. Recently, gaseous reagents, including HF, have been used for cleavage of compounds from benzyl ester linkages [22]. An alternative cleavage reagent is TFMSA [28]; however, this reagent has a high boiling point and isolation of the desired product(s) can be more difficult.

For the synthesis of long peptides using the Boc/Bn protection strategy on Merrifield resin, the stability of the benzyl ester towards 50% TFA in DCM is not sufficient, even though only approximately 1–2% of the peptide is cleaved during each cycle.

Using the PAM linker **2.1** [77], the stability of the ester bond towards treatment with 50% TFA in DCM is increased 100-fold when compared with the original Merrifield resin **2.11** [55].

Recently, 4-(3-hydroxy-4-methylpentyl)phenylacetic acid was used to replace the PAM linker [78]. This linker is more stable and it survives cleavage of side-chain protecting groups by a "low" TFMSA/DMS/TFA mixture [21], thus allowing the side-chain cleavage mixture to be removed before the peptide is cleaved from the support using a high concentration of HF. In addition, the authors suggested that the linker has self-scavenging properties in that the carbocation formed on cleavage of the peptide from the support is stabilized by an intramolecular cyclization.

To facilitate cleavage under milder conditions, the linker was further fine-tuned by substitutions on the aromatic ring. An electron-donating alkoxy group increases the acid lability of the ester bond. This approach led to the development of the Wang linker **2.2** [56], the most often used TFA-cleavable linker for the preparation of carboxylic acids.

Esters of 3-(4-hydroxymethyl-phenoxy)-propionic acid (HMPP) **2.7** [79] are cleaved 2–3 times faster than 2-(4-hydroxymethyl-phenoxy)-acetic acid (HMPA) esters **2.6**.

Addition of a methoxy group to the aromatic ring further increases the acid lability [80].

SASRIN (super acid-sensitive resin) resin **2.3** is reported in the literature as a super acid-sensitive resin [57, 67, 81]. Extreme acid lability is achieved also by the hypersensitive acid-labile (HAL) linker **2.4** [58], where acids can be cleaved by as low as 0.05–0.1% TFA.

The Wang linker **2.2** [56] attached to polystyrene resin, referred to as Wang resin, is currently the most frequently used acid-cleavable ester linker. Most peptide acids are prepared on the Wang resin using the Fmoc/*t*Bu protection strategy. The base-labile Fmoc group is removed from the *N*-α-amino group by treatment with piperidine in DMF and the *t*Bu-based side-chain protecting groups are cleaved by TFA at the same time the peptide is cleaved from the linker.

**Benzhydrol esters** are more acid labile than benzyl esters. Esters formed from the benzhydrol linker are cleaved with 1% TFA or 0.2 M HCl in DCM [82]. Similar acid lability to the HAL linker was achieved by the Rink ester linker **2.5** [59], which is cleavable by 10% AcOH or 0.2% TFA.

**Trityl esters** are among the most acid labile linkages in this group. The trityl linker was introduced by Leznoff [6, 83] and Fréchet [4, 5] as a polymer-supported protecting group and later applied to peptide synthesis using the Fmoc/*t*Bu protection strategy by Barlos [84]. The acid lability can be reduced, and currently the trityl linker of choice is the 2-chlorotrityl derivative **2.14**. This linker is more stable towards acids when compared to the unsubstituted trityl [74, 85]. Cleavage of acids from the 2-chlorotrityl linker was achieved by treatment with AcOH/trifluoroethanol/DCM (1 : 1 : 8) [74, 85]. Typical conditions for cleavage of peptide from the various resins are summarized in Table 8.4.

**Table 8.4** Comparison of the acid lability of ester linkers.

| Linker | Structure | Cleavage | Reference |
|---|---|---|---|
| PAM | 2.1 | HF or TFMSA | [77] |
| Wang | 2.2 | 50 or 95% TFA | [56] |
| SASRIN | 2.3 | 1% TFA | [57, 67] |
| HAL | 2.4 | 0.1% TFA | [58] |
| Rink | 2.5 | 0.2% TFA | [59] |
| HMPA | 2.6 | 90%TFA | [60] |
| HMPP | 2.7 | 0.2 M EDT and 0.1 M TMSBr in TFA | [86] |
| HMBA | 2.8 | acidic stable | [87] |
| Phenylsulfanyl | 2.9 | acidic stable | [63] |
| Benzenesulfonyl | 2.10 | acidic stable | [63] |
| Merrifield | 2.11 | HF or TFMSA | [88] |
| Nitro-halomethyl | 2.12 | acidic stable | [72] |
| 2-Chlorotrityl | 2.14 | AcOH/trifluoroethanol/DCM (1 : 1 : 8) | [74] |

EDT, 1,2-ethanedithiol; TMSBr, bromotrimethylsilane.

**Hydrolysis** Benzyl ester linkers can also be cleaved by treatment with a hydroxide to release the peptide with a terminal carboxyl group. Whereas acid induced cleavage is accelerated by the introduction of electron-donating substituents into the linker to stabilize the carbocation formed, nucleophilic attack is accelerated by the presence of electron-withdrawing substituents that increase the electrophilicity of the carbonyl carbon. The presence of electron-withdrawing substituents presumably also stabilizes the anion that is formed on hydrolysis or aminolysis of the ester bond. The reactivity is in a reverse order to that of acid-induced cleavage, with esters formed from the Merrifield resin being the most labile, whereas esters formed using the HAL linker **2.4** are the most stable. Saponification of the benzyl ester from nitrated polystyrene–divinylbenzene copolymer (PS-DVB) resin was used by Merrifield [1] to release the first peptide synthesized on solid phase. Commonly used aqueous saponification solutions, dilute (0.1 M) alkali in water, are compatible with hydrophilic solid supports (TentaGel, ArgoGel). However, the solutions are not compatible with the hydrophobic character of polystyrene resin. Numerous conditions for the saponification of esters bound to polystyrene resin have been reported, including 1 M aqueous NaOH/tetrahydrofuran (THF) (7 : 3) [89] and 1 M aqueous KOH/MeOH/dioxane [90]. Cleavage under these conditions is substantially slower when compared to the solution rate. The use of trimethyltin hydroxide has also been reported for the selective cleavage of resin-bound benzyl esters [91].

**Ammonolysis** Peptide amides have been synthesized by ammonolysis of polymer-bound benzyl esters prepared from nitrated Merrifield linker **2.12** [92] or 4-hydroxymethyl-benzoylaminomethyl resin (HMBA) linker **2.8** [87]. The sensitivity towards nucleophile-mediated cleavage was increased by introducing an electron-withdrawing group on the aromatic ring of the benzyl ester. Nitration of the Merrifield resin facilitates the release of resin-bound deamino-oxytocin using a saturated solution of ammonia in MeOH at 0 °C [92]. Cleavage from the HMBA linker **2.8** by ammonolysis was achieved using a solution of ammonia in 2-propanol. The secondary alcohol was used rather than MeOH to avoid transesterification [93]. Ammonolysis of peptides from linker **2.8** by ammonia vapors was also reported [94–96].

The ester linker **2.9** attached to solid support via a thioether bridge has been used as a safety-catch linker. The sulfur was oxidized to sulfone **2.10** [63] to enhance its sensitivity to nucleophilic attack. This linker has been used for the synthesis of cyclic peptides [97].

**Enzymatic Cleavage** Highly selective reagents for the cleavage of peptides from the linker have been achieved under very mild conditions with use of suitable enzymes [98]. New resins have been developed that allow penetration of enzyme inside resin beads. The smooth, specific splitting off of nonapeptide from the resin by treatment with protease Subtilisin during 30 min was demonstrated with use of polyethylenglycol (PEG) resin **2.15** named SPOCC (superpermeable organic combinatorial chemistry resin) [99] (Scheme 8.7).

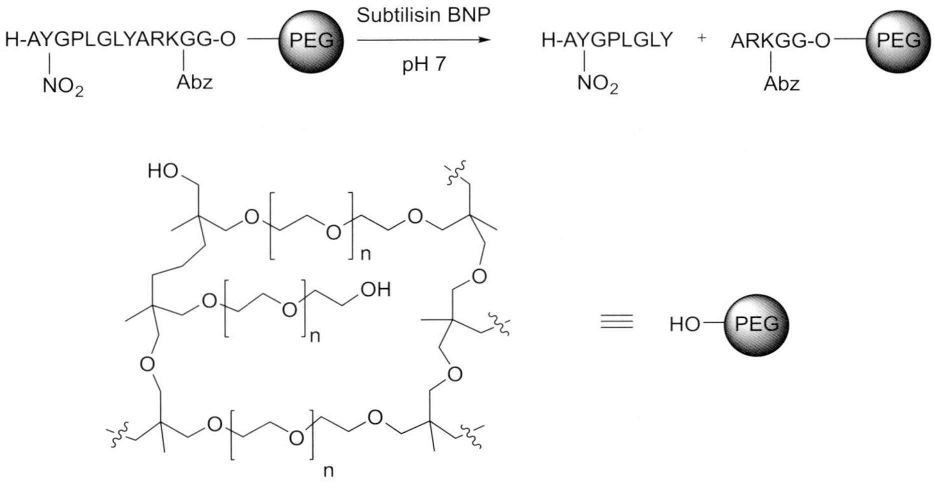

Scheme 8.7 Enzymatic cleavage of peptide from PEG resin.

In 1992, Elmore developed a phosphodiester type of linker **2.16** that facilitates direct cleavage by phosphodiesterase (Scheme 8.8) [100].

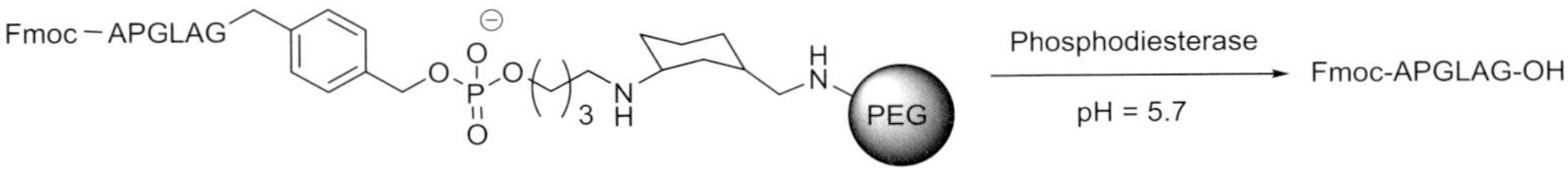

Scheme 8.8 Enzymatic cleavable phosphodiester linker.

**Other Cleavage Conditions** Typical protocols for cleavage of side-chain protecting groups include acidic or basic cleavage cocktails. To prevent potential cleavage of the peptide from the resin, new linkers cleavable by orthogonal cleavage reagents have been developed including 4-(*N*-[1-(4,4-dimethyl-2,6-dioxocyclohexylidene)-3-methylbutyl]amino)benzyl ester (*O*Dmab) **2.17** [101], where cleavage of the peptide from the linker is performed by hydrazine, and it is suitable for Fmoc/*t*Bu SPPS (Scheme 8.9).

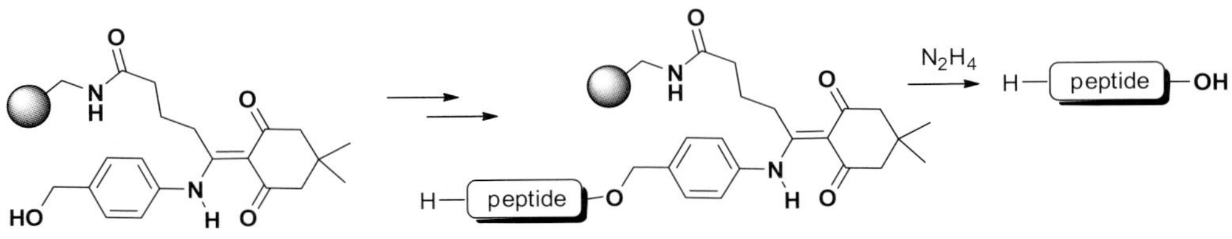

Scheme 8.9 Structure of the *O*Dmab linker and cleavage of the peptide.

Another example is the linker **2.18** derived from *p*-nitromandelic acid and reported in 2007, designed as a safety-catch linker for the Boc/Bn protection strategy. After acid-mediated cleavage of protecting groups, release of peptide from the resin was realized by reduction of the nitro group by $SnCl_2$ followed by microwave irradiation in 5% TFA [102] (Scheme 8.10).

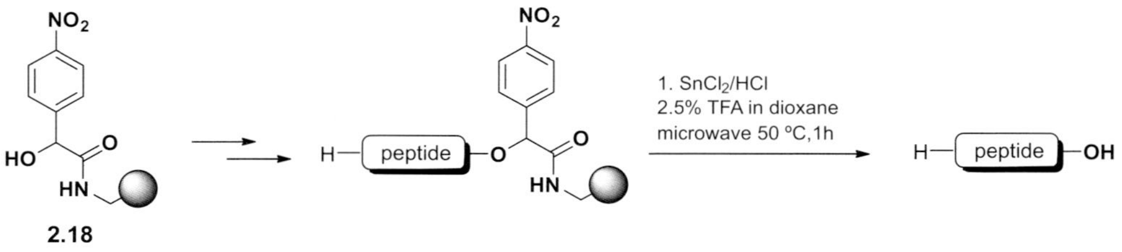

**Scheme 8.10** Use of *p*-nitromandelic acid linker for SPPS.

### 8.2.2
### Amides

Since numerous natural peptides are C-terminal amides (oxytocin, vasopressin, luteinizing-hormone releasing hormone, calcitonin, etc.), amide linkers have been studied very intensively and a rich variety of linkers with fine-tuned properties have been reported. Many of these linkers are now commercially available. Typically, benzylamine-, benzhydrylamine-, and tritylamine-type linkers are used for the preparation of peptide primary amides (Figure 8.6).

**Benzylamines** require the presence of electron donating groups on the aromatic nucleus to increase the acid lability. For example, the amine linker analogous to the SASRIN linker **2.19** [103], activated by the presence of two alkoxy groups, releases the corresponding amide in 70% TFA in DCM (ambient temperature, 2 h) in only 24% yield. The addition of 5% TFMSA to TFA cleavage cocktail is necessary to elevate the yield to 80%. Increased acid lability can be achieved for *N*-benzylamides by introducing three electron-donating substituents onto the phenyl ring. The peptide amide linker (PAL) linker **2.20** reported by Albericio *et al.* [103, 104], was developed for the synthesis of peptide amides and it is sufficiently acid sensitive to allow the products to be cleaved from the support with TFA.

**Benzhydrylamines** are the most frequently used linkers to prepare primary amides. The linker **2.21** described by Pietta and Marshall [105] was the first linker used for the synthesis of peptide amides by acidic cleavage. Matsueda and Steward [106] added a methyl group to the aromatic ring and prepared a more acid-labile MeBHA linker **2.22**, which is the most commonly used linker for peptide amides synthesis using the Boc/Bn protection strategy. Both linkers require cleavage by HF or TFMSA. To further increase the acid sensitivity of the benzhydrylamine linkers and to allow peptide synthesis using the Fmoc/*t*Bu strategy, Rink [59] and Knorr introduced linkers **2.23** and **2.24** with electron-donating methoxy groups. The

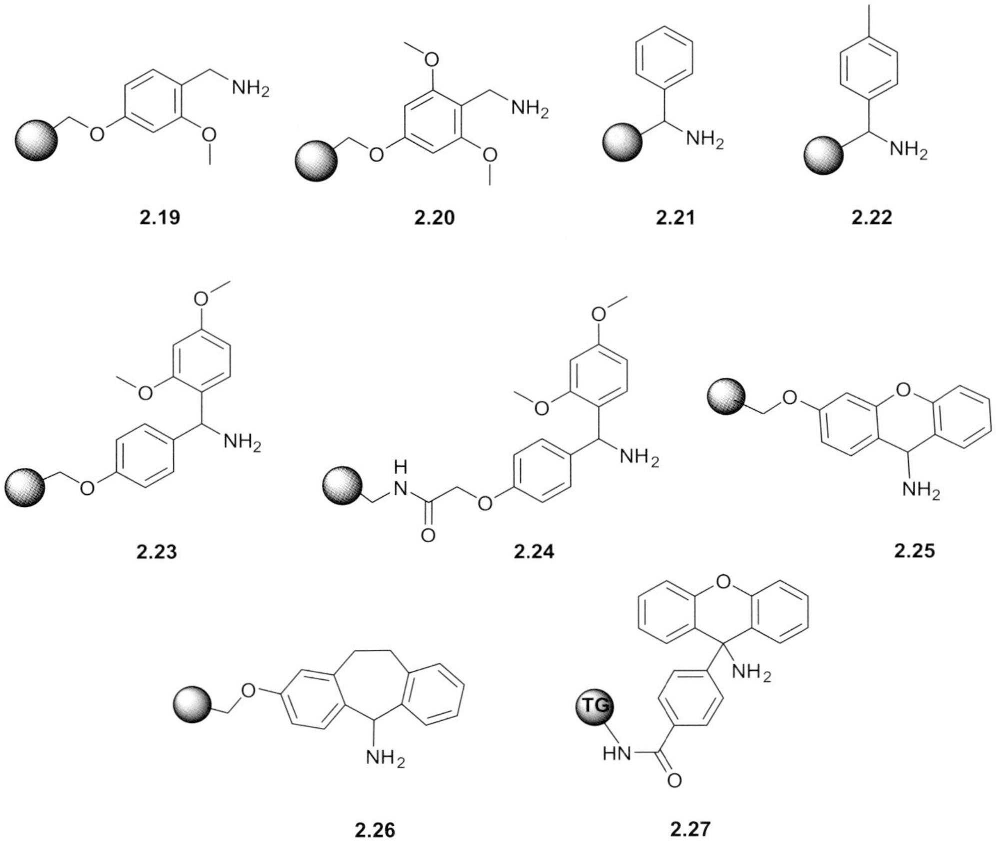

**Figure 8.6** Benzylamine, benzhydrylamine, and tritylamine linkers.

difference between the Rink [59] and Knorr linkers is the method of attachment of the linker to the resin. The Rink linker uses a benzyl ether linkage, which can be partially cleaved at high TFA concentrations. The Knorr linker is attached by a more acid stable amide; however, the bond between the linker and the target molecule is also more acid stable and requires higher TFA concentrations to affect cleavage of the target molecule from the linker. When using TFA for cleavage, the Rink amide linker **2.23** is the most common linker for the preparation of primary amides. Several other variations on the benzhydryl type linker have been reported. Sieber's xanthenyl linker **2.25** [107] provides amides under very mild cleavage conditions (2% TFA), while Ramage [108, 109] has reported a linker based on dibenzocyclohepta-1,4-diene **2.26** and a similar linker was described by Nokihara [110].

**Tritylamine** linker **2.27** was described by Henkel and used for SPPS synthesis on TentaGel resin [111]. The linker is based on a xanthone moiety and loading of the first amino acid required an 8-fold excess of Fmoc amino acid fluoride. The final cleavage was carried out with TFA containing the scavengers phenol, ethanedithiol, thioanisole, and water for 1 h.

Also, a number of examples of the preparation of secondary amides by acylation of secondary amine functionalized benzhydryl-type linkers have been reported (Scheme 8.11). For example, preparation of the polymer-bound secondary amine **2.28** was accomplished by treating chloro-Rink resin **2.13** with a primary amine [73, 112]. Using an analogous procedure, a resin-bound secondary amine was prepared from MAMP (Merrifield α-methoxyphenyl resin) resin [113]. Alternative routes include reductive alkylation of the Rink amine linker **2.23** [114].

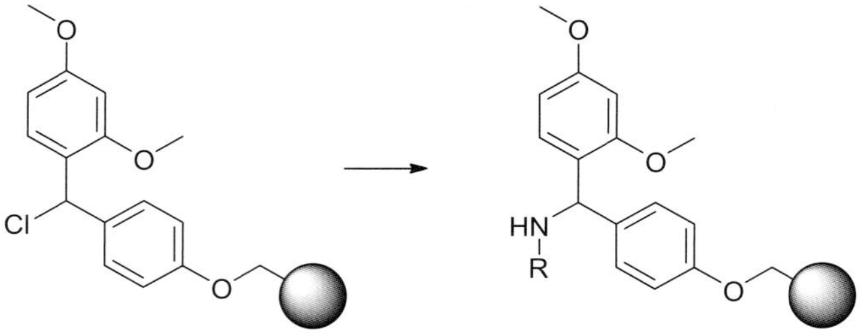

**2.13: X = Cl**
**2.23: X = $NH_2$**

**2.28**

**Scheme 8.11** Preparation of linker **2.28** for the synthesis of peptide secondary amides.

Acylation of amine linkers using *N*-protected amino acids is carried out under typical conditions used for peptide synthesis. The cleavage of the final peptides is accomplished under acidic conditions. Examples of the cleavage conditions for individual linkers are summarized in Table 8.5.

**Table 8.5** Cleavage from individual amide linkers.

| Linker | Structure | Cleavage | Reference |
|---|---|---|---|
| SASRIN-$NH_2$ | **2.19** | 5% TFMSA in TFA | [103] |
| PAL | **2.20** | 70% TFA | [103, 104] |
| BHA | **2.21** | HF or TFMSA | [105] |
| MeBHA | **2.22** | HF or TFMSA | [106] |
| Rink | **2.23** | 10–95% TFA | [59] |
| Knorr | **2.24** | 95% TFA | [115] |
| Xanthenyl | **2.25** | 2% TFA | [107] |
| Ramage | **2.26** | 3% TFA | [108, 109] |
| Tritylamine | **2.27** | 10% TFA | [111] |

## 8.2.3
## Hydrazides

The hydrazide linker **2.29** was introduced at the end of the last century, first as the tool for peptide ester synthesis [116], and later also for amides [117–119] and several other derivatives, including lipidated peptides [120] and cyclic depsipeptides [121]. Applications of this linker for synthesis of modified peptides have been compiled in a review article [122].

The key transformation of this linker is formation of a hydrazide in reaction with a protected amino acid according to Scheme 8.12. The coupling conditions include the standard Fmoc protocol. Conversion could be accelerated with use of microwave irradiation, but the yield varies only between 72 and 86% [123].

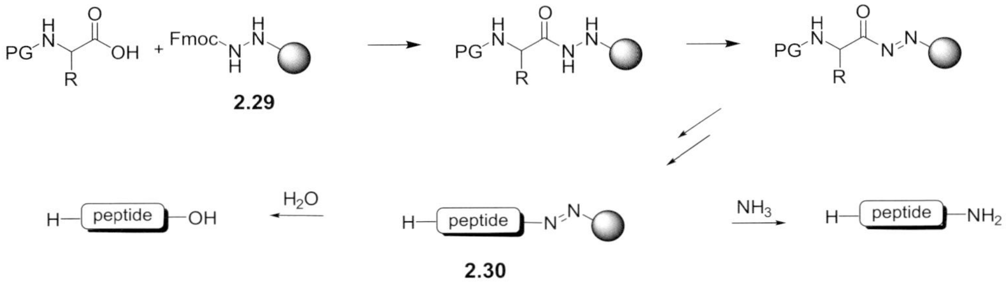

**Scheme 8.12** Hydrazide linker.

To facilitate release of the target peptide, the hydrazide was oxidized to an acyl diazene derivative **2.30** by oxygen in the presence of Cu(II) or with *N*-bromosuccinimide (NBS). Nucleophilic release from the diazene **2.30** was carried out with ammonia and water, and afforded corresponding amides and carboxylic acids, respectively.

This linker was also applied for synthesis of functional derivatives such as *N*-(di) substituted amides (aliphatic as well as aromatic) and esters [122].

A considerable advantage of the hydrazine linker is its stability towards standard deprotection conditions, thus enabling smooth Boc- and Fmoc-based SPPS and, as reported recently [118, 124], without racemization of amino acids.

## 8.2.4
## Oximes

Oxime linkers for peptide synthesis were developed in the early eighties of the last century as a tool for synthesis of amides. Substituted benzophenones, described by Kaiser and De Grado [125], were applied as linkers for acid immobilization through ester formation. *p*-Nitrobenzophenone was advocated as the most useful derivative in this class (Scheme 8.13).

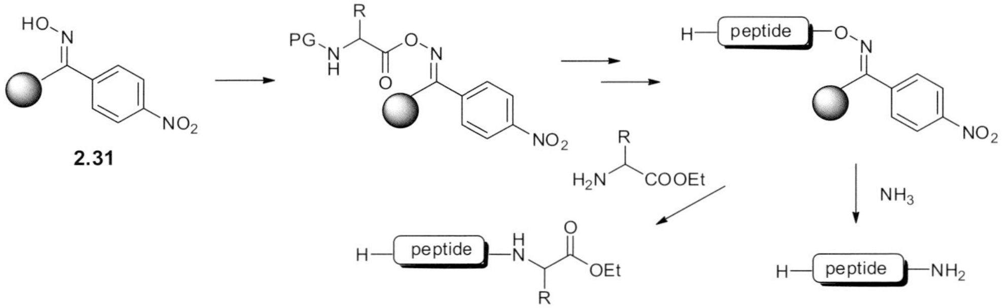

Scheme 8.13 Kaiser oxime linker in SPPS.

The linker **2.31**, commonly referred to as Kaiser's linker or Kaiser oxime linker, was stable to acid-mediated cleavage up to 25% TFA, but it was very sensitive to nucleophilic attack. It was cleaved by ammonia at the end of the synthesis to afford the appropriate amide or by the amino group of an amino acid ester to elongate the final peptide by one unit [126]. In this synthesis the effects of $Ca^{2+}$ or $Eu^{3+}$ salts as cleaving agents in very mild conditions were studied [127].

### 8.2.5 Thioesters

Thioester linkers, introduced in 1991 [128], have been used particularly for formation of peptide dendrimers [129], cyclic peptides [130, 131], or for synthesis of peptides with a terminal thioester group. The last mentioned derivatives can serve as substrates for native chemical ligation (NCL), leading to very long peptide chains [132] (usually over 60 amino acids), when the optimization of α-*N*-deprotection and subsequent amide bond formation as well as purification of final peptide is too difficult. Therefore, these peptides are built effectively by NCL according to Scheme 8.14 [133].

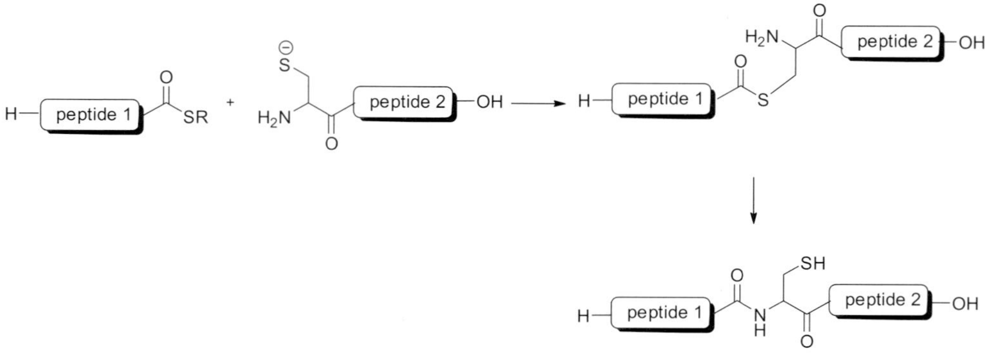

Scheme 8.14 Preparation of long peptide chain by NCL.

The thioester linkers are stable to TFA treatment used in the standard Boc/Bn protection strategy. The cleavage is performed with strong acid, usually HF, to afford a peptide thioester, or by nucleophilic attack for the preparation of peptides with a modified terminal carboxyl.

The immobilization of the first amino acid is described in Scheme 8.15. Diphenyl-methanethiol linker **2.32** [134, 135] is acylated by the amino acid in solution, followed by immobilization to the resin by standard amide-coupling methods.

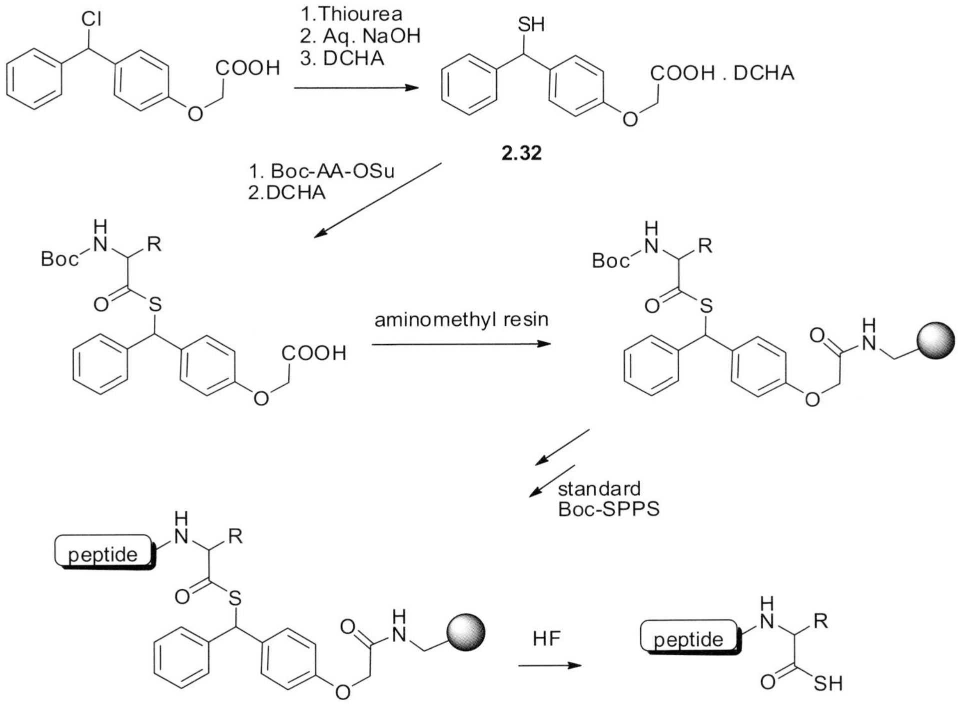

**Scheme 8.15** Immobilization of amino acid-bound thioester linker complex to solid support.

The first amino acid can also be bound directly to alkylthio resin, as described in the case of 3-mercaptopropionyl **2.33** and 2-mercaptoacetyl **2.34** resins. Boc-amino acids were preactivated by 7-azabenzotriazol-1-yloxy-tris-(pyrrolidino)phosphonium hexafluorophosphate (PyAOP) and *N,N*-diisopropylethylamine (DIEA) [131]. Immobilized peptides were successfully used for synthesis of hydroxamic acids by nucleophilic attack with $NH_2OTMS$ [136] (Scheme 8.16)

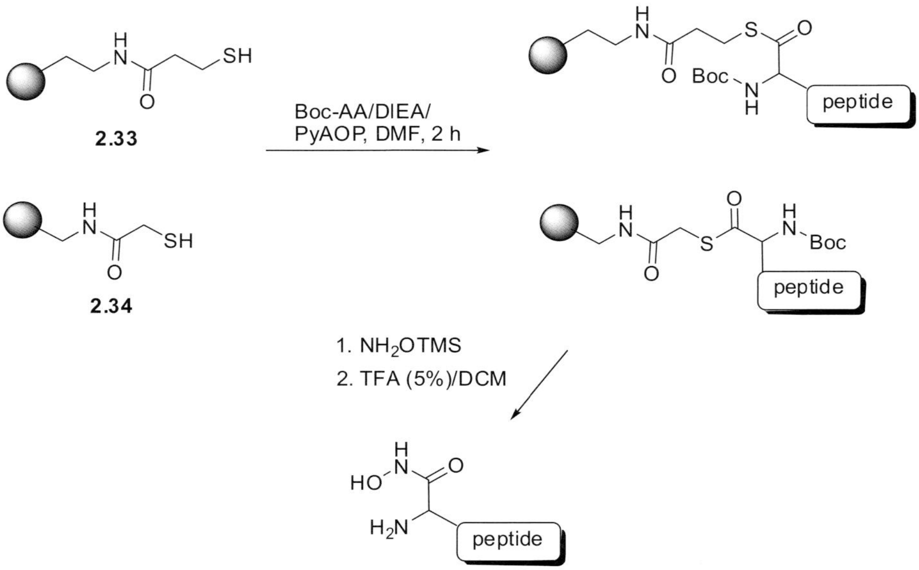

Scheme 8.16 Immobilization of amino acid to thioester resin.

## 8.3
## Immobilization via Amino Group

The stepwise SPPS from the N → C terminus represents a substantially less-frequently used strategy when compared to the C-terminus immobilization. The main reason is on-resin activation of the carboxylic acid. Apart from potentially compromising of the optical integrity, the activated species that is prone to side-reactions is resin-bound. On the other hand, the advantage of building the peptide from the N- to C-terminus is direct generation of C-terminal modified peptides. The commonly used method for immobilization of amino acid via its N-terminus consists of formation of carbamates – the most frequently used protecting group for amines. The first reported use of carbamates for amino acids immobilization was described by Letsinger and Kornet [18]. The most frequently used carbamate linker is derived from the Wang resin **2.2**. The carbamate linkage is typically prepared by reaction with phosgene [137], carbonyldiimidazole [137, 138], or *p*-nitrophenylchloroformate [139–143], followed by reaction with an excess of C-protected amino acid (Scheme 8.17). After completion of the synthesis, the peptides are cleaved from the support by treatment with TFA/DCM (1 : 1) to yield the peptides as trifluoroacetic salts. The carbamate protection has also been applied to hydroxymethyl polystyrene [144], hydroxy TentaGel [137], and the Sheppard linker [140, 145, 146]. The use of a polymer-supported *tert*-alkoxycarbonyl group was also reported for the solid-supported synthesis of peptides [38]. In this case cleavage was achieved by reaction with 10% TFA.

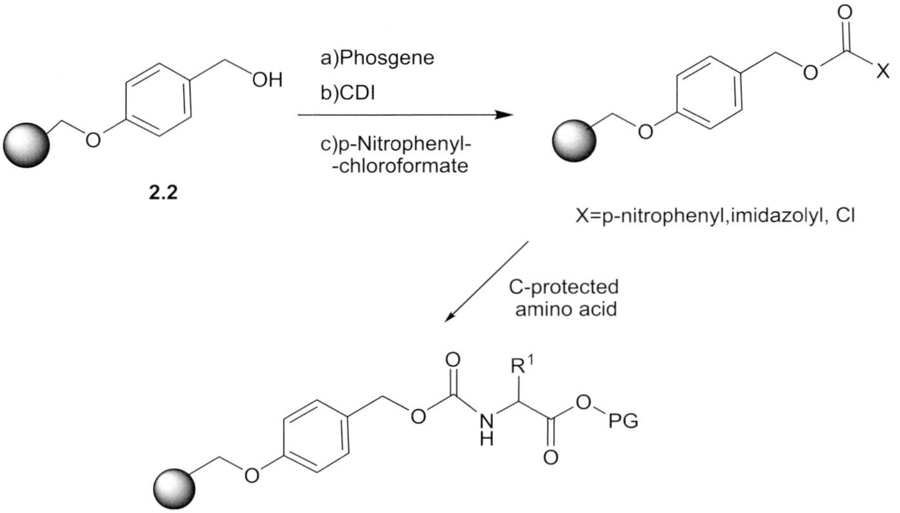
a)Phosgene
b)CDI
c)p-Nitrophenyl- -chloroformate
OH
O
X
2.2
X=p-nitrophenyl,imidazolyl, Cl
C-protected amino acid
R1
N H
PG

**Scheme 8.17** Peptide synthesis using Wang-derived carbamate linker.

Lipshutz and Shin introduced a silyl linker **3.1** applicable for reversed peptide synthesis. The first amino acid was anchored via its N-terminus through the carbamate–silyl bond [147]. The target peptides were cleaved from the polymer support by 49% aqueous HF solution, leaving the linker recyclable for repeated use (Scheme 8.18).

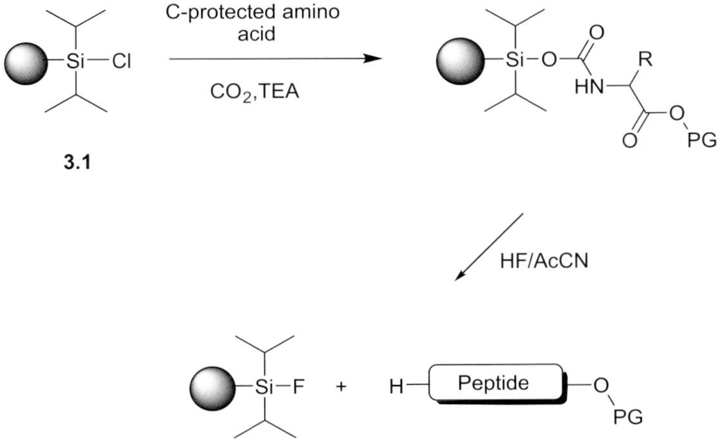
Si—Cl
3.1
C-protected amino acid
CO2,TEA
Si–O
HN
R
PG
HF/AcCN
Si–F
+
H
Peptide
O
PG

**Scheme 8.18** Carbamate-silyl anchoring in reversed SPPS.

Another example of a protecting group that can serve as a linker for immobilization of amines is trityl. The sensitivity of the trityl protective group towards acidic cleavage

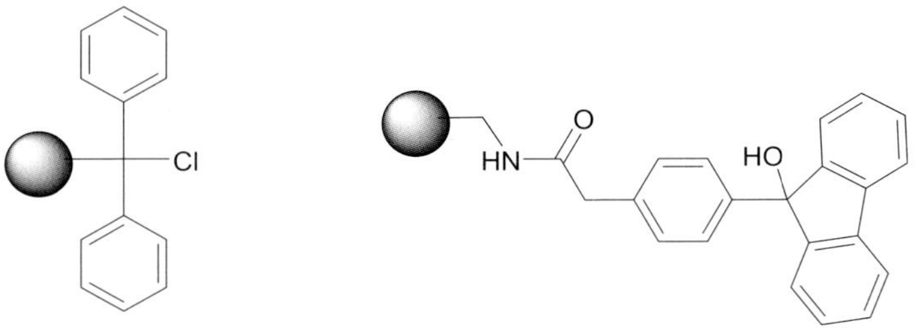

3.2

3.3

Figure 8.7 Trityl chloride resin and PhFl resin.

allows its use for amino acid immobilization, whereas benzylamines are not suitable for these purposes due to their acidic stability that necessitates too harsh cleavage conditions. Support-bound trityl chloride **3.2** was used for immobilization of amino acids via the N-terminus and subsequent peptide synthesis [84]. The cleavage of the final product was carried out with 2% TFA in DCM. Analogously, 2-chlorotrityl resin **2.14** [148] and 9-phenylfluoren-9-yl polystyrene based resin (PhFI) **3.3** [37] were applied for reversed peptide synthesis (Figure 8.7). Before reaction with nucleophiles the PhFI resin **3.3** has to be activated with 20% of acetyl chloride in DCM. The last two mentioned resins differ in acidic stability: peptides from PhFI resin **3.3** were cleaved with 95% TFA whereas the cleavage from 2-chlorotrityl resin **2.14** was performed with 10% TFA in DCM.

## 8.4 Backbone Immobilization

In the mid-1990s, several research groups independently described novel linkers for immobilization of secondary amides [149–158]. These linkers shared a common structural motif – an electron-rich benzyl group attached to the amide nitrogen. After a synthesis is completed on this kind of linker, acid-mediated cleavage affords secondary amides and enables synthesis of peptides linked to the polymeric support by the backbone amide. The backbone amide linker (BAL) strategy has become an indispensable tool for syntheses of secondary amides and numerous other compound classes such as sulfonamides, ureas and even heterocycles [159]. Not surprisingly, numerous variations of the original BAL concept later appeared, documenting the potential of this linker strategy. Diverse applications of BAL have been recently reviewed [160].

The most common method for accessing the resin-bound secondary amide is based on reductive amination of a resin-bound aldehyde linker followed by acylation. According to the structure of the aromatic aldehyde, the linkers can be divided into several groups: benzaldehyde, indole, naphthalene, thiophene, safety-catch linkers, and photolabile aldehyde linkers (Table 8.6). Individual linkers mainly differ by stability under acidic conditions.

Table 8.6 Heterocyclic-related, safety-catch, and photolabile linkers for the BAL strategy.

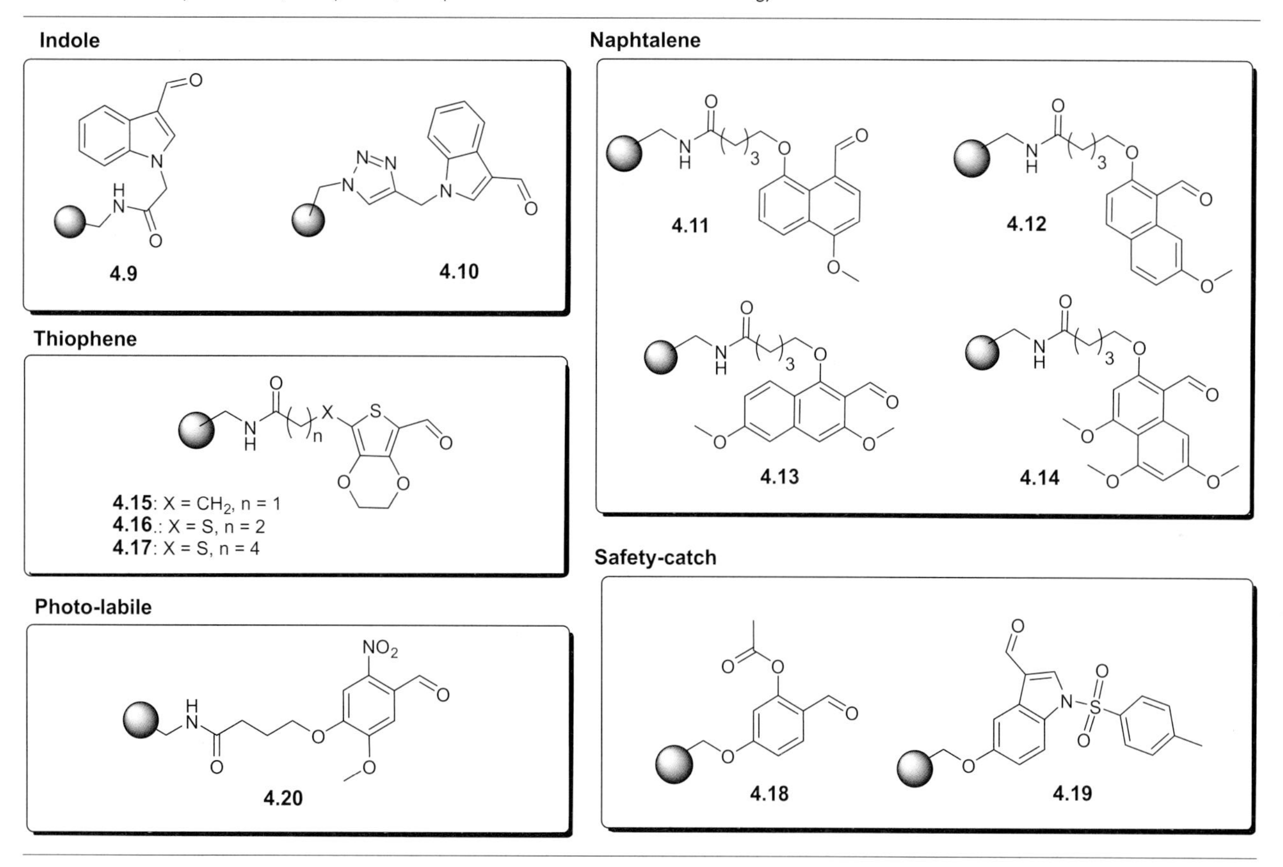

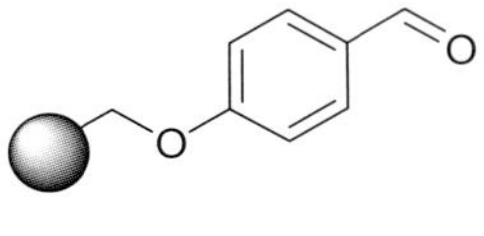

4.1 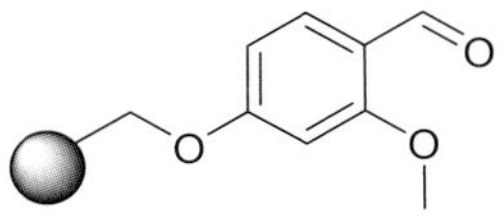 4.2 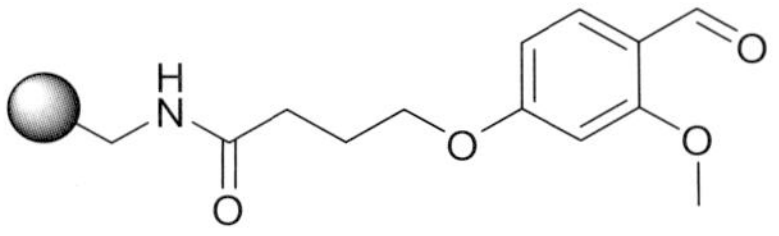 4.3

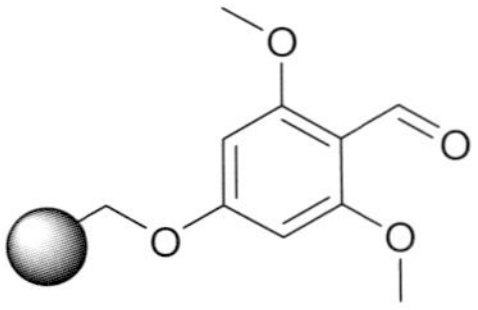

4.4 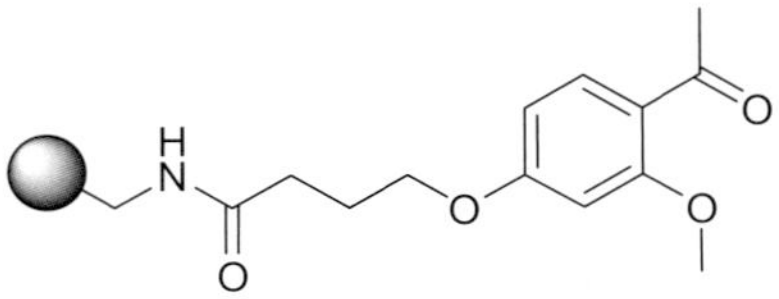 4.5 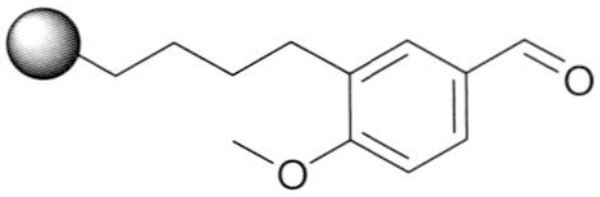 4.6

**Figure 8.8** Benzaldehyde-related linkers.

## 8.4.1
## Benzaldehyde-Based Linkers

Benzaldehyde-based linkers are the most commonly used BAL (Figure 8.8 and Scheme 8.19) [149–158]. The linkers **4.1** [154, 155, 157] and **4.2** [154], referred to as AMEBA (acid-sensitive methoxybenzaldehyde) [151] or MALDRE (2-methoxy-4-benzyloxy-polystyrene aldehyde resin) [153], are attached to polystyrene [151, 153] or ArgoGel [154] resins via an ether [154] or amide bond [157]. Immobilization of the linker into a support via an amide bond (linker **4.3**) results in the formation of a more stable attachment to the support. The more acid-labile BAL **4.4** was described by the Barany group [152] for synthesis of peptides attached to the polymer support via the nitrogen of an amide bond. This attachment leaves the carboxyl group of amino acids available for further transformations [161]. An *o*-BAL linker that was linked to the solid support via the alkyl group in the *ortho* position to the aldehyde function was also reported [162–166]. Mono and dimethoxy linkers attached to SynPhase Crowns showed no differences in purity or yield of products, other than the greater lability of the amides on the BAL linker [167]. The linker was also useful for the synthesis of peptides with a carboxy terminal amino acid prone to racemization (His), as no activation of the carboxylic acid was needed. Furthermore, this linker is ideal for peptides with a C-terminal dipeptide sequence that is prone to diketopiperazine

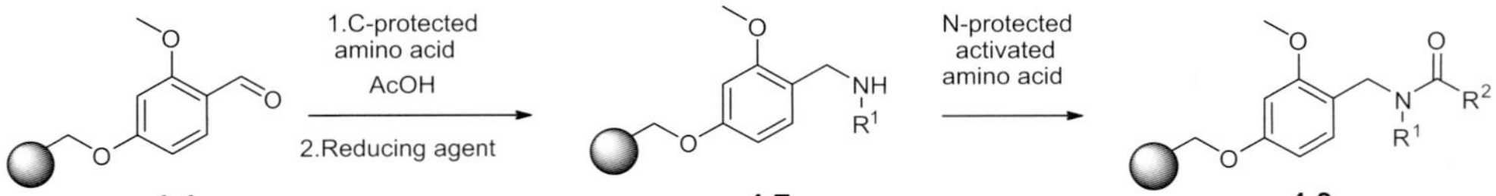

1.C-protected amino acid AcOH; 2.Reducing agent; N-protected activated amino acid

4.4 → 4.7 → 4.8

**Scheme 8.19** Preparation of peptide amides with use of benzaldehyde linker.

(DKP) formation (e.g., proline-containing dipeptides), as no benzyl ester of a dipeptide is present during synthesis [168]. The length of the spacer between the linker and the resin, as in linkers **4.2** and **4.3**, can significantly influence the yield of isolated products, with a longer spacer having a favorable effect [169]. Maeji's group introduced acetophenone-based linker **4.5** [170], and Gu with Silverman extended the group of benzaldehyde linkers with the stable linker **4.6** cleavable under harsh conditions (refluxing TFA) [171].

Loading of aldehyde and acetophenone linkers is accomplished by a reductive amination with primary amines and provided resin-bound secondary amines **4.7**. The reductive amination conditions were optimized on numerous examples. The reaction is usually performed at room temperature under acidic conditions (typically 1–10% AcOH). If the starting amino acid is used as a HCl salt, the acidic environment is not necessary. The most suitable solvents are MeOH, DMF, and *N*-methylpyrrolidone (NMP). The reaction time can be effectively shortened with the help of microwave heating [172]. The reduction step has been performed in one-pot fashion without the isolation of the imine intermediate and has been accomplished with various reducing agents, most commonly $NaBH_3CN$ or $NaBH(OAc)_3$. Alternatively, the reductive alkylation has been performed in solution and the product (after protection of the amine by Fmoc) attached to the resin [173]. The secondary amine is then acylated to produce amide **4.8** (Scheme 8.19). The acylation is usually performed with a symmetrical *N*-protected amino acid anhydride generated *in situ* using 2-(1*H*-9-azobenzotriazole-1-yl)-1,1,3,3-tetramethylaminium hexafluorophosphate (HATU) in a mixture of DCM/DMF (usually 9 : 1) [160].

Cleavage of amides from linker **4.1** was slow with TFA and was often more effectively carried out by treatment with HF. Cleavage from dialkoxy linker **4.3** was affected with TFA containing 2.5% $Et_3SiH$ in 15 min [154], while amides synthesized on trialkoxy linker **4.4** were cleaved using 5% TFA/DCM [151].

### 8.4.2 Indole Aldehyde Linkers

Indole aldehyde linkers, heterocyclic analogs of the BAL linker, were introduced at the end of the 1990s. An indole aldehyde linker **4.9** reported by Estep *et al.* [156] served an analogous purpose to that of the BAL linker and was later further studied [174, 175]. After derivatization of the linker by reductive amination and acylation with a carboxylic acid, the resulting amide derivatives were released by treatment with 2–50% TFA. A similar indole linker **4.10** anchored to resin via a triazole spacer was used for the synthesis of dopamine receptor ligands [176].

### 8.4.3 Naphthalene Aldehyde Linkers (NALs)

NALs were developed as less or more acid stable alternatives to benzaldehyde linkers and have been successfully used for peptide synthesis. The acid-lability of the NAL1 **4.11** linker is similar to BAL (50% TFA/DCM) whereas the cleavage from NAL2 **4.12** requires more highly concentrated TFA (95% TFA/$H_2O$) [160, 177]. Subsequently, NAL3 **4.13** and NAL4 **4.14** with increased acidic lability were reported. Cleavage of the peptides from the extremely labile NAL4 **4.14** was accomplished with only 0.5% TFA/DCM which did not affect *t*Bu esters and *t*Bu ethers [178].

### 8.4.4
### Thiophene Aldehyde Linkers (T-BALs)

T-BALs **4.15–4.17** represent a new class of highly acid-labile linkers for solid-phase synthesis of peptide amides [160, 179]. Their structure is based on the 3,4-ethylenedioxythiophene (EDOT) skeleton. In comparison to benzaldehyde linkers **4.3** or **4.4**, the electron-richness of EDOT allows the cleavage of the target amides from T-BALs with use of very low acid concentrations (only 1% TFA/DCM).

### 8.4.5
### Safety-Catch Aldehyde Linkers

Safety-catch aldehyde linkers with a reversed acid stability have also been described. Safety-catch acetoxyalkoxybenzyl BAL linker **4.18** was prepared by acetylation of the corresponding resin-bound hydroxy-benzaldehyde linker. The linker is stable towards 95% TFA and thus it is compatible with the Boc protection strategy [180]. After the cleavage of the acetyl group with piperidine, the resulting hydroxybenzylaldehyde linker is acid labile and the product was cleaved with 95% TFA in water. Also, safety-catch indole linker **4.19** was developed [181]. Tosylation of the indole nitrogen decreased the acid lability (no cleavage detected with 50% TFA in DCM after 2 h). After the tosyl group was removed with tetrabutylammonium fluoride (TBAF), the amide was cleaved in 1% TFA in DCM within 1 h.

### 8.4.6
### Photolabile Aldehyde Linker (PhoB)

Photolabile aldehyde linker (PhoB) based on nitro-substituted alkoxybenzaldehyde **4.20** was introduced by Minkwitz and Meldal [182]. The linker was synthesized from inexpensive starting materials in three reaction steps and allowed mild orthogonal cleavage of the substrate with use of long-wave UV light (365 nm).

## 8.5
## Immobilization via Amino Acid Side-Chain

This scenario belongs to rather special cases since only amino acids with side-chains suitable for immobilization can be used. The immobilization via side-chain usually requires both N- and C-protections of the first amino acid. After the amino acid side-chain immobilization, the peptides are synthesized in both N $\rightarrow$ C and C $\rightarrow$ N directions (after selective deprotection of the corresponding amino acid terminus). According to the character of the side-chain, different types of linkers are used to immobilize amino acids via the carboxyl group (Asp, Glu), amino group (Lys, Orn, His, Arg), hydroxy group (Ser, Thr, Tyr), sulfanyl group (Cys), or aromatic ring (Phe).

### 8.5.1 Carboxyl Group

Amino acids that contain a carboxylate in their side-chains have been immobilized via ester or amide formation using strategies and linkers described in Section 8.2. Asp and Glu, protected at both termini, were used to acylate Wang **2.2** [183] and Rink amide **2.22** [183–185] resins (Scheme 8.20). After completion of the synthesis, the acid-mediated cleavage afforded final peptides containing Asp and Glu in the case of Wang resin and Asn and Gln in the case of Rink amide resin.

N,C-protected activated Asp or Glu

Wang resin: X = H, Y = OH
Rink amide resin: X = 2,4-methoxyphenyl, Y = $NH_2$

Wang resin: X = H, Y = O
Rink amide resin: X = 2,4-methoxyphenyl, Y = NH
n = 1,2

**Scheme 8.20** Immobilization of amino acids via side-chain carboxylate on Wang and Rink amide resin.

Also, immobilization of Asp and Glu on bromomethyl resin **5.1** [186], HMPA resin **2.6** [186], PAL linker **2.18** [187, 188], hydroxymethyl resin **5.2** [189–191], and tris (methoxy)benzhydrylamine (TMBPA) linker **5.3** [192] have been described (Figure 8.9).

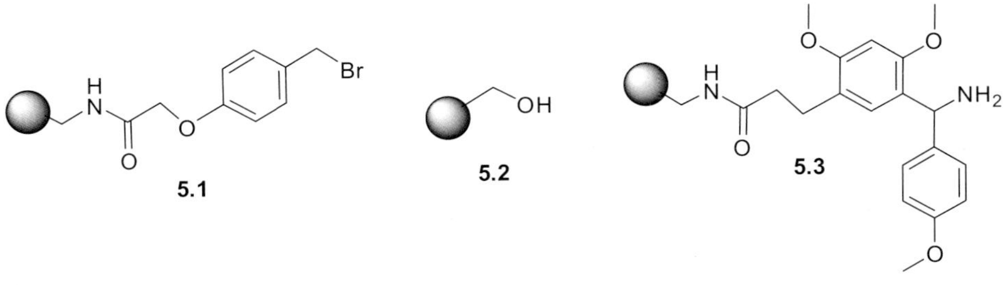

5.1 5.2 5.3

**Figure 8.9** Other resins used for Glu and Asp side-chain immobilization.

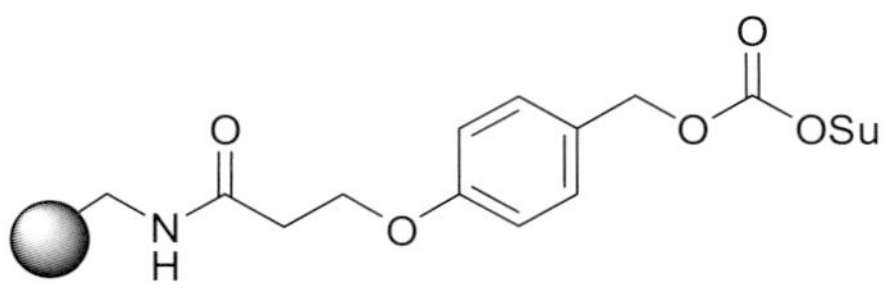

5.4

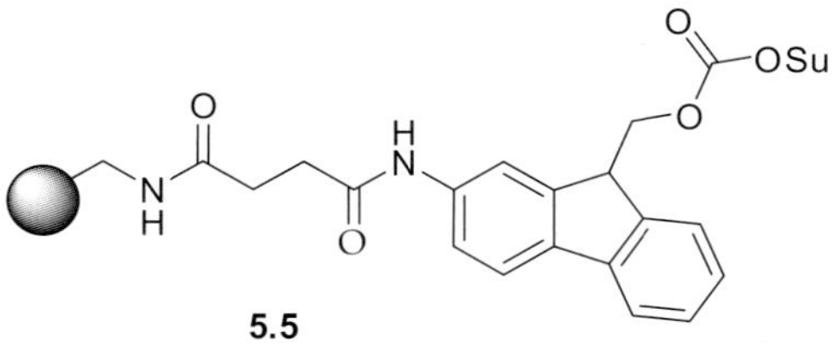

5.5

Figure 8.10 Linkers for side-chain immobilization of Lys.

## 8.5.2
## Amino and Other Nitrogen-Containing Groups

### 8.5.2.1 Lys

Side-chain immobilization of Lys (and any other amino acids with amino groups in their side-chains, such as ornithine, homolysine, etc.) has been carried out using linkers described in Section 8.3. Two carbonate linkers were used for side-chain anchoring of Lys (Figure 8.10) [193]. Polymer supported peptides were subsequently on-resin cyclized "head-to-tail" [191]. The cleavage from linker **5.4** was carried out with TFA/thioanisole/ethanedithiol/anisole (95 : 5 : 3 : 2). The second linker **5.5** is base labile and cleavable with secondary amines (piperidine or morpholine) [194].

The versatile side-chain anchoring strategy has also been described with use of chlorotritylchloride linker **2.14**. The resin was used for side-chain immobilization of various amino acids (such as Lys, Cys, His, Trp and Tyr) [195]. The target peptides were immobilized via side-chain of Lys and were cleaved with 95% TFA.

### 8.5.2.2 His

His-containing peptides have been synthesized after immobilization of His via its imidazole nitrogen to the dinitrofluorobenzene linker **5.6** (Scheme 8.21) [196]. Target peptides were released with 2-mercaptoethanol and triethylamine (TEA) in DMF (1 : 0.1 : 20). Also, immobilization of His on chlorotritylchloride resin **2.14** was described with prolonged acid-mediated cleavage (95% TFA, 2 h) [195].

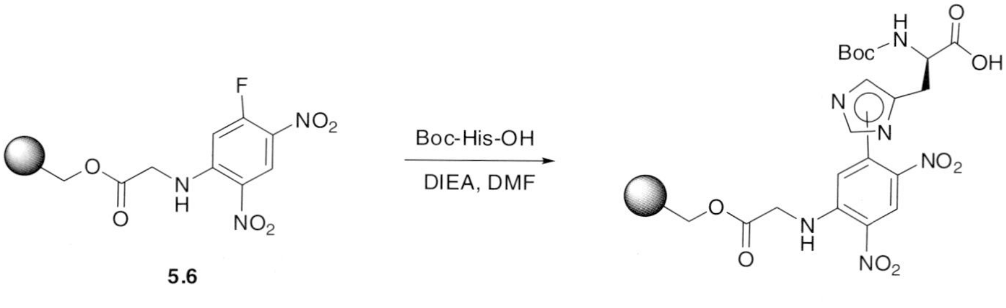

Boc-His-OH

DIEA, DMF

5.6

Scheme 8.21 Immobilization of histidine via imidazole skeleton.

#### 8.5.2.3 **Arg**

Arg and other guanidine-containing compounds have been immobilized to benzopyran-derived sulfonamide linker **5.7** (Scheme 8.22) [197]. The linker is compatible with the Fmoc/*t*Bu protection strategy, cleavage of the final peptide was performed with 90% TFA in water.

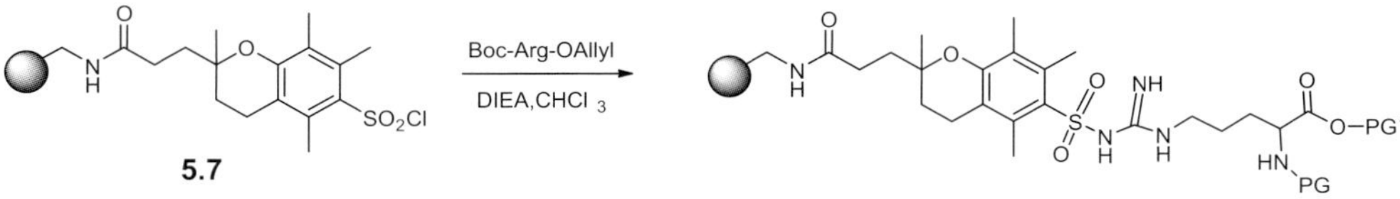

**Scheme 8.22** Benzopyran linker for side-chain immobilization of arginine.

Also, the BAL strategy was applied for side-chain anchoring of Arg (Scheme 8.23). Indole BAL **4.9** was used for immobilization of Orn via its side-chain amino group and the resulting secondary amine was reacted with *N*,*N*′-bis-Boc-thiourea to form a guanidine [175]. The final cleavage was performed with 50% TFA in DCM.

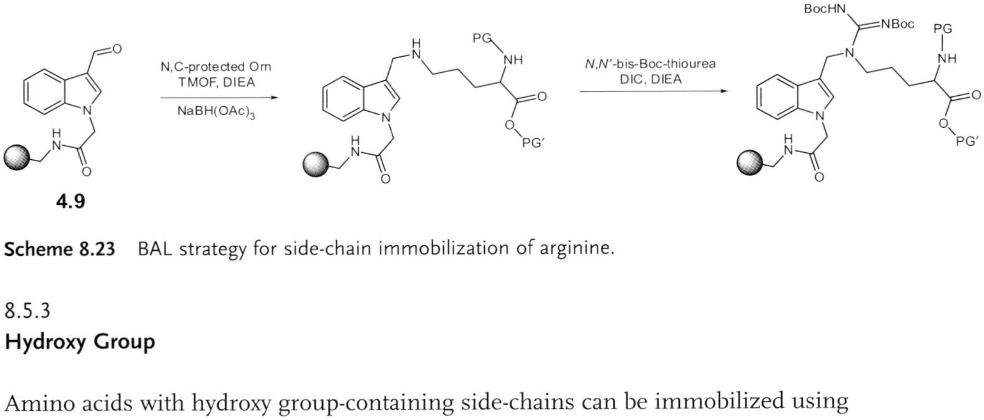

**Scheme 8.23** BAL strategy for side-chain immobilization of arginine.

### 8.5.3 Hydroxy Group

Amino acids with hydroxy group-containing side-chains can be immobilized using traditional linkers for solid-phase synthesis of alcohols (Scheme 8.24). Ser, Thr, and Tyr were immobilized on bromo(4-methoxyphenyl)methyl resin **5.8** and used for synthesis of peptides thioesters [183]. The cleavage of the final peptides was performed with TFA, thioanisole, triisopropylsilane, and $H_2O$ (85 : 5 : 5 : 5). Polymer-bound *p*-alkoxybenzyl trichloroacetimidate **5.9** immobilized hydroxy amino acids in the presence of $BF_3{\cdot}OEt_2$ [198]. Tyr was anchored to Wang **2.2** and SASRIN **2.3** resins via Mitsunobu coupling [71]. Hydroxy amino acid derivatives were also immobilized as benzhydrol ethers using polymeric diphenyldiazomethane, and mild cleavage conditions with 1–2% TFA/DCM allowed synthesis of fully protected peptide alcohols [199]. Rink chloride **2.13** was converted to the Rink ether derivative **5.10** and cleaved to yield the required peptide-alcohol with 5% TFA/DCM. A potential

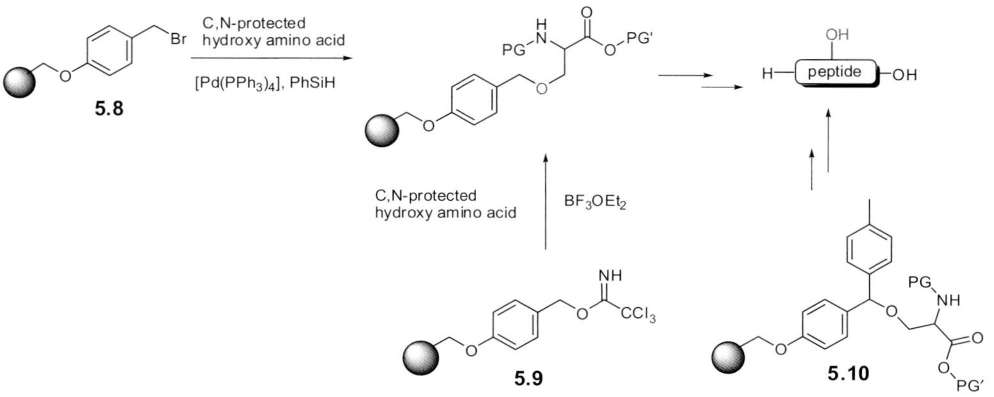

**Scheme 8.24** Immobilization of amino acids via side-chain hydroxy group (demonstrated on Ser).

byproduct from this route was the TFA ester [200, 201]; however, the ester was readily hydrolyzed in aqueous sodium carbonate solution or simply by exposure to MeOH. Alternatively, 0.2 M HCl in DCM was used to cleave alcohols from the ether linker [82].

### 8.5.4
### Sulfanyl Group

Sulfanyl groups smoothly undergo oxidation to disulfides or nucleophilic substitution to thioethers. Both reactions have been applied for attachment to the polymer support. However, the sensitivity of the sulfanyl group towards oxidation usually requires carrying out the cleavage under inert gas to eliminate potential thiol dimerization.

The disulfide linker **5.11** [202] takes advantage of disulfide bond formation. The linker was used for immobilization of cysteine and subsequent peptide synthesis in the N → C direction. The cleavage was performed with 2 equiv. dithiothreitol in a mixture of *i*PrOH/$H_2O$ (1 : 1). Immobilization of cysteine via dinitrofluorobenzene linker **5.6** was used for bidirectional synthesis of peptides [196]. The cleavage was performed under mild conditions with 2-mercaptoethanol in DMF. The cysteine-preformed handle **5.12** based on the xanthenyl (XAL) moiety was introduced by Barany *et al.* and was used in connection with PEG-containing resin [203]. The linker allowed syntheses of cyclic peptide disulfides (Scheme 8.25) (oxidative iodine cleavage) or peptide thiols (acidic TFA cleavage).

Bromo(4-methoxyphenyl)methyl resin **5.8** was used for attachment of cysteine via thioether [204]. The growing peptide chain was terminated with cysteine, which allowed reductive cyclative cleavage (triggered by *N*-chlorosuccinimide and DMS) leading to formation of peptide disulfides. The linker was used for synthesis of peptide mimetics with a single disulfide bridge. The cyclic peptide disulfides were also synthesized after side-chain immobilization of cysteine on chlorotrityl resin **2.14** [205]. The final oxidative cleavage was performed with iodine.

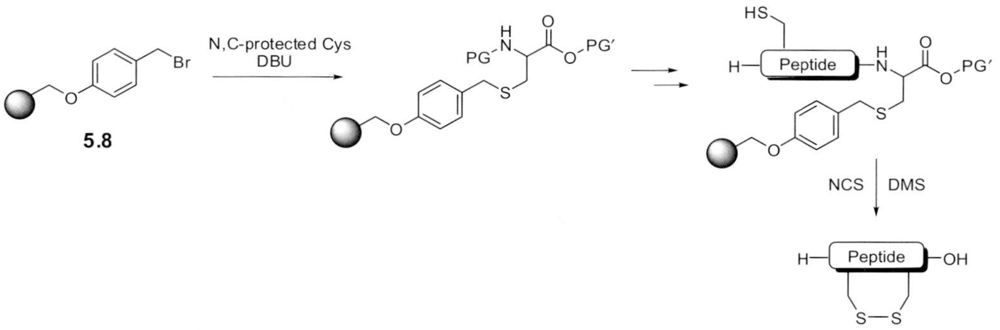


**Scheme 8.25** Immobilization of cysteine via thioether and cyclic peptide synthesis.

Cysteine was also immobilized on poly(*N*-acrylylpyrrolidine) resin via thiocarbamate (Figure 8.11) [206]. The cleavage of the target peptides was performed with 0.1 M NaOH in *n*-BuOH/$H_2O$ (9 : 1).

### 8.5.5
### Aromatic Ring

The immobilization of amino acids via side-chain aromatic ring was described by Lee and Silverman (Figure 8.12). Phenylalanine and phenyl-β-alanine were immobilized

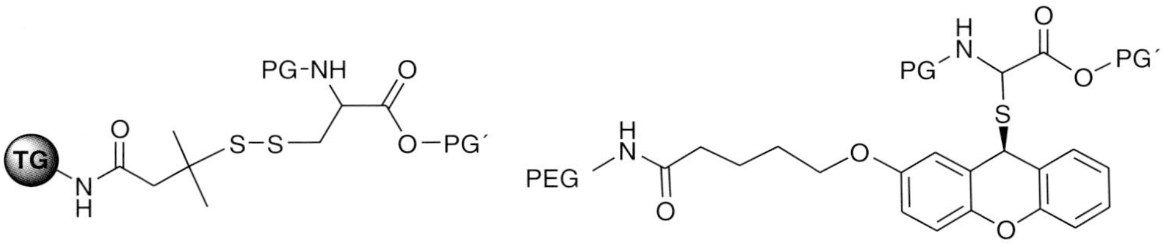


**5.11** **5.12**

**Figure 8.11** Linkers for side-chain immobilization of Cys.

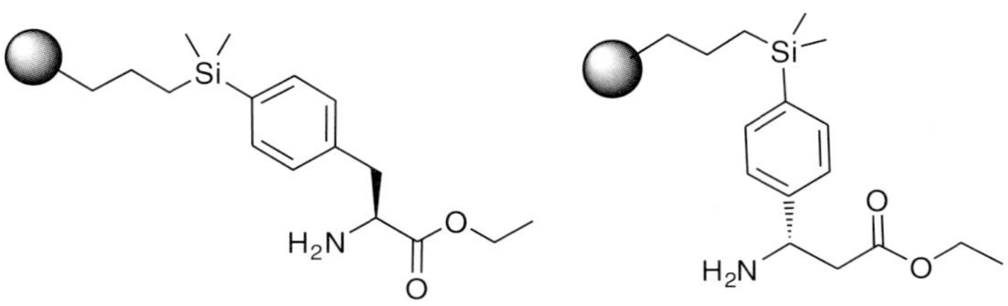


**5.13** **5.14**

**Figure 8.12** Linkers with amino acids immobilized via aromatic skeleton.

on silicon-based linkers **5.13** and **5.14** [207, 208]. The final peptides were cleaved with 50% TFA in DCM.

## 8.6 Conclusions

The ever-growing number of linkers and chemistries used to immobilize amino acids has documented the power of Merrifield SPPS. There are three major factors that contribute to a successful peptide synthesis: selection of resin, linker, and protection strategy. This chapter covered linkers. Our strategy in selecting the most suitable linker follows reverse order when compared to the practical synthesis.

i) What reagents and reaction conditions applied for peptide release will the target peptide tolerate?
ii) What reagent will be used to remove temporary protection of amino/carboxyl groups?
iii) What is the structure of the peptide?
iv) Is the resin with a linker attached to it commercially available from a reliable source? (The authors experienced problems with several resin batches in the not very distant past.) [17].
v) Irrespective of the origin of the resin-bound linker and its loading, it is imperative to carry out quantitative analysis of the first amino acid loading.

The correct answers to these questions play crucial roles in the outcome of the synthesis, particularly for long-chain peptides where small discrepancies in reaction sequences can have a profound effect on the outcome of the synthetic effort.

Current increased interest in peptide synthesis, caused particularly by its successful application in drug discovery, has attracted chemists to find new linkers with ever-improving parameters. Their effective application has been demonstrated especially in combinatorial synthesis of natural peptides, mimetics, peptide derivatives, and various conjugates, representing an endless number of variations. Although anyone can buy a plethora of resins with bound linker and the first amino acid or small oligopeptide as a starting material, the knowledge of the linker properties is one of the most important requirements for their effective use in the synthesis of a desired peptide, regardless of whether a manual or automatic process is applied.

## References


1 Merrifield, R.B. (1963) *Journal of the American Chemical Society*, **85**, 2149–2154.
2 Leznoff, C.C. and Wong, J.Y. (1972) *Canadian Journal of Chemistry*, **50**, 2892–2893.
3 Wong, J.Y. and Leznoff, C.C. (1973) *Canadian Journal of Chemistry*, **51**, 2452–2456.
4 Frechet, J.M.J. and Haque, K.E. (1975) *Tetrahedron Letters*, **16**, 3055–3056.

5 Frechet, J.M.J. and Nuyens, L.J. (1976) *Canadian Journal of Chemistry*, **54**, 926–934.

6 Fyles, T.M. and Leznoff, C.C. (1976) *Canadian Journal of Chemistry*, **54**, 935–942.

7 Leznoff, C.C. and Dixit, D.M. (1977) *Canadian Journal of Chemistry*, **55**, 3351–3355.

8 Leznoff, C.C. and Wong, J.Y. (1973) *Canadian Journal of Chemistry*, **51**, 3756–3764.

9 Leznoff, C.C. and Greenberg, S. (1976) *Canadian Journal of Chemistry*, **54**, 3824–3829.

10 Xu, Z.H., McArthur, C.R., and Leznoff, C.C. (1983) *Canadian Journal of Chemistry*, **61**, 1405–1409.

11 Leznoff, C.C. and Goldwasser, J.M. (1977) *Tetrahedron Letters*, **18**, 1875–1878.

12 Goldwasser, J.M. and Leznoff, C.C. (1978) *Canadian Journal of Chemistry*, **56**, 1562–1568.

13 Dixit, D.M. and Leznoff, C.C. (1978) *Israel Journal of Chemistry*, **17**, 248–252.

14 James, I.W. (1999) *Tetrahedron*, **55**, 4855–4946.

15 Guillier, F., Orain, D., and Bradley, M. (2000) *Chemical Reviews*, **100**, 2091–2158.

16 Comely, A.C. and Gibson, S.E. (2001) *Angewandte Chemie, International Edition*, **40**, 1012–1032.

17 Bouillon, I., Soural, M., Miller, M.J., and Krchnak, V. (2009) *Journal of Combinatorial Chemistry*, **11**, 213–215.

18 Letsinger, R.L. and Kornet, J. (1963) *Journal of the American Chemical Society*, **85**, 3045–3046.

19 Letsinger, R.L., Kornet, M.J., Mahadevan, V., and Jerina, D.M. (1964) *Journal of the American Chemical Society*, **86**, 5163–5165.

20 Sakakibara, S. and Shimonishi, Y. (1965) *Bulletin of the Chemical Society of Japan*, **38**, 1412–1413.

21 Tam, J.P., Heath, W.F., and Merrifield, R.B. (1983) *Journal of the American Chemical Society*, **105**, 6442–6455.

22 Kerschen, A., Kaniszai, A., Botros, I., and Krchnak, V. (1999) *Journal of Combinatorial Chemistry*, **1**, 480–484.

23 Darling, G.D. and Frechet, J.M. (1986) *The Journal of Organic Chemistry*, **51**, 2270–2276.

24 Stranix, B.R., Gao, J.P., Barghi, R., Salha, J., and Darling, G.D. (1997) *The Journal of Organic Chemistry*, **62**, 8987–8993.

25 Stranix, B.R. and Darling, G.D. (1997) *The Journal of Organic Chemistry*, **62**, 9001–9004.

26 Sylvain, C., Wagner, A., and Mioskowski, C. (1998) *Tetrahedron Letters*, **39**, 9679–9680.

27 Sakakibara, S., Shimonishi, Y., Kishida, Y., Okada, M., and Sugihara, H. (1967) *Bulletin of the Chemical Society of Japan*, **40**, 2164–2167.

28 Yajima, H., Fujii, N., Ogawa, H., and Kawatani, H. (1974) *Journal of the Chemical Society, Chemical Communications*, 107.

29 Fujii, N., Otaka, A., Ikemura, O., Akaji, K., Funakosho, S., Hayashi, Y., Kuroda, Y., and Yajima, H. (1987) *Journal of the Chemical Society, Chemical Communications*, 274–275.

30 Marshall, G.R. and Merrifield, R.B. (1965) *Biochemistry*, **4**, 2394–2401.

31 Larsen, B.D. and Holm, A. (1994) *International Journal of Peptide and Protein Research*, **43**, 1–9.

32 Meldal, M., Bielfeldt, T., Peters, S., Jensen, K.J., Paulsen, H., and Bock, K. (1994) *International Journal of Peptide and Protein Research*, **43**, 529–536.

33 Li, X., Kawakami, T., and Aimoto, S. (1998) *Tetrahedron Letters*, **39**, 8669–8672.

34 Wade, J.D., Bedford, J., Sheppard, R.C., and Tregear, G.W. (1991) *Peptide Research*, **4**, 194–199.

35 Liebe, B. and Kunz, H. (1997) *Angewandte Chemie International Edition*, **36**, 618–621.

36 Dölling, R., Beyermann, M., Haenel, J., Kernchen, F., Krause, E., Franke, P., Brudel, M., and Bienert, M. (1994) *Journal of the Chemical Society, Chemical Communications*, 853–854.

37 Bleicher, K.H. and Wareing, J.R. (1998) *Tetrahedron Letters*, **39**, 4587–4590.

38 Leger, R., Yen, R., She, M.W., Lee, V.J., and Hecker, S.J. (1998) *Tetrahedron Letters*, **39**, 4171–4174.

39 Bordusa, F., Ullmann, D., and Jakubke, H.-D. (2009) *Angewandte Chemie International Edition*, **36**, 1099–1101.

40 Felix, A.M. and Merrifield, R.B. (1970) *Journal of the American Chemical Society*, **92**, 1385–1391.

41 Abraham, N.A., Fazal, G., Ferland, J.M., Rakhit, S., and Gauthier, J. (1991) *Tetrahedron Letters*, **32**, 577–580.

42 Sharma, R.P. (2006) Inverse solid phase peptide synthesis with additional capping step. US 2009099307; (2009) *Chemical Abstracts*, **145**, 336323.

43 Quibell, M., Owen, D., Packman, L.C., and Johnson, T. (1994) *Journal of Chemical Society, Chemical Communication*, **20**, 2343–2344.

44 Maffre-Lafon, D., Escale, R., Dumy, P., Vidal, J.-P., and Girard, J.-P. (1994) *Tetrahedron Letters*, **35**, 4097–4098.

45 Dixon, M.J., Nathubhai, A., Andersen, O.A., van Aalten, D.M.F., and Eggleston, I.M. (2009) *Organic & Biomolecular Chemistry*, **7**, 259–268.

46 Loffet, A. and Zhang, H.X. (1993) *International Journal of Peptide and Protein Research*, **42**, 346–351.

47 Novabiochem (2009) *Guide to the Selection of Building Blocks for Peptide Synthesis*. Merck Biosciences, Läufingen.

48 Mergler, M., Nyfeler, R., and Gosteli, J. (1989) *Tetrahedron Letters*, **30**, 6741–6744.

49 Collini, M.D. and Ellingboe, J.W. (1997) *Tetrahedron Letters*, **38**, 7963–7966.

50 Nugiel, D.A., Wacker, D.A., and Nemeth, G.A. (1997) *Tetrahedron Letters*, **38**, 5789–5790.

51 Raju, B. and Kogan, T.P. (1997) *Tetrahedron Letters*, **38**, 4965–4968.

52 Ngu, K. and Patel, D.V. (1997) *Tetrahedron Letters*, **38**, 973–976.

53 Bhalay, G. and Dunstan, A.R. (1998) *Tetrahedron Letters*, **39**, 7803–7806.

54 Phoon, C.W., Oliver, S.F., and Abell, C. (1998) *Tetrahedron Letters*, **39**, 7959–7962.

55 Mitchell, A.R., Erickson, B.W., Ryabtsev, M.N., Hodges, R.S., and Merrifield, R.B. (1976) *Journal of the American Chemical Society*, **98**, 7357–7362.

56 Wang, S.-S. (1973) *Journal of the American Chemical Society*, **95**, 1328–1333.

57 Mergler, M., Tanner, R., Gosteli, O., and Grogg, P. (1988) *Tetrahedron Letters*, **29**, 4005–4008.

58 Albericio, F. and Barany, G. (1991) *Tetrahedron Letters*, **32**, 1015–1018.

59 Rink, H. (1987) *Tetrahedron Letters*, **28**, 3787–3790.

60 Cardno, M. and Bradley, M. (1996) *Tetrahedron Letters*, **37**, 135–138.

61 Alsina, J., Rabanal, F., Chiva, C., Giralt, E., and Albericio, F. (1998) *Tetrahedron*, **54**, 10125–10152.

62 St. Hilaire, P.M., Willert, M., Juliano, M.A., Juliano, L., and Meldal, M. (1999) *Journal of Combinatorial Chemistry*, **1**, 509–523.

63 Marshall, D.L. and Liener, I.E. (1970) *The Journal of Organic Chemistry*, **35**, 867–868.

64 Blankenmeyer-Menge, B., Nimtz, M., and Frank, R. (1990) *Tetrahedron Letters*, **31**, 1701–1704.

65 Sieber, P. (1987) *Tetrahedron Letters*, **28**, 6147–6150.

66 Atherton, E., Benoiton, N.L., Brown, E., Sheppard, R.C., and Williams, B.J. (1981) *Journal of the Chemical Society, Chemical Communications*, 336–337.

67 Mergler, M., Nyfeler, R., Tanner, R., Gosteli, J., and Grogg, P. (1988) *Tetrahedron Letters*, **29**, 4009–4012.

68 Grandas, A., Jorba, X., Giralt, E., and Pedroso, E. (1989) *International Journal of Peptide and Protein Research*, **33**, 386–390.

69 Van Nispen, J.W., Polderdijk, J.P., and Greven, H.M. (1985) *Recueil des Travaux Chimiques des Pays-Bas*, **104**, 99–100.

70 Granitza, D., Beyermann, M., Wenschuh, H., Haber, H., Carpino, L.A., Truran, G.A., and Bienert, M. (1995) *Journal of the Chemical Society, Chemical Communications*, **21**, 2223–2224.

71 Cabrele, C., Langer, M., and Beck-Sickinger, A.G. (1999) *The Journal of Organic Chemistry*, **64**, 4353–4361.

72 Gisin, B.F. (1973) *Helvetica Chimica Acta*, **56**, 1476–1482.

73 Garigipati, R.S. (1997) *Tetrahedron Letters*, **38**, 6807–6810.

74 Barlos, K., Chatzi, O., Gatos, D., and Stavropoulos, G. (1991) *International Journal of Peptide and Protein Research*, **37**, 513–520.

75 Bayer, E., Goldammer, C., and Zhang, L. (1993) *Chemistry of Peptides and Proteins*, **5/6**, 3–8.

76 Merrifield, R.B. (1964) *Biochemistry*, **3**, 1385–1390.
77 Mitchell, A.R., Kent, S.B.H., Engelhard, M., and Merrifield, R.B. (1978) *The Journal of Organic Chemistry*, **43**, 2845–2852.
78 Rosenthal, K., Erlandsson, M., and Unden, A. (1999) *Tetrahedron Letters*, **40**, 377–380.
79 Albericio, F. and Barany, G. (1985) *International Journal of Peptide and Protein Research*, **26**, 92–97.
80 Sheppard, R.C. and Williams, B.J. (1982) *Journal of the Chemical Society, Chemical Communications*, 587–588.
81 Katritzky, A.R., Toader, D., Watson, K., and Kiely, J.S. (1997) *Tetrahedron Letters*, **38**, 7849–7850.
82 Atkinson, G.E., Fischer, P.M., and Chan, W.C. (2000) *The Journal of Organic Chemistry*, **65**, 5048–5056.
83 Leznoff, C.C. (1985) In *Biotechnology: Applications and Research* (ed. P.N. Cheremisinoff), Technomie, Westport, CT, pp. 586–593.
84 Barlos, K., Gatos, D., Kallitsis, I., Papaioannou, D., and Sotiriou, P. (1988) *Liebigs Annalen der Chemie*, 1079–1081.
85 Barlos, K., Gatos, D., Kapolos, S., Papaphotiu, G., Schafer, W., and Wenqing, Y. (1989) *Tetrahedron Letters*, **30**, 3947–3950.
86 Erlandsson, M. and Undén, A. (2006) *Tetrahedron Letters*, **47**, 5829–5832.
87 Atherton, E., Logan, C.J., and Sheppard, R.C. (1981) *Journal of the Chemical Society, Perkin Transactions 1*, 538–546.
88 Urban, F.J. and Moore, B.S. (1992) *Journal of Heterocyclic Chemistry*, **29**, 431–438.
89 Fancelli, D., Fagnola, M.C., Severino, D., and Bedeschi, A. (1997) *Tetrahedron Letters*, **38**, 2311–2314.
90 Bilodeau, M.T. and Cunningham, A.M. (1998) *The Journal of Organic Chemistry*, **63**, 2800–2801.
91 Furlan, R.L.E., Mata, E.G., Mascaretti, O.A., Pena, C., and Coba, M.P. (1998) *Tetrahedron*, **54**, 13023–13034.
92 Takashima, H., du Vigneaud, V., and Merrifield, R.B. (1968) *Journal of the American Chemical Society*, **90**, 1323–1325.
93 Story, S.C. and Aldrich, J.V. (1992) *International Journal of Peptide and Protein Research*, **39**, 87–92.
94 Bray, A.M., Maeji, N.J., Jhingran, A.G., and Valerio, R.M. (1991) *Tetrahedron Letters*, **32**, 6163–6166.
95 Bray, A.M., Valerio, R.M., and Maeji, N.J. (1993) *Tetrahedron Letters*, **34**, 4411–4414.
96 Andersson, L., Blomberg, L., Flegel, M., Lepsa, L., Nilsson, B., and Verlander, M. (2000) *Biopolymers*, **55**, 227–250.
97 Flanigan, E. and Marshall, G.R. (1970) *Tetrahedron Letters*, **11**, 2403–2406.
98 Reents, R., Jeyaraj, D.A., and Waldmann, H. (2002) *Drug Discovery Today*, **7**, 71–76.
99 Rademann, J., Grotli, M., Meldal, M., and Bock, K. (1999) *Journal of the American Chemical Society*, **121**, 5459–5466.
100 Elmore, D.T., Guthrie, D.J.S., Wallace, A.D., and Bates, S.R.E. (1992) *Journal of the Chemical Society, Chemical Communications*, 1033.
101 Chhabra, S.R., Parekh, H., Khan, A.N., Bycroft, B.W., and Kellam, B. (2001) *Tetrahedron Letters*, **42**, 2189–2192.
102 Isidro-Llobet, A., Alvarez, M., Burger, K., Spengler, J., and Albericio, F. (2007) *Organic Letters*, **9**, 1429–1432.
103 Albericio, F. and Barany, G. (1987) *International Journal of Peptide and Protein Research*, **30**, 206–216.
104 Albericio, F., Kneib-Cordonier, N., Biancalana, S., Gera, L., Masada, R.I., Hudson, D., and Barany, G. (1990) *The Journal of Organic Chemistry*, **55**, 3730–3743.
105 Pietta, P.G. and Marshall, G.R. (1970) *Journal of the Chemical Society D, Chemical Communications*, 650.
106 Matsueda, G.R. and Stewart, J.M. (1981) *Peptides*, **2**, 45–50.
107 Sieber, P. (1987) *Tetrahedron Letters*, **28**, 2107–2110.
108 Ramage, R., Irving, S.L., and Mcinnes, C. (1993) *Tetrahedron Letters*, **34**, 6599–6602.
109 Patterson, J.A. and Ramage, R. (1999) *Tetrahedron Letters*, **40**, 6121–6124.
110 Noda, M., Yamaguchi, M., Ando, E., Takeda, K., and Nokihara, K. (1994) *The Journal of Organic Chemistry*, **59**, 7968–7975.
111 Henkel, B., Zeng, W., and Bayer, E. (1997) *Tetrahedron Letters*, **38**, 3511–3512.

112 Edvinsson, K.M., Herslöf, M., Holm, P., Kann, N., Keeling, D.J., Mattsson, J.P., Norden, B., and Shcherbukhin, V. (2000) *Bioorganic & Medicinal Chemistry Letters*, **10**, 503–507.

113 Brown, D.S., Revill, J.M., and Shute, R.E. (1998) *Tetrahedron Letters*, **39**, 8533–8536.

114 Brown, E.G. and Nuss, J.M. (1997) *Tetrahedron Letters*, **38**, 8457–8460.

115 Bernatowicz, M.S., Daniels, S.B., and Koster, H. (1989) *Tetrahedron Letters*, **30**, 4645–4648.

116 Semenov, A.N. and Gordeev, K.Y. (1995) *International Journal of Peptide and Protein Research*, **45**, 303–304.

117 Albrecht, M., Stortz, P., Runsink, J., and Weis, P. (2004) *Chemistry – A European Journal*, **10**, 3657–3666.

118 Camarero, J.A., Hackel, B.J., de Yoreo, J.J., and Mitchell, A.R. (2004) *The Journal of Organic Chemistry*, **69**, 4145–4151.

119 Millington, C.R., Quarrell, R., and Lowe, G. (1998) *Tetrahedron Letters*, **39**, 7201–7204.

120 Lumbierres, M., Palomo, J.M., Kragol, G., Roehrs, S., Mueller, O., and Waldmann, H. (2005) *Chemistry – A European Journal*, **11**, 7405–7415.

121 Shigenaga, A., Moss, J.A., Ashley, F.T., Kaufmann, G.F., and Janda, K.D. (2006) *Synlett*, **4**, 551–554.

122 Woo, Y.-H., Mitchell, A.R., and Camarero, J.A. (2007) *International Journal of Peptide Research and Therapeutics*, **13**, 181–190.

123 Lindquist, C., Tedebark, U., Ersoy, O., and Somfai, P. (2003) *Synthetic Communications*, **33**, 2257–2262.

124 Kwon, Y., Welsh, K., Mitchell, A.R., and Camarero, J.A. (2004) *Organic Letters*, **6**, 3801–3804.

125 Garigipati, R.S. (1998) Rink-chloride linker for solid phase organic synthesis of organic molecules. WO 9844329; (1998) *Chemical Abstracts*, **129**, 289927.

126 Siemens, L.M., Rottnek, F.W., and Trzupek, L.S. (1990) *The Journal of Organic Chemistry*, **55**, 3507–3511.

127 Moraes, C.M., Bemquerer, M.P., and Miranda, M.T.M. (2000) *The Journal of Peptide Research*, **55**, 279–288.

128 Hojo, H. and Aimoto, S. (1991) *Bulletin of the Chemical Society of Japan*, **64**, 111–117.

129 Zhang, L. and Tam, J.P. (1997) *Journal of the American Chemical Society*, **119**, 2363–2370.

130 Zhang, L. and Tam, J.P. (1997) *Tetrahedron Letters*, **38**, 4375–4378.

131 Zhang, L. and Tam, J.P. (1999) *Journal of the American Chemical Society*, **121**, 3311–3320.

132 Dawson, P.E., Muir, T.W., Clark-Lewis, I., and Kent, S.B. (1994) *Science*, **266**, 776–779.

133 Clippingdale, A.B., Barrow, C.J., and Wade, J.D. (2000) *Journal of Peptide Science*, **6**, 225–234.

134 Canne, L.E., Walker, S.M., and Kent, S.B.H. (1995) *Tetrahedron Letters*, **36**, 1217–1220.

135 Gaertner, H., Villain, M., Botti, P., and Canne, L. (2004) *Tetrahedron Letters*, **45**, 2239–2241.

136 Zhang, W., Zhang, L., Li, X., Weigel, J.A., Hall, S.E., and Mayer, J.P. (2001) *Journal of Combinatorial Chemistry*, **3**, 151–153.

137 Hauske, J.R. and Dorff, P. (1995) *Tetrahedron Letters*, **36**, 1589–1592.

138 Munson, M.C., Cook, A.W., Josey, J.A., and Rao, C. (1998) *Tetrahedron Letters*, **39**, 7223–7226.

139 Dressman, B.A., Spangle, L.A., and Kaldor, S.W. (1996) *Tetrahedron Letters*, **37**, 937–940.

140 Marsh, I.R., Smith, H., and Bradley, M. (1996) *Journal of the Chemical Society, Chemical Communications*, 941–942.

141 Zaragoza, F. (1995) *Tetrahedron Letters*, **36**, 8677–8678.

142 Tomasi, S., Le Roch, M., Renault, J., Corbel, J.-C., Uriac, P., Carboni, B., Moncoq, D., Martin, B., and Delcros, J.-G. (1998) *Bioorganic & Medicinal Chemistry Letters*, **8**, 635–640.

143 Kim, S.W., Hong, C.Y., Lee, K., Lee, E.J., and Koh, J.S. (1998) *Bioorganic & Medicinal Chemistry Letters*, **8**, 735–738.

144 Burdick, D.J., Struble, M.E., and Burnier, J.P. (1993) *Tetrahedron Letters*, **34**, 2589–2592.

145 Page, P., Burrage, S., Baldock, L., and Bradley, M. (1998) *Bioorganic & Medicinal Chemistry Letters*, **8**, 1751–1756.

146 Kaljuste, K. and Unden, A. (1995) *Tetrahedron Letters*, **36**, 9211–9214.

**147** Lipshutz, B.H. and Shin, Y.-J. (2001) *Tetrahedron Letters*, **42**, 5629–5633.

**148** Thieriet, N., Guibé, F., and Albericio, F. (2000) *Organic Letters*, **2**, 1815–1817.

**149** Boojamra, C.G., Burow, K.M., and Ellman, J.A. (1995) *The Journal of Organic Chemistry*, **60**, 5742–5743.

**150** Jensen, K.J., Songster, M.F., Vagner, J., Alsina, J., Albericio, F., and Barany, G. (1996) In *Peptides: Chemistry, Strucure and Biology* (eds P.T.P. Kaumaya and R.S. Hodges), Mayflower, Birmingham, p. 30.

**151** Fivush, A.M. and Willson, T.M. (1997) *Tetrahedron Letters*, **38**, 7151–7154.

**152** Jensen, K.J., Alsina, J., Songster, M.F., Vagner, J., Albericio, F., and Barany, G. (1998) *Journal of the American Chemical Society*, **120**, 5441–5452.

**153** Sarantakis, D. and Bicksler, J.J. (1997) *Tetrahedron Letters*, **38**, 7325–7328.

**154** Swayze, E.E. (1997) *Tetrahedron Letters*, **38**, 8465–8468.

**155** Bourne, G.T., Meutermans, W.D.F., Alewood, P.F., McGeary, R.P., Scanlon, M., Watson, A.A., and Smythe, M.L. (1999) *The Journal of Organic Chemistry*, **64**, 3095–3101.

**156** Estep, K.G., Neipp, C.E., Stramiello, L.M.S., Adam, M.D., Allen, M.P., Robinson, S., and Roskamp, E.J. (1998) *The Journal of Organic Chemistry*, **63**, 5300–5301.

**157** Bourne, G.T., Meutermans, W.D.F., and Smythe, M.L. (1999) *Tetrahedron Letters*, **40**, 7271–7274.

**158** Jin, J., Graybill, T.L., Wang, M.A., Davis, L.D., and Moore, M.L. (2001) *Journal of Combinatorial Chemistry*, **3**, 97–101.

**159** Krchnak, V. and Holladay, M.W. (2002) *Chemical Reviews*, **102**, 61–92.

**160** Boas, U., Brask, J., and Jensen, K.J. (2009) *Chemical Reviews*, **109**, 2092–2118.

**161** Alsina, J., Yokum, T.S., Albericio, F., and Barany, G. (1999) *The Journal of Organic Chemistry*, **64**, 8761–8769.

**162** Guillaumie, F., Kappel, J.C., Kelly, N.M., Barany, G., and Jensen, K.J. (2000) *Tetrahedron Letters*, **41**, 6131–6135.

**163** Olsen, J.A., Jensen, K.J., and Nielsen, J. (2000) *Journal of Combinatorial Chemistry*, **2**, 143–150.

**164** Brask, J. and Jensen, K.J. (2001) *Bioorganic & Medicinal Chemistry Letters*, **11**, 697–700.

**165** Tolborg, J.F. and Jensen, K.J. (2000) *Chemical Communications*, 147–148.

**166** Boas, U., Brask, J., Christensen, J.B., and Jensen, K.J. (2002) *Journal of Combinatorial Chemistry*, **4**, 223–228.

**167** Bui, C.T., Bray, A.M., Pham, Y., Campbell, R., Ercole, F., Rasoul, F.A., and Maeji, N.J. (1998) *Molecular Diversity*, **4**, 155–163.

**168** Alsina, J., Yokum, T.S., Albericio, F., and Barany, G. (2000) *Tetrahedron Letters*, **41**, 7277–7280.

**169** Bui, C.T., Rasoul, F.A., Ercole, F., Pham, Y., and Maeji, N.J. (1998) *Tetrahedron Letters*, **39**, 9279–9282.

**170** Bui, C.T., Bray, A.M., Ercole, F., Pham, Y., Rasoul, F.A., and Maeji, N.J. (1999) *Tetrahedron Letters*, **40**, 3471–3474.

**171** Gu, W. and Silverman, R.B. (2003) *Organic Letters*, **5**, 415–418.

**172** Brandt, M., Madsen, J.C., Bunkenborg, J., Jensen, O.N., Gammeltoft, S., and Jensen, K.J. (2006) *Chemistry and Biochemistry*, **7**, 623–630.

**173** Austin, R.E., Waldraff, C.A., and Al-Obeidi, F. (2002) *Tetrahedron Letters*, **43**, 3555–3556.

**174** Yan, B. and Yan, H. (2001) *Journal of Combinatorial Chemistry*, **3**, 78–84.

**175** Beythien, J., Barthélémy, S., Schneeberger, P., and White, P.D. (2006) *Tetrahedron Letters*, **47**, 3009–3012.

**176** Bettinetti, L., Löber, S., Hübner, H., and Gmeiner, P. (2005) *Journal of Combinatorial Chemistry*, **7**, 309–316.

**177** Boas, U., Christensen, J.B., and Jensen, K.J. (2004) *Journal of Combinatorial Chemistry*, **6**, 497–503.

**178** Pittelkow, M., Boas, U., and Christensen, J.B. (2006) *Organic Letters*, **8**, 5817–5820.

**179** Jessing, M., Brandt, M., Jensen, K.J., Christensen, J.B., and Boas, U. (2006) *The Journal of Organic Chemistry*, **71**, 6734–6741.

**180** Okayama, T., Burrit, A., and Hruby, V.J. (2000) *Organic Letters*, **2**, 1787–1790.

**181** Scicinski, J.J., Congreve, M.S., and Ley, S.V. (2006) *Journal of Combinatorial Chemistry*, **6**, 375–384.

**182** Minkwitz, R. and Meldal, M. (2005) *QSAR & Combinatorial Science*, **24**, 343–353.

**183** Ficht, S., Payne, R.J., Guy, R.T., and Wong, Ch. (2008) *Chemistry – A European Journal*, **14**, 3620–3629.

**184** Woon, E.C.Y., Arcieri, M., Wilderspin, A.F., Malkinson, J.P., and Searcey, M. (2007) *The Journal of Organic Chemistry*, **72**, 5146–5151.

**185** Malkinson, J.P., Zloh, M., Kadom, M., Errington, R., Smith, P.J., and Searcey, M. (2003) *Organic Letters*, **5**, 5051–5054.

**186** Valero, M.-L., Giralt, E., and Andreu, D. (1996) *Tetrahedron Letters*, **37**, 4229–4232.

**187** Albericio, F., Van Abel, R., and Barany, G. (1990) *International Journal of Peptide and Protein Research*, **35**, 284–286.

**188** Kates, S.A., Sole, N.A., Johnson, C.R., Hudson, D., Barany, G., and Albericio, F. (1993) *Tetrahedron Letters*, **34**, 1549–1552.

**189** Spatola, A.F., Darlak, K., and Romanovskis, P. (1996) *Tetrahedron Letters*, **37**, 591–594.

**190** Spatola, A.F., Crozet, Y., deWit, D., and Yanagisawa, M. (1996) *Journal of Medicinal Chemistry*, **39**, 3842–3846.

**191** Romanovskis, P. and Spatola, A.F. (1998) *The Journal of Peptide Research*, **52**, 356–374.

**192** Breipohl, J., Knolle, J., and Stueber, W. (1990) *International Journal of Peptide and Protein Research*, **35**, 281–283.

**193** Alsina, J., Rabanal, F., Giralt, E., and Albericio, F. (1994) *Tetrahedron Letters*, **35**, 9633–9636.

**194** Rabanal, F., Giralt, E., and Albericio, F. (1992) *Tetrahedron Letters*, **33**, 1775–1778.

**195** Bernhardt, A., Drewello, M., and Schutkowski, M. (1997) *The Journal of Peptide Research*, **50**, 143–152.

**196** Glass, J.D., Schwartz, I.L., and Walter, R. (1972) *Journal of the American Chemical Society*, **94**, 6209–6211.

**197** García, O., Nicolás, E., and Albericio, F. (2003) *Tetrahedron Letters*, **44**, 5319–5321.

**198** Hanessian, S. and Xie, F. (1998) *Tetrahedron Letters*, **39**, 733–736.

**199** Mergler, M., Dick, F., Gosteli, J., and Nyfeler, R. (1999) *Tetrahedron Letters*, **40**, 4663–4664.

**200** Leznoff, C.C. and Fyles, T.M. (1976) *Journal of the Chemical Society, Chemical Communications*, 251.

**201** Deegan, T.L., Gooding, O.W., Baudart, S., and PorcoJr, J.A. (1997) *Tetrahedron Letters*, **38**, 4973–4976.

**202** Tegge, W., Bautsch, W., and Frank, R. (2007) *Journal of Peptide Science*, **13**, 693–699.

**203** Barany, G., Han, Y., Hargittai, B., Liu, R.-Q., and Varkey, J.T. (2003) *Biopolymers*, **71**, 652–666.

**204** Zoller, T., Ducep, J.-B., Tahtaoui, C., and Hibert, M. (2000) *Tetrahedron Letters*, **41**, 9989–9992.

**205** Rietman, B.H., Smulders, R.H.P.H., Eggen, I.F., Van Vliet, A., van de Werken, G., and Tesser, G.I. (1994) *International Journal of Peptide and Protein Research*, **44**, 199–206.

**206** Stahl, G.L., Walter, R., and Smith, C.W. (1979) *Journal of the American Chemical Society*, **101**, 5383–5394.

**207** Lee, Y. and Silverman, R.B. (1999) *Journal of the American Chemical Society*, **121**, 8407–8408.

**208** Lee, Y. and Silverman, R.B. (2001) *Organic Letters*, **2**, 303–306.

# 9
# Orthogonal Protecting Groups and Side-Reactions in Fmoc/*t*Bu Solid-Phase Peptide Synthesis

*Stefano Carganico and Anna Maria Papini*

## 9.1
## Orthogonal Protecting Groups in Fmoc/*t*Bu Solid-Phase Peptide Synthesis

### 9.1.1
### Arg

During solid-phase peptide synthesis (SPPS) the three nitrogen atoms of the guanidine group of Arg (Figure 9.1), being strongly nucleophilic, are prone to alkylation and subsequent Orn (ornithine) formation upon base-mediated decomposition [1], and therefore need to be protected. However, in common practice, most protecting groups block only the ω-nitrogen. In addition, free unprotected Arg residues tend to cyclize upon activation of the α-carboxylic group to form δ-lactams.

In the fluorenylmethoxycarbonyl (Fmoc)/*tert*-butyl (*t*Bu) strategy the most commonly used protecting groups of Arg are the arylsulfonyl-based derived from the tosyl group (Tos) [2] such as the 4-methoxy-2,3,6-trimethylbenzenesulfonyl (Mtr) group **1**, now superseded by the two cyclic ether derivates 2,2,5,7,8-pentamethylchroman-6-sulfonyl (Pmc) **2** and 2,2,4,6,7-pentamethyldihydrobenzofuran-5-sulfonyl (Pbf) **3**. Mtr removal requires several hours of trifluoroacetic acid (TFA) treatment and often causes sulfonation of Trp residues [3], which can be avoided using 1 M bromotrimethylsilane (TMSBr) in TFA [4]. Moreover, long TFA treatment can cause *O*-sulfonation of Ser and Thr, which can be suppressed by adding thiocresol to the cleavage cocktail [5]. The Pmc group, being much more acid sensitive than Mtr, can be removed faster and the Trp/Tyr modifications are less pronounced [6], and can be overcome if the Trp indole ring is Boc protected [7]. The Pbf group [8], the dihydrofuran analog of the Pmc group, is at present the most widely used Arg-protecting group. It has proved to be more acid labile than Pmc (its removal is 1–2 times faster) and generates less alkylation than the other arylsulfonyl-protecting groups.

*Amino Acids, Peptides and Proteins in Organic Chemistry.*
*Vol.3 – Building Blocks, Catalysis and Coupling Chemistry.* Edited by Andrew B. Hughes

ISBN: 978-3-527-32102-5

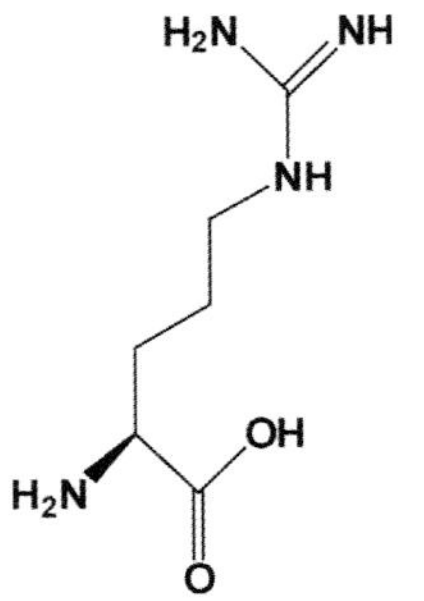

**Figure 9.1** The guanidino group of Arg.

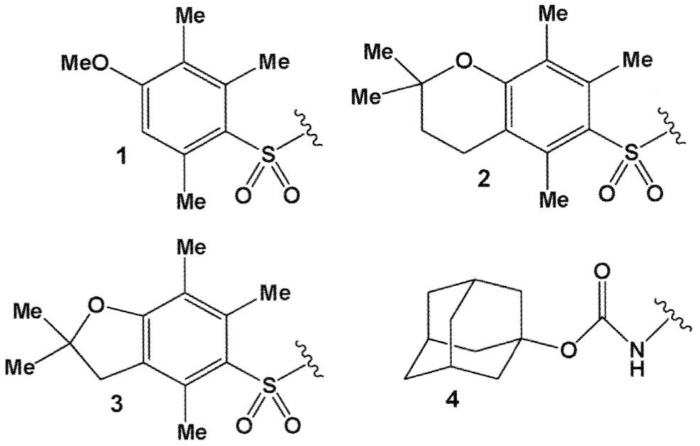

Other reported protecting strategies have not gained popularity due to several different shortcomings. For example, the nitro group ($NO_2$) is used to protect the ω-nitrogen and can be removed with $H_2$/Pd or with hydrazinium monoformate and magnesium [9], but it is not completely stable under coupling conditions [10]. It generally requires long deprotection times and it is prone to generate partially reduced products [11]. The urethane protections, ω-Boc [12] and δ,ω-bis-adamantyloxycarbonyl $(Adoc)_2$ [13] **4**, are unable to completely suppress the guanidine group nucleophilicity. On the other hand bis-urethane Arg derivatives with both ω and ω′-nitrogens Boc protected [14] do not show any side-reaction on deprotection, but are highly hindered adducts and the coupling time of these materials needs to be extended. The trityl (Trt) group is not normally used for Arg protection because it gives an adduct with poor solubility in *N,N*-dimethylformamide (DMF)/dichloromethane (DCM) [15].

A completely different approach to the problem of Arg protection is the use of a suitably protected Orn residue that can be converted into Arg at the end of the synthesis by guanylation with reagents such as 1*H*-pyrazole-1-carboxamide hydrochloride [16]. An additional level of orthogonality is needed for the Orn δ-amino group, which has to be selectively deprotected before the conversion. Protecting groups such as the 1-(4-methoxyphenyl)ethyloxycarbonyl (Mpeoc) [17] **4a**, cleavable under mild acidic conditions, have been specifically developed for this application.

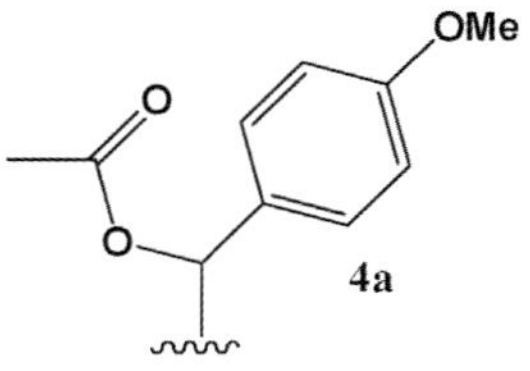

### 9.1.2
### Asn and Gln

Asn and Gln could in principle be incorporated into peptides without protection, but the unprotected derivatives display a low solubility in solvents commonly used in peptide synthesis and have low coupling rates. In addition, the amide side-chain, especially that of Asn, can suffer partial dehydration on activation [18]. Finally, although more common for the Boc/Bzl chemistry, Gln in the N-terminal position can undergo weak acid-catalyzed cyclization forming pyroglutamyl residues that cause the truncation of the peptide sequence [19].

Side-chain protection prevents all these undesired reactions and, in addition, inhibits hydrogen bond interactions of the amide, which stabilize secondary structures causing incomplete deprotection and reduced coupling rate.

The most common amide-protecting group for Fmoc/*t*Bu SPPS is the triphenylmethyl group (Trt) [20] **5**, which requires care in the choice of the cleavage scavengers since it generates stable carbocations that tend to alkylate Trp. Such alkylation is reduced by using the 9-xanthenyl (Xan) group **6** and its 2-methoxy derivative (2-Moxan) [21] **7**; however, it generates less-soluble derivatives.

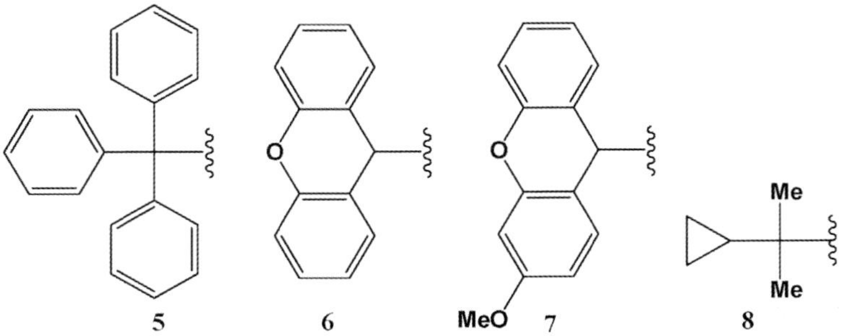

When Asn is the N-terminal amino acid, Trt deprotection is slower, due to the vicinity of the free α-amino group [22]. In this case cleavage time needs to be extended or methyltrityl protection can be used instead of Trt.

The *N*-dimethylcyclopropylmethyl (Dcmp) group **8** represents a convenient alternative to Trt with several advantages like rapid removal (even at the N-terminal

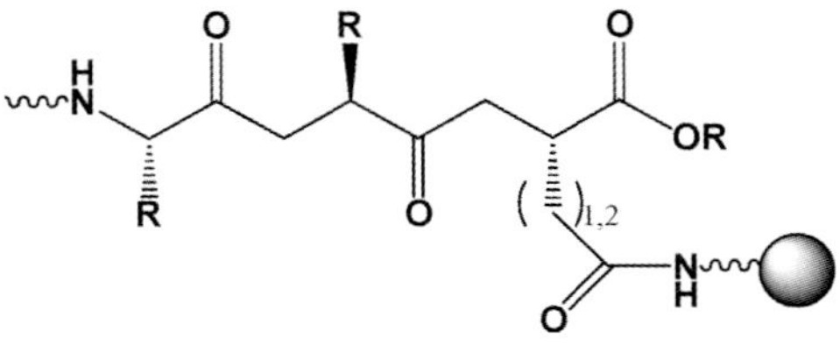

Figure 9.2 Side-chain anchoring of C-terminal Asn- and Gln-containing peptides.

position), faster coupling rate (due to the minor steric hindrance) and better solubility in DMF [23].

To overcome the problem of the slow and troublesome attachment to most resins the resin can be linked to the amide side-chain [24] (Figure 9.2) instead of to the carboxylic function (that needs to be protected during the synthesis).

More used in the past than now, before Trt protection gained popularity, are the 2,4,6-trimethoxybenzyl group (Tmob) [25] **9** and the 4,4′-dimethoxybenzhydryl group (Mbh) [26] **10**, both less acid-labile and soluble than Trt [20].

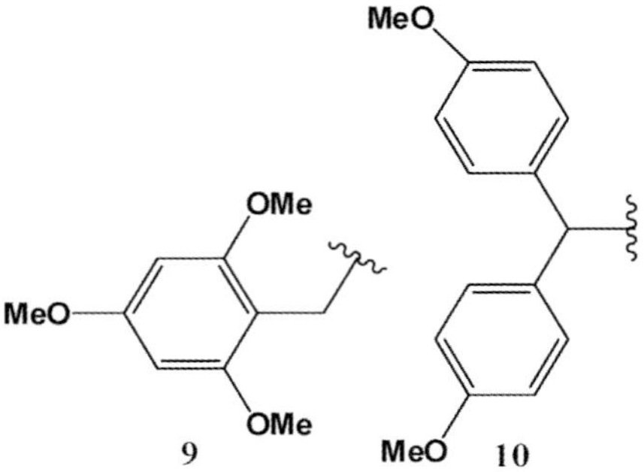

### 9.1.3
### Asp and Glu

The carboxylic acid side-chains of Asp and Glu need to be protected during peptide synthesis in order to prevent amide bond formation with incoming amino acids and, as a consequence, branching of the peptide [27].

Although protected, Asp and Glu residues could still be affected from side-reactions, particularly acid/base (particularly in Fmoc/*t*Bu chemistry)-catalyzed cyclization to form aspartimides and glutarimides, respectively [28]. Subsequent hydrolysis of the imide-containing peptides leads to a mixture of the desired peptide along with a product, called a β-peptide in which the side-chain carboxylic group forms part of the backbone and a β-piperidide adduct [29].

The reaction occurs less often with Glu [30] and is highly sequence dependent. Susceptible sequences are Asp–Xxx with Xxx being Gly, Asn, Ala, and Gln [31].

Since the beginning of Fmoc/*t*Bu SPPS, Asp and Glu have been successfully protected by the *t*Bu **11** group, which is base stable and TFA labile [32].

Equally favorable properties are displayed by 1-adamantyl (1-Ada) protection [33] **12**. Both *t*Bu and 1-Ada minimize piperidine-catalyzed aspartimide formation although several bulky *t*Bu derivatives give better results in this respect like the 3-methylpent-3-yl (Mpe) [34] group **13** and β-2,4-dimethyl-3-pentyl (Dmp) [35] group **14**.

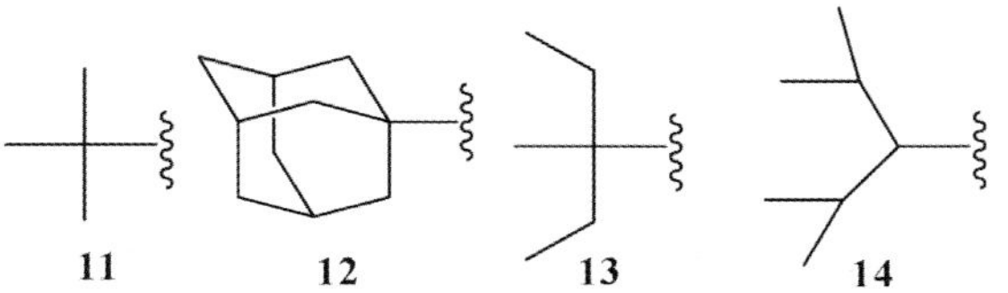

Aspartimide formation is also greatly reduced by adding either 1-hydroxybenzotriazole (HOBt) or 2,4-dinitrophenol to the piperidine deprotection solution [36], but can be completely eliminated, especially with susceptible sequences in long peptide sequences, only by employing amide-backbone protection for the introduction of residues preceding Asp. This is achieved using the 2-hydroxy-4-methoxybenzyl (Hmb) [37] **15** or the 2,4-dimethoxybenzyl (Dmb) [38] **16** (only applied to glycine residues) amide-protecting groups. Hmb and Dmb, removed at the same time as the final cleavage, prevent undesired side-reactions and suppress aggregation during chain extension.

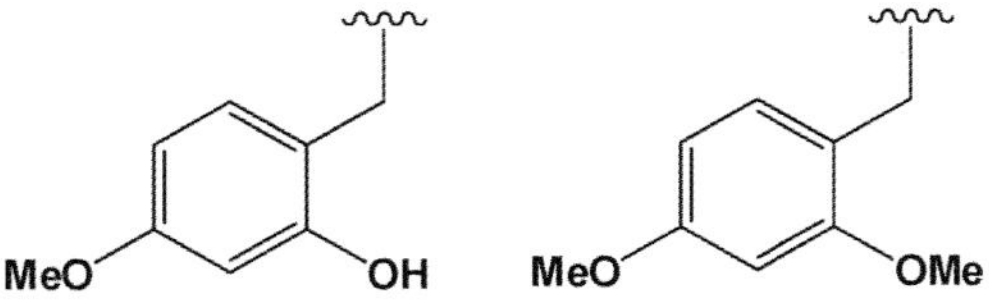

When an additional degree of orthogonality is required, such as in the case of lactam-bridged peptide synthesis, a number of different Asp and Glu protecting groups exist. For example, the benzyl group (Bzl) [39] **17**, the 2-amantyl (2-Ada) [40] **18**, and allyl esters [41] (thought more prone to imide formation than *t*Bu) all are removed by palladium-catalyzed transfer to a suitable nucleophile. Alternatively super-acid-labile groups can be used such as the 2-phenyl isopropyl (Pp) [42] group **19**, removable in the presence of *t*Bu/Boc with 1% TFA in DCM and the phenyl-3,4-ethylenedioxy-2-thienyl (EDOT-Ph) [43] **20** cleaved by 0.1–0.5% TFA in DCM.

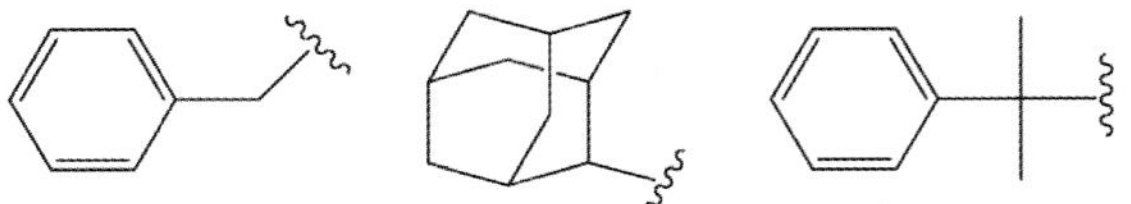

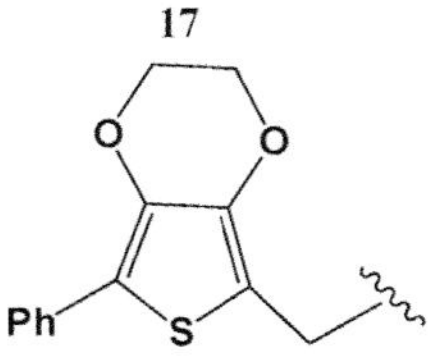

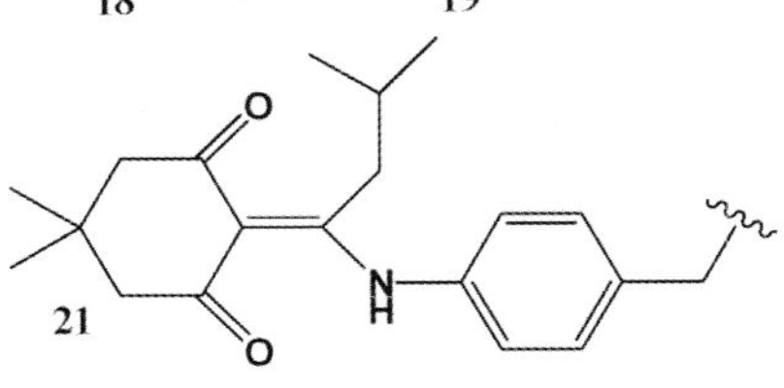

Another orthogonal protecting group is the 4-[*N*-[1-(4,4′-dimethyl-2,6-dioxocyclohexylidene)-3-methylbutyl]aminobenzyl group (Dmab) [44] **21**, removed with 2% hydrazine in DMF. With aspartimide-susceptible sequences it is recommended to use Dmab along with additional precautions such as backbone-amide protection [45].

For applications in native chemical a new photo-labile protecting group, {7-[bis(carboxymethyl)amino]coumarin-4-yl}methyl (BCMACM) [46] group **22** ligation has recently been described, removed with UV irradiation at 405 nm.

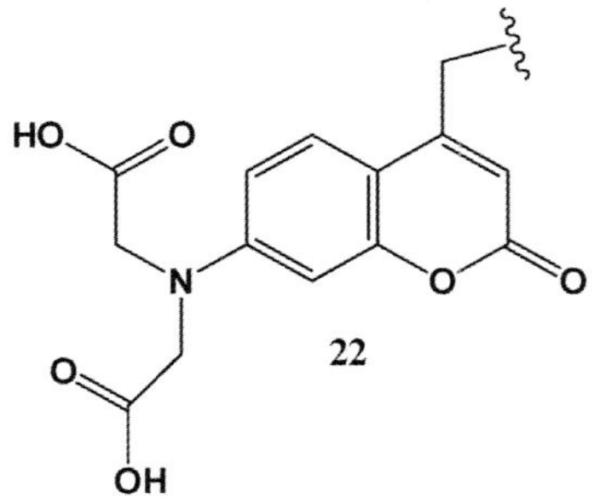

**22**

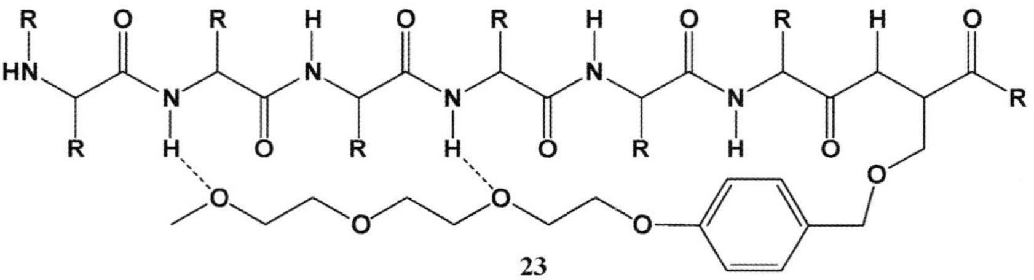

**23**

A special carboxylic protecting group, 4-(3,6,9-trioxadecyl)oxybenzyl (TEGBz) [47] **23**, has been developed to suppress aggregation of those "difficult sequences," in which intermediate resin-bound peptide chains associate into extended β-sheet-type structures. TEGBz forms hydrogen bonds with the backbone amino groups, enabling the so-called "internal solvation" that inhibits aggregation by enhancing backbone linearity.

### 9.1.4
### Cys

In SPPS, protection of the Cys side-chain sulfhydryl group is mandatory, otherwise it would easily undergo alkylation and acylation. Free Cys residues are also prone to oxidation, even by atmospheric oxygen, to form intra- and inter-molecular disulfide bonds.

The chemistry of Cys-protecting groups is particularly rich due to the key importance of this amino acid in forming inter- and intra-molecular disulfide bridges, and in consideration of the versatile reactivity of the thiol group.

With the Fmoc/*t*Bu strategy trityl (Trt) [48] **24** is the most used Cys protection. Since acid *S*-detritylation is an equilibrium reaction it needs to be driven to completion by capture of the forming carbocation. This can be achieved by adding $H_2O$, thiols and especially silanes to the reaction mixture. Triisopropylsilane (TIS), in particular, is preferred over triethylsilane (TES) [49] which can lead to reduction of free Trp residues. In spite of its popularity, Cys(Trt) is prone to racemization (up to 10–20%) during peptide coupling by base mediated *in situ* activation [50], especially with the 2-(1*H*-benzotriazol-1-yl)-1,1,3,3,-tetramethyluronium tetrafluoroborate (TBTU)/*N,N*-diisopropylethylamine (DIPEA) system. The use of 2-(1*H*-benzotriazol-1-yl)-1,1,3,3-tetramethyluronium hexafluotriphosphate (HBTU), benzotriazol-1-yl-oxytripyrrolidinephosphonium hexafluorophosphate (PyBOP)/HOBt, or preactivated reagents such as symmetrical anhydrides, OPfp esters, and diisopropylcarbodiimide (DIPCDI)/HOBt minimize this problem [51]. Enantiomerization occurs also with the attachment of Cys(Trt) to Wang type resins and during chain extension when Cys (Trt) is the C-terminal residue. The use of chloro-trityl resins is recommended [52].

In alternatives to Trt, other protecting groups removed with concentrated TFA [53] 1,4,6-trimethoxybenzyl (Tmob) **25** and 9-phenylxanthen-9-yl (pixyl) **26**, whereas monomethoxytrityl (Mmt) **27**, 9*H*-xanthen-9-yl (xanthyl) **28**, and 2-methoxy-9*H*-xanthen-9-yl (2-Moxan) **7** are more acid-labile [54], and can be selectively cleaved in the presence of *t*Bu groups.

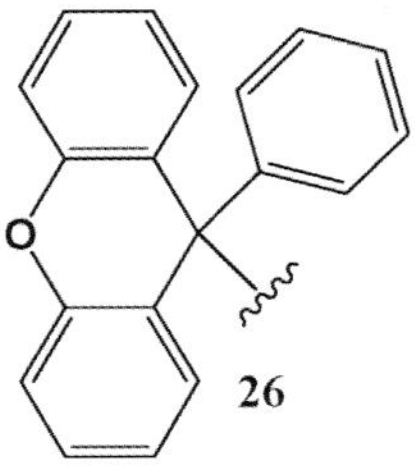

O

26

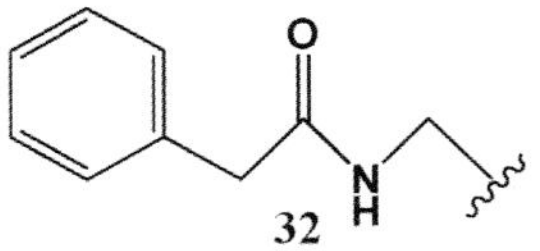

O

N
H

32

The *t*Bu **29**, *S*-(1-adamantyl) (1-Ada) **30**, acetamidomethyl (Acm) **31**, trimethylacetamidomethyl (Tacm) **32**, and phenylacetamidomethy (Phacm) **33** groups (the last two developed to avoid formation of thiazolidine-2-carboxylic acid) are stable to acid and compatible with both Boc and Fmoc SPPS strategies. Those groups can be removed in several ways (see Table 9.1) enabling concomitant disulfide bridge formation even multiple, selective cyclization if used in combination [55]. Phacm has an additional level of orthogonality since it is enzymatically cleavable by penicillin G acylase [56].

A different protecting approach is to use mixed disulfides such as the *S-tert*-Butylsulfanyl (S*t*Bu) group **34**, which is stable to TFA and is removed with thiols [57] or tributylphoshine [58]. Coupling efficiency is reported to be highly sequence dependent [59]. Allyl-based Cys-protecting groups are base-labile and therefore cannot be used in Fmoc/*t*Bu SPPS [60].

Finally, several non-conventional protecting groups have been developed for those chemical ligations based on Cys. The thiazolidine (Thz) [61] **35** protection has found a special application allowing masking of N-terminal Cys during the tandem native chemical ligation (TNCL) reaction. It simultaneously protects the α-amino and the side-chain thiol groups of protected N-terminal free Cys and is stable to acids, and can be removed in aqueous conditions in the presence of methoxylamine. The thiosul-

**Table 9.1** Deprotection and deprotection/oxidation conditions of the most common Cys-protecting groups (concomitant disulfide formation in green).

| Protecting group | Structure | Removed by | Stable to |
|---|---|---|---|
| Trityl (Trt) | 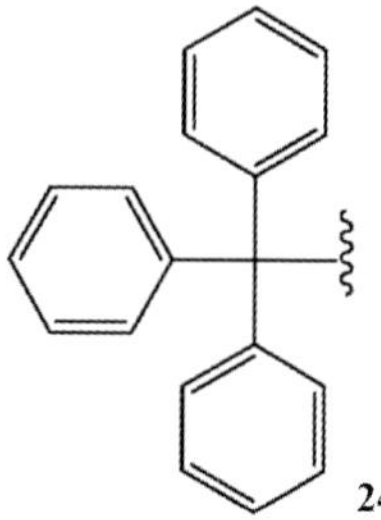 24 | dilute TFA/scavengers, Ag(I) [63], Hg(II) [64], RSCl, $I_2$ [65], Tl(III)-trifluoracetate [78] TFA/DMSO/anisole [66] | base, nucleophiles, RSH |
| 1,4,6-Trimethoxybenzyl (Tmob) | 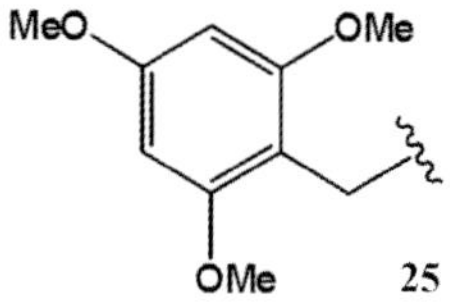 MeO, OMe, OMe 25 | dilute TFA/DCM/silanes [53] $I_2$ [67], Tl(III)-trifluoracetate | base, nucleophiles |
| Monomethoxytrityl (Mmt) | 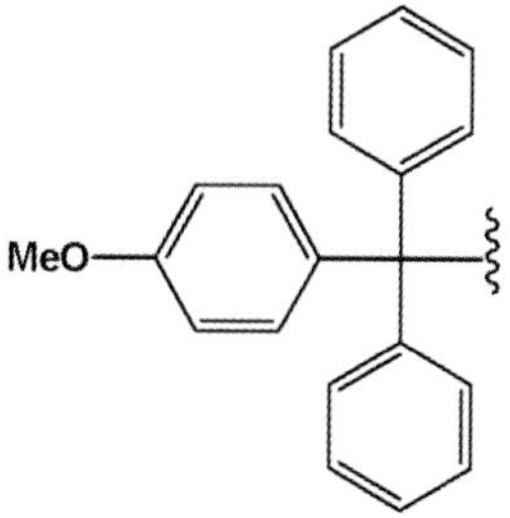 MeO 27 | dilute TFA/DCM/TIS [49], AcOH/TFE/DCM (1: 2: 7) [68], $I_2$ [69] | base, nucleophiles, RSH |

| | | | |
|---|---|---|---|
| 9*H*-Xanthen-9-yl (Xan) | 28 | dilute TFA/TIS [54] $I_2$, Tl(III)-trifluoracetate [70] | base, nucleophiles |
| *t*Bu | 29 | TFMSA [71], TMSBr/TFA/RSH [72], tetrafluoroboric acid [73], Hg(II) acetate [74], Tl(III)-trifluoracetate [78], $MeSiCl_3/Ph_2SO$/TFA [75], TFA/DMSO/Anisole | TFA, Ag(I), base, $I_2$, RSH |
| 1-Adamantyl (1-Ada) | 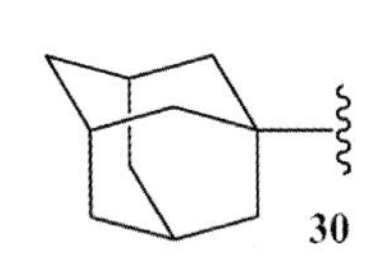 | TfOH/TFA/RSH [76], Hg(II) acetate [75], Tl(III)-trifluoracetate [77] | TFA, Ag(I), base, $I_2$ |
| *S*-*tert*-Butylsulfanyl (S*t*Bu), | 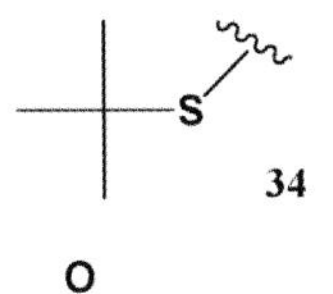 | RSH [78], $NaBH_3$ [79], phosphines [80] | TFA, TFMSA, base, RSCl |
| Acetamidomethyl (Acm) | 31 | Hg(II) acetate [81], Ag(I)/TFA [82], $I_2$ [83], Tl(III)-trifluoracetate [78], $AgBF_4$/TFA [84], AgTMS/DMSO-aq. HCl [85] | TFA, TFMSA, base, RSH |

(*Continued*)

**Table 9.1** *(Continued)*

| Protecting group | Structure | Removed by | Stable to |
|---|---|---|---|
| Trimethylacetamidomethyl (Tacm) | O; N H; 32 | $AgBF_4$/TFA [85], Hg(II) acetate [86], $I_2$ [87], Tl(III)-trifluoracetate | TFA, TFMSA, base, RSH |
| Thiazolidine (Thz) | O; S; OH; NH; 35 | methoxylamine/$H_2O$ [88] | TFA, base, RSH |

fonate group ($S_2O_3$) [62] can be used as well for Cys protection during TNCL. It is introduced with sodium tetrathionate ($Na_2S_4O_6$) in a solvent, which can then be removed by treatment with dithiothreitol (DTT).

## 9.1.5
## His

Under standard SPPS conditions the two imidazole nitrogens of unprotected His react with electrophiles such *N,N′*-dicyclohexylcarbodiimide [89], catalyze acyl-transfer reactions [90], and, above all, promote the racemization of the chiral α-carbon [91].

To prevent both alkylation and racemization the π-nitrogen of the imidazole ring must be protected or rather the protection can be located on the τ-position, reducing the nucleophilicity of the π-nitrogen by inductive effects. Between the two non-equivalent nitrogen atoms (π and τ), of the imidazole ring a rapid proton exchange takes place and makes the two tautomers inseparable (Figure 9.3).

The two positions have almost the same basicity; however, their nucleophilicity is significantly different and upon reaction with electrophiles the N-τ product is usually the major one. Thus regiospecific protection of the π-position first requires an orthogonal protection of the τ-nitrogen and counts for the fact that the synthesis of π-products is often troublesome.

The trityl group (Trt) **36** on the τ-position is the most commonly used protecting group of His for SPPS with the Fmoc strategy [92]. It is stable, commercially available at an affordable price, and its mild acidolytic deprotection is fast and smooth. Despite early cautious claims, Trt protection keeps racemization at a very low rate under normal SPPS conditions and it is regarded to be an exception to the rule that τ-located protection does not completely suppress racemization [93]. Except in those cases where significant steric hindrance is displayed, as for the coupling of His to Pro, in which case a small amount of enantiomerization (5%) occurs [94]. However, racemization becomes a serious issue upon esterification of the His carboxylic group (e.g., when hydroxyl-resins are used and His is the first amino acid of a sequence). In this case enantiomerization can be reduced using the Trt group for the α-nitrogen protection as well [95]. The best solution to this problem is the use of 2-chlorotrityl resin, which can be esterified without racemization [96]. The protection with the super acid-labile methyltrityl (Mtt) **37** and momomethoxytrityl (Mmt) **38**

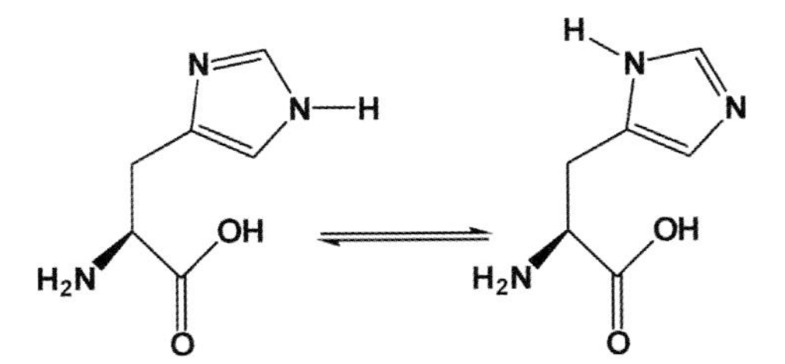


**Figure 9.3** Tautomeric equilibrium of His side-chain imidazole ring.

groups has also been described [97].

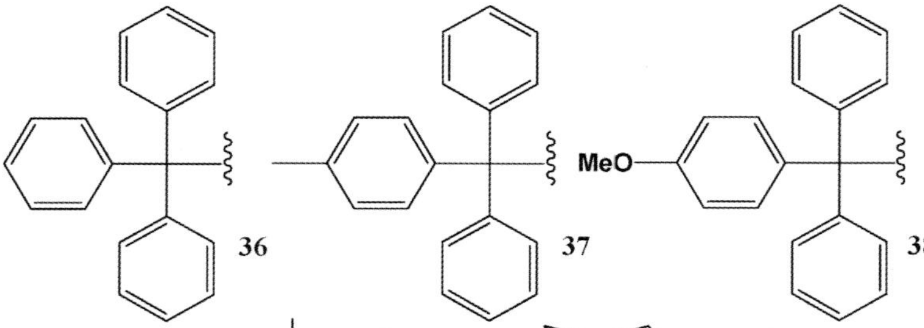

MeO

36 37 38

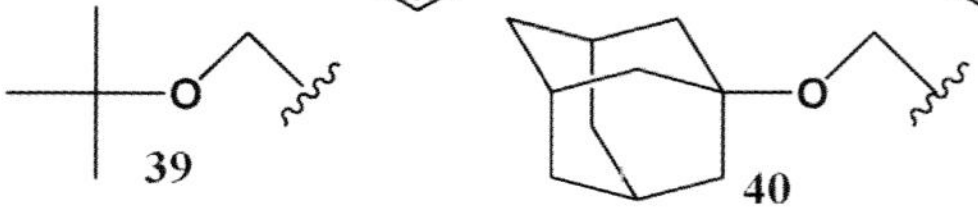

39 40

In the case of His-rich peptides the *tert*-butoxymethyl (Bum) **39** protection of the π-imidazole position is recommended instead of τ-Trt because it minimizes racemization during peptide coupling reactions [98]. A significant drawback, common to all π-protected derivatives, is their difficult synthesis due to the presence of the more reactive τ-position. Bum cleavage with TFA requires slightly longer times compared to *t*Bu, Boc, and Trt. In the case of sequences containing N-terminal Cys, methoxylamine should be added to the cleavage mixture as a scavenger since Bum deprotection generates formaldehyde that can mask Cys as a thiazolidine (Thz) giving an adduct with a 12 mass unit difference [95, 99].

Another protection of the π imidazole position is the 1-adamantyloxymethyl (1-Adom) **40**, whose derivatives are more soluble in organic solvents than Bum ones and give better synthetic yields [100].

### 9.1.6 Lys

Lys side-chain protection is a must in SPPS, otherwise the ε-amino group would react with acylating agents leading to uncontrolled peptide branching.

The standard protection in the Fmoc strategy is the *tert*-butyloxycarbonyl (Boc) **41**, cleaved with concentrated TFA, which represents a perfect combination within Fmoc/*t*Bu SPPS [101].

Lys residues are often postsynthetically modified in modern peptide synthesis of bioactive and modified peptides. Usually the goal is to introduce, on a given sequence, post-translational modifications such as glycations [102] and glycosylations [103] or taking advantage of the ability to form amide bonds between Lys side-chains and molecular devices bearing carboxylic groups [104]. This latter application applies to a broad range of peptide modifications, like linking to chromophores, chelating agents, radioactive molecules, and many more substituents each conferring a specific property to the modified peptide. The amide bond formation is a straightforward one and highly compatible with on-resin peptide synthesis. For all those applications of selective modifications a broad panel of orthogonal Lys side-chain protecting groups has been developed.

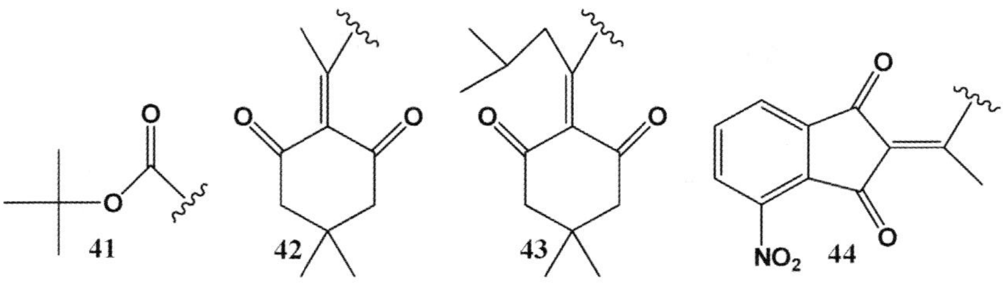

41 42 43 44

Particularly popular among orthogonal Lys side-chain protection is the 1-(4,4′-dimethyl-2,6-dioxocyclohexylidene)ethyl group (Dde) **42** removed with 2% hydrazine in DMF [105]. However, Dde has some limitations, it is partially labile to piperidine (once removed it can migrate to other unprotected Lys), and then hydrazine partially removes the $N^{\alpha}$-Fmoc [106]. In addition, hydrazine can also reduce Alloc protection preventing its subsequent removal [107]. In order to prevent that, allyl alcohols need to be added as scavengers when Dde is deprotected.

Several hindered Dde variants, in particular 1-(4,4′-dimethyl-2,6-dioxocyclohexylidene)-3-methylbutyl (ivDde) [108] **43** and the structurally similar 2-acetyl-4-nitroindane-l,3-dione (Nde) [109] **44**, completely overcome these side-reactions. Dde, ivDde and Nde deprotection can be monitored both spectrophotometrically at various UV wavelengths and by mass spectrometry because of the different hydrazine adducts **42a** (Scheme 9.1). Nde removal can also be followed by a change of color of the resin and solution.

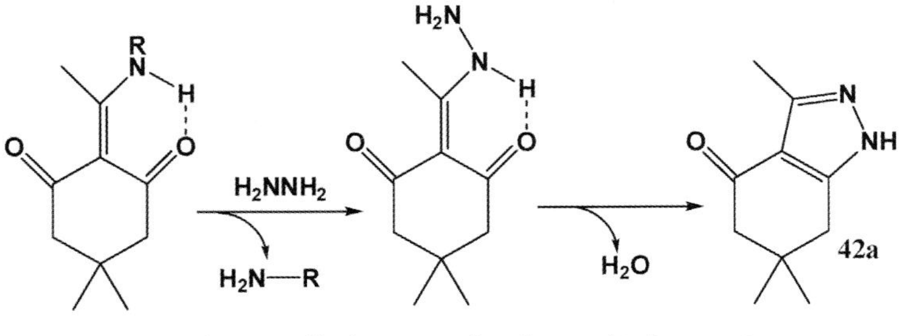

$H_2NNH_2$ $H_2N—R$ $H_2O$ 42a

**Scheme 9.1** Mechanism of hydrazine-mediated Lys(Dde) deprotection.

The acid sensitivity of the trityl group, removed with 20% TFA in DCM, is increased by the introduction of electron-donating substituents leading to super-acid-labile groups as 4-methyltrityl (Mtt) [110] **45** monomethoxytrityl (Mmt) **46**, and dimethoxytrityl (Dmt) [111] **47**.

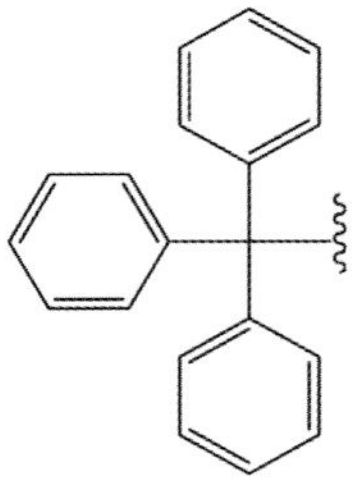

Mtt and Dmt deprotection occurs with 1% TFA in DCM or DCM/hexafluoroisopropanol (HFIP)/trifluoroethanol (TFE)/TES (6.5 : 2 : 1 : 0.5) allowing selective removal in the presence of other acid-labile protecting groups, like *t*Bu and Boc, that are cleaved by concentrated TFA [41]. Mmt deprotections take place at even milder conditions, with AcOH/TFE/DCM (1 : 2 : 7) and even in the presence of hydrophilic resins such as TentaGel and cellulose that display a quenching effect for Mtt and Dmt removal.

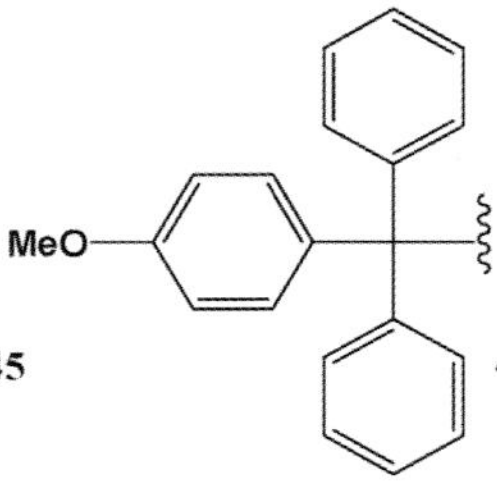

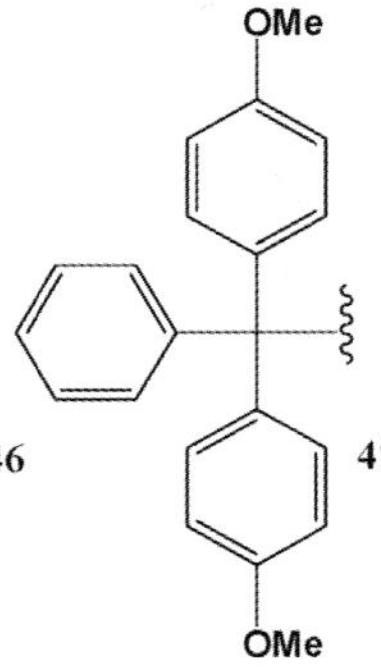

The ε-N function in Lys can bear the allyloxycarbonyl (Alloc) protection **48**, compatible with both Boc/Bzl and Fmoc/*t*Bu SPPS strategies and can be removed by palladium catalyst in the presence of nucleophiles like *N*-methyl morpholine (NMM) [112]. Other groups removed by palladium catalyzed hydrogenolysis, like benzyloxycarbonyl (Z or Cbz) **49** and 2-chlorobenzyloxycarbonyl (2-Cl-Z) [113] **50** or the benzyl group (Bzl or Bn) [114] **51** are also used.

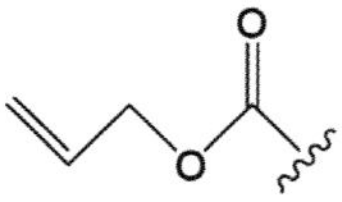

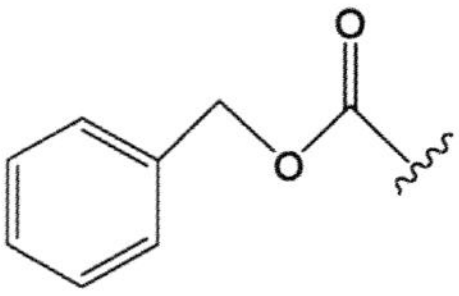

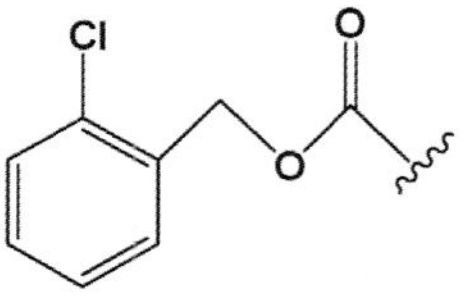

The trifluoroacetyl group (Tfa) **52**, stable to both acid and mild base, is removed by strong alkaline aqueous solutions or sodium borohydride. Although its use is more common in the Boc strategy, it has found some applications in Fmoc chemistry as well [115].

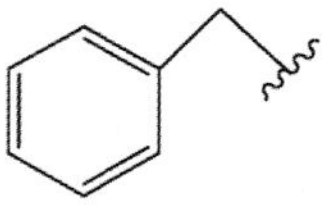

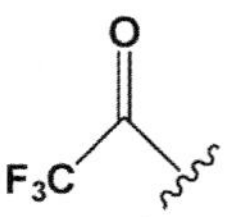

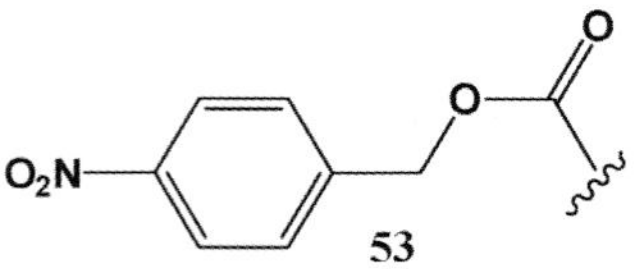

Another semi-permanent side-chain protection of Orn and Lys is *p*-nitrobenzyloxycarbonyl (pNZ) **53**. For Fmoc/*t*Bu chemistry it does not result in partial removal of the $N^{\alpha}$-Fmoc that occurs when groups such as Alloc are used for the same application due to the formation of highly basic free amine [116]. Furthermore, pNZ, removed by 6 M $SnCl_2$ and 1.6 mM HCl/dioxane in DMF, can be used in conjunction with *p*-nitrobenzyl ester (pNB) to prepare cyclic peptides [117].

Finally, the Fmoc protection of both α- and ε-nitrogen represents a specialized derivative for simultaneous multiple peptide synthesis, where side-chain branching is desired after α-amino deprotection [118].

### 9.1.7 Met

The thioether function of Met can undergo acid-catalyzed oxidation forming methionyl sulfoxide and can also be alkylated. In standard Fmoc-based peptide synthesis it is common practice to use unprotected Met because the use of the proper scavengers like thioanisole [119] in the cleavage mixture greatly reduces both side-reactions.

Sulfoxides are also generated by prolonged air exposure but can be reduced back to the sulfide in several ways such as $NH_4I/Me_2S$ in TFA at low temperatures [120] (Scheme 9.2) or $NaI/CH_3COCl$ in DMF [121].

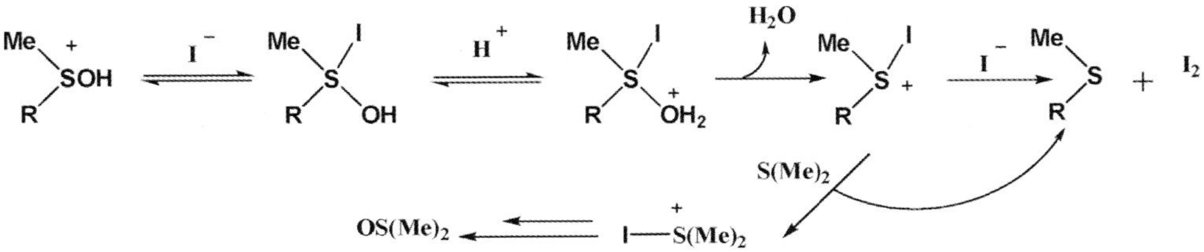

**Scheme 9.2** Acid $NH_4I/Me_2S$-mediated Met(O) reduction mechanism.

Another approach to the oxidation problem, although more common of the Boc strategy, is introducing Met as the sulfoxide derivate Met(O) [122] prepared by treatment with $H_2O_2$ and then reducing it at a convenient time in the synthesis.

### 9.1.8 Ser and Thr

The side-reactions affecting the unprotected hydroxyl group of Ser, Thr and Tyr side-chains, under the standard synthetic conditions used in peptide chemistry, are less severe compared to the amino and carboxylic functions. As a consequence, a number of syntheses describing unprotected amino acid incorporation have been reported. However, unprotected hydroxyl groups can suffer acylation and dehydration, and it is therefore normal practice to protect them.

Ser and Thr have very similar behavior and characteristics, but the minor steric hindrance of the former's hydroxyl makes it more easy to protect but also more reactive toward acylating reagents.

The classical protection of such amino acids is *t*Bu removed under strong acidic conditions [32]. Selective deprotection of hydroxyl side-chains is often performed during postsynthetic modification of Ser and Thr, especially with phosphorylation and glycosylation. This can be achieved by the trityl protection, removed orthogonally from *t*Bu and Boc under mild acidic conditions, 1% TFA and 5% TIS in DCM [123] or 20% dichloroacetic acid in DCM [124]. Another acid-labile group is *tert*-butyldimethylsilyl (TBDMS) **54** which can be removed with AcOH/THF/$H_2O$ (3 : 1 : 1) [125]. The photocleavable 4,5-dimethoxy-2-nitrobenzyl group [126] **55** has been used for Ser protection. Finally, Ser and Thr side-chains can be protected as benzyl ethers, removed with TFA.

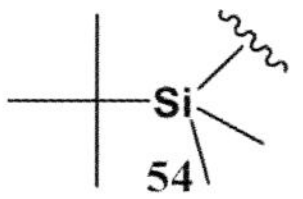

Si

54

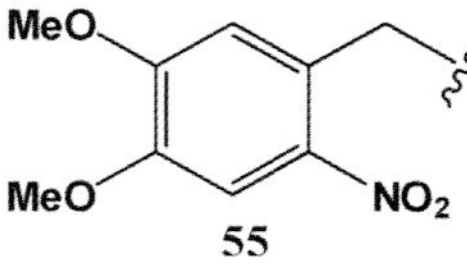

MeO

MeO

$NO_2$

55

## 9.1.9 Trp

The two main side-reactions affecting Trp residues in Fmoc chemistry are oxidation and alkylation of the indole ring by carbonium ions generated during the TFA cleavage [127]. Alkylation of unprotected Trp could be kept under control by using 1,2-ethanedithiol (EDT) as a cleavage mixture scavenger [128]. However, sulfonation by the byproducts of Mtr-, Pmc-, and Pbf-protected Arg cannot be completely eliminated. This problem has been solved by developing Boc-protected Trp [129], which generates, after the cleavage step, a TFA-stable-*N*-carboxy indole intermediate, capable of reducing the susceptibility of the heterocyclic ring to electrophilic attack [130] (Scheme 9.3). The carbamic acid derivative, associated with a 44 mass unit gain, is not stable in solution and decomposes slowly in water during the routine work-up stages, leaving the indole ring free.

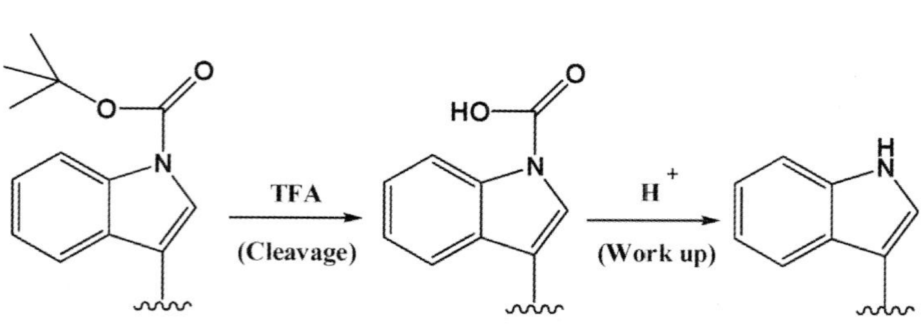

O O N TFA (Cleavage) HO O N $H^+$ (Work up) H N

**Scheme 9.3** Mechanism of Trp(Boc) deprotection.

Trp can also suffer partial reduction by TES used as a scavenger of trityl groups [49] and is affected by the presence of silver salts (used for removal of several Cys-protecting groups). The latter side effect can be avoided by adding an excess of free Trp in the deprotection mixture [131].

Also allyloxycarbonyl (Alloc) protection eliminates the oxidation-alkylation problems and, although unstable to 20% piperidine, it is stable to 1,8-diazabicyclo[5.4.0.] undec-7-ene (DBU) that needs to be used for Fmoc removal [132].

Recently a new TFA stable Trp protection has been proposed, 4-(*N*-methylamino) butanoyl (Nmbu), whose function is to improve the solubility of the peptides in view of high-performance liquid chromatography (HPLC) purification [133]. Nmbu is stable to TFA cleavage and on treatment of the purified peptide at pH 9.5 it undergoes an intramolecular cyclization reaction (Scheme 9.4) that results in the fully deprotected peptide and *N*-methylpyrrolidone.

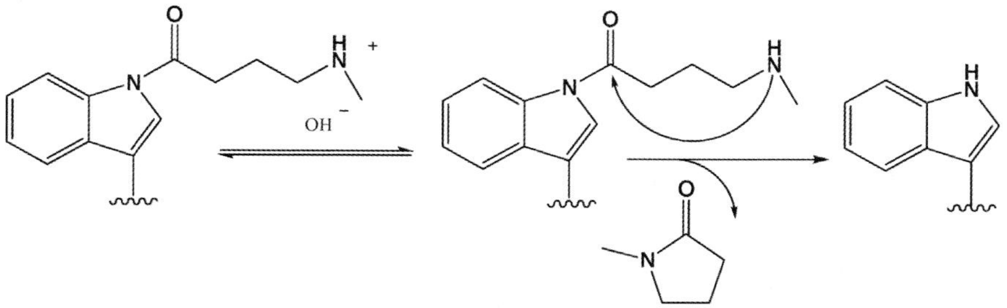

**Scheme 9.4** Mechanism of Trp(Nmbu) deprotection.

## 9.1.10
## Tyr

The Tyr side-chain, if left unprotected during the peptide synthesis, is prone to *O*-acylation because the basic conditions of Fmoc removal generates the phenolate anion which is acylated due to its strong nucleophilicity. Tyr is commonly protected with a *t*Bu group [134], which generates very little of the 3-alkylated product (see side-reaction section).

The TBDMS ether protection **56**, although less acid-labile than the corresponding analogs of Ser and Thr, can nevertheless be removed selectively using tetrabutylammonium fluoride (TBAF) [125].

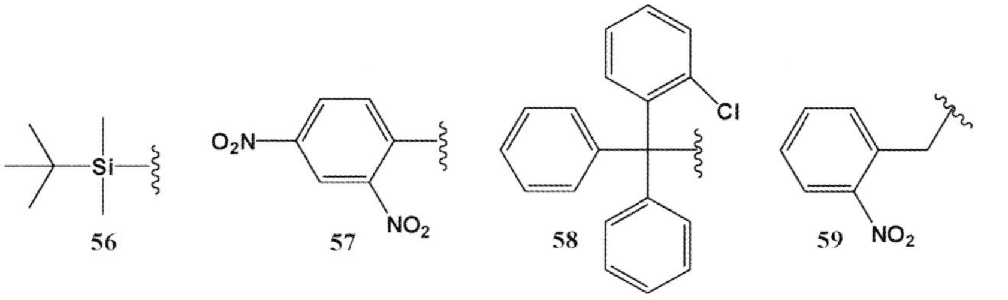

Despite some misunderstandings [135], the 2,4-dinitrophenyl (Dnp) group **57**, deprotected with 2-thiophenol/pyridine/DMF (2 : 1 : 10), is another suitable choice for Tyr selective protection as for example during on-resin postsynthetic modifications (such as phosphorylation or glycosylation). However, since Dnp is readily cleaved by 20% piperidine or 2% DBU in DMF [136], Dnp-protected Tyr should be employed as the N-terminal residue or could be immediately modified after incorporation. Another possibility is using the more flexible 2-Cl-Trt group **58** that can be selectively removed at any time in the synthesis with 1% TFA/5% TIS in DMF [123].

Some photo-cleavable Tyr protecting groups have been reported, particularly useful for the synthesis of those molecular devices whose activity is controlled by light (caged compounds). One example is the 2-nitrobenzyl group (NB) **59**, removed by UV light [137] (Scheme 9.5). Upon irradiation the excited nitro compound abstracts a hydrogen from the benzylic position and the intermediate rapidly rearranges into a nitroso hemiacetal.

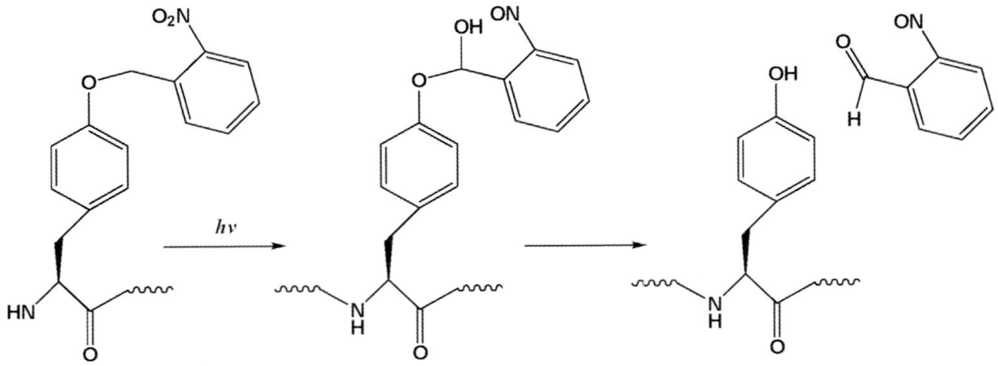

**Scheme 9.5** Mechanism of Tyr(NB) deprotection.

## 9.1.11 Conclusions

In conclusion of this section, we reported those amino acid-protecting groups, already accepted for their possible use in multidimensional orthogonal protecting schemes for SPPS. We focused on the most popular protecting groups, selecting them on the base of their commercial availability, stability in the commonly used reaction conditions, limited side-reactions effects, and wide application in the synthesis of complex peptides. It was beyond the scope of this chapter to mention those protecting groups that, even if promising, have not been yet widely validated.

# 9.2 Side-Reactions in Fmoc/*t*Bu Solid-Phase Peptide Synthesis

In this section we propose a compendium of the best-known side-reactions occurring in SPPS using the Fmoc/*t*Bu strategy (Table 9.2). Most reactions can be avoided with

**Table 9.2** The most frequent side-reactions occurring in SPPS using the Fmoc/*t*Bu strategy.

| Side-reaction | Origin | Peptide byproduct characterization |
|---|---|---|
| Racemization of His (Section 9.2.1) | alkylation of His imidazole ring | same mass, different retention times (diastereomeric peptides) |
| O-to-N acyl transfer (Section 9.2.2) | unprotected hydroxyl groups | higher molecular weight (double or triple coupling) |
| Met oxidation (Section 9.2.3) | air exposure | two peaks with mass of +16 units and different retention times (diastereomeric sulfoxide peptides) |
| Asn and Gln dehydration (Section 9.2.4) | unprotected amide side-chain function | molecular weight of −18 mass units (cyano-containing peptides) |
| Aspartimide formation (Section 9.2.5) | susceptible sequences such as Asp–Gly, Asp–Asn, Asp–Ala, Asp–Gln | molecular weight of −18 units (aspartimide derivate), +67 units (piperidide adducts), and same mass (β-peptide) |
| Formation of diketopiperazines (Section 9.2.6) | stop of peptide elongation because of the resin cleavage after the first two C-terminal amino acid residues | from reduced yield to no desired peptide formation |
| Protected Cys oxidation (Section 9.2.7) | air exposure | molecular weight of +16 (sulfoxide), −2 (undesired intramolecular disulfide bridges), double mass, and −2 units (undesired intermolecular disulfide bridges) |
| Protected Cys β-elimination (Section 9.2.7) | base-catalyzed $C^{\alpha}$ carbon abstraction | molecular weight of +51 (β-(1-piperidyl)Ala adduct) |
| Racemization of protected Cys (Section 9.2.7) | base-catalyzed racemization upon coupling | same molecular weight and different retention times (diastereomeric peptides) |
| Deletion peptide sequences (Section 9.2.8) | incomplete coupling or Fmoc deprotection | lower molecular weight (peptides lacking in one or more amino acid residues) |
| Truncated sequences (Section 9.2.8) | long or difficult sequences | synthesis cannot proceed |
| Multiple addition (Section 9.2.8) | partial undesired Fmoc deprotection and impurities of Fmoc-dipeptides in commercially available protected amino acids | higher molecular weight (peptides with more residues than expected) |

(*Continued*)

**Table 9.2** (*Continued*)

| Side-reaction | Origin | Peptide byproduct characterization |
|---|---|---|
| Guanidino capping (Section 9.2.9) | uronium/guanidinium salt exposure of free amino groups in the absence of incoming amino acids | truncated adduct at the position of capping with a molecular weight of + 98 |
| Arg conversion into Orn (Section 9.2.10) | acylation of urethane-protected Arg guanidino group | molecular weight of −42 |

careful choice of side-chain protecting groups and synthetic protocols. Moreover, it should be remembered that commercially available products used in standard SPPS (i.e., activating agents and protected amino acids) have to be tested for their high purity. In fact, the presence of impurities, even if in small amounts, can strongly limit purity of final peptides.

### 9.2.1 Imidazole Ring-Mediated Racemization of Chiral α-Carbon

Under the standard conditions of SPPS the two imidazole nitrogens of unprotected His can catalyze acyl-transfer reactions [90] and, above all, promote the racemization of the chiral α-carbon [91]. This enantiomerization is a unique case among the natural amino acids and happens regardless of having the α-nitrogen protected as a urethane. Such behavior can be explained by two different pathways as reported in Figure 9.4 [138].

In the first mechanism (*A*), the π nitrogen acts as a nucleophile, attacking the activated carbonyl group intramolecularly, forming cation **60**. Further enolization will lead to loss of optical activity during the coupling with the liberated peptidyl $N^{\alpha}$-nitrogen. In the second mechanism (*B*), the π-nitrogen acts as a base leading directly to the zwitterionic enolate **61**, which also brings about enantiomerization upon coupling. Since in both cases the π-nitrogen is involved, His racemization can be suppressed by blocking the π-position [3].

### 9.2.2 Hydroxyl-Mediated *O* → *N* Acyl Transfer

Hydroxyl-containing amino acids such as Ser, Thr, Tyr, and Hyp (hydroxyproline) undergo a small degree of acylation during standard coupling conditions, and are therefore usually protected.

In fact, acylation can lead to an undesired acyl-transfer reaction, especially in the case of Ser, which is the most reactive on the account of being the least

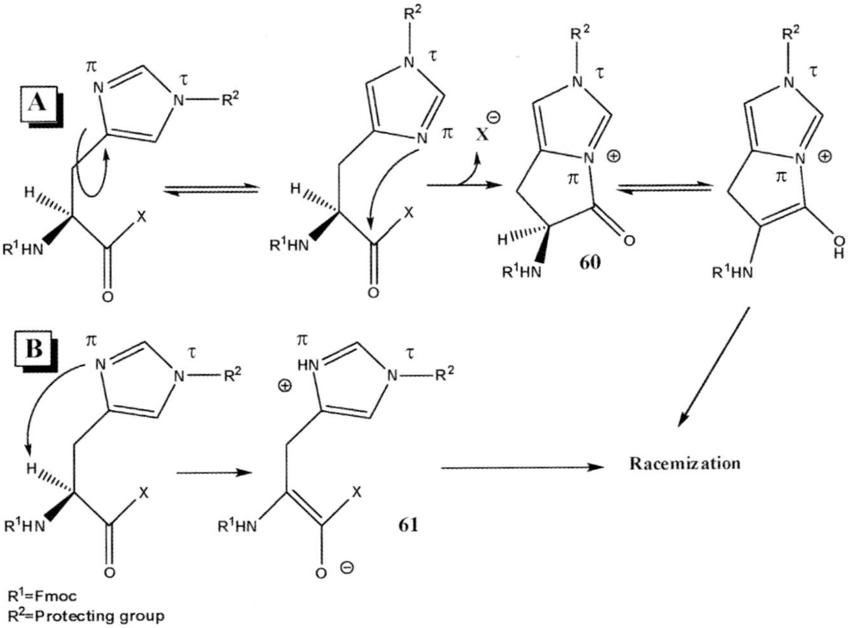

Figure 9.4 π Nitrogen-mediated His racemization.

hindered and bearing a primary, more nucleophilic hydroxyl function [139]. The free hydroxyl group competes with the unprotected α-nitrogen in the reaction with the activated carboxylic group of the incoming amino acid. After Fmoc deprotection the undesired branch elongates and the N-terminal nitrogen reacts intramolecularly with the newly formed ester (Figure 9.5). The result is an $O \rightarrow N$ acyl migration of a mono- or dimeric amino acid unit leading to peptides with higher molecular weight with respect to the desired product.

*O*-Acylation of Tyr the side-chain is a definitely much more severe side-reaction than that occurring with Ser and Thr hydroxyl functions because the basic conditions for Fmoc removal generate the phenolate anion, which is acylated because of its strong nucleophilicity. However, Tyr once acylated does not display the $O \rightarrow N$ acyl transfer, possibly because of the steric hindrance of the aromatic ring.

However, in spite of this minor side-reaction, the so-called "minimum protection strategy" makes use of unprotected hydroxyl groups (with the exception of Tyr), producing more soluble peptides and requiring less harsh deprotection conditions [140]. On the other hand, the "global protection strategy," which applies fully protected amino acids, will avoid many side-reactions generated by unprotected side-chains and it is usually preferred.

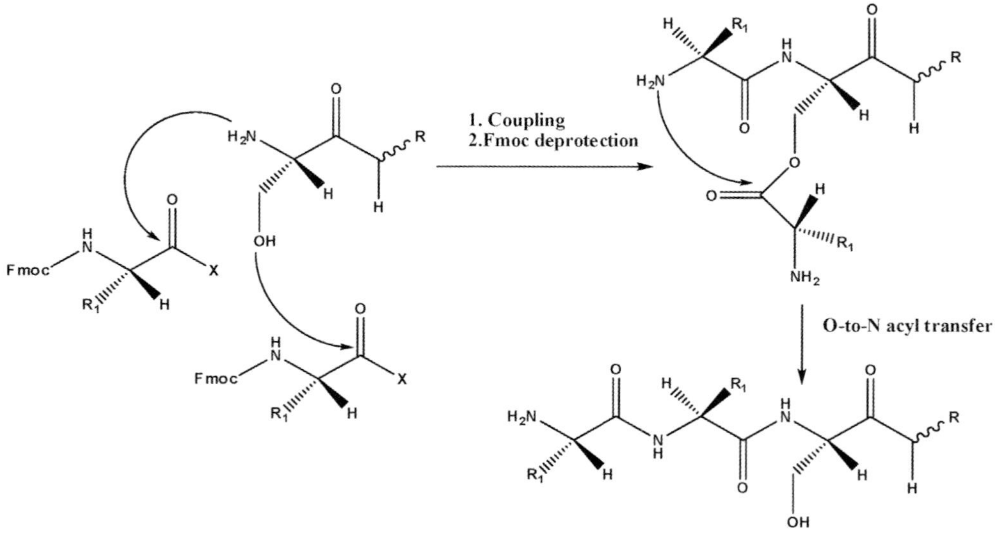

Figure 9.5 Acylation of the unprotected side-chain of Ser upon coupling followed by $O \rightarrow N$ acyl transfer.

### 9.2.3
### Met Oxidation to Methionyl Sulfoxide

In the Fmoc/*t*Bu strategy, Met can undergo partial oxidation to the corresponding sulfoxide, Met(O), if scavengers such as TMSBr and EDT are not used during the acidic TFA cleavage [141]. Considerable amounts of oxidation product are also formed by prolonged air exposure especially during automatic multistep peptide synthesis. Although reversible, this side-reaction causes problems at the purification stage because the sulfoxide is chiral and generates a mixture of different peptides, three in the simplest case of only one Met residue [3], namely the unoxidized form plus the two diastereomeric oxidized forms (Figure 9.6), which usually show a shorter retention time in reverse-phase HPLC.

Several methods have been proposed for reduction of the Met sulfoxide after cleavage, including the use of *N*-methylsulfanylacetamide in 10% aqueous acetic acid [142], thioglycolic acid [143], $NH_4I$/TFA [144], $SO_3$/EDT in 20% piperidine/DMF [145], and $TiCl_4$/NaI in MeOH/acetonitrile/DMF (5 : 5 : 4) [146].

### 9.2.4
### Dehydration of Asn and Gln Amide Side-Chain

The major side-reaction during coupling of Asn (and Gln to a lesser extent) in both Boc and Fmoc peptide synthesis is dehydration of the amide side-chain to the corresponding nitrile, β-cyano-alanine [18]. Dehydration occurs during carboxyl activation of the $N^{\alpha}$-protected Asn (Figure 9.7).

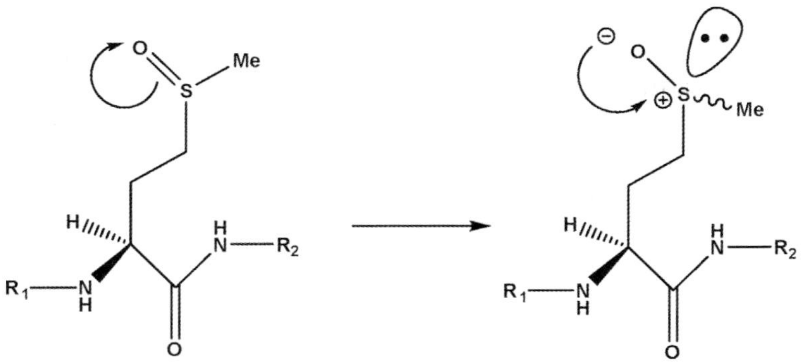

Figure 9.6 Mechanism of Met sulfoxide generating two diastereomeric peptides.

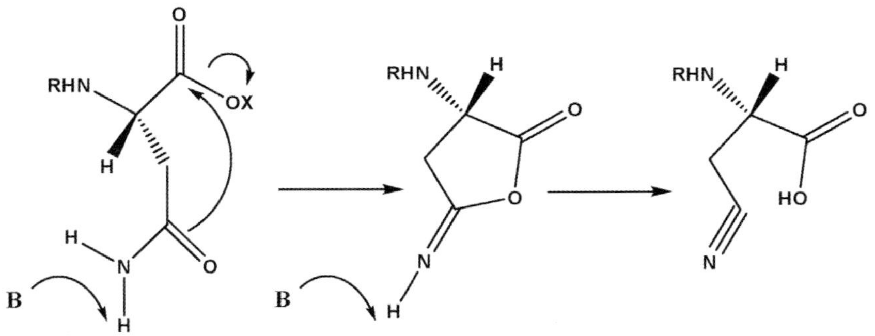

Figure 9.7 Dehydration of the amide function of Asn on activation.

The nitrile-bearing amino acid can then be activated again and become incorporated into the peptide chain, and the resulting cyano-containing peptide (with a molecular weight of −18 mass units) is difficult to separate from the desired one. This side-reaction is usually not noticed in Boc chemistry because the final HF cleavage reverses nitrile formation [147]. However, in Fmoc chemistry, the conditions of final cleavage and deprotection are too mild to rehydrate the nitrile [148]. Amide side-chain protection effectively prevents this dehydration.

The amide side-chain group of N-terminal Gln has a pronounced tendency towards cyclization to form a pyroglutamyl residue [19]. This side-reaction is not normally an issue in Fmoc chemistry since the reaction is catalyzed by weak acids, but not by strong acids such as the ones required for final cleavage and deprotection. However, weak acids are used in the Fmoc/*t*Bu strategy in the case of super acid-labile group deprotection or for peptide cleavage from highly acid-sensitive resins, such as 2-Cl-Trt and SASRIN (super acid-sensitive resin) [110]. Whatever the case, pyroglutamyl formation is efficiently prevented by Gln side-chain protection.

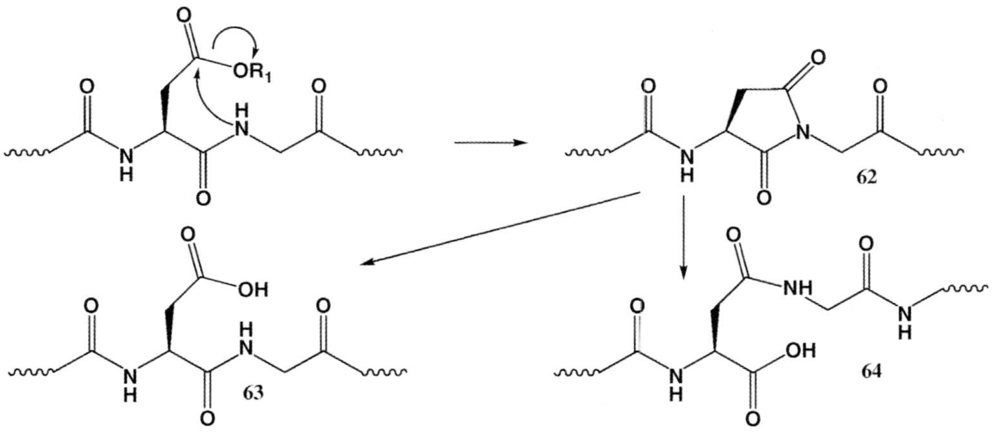

Figure 9.8 Aspartimide formation in Asp–Gly sequences and subsequent hydrolysis.

### 9.2.5
### Aspartimide Formation

The most common side-reaction concerning protected Asp and Glu residues in Fmoc/*t*Bu SPPS is an acid/base-catalyzed cyclization forming aspartimides and glutamides, respectively (Figure 9.8). Subsequent hydrolysis of the imide-containing peptides **62** leads to a mixture of the desired peptide **63** (termed the α-peptide) along with product **64**, termed β-peptide, in which the side-chain carboxylic group forms part of the backbone.

The aspartimide ring can be opened by nucleophilic attack of piperidine on either of the two carbonyl groups to give a β-piperidide **66** and α-piperidide **67** adducts, respectively (Figure 9.9) [29].

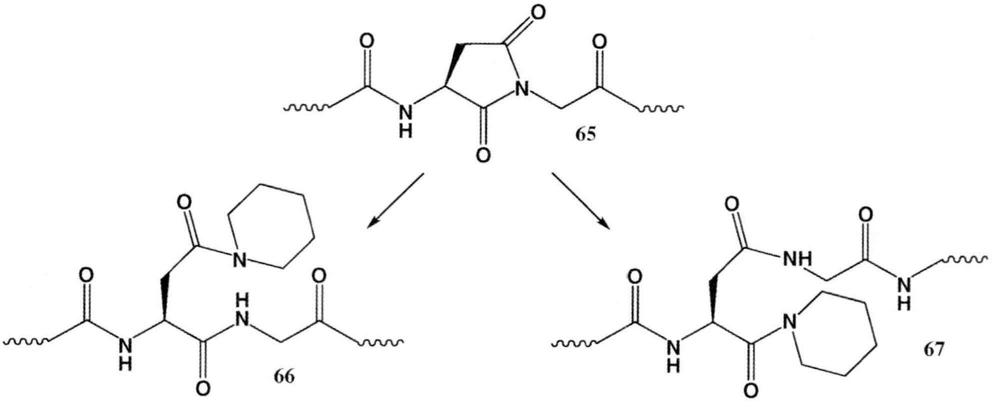

Figure 9.9 Piperidine opening of the aspartimide ring.

In addition, they are readily susceptible to base-catalyzed epimerization, presumably through an α-proton abstraction mechanism [28].

Regarding the differences in mass with respect to the desired peptide the values found are:

- For the corresponding aspartimide derivate −18 mass units.
- For both the piperidide adducts +67 mass units.
- For the β-peptide the same mass value.

The reaction occurs less often with Glu [30] and it is perhaps the most characteristic example of a sequence-dependent undesirable process. Susceptible sequences are, with piperidine adduct formation, Asp–Gly and Asp–Asn, followed by less-susceptible Asp–Ala and Asp–Gln, both without piperidine adduct formation [31]. With the other amino acids the cyclization does not occur because of steric hindrance.

Fmoc removal with DBU also promotes aspartimide formation [149].

All together these side-reactions give rise to a heterogeneous mixture that apart from reducing the yield of the desired peptide may lead to difficulties in separation of the different adducts, which frequently coelute in reverse-phase chromatographies.

Several strategies can be applied to avoid aspartimide formation such as:

- Addition of HOBt or dinitrophenol to the piperidine/DMF deprotection solution [150].
- Aspartyl backbone amide protection with the 2-hydroxy-4-methoxybenzyl (Hmb) [37] or the 2,4-dimethoxybenzyl (Dmp) [151] groups.
- The use of the dipeptide Fmoc-Asp(O*t*Bu)–(Dmp/Hmb)Gly-OH [77]. The Dmp derivatives, being less hindered, display a better reactivity than the corresponding Hmb ones.

## 9.2.6 Formation of Diketopiperazines

The solid-phase coupling of the third amino acid of a given peptide sequence requires special attention since a resin-bound dipeptide with the N-terminal amino group deprotected is susceptible to cleavage from the solid support as a consequence of an intermolecular cyclization to form a diketopiperazines (Figure 9.10) [152].

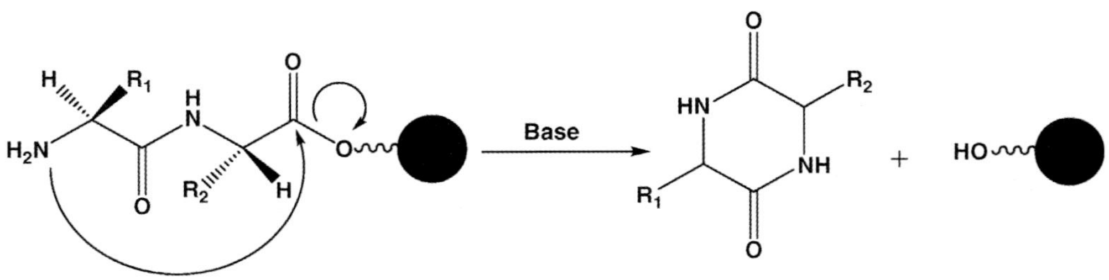

**Figure 9.10** Diketopiperazine formation.

The reaction may be acid- [153] or base-catalyzed [154] and is driven by multiple factors:

- The nature of the anchorage between peptide and resin.
- The type of amino acid in the dipeptide.
- The N-terminal Fmoc deprotection protocols.
- The third amino acid.

In the first place, the cyclization is favored by the presence of good leaving groups at the peptide-resin anchorage such as benzyl resins [155]. For the same reason, C-terminal amide peptides are not generally prone to diketopiperazine formation. The presence of a bulky environment at the place of anchorage in resins such as the trityl ones reduces the cyclization [156]. Amino acids either in the first, second and in both positions, that can adopt easily a *cis*-amide bond configuration, such as the ones bearing an alkyl side-chain or Pro, favor the cyclization [157]. Other factors that contribute to this side-reaction are longer reaction times for the N-terminal Fmoc deprotection [158], slow coupling of the third amino acid (usually bulky amino acids), and the presence of one L- and one D-amino acid in the dipeptide sequence [159].

### 9.2.7 Side-Reactions Affecting Protected Cys

*S*-Protected cysteine residues can undergo partial oxidation by atmospheric oxygen to give the corresponding sulfoxides similarly to the case of Met residues, but to a lesser extent [160]. Under strong acidic conditions, more common in the Boc/Bzl than in the Fmoc/*t*Bu strategy, those sulfoxides react with other Cys, even protected, to form disulfide bonds (Figure 9.11) [161].

It has also been noticed that peptides containing a C-terminal Cys protected with Trt, Acm, and especially S-S*t*Bu [162], can be affected by base-catalyzed β-elimination of the *S*-protecting group with formation of dehydroalanine **68** (Figure 9.12), followed by generation of the corresponding piperidine adduct β-(1-piperidyl)Ala (**69**) (Figure 9.13) [163]. This side-reaction proceeds with epimerization at the $C^{\alpha}$-carbon and generates byproducts with a molecular weight of +51 mass units.

Finally, Cys-protected active ester derivatives have a partial tendency toward racemization upon coupling and anchoring to the resin [50]. This happens via a base-induced $C^{\alpha}$-proton abstraction leading to a carbanion and its stabilization by enolization of the carboxyl function (Figure 9.13).

When a Cys residue is involved, it is advisable to use a lower excess of base (i.e., DIPEA) [164], during the coupling and to reduce the time of piperidine-mediated Fmoc deprotection of the $N^{\alpha}$-amino function [165].

### 9.2.8 Deletion Peptides, Truncated Sequences, and Multiple Additions

At the end of most SPPS a certain number of peptides lacking in one or more amino acids are usually formed alongside the desired product (which should be

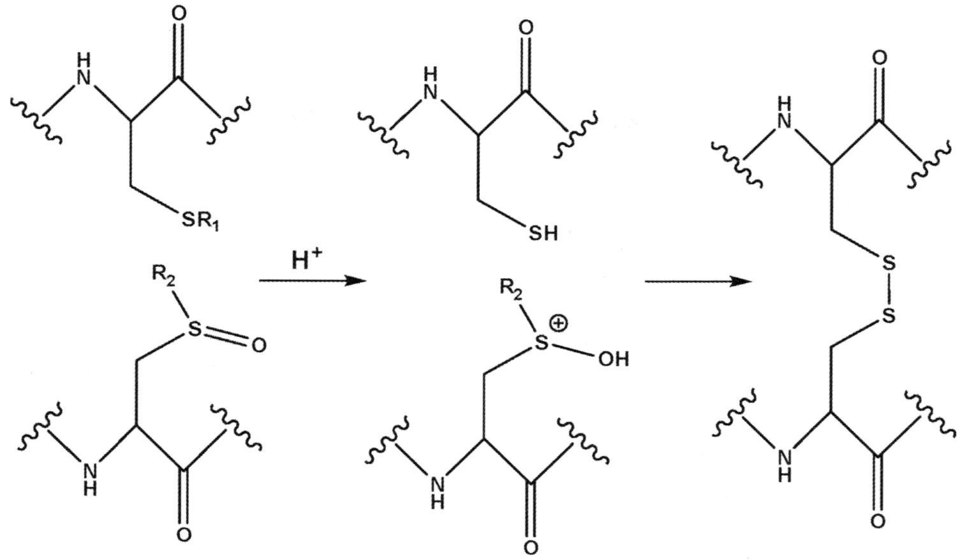

Figure 9.11 Sulfoxide-mediated disulfide formation.

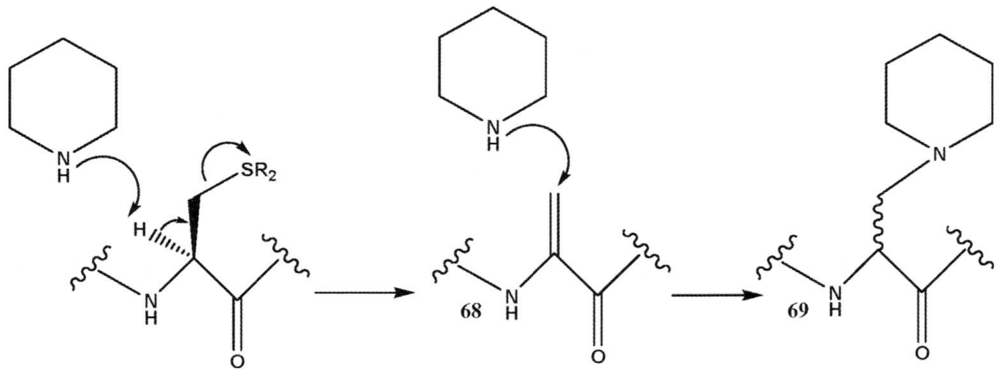

Figure 9.12 Formation of a β-(1-piperidyl)Ala adduct as a consequence of β-elimination of S-protected Cys.

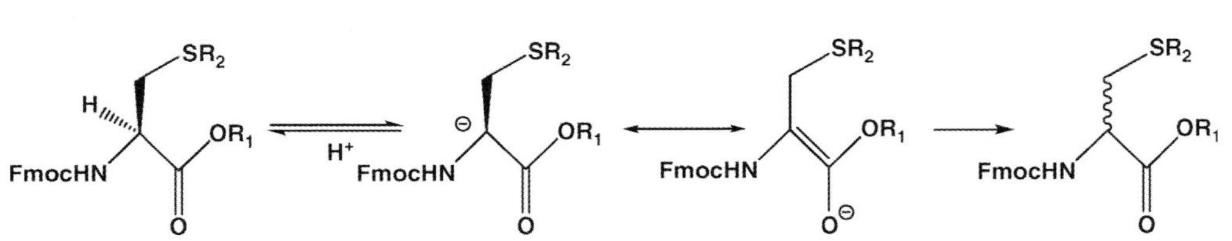

Figure 9.13 Base induced racemization of S-protected Cys residues upon coupling.

the main product in the case of a successful synthesis). Those undesired peptides are called "deletion peptides." In the Fmoc/*t*Bu strategy they form either because some of the $N^{\alpha}$-amino groups remain Fmoc-protected after piperidine treatment or because not all the liberated $N^{\alpha}$-amino groups react with the incoming amino acid.

Truncated sequences are usually generated by steric hindrance around the N-terminal α-amino acid of the peptide as, for example, in the case of long sequences (longer than 40 amino acid residues). Difficult sequences can also generate both deletion and truncated peptides. The coupling rate is slower for bulky amino acids such as Val and Leu, which have a branching at the β-carbon atom. Similar considerations, of slow coupling induced by steric hindrance, apply to unusual amino acids, those carrying special modifications (e.g., post-translational modifications), or chromophore groups [102–104].

Finally, truncated products may arise from side-reactions such as guanidino capping (see Section 9.2.9) or undesired amide bond formation (in the presence of carboxylic groups other than the ones of the incoming amino acid).

A multiple coupling cycle is a special case of an incomplete one and produces peptides with one or more residues than needed. This is generally due to partial Fmoc removal during the coupling step, usually when the bases such DIPEA are used in high excess and with reduced reaction volumes [166]. Multiple addition can also occur if the incoming amino acid building block used in the coupling cycle is partially Fmoc-deprotected either because of impurities or degradation. The formation of dipeptides side-products upon Fmoc protection of amino acids using Fmoc-Cl has also been reported [167]. Regarding the latter case, it should be remembered that Fmoc-amino acids are sensitive to heating and moisture.

### 9.2.9
### Uronium/Guanidinium Salts-Induced Guanidino Capping

Unlike phoshonium salts (i.e., HOBt and HOAt) commonly used in SPPS as carboxylic activating agents, uronium and guanidinium salts, that is, HBTU, TBTU, and 2-(1*H*-7-azabenzotriazol-1-yl)-1,1,3,3,-tetramethylurinium hexafluorophosphate (HATU), can react with liberated N-terminal amino groups for guanidino-derivative adducts [168] (Figure 9.14).

The reaction occurs if the coupling agent is added in solution to the resin before the incoming amino acid or if the coupling agent is used in excess with respect to the amino acid. The guanidinylation of the N-terminal amino acid is a form of peptide capping because it is stable to the synthetic conditions and survives the final cleavage as well. The result is a peptide byproduct truncated at the position of capping with a molecular weight of +98 mass units. This side-reaction is usually avoided under standard Fmoc/*t*Bu SPPS protocols that make use of commercially available protected amino acids. However, it could become an issue in the case of the synthesis of peptides carrying special modifications requiring the use of a minimum amount of specific modified building blocks often troublesome to synthesize.

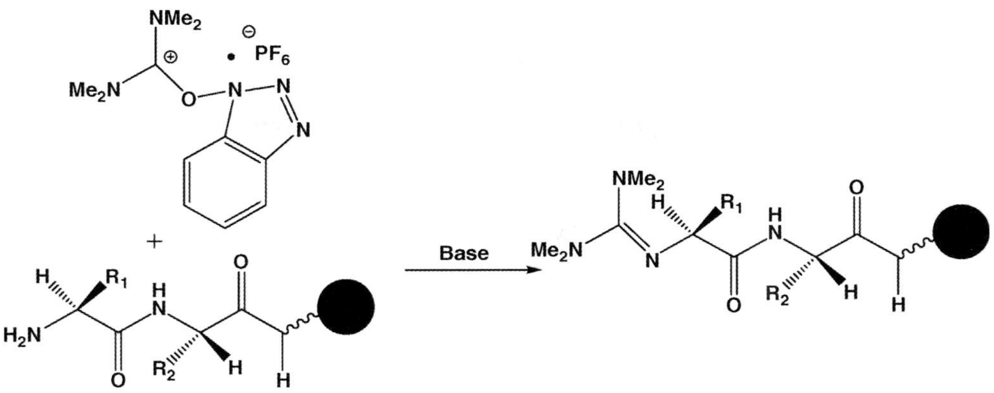

Figure 9.14 HBTU-mediated guanidino capping of the liberated N-terminal amino group.

## 9.2.10 Arg Cyclization and Arg Conversion into Orn

The main side-reaction affecting Arg residues is cyclization during the coupling step under basic conditions when the δ-guanidino nitrogen may lead a nucleophilic attack on the activated carboxylic group of the incoming amino acid [169]. The product formed is a δ-lactam derivative of 2-piperidone (Figure 9.15).

This reaction occurs particularly if the guanidino group is left unprotected and it is not completely eliminated by protection on the ω position whereas the di-ω- or δ-protections efficiently suppress it [14].

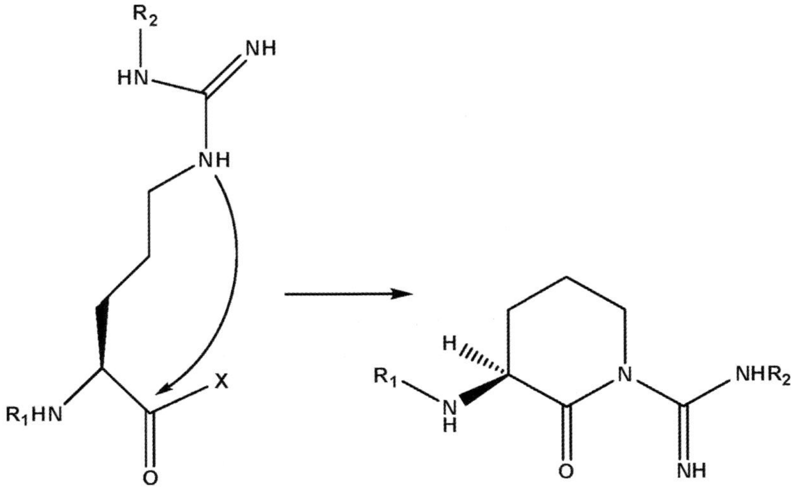

Figure 9.15 δ-Lactam formation from Arg derivates upon coupling.

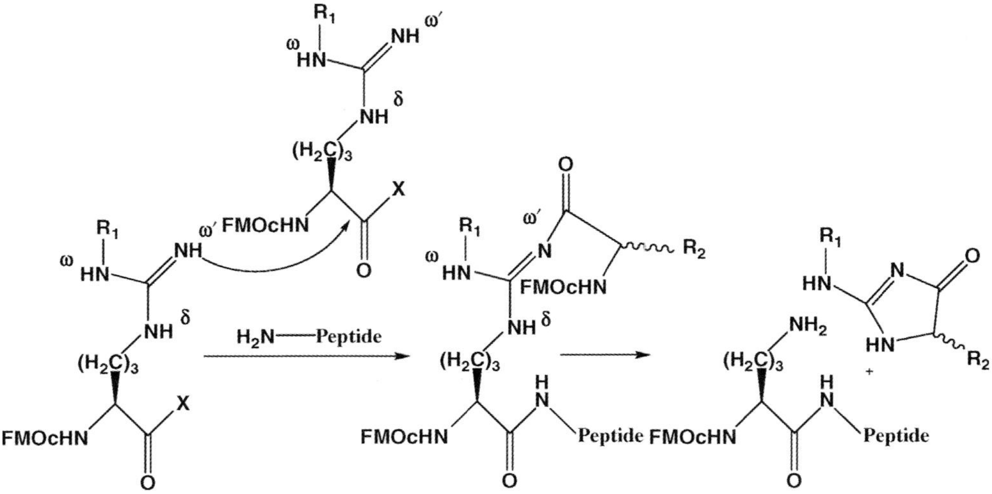

Figure 9.16 Conversion of Arg residues into Orn residues.

Furthermore, when Arg residues are protected with urethane-type groups, such as Boc and Alloc, they suffer partial conversion into the corresponding Orn residues, especially in the Boc/Bzl strategy but also in the Fmoc/*t*Bu technique [1]. This undesired conversion is caused by guanidino acylation during the coupling step followed by intermolecular decomposition under the strong basic conditions required for Fmoc removal. The overall process generates a product with a difference of −42 mass units (Figure 9.16).

## 9.2.11
## Conclusions

Since its early start in the 1970s, the main operations in Fmoc/*t*Bu SPPS (i.e., deprotections, couplings, cleavage) have reached an exceptionally good degree of efficiency. However, a certain number of side-reactions are still recognized. As a consequence of these side-reactions a peptide synthesis can lead to mixtures of peptides having, at best, the desired product as the main element. Only a precise design of the synthetic strategy to limit the possibility of having side-reactions can lead to the purest crude optimized for the final purification step. In fact, the principal weak point of SPPS often lies with the purification step. Isolation of the desired material from a mixture of different peptide byproducts can be troublesome and time consuming. A careful choice of protecting groups, synthetic conditions, *ad hoc* scavenger cocktail for cleavage from the resin, together with a detailed knowledge of the possible side-reactions is paramount for achieving a successful synthesis. Moreover, extensive washing of the resin during all the synthetic steps is a simple operation to avoid mixtures of the final cleaved peptide with those side-products that were incorporated into the resin. Finally, it should be remembered that commercially available reagents for SPPS, such as protected amino acids and activating agents, should not contain impurities, even if in small amounts. Moreover, those materials could be a source of other uncontrolled side-products, being also susceptible to degradation if not carefully stored.

## References


1 Rink, H., Sieber, P., and Raschdorf, F. (1984) *Tetrahedron Letters*, **25**, 621–624.
2 Ramage, R., Green, J., and Blake, A.J. (1991) *Tetrahedron*, **47**, 6353–6370.
3 Sieber, P. (1987) *Tetrahedron Letters*, **28**, 1637–1640.
4 Fujii, N., Otaka, A., Sugiyama, N., Hatano, M., and Yajima, H. (1987) *Chemical and Pharmaceutical Bulletin*, **35**, 3880–3883.
5 Jaeger, E., Remmer, H., Jung, G., Metzger, J., Oberthür, W., Rücknagel, K.P., Schäfer, W., Sonnenbichler, J., and Zetl, I. (1993) *Biological Chemistry Hoppe Seyler*, **5**, 349–362.
6 Green, J., Ogunjobi, O.M., Ramage, R., Stewart, A.S.J., McCurdy, S., and Noble, R. (1988) *Tetrahedron Letters*, **29**, 4341–4344.
7 White, P. (1992) In *Peptides: Chemistry, Structure & Biology, Proceedings of the 12th American Peptide Symposium* (eds J.A. Smith and J.E. Rivier), ESCOM, Leiden, p. 537.
8 Carpino, L.A., Shroff, H., Triolo, S.A., Mansour, E.M.E., Wenschuh, H., and Albericio, F. (1993) *Tetrahedron Letters*, **34**, 7829–7832.
9 Gowda, D.C. (2002) *Tetrahedron Letters*, **43**, 311–313.
10 Wunsch, E. (1974) *Houben-Weyls Methods der Organischen Chemie*, **15**, Thieme, Stuttgart, Parts 1 and 2.
11 Turàn, A., Patthy, A., and Bajusz, S. (1975) *Acta Chimica Academiae Scientiarum Hungaricae*, **85**, 327–332.
12 Gronvald, F.C., Johansen, N.L., and Lundt, F.G. (1981) In *Peptides 1980* (ed. K. Brunfeldt), Scriptor, Copenhagen, p. 111.
13 Presentini, R. and Antoni, G. (1986) *International Journal of Peptide and Protein Research*, **27**, 123–126.
14 Verdini, A.S., Lucietto, P., Fossati, G., and Giordani, C. (1992) *Tetrahedron Letters*, **33**, 6541–6542.
15 Caciagli, V. and Verdini, A.S. (1988) In *Peptide Chemistry 1987* (eds T. Shiba and S. Sakakibara), Protein Research Foundation, Osaka, p. 283.
16 Bernatowicz, M.S., Wu, Y., and Matsueda, G.R. (1992) *The Journal of Organic Chemistry*, **57**, 2497–2502.
17 Bernatowicz, M.S. and Matsueda, G.R. (1994) In *Peptides: Chemistry, Structure & Biology, Proceedings of the 13th American Peptide Symposium* (eds R.S. Hodges and J.A. Smith), ESCOM, Leiden, p. 107.
18 Gausepohl, H., Kraft, M., and Frank, R.W. (1989) *International Journal of Peptide and Protein Research*, **34**, 287–294.
19 Dimarchi, R.D., Tam, J.P., Kent, S.B.H., and Merrifield, R.B. (1982) *The Journal of Peptide Research*, **19**, 88–93.
20 Sieber, P. and Riniker, B. (1991) *Tetrahedron Letters*, **32**, 739–742.
21 Han, Y., Solè, N.A., Tejbrant, J., and Barany, G. (1996) *Peptide Research*, **9**, 166–173.
22 Friede, M., Denery, S., Neimark, J., Kieffer, S., Gausepohl, H., and Briand, J.P. (1992) *Peptide Research*, **5**, 145–147.
23 Carpino, L.A., Chao, H.G., Ghassemi, S., Mansour, E.M.E., Riemer, C., Warass, R., Sadat-Aulace, D., Truran, G.A., Imazumi, H., El-Faham, A., Ionescu, D., Ismail, M., Kowaleski, T.L., Han, C.H., Wenschuh, H., Beyermann, M., Bienert, M., Shroff, H., Albericio, F., Triolo, S.A., Sole, N.A., and Kates, S.A. (1995) *The Journal of Organic Chemistry*, **60**, 7718–7719.
24 Breipohl, G., Knolle, J., and Stuber, W. (1990) *International Journal of Peptide and Protein Research*, **35**, 281–283.
25 Weygand, F., Steglich, W., Bjarnason, J., Ahktar, R., and Chytil, N. (1968) *Chemische Berichte*, **101**, 3623–3641.
26 König, W. and Geiger, R. (1970) *Chemische Berichte*, **103**, 2041–2051.

27 Natarajan, S. and Bodanszky, M. (1976) *The Journal of Organic Chemistry*, **41**, 1269–1272.
28 Tam, J.P., Riemen, M.W., and Merrifield, R.B. (1988) *Peptide Research*, **1**, 6–18.
29 Dölling, R., Beyermann, M., Haenel, J., Kernchen, F., Krause, E., Franke, P., Brudel, M., and Bienert, M. (1994) *Journal of the Chemical Society, Chemical Communications*, 853–854.
30 Kates, S.A. and Albericio, F. (1994) *Letters in Peptide Science*, **1**, 213–220.
31 Yang, Y., Sweeney, W.V., Schneider, K., Thörnqvist, S., Chait, B.T., and Tam, J.P. (1994) *Tetrahedron Letters*, **35**, 9689–9692.
32 Chang, C.-D., Waki, M., Ahmad, M., Meienhofer, J., Lundell, E.O., and Huang, J.D. (1980) *International Journal of Peptide and Protein Research*, **15**, 59–66.
33 Okada, Y., Iguchi, S., and Kawasaki, K. (1987) *Journal of the Chemical Society, Chemical Communications*, 1532–1534.
34 Karlström, A. and Unden, A. (1996) *Tetrahedron Letters*, **37**, 4243–4246.
35 Karlström, A. and Unden, A. (1995) *Tetrahedron Letters*, **36**, 3909–3912.
36 Martinez, J. and Bodanszky, M. (1978) *International Journal of Peptide and Protein Research*, **12**, 277–283.
37 Quibell, M., Owen, D., Packmann, L.C., and Johnson, T. (1994) *Journal of the Chemical Society, Chemical Communications*, 2343–2344.
38 Zahariev, S., Guarnaccia, C., Pongor, C.I., Quaroni, L., Cemazar, M., and Pongora, S. (2006) *Tetrahedron Letters*, **47**, 4121–4124.
39 Benoiton, L. (1962) *Canadian Journal of Chemistry*, **40**, 570–572.
40 Okada, Y. and Iguchi, S. (1988) *Journal of the Chemical Society, Perkin Transactions I*, 2129–2136.
41 Barlos, K., Gatos, D., Chatzi, O., Koutsogianni, S., and Schaefer, W. (1993) In *Peptides 1992* (eds C.H. Schneider and A.N. Eberle), ESCOM, Leiden, p. 283.
42 Kunz, H., Waldmann, H., and Unverzagt, C. (1985) *International Journal of Peptide and Protein Research*, **26**, 493–497.
43 Isidro-Llobet, A., Alvarez, M., and Albericio, F. (2008) *Tetrahedron Letters*, **49**, 3304–3307.
44 Chan, W.C., Bycroft, B.W., Evans, D.J., and White, P.D. (1995) *Journal of the Chemical Society, Chemical Communications*, 2209–2210.
45 Ruczynski, J., Lewandowska, B., Mucha, P., and Rekowski, P. (2008) *Journal of Peptide Science*, **14**, 335–341.
46 Briand, B., Kotzur, N., Hagen, V., and Beyermann, M. (2008) *Tetrahedron Letters*, **49**, 85–87.
47 Kocsis, L., Bruckdorfer, T., and Orosz, G. (2008) *Tetrahedron Letters*, **49**, 7015–7017.
48 Fujii, N., Otaka, A., Funakoshi, S., Bessho, K., and Yajima, H. (1987) *Journal of the Chemical Society, Chemical Communications*, 163–164.
49 Pearson, D.A., Blanchette, M., Baker, M.L., and Guindon, C.A. (1989) *Tetrahedron Letters*, **30**, 2739–2742.
50 Kaiser, T., Nicholson, G.J., Kohlbau, H.J., and Voelter, W. (1996) *Tetrahedron Letters*, **37**, 1187–1190.
51 Angell, Y.M., Alsina, J., Albericio, F., and Barany, G. (2002) *The Journal of Peptide Research*, **60**, 292–299.
52 Fujiwara, Y., Akaji, K., and Kiso, Y. (1994) *Chemical and Pharmaceutical Bulletin*, **42**, 724–726.
53 Munson, M.C., Garcia-Echeverria, C., Albericio, F., and Barany, G. (1992) *The Journal of Organic Chemistry*, **57**, 3013–3018.
54 Han, Y. and Barany, G. (1997) *The Journal of Organic Chemistry*, **62**, 3841–3848.
55 Albericio, F. (2000) In *Fmoc Solid Phase Peptide Synthesis* (eds W.C. Chan and P.D. White), Oxford University Press, Oxford, p. 77.
56 Greiner, G. and Hermann, P. (1991) In *Peptides 1990* (eds E. Giralt and D. Andreu), ESCOM, Leiden, p. 277.
57 Weber, U. and Hartter, P. (1970) *Hoppe-Seyler's Zeitschrift fur Physiologische Chemie*, **351**, 1384–1388.
58 Beekman, N.J.C.M., Schaaper, W.M.M., Tesser, G.I., Dalsgaard, K., Kamstrup, S., Langeveld, J.P.M., Boshuizen, R.S., and Meloen, R.H. (1997) *The Journal of Peptide Research*, **50**, 357–364.
59 Berangere, D. and Trifilieff, E. (2000) *Journal of Peptide Science*, **6**, 372–377.

60 Loffet, A. and Zhang, H.X. (1993) *International Journal of Peptide and Protein Research*, **42**, 346–351.
61 Bang, D. and Kent, S.B.H. (2004) *Angewandte Chemie International Edition*, **43**, 2534–2538.
62 Sato, T. and Aimoto, S. (2003) *Tetrahedron Letters*, **44**, 8085–8087.
63 Zervas, L. and Photaki, I. (1962) *Journal of the American Chemical Society*, **84**, 3887–3897.
64 Photaki, I., Taylor-Papadimitriou, J., Sakarellos, C., Mazarakis, P., and Zervas, L. (1970) *Journal of the Chemical Society, Perkin Transactions 1*, 2683–2687.
65 Kamber, B. and Rittel, W. (1968) *Helvetica Chimica Acta*, **51**, 2061–2064.
66 Otaka, A., Koide, T., Shide, A., and Nobutaka, F. (1991) *Tetrahedron Letters*, **32**, 1223–1226.
67 Munson, M.C. and Barany, G. (1993) *Journal of the American Chemical Society*, **115**, 10203–10210.
68 Barlos, K., Gatos, D., Kallitsis, J., Papaphotiu, G., Sotitiu, P., Wenqing, Y., and Schäfer, W. (1989) *Tetrahedron Letters*, **30**, 3943–3946.
69 Barlos, K., Gatos, D., Kutsogianni, S., Papaphotiu, G., Poulos, C., and Tsegenidis, T. (1991) *International Journal of Peptide and Protein Research*, **38**, 562–568.
70 Hargittai, B. and Barany, G. (1999) *The Journal of Peptide Research*, **54**, 468–479.
71 McCurdy, S. (1989) *Peptide Research*, **2**, 147–152.
72 Wang, H., Miao, Z., Lai, L., and Xu, X. (2000) *Synthetic Communications*, **30**, 727–735.
73 Akaji, K., Yoshida, M., Tatsumi, T., Kimura, T., Fujiwara, Y., and Kiso, Y. (1990) *Journal of the Chemical Society, Chemical Communications*, 288–290.
74 Atherton, E., Pinori, M., and Sheppard, R.C. (1985) *Journal of the Chemical Society, Perkin Transactions I*, 2057–2064.
75 Akaji, K., Tatsumi, T., Yoshida, M., Kimura, T., Fujiwara, Y., and Kiso, Y. (1991) *Journal of the Chemical Society, Chemical Communications*, 167–168.
76 Fujii, N., Otaka, A., Funakoshi, S., Bessho, K., Watanabe, T., Akaji, K., and Yajima, H. (1987) *Chemical and Pharmaceutical Bulletin*, **35**, 2339–2347.
77 Yajima, H., Fujii, N., Funakoshi, S., Watanabe, T., Murayama, E., and Otaka, A. (1988) *Tetrahedron*, **44**, 805–819.
78 Threadgill, M.D. and Gledhill, A.P. (1989) *The Journal of Organic Chemistry*, **54**, 2940–2949.
79 Wunsch, E. (1974) In *Houben-Weyls Methods der Organischen Chemie*, **15**, Part 1 (ed. E. Muller), Thieme, Stuttgart, p. 789.
80 Moroder, J., Gemeiner, M., Gohring, W., Jaeger, E., and Wunsch, E. (1981) In *Peptides 1980* (ed. K. Brunfeldt), Scriptor, Copenhagen, p. 121.
81 Sakakibara, S. (1995) *Biopolymers*, **37**, 17–28.
82 Fujii, N., Otaka, A., Watanabe, T., Okamachi, A., Tamamura, H., Yajima, H., Inagaki, Y., Nomizu, M., and Asano, K. (1989) *Journal of the Chemical Society, Chemical Communications*, 283–284.
83 Kamber, B. (1971) *Helvetica Chimica Acta*, **54**, 927–930.
84 Yoshida, M., Akaji, K., Tatsumi, T., Iinuma, S., Fujiwara, Y., Kimura, T., and Kiso, Y. (1990) *Chemical and Pharmaceutical Bulletin*, **38**, 273–275.
85 Tamamura, H., Otaka, A., Nakamura, J., Okubo, K., Koide, T., Ikeda, K., Ibuka, T., and Fujii, N. (1995) *International Journal of Peptide and Protein Research*, **45**, 312–319.
86 Xu, Y. and Wilcox, D.E. (1998) *Journal of the American Chemical Society*, **120**, 7375–7276.
87 Kiso, Y., Yoshida, M., Kimura, T., Fujiwara, Y., and Shimokura, M. (1989) *Tetrahedron Letters*, **30**, 1979–1982.
88 Wu, B., Warren, J.D., Chen, J., Chen, G., Hua., Z., and Danishefskya, S.J. (2006) *Tetrahedron Letters*, **47**, 5219–5223.
89 Rink, H. and Riniker, B. (1974) *Helvetica Chimica Acta*, **57**, 831–835.
90 Bodanszky, M., Fink, M.L., Klausner, Y.S., Natarajan, S., Tatemoto, K., Yiotakis, A.E., and Bodanszky, A. (1977) *The Journal of Organic Chemistry*, **42**, 149–152.
91 Jones, J.H. and Ramage, W.I. (1978) *Journal of the Chemical Society, Chemical Communications*, 472–473.

92 Barlos, K., Papaioannou, D., and Theodoropoulos, D. (1982) *The Journal of Organic Chemistry*, **47**, 1324–1326.

93 Harding, S., Heslop, I., Jones, J., and Wood, M. (1992) In *Peptides 1994* (ed. H. Maia), ESCOM, Leiden, p. 641.

94 Mergler, M., Dick, F., Sax, B., Shwindling, J., and Vorherr, Th. (2001) *Journal of Peptide Science*, **7**, 502–510.

95 Sieber, P. and Riniker, B. (1987) *Tetrahedron Letters*, **28**, 6031–6034.

96 Barlos, K., Chatzi, O., Gatos, D., and Stravropoulos, G. (1991) *International Journal of Peptide and Protein Research*, **37**, 513–520.

97 Barlos, K., Chatzi, O., Gatos, D., Stravropoulos, G., and Tsegenidis, T. (1991) *Tedrahedron Letters*, **32**, 475–478.

98 Colombo, R., Colombo, F., and Jones, J.H. (1984) *Journal of the Chemical Society, Chemical Communications*, 292–293.

99 Gesquiere, J., Najib, J., Diesis, E., Barbry, D., and Tartar, A. (1992) In *Peptides: Chemistry, Structure & Biology, Proceedings of the 12th American Peptide Symposium* (eds J.A. Smith and J.E. River), ESCOM, Leiden, p. 641.

100 Okada, Y., Wang, J., Yamamoto, T., and Mu, Y. (1996) *Chemical and Pharmaceutical Bulletin*, **44**, 871–873.

101 Schwyzer, R. and Rittel, W. (1961) *Helvetica Chimica Acta*, **44**, 159–169.

102 Carganico, S., Rovero, P., Halperin, J.A., Papini, A.M., and Chorev, M. (2009) *Journal of Peptide Science*, **15**, 67–71.

103 Paolini, I., Nuti, F., Pozo-Carrero, M., de la, C., Barbetti, F., Kolesinska, B., Kaminski, Z.J., Chelli, M., and Papini, A.M. (2007) *Tetrahedron Letters*, **48**, 2901–2904.

104 Grandjean, C., Rommens, C., Gras-Masse, H., and Melnyk, O. (1999) *Tetrahedron Letters*, **40**, 7235–7238.

105 Rohwedder, B., Mutti, Y., Dumy, P., and Mutter, M. (1998) *Tetrahedron Letters*, **39**, 1175–1178.

106 Augustyns, K., Kraas, W., and Jung, G. (1998) *The Journal of Peptide Research*, **51**, 127–133.

107 Eichler, J., Lucka, A.W., and Houghten, R.A. (1994) *Peptide Research*, **7**, 300–307.

108 Chhabra, S.R., Hothi, B., Evans, D.J., White, P.D., Bycroft, B.W., and Chan, W.C. (1998) *Tetrahedron Letters*, **39**, 1603–1606.

109 Kellam, B., Bycroft, B.W., Chan, W.C., and Chhabra, S.R. (1998) *Tetrahedron*, **54**, 6817–6832.

110 Aletras, A., Barlos, K., Gatos, D., Koutsogianni, S., and Mamos, P. (1995) *International Journal of Peptide and Protein Research*, **45**, 488–496.

111 Matysiak, S., Böldicke, T., Tegge, W., and Frank, R. (1998) *Tetrahedron Letters*, **39**, 1733–1734.

112 Lyttle, M. and Hudson, D. (1992) In *Peptides: Chemistry, Structure & Biology, Proceedings of the 12th American Peptide Symposium* (eds J.A. Smith and J.E. River), ESCOM, Leiden, p. 583.

113 Erickson, B.W. and Merrifield, R.B. (1973) *Journal of the American Chemical Society*, **95**, 3757–3763.

114 Huang, Z.-P., Su, X.-Y., Du, J.-T., Zhao, Y.-F., and Li, Y.-M. (2006) *Tetrahedron Letters*, **47**, 5997–5999.

115 Stetsenko, D.A. and Gait, M.J. (2001) *Bioconjugate Chemistry*, **12**, 576–586.

116 Farrera-Sinfreu, J., Royo, M., and Albericio, F. (2002) *Tetrahedron Letters*, **43**, 7813–7815.

117 Isidro-Llobet, A., Alvarez, M., and Albericio, F. (2005) *Tetrahedron Letters*, **46**, 7733–7736.

118 Tam, J.P. (1988) *The Proceedings of the National Academy of Sciences of the United States of America*, **85**, 5409–5413.

119 Yajima, H., Kanaki, J., Kitajima, M., and Funakoshi, S. (1980) *Chemical & Pharmaceutical Bulletin*, **28**, 1214–1218.

120 Vilaseca, M., Nicolas, E., Capdevila, F., and Giralt, E. (1998) *Tetrahedron*, **54**, 15273–15286.

121 Norris, K., Halstrom, J., and Brunfeldt, K. (1971) *Acta Chemica Scandinavica*, **25**, 945–954.

122 Iselin, B. (1961) *Helvetica Chimica Acta*, **44**, 61–78.

123 Barlos, K., Gatos, D., and Koutsogianni, S. (1998) *The Journal of Peptide Research*, **51**, 194–200.

124 Coba, M.P., Turyn, D., and Pena, C. (2003) *The Journal of Peptide Research*, **61**, 17–23.

125 Fisher, P. (1992) *Tetrahedron Letters*, **33**, 7605–7608.

126 Pirrung, M.C. and Nunn, D.S. (1992) *Bioorganic and Medicinal Chemistry Letters*, **2**, 1489–1492.

127 Fields, C.G. and Fields, G.B. (1993) *Tetrahedron Letters*, **34**, 6661–6664.

128 Fields, G.B. and Noble, R.L. (1990) *International Journal of Peptide and Protein Research*, **35**, 161–214.

129 White, P. (1992) In *Peptides: Chemistry, Structure & Biology, Proceedings of the 12th American Peptide Symposium* (eds J.A. Smith and J.E. Rivier), ESCOM, Leiden, p. 537.

130 Franzen, H., Grehn, L., and Ragnarsson, U. (1984) *Journal of the Chemical Society, Chemical Communications*, 1699–1700.

131 Najib, J., Letailleur, T., Gesquiere, J.-C., and Tartar, A. (1996) *Journal of Peptide Science*, **2**, 309–317.

132 Vorherr, T., Trzeciak, A., and Bannwarth, W. (1996) *International Journal of Peptide and Protein Research*, **48**, 553–558.

133 Wahlström, K. and Undén, A. (2009) *Tetrahedron Letters*, **50**, 2976–2978.

134 Adamson, J.G., Blaskovich, M.A., Groenevelt, H., and Lajoie, G.A. (1991) *The Journal of Organic Chemistry*, **56**, 3447–3449.

135 Doherty-Kirby, A. and Lajoie, G. (2000) In *Solid-Phase Synthesis: A Practical Guide* (eds S.A. Kates and F. Albericio), Dekker, New York, p. 148.

136 Philosof-Oppenheimer, R., Pecht, I., and Fridkin, M. (1995) *International Journal of Peptide and Protein Research*, **45**, 116–121.

137 Tatsu, Y., Shigeri, Y., Sogabe, S., Yumoto, N., and Yoshikawa, S. (1996) *Biochemical and Biophysical Research Communications*, **227**, 688–693.

138 Lloyd-Williams, P., Albericio, F., and Giralt, E. (1997) *Chemical Approaches to the Synthesis of Peptides Protein*, CRC Press, Boca Raton, FL, p. 29.

139 Doherty-Kirby, A. and Lajoie, G. (2000) In *Solid-Phase Synthesis: A Practical Guide* (eds S.A. Kates and F. Albericio), Marcel Dekker, New York, p. 137.

140 Reissmann, S., Steinmetzer, T., Greiner, G., Seyfarth, L., Besser, D., and Schumann, C. (2004) In *Houben-Weyls Methods der Organischen Chemie*, **E22**, Thieme, Stuttgart, p. 347.

141 Beck, W. and Jung, G. (1994) *Letters in Peptide Science*, **1**, 31–37.

142 Houghten, R.A. and Li, C.H. (1979) *Analytical Biochemistry*, **98**, 36–46.

143 Hofmann, K.H., Finn, F.M., Limetti, M., Montibeller, J., and Zanetti, G. (1966) *Journal of the American Chemical Society*, **88**, 3633–3639.

144 Nicolas, E., Vilaseca, M., and Giralt, E. (1995) *Tetrahedron*, **51**, 5701–5710.

145 Futaki, S., Taike, T., Yagami, T., Akita, T., and Kitagawa, K. (1989) *Tetrahedron Letters*, **30**, 4411–4412.

146 Pennington, M.W. and Byrnes, M. (1995) *Peptide Research*, **8**, 39–43.

147 Mojsov, S., Mitchell, A.R., and Merrifield, R.B. (1980) *The Journal of Organic Chemistry*, **45**, 555–560.

148 Carpino, L.A. and Han, G.Y. (1972) *The Journal of Organic Chemistry*, **37**, 3404–3409.

149 Kitas, E., Knorr, R., Trzeciak, A., and Bannwarth, W. (1991) *Helvetica Chimica Acta*, **74**, 1314–1328.

150 Merrifield, R.B., Vizoli, L.D., and Boman, H.G. (1982) *Biochemistry*, **21**, 5020–5031.

151 El Haddadi, M., Cavelier, F., Vives, E., Azmani, A., Verducci, J., and Martinez, J. (2000) *Journal of Peptide Science*, **6**, 560–570.

152 Gisin, B.F. and Merrifield, R.B. (1972) *Journal of the American Chemical Society*, **94**, 3102–3106.

153 Giralt, E., Eritja, R., and Pedroso, E. (1981) *Tetrahedron Letters*, **22**, 3779–3782.

154 Barany, G. and Albericio, F. (1985) *Journal of the American Chemical Society*, **107**, 4936–4942.

155 Lloyd-Williams, P., Albericio, F., and Giralt, E. (1997) *Chemical Approaches to the Synthesis of Peptides Protein*, CRC Press, Boca Raton, FL, p. 59.

156 Akaji, K., Kiso, Y., and Carpino, L.A. (1990) *Journal of the Chemical Society, Chemical Communications*, 584–586.

157 Khosla, M.C., Smeby, R.R., and Bumpus, F.M. (1972) *Journal of the American Chemical Society*, **94**, 4721–4724.

158 Pedroso, E., Gras, A., de Las Heras, X., Eritja, R., and Giralt, E., (1986) *Tetrahedron Letters*, **27**, 743–746.

**159** Gairì, M., Lloyd-Williams, P., Albericio, F., and Giralt, E. (1990) *Tetrahedron Letters*, **31**, 7363–7366.

**160** Yajima, H., Akaji, K., Funakoshi, S., Fuji, N., and Irie, H. (1980) *Chemical Pharmaceutical Bulletin*, **28**, 1942–1945.

**161** Fujii, N., Otaka, A., Watanabe, T., Arai, H., Funakoshi, S., Bessho, K., and Yajima, H. (1987) *Journal of the Chemical Society, Chemical Communications*, **21**, 1676.

**162** Eritja, R., Ziehler-Martin, J.P., Walker, P.A., Lee, T.D., Legesse, K., Albericio, F., and Kaplan, B.E. (1987) *Tetrahedron*, **43**, 2675–2680.

**163** Lukszo, J., Patterson, D., Albericio, F., and Kates, S. (1996) *Letters in Peptide Science*, **3**, 157–166.

**164** Behrendt, R., Sehenk, M., Musiol, H.-J., and Moroder, L. (1999) *Journal of Peptide Science*, **5**, 519–529.

**165** Atherton, E., Benoiton, N.L., Brown, E., Sheppard, R.C., and Williams, B.J. (1981) *Journal of the Chemical Society, Chemical Communications*, 336–337.

**166** Atherton, E. and Wellings, D.A. (2004) In *Houben-Weyls Methods der Organischen Chemie*, **E22**, Thieme, Stuttgart, p. 334.

**167** Isidro-Llobet, A., Just-Baringo, X., Ewenson, A., Alvarez, M., and Albericio, F. (2007) *Biopolymers*, **88**, 733–737.

**168** Gausepohl, H., Pieles, U., and Frank, R.W. (1992) In *Peptides: Chemistry, Structure & Biology, Proceedings of the 12th American Peptide Symposium* (eds J.A. Smith and J.E. River), ESCOM, Leiden, p. 523.

**169** Zervas, L., Winitz, M., and Greenstein, J.P. (1957) *The Journal of Organic Chemistry*, **22**, 1515–1521.

# 10
# Fmoc Methodology: Cleavage from the Resin and Final Deprotection

*Fernando Albericio, Judit Tulla-Puche, and Steven A. Kates*

## 10.1
## Introduction

The fluorenylmethoxycarbonyl (Fmoc)/*tert*-butyl (*t*Bu) solid-phase method is categorically the present strategy of choice for the preparation of peptides for both research and industrial scales [1–5]. The main advantage of the Fmoc/*t*Bu approach compared to the earlier *tert*-butyloxycarbonyl (Boc)/benzyl (Bzl) strategy is the conditions used to remove the temporal protecting group (Fmoc) and the permanent side-chain protecting groups as well as the separation/cleavage of the peptide from the resin. For the Boc/Bzl strategy, each removal of the temporal protecting group (Boc) requires treatment with trifluoroacetic acid (TFA) and the final step – the cleavage of the peptide from the resin – ideally requires the concourse of HF. Although HF provokes very clean chemical reactions and is an optimal reagent to incorporate into a synthetic strategy, its use requires special Teflon reservoirs and may be forbidden in some geographical areas by local safety departments. These drawbacks are further magnified at an industrial scale. Conversely, the Fmoc/*t*Bu strategy requires typically one or a maximum of two acid treatments.

The absence of TFA for the release of the temporal protecting group in the Fmoc/*t*Bu strategy allows a gradual removal/cleavage of the remaining protecting groups. A final treatment of the peptide resin with a high concentration of TFA (approximately 90%) will render the totally free unprotected peptide. Alternatively, after choosing the proper resin, it is possible first to remove the protected peptide from the resin (1–5% TFA) and then to obtain the free peptide by treatment with a high concentration of TFA (Figure 10.1). This two-step cleavage/deprotection strategy allows modulation of the TFA treatment to minimize side-reactions and therefore increase the quality of the final product.

Conversely, the protected peptide may be further elongated in solution by reaction with other protected peptides also prepared by solid-phase to accomplish the target molecule (convergent strategy, Figure 10.2) [6, 7].

Although there has been continuous development of new resins, linkers, protecting groups, and coupling reagents (see Chapter 12), the treatment of the protected peptide(-resin) with a high concentration of TFA is perhaps the key step for obtaining

*Amino Acids, Peptides and Proteins in Organic Chemistry.*
*Vol.3 – Building Blocks, Catalysis and Coupling Chemistry.* Edited by Andrew B. Hughes

ISBN: 978-3-527-32102-5

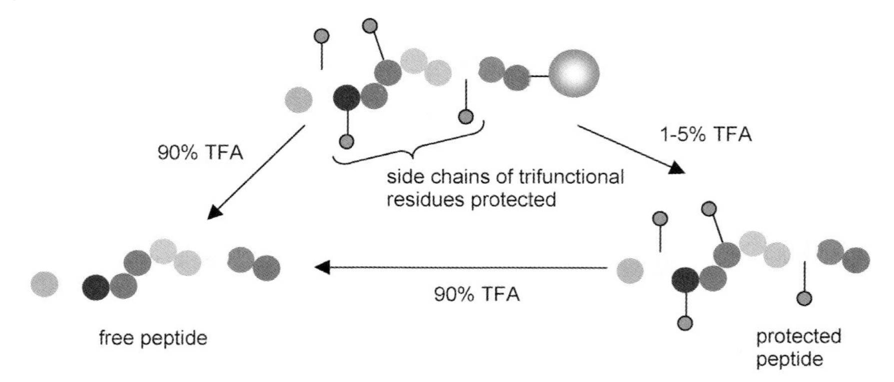


**Figure 10.1** The versatility of the Fmoc/*t*Bu strategy can allow the isolation of the free peptide in one or two steps.

a high-quality final peptide. This final deprotection step can ruin all the synthetic work previously carried out and jeopardize the purification of the crude product. Both of these outcomes will translate into a peptide with less purity and/or a greater production cost.

This chapter is organized into two main sections devoted to cleavage from the resin with a low content of TFA that renders the protected peptide and treatment with a high content of TFA that leads to the final peptide. (For cyclic Cys-containing peptides (disulfides), the final oxidation is carried out typically after the high TFA content treatment to render the heterodetic peptide.) A large number of linkers/resins and their corresponding cleavage methods for the preparation of peptides, many

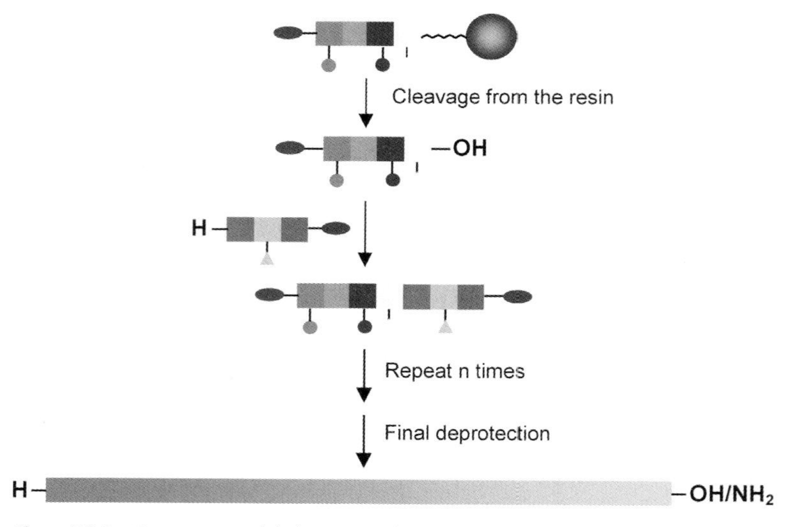


**Figure 10.2** Convergent solid-phase peptide synthesis.

of which have use for particular sequences, have been reported. This chapter will only discuss the conventional and standard methods that have gained broad acceptance by the scientific community. Special attention to the potential side-reactions involved in the TFA treatments will also be described.

## 10.2 "Low" TFA-Labile Resins

### 10.2.1 Cleavage

The use of TFA-labile resins allows for the isolation of protected peptides. Once in solution these protected fragments can be further treated with high concentrations of TFA or subjected to the coupling with other protected peptides prior to the final TFA step.

Natural peptides and synthetic analogs normally possess their *C*-terminal function as free carboxylic acids or amides. For the preparation of a *C*-terminal peptide acid, there are three commercially available resins, which are broadly used and are labile to a low concentration of TFA. The most widely used polymeric supports are the chlorotrityl chloride (CTC, Barlos) resin [8–10], followed by SASRIN (super acid-sensitive resin) resin [11–14], and the *p*-methylbenzhydrylbromide (bromide) resin (Figure 10.3) [15]. Their lability to acid mimics this pattern as well; CTC and *p*-methylbenzhydrylbromide (bromide) resin are the most and least labile, respectively. The conditions for final deprotection to produce a *C*-terminal peptide amide require stronger conditions compared to those used for the preparation of its

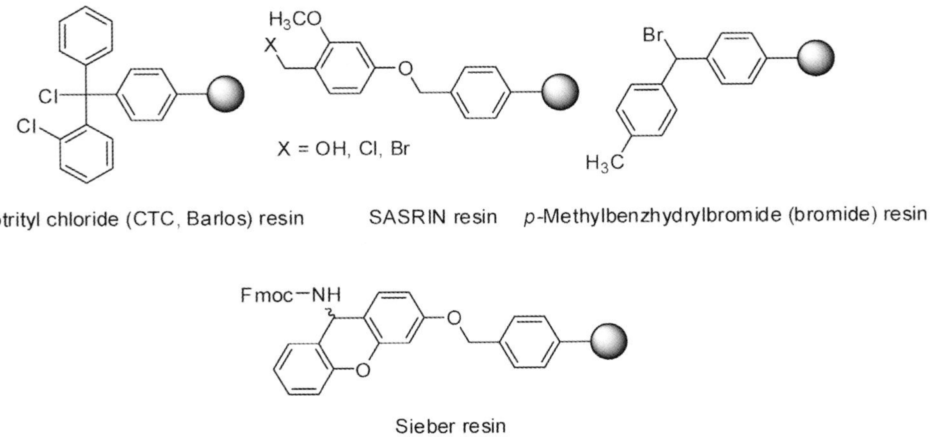

Cl
Cl
$H_3CO$
X
O
X = OH, Cl, Br
Br
$H_3C$

Chlorotrityl chloride (CTC, Barlos) resin SASRIN resin *p*-Methylbenzhydrylbromide (bromide) resin

Fmoc–NH
O
O

Sieber resin

**Figure 10.3** Resins used for the preparation of *C*-terminal peptide acids and amides using low TFA concentration solutions for release from the solid support.

corresponding *C*-terminal peptide acid. The Sieber resin is the most widely used support for the release of *C*-terminal peptide amides from the solid support using low TFA concentrations (Figure 10.3) [16, 17]. Sieber resin and the bromide resin possess a similar TFA lability.

Cleavage of a protected peptide from CTC resins is accomplished by three treatments of alternating washes with TFA/$CH_2Cl_2$ (1 : 99) (1 min) and $CH_2Cl_2$ (30 s) [7–10]. At research scale, this cleavage is carried out at room temperature. For large-scale synthesis, the cleavage is conducted in one step for 45–60 min at 0–5 °C [18]. For peptides containing His(Trt), the solution filtrate is collected over $CH_3OH$/pyridine (9 : 1). For all cleavage reactions, the filtrates are concentrated to at least one-third of the original volume and the peptide is precipitated with $H_2O$ with an amount that is usually about one-half of the organic volume. The precipitate is filtered off and dried under reduced pressure over $P_2O_5$. All common protecting groups for the Fmoc/*t*Bu strategy, including His(Trt), are stable under these conditions.

Protected peptides are also quantitatively detached from CTC resin using hexafluoroisopropanol (HFIP)/$CH_2Cl_2$ (1 : 4) (3 min) [19]. Neutralization of the effluents with $CH_3OH$/pyridine (9 : 1) is recommended because several protecting groups are not completely stable to these cleavage conditions. Cleavage is also performed using trifluoroethanol (TFE)/HOAc/$CH_2Cl_2$ (1 : 1.3) (15–60 min) [20, 21], followed by a similar work-up as described below. If protected peptides are to be used further in a convergent strategy, the HOAc has to be removed completely because its presence will lead to capping of the resin.

The use of scavengers is not mandatory since the amount of acid is maintained at a low concentration and carbocations are not generated in solution. However, oxidation of Met or Cys(Me) has been reported and may be avoided by adding $H_2O$ (up to 20% volume) in the cleavage mixture [22, 23]. The presence of other scavengers (1%) such as triethylsilane (TES) or triethylisopropylsilane (TIS) for Trp and 1,2-ethanedithiol (EDT) for Cys can accelerate the cleavage from the resin as well as the removal of side-chain protecting groups.

Cleavage of protected peptides from SASRIN, bromide and Sieber resins is carried out similarly to those carried out from CTC resin except that a larger concentration of TFA is used in the cleavage cocktail. A 1–2% TFA concentration is recommended for SASRIN resin [11–14]. Cleavage conditions for bromide and Sieber resins should implement up to 5% TFA conducted at 5 °C.

Polymeric supports that are composed of polyethyleneglycol (PEG) such as TentaGel or ChemMatrix require a greater amount of TFA since the PEG units act as a sponge and absorb the acid. It was reported that this phenomenon is more significant with the PEG-based resin ChemMatrix [24].

### 10.2.2
### Choice of Resin for the Preparation of Peptide Acids

The advantages and disadvantages of CTC, SASRIN and bromide resins are discussed in relation to their lability/stability, the method of incorporation of the first amino acid, and side-reactions associated to their use.

#### 10.2.2.1 **CTC Resin**

The CTC resin is the most acid-labile support and therefore is the most secure for maintaining all protecting groups attached to their corresponding side-chains. Unfortunately this lability is also the main drawback to this resin. Premature cleavage of the protected peptide can be detected by washings with acidic compounds such as the coupling additive 7-azabenzotriazole (HOAt). Interestingly, the use of the Kessler method [25] for solid-phase *N*-methylation of the first amino acid anchored to the CTC resin cannot be achieved [26]. In the synthesis of the natural peptide analog *N*Me-azathiocoraline, Boc-D-diaminopropionic acid (Dap)(Fmoc)-OH was loaded to the CTC resin. Following Fmoc removal, the amino group was protected with the *o*-nosyl (NBS) group, and then *N*-methylation was carried out using triphenylphosphine ($PPh_3$), $CH_3OH$, and diisopropyl azodicarboxylate (DIAD) in tetrahydrofuran (THF). Finally, when the *o*-NBS group was removed with β-mercaptoethanol and 1,8-diazabicyclo[5,4,0]undec-7-ene (DBU) in *N*,*N*-dimethylformamide (DMF), a significant loss of the peptide from the resin was detected. The premature peptide cleavage was due to the instability of the ester of an *o*-NBS-amino acid to the nucleophilic conditions involved under these reaction conditions. Presumably the electron-withdrawing effect of the *o*-NBS group is responsible for this premature release of the peptide (Figure 10.4) since the ester bond was completely stable to the same conditions when *N*-methylation was carried out at a residue positioned not at the *C*-terminus [26].

CTC resin is also not compatible with the use of the *p*-nitrobenzyloxycarbonyl (pNZ) protecting group for an α-amino function. The removal of the pNZ group

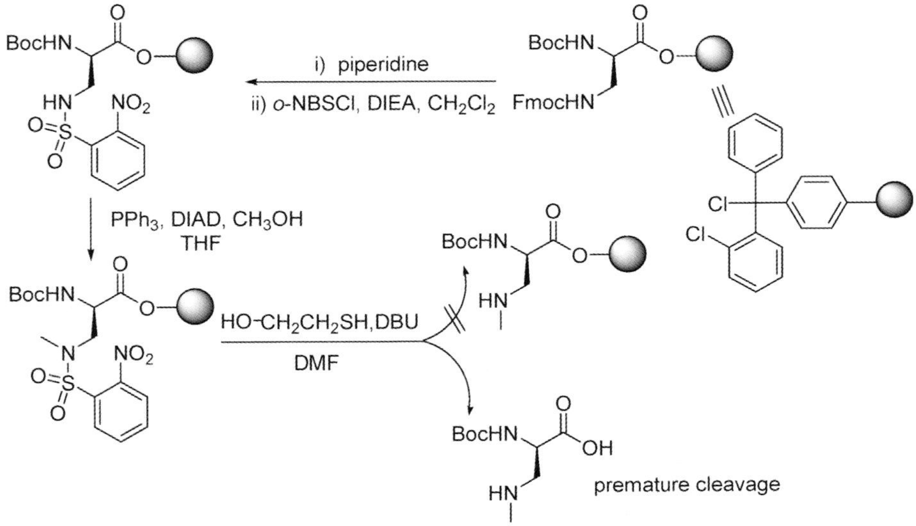

**Figure 10.4** Premature cleavage of peptide from the CTC resin during the removal of the *o*-NBS group.

requires $SnCl_2$, which provokes premature release (bleeding) of the peptide from the resin [22].

The synthesis of peptides larger than 15–18 amino acids using CTC resin is not recommended due to the stability and properties of the polymeric support. To improve the purity of the peptides synthesized with the CTC resin, it is strongly recommended to reduce the initial loading of the support. The initial loading of the commercial resin is typically 1.5 mmol/g. The first amino acid incorporation should be conducted using 0.7 equiv. of reagent to reduce the resin loading to 0.6–0.8 mmol/g [27].

The incorporation of the first amino acid can be considered racemization-free since it occurs via the carboxylate of the protected amino acid derivative by a nucleophilic displacement to a halogen. This is an advantageous feature for the *C*-terminal incorporation of Cys and His derivatives, which show a marked tendency to undergo racemization.

For Fmoc/*t*Bu synthesis of *C*-terminal Cys peptide acids, racemization can also be caused by piperidine during the Fmoc removal step. This side-reaction has been extensively studied and is particularly significant when 4-hydroxymethylphenoxy-type resins are used in conjunction with acetamidomethyl (Acm) protection of the Cys unit (Figure 10.5, top) [28]. Racemization is almost completely avoided when CTC resin is used presumably due to the high degree of steric hindrance. Another side-reaction associated with the synthesis of *C*-terminal Cys peptide acids is the formation of a 3-(piperidin-1-yl)Ala residue at the *C*-terminus [29, 30]. The mechanism of this side-reaction is related to racemization. The *C*-terminal Cys peptide undergoes a base-catalyzed β-elimination of the protected sulfur of the side-chain to give a dehydroalanine intermediate which is trapped by nucleophilic addition of piperidine (Figure 10.5, bottom). The existence of the side-reaction is diagnosed by the observation in mass spectrometry of a peak that is 51 Da higher then the desired mass of the synthesized peptide. This side-reaction is minimized with CTC resin, but unfortunately still prevalent with the nonproteinogenic Cys(Me) residue [23].

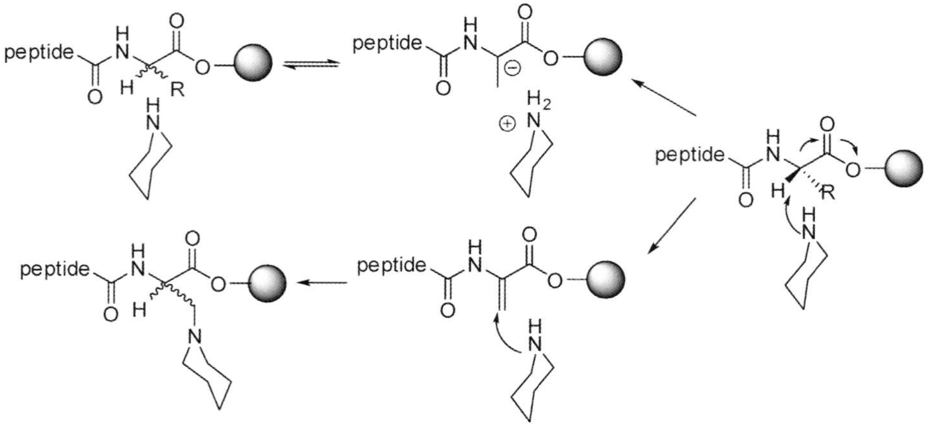

**Figure 10.5** Racemization (top) and formation of 3-(piperidin-1-yl)Ala residue (bottom) with *C*-terminal Cys.

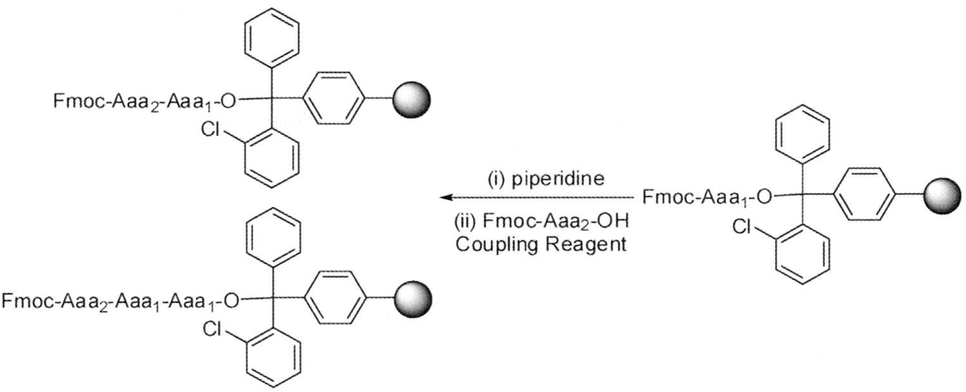

Figure 10.6 Double incorporation of the first amino acid during the coupling of the second amino acid with a CTC resin.

An intriguing side-reaction has been reported during the incorporation of the second Fmoc-amino acid to the CTC resin, which results in a double incorporation of the first amino acid (Figure 10.6). This undesired reaction can be interpreted by cleavage of the unprotected first amino acid and subsequent coupling to the incoming activated second Fmoc-amino acid.

The formation of diketopiperazines (DKPs) is a well-known and important side-reaction in peptide synthesis (Figure 10.7). DKP formation may occur during removal of the α-amino protecting group of the second residue. Due to the steric hindrance of the CTC resin, this side-reaction is minimized. DKP formation with CTC resin is detected when two *N*-methylamino acids are at the *C*-terminal position [22, 27, 31].

#### 10.2.2.2 **SASRIN Resin**

Premature release of the peptide typically does not occur when SASRIN resin is used for solid-phase chain elongation. Conversely, the incorporation of the first amino acid to the hydroxymethyl SASRIN resin occurs with racemization due to the use of a base catalyst [13]. DKP formation and side-reactions associated to the presence of Cys as the *C*-terminal residue also occur with the SASRIN resin, limiting its use for sequences prone to these phenomena.

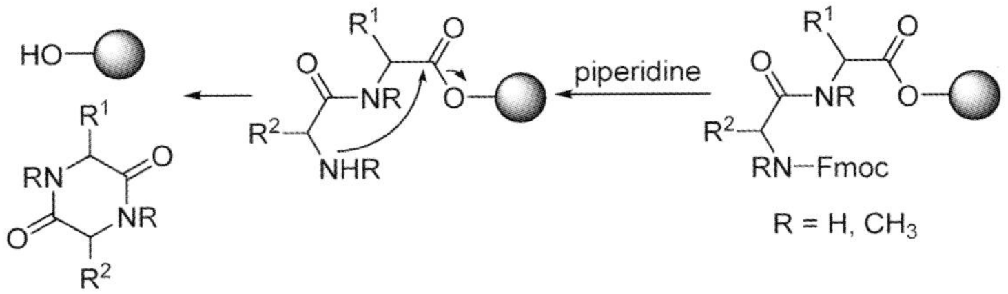

Figure 10.7 DKP formation during the synthesis of *C*-terminal peptide acids.

#### 10.2.2.3 Bromide Resin

Cleavage from bromide resin for the preparation of protected peptides has to be carried out carefully and at low temperatures due to its average sensitivity/lability to acid. Alternatively, the bromide resin represents a compromise between CTC and SASRIN resin as it relates to undesired side-reactions. Bromide resin is compatible with pNZ protection and solid-phase *N*-methylation of the first residue. Bromide resin is an excellent alternative to Wang resin for the preparation of unprotected peptides but may not always be a better replacement to CTC resin.

### 10.2.3 Final Deprotection

Final deprotection is referred to as the treatment with a high TFA content in the presence of scavengers of protected peptides, which are either anchored to the resin or in solution and render the free unprotected peptide. (For cyclic Cys-containing peptides (disulfides), the final oxidation is carried out typically after the high TFA content treatment to render the heterodetic peptide.) Arguably, this is one of the most crucial steps in the peptide synthesis process and can jeopardize all efforts used in the peptide elongation. The treatment of a peptide resin with a cleavage solution is not one simple reaction, but a series of competing reactions, which can lead to a peptide irreversibly modified or damaged. Final deprotection should be performed as quickly as possible to minimize the exposure of the peptide to the cleavage reagent. It is important that the step should promote the cleavage of the peptide from the resin and remove the protecting groups. The 2,2,4,6,7-pentamethyl-dihydrobenzofuran-5-sulfonyl group (Pbf) for the side-chain of Arg is the most demanding protecting group since it is the least labile and requires either a greater concentration of TFA, a longer reaction time or higher temperature for complete removal [32]. Furthermore, the cleaved sulfonyl derivative can react with some residues such as the indole of a Trp residue and provide back alkylation side products. Other residues that require special attention are Cys and Met. Thus, peptides containing Arg, Trp, Cys, and Met should be treated carefully during final deprotection.

Prior to conducting the final deprotection of a full batch of protected peptide (-resin), it is recommended to optimize the deprotection conditions with small aliquots of the batch. In a production campaign, it is advisable to perform a design of experiments (DoE) process before carrying out the final deprotection at scale.

## 10.3 "High" TFA-Labile Resins

Resins labile to high TFA concentration are based on benzyl and benzhydryl linkers containing electron-donating substituents such as methoxy groups: Wang-type resins for the preparation of *C*-terminal acid peptides [33], and Rink [34] and peptide amide linker (PAL)/backbone amide linker (BAL) resins [35–37] for *C*-terminal amide peptides. These linkers are usually incorporated to the solid support via a

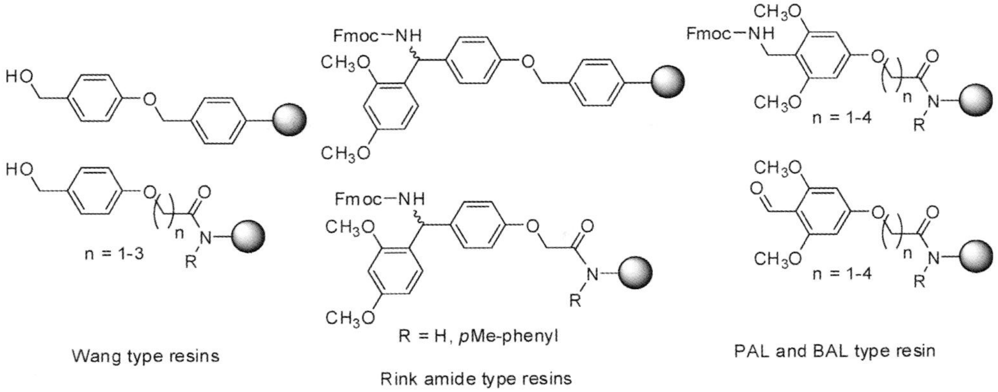

Figure 10.8 The most commonly used resins cleaved by high concentration of TFA.

phenyl benzyl ether or an amide bond. For the latter, the linkers contain a carboxylic acid, which is coupled to a solid support containing an amine. The most commonly used resins that release free peptides through a high TFA concentration cocktail are displayed in Figure 10.8.

### 10.3.1
### Cleavage

Cleavage solutions are commonly referred to as cleavage "cocktails." Similar to "bar cocktails," each researcher/laboratory has their "best" recipe for carrying out the procedure. The cleavage cocktails contain TFA in addition to scavengers with a dual chemical function. The TFA content is in all cases greater than 70%, and many times is between 85 and 95%. Through a push–pull mechanism, the cleavage cocktail can accelerate release of the peptide from the resin with concomitant removal of the side-chain protecting groups, but also will react with the carbocation and/or sulfonyl species formed to minimize back-reactions to the desired product. Some common guidelines are described, but it is advisable to optimize the cleavage cocktail for each peptide. Table 10.1 represents some TFA-based cocktails broadly used for the final cleavage/deprotection of peptide resins.

Although there is a tendency to prepare complex scavenger mixtures, for many peptides that do not contain Arg, the simple addition of $H_2O$ in the cleavage cocktail has been shown to render good results and peptide quality similar to multiple component cocktails. Silane-based cocktails (TES, TIS) generally provide good results and have additional benefits being nonodorous and less toxic than thiol-based cocktails [38]. TES has better scavenging properties then TIS. However, the latter (TIS) is recommended for Trp-containing sequences since the former (TES) has the ability to reduce the indole function of this residue [38]. The use of silanes is also not advisable for Cys(Acm)-containing peptides due to their ability to cleave the Acm group which can subsequently alkylate various peptide sites [44]. Other simple cleavage mixtures include Reagents P, In, and A which contain phenol, indole, and

**Table 10.1** High concentration of TFA-based cleavage cocktails.[a)].

| Reagent | Composition | Reference/comments |
|---|---|---|
| W | TFA/$H_2O$ 95 : 5 | [36] no: Trp; 20% $H_2O$ to reduce oxidation of Cys(Me) [22] |
| T | TFA/TES 95 : 5 | [38, 39] no: Trp, Acm. |
| T′ | TFA/TIS 95 : 5 | [38] no: Acm; acceptable: Trp |
| WT(′) | TFA/TES–TIS/$H_2O$ 90 : 5 : 5 | combination of W and T(′) |
| P | TFA/phenol 95 : 5 | [36] |
| $P^+$ | TFA/phenol/methanesulfonic acid 95 : 2.5 : 2.5 | 15-min reaction |
| In | TFA/$CH_2Cl_2$/indole 70 : 28 : 2 | [36] no: Trp |
| A | TFA/$CH_2Cl_2$/DMS 70 : 25 : 5 | [36] classical cocktail in which DMS has a push–pull effect |
| R | TFA/thioanisole/EDT/anisole 90 : 5 : 3 : 2 | [36] developed for accelerating Arg deprotection |
| K | TFA/phenol/$H_2O$/thioanisole/EDT 82.5 : 5 : 5 : 5 : 2.5 | [40] acceptable: Arg, Met, Trp |
| K′ | TFA/phenol/$H_2O$/thioanisole/1-dodecanethiol 82.5 : 5 : 5 : 5 : 2.5 | similar to K, but less odorous. |
| B | TFA/phenol/$H_2O$/TIS 88 : 5 : 5 : 2 | [41] similar results to K, but more user friendly |
| L | TFA/DTT/$H_2O$/TIS 88 : 5 : 5 : 2 | similar to B, but better for Cys |
| | TFA/thioanisole/phenol/TIS/DTT/$H_2O$/$CH_2Cl_2$ 60 : 5 : 5 : 1 : 2.5 : 5 : 21.5 | [46] acceptable: multi Arg, Trp |
| I | TFA/DMB/TIS 92.5 : 5 : 2.5 | [45] to prevent alkylation of linker fragments |
| S | TFA/$CH_2Cl_2$/BME/anisole 50 : 45 : 3 : 2 | [17] cleavage with Sieber resin to minimize loss of sulfate |
| H | TFA/phenol/thioanisole/EDT/$H_2O$/DMS-$NH_4I$ 81 : 5 : 5 : 2.5 : 3 : 2 : 1.5 | [42] concomitant reduction of Met; disulfide formation depending on time |
| N | $NH_4Br$ (40 equiv.) TFA/thioanisole/ EDT/anisole 90 : 5 : 3 : 2 | [43] concomitant reduction of Met |

a) This is a representative and not an exhaustive list.
CAUTION: The preparation of cleavage cocktails and the cleavage/deprotection reaction must be conducted in a well-vented fume hood. Reagents must be freshly prepared.

dimethylsulfide (DMS), respectively, as a scavenger [36]. Reagents In and A contain a large amount of $CH_2Cl_2$ to promote swelling of the peptide resin. Although methanesulfonic acid is an extremely corrosive reagent and requires careful handling, its addition to Reagent P vastly accelerates the cleavage/deprotection step.

Reagent R (TFA/thioanisole/EDT/anisole; 90 : 5:3 : 2) is arguably the first mixture developed following a rational screening of cocktail mixtures using model sequences containing two Trp residues. Reagent R accelerates cleavage of the peptide from the resin and deprotection of Arg, but does not prevent back-alkylation of Trp by fragments generated from handles (see Section 10.3.3) [36].

Reagent K (TFA/phenol/$H_2O$/thioanisole/EDT; 82.5 : 5 : 5 : 5 : 2.5) evolved from a DoE process involving a combination of scavengers and has been shown to be very effective avoiding side-reactions to Trp [40]. EDT competes favorably with Pbf and *t*Bu protecting groups to prevent re-attachment of side-products to Trp, Tyr, or Met. Unfortunately, EDT reacts with TFA to form a TFA–EDT adduct that subsequently can modify a Trp residue. $H_2O$ is added to the cleavage mixture since it inhibits the formation of the TFA–EDT adduct and ensuing Trp modification (see Figure 10.11). Thioanisole acts as a scavenger, prevents oxidation of Met, and accelerates Pbf removal. Phenol is added to accelerate protecting group removal.

Reagent B (TFA/phenol/$H_2O$/TIS; 88 : 5 : 5 : 2) performs similar to Reagent K and is recommended primarily due to the absence of malodorous components in its formulation and use [41].

Dimethoxybenzyl (DMB) (Reagent I, TFA/DMB/TIS; 92.5 : 5 : 2.5) has been shown to avoid side-reactions provoked by linker fragments (see Section 10.3.3) [45].

The rich Arg- and Trp-containing peptide resin [Boc-Ile-Leu-Arg(Pbf)-Trp(Boc)-Pro-Trp(Boc)-Trp(Boc)-Pro-Trp(Boc)-Arg(Pbf)-Arg(Pbf)-Lys(CTC resin)-$NH_2$] has been successfully obtained in a multigram scale using TFA/thioanisole/phenol/TIS/dithiothreitol (DTT)/$H_2O$/$CH_2Cl_2$ (60 : 5 : 5 : 1 : 2.5 : 5 : 21.5) [46].

Cleavage conditions for sequences containing modified Tyr residues have been examined. For minimizing the loss of sulfate in Tyr($SO_3Na$)-containing peptides prepared with Sieber-type resins, a 15-min cleavage at room temperature with Reagent S (TFA/$CH_2Cl_2$/β-mercaptoethanol (BME)/anisole; 50 : 45 : 3 : 2) has been recommended [17]. Other authors have suggested conducting the cleavage at 4 °C in the absence of thiol scavengers and using *m*-cresol and 2-methylindole [47]. Tyr($OPO_3H$)-containing peptides do not require special attention to avoid loss of the phosphorylation site [48].

Reaction cleavage times are typically 1–2 h and as long as 4 h for those containing several Arg residues. Control of the temperature can also be useful for minimizing side-reactions. Specific conditions are always dependent upon the residues present in a sequence.

### 10.3.2 Final Deprotection of Protected Peptides in Solution

Final deprotection of protected peptides in solution does not require the demanding reaction conditions of peptide removal from the resin, which affect the global yield and generate reactive carbocations from the linker (see Section 10.3.3). Protected peptides prepared by cleavage from CTC resin or a similar polymeric support, or from solution coupling of protected peptides are deprotected using similar cleavage cocktails to those mentioned in the previous section. However, the amount of scavengers and time of reaction can be reduced. For example, pramlintide, is a 37-amino-acid peptide that contains an Arg residue and disulfide bridge, but no Trp. Following disulfide bond formation, the protected peptide was deprotected at multi-gram scale with TFA/phenol/TIS (95 : 2.5 : 2.5) for 4 h at 20 °C [49].

Work-up of the protected peptide following treatment with a cleavage mixture generally involves precipitation with ether (methyl-*tert*-butylether or diisopropyl-ether) of the filtrates. For some peptides, the filtrates can be concentrated but is dependent upon potential side-reactions (see Section 10.3.3).

## 10.3.3 Side-Reactions

### 10.3.3.1 Linker/Resin

Linkers containing electron-rich aromatic rings can be cleaved at different positions (Figure 10.9). The desired cleavage position for the Wang, Rink, and PAL/BAL resins to produce the *C*-terminal peptide acid or amide is at Site #1. For these resins, additional cleavage sites are possible. Cleavage at Site #2 in the Wang resin will

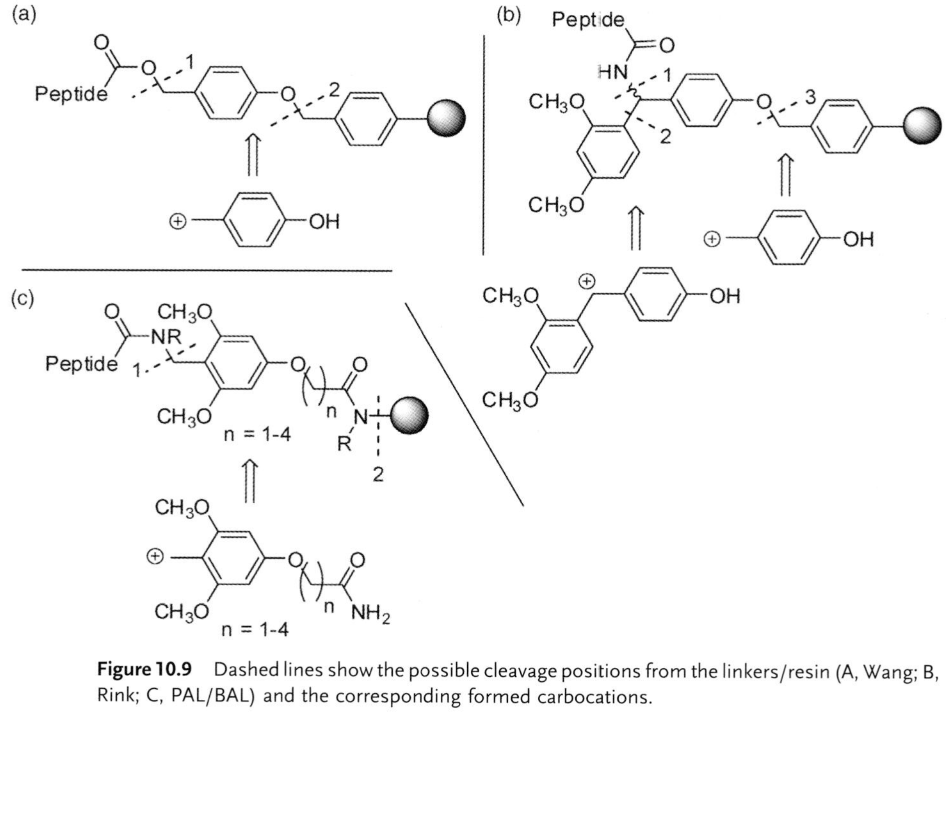

**Figure 10.9** Dashed lines show the possible cleavage positions from the linkers/resin (A, Wang; B, Rink; C, PAL/BAL) and the corresponding formed carbocations.

liberate a 4-hydroxybenzyl carbocation. In the Rink resin, two additional cleavage sites are possible. Cleavage at Site #2 and the ether at Site #3 will liberate a dimethoxyphenyl and 4-hydroxybenzyl carbocation, respectively. For PAL/BAL resin, cleavage can occur at the amide at Site #2 to generate a dimethoxybenzyl carbocation.

The 4-hydroxybenzyl carbocation formed from the anomalous cleavage at Site #2 in Wang resin is able to alkylate the indole function of Trp as well as other peptide sites in the sequence [50–52]. The existence of this side-reaction is diagnosed by a + 106 Da peak in the mass spectrum. The same carbocation may be formed from Rink resin cleavage at Sites #1, #2, and #3. Cleavage at Sites #2 and #3 of the Rink resin produce a *C*-terminal *N*-alkylated peptide amide. Similarly, cleavage at only Site # 3 yields a *C*-terminal *N*-alkylated peptide amide from the benzhydryl scaffold. This carbocation has the potential to alkylate other sites in the peptide chain [45, 53]. Cleavage at Site #2 of the BAL/PAL resin forms a tris-alkoxybenzyl carbocation capable of producing a *C*-terminal *N*-alkylated peptide amide or alkylating other sites in the peptide chain [36, 53].

These side-reactions can be minimized by using appropriate cleavage cocktails such as DMB [45]. The connectivity of the linker to the resin can also improve the stability to cleavage conditions. PAL/BAL, Wang, or Rink supports should incorporate aminomethyl (R = H, Figure 10.8) as opposed to *p*-benzhydrylamine (R = *p*-methylphenyl, Figure 10.8) resin for the amide bond linkage to the carboxylic acid function of the linker since the former is more stable to TFA [36, 53].

The use of nondegradable linkers/resins has been proposed as alternative handles but are not commercially available (Figure 10.10) [54, 55].

#### 10.3.3.2 **Trp and Tyr Modification**

The indole ring of a Trp residue is the peptide site that is most likely to be modified by alkylation. Figure 10.11 displays the most important side-reactions that can affect the indole function [56], but these carbocations also have the ability to modify other peptide sites.

Protection of the indole nitrogen of Trp with a Boc group minimizes these side-reactions. Optimal removal of the Boc group from Trp is a two-step process. Initial TFA treatment leads to a carbamic acid derivative followed by treatment with aqueous $NH_4OH$ to render the free indole peptide. This two-step process is the preferred

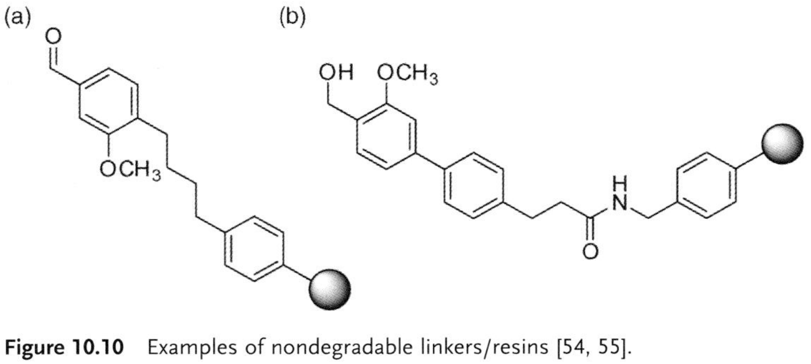

(a) O OCH3

(b) OH OCH3 H N O

**Figure 10.10** Examples of nondegradable linkers/resins [54, 55].

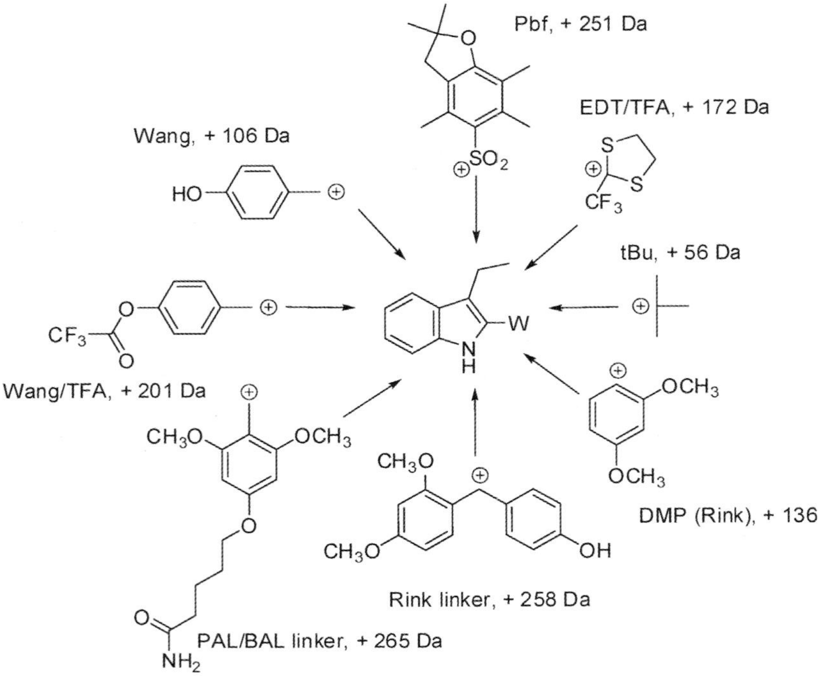

**Figure 10.11** Modifications than can occur at position 2 of the indole of Trp from treatment with a TFA-based cleavage cocktail [56].

method because during the cleavage/deprotection with a TFA cocktail, the Trp residue is still protected with the carbamic acid. In several peptides, the use of Reagent K allows the isolation of the carbamic peptide (Paradis and Albericio, unpublished results).

Similar alkylations, albeit less severe, can occur to Tyr at the aromatic ring. Tyr may also be alkylated by the Acm group that is used to protect Cys (see Section 10.3.3.3).

#### 10.3.3.3 **Sulfur-Containing Residues: Cys and Met**

Cys residues containing TFA-labile side-chain protecting groups can be oxidized during the cleavage/deprotection step directly rendering a disulfide bridge or a sulfonic acid derivative.

The Acm protecting group of the Cys thiol has been shown to be unstable to the acid conditions used for the cleavage/deprotection step. The Acm group can alkylate the *ortho* position of the hydroxyl function of Tyr [57]. This side-reaction can be suppressed by carrying out the cleavage/deprotection at high dilution and adding phenol as a scavenger in the cleavage cocktail. For final deprotection in solution with HCl treatment, transfer of Acm from a Cys residue to the amide of Asn has been detected [58].

During solid-phase assembly, oxidation of Met to its corresponding sulfoxide as well as sulfonium derivative has been observed [59]. The sulfoxide can be reduced during the cleavage/deprotection step with Reagent B followed by the addition of bromotrimethylsilane (TMSBr) and EDT to the cocktail mixture for the final 15 min of reaction time [60, 61]. Work-up is carried out using the standard cleavage/deprotection procedure. Less aggressive conditions for Met sulfoxide reduction use Reagents H (TFA/phenol/thioanisole/EDT/$H_2O$/DMS/$NH_4I$; 81 : 5 : 5 : 2.5 : 3 : 2 : 1.5) [42] and N ($NH_4Br$ (40 equiv.) TFA/thioanisole/EDT/anisole; 90 : 5 : 3 : 2) [43]. Direct formation of a disulfide bridge can occur when using both Reagent H and N.

Met is susceptible to *tert*-butylation during cleavage/deprotection. Des-*tert*-butylation is carried out dissolving the peptide in 4% HOAc–$H_2O$ in a water bath at 60–65 °C for 2 h.

#### 10.3.3.4 Ser and Thr, *N* → O Migration

*N* → *O* acyl migration in Ser- and Thr-containing peptides is known to occur upon exposure to strong acids such as HF used in the Boc/Bzl-based cleavage/deprotection step. It has been described that this side-reaction can also be catalyzed by TFA during the cleavage/deprotection step (Figure 10.12) [62]. The extent of this side-reaction is strongly dependent on the sequence and can be clearly influenced by the introduction of a D-amino acid in the peptide chain. *N* → *O* acyl migration can be reversed by treatment of the undesired desipeptide with a basic pH solution (0.1 M $Na_2CO_3$, pH 8.5, 1 h or dilute aqueous $NH_4OH$ [63], phosphate-buffered saline (PBS), pH 7.4, 37 °C for 2 h [64]).

An *N* → *O* acyl migration has been detected at Thr residues during Fmoc-based solid-phase synthesis of phospho-Tyr-containing peptides. The most prone sequence to undergo this side-reaction occurs if a phospho-Tyr residue is located at position +2 with respect to Thr [65].

$N^{\alpha}$-Trifluoroacetylation of *N*-terminal Ser or Thr has been detected. The reaction proceeds prevalently via the formation of a trifluoroacetyl ester of the *N*-terminal hydroxyl amino acid, followed by an *O* → *N* shift based upon a hydroxyoxazolidine intermediate [66].

#### 10.3.3.5 Asp and Asn

In the Fmoc/*t*Bu method, the formation of aspartimide occurs primarily during piperidine treatment used to remove the $N^{\alpha}$-Fmoc group. This undesired side-reaction has been observed during the acid-catalyzed cleavage/protection step [67–69]. Following aspartimide formation, the ring is opened during reaction work-up, and a mixture of α- and β-peptides is obtained (Figure 10.13).

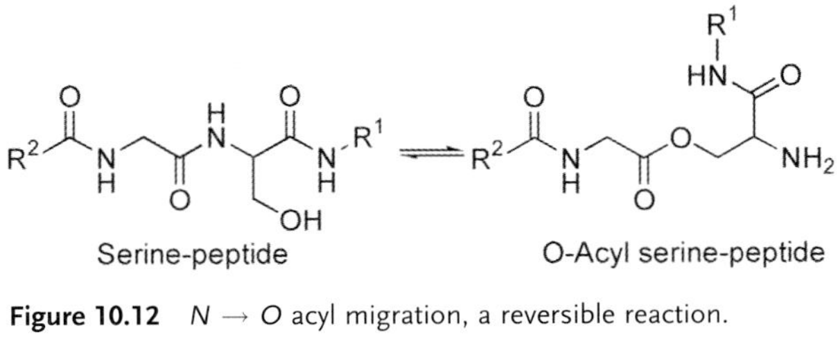

Serine-peptide ⇌ O-Acyl serine-peptide

**Figure 10.12** *N* → *O* acyl migration, a reversible reaction.

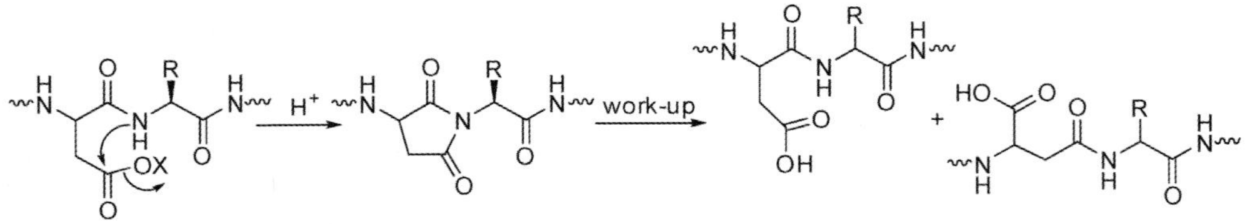

Figure 10.13 Acid-catalyzed aspartimide formation.

It has been proposed that a two-step procedure circumvents this problem. Initial treatment of the resin with a low TFA content to produce the protected peptide followed by removal of the side-chain protecting groups with aqueous HCl minimizes aspartimide formation [70].

Deprotection of Trt-protected *N*-terminal Asn is incomplete under standard conditions [71]. This phenomenon does not occur if the Asn residue is internally positioned in the sequence and for Trt-protected *N*-terminal Gln. Presumably, the incomplete deprotection is due to the extremely slow removal of the Trt group proximal to an amino group. The use of the new methyltrityl protecting group overcomes this problem resulting in rapid and complete deprotection.

#### 10.3.3.6 Arg

During cleavage/deprotection step, Arg rich peptides can form sulfonated Arg peptides [72]. Maximum amounts of sulfonated peptides were observed with

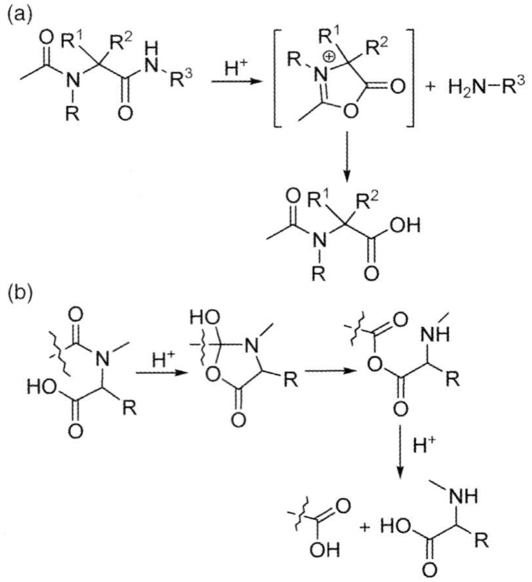

Figure 10.14 TFA-based catalyzed fragmentations can occur in *N*-methylamino acid-containing peptides at the *N*- and *C*-termini. (A) Loss of the *N*-terminal acetylated amino acid where $R^3$ represents the peptide chain [75]. (B) Cleavage of the *C*-terminal amino acid residue [75, 77].

thioanisole or anisole used as stand-alone scavengers. The best scavenger combination is thiocresol/thioanisole (1 : 1).

#### 10.3.3.7 *N*-Alkylamino Acids

Solid-phase synthesis of *N*-methylamino acid-rich peptides constitutes a real challenge to prepare in high purity. Coupling of protected *N*-methylamino acids to *N*-methylamino acids proceeds with low yields [73]. In addition, several side-reactions occur during the cleavage/deprotection step with TFA [73–76]. Fragmentation of Ac-*N*-methylamino acids at the *N*-terminus and cleavage of the *N*-methylamino acid at the *C*-terminus can occur during TFA cleavage of the peptide from the resin (Figure 10.14) [75, 77]. Fragmentation between consecutive *N*-methylamino acids during TFA cleavage is possible via a diketopiperazinium [75, 76] or an oxazolone intermediate mechanism (Figure 10.15) [74, 75]. Concentration of TFA, reaction time, and temperature are important parameters to optimize in order to minimize these side-reactions.

(a)

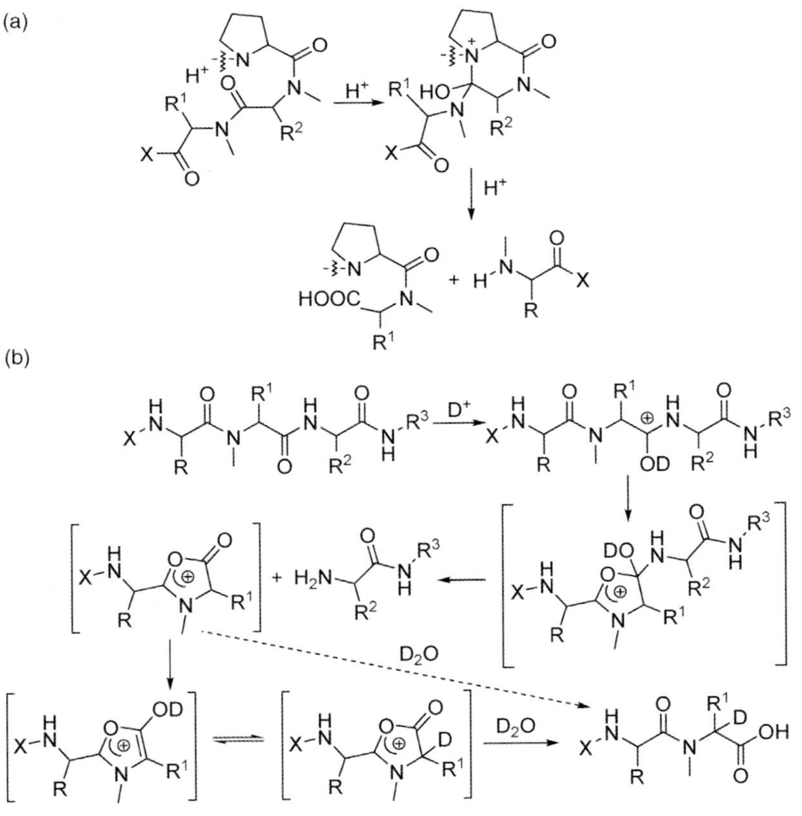

(b)

**Figure 10.15** TFA-based-catalyzed fragmentations between consecutive *N*-methylamino acids via (A) a diketopiperazinium ion intermediate (X: OH, $NH_2$, or resin) [75, 76] and (B) an oxazolone intermediate [74, 75].

#### 10.3.3.8 Work-Up

Methyl *tert*-butylether and diethyl ether are regarded as interchangeable solvents for the "cold ether" work-up following the cleavage/deprotection step. However, the use of methyl *tert*-butylether to precipitate peptides from acid solutions was shown to produce *t*Bu alkylation byproducts. Thus, diethyl ether work-up is advisable, as it consistently leads to cleaner products [78].

## 10.4 Final Remarks

The cleavage/deprotection step is crucial for obtaining peptides with both high purity and yield. It is the final procedure following a synthetic route that often is quite time-consuming and expensive. For complicated peptides and/or industrial synthesis, a DoE process is recommended for optimizing the best cleavage mixtures (proper addition of scavengers), reaction time, and temperature. The work-up following acid treatment also has to be controlled to avoid further modification of the peptide.

## Acknowledgments

Work in the authors laboratory (Spain) was supported by funds from CICYT (CTQ2009-07758). The Generalitat de Catalunya (2009SGR 1024), the Institute for Research in Biomedicine, and the Barcelona Science Park. J.T.-P. is a Juan de la Cierva fellow (MICINN, Spain).

## References

1 Lloyd-Williams, P., Albericio, A., and Giralt, E. (1997) *Chemical Approaches to the Synthesis of Peptides and Proteins*, CRC, Boca Raton, FL.

2 Bruckdorfer, T., Marder, O., and Albericio, F. (2004) *Current Pharmaceutical Biotechnology*, **5**, 29–43.

3 Cupido, T., Tulla-Puche, J., Spengler, J., and Albericio, F. (2007) *Current Opinion in Drug Discovery & Development*, **10**, 768–783.

4 Tulla-Puche, J. and Albericio, F. (2008) *The Power of Functional Resins in Organic Chemistry*, Wiley-VCH Verlag GmbH, Weinheim.

5 Zompra, A.A., Galanis, A.S., Werbitzky, O., and Albericio, F. (2009) *Future Medicinal Chemistry*, **1**, 361.

6 Lloyd-Williams, P., Albericio, F., and Giralt, E. (1993) *Tetrahedron*, **49**, 11065–11133.

7 Barlos, K. and Gatos, D. (1999) *Biopolymers*, **51**, 266–278.

8 Barlos, K., Gatos, D., Hondrelis, J., Matsoukas, J., Moore, G.J., Schaefer, W., and Sotiriou, P. (1989) *Liebigs Annalen der Chemie*, 951–955.

9 Barlos, K., Gatos, D., Kallitsis, J., Papaphotiu, G., Sotiriu, P., Wengqing, Y., and Schäfer, W. (1989) *Tetrahedron Letters*, **30**, 3943–3946.

10 Barlos, K., Gatos, D., and Schäfer, W. (1991) *Angewandte Chemie International Edition*, **30**, 590–593.

11 Mergler, M., Tanner, R., Gosteli, J., and Grogg, P. (1988) *Tetrahedron Letters*, **29**, 4005–4008.

12 Mergler, M., Nyfeler, R., Tanner, R., Gosteli, J., and Grogg, P. (1988) *Tetrahedron Letters*, **29**, 4009–4012.

13 Mergler, M., Nyfeler, R., and Gosteli, J. (1989) *Tetrahedron Letters*, **30**, 6741–6744.

14 Mergler, M., Nyfeler, R., Gosteli, J., and Tanner, R. (1989) *Tetrahedron Letters*, **30**, 6745–6748.

15 Barlos, K., Gatos, D., Kalaitzi, V., Katakalou, C., Scariba, E., Mourtas, S., Karavoltsos, M., and Midard, G. (2002) In *Peptides 2002; Proceedings of the European Peptide Symposium* (eds E. Benedetti and C. Pedone), Edizioni Ziino, Castellammare di Stabia, p. 38.

16 Sieber, P. (1987) *Tetrahedron Letters*, **28**, 2107–2110.

17 Han, Y., Bontems, S.L., Hegyes, P., Munson, M.C., Minor, C.A., Kates, S.A., Albericio, F., and Barany, G. (1996) *The Journal of Organic Chemistry*, **61**, 6326–6339.

18 Zhang, H., Schneider, S.E., Bray, B.L., Friedrich, P.E., Tvermoes, N.A., Mader, C.J., Whight, S.R., Niemi, T.E., Silinski, P., Picking, T., Warren, M., and Wring, S.A. (2008) *Organic Process Research & Development*, **12**, 101–110.

19 Bollhagen, R., Schmiedberger, M., Barlos, K., and Grell, E. (1994) *Journal of the Chemical Society, Chemical Communications*, 2559–2560.

20 Barlos, K., Chatzi, O., Gatos, D., and Stavropoulos, G. (1991) *International Journal of Peptide and Protein Research*, **37**, 513–520.

21 Athanassopoulos, P., Barlos, K., Gatos, D., Hatzi, O., and Tzavara, C. (1995) *Tetrahedron Letters*, **36**, 5645–5648.

22 Tulla-Puche, J., Bayo-Puxan, N., Moreno, J.A., Francesch, A.M., Cuevas, C., Alvarez, M., and Albericio, F. (2007) *Journal of the American Chemical Society*, **129**, 5322–5323.

23 Bayo-Puxan, N., Tulla-Puche, J., and Albericio, F. (2009) *European The Journal of Organic Chemistry*, 2957–2974.

24 Garcia-Martin, F., Quintanar-Audelo, M., Garcia-Ramos, Y., Cruz, L.J., Gravel, C., Furic, R., Cote, S., Tulla-Puche, J., and Albericio, F. (2006) *Journal of Combinatorial Chemistry*, **8**, 213–220.

25 Biron, E. and Kessler, H. (2005) *The Journal of Organic Chemistry*, **70**, 5183–5189.

26 Tulla-Puche, J., Marcucci, E., Prats-Alfonso, E., Bayo-Puxan, N., and Albericio, F. (2009) *Journal of Medicinal Chemistry*, **52**, 834–839.

27 Chiva, C., Vilaseca, M., Giralt, E., and Albericio, F. (1999) *Journal of Peptide Science*, **5**, 131–140.

28 Atherton, E., Hardy, P.M., Harris, D.E., and Matthews, B.H. (1991) In *Peptides 1990: Proceedings of the European Peptide Symposium* (eds E. Giralt and D. Andreu), ESCOM, Leiden, p. 243.

29 Eritja, R., Ziehler-Martin, J.P., Walker, P.A., Lee, T.D., Legesse, K., Albericio, F., and Kaplan, B.E. (1987) *Tetrahedron*, **43**, 2675–2680.

30 Lukszo, J., Patterson, D., Albericio, F., and Kates, S.A. (1996) *Letters in Peptide Science*, **3**, 157–166.

31 Rovero, P., Vigano, S., Pegoraro, S., and Quartara, L. (1996) *Letters in Peptide Science*, **2**, 319–323.

32 Carpino, L.A., Shroff, H., Triolo, S.A., Mansour, E.M.E., Wenschuh, H., and Albericio, F. (1993) *Tetrahedron Letters*, **34**, 7829–7832.

33 Albericio, F. and Barany, G. (1985) *International Journal of Peptide and Protein Research*, **26**, 92–97.

34 Rink, H. (1987) *Tetrahedron Letters*, **28**, 3787–3790.

35 Albericio, F. and Barany, G. (1987) *International Journal of Peptide and Protein Research*, **30**, 206–216.

36 Albericio, F., Kneib-Cordonier, N., Biancalana, S., Gera, L., Masada, R.I., Hudson, D., and Barany, G. (1990) *The Journal of Organic Chemistry*, **55**, 3730–3743.

37 Jensen, K.J., Alsina, J., Songster, M.F., Vagner, J., Albericio, F., and Barany, G. (1998) *Journal of the American Chemical Society*, **120**, 5441–5452.

38 Pearson, D.A., Blanchette, M., Baker, M.L., and Guindon, C.A. (1989) *Tetrahedron Letters*, **30**, 2739–2742.

39 Mehta, A., Jaouhari, R., Benson, T.J., and Douglas, K.T. (1992) *Tetrahedron Letters*, **33**, 5441–5444.

40 King, D.S., Fields, C.G., and Fields, G.B. (1990) *International Journal of Peptide and Protein Research*, **36**, 255–266.

41 Sole, N.A. and Barany, G. (1992) *The Journal of Organic Chemistry*, **57**, 5399–5403.

42 Huang, H. and Rabenstein, D.L. (1999) *The Journal of Peptide Research*, **53**, 548–553.

43 Taboada, L., Nicolas, E., and Giralt, E. (2001) *Tetrahedron Letters*, **42**, 1891–1893.

44 Cuthbertson, A. (2001) PCT Int. Appl. WO 2001005757, *Chemical Abstracts*, **134**, 101196.

45 Stathopoulos, P., Papas, S., and Tsikaris, V. (2006) *Journal of Peptide Science*, **12**, 227–232.

46 Giraud, M., Albericio, F., Quattrini, F., Werbitzky, O., Senn, K., and Williner, M. (2008) PCT Int. Appl. WO 2008040536, *Chemical Abstracts*, **148**, 379970.

47 Yagami, T., Shiwa, S., Futaki, S., and Kitagawa, K. (1993) *Chemical & Pharmaceutical Bulletin*, **41**, 376–380.

48 Perich, J.W. (1992) *International Journal of Peptide and Protein Research*, **40**, 134–140.

49 Brunner, A., Werbitzky, O., Varray, S., Quattrini, F., Hermann, H., Strong, A., Albericio, F., Tulla-Puche, T., and Garcia-Ramos, Y. (2009) PCT Int. Appl. WO 2009003666, *Chemical Abstracts*, **150** 98664.

50 Rovero, P., Pegoraro, S., Vigano, S., Bonelli, F., and Triolo, A. (1994) *Letters in Peptide Science*, **1**, 149–155.

51 Giraud, M., Cavelier, F., and Martinez, J. (1999) *Journal of Peptide Science*, **5**, 457–461.

52 Cironi, P., Tulla-Puche, J., Barany, G., Albericio, F., and Álvarez, M. (2004) *Organic Letters*, **6**, 1405–1408.

53 Yraola, F., Ventura, R., Vendrell, M., Colombo, A., Fernàndez, J.-C., de la Figuera, N., Fernández-Forner, D., Royo, M., Forns, P., and Albericio, F. (2004) *QSAR & Combinatorial Science*, **23**, 145–152.

54 Gu, W. and Silverman, R.B. (2003) *Organic Letters*, **5**, 415–418.

55 Colombo, A., de la Figuera, N., Fernandez, J.C., Fernandez-Forner, D., Albericio, F., and Forns, P. (2007) *Organic Letters*, **9**, 4319–4322.

56 Guy, C.A. and Fields, G.B. (1997) *Methods in Enzymology*, **289**, 67–83.

57 Engebretsen, M., Agner, E., Sandosham, J., and Fischer, P.M. (1997) *The Journal of Peptide Research*, **49**, 341–346.

58 Mendelson, W.L., Tickner, A.M., Holmes, M.M., and Lantos, I. (1990) *International Journal of Peptide and Protein Research*, **35**, 249–257.

59 Nokihara, K., Yasuhara, T., Blahunka, A., Kasama, T., and Wray, V. (1999) *Peptide Science* **36**, 101.

60 Beck, W. and Jung, G. (1994) *Letters in Peptide Science*, **1**, 31–49.

61 Teixido, M., Altamura, M., Quartara, L., Giolitti, A., Maggi, C.A., Giralt, E., and Albericio, F. (2003) *Journal of Combinatorial Chemistry*, **5**, 760–768.

62 Carpino, L.A., Krause, E., Sferdean, C.D., Bienert, M., and Beyermann, M. (2005) *Tetrahedron Letters*, **46**, 1361–1364.

63 Coin, I., Dölling, R., Krause, E., Bienert, M., Beyermann, M., Sferdean, C.D., and Carpino, L.A. (2006) *The Journal of Organic Chemistry*, **71**, 6171–6177.

64 Taniguchi, A., Sohma, Y., Kimura, M., Okada, T., Ikeda, K., Hayashi, Y., Kimura, T., Hirota, S., Matsuzaki, K., and Kiso, Y. (2006) *Journal of the American Chemical Society*, **128**, 696–697.

65 Eberhard, H. and Seitz, O. (2008) *Organic & Biomolecular Chemistry*, **6**, 1349–1355.

66 Huebener, G., Goehring, W., Musiol, H.J., and Moroder, L. (1992) *Peptide Research*, **5**, 287–292.

67 Tam, J.P., Riemen, M.W., and Merrifield, R.B. (1988) *Peptide Research*, **1**, 6–18.

68 Nicolas, E., Pedroso, E., and Giralt, E. (1989) *Tetrahedron Letters*, **30**, 497–500.

69 Stathopoulos, P., Papas, S., Kostidis, S., and Tsikaris, V. (2005) *Journal of Peptide Science*, **11**, 658–664.

70 Dixon, M.J., Nathubhai, A., Andersen, O.A., van Aalten, D.M.F., and Eggleston, I.M. (2009) *Organic & Biomolecular Chemistry*, **7**, 259–268.

71 Friede, M., Denery, S., Neimark, J., Kieffer, S., Gausepohl, H., and Briand, J.P. (1992) *Peptide Research*, **5**, 145–147.

72 Beck-Sickinger, A.G., Schnorrenberg, G., Metzger, J., and Jung, G. (1991) *International Journal of Peptide and Protein Research*, **38**, 25–31.

73 Carpino, L.A., El-Faham, A., Minor, C.A., and Albericio, F. (1994) *Journal of the*

*Chemical Society, Chemical Communications*, 201–203.
**74** Urban, J., Vaisar, T., Shen, R., and Lee, M.S. (1996) *International Journal of Peptide and Protein Research*, **47**, 182–189.
**75** Teixido, M., Albericio, F., and Giralt, E. (2005) *The Journal of Peptide Research*, **65**, 153–166.
**76** Anteunis, M.J.O. and Van der Auwera, C. (1988) *International Journal of Peptide and Protein Research*, **31**, 301–310.
**77** Van der Auwera, C. and Anteunis, M.J.O. (1988) *International Journal of Peptide and Protein Research*, **31**, 186–191.
**78** de La Torre, B.G. and Andreu, D. (2008) *Journal of Peptide Science*, **14**, 360–363.

# 11
# Strategy in Solid-Phase Peptide Synthesis

*Kleomenis Barlos and Knut Adermann*

## 11.1
## Synthetic Strategies Utilizing Solid-Phase Peptide Synthesis Methods

Peptides and proteins are involved in the regulation of all body functions in living organisms. A large number of structurally diverse peptides and proteins are required to study their function. The goal of such research is not only to understand their biological role, but also the development of selectively acting pharmaceuticals influencing or replacing their functions. The large number of peptides required for this purpose cannot be obtained in a reasonable time applying the traditional solution-phase or biotechnological methods for peptide synthesis.

Chemical peptide synthesis was considerably improved with the introduction of solid-phase peptide synthesis (SPPS) [1]. The initial concept of SPPS includes the attachment of the C-terminal amino acid of the required peptide onto an insoluble polymeric support (resin). Subsequently, the peptide chain is elongated in the N-terminal direction by the stepwise addition of individual amino acids using appropriately protected and activated derivatives followed by reactions to cleave the peptide–resin bond and remove side-chain-protecting groups (Scheme 11.1). The main advantage of SPPS, compared to traditional solution-phase techniques, is the opportunity to apply an excess of reagents in all reaction steps, thereby driving the reactions fast to completion. Excessive reagents are easily removed by washing of the support and filtration, yielding quantitatively the elongated resin-bound peptide. Importantly, this method is well suited for use in fully automated peptide synthesizers capable of the simultaneous production of hundreds of peptides per day.

Originating from the basic step-by-step (SBS) method, two technologies combining the advantages of both solid-phase and solution chemistry have been developed: the condensation of protected peptide fragments synthesized by the SBS method [2–7] and the chemical ligation approach [8]. Among different ligation strategies, native chemical ligation (NCL) is the most widely used today. By this technology peptides and proteins are synthesized through the ligation of deprotected and purified solid-phase-synthesized peptide thioesters with a peptide containing Cys

*Amino Acids, Peptides and Proteins in Organic Chemistry.*
*Vol.3 – Building Blocks, Catalysis and Coupling Chemistry.* Edited by Andrew B. Hughes

ISBN: 978-3-527-32102-5

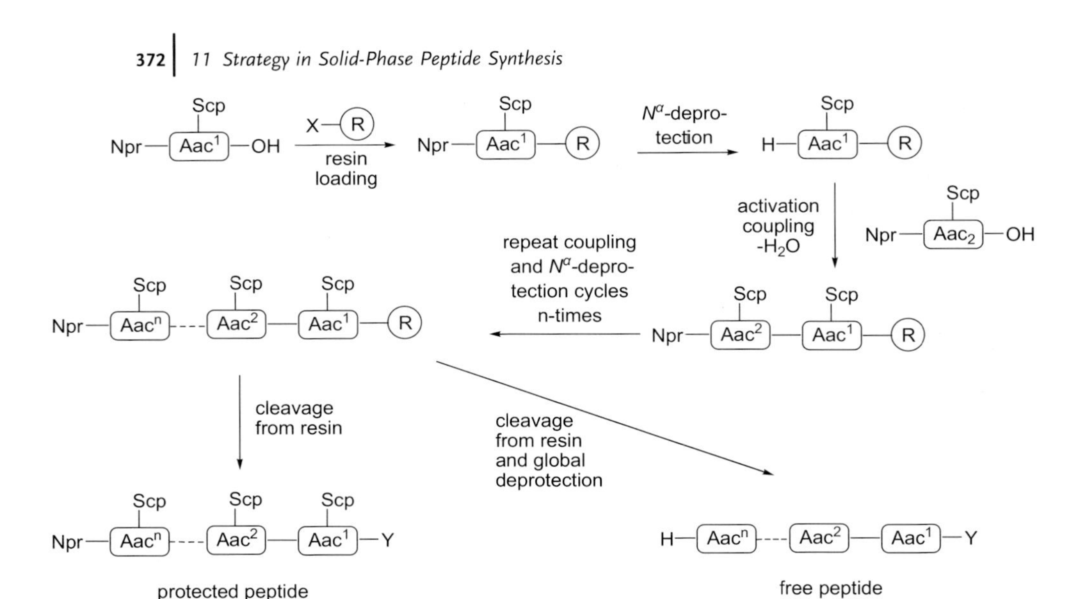


**Scheme 11.1** SBS SPPS (linear route). Aac = amino acid; Npr = $N^{\alpha}$-protecting group; Scp = side-chain-protecting group; X = resin functional group; Y = OH, $NH_2$; R = resin.

at its N-terminus [9]. The condensation of protected segments can be performed on the solid support or in solution, referred to as "convergent synthesis" [10] or the "hybrid method," respectively. In comparison to the traditional SBS method where the peptide chain grows steadily from the C- to the N-terminus, the fragment condensation method allows the peptide to be attached also through a side-chain functional group or through its backbone amide groups to a suitable resin. The peptide chain can be extended towards both the N- and the C-terminal directions (two-directional synthesis) [11] or modified, such as by cyclization (Scheme 11.2). Depending on the complexity of the target peptide, the growing peptide chain can be assembled in part on a solid support, cleaved from the resin and modified in solution, and subsequently reattached onto a resin for the continuation of the synthesis (phase-change synthesis) [12]. In a further variation of NCL and convergent synthesis, the segment condensation can be performed on a side-chain hydroxy function followed by a rearrangement to a neighboring *N*-terminal side-chain function providing the native sequence in a second step (isopeptide or depsipeptide method) [13, 14].

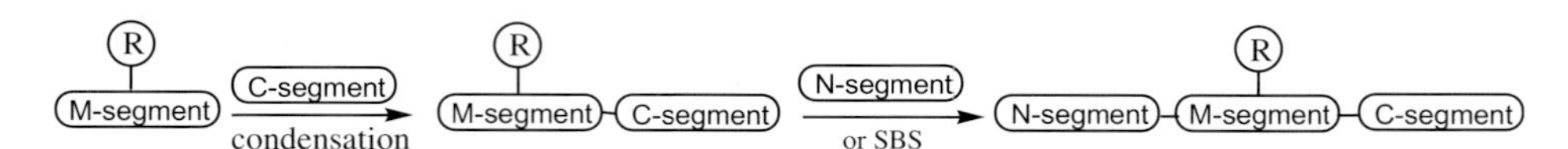


**Scheme 11.2** SPPS with extension in both the N- and C-terminal directions. M-segemnt = middle peptide segment; C-segment = C-terminal peptide segment; N-segment; N-terminal peptide segment; R= resin.

For evaluation of their biological activity peptides are usually required in milligram to gram quantities. For their pharmaceutical, nutritional, or cosmetic applications, amounts of up to hundreds of kilograms to tons of high-purity peptide material must be produced by robust processes. In order to economically produce a peptide at this scale, a synthetic strategy must be determined by careful analysis of all available technologies and parameters influencing the performance of the peptide synthesis and purification.

No other field of synthetic organic chemistry is so complex and challenging like peptide chemistry. More or less all scientific issues of peptide chemistry are controversially discussed, but it is still impossible to establish precise instructions or even unambiguous rules allowing the performance of an optimum synthesis without considerable experimental efforts. Here, we describe general chemical strategies for SPPS of structurally diverse peptides in the laboratory and on a large scale.

## 11.2 Solid Support: Resins and Linkers

Two major procedures are used in SPPS: batch synthesis and continuous-flow synthesis [15, 16], differing in the way solvents and reagents are added. In batch synthesis, the resin is placed into a solid-phase reactor, usually a fritted vessel, and reagents and solvents are added batchwise. After finishing the individual operation, the resin is filtered, and synthesis is continued by the addition of reagents and solvents required for the next operation. In continuous-flow synthesis, the resin is placed into a fritted column, and the reagent solutions and solvents are steadily pumped through the column. For application in these methods, two types of resins have been developed. In batch synthesis, cross-linked polystyrene (PS) and its derivatives are used (Scheme 11.3). The advantages of PS resins are low cost and

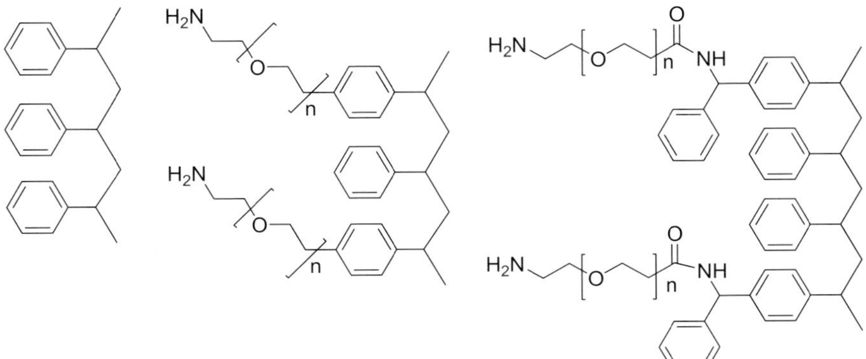

Polystyrene (PS) — Tentagel — PEG-PS

**Scheme 11.3** Basic resins used in SPPS.

good swelling properties in solvents suitable for peptide synthesis, such as dichloromethane (DCM), *N,N*-dimethylformamide (DMF), *N*-methylpyrrolidone (NMP), and dimethylsulfoxide (DMSO). Additionally, PS is inert to all reagents used during peptide synthesis. For continuous-flow syntheses, the resins consist partly of macroporous PS, methacrylates, or kieselguhr. Within the pores of these materials acrylamides are polymerized. The synthesis is then performed on the functionalized polyacrylamide. In the large majority of SPPS PS-type resins are utilized.

Resins must be properly functionalized in order to bind the first amino acid and to allow the cleavage of the peptide from the resin after the completion of the synthesis. Taking these requirements into consideration, two routes are followed to introduce suitable functional groups to the solid support. In the first case, the aromatic rings of PS are properly modified. Examples of such resins are Merrifield's resin, methoxybenzhydrylamine (MBHA), MBH-bromide, and 2-chlorotrityl resin (CLTR). In the second case, an additional bifunctional reagent (linker or handle) combines the amino acid with the functional group of the basic resin. Examples of such linker–resin combinations are the Wang and the SASRIN (super acid-sensitive resin) resins where the Merrifield resin is combined with phenoxybenzyl alcohols. 4-Hydroxymethyl-2,6-di-*tert*-butylphenol (HMBP), Rink amide, peptide amide linker (PAL), Sieber amide and trityl resins (Scheme 11.4) contain corresponding linkers that are combined with alkylamino resins such as MBHA and resins of the TentaGel [35] and polyethyleneglycol (PEG)-PS type.

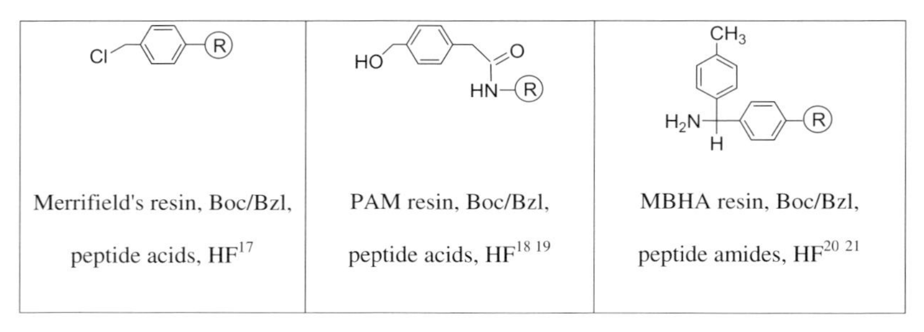


**Scheme 11.4** Resins used in SPPS, protection strategy, and conditions for the peptide cleavage from the resin.

## 11.3
## Developing the Synthetic Strategy: Selection of the Protecting Group Scheme

The development of a peptide synthesis strategy begins always with the selection of the most suitable resin. This determines which protecting group scheme is required for peptide synthesis. The two schemes that are most frequently used are the Boc/Bzl and the Fmoc/*t*Bu protection schemes. If Boc/Bzl is selected, HF-labile Merrifield or 4-hydroxymethyl-phenylacetamidomethyl (PAM) resins are used for the synthesis of

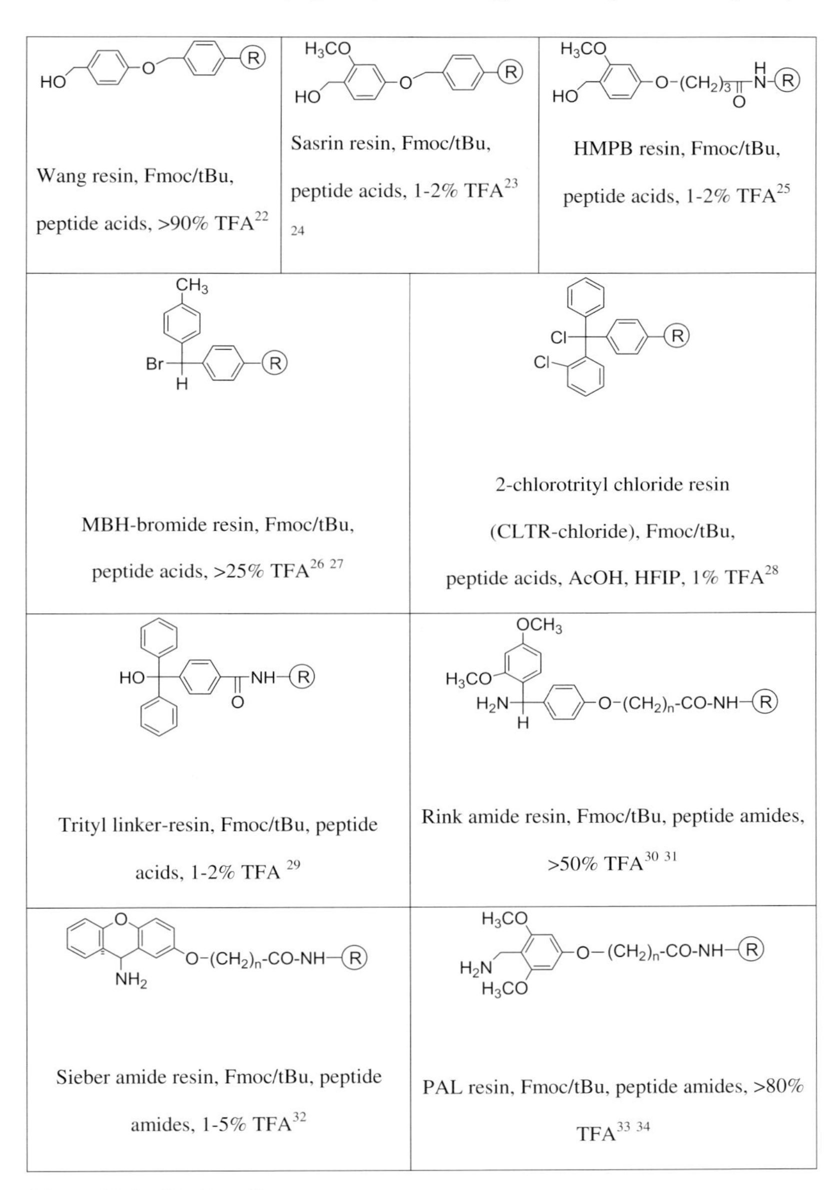


**Scheme 11.4** *(Continued)*

peptide acids, while MBHA resins are used for peptide amides. In the Fmoc/*t*Bu strategy for SPPS of peptide acids, trifluoroacetic acid (TFA)-labile resins of the benzyl type such as Wang, SASRIN, or HMBP resins, resins of the benzhydryl type such as the 4-methyl benzhydryl, or trityl-type resins are used. For the synthesis of peptide amides following the Fmoc/*t*Bu protection strategy, resins of the trialkoxybenzhydryl type such as the Rink amide resin, the Sieber amide resin, and the PAL resin are preferred (Scheme 11.4). Peptide amides are also synthesized by aminolysis of peptides produced on Merrifield's resin or following the hybrid strategy by amidation of a suitably protected peptide fragment, or the condensation with a peptide or amino acid amide [36, 37].

The main advantage of the Fmoc/*t*Bu strategy are the mild conditions used for deprotection and cleavage of the peptide from the resin. In addition, the electrophilic *t*Bu species formed during the global deprotection with TFA solutions are less reactive with nucleophilic side-chains of the amino acids and easier to scavenge than benzyl cations generated in Boc/Bzl chemistry during the removal of protecting groups and the resin cleavage with HF. An additional problem related to Boc/Bzl protection schemes is the repeated use of hazardous TFA during Boc removal as well as the use of toxic and highly metal- and glass-corrosive HF required for global deprotection. Although TFA can be handled with proper safety in small-scale syntheses, it represents a critical issue for large-scale syntheses. A benefit of the Boc/Bzl strategy is the often observed higher level of completion of the Boc group removal in difficult regions by TFA treatment compared to the corresponding Fmoc group removal in the same sequence using piperidine.

For SPPS of protected fragments required for the convergent and hybrid strategy, resins are used that can be cleaved under orthogonal conditions to the N-terminal and side-chain protection. Most frequently, HMBP, SASRIN, and CLTR are used for the preparation of protected peptide acids, while the Sieber amide resin is used for the preparation of protected peptide amides.

## 11.4 Resin Loading

The loading of the first amino acid onto the resin must be performed under conditions that prevent racemization of this amino acid and to minimize other possible side-reactions. The ideal loading degree (i.e., moles of amino acid per gram of resin) depends on the molecular weight of the targeted final protected peptide before its cleavage from the resin. Fast coupling and deprotection rates during SBS are usually observed when the obtained peptide/resin mass ratio does not exceed 1: 1 [38]. Using resins with a higher loading, the peptide-resin loses its resin-like properties with the progression of the synthesis due to its high and growing peptide content. It must be noted that the flow of the dissolved activated amino acids and reagents through the resin channels, pores, and membranes occurs in a manner similar to that known from chromatographic separations. Therefore, the speed of the penetration of the beads changes steadily as the peptide chain changes during assembly. Sequence regions where the peptide assembly becomes very difficult are indicated by an abrupt deswelling (collapse) of the resin. This can occur particularly when β-sheets form between two resin-bound chains, resulting in an increase of the degree of cross-linking of the resin. Consequently, increasing the resin loading increases the risk of aggregation, which renders the continuation of a peptide synthesis more difficult and in some cases even impossible.

Usually the C-terminal amino acid of the required peptide is attached to the functional group of the resin. Depending on the resin used for the synthesis of peptide acids, *N*-protected amino acids must be activated nucleophilically or electrophilically in order to be attached onto the resin. Benzylalcohol-type resins such as Wang, SASRIN, HMBP, and PAM resins require electrophilic activation of the amino acid. As a result, sensitive amino acids such as Cys, His, and Phe are susceptible to racemization [39, 40]. Several methods have been proposed and tested for suppression of racemization, including modest activation of the carboxyl function [41] and the Mitsunobu reaction [42]. In the latter case, an alcohol function of the resin is activated rather than the carboxyl group of the amino acid. Resins of the alkyl halide type such as Merrifield's [43], MBH-bromide and CLTR-chloride [44] require nucleophilic activation of the amino acids. As a result, racemization during the esterification step is almost completely suppressed.

The linkage of the first amino acid can also be achieved through a reactive side-chain function [45], a backbone amide [46], or the N-terminal function [47]. Such linkages are usually used for the preparation of modified peptides. In contrast to the synthesis of peptide acids, resins for the SPPS of peptide amides can be loaded without racemization of the free amino function to the resin applying identical protocols used for the coupling of the amino acids during SPPS.

## 11.5 SBS Peptide Chain Elongation: Coupling and Activation

After loading of the first amino acid onto the resin, the *N*-terminal protecting group of the amino acid must be removed by treatment with TFA in Boc/Bzl protection schemes or piperidine in Fmoc/*t*Bu schemes. The next amino acid is then coupled to the liberated amino function. These two chemical operations, a cycle, are repeated until the entire peptide chain assembly has been completed. To achieve acceptable product purity and yields it is essential to drive coupling and *N*-deprotection steps to completion. If this is not accomplished, the final peptide is contaminated with failure peptides, that is, peptides lacking one or more residues, which are difficult to separate by chromatographic techniques [48].

To perform coupling reactions effectively, amino acids are activated at their α-carboxyl function by reaction with mild activating reagents such as *N,N'*-diisopropylcarbodiimide (DIC) or uronium-type reagents such as 2-(1*H*-benzotriazol-1-yl)-1,1,3,3-tetramethyluronium hexafluorotriphosphate (HBTU) [49, 50] and 2-(1*H*-7-azabenzotriazol-1-yl)-1,1,3,3,-tetramethyluronium hexafluorophosphate (HATU) [51]. Furthermore, catalysts such as 1-hydroxybenzotriazole (HOBt) [52] and 1-hydroxy-7-azabenzotriazole (HOAt) [51] are usually added for the suppression of racemization (Scheme 11.5). The solvent used for the coupling also plays an important role in suppressing racemization. DCM was found to be superior over polar solvents such as DMF in the suppression of racemization [53]. The extent of

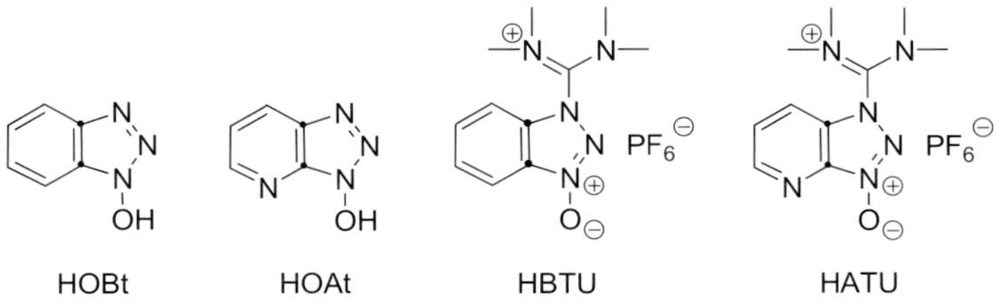

HOBt HOAt HBTU HATU

**Scheme 11.5** Additives and coupling reagents preferably used in SPPS.

racemization depends on the time a protected amino acid remains activated. In the cases of difficult couplings, in particular when amino acids susceptible to racemization are used (Cys, His, and Phe), double coupling cycles should be preferably applied rather than the extension of the coupling time. In addition to racemization, prolonged coupling times can cause side-reactions such as double coupling and chain termination due to the reaction of the free amino group with carboxyl activators such as DIC and especially uronium-type reagents [54].

In some cases, coupling reactions proceed sluggishly and remain incomplete even after repeated acylations. The remaining unreacted amino functions must be capped to reduce the generation of failure peptides. The capping reagent must be selected carefully in order to provide advantageous chromatographic properties of the terminated shorter peptides to facilitate high-performance liquid chromatography (HPLC) separation from the target product. Acetylation or benzoylation are typically employed for capping, both increasing lipophilicity of the terminated peptide chain. Unfortunately, capping without the consideration of the HPLC separation criteria is included in many protocols as a routine procedure to avoid the elongation of truncated peptide, resulting in many cases in difficult chromatographic purifications.

## 11.6 Piperazine Formation

There is a high risk of losing a significant quantity of the resin-bound peptide during chain assembly and removal of the *N*-protecting group of the second amino acid of the C-terminal dipeptide. Often the liberated amino function of the second amino acid intramolecularly attacks the ester bond of the first amino acid to the resin, resulting in the formation of a sterically favored, six-membered diketopiperazine ring. This can even result in the complete cleavage of the peptide from the resin within a few minutes. Peptides that are especially susceptible to piperazine formation contain cyclic amino acids such as Pro and Tic, Gly, D-, *N*-alkyl and $C^{\alpha}$-alkyl amino acids in the C-terminal dipeptide [55]. The problem of piperazine formation is not only the reduction in yield, but also the formation of truncated peptides. These are generated by the reloading and coupling of the liberated hydroxyl functions of the resin with new incoming amino acids. Piperazine formation can be effectively suppressed by using bulky and electron-withdrawing resins such as CLTR. Piperazine formation is more likely to occur when resins are used that electrophilically activate the C-terminal amino acid, as is the case with thioester-generating or oxime resins. Peptide amides are less susceptible to this side-reaction due to the electrophilic deactivation of the C-terminal carbonyl function by the amino group.

## 11.7 Solid-Phase Synthesis of Protected Peptide Segments

Protected peptide fragments are required for application in fragment condensations and for the synthesis of modified peptides. Typically, Fmoc/*t*Bu-protected segments are used in such applications. In some special cases, the use of Fmoc/Bzl-type protection has been proved advantageous [56]. In most cases, the segments are synthesized in a SBS manner and cleaved from the resin by mild acidic treatment so that side-chain protection of the *t*Bu type remains intact. Resins and linkers of the trityl type and of the dialkoxybenzyl type are preferentially used as solid support. In many applications protected fragments with Gly, Pro, or pseudo-Pro residues at their C-terminus are required because of their stereochemical stability. The use of trityl-type resins is especially advantageous due to its minor susceptibility to piperazine formation. The protected peptides are obtained by resin cleavage with mild acidic compounds. For example, mixtures containing acetic acid and fluorinated alcohols, 1% TFA or mixtures containing hexafluoroisopropanol (HFIP) [57] are used for cleaving protected peptides from trityl-resins. Cleavage of the protected peptides from the resin and simultaneous disulfide bond formation occurs usually in high yield [58]. Protected peptides synthesized on SASRIN and HMBP resins can be cleaved by treatment with 1–2% TFA. Mixtures containing fluorinated alcohols not only cleave the peptide from the resin, but also effectively solubilize most protected peptides. In most cases, protected peptides of 5–15 amino acids in length are obtained in high purity and do not require further purification.

## 11.8 Fragment Condensation Approach: Convergent and Hybrid Syntheses

While SBS synthesis is typically considered to be a straightforward process, several problems can be encountered during the peptide assembly. Problems originate mainly from incomplete coupling and deprotection steps, and racemization of the activated amino acids. Consequently, the number and quantity of peptides with failure sequences increases with the length of the peptide. Failure sequence peptides display very similar chromatographic behavior compared to the target peptide and therefore their separation, if possible at all, causes severe losses in yield. Mass spectrometric techniques and sophisticated instrumentation must be applied in order to identify and quantify these impurities, which often elute close to or even under the main peak during HPLC.

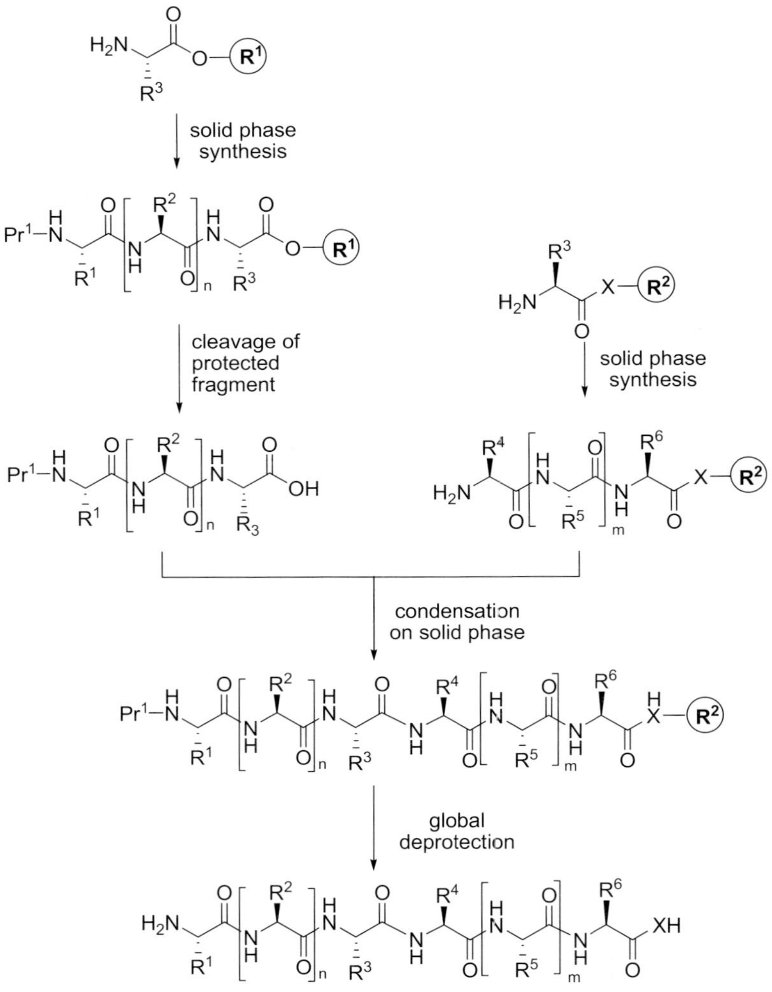

solid phase synthesis

cleavage of protected fragment

solid phase synthesis

condensation on solid phase

global deprotection

**Scheme 11.6** Schematic representation of convergent peptide synthesis.

To overcome the problem of failure sequences, suitably protected peptide fragments can be condensed on the solid phase (convergent synthesis (Scheme 11.6)) [36] or in solution (hybrid method (Scheme 11.7) or by the thioester method (Scheme 11.8)). The required protected peptide segments are favorably synthesized by SBS solid-phase synthesis. As the protected peptides are short, usually with 5–20 amino acid residues, their synthesis can be easily controlled and optimized. In most cases, peptide fragments with maximum protection including the β-turn/sheet-inducing Asn and Gln are used. Several complex peptides were synthesized by this method [59, 60]. This approach has been proven to be

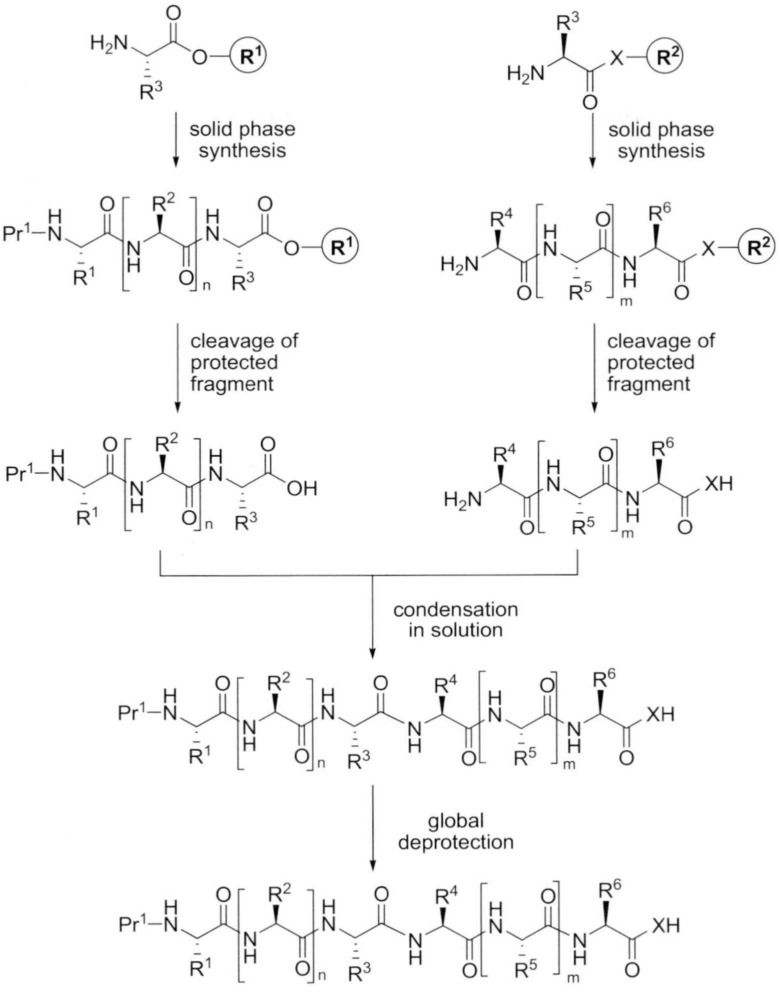

**Scheme 11.7** Schematic representation of hybrid (solid-phase/solution) peptide synthesis.

more suitable for large-scale peptide synthesis than the linear SBS route (see below). It has been adapted for the production of hundreds of kilograms to tons of pharmaceutical peptides, including the anti-HIV-peptide enfurvirtide and several glucagon-like peptide-1 peptides [37, 61]. The main drawback of

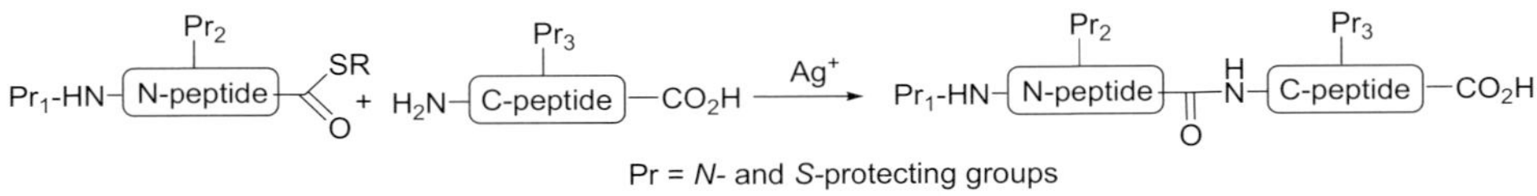

Pr = *N*- and *S*-protecting groups

**Scheme 11.8** The thioester condensation method catalyzed with silver ions.

convergent and hybrid synthesis is that fragment condensation using fragments with different amino acids than Gly and Pro as their C-terminal amino acid requires a challenging optimization to avoid substantial racemization. This disadvantage is reduced considerably when pseudo-Pro residues are used, extending racemization-free condensations to peptides with Ser and Thr as the C-terminal amino acid.

An additional, effective fragment condensation approach is the use of peptide thiols or thioesters as activated *N*-components. Apart from the thiol group of Cys only the $N^{\alpha}$-function of the *N*-component and the $N^{\varepsilon}$-functions of lysine side-chains require protection during this condensation. The activation of the thioester is achieved by treatment with silver ions using $AgNO_3$ or AgCl (Scheme 11.8). Very acid-sensitive sulfated peptides of the cholecystokinin/gastrin family have been synthesized according to this method [62].

During the elongation of the peptide chain in stepwise SPPS the rapid completion of coupling and deprotection steps during the assembly of certain regions can be very difficult or even impossible. This is often a result of the formation of β-turns and -sheets of the resin-bound peptide chains, leading to extensive aggregation [63]. Successful chain assembly of peptides containing such regions remains one of the major unsolved problems in peptide synthesis. In Fmoc SPPS, several tools have been developed in order to overcome this problem. Pseudoprolines can be introduced into the peptide chains. Similar to Pro, their application improves the solubility properties of the peptides and of the corresponding protected peptide segments [64] by limiting β-sheet and also helix conformation of the peptide chain, thus preventing aggregation [65]. In a similar way benzyl-type groups at the backbone have been used to effectively interrupt β-sheet formation [66].

Applying convergent, hybrid, or phase-change synthetic strategies, peptide aggregation can also be circumvented. Generally, properly protected fragments must be selected for the condensation reactions to avoid the formation of β-turn and -sheet structural elements at both the C- and N-terminal regions [67]. In most cases, if the peptide chain is extended farther from the difficult region, peptide synthesis proceeds without problems (Scheme 11.9).

In a similarly effective way, the β-sheet/turn-forming regions can be interrupted by a fragment condensation or SBS assembly of a peptide chain on a hydroxy group of a serine residue. This “isopeptide or depsipeptide” technique (Scheme 11.10) was successfully utilized for the synthesis of the very challenging amyloid peptides [68].

## 11.9
## Cleavage from the Resin and Global Peptide Deprotection

After completion of the chain assembly either by SBS synthesis or by fragment condensation, the side-chain-protected peptide is cleaved from the resin and deprotected. Depending on the resin functional group and the protection scheme, the

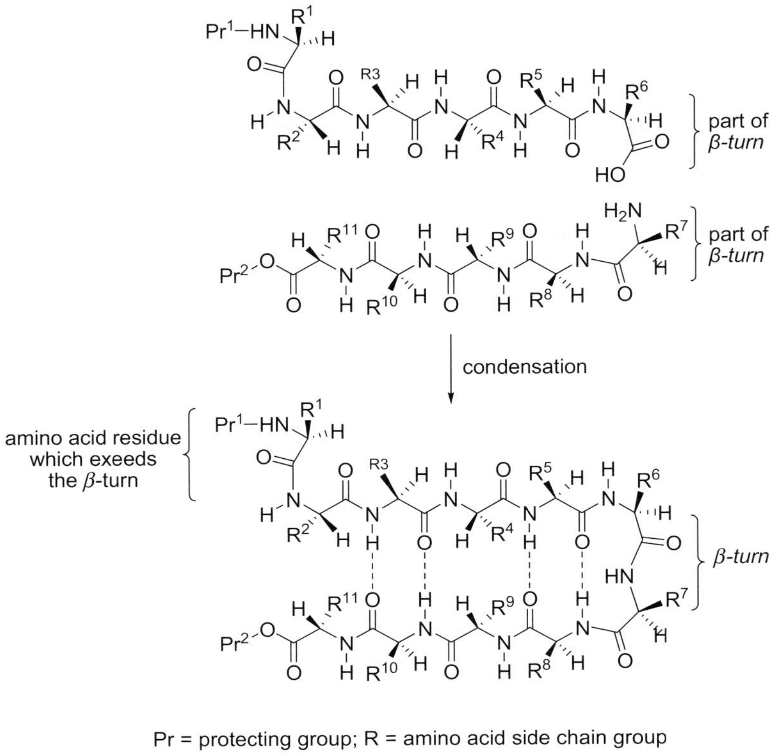

Pr = protecting group; R = amino acid side chain group

**Scheme 11.9** β-Turn exceeding by fragment condensation.

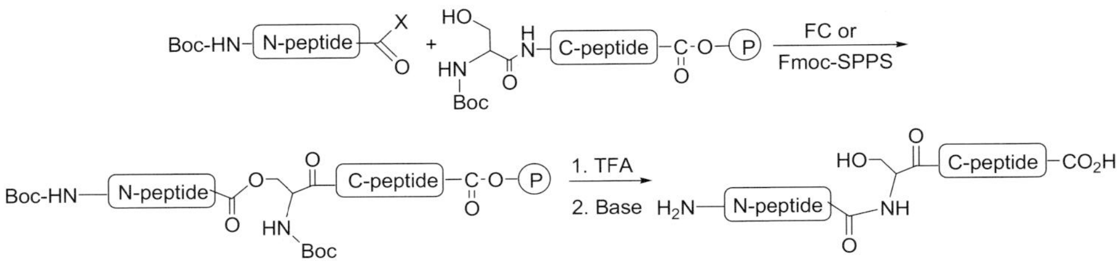

**Scheme 11.10** β-Turn/sheet avoidance using the isopeptide method.

covalent bond of the peptide to the resin and the deprotection are performed by treatment with acids of suitable concentration. The conditions of the cleavage and deprotection must be designed as mild as possible. Otherwise sensitive peptide side-chains can be modified by cationic species formed during the removal of either Bzl- or *t*Bu-type protecting groups. Resin-bound cationic species formed during the acidic treatment can irreversibly bind the peptide through one of its nucleophilic side-chains to the resin. For the effective cleavage of the peptides from the resin and side-chain deprotection, several methods and mixtures of reagents called cocktails have been developed. Most cocktails contain, in addition to the acid, HF in Boc/Bzl chemistry and TFA in Fmoc/*t*Bu chemistry, a number of nucleophiles such as thioanisole, dimethylsulfide (DMS), ethanedithiol (EDT), dithiothreitol (DTT), and silanes. Their function is to scavenge the resin-bound cations formed during the acidolytic cleavage of the peptide from the resin and the cations formed during the side-chain deprotection. Depending on the amino acid composition, the peptide sequence, and the nature of the protecting groups used, global deprotection and resin cleavage must be carefully optimized for any individual peptide. In addition to the cocktail composition, parameters such as temperature, concentration of the peptide, concentration of the cations formed in the deprotection mixture, and duration of the deprotection are important to obtain the products in optimum yield and purity.

## 11.10
## Disulfide Bond-Containing Peptides

Peptides and proteins containing one or more disulfide bonds are widely distributed in nature, playing crucial biological roles. The participating thiol, Cys, is the most nucleophilic group contained in peptides. In addition protected thiol groups are good leaving groups adding to Cys an additional danger of racemization through an elimination/Michael addition mechanism. This occurs during the esterification of Cys-derivatives onto the resin, during the activation and coupling of Cys and during the basic treatments by the normal process of SPPS. Usually protecting groups of the benzyl and trityl type are used that inhibit both the nucleophilic, by blocking the ionization possibility, and with their electron-withdrawing character the leaving group properties of the thiol functions as well. In addition, the very acid-stable Acm and the thiol-sensitive S*t*Bu protecting groups are frequently used, and are orthogonal to benzyl- and trityl-type protection.

The SPPS of Cys-rich peptides, in particular those with two or three disulfide bonds, is very challenging because the disulfide bonds must be introduced regioselectively. There are three general strategies for the synthesis of multiple disulfide-bonded peptides: (i) the oxidative folding of a linear totally deprotected precursor (Scheme 11.11) – this method takes advantage of the structural information retained within the linear peptide chain controlling the folding process, (ii) an oxidation of four free thiol functions in the presence of a pair of protected Cys that are oxidized regioselectively after the oxidation of the free thiol groups (semiselective oxidation; in Scheme 11.12 the method is illustrated with the example of the convergent synthesis of the proteinase inhibitor hirudin), and (iii) of the totally regioselective formation of all disulfide bonds using three orthogonally protected pairs of Cys residues. In the latter case the disulfide bonds are generated in a hierarchical manner by pairwise deprotecting and/or selective oxidation of the Cys residues containing orthogonal protecting groups on their thiol functions (Scheme 11.13).

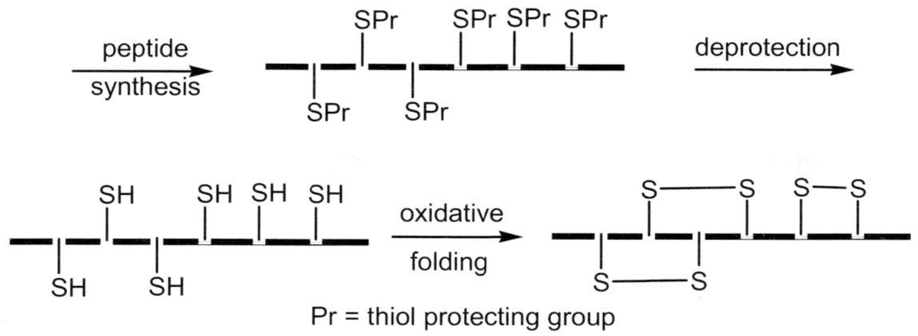

peptide synthesis → SPr SPr SPr SPr / SPr SPr → deprotection

SH SH SH SH / SH SH → oxidative folding → S—S S—S / S—S

Pr = thiol protecting group

**Scheme 11.11** Schematic representation of oxidative folding.

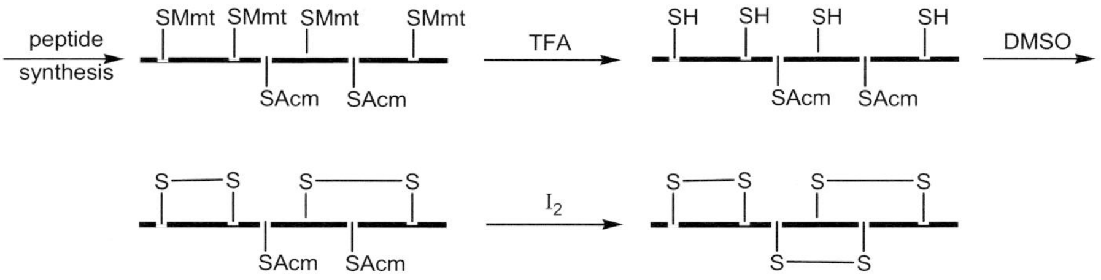

peptide synthesis → SMmt SMmt SMmt SMmt / SAcm SAcm → TFA → SH SH SH SH / SAcm SAcm → DMSO

S—S S—S / SAcm SAcm → $I_2$ → S—S S—S / S—S

**Scheme 11.12** Schematic representation of the semiselective synthesis of hirudin.

In Boc/Bzl schemes the hierarchical orthogonal deprotection conditions can be achieved with the application of Bzl(4-Me)/Acm/S*t*Bu while in Fmoc/*t*Bu schemes with Trt/Acm/S*t*Bu groups are preferably used. Although this approach is chemoselective, the yield of such syntheses is often unacceptably low due to the three chemical steps required and the corresponding HPLC purifications. An example for an Fmoc-based regioselective disulfide bond formation is shown in Scheme 11.13,

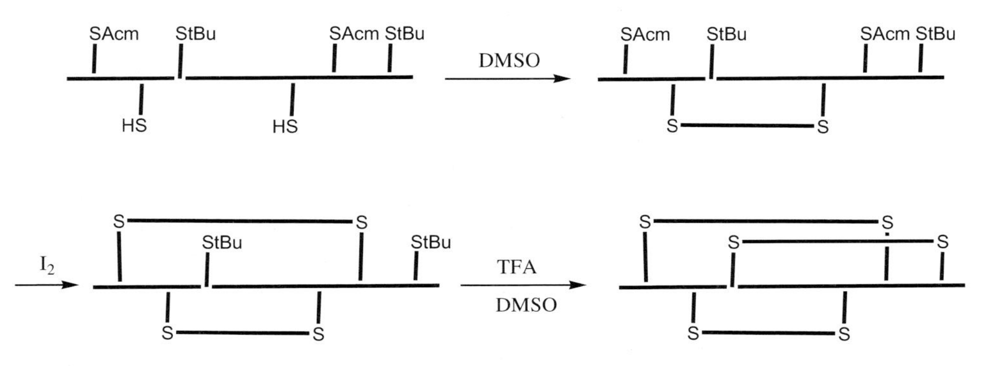


**Scheme 11.13** Directed synthesis of defensins.

which describes the synthesis of β-defensins [69]. It should be noted that the optimum positioning of the three pairs of Cys-protecting groups cannot be predicted, but has to be determined experimentally.

Even more challenging is the synthesis of single-, double-, and multiple-stranded Cys-rich peptides such as insulin and relaxin. To achieve the intermolecular disulfide bond formation, methods were developed where the first interchain disulfide bond is formed between an activated pyridylthiol function of the first chain with a free thiol function of the second chain. The additional disulfide bonds are introduced by the methods used for intramolecular disulfide bond formation. In Scheme 11.14 this strategy is schematically described in the synthesis of insulin-like peptide-5.

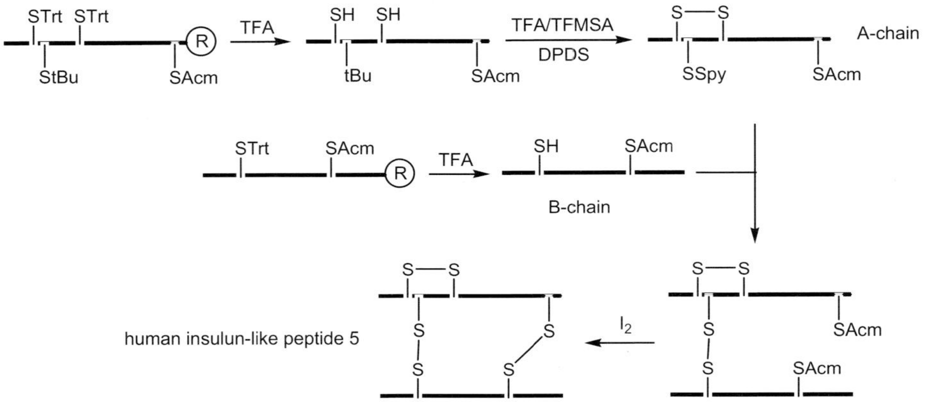

**Scheme 11.14** Schematic representation of the site-directed strategy followed for the hybrid synthesis of insulin-like peptide-5.

## 11.11 Native Chemical Ligation (NCL)

In ligation strategies, two unprotected peptide chains are selectively combined through two of their functional groups, thereby yielding a larger polypeptide. Although many variations have been developed [8], the most effective ligation strategy is NCL [70, 71]. NCL is a synthetic methodology, which provides straightforward access to Cys-containing large polypeptides and proteins with a chain length exceeding 100 amino acids. In NCL, unprotected peptides are selectively condensed in an appropriate aqueous buffer [72]. This approach proceeds initially in a selective reaction between a peptide thioester, the N-terminal peptide (*N*-peptide), with a peptide containing an N-terminal Cys residue, the carboxyl component (*C*-peptide). The reaction starts with the replacement of the thioalkyl group of the *N*-peptide by the thiol function of the N-terminal Cys of the *C*-peptide (Scheme 11.15). The interme-

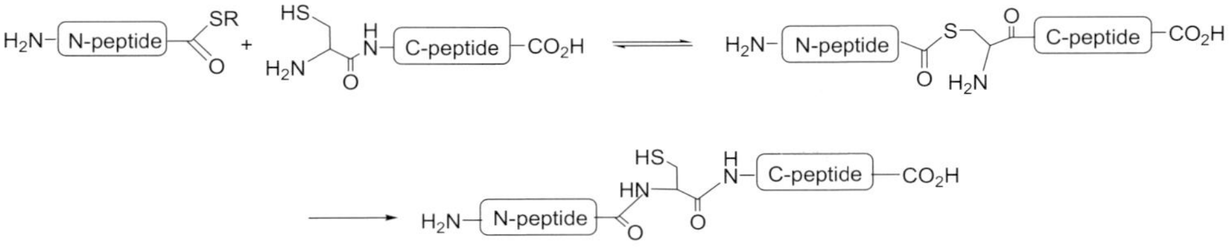

Scheme 11.15 Native chemical ligation.

diate thioester formed, subsequently undergoes a fast $S \rightarrow N$-acyl transfer resulting in a native peptide amide bond at the site of ligation [73]. During ligation, the *trans*-thioesterification reaction is reversible, creating an equilibrium between all thiols of the participating peptides and of thiols (e.g., thiophenol) added to catalyze the condensation [74].

The methods employed for the synthesis of required peptide thioesters comprise Boc as well as Fmoc chemistry (Scheme 11.16). Using Boc chemistry, the linear peptide thioester is assembled on a resin loaded with a thioacid derivative. Peptide thioester release from the resin is achieved using anhydrous HF. The most frequently used resins to assemble peptide thioesters are PAM and MBHA resins. Chain elongation is carried out with *in situ* neutralization methodologies [75]. Alternatively, the peptide thioester can be released from a safety-catch sulfonamide resin by activation of the peptide resin, followed by treatment with thiols [76].

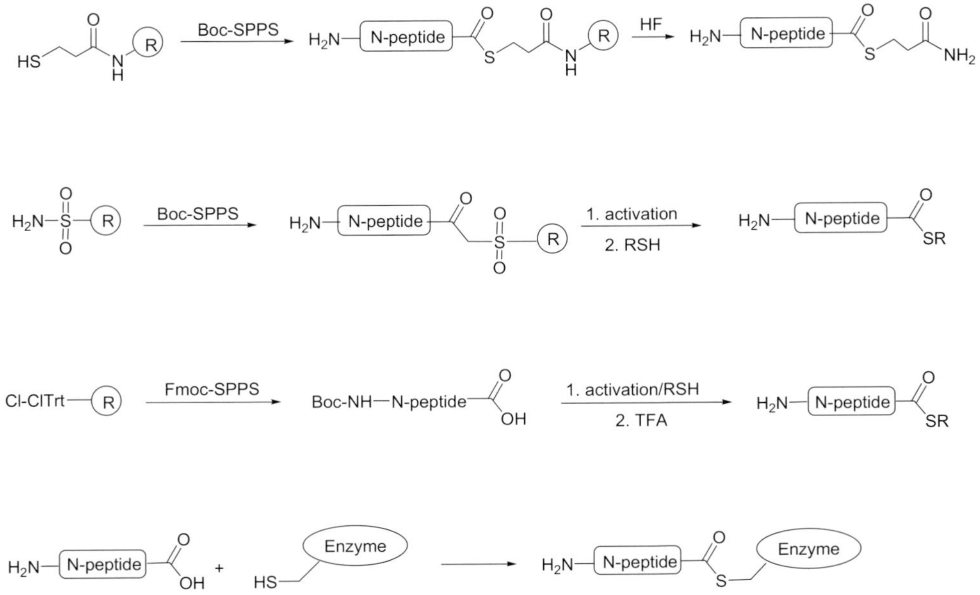

Scheme 11.16 SPPS of peptide thioesters.

Peptide thioesters can also be obtained in solution by thioesterification of a protected peptide segment. Resins of the trityl type such as CLTR [77] and trityl-TGT resins [78] are more frequently used for the preparation of peptide thioesters following this methodology. By combination of synthetic peptide chemistry and enzymatic catalysis, thioesters can also be produced by treatment of peptides with suitable enzymes [79].

The value of NCL for the synthesis of proteins and complex and modified polypeptides is further enhanced by the fact that internal Cys ligation still proceeds fast if the Cys residue is positioned two to six amino acid residues from the N-terminus [80]. Homo-Cys can also be used as the N-terminal amino acid of the *N*-peptide. Methylation of the thiol group results in proteins containing a Met residue at the ligation site [81]. Finally, NCL can be performed repeatedly to assemble larger polypeptides from more than two unprotected fragments [82].

An additional advantage of NCL is the mild electrophilic activation of the *N*-peptide as a thioester during the condensation with the *C*-peptide. As a result, less racemization is observed compared to protected fragments activated by carbodiimides or uronium salts used in convergent synthesis. Nevertheless, the ligation site, if possible, must be selected carefully. Pro and sterically hindered amino acids such as Val and Ile as the thioester component ligate sluggishly [83]. Racemization-prone amino acids such as His and Phe should be avoided as the C-terminal residue of the *N*-peptide. In the SPPS of peptide thioesters, the risk of diketopiperazine formation is higher due to the electrophilic activation of the C-terminal amino acid thioester. This may result in the formation of truncated peptides lacking the C-terminal dipeptide and larger C-terminal sequences, especially in cases where Gly and Pro are contained in the C-terminal dipeptide.

## 11.12
## SPPS of Peptides Modified at their C-Terminus

Besides the commonly synthesized peptide acids and amides, a large number of C-terminally modified peptides are required for studies as potential pharmaceuticals and in structure–activity relationship (SAR) analyses. Such modified peptides are produced by solid-phase synthesis using suitably loaded resins. Of great importance are the peptide amino alcohols (peptaibols), which are widely distributed in nature. A prominent example among the ion channel and antibiotic peptaibols is alamethicin [84]. Peptaibols are synthesized by SBS methods on suitable resins. Naturally occurring peptaibols contain bulky α-aminoisobutyric acid (Aib) residues and are therefore difficult to assemble. The amino acids used are therefore strongly activated (e.g., as amino acyl fluorides) in order to achieve acceptable coupling yields for those regions containing Aib residues [85].

In most cases, peptides functionalized at their C-terminus are synthesized by introducing the required functionality onto a suitable resin. Scheme 11.17 illustrates some important compound classes, which are obtained by SBS synthesis using CLTR. Besides peptaibols, peptide amino thiols [86] and peptide aminoalkyl

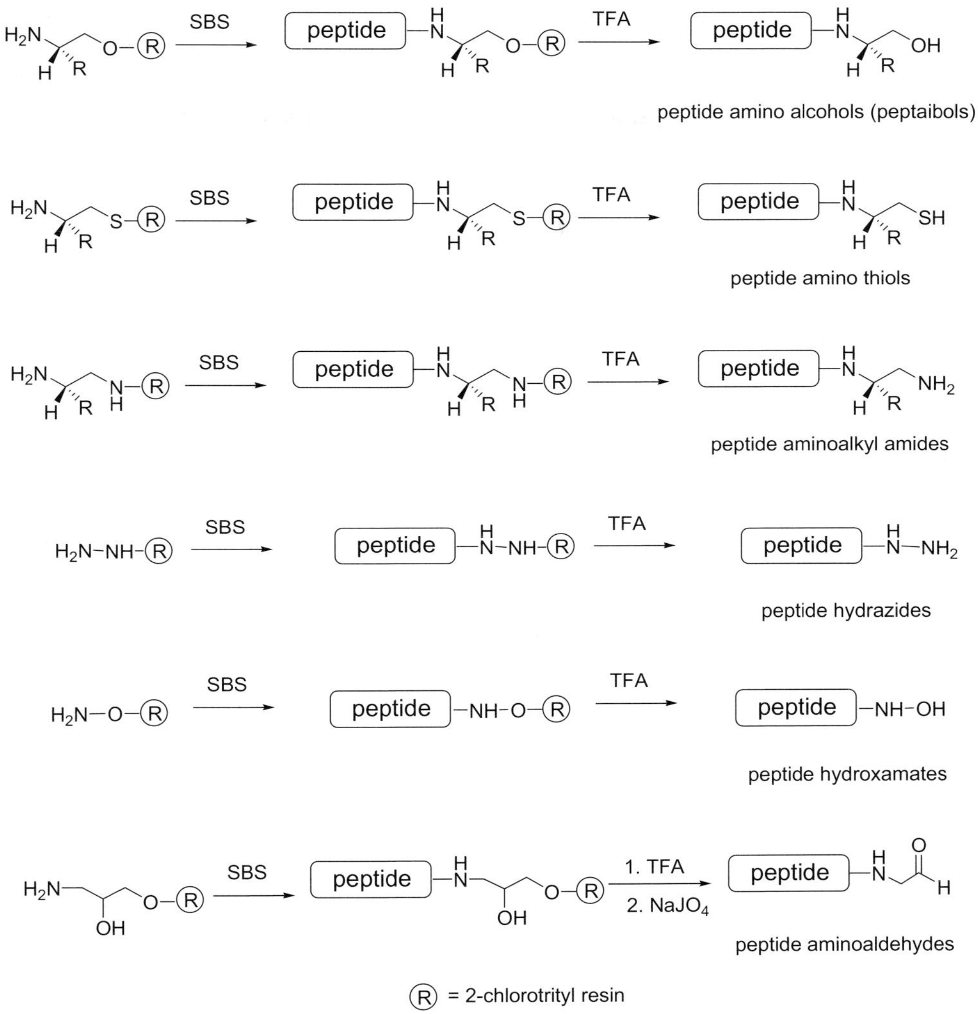

SBS TFA

peptide amino alcohols (peptaibols)

SBS TFA

peptide amino thiols

SBS TFA

peptide aminoalkyl amides

$H_2N$-NH-(R) SBS peptide -NH-NH-(R) TFA peptide -NH-$NH_2$

peptide hydrazides

$H_2N$-O-(R) SBS peptide -NH-O-(R) TFA peptide -NH-OH

peptide hydroxamates

SBS 1. TFA 2. $NaJO_4$

peptide aminoaldehydes

(R) = 2-chlorotrityl resin

**Scheme 11.17** Solid-phase synthesis of peptides modified at their C-terminus.

amides [87] including spermine derivatives [88], peptide hydroxamates [89], peptide hydrazides [90], and peptide aldehydes are frequently required derivatives [91]. Peptide hydroxamates can be also obtained using SASRIN resin [92], which is also suitable to prepare aminothiol peptides. Peptide aldehydes, used extensively for the introduction of reduced peptide bonds into a peptide chain and in chemoselective ligation, are routinely produced using resins loaded with derivatives of Weinreb amide. The peptide aldehydes are released from these resins upon treatment with aluminum hydrides [93].

## 11.13
## Side-Chain-Modified Peptides

A large number of peptides and proteins with modified side-chains occurs naturally and are of high biological importance. The most essential among them are glycosylated, phosphorylated, sulfated and lipidated peptides. Examples for side-chain modifications are peptides carrying a biotinyl or fluorescent label, branched [94], multiple antigenic peptides (MAPs) [95], and multi-branched peptides (e.g., dendrimers) [96], which are of significance as biochemical research tools. Side-chain-modified peptides are synthesized mainly via two solid-phase routes. The first route utilizes modified amino acid building blocks, which already contain a proper modifying group such as a carbohydrate [97], protected phosphoryl group [98], sulfate group [99], affinity tag [100], PEG chain [101], or a chromophore at the respective side-chain. In the second general approach, a peptide containing a semipermanent protecting group is first assembled on a suitable resin. In Boc/Bzl synthesis, *t*Bu-type protecting groups can be used that are orthogonal to the protection schemes and resins used. For this approach Fmoc chemistry is preferred over Boc chemistry, and the most frequently applied semipermanent protecting groups are Alloc [102], Mtt [103], and Dde for the side-chain of Lys and Orn [104], the Allyl [102] and Dmb groups for Glu and Asp [105], the Trt group [106] for Ser, Thr, and Hse, the 2-Cl-Trt group for Tyr, and the Mmt group for Cys [107] and His [108] (Scheme 11.18). The removal of the acid-sensitive groups of the Trt type and Dmb is not compatible with highly acid-sensitive resins such as the CLT, SASRIN and the Sieber amide resin. More acid-stable resins, such as the Wang and the Rink amide resin have to be used in these cases. After the selective removal of the side-chain protection, the modifying groups are introduced by appropriate specific methods, followed by the cleavage of the peptide from the resin and its deprotection (Scheme 11.19).

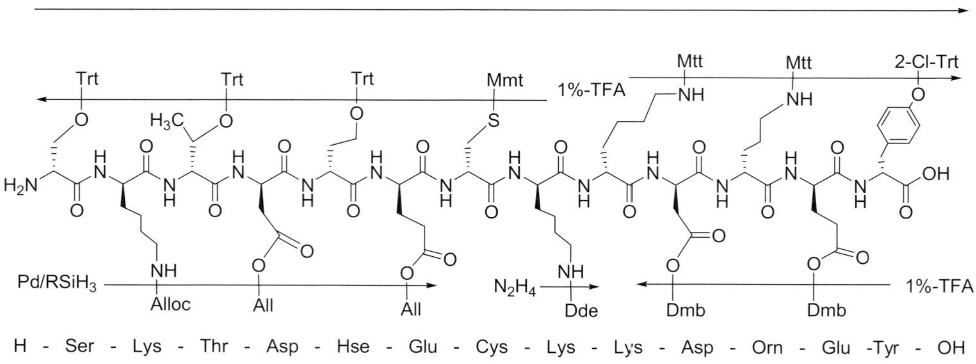

**Scheme 11.18** Selective side-chain deprotection of varicus trifunctional amino acids.

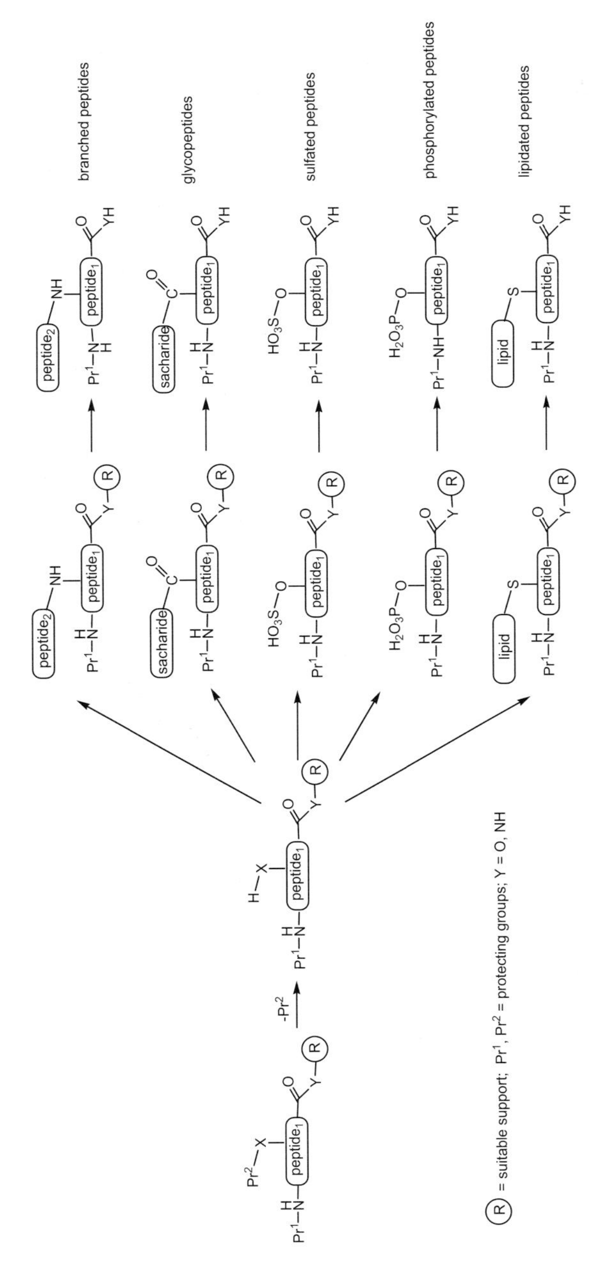


**Scheme 11.19** Solid-phase synthesis of side-chain-modified peptides.

## 11.14
## Cyclic Peptides

Several cyclic peptides and depsipeptides display various important biological activities. Cyclosporine is the most prominent example of such peptides. Cyclosporine, a potent immunosuppressant, has considerably boosted the efforts to improve the technology of the synthesis of cyclic peptides. Apart from the long known antibiotic peptides, the extremely cystine-rich cyclotides are a recently discovered class of highly biologically active cyclic peptides [109]. The interest in cyclotides is increasing because, in addition to their biological activity, they show considerable resistance against enzymatic degradation. Enzymatic degradation is still a major drawback for the pharmaceutical application of peptides. The cyclization of peptides leads generally to the restriction of their conformational freedom. Therefore, cyclic peptides often interact more favorably with their receptors or binding partners as compared to the corresponding open-chain peptides.

The general strategy utilized for the synthesis of head-to-tail cyclic peptides is based on the assembly of the peptide chain on a suitable resin. Two alternative routes are available to achieve cyclization: the on-resin cyclization and the cyclization of a suitably protected peptide including the cyclization of a peptide thioester [110].

For the on-resin cyclization, two synthetic strategies are applicable (Scheme 11.20). In the first route, the peptide is linked to a resin such as an oxime resin [111] or a safety-catch resin [112], which act as electrophilic activators of the C-terminal carboxyl group of the resin-bound peptide. In these cases, cyclization occurs after deprotection of the N-terminus by treatment with bases. In the second approach the peptide is bound to a suitable resin through a side-chain functional group or through its backbone [113]. The side-chain functional groups mainly used for this approach are carboxylic groups of Asp [114] and Glu [115], basic side-chain functions of Lys [116] and His [117], hydroxyl and phenoxy amino acids [118, 119], and the thiol group of Cys [11]. Cyclization is then carried out on-resin by similar methods applied for fragment condensation in solution (Scheme 11.20). It is believed that although the

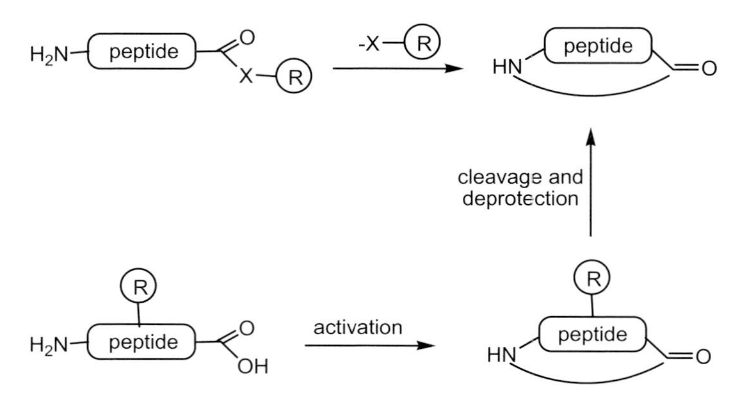


**Scheme 11.20** General routes for the solid-phase synthesis of head-to-tail cyclic peptides.

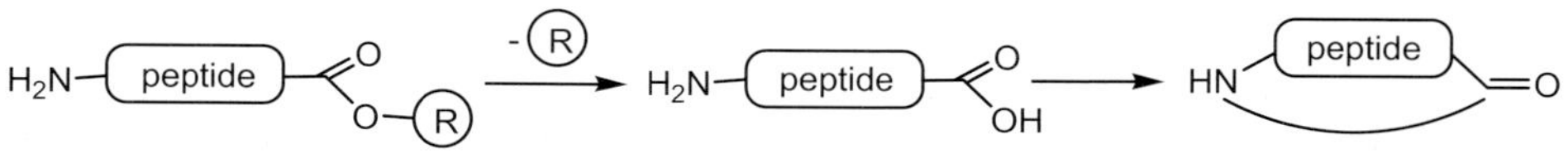

Scheme 11.21 Hybrid synthesis of head-to-tail cyclic peptides.

peptide is present in a high concentration on the resin-beads, a pseudo-dilution effect reduces the oligomerization and polymerization risk during the cyclization reaction [120]. Nevertheless, in comparative studies cyclization in solution under high dilution was proved superior over on-resin cyclization [121]. The chemical parameters for cyclization reactions such as the type of activation reagent and solvent were investigated in several studies [122]. However, a general trend or even a rule for optimum cyclization conditions is not obvious and it seems that cyclization reactions have to be optimized for any individual peptide.

Cyclization in solution (Scheme 11.21) can be performed under high dilution reducing the risk of oligomerization. In general, similar methods are used for the preparation of protected peptides as precursors of cyclic peptides compared with those applied for the synthesis of protected peptide fragments. If possible, Gly or Pro are selected as the C-terminal amino acid due to their racemization safety. Usually resins of the trityl type [123] or dialkoxybenzyl type [124] are applied for the preparation of the required protected linear peptides [97, 125]. In an interesting variation, peptide thioesters are synthesized, and the selective cyclization is carried out by thioesterase catalysis [126] and NCL methodology [127]. The depsipeptide method can also utilized for the preparation of cyclic peptides in solution [128].

For the preparation of cyclic peptides where side-chain functions contribute to the cyclization, selectively removable, semipermanent protecting groups are required. For example, the side-chain to side-chain lactamization (Scheme 11.22) requires a protected peptide, selectively deprotected at the side-chain of a diamino (e.g., Lys) and of a diacidic (e.g., Asp, Glu) amino acid. A frequently used pair of protecting groups allowing selective removal in the presence of Fmoc/*t*Bu- and Boc/Bzl-type protection and various resins is the pair is All/Alloc, which can be cleaved by palladium-catalyzed hydrogenation in the presence of silanes [129]. Other semipermanent pairs of protecting groups have also been suggested for the preparation of cyclic peptides and depsipeptides [130]. Side-chain-to-tail [131] and head-to-tail cyclic (Scheme 11.23) peptides are prepared by similar methods.

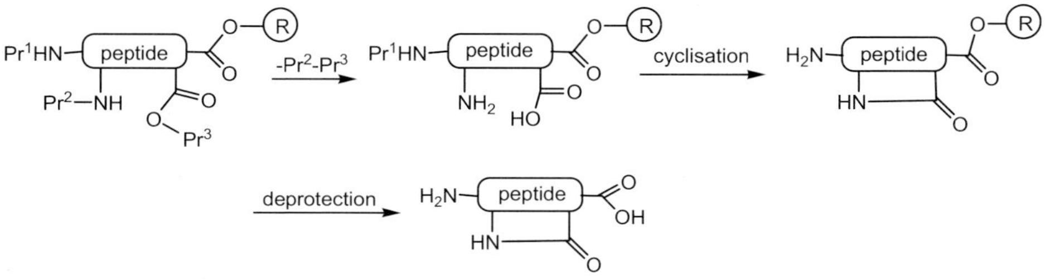

Scheme 11.22 Solid-phase synthesis of side-chain-to-side-chain cyclic peptides.

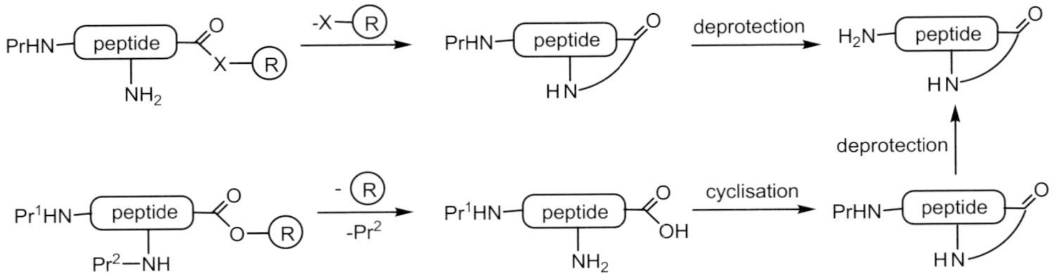

**Scheme 11.23** Solid-phase and hybrid synthesis of head-to-tail cyclic peptides.

## 11.15
## Large-Scale Solid-Phase Synthesis

After the identification of a new peptide pharmaceutical, a generally pursued strategy applied for the production of clinical material and subsequent commercial material is the following. First, a laboratory scale synthesis is developed to quickly supply small amounts of material for SAR, biological activity, and toxicology. Next, the development of an improved process for material used in early clinical phase I trials and then for phase II and III clinical supplies is implemented, which is finally capable of producing the necessary commercial quantities.

In the first instance, a main consideration for such production processes is the time in which the peptide can be obtained in sufficient amounts. Then, for commercial processes several other criteria are important, notably cost and quality of goods, environmental considerations, and the ability to produce estimated commercial product volumes. In the case of a moderately or highly complex peptide requiring a large volume of material (e.g., the 36-residue enfuvirtide manufactured by a convergent synthesis route using three fragments), the production costs can become critical for the commercial success of the peptide in the market. Therefore, the synthetic strategy for the commercial manufacturing process has to be carefully evaluated and tested.

The first decision to be made during the development of a commercial process for large quantities of a peptide is the strategy of synthesis. The selection of the methods depends predominantly on the peptide's length and complexity. Peptides of small size up to 15 amino acid residues without the need of a post-synthetic modification can be synthesized following the linear SBS route. Peptides of medium size, containing 15–70 amino acid residues in their peptide chain, are more economically synthesized by the hybrid method applying protected peptide fragments. For larger peptides and proteins containing Cys, a mixed ligation–hybrid process may be the method of choice.

The main factors determining the cost of the SPPS are amino acid derivatives, solvents, resins, coupling and deprotection reagents, and materials applied in the peptide purification. The solvent requirements become higher with a decrease in the loading of the resin. For example, in the linear synthesis of a 30mer peptide, a resin loading of 0.33 mmol/g for the first amino acid will ensure that the peptide/resin mass ratio does not differ significantly from 1:1 at the end of the synthesis and adverse resin conditions will be avoided [132]. If the content of the peptide becomes higher, the resin loses much of its desired properties and the synthesis can become problematic or even impossible.

Comparing the linear synthesis of a given 30mer peptide with a three-fragment hybrid synthesis, a higher resin loading of 1.0 mmol/g can be used to manufacture the intermediate fragments. This higher resin loading reduces the solvent requirements by approximately one-third while resin usage is equal. In addition, the concentration of the applied activated amino acids is 3 times greater in the case of the hybrid route than of the linear route, by using the same quantity of amino acid. The hybrid route requires therefore lower excess of reagents and shorter reaction times for each coupling and deprotection step. Consequently, the material costs of the hybrid route will be almost 50% less as compared with the linear route. This difference increases with the length of the peptide chain.

For the effective Good Manufacturing Practice production of long-chain pharmaceutical peptides (longer than 50 amino acids) the ligation and hybrid methods should be compared. The main advantage of the ligation method over the hybrid technology is that HPLC-purified peptides are ligated. Therefore, the final crude product obtained is usually of higher purity than the product obtained by the hybrid method. However, this advantage in purity comes with higher manufacturing costs. If a peptide is synthesized by $n$ ligation steps, then $2n + 1$ purifications and isolations including costly lyophilization steps are needed, considerably increasing the production cost of the large peptide. An additional cost disadvantage of NCL is connected to the synthesis of the peptide thioesters. Usually the synthesis of the peptide thioesters requires handling HF, which adds considerable costs by the required special manufacturing equipment, processing costs, and expenses related to HF safety and environmental issues. In summarizing, there are many chemical parameters as well as safety and environmental considerations that must be carefully analyzed in designing a cost-effective commercially viable manufacturing process for peptides.

## 11.16 Conclusions

Peptide synthesis, especially at larger scale, is possibly one of the most challenging chemical problems faced today by synthetic chemists and the pharmaceutical industry. Many peptide and protein pharmaceuticals have never reached the market because their synthesis was considered impossible or too costly. Despite the recent advances in peptide synthetic methodology, there is still great need for considerable improvements. The ultimate goal to reach Nature's perfection in synthesizing peptides is still far away.

## References


1 Merrifield, R.B. (1985) Solid phase synthesis. *Angewandte Chemie*, **97**, 801–812; *Angewandte Chemie (International Edition in English)*, **24**, 799–810.

2 Hirschmann, R., Nutt, R.F., Veber, D.F., Vitali, R.A., Varga, S.L., Jacob, T.A., Holly, F.W., and Denkewalter, R.G. (1969) Total synthesis of an enzyme. V. Preparation of enzymatically active material. *Journal of the American Chemical Society*, **91**, 507–508.

3 Kaiser, E.T. (1989) Synthetic approaches to biologically active peptides and proteins including enzymes. *Accounts of Chemical Research*, **22**, 47–54.

4 Aimoto, S., Mizoguchi, N., Hojo, H., and Yoshimura, S. (1989) Development of a facile method for polypeptide synthesis. Synthesis of bovine pancreatic trypsin inhibitor (BPTI). *Bulletin of the Chemical Society of Japan*, **62**, 524–531.

5 Romovacek, H., Dowd, S.R., Kawasaki, K., Nishi, N., and Hofmann, K. (1979) Studies on polypeptides. 54. The synthesis of a peptide corresponding to position 24–104 of the peptide chain of ribonuclease T1. *Journal of the American Chemical Society*, **101**, 6081–6091.

6 Meienhofer, J. (1981) Why peptide synthesis continues to remain a formidable challenge. *Biopolymers*, **20**, 1761–1784.

7 Yajima, H. and Fujii, N. (1981) Studies on peptides. 103. Chemical synthesis of a crystalline protein with the full enzymic activity of ribonuclease A. *Journal of the American Chemical Society*, **103**, 5867–5871.

8 Tam, J.P., Xu, J., and Eom, K.D. (2001) Methods and strategies of peptide ligation. *Biopolymers*, **60**, 194–205.

9 Dawson, P.E. and Kent, S.B.H. (2000) Synthesis of native proteins by chemical ligation. *Annual Review of Biochemistry*, **69**, 923–960.

10 Albericio, F. (2004) Developments in peptide and amide synthesis. *Current Opinion in Chemical Biology*, **8**, 211–221.

11 Barlos, K. and Gatos, D. (2000) In *Fmoc Solid Phase Synthesis: A Practical Approach* (eds W.C. Chan and P.D. White), Oxford University Press, New York, pp. 215–227.

12 Athanassopoulos, P., Barlos, K., Gatos, D., Hatzi, O., and Tzavara, C. (1995) Application of 2-chlorotrityl chloride in convergent peptide synthesis. *Tetrahedron Letters*, **36**, 5645–5648.

13 Sohma, Y., Chiyomori, Y., Kimura, M., Fukao, F., Taniguchi, A., Hayashi, Y., Kimura, T., and Kiso, Y. (2005) "*O*-Acyl isopeptide method" for the efficient preparation of amyloid β peptide 1–42 mutants. *Bioorganic and Medicinal Chemistry*, **13**, 6167–6174.

14 Sohma, Y. and Kiso, Y. (2006) "Click Peptides"-chemical biology-oriented synthesis of Alzheimer's disease-related amyloid β peptide (Aβ) analogues based on the "*O*-acyl isopeptide method". *ChemBioChem*, **7**, 1549–1557.

15 Atherton, E., Brown, E., Sheppard, R.C., and Rosevear, A. (1981) A physically supported gel polymer for low pressure, continuous flow solid phase reactions. Application to solid phase peptide synthesis. *Journal of the Chemical Society, Chemical Communications*, 1151–1152.

16 Atherton, E. and Sheppard, R.C. (1989) *Solid Phase Peptide Synthesis: A Practical Approach*, IRL Press, Oxford, p. 25.

17 Merrifield, R.B. (1964) Solid phase synthesis. III. An improved synthesis of bradykinin. *Biochemistry*, **3**, 1385–1390.

18 Mitchell, A.R., Kent, S.B.H., Engelhard, M., and Merrifield, R.B. (1978) A new synthetic route to *tert*-butyloxycarbonylaminoacyl-4-(oxymethyl)phenylacetamidomethyl-resin, an improved support for solid-phase peptide synthesis. *The Journal of Organic Chemistry*, **43**, 2845–2852.

19 Tam, J.P., Kent, S.B.H., Wong, T.W., and Merrifield, R.B. (1979) Improved synthesis of 4-(Boc-aminoacyloxymethyl) phenylacetic acids for use in solid-phase peptide synthesis. *Synthesis*, 955–957.

20 Matsueda, G.R. and Stewart, J.M. (1981) A *p*-methylbenzhydrylamine resin for

improved solid-phase synthesis of peptide amides. *Peptides*, **2**, 45–50.

21 Pietta, P.G. and Marshall, G.R. (1970) Amide protection and amide supports in solid-phase peptide synthesis. *Journal of the Chemical Society, Chemical Communications*, 650–651.

22 Wang, S.-S. (1973) p-Alkoxybenzyl alcohol resin and *p*-alkoxybenzyloxycarbonylhydrazide resin for solid phase synthesis of protected peptide fragments. *Journal of the American Chemical Society*, **95**, 1328–1333.

23 Mergler, M., Tanner, R., Gosteli, J., and Grogg, P. (1988) Peptide synthesis by a combination of solid-phase and solution methods I: a new very acid-labile anchor group for the solid phase synthesis of fully protected fragments. *Tetrahedron Letters*, **29**, 4005–4008.

24 Mergler, M., Nyfeler, R., and Gosteli, J. (1989) Peptide synthesis by a combination of solid-phase and solution methods III: resin derivatives allowing minimum-racemization coupling of $N^{\alpha}$-protected amino acids. *Tetrahedron Letters*, **30**, 6741–6744.

25 Riniker, B., Flörsheimer, A., Fretz, H., Sieber, P., and Kamber, B. (1993) A general strategy for the synthesis of large peptides: the combined solid-phase and solution approach. *Tetrahedron*, **49**, 9307–9320.

26 Barlos, K., Mamos, P., Papaioannou, D., Patrianakou, S., Sanida, C., and Schafer, W. (1987) Application of the Trt and Fmoc groups for the protection of polyfunctional α-amino acids. *Liebigs Annalen der Chemie*, 1025–1030.

27 Barlos, K., Gatos, D., Hondrelis, J., Matsoukas, J., Sotiriou, P., Moore, G.J., and Schäfer, W. (1989) Preparation of new acid-labile resins of the sec-alcohol type and their application in peptide synthesis. *Liebigs Annalen der Chemie*, 951–955.

28 Barlos, K., Gatos, D., Kallitsis, J., Papaphotiu, G., Sotiriu, P., Wenqing, Y., and Schäfer, W. (1989) Synthesis of protected peptide-fragments using substituted triphenylmethyl resins. *Tetrahedron Letters*, **30**, 3943–3946.

29 Bayer, E. and Goldammer, C. (1994) Solid-phase system containing a trityl group, process for its preparation and its use in solid-phase reactions. Patent DE4306839 (A1).

30 Rink, H. (1987) Solid-phase synthesis of protected peptide fragments using a trialkoxy-diphenyl-methylester resin. *Tetrahedron Letters*, **28**, 3787–3790.

31 Bernatowicz, M.S., Daniels, S.B., and Köster, H. (1989) A comparison of acid labile linkage agents for the synthesis of peptide C-terminal amides. *Tetrahedron Letters*, **30**, 4645–4648.

32 Sieber, P. (1987) A new acid-labile anchor group for the solid-phase synthesis of C-terminal peptide amides by the Fmoc method. *Tetrahedron Letters*, **28**, 2107–2110.

33 Albericio, F. and Barany, G. (1987) An acid-labile anchoring linkage for solid-phase synthesis of C-terminal peptide amides under mild conditions. *International Journal of Peptide and Protein Research*, **30**, 206–216.

34 Albericio, F., Kneib-Cordonier, N., Biancalana, S., Gera, L., Masada, R.I., Hudson, D., and Barany, G. (1990) Preparation and application of the 5-(4-(9-fluorenylmethyloxycarbonyl) aminomethyl-3,5-dimethoxyphenoxy)-valeric acid (PAL) handle for the solid-phase synthesis of C-terminal peptide amides under mild conditions. *The Journal of Organic Chemistry*, **55**, 3730–3743.

35 Bayer, E. and Rapp, W. (1990) Graft copolymers of crosslinked polymers and polyoxyethylene, processes for their production, and their usage. US Patent 4908405.

36 Barlos, K., Gatos, D., Papaphotiou, G., and Schäfer, W. (1993) Synthesis of Calcitonin derivatives by fragment condensation in solution and on 2-chlorotrityl resin. *Liebigs Annalen der Chemie*, 215–220.

37 Bray, B.L. (2003) Large-scale manufacture of peptide therapeutics by chemical synthesis. *Nature Reviews Drug Discovery*, **2**, 587–593.

38 Sarin, V.K., Kent, S.B.H., Mitchell, A.R., and Merrifield, R.B. (1984) A general

approach to the quantitation of synthetic efficiency in solid-phase peptide synthesis as a function of chain length. *Journal of the American Chemical Society*, **106**, 7845–7850.

39 Atherton, E., Benoiton, N.L., Brown, E., Sheppard, R.C., and Williams, B.J. (1981) Racemization of activated, urethane-protected amino-acids by *p*-dimethyl-aminopyridine. Significance in solid-phase peptide synthesis. *Journal of the Chemical Society, Chemical Communications*, 336–337.

40 Benoiton, N.L. (1996) 2-Alkoxy-5(4*H*)-oxazolones and the enantiomerization of *N*-alkoxycarbonylamino acids. *Biopolymers*, **40**, 245–254.

41 Benoiton, N.L. (2005) *Chemistry of Peptide Synthesis*, CRC Press, Boca Raton, FL.

42 Barlos, K., Gatos, D., Kallitsis, J., Papaioannou, D., Sotiriou, P., and Schäfer, W. (1987) Anchoring of amino-acids on hydroxyl groups containing resins and their application to peptide-synthesis using *N*-tritylamino acid 1-benzotriazolylesters. *Liebigs Annalen der Chemie*, 1031–1035.

43 Gisin, B.F. (1973) The preparation of Merrifield-resins through total esterification with cesium salts. *Helvetica Chimica Acta*, **56**, 1476–1482.

44 Barlos, K., Chatzi, O., Gatos, D., and Stavropoulos, G. (1991) 2-Chlorotrityl chloride resin. Studies on anchoring of Fmoc-amino acids and peptide cleavage. *International Journal of Peptide and Protein Research*, **37**, 513–520.

45 Barlos, K. and Gatos, D. (2000) In *Fmoc Solid Phase Peptide Synthesis: A Practical Approach* (eds W.C. Chan and P.D. White), Oxford University Press, New York, pp. 215–227.

46 Jensen, K.J., Alsina, J., Songster, M.F., Vagner, J., Albericio, F., and Barany, G. (1998) Backbone amide linker strategy for solid-phase synthesis of C-terminal-modified and cyclic peptides. *Journal of the American Chemical Society*, **120**, 5441–5452.

47 Barlos, K., Gatos, D., Kallitsis, I., Papaioannou, D., and Sotiriou, P. (1988) Application of 4-polystyryltriphenyl-methyl chloride to the synthesis of peptides and amino-acid derivatives. *Liebigs Annalen der Chemie*, 1079–1081.

48 Barany, G. and Merrifield, R.B. (1980) In *The Peptides: Analysis, Synthesis, Biology* (eds E. Gross and E. Meienhofer), Academic Press, London, pp. 159–163.

49 Dourtoglou, V., Ziegler, J.-C., and Gross, B. (1978) L'hexafluorophosphate de *O*-benzotriazolyl-*N,N*-tetramethyluronium: un reactif de couplage peptidique nouveau et efficace. *Tetrahedron Letters*, **19**, 1269–1272.

50 Dourtoglou, V., Gross, B., Lambropoulou, V., and Zioudrou, C. (1984) O-benzotriazolyl-*N,N,N′,N′*-tetramethyluronium hexafluoro-phosphate as coupling reagent for the synthesis of peptides of biological interest. *Synthesis*, 572–574.

51 Carpino, A.L. (1993) 1-Hydroxy-7-azabenzotriazole. An efficient peptide coupling additive. *Journal of the American Chemical Society*, **115**, 4397–4398.

52 König, W. and Geiger, R. (1970) New method for the synthesis of peptides: activation of carboxyl group with dicyclohexylcarbodiimide using 1-hydroxybenzotriazoles as additives. *Chemische Berichte*, **103**, 788–798.

53 (a) Benoiton, N.L. and Kuroda, K. (1981) Studies on racemization during couplings using a series of model tripeptides involving activated residues with unfunctionalized side-chains. *International Journal of Peptide and Protein Research*, **17**, 197–204; (b) Benoiton, N.L., Kuroda, K., and Chen, F.M.F. (1979) Series of lysyldipeptide derivatives for racemization studies in peptide-synthesis. *International Journal of Peptide and Protein Research*, **13**, 403–408; (c) Griehl, C., Kolbe, A., and Merkel, S. (1996) Quantitative description of epimerization pathways using the carbodiimide method in the synthesis of peptides. *Journal of the Chemical Society, Perkin Transactions 2*, 2525–2529.

54 (a) Story, S.C. and Aldrich, J.V. (1994) Side-product formation during cyclization with HBTU on a solid support. *International Journal of Peptide and Protein Research*, **43**, 292–296; (b) Hachmann, J. and Lebl, M. (2006) Search

for optimal coupling reagent in multiple peptide synthesizer. *Biopolymers*, **84**, 340–347.

**55** Fischer, P.M. (2003) Diketopiperazines in peptide and combinatorial chemistry. *Journal of Peptide Science*, **9**, 9–35.

**56** Bodi, J., Süli-Vargha, H., Ludhnyi, K., Vekey, K., and Orosz, G. (1997) New strategy for the synthesis of large peptides as applied to the C-terminal cysteine-rich 41 amino acid fragment of the mouse agouti protein. *Tetrahedron Letters*, **38**, 3293–3296.

**57** Bollhagen, R., Schmiedberger, M., Barlos, K., and Grell, E. (1994) A new reagent for the cleavage of fully protected peptides synthesised on 2-chlorotrityl chloride resin. *Journal of the Chemical Society, Chemical Communications*, 2559–2560.

**58** Barlos, K., Gatos, D., Kutsogianni, S., Papaphotiou, G., Poulos, C., and Tsegenidis, T. (1991) Solid-phase synthesis of partially protected and free peptides containing disulfide bonds by simultaneous cysteine oxidation-release from 2-chlorotrityl resin. *International Journal of Peptide and Protein Research*, **38**, 562–568.

**59** Chiva, C., Barthe, P., Codina, A., Gairi, M., Molina, F., Granier, C., Pugniere, M., Inui, T., Nishi, H., Nishiuchi, Y., Kimura, T., Sakakibara, S., Albericio, F., and Giralt, E. (2003) Synthesis and NMR structure of P41icf, a potent inhibitor of human cathepsin L. *Journal of the American Chemical Society*, **125**, 1508–1517.

**60** (a) Makino, T., Matsumoto, M., Suzuki, Y., Kitajima, Y., Yamamoto, K., Kuramoto, M., Minamitake, Y., Kangawa, K., and Yabuta, M. (2005) Semisynthesis of human ghrelin: condensation of a Boc-protected recombinant peptide with a synthetic *O*-acylated fragment. *Biopolymers*, **79**, 238–247; (b) Krambovitis, E., Hatzidakis, G., and Barlos, K. (1998) Preparation of MUC-1 oligomers using an improved convergent solid-phase peptide synthesis. *The Journal of Biological Chemistry*, **273**, 10874–10879; (c) Felix, A.M., Zhao, Z., Lambros, T., Ahmad, M., Liu, W., Daniewski, A., Michalewsky, J., and Heimer, E.P. (1998) Combined solid-phase/solution synthesis of a 31-residue vasoactive intestinal peptide analog: general method for repetitive coupling of fragments without isolation and purification of intermediates. *The Journal of Peptide Research*, **52**, 155–164; (d) Barlos, K., Gatos, D., and Schäfer, W. (1991) Synthesis of prothymosin α (Prot-α): a protein consisting of 109 amino acid residues. *Angewandte Chemie (International Edition in English)*, **30**, 590–593; (e) Goulas, S., Gatos, D., and Barlos, K. (2006) Convergent solid-phase synthesis of hirudin. *Journal of Peptide Science*, **12** 116–123.

**61** Ewing, W.R., Mapelli, C., Sulsky, R.B., Haque, T., Lee, V.G., Riexinger, D.J., Martinez, R.L., and Zhu, Y. (2008) Human glucagon-like-peptide-1 modulators and their use in treatment of diabetes and related conditions. US Patent 7417028.

**62** (a) Blake, J. and Li, C.H. (1981) New segment-coupling method for peptide synthesis in aqueous solution: application to synthesis of human [Gly17]-β-endorphin. *Proceedings of the National Academy of Sciences of United States of America*, **78**, 4055–4058; (b) Kitagawa, K., Adachi, H., Sekigawa, Y., Yagami, T., Futaki, S., Gu, Y.J., and Inoue, K. (2004) Total chemical synthesis of large CCK isoforms using a thioester segment condensation approach. *Tetrahedron*, **60**, 907–918; (c) Kitagawa, K., Aida, C., Fujiwara, H., Yagami, T., Futaki, S., Kogire, M., Ida, J., and Inoue, K. (2001) Facile solid-phase synthesis of sulfated tyrosine-containing peptides: total synthesis of human big gastrin-II and cholecystokinin (CCK)-39. *The Journal of Organic Chemistry*, **66**, 1–10; (d) Aimoto, S. (1999) Polypeptide synthesis by the thioester method. *Biopolymers*, **51** 247–265.

**63** Toniolo, C., Bonora, G.M., Mutter, M., and Pillai, V.N.R. (1981) Linear oligopeptides. 78. The effect of the insertion of a proline residue on the solution conformation of host peptides. *Makromolekulare Chemie*, **182**, 2007–2014.

64 King, B.T., Bury, P.A., Gabel, R.A., Crider, J.E., Carr, R.T. II, DeHoff, and Bradley, S. (2009) Insulinotropic peptide synthesis using solid and solution phase combination techniques. US Patent Application 0149628.

65 (a) Mutter, M., Nefzi, A., Sato, T., Sun, X., Wahl, F., and Wuhl, T. (1995) Pseudo-prolines (ψPro) for accessing "inaccessible" peptides. *Peptide Research*, **8**, 145–153; (b) Wöhr, T., Wahl, F., Nefzi, A., Rohwedder, B., Sato, T., Sun, X.C., and Mutter, M. (1996) Pseudo-prolines as a solubilizing, structure-disrupting protection technique in peptide synthesis. *Journal of the American Chemical Society*, **118**, 9218–9227.

66 (a) Johnson, T., Quibell, M., Owen, D., and Sheppard, R.C. (1993) A reversible protecting group for the amide bond in peptides – use in the synthesis of difficult sequences. *Journal of the Chemical Society, Chemical Communications*, 369–372; (b) Hyde, C., Johnson, T., Owen, D., Quibell, M., and Sheppard, R.C. (1994) Some difficult syntheses made easy. A study of interchain association in solid-phase peptide synthesis. *International Journal of Peptide and Protein Research*, **43**, 431–440.

67 Gatos, D., Patrianakou, S., Hatzi, O., and Barlos, K. (1997) Comparison of the stepwise and convergent approaches in solid-phase synthesis of rat $Tyr^0$-atriopeptin II. *Letters in Peptide Science*, **4**, 177–184.

68 (a) Coin, I., Schmieder, P., Bienert, M., and Beyermann, M. (2008) The depsipeptide technique applied to peptide segment condensation: scope and limitations. *Journal of Peptide Science*, **14**, 299–306; (b) Barlos, K. and Gatos, D. (1999) 9-Fluorenylmethyloxycarbonyl/*t*Butyl-based convergent protein synthesis. *Biopolymers*, **51**, 266–278.

69 (a) Schulz, A., Klüver, E., Schulz-Maronde, S., and Adermann, K. (2005) Engineering disulfide bonds of the novel human β-defensins hBD-27 and hBD-28: differences in disulfide formation and biological activity among human β-defensins. *Biopolymers*, **80**, 34–49; (b) Klüver, E., Schulz-Maronde, S., Scheid, S., Meyer, B. Forssmann, W.-G., and Adermann, K. (2005) Structure–activity relation of human β-defensin 3: influence of disulfide bonds and cysteine substitution on antimicrobial activity and cytotoxicity. *Biochemistry*, **44**, 9804–9816.

70 (a) Kent, S.B.H. (2009) Total chemical synthesis of proteins. *Chemical Society Reviews*, **38**, 338–351; (b) Hackenberger, C.P.R. and Schwarzer, D. (2008) Chemoselective ligation and modification strategies for peptides and proteins. *Angewandte Chemie International Edition*, **47**, 10030–10074; (c) David, R., Richter, M.P.O., and Beck-Sickinger, A.G. (2004) Expressed protein ligation. *European Journal of Biochemistry*, **271**, 663–677.

71 (a) Dawson, P.E. and Kent, S.B.H. (2000) Synthesis of native proteins by chemical ligation. *Annual Review of Biochemistry*, **69**, 923–960; (b) Schnolzer, M. and Kent, S.B. (1992) Constructing proteins by dovetailing unprotected synthetic peptides: backbone-engineered HIV protease. *Science*, **256**, 221–225.

72 (a) Dawson, P.E., Muir, T.W., Clark-Lewis, I., and Kent, S.B.H. (1994) Synthesis of proteins by native chemical ligation. *Science*, **266**, 776–779; (b) Tam, J.P., Lu, Y.A., Liu, C.-F., and Shao, J. (1995) Peptide synthesis using unprotected peptides through orthogonal coupling methods. *Proceedings of the National Academy of Sciences of United States of America*, **92**, 12485–12489.

73 Wieland, T., Bokelmann, E., Bauer, L., Lang, H.U., and Lau, H. (1953) Uber peptidsynthesen. 8. Mitteilung bildung von s-haltigen peptiden durch intramolekulare wanderung von aminoacylresten. *Liebigs Annalen der Chemie*, **583**, 129–149.

74 Dawson, P.E., Churchill, M.J., Ghadiri, M.R., and Kent, S.B.H. (1997) Modulation of reactivity in native chemical ligation through the use of thiol additives. *Journal of the American Chemical Society*, **119**, 4325–4329.

75 Schnölzer, M., Alewood, P., Jones, A., Alewood, D., and Kent, S.B.H. (1992) *In situ* neutralization in Boc-chemistry solid phase peptide synthesis. Rapid, high

yield assembly of difficult sequences. *International Journal of Peptide and Protein Research*, **40**, 180–193.

**76** Kenner, G.W., McDermott, J.R., and Sheppard, R.C. (1971) The safety catch principle in solid phase peptide synthesis. *Journal of the Chemical Society, Chemical Communications*, 636–637.

**77** Futaki, S., Sogawa, K., Maruyama, J., Asahara, T., Niwa, M., and Hojo, H. (1997) Preparation of peptide thioesters using Fmoc-solid-phase peptide synthesis and its application to the construction of a template-assembled synthetic protein (TASP). *Tetrahedron Letters*, **38**, 6237–6240.

**78** (a) Biancalana, S., Hudson, D., Songster, M.F., and Thompson, S.A. (2001) Fmoc chemistry compatible thio-ligation assembly of proteins. *Letters in Peptide Science*, **7**, 291–297; (b) von Eggelkraut-Gottanka, R., Klose, A., Beck-Sickinger, A.G., and Beyermann, M. (2003) Peptide a-thioester formation using standard Fmoc-chemistry. *Tetrahedron Letters*, **44**, 3551–3554; (c) Briand, B., Kotzur, N., Hagen, V., and Beyermann, M. (2008) A new photolabile carboxyl protecting group for native chemical ligation. *Tetrahedron Letters*, **49**, 85–87; (d) Hackenberger, C.P.R. (2006) The reduction of oxidized methionine residues in peptide thioesters with $NH_4I$–$Me_2S$. *Organic & Biomolecular Chemistry*, **4**, 2291–2295; (e) Mezo, A.R., Cheng, R.P., and Imperiali, B. (2001) Oligomerization of uniquely folded mini-protein motifs: development of a homotrimeric betabetaalpha peptide. *Journal of the American Chemical Society*, **123**, 3885–3891.

**79** (a) Tsukiji, S. and Nagamune, T. (2009) Sortase-mediated ligation: a gift from gram-positive bacteria to protein engineering. *ChemBioChem*, **10**, 787–798; (b) Perler, F.B., Xu, M.-Q., and Paulus, H. (1997) Protein splicing and autoproteolysis mechanisms. *Current Opinion in Chemical Biology*, **1**, 292–299; (c) Muir, T.W., Sondhi, D., and Cole, P.A. (1998) Expressed protein ligation: a general method for protein engineering. *Proceedings of the National Academy of Sciences of the United States of America*, **95**, 6705–6710; (d) Evans, J.T.C., Benner, J., and Xu, M.Q. (1998) Semisynthesis of cytotoxic proteins using a modified protein splicing element. *Protein Science*, **7**, 2256–2264; (e) Muralidharan, V. and Muir, T.W. (2006) Protein ligation: an enabling technology for the biophysical analysis of proteins. *Nature Methods*, **3**, 429–438; (f) Xu, M.-Q. and Evans, T.C. (2005) Recent advances in protein splicing: manipulating proteins *in vitro* and *in vivo*. *Current Opinion in Biotechnology*, **16**, 440–446.

**80** Haase, C. and Seitz, O. (2009) Internal cysteine accelerates thioester-based peptide ligation. *European Journal of Organic Chemistry*, 2096–2101.

**81** Tam, J.P. and Yu, Q. (1998) Methionine ligation strategy in the biomimetic synthesis of parathyroid hormones. *Biopolymers*, **46**, 319–327.

**82** (a) Bang, D. and Kent, S.B.H. (2005) His6 tag-assisted chemical protein synthesis. *Proceedings of the National Academy of Sciences of United States of America*, **102**, 5014–5019; (b) Durek, T., Torbeev, V.Y., and Kent, S.B.H. (2007) Convergent chemical synthesis and high-resolution x-ray structure of human lysozyme. *Proceedings of the National Academy of Sciences of United States of America*, **104**, 4846–4851.

**83** Hackeng, T.M., Griffin, J.H., and Dawson, P.E. (1999) Protein synthesis by native chemical ligation: expanded scope by using straightforward methodology. *Proceedings of the National Academy of Sciences of the United States of America*, **96**, 10068–10073.

**84** Leitgeb, B., Szekeres, A., Manczinger, L., Vagvölgyi, C., and Kredics, L. (2007) The history of alamethicin: a review of the most extensively studied peptaibol. *Chemistry & Biodiversity*, **4**, 1027–1051.

**85** Wenschuh, H., Beyermann, M., Rothemund, S., Carpino, L.A., and Bienert, M. (1995) Multiple solid phase synthesis via Fmoc-amino acid fluorides. *Tetrahedron Letters*, **36**, 1247–1250.

**86** Mourtas, S., Katakalou, C., Nicolettou, A., Tzavara, C., Gatos, D., and Barlos, K.

(2003) Resin-bound aminothiols: synthesis and application. *Tetrahedron Letters*, **44**, 179–182.

87 Karavoltsos, M., Mourtas, S., Gatos, D., and Barlos, K. (2003) Solid phase insertion of diamines into peptide chains. *Tetrahedron Letters*, **44**, 3979–3982.

88 Nash, I.A., Bycroft, B.W., and Chan, W.C. (1996) Dde – a selective primary amine protecting group: a facile solid phase synthetic approach to polyamine conjugates. *Tetrahedron Letters*, **37**, 2625–2628.

89 Mellor, S.L., McGuire, C., and Chan, W.C. (1997) N-Fmoc-aminooxy-2-chlorotrityl polystyrene resin: a facile solid-phase methodology for the synthesis of hydroxamic acids. *Tetrahedron Letters*, **38**, 3311–3314.

90 Stavropoulos, G., Gatos, D., Magafa, V., and Barlos, K. (1995) Preparation of polymer-bound trityl-hydrazines and their application in the solid phase synthesis of partially protected peptide hydrazides. *Letters in Peptide Science*, **2**, 315–318.

91 Zhang, L., Torgerson, T.R., Liu, X.-Y., Timmons, S., Colosia, A.D., Hawiger, J., and Tam, J.P. (1998) Preparation of functionally active cell-permeable peptides by single-step ligation of two peptide modules. *Proceedings of the National Academy of Sciences of the United States of America*, **95**, 9184–9189.

92 Alsina, J. and Albericio, F. (2003) Solid-phase synthesis of C-terminal modified peptides. *Biopolymers*, **71**, 454–477.

93 Dinh, T.Q. and Armstrong, R.W. (1996) Synthesis of ketones and aldehydes *via* reactions of Weinreb-type amides on solid support. *Tetrahedron Letters*, **37**, 1161–1164.

94 Jung, J.E., Wollscheid, H.-P., Marquardt, A., Manea, M., Scheffner, M., and Przybylski, M. (2009) Functional ubiquitin conjugates with lysine–ε-amino-specific linkage by thioether ligation of cysteinyl-ubiquitin peptide building blocks. *Bioconjugate Chemistry*, **20**, 1152–1162.

95 Tam, J.P. (1996) Recent advances in multiple antigen peptides. *Journal of Immunological Methods*, **196**, 17–32.

96 (a) Maillard, N., Clouet, A., Darbre, T., and Reymond, J.-L. (2009) Combinatorial libraries of peptide dendrimers: design, synthesis, on-bead high-throughput screening, bead decoding and characterization. *Nature Protocols*, **4**, 132–142; (b) Crespo, L., Sanclimens, G., Pons, M., Giralt, E., Royo, M., and Albericio, F. (2005) Peptide and amide bond-containing dendrimers. *Chemical Reviews*, **105**, 1663–1681; (c) Niederhafner, P., Sebestik, J., and Jezek, J. (2005) Peptide dendrimers. *Journal of Peptide Science*, **11** 757–788.

97 (a) Hu, H., Xue, J., Swarts, B.M., Wang, Q., Wu, Q., and Guo, Z. (2009) Synthesis and antibacterial activities of *N*-glycosylated derivatives of tyrocidine A, a macrocyclic peptide antibiotics. *Journal of Medicinal Chemistry*, **52**, 2052–2059; (b) Baumann, K., Kowalczyk, D., Gutjahr, T., Pieczyk, M., Jones, C., Wild, M.K., Vestweber, D., and Kunz, H. (2009) Sulfated and non-sulfated glycopeptide recognition domains of P-selectin glycoprotein ligand 1 and their binding to P- and E-selectin. *Angewandte Chemie International Edition*, **48**, 3174–3178; (c) Gamblin, D.P., Scanlan, E.M., and Davis, B.G. (2009) Glycoprotein synthesis: an update. *Chemical Reviews*, **109**, 131–163; (d) Warren, J.D., Miller, J.S., Keding, S.J., and Danishefsky, S.J. (2004) Toward fully synthetic glycoproteins by ultimately convergent routes: a solution to a long-standing problem. *Journal of the American Chemical Society*, **126**, 6576–6578; (e) Hackenberger, C.P.R., Friel, C.T., Radford, S.E., and Imperiali, B. (2005) Semisynthesis of a glycosylated Im7 analogue for protein folding studies. *Journal of the American Chemical Society*, **127**, 12882–12889; (f) Shin, Y., Winans, K.A., Backes, B.J., Kent, S.B.H., Ellman, J.A., and Bertozzi, C.R. (1999) Fmoc-based synthesis of peptide-α thioesters: application to the total chemical synthesis of a glycoprotein by native chemical ligation. *Journal of the American Chemical Society*, **121**, 11684–11689.

98 (a) Attard, T.J., O'Brien-Simpson, N., and Reynolds, E.C. (2007) Synthesis of

phosphopeptides in the Fmoc mode. *International Journal of Peptide. Research and Therapeutics*, **13**, 447–468; (b) McMurray, J.S., Coleman, D.R., Wang, W., and Campbell, M.L. (2001) The synthesis of phosphopeptides. *Biopolymers*, **60**, 3–31.

99 Seibert, C. and Sakmar, T.P. (2007) Toward a framework for sulfoproteomics: synthesis and characterization of sulfotyrosine-containing peptides. *Biopolymers*, **90**, 459–477.

100 Xie, N., Elangwe, E.N., Asher, S., and Zheng, Y.G. (2009) A dual-mode fluorescence strategy for screening HAT modulators. *Bioconjugate Chemistry*, **20**, 360–366.

101 (a) Krüger, J., Minuth, T., Schröder, W., and Werwath, J. (2008) Manufacturing and PEGylation of a dual-acting peptide for diabetes. *European Journal of Organic Chemistry*, 5936–5945; (b) Veronese, F.M., Mammucari, C., Caliceti, P., Schiavon, O., and Lora, S. (1999) Influence of PEGylation on the release of low and high molecular-weight proteins from PVA matrices. *Journal of Bioactive and Compatible Polymers*, **14**, 315–330.

102 Loffet, A. and Zhang, H.X. (1993) Allyl-based groups for side-chain protection of amino-acids. *International Journal of Peptide and Protein Research*, **42**, 346–351.

103 Aletras, A., Barlos, K., Gatos, D., Koutsogianni, S., and Mamos, P. (1995) Preparation of the very acid-sensitive Fmoc-Lys(Mtt)-OH. Application in the synthesis of side-chain to side-chain cyclic peptides and oligolysine cores suitable for the solid-phase assembly of MAPs and TASPs. *International Journal of Peptide and Protein Research*, **45**, 488–496.

104 (a) Bycroft, B.W., Chan, W.C., Chhabra, S.R., and Hone, N.D. (1993) A novel lysine-protecting procedure for continous flow solid phase synthesis of branched peptides. *Journal of the Chemical Society, Chemical Communications*, 778–779; (b) Kellam, B., Bycroft, B.W., and Chhabra, S.R. (1997) Solid phase applications of Dde and the analogue Nde: synthesis of trypanothione disulphide. *Tetrahedron Letters*, **38**, 4849–4852.

105 McMurray, J.S. (1991) Solid phase synthesis of a cyclic peptide using Fmoc chemistry. *Tetrahedron Letters*, **32**, 7679–7682.

106 (a) Barlos, K., Mamos, P., Papaioannou, D., Patrianakou, S., Sanida, C., and Schaefer, W. (1987) Application of the Trt and Fmoc groups for the protection of α-amino acids. *Liebigs Annalen der Chemie*, 1025–1030; (b) Barlos, K., Gatos, D., Koutsogianni, S., Schäfer, W., Stavropoulos, G., and Wenging, Y. (1991) Preparation and use of *N*-Fmoc-*O*-Trt-hydroxyamino acids for solid-phase synthesis of peptides. *Tetrahedron Letters*, **32**, 471–474; (c) Barlos, K., Gatos, D., and Koutsogianni, S. (1998) Fmoc/Trt amino acids: comparison to Fmoc/*t*Bu amino acids in peptide synthesis. *The Journal of Peptide Research*, **51**, 194–200.

107 Barlos, K., Gatos, D., Hatzi, O., Koch, N., and Koutsogianni, S. (1996) Synthesis of the very acid-sensitive Fmoc-Cys(Mmt)-OH and its application in solid-phase peptide synthesis. *International Journal of Peptide and Protein Research*, **47**, 148–153.

108 Barlos, K., Chatzi, O., Gatos, D., Stavropoulos, G., and Tsegenidis, T. (1991) Fmoc-His(Mmt)-OH and Fmoc-His(Mtt)-OH. Two new histidine derivatives Nim-protected with highly acid-sensitive groups. Preparation properties and use in peptide synthesis. *Tetrahedron Letters*, **32**, 475–478.

109 Craik, D.J., Cemazar, M., Wang, C.K.L., and Daly, N.L. (2006) The cyclotide family of circular miniproteins: nature's combinatorial peptide template. *Biopolymers*, **84**, 250–266.

110 (a) Richter, L.S., Tom, J.Y.K., and Burnier, J.P. (1994) Peptide-cyclizations on solid support: a fast and efficient route to small cyclopeptides. *Tetrahedron Letters*, **35**, 5547–5550; (b) Zhang, L. and Tam, J.P. (1999) Lactone and lactam library synthesis by silver ion assisted orthogonal cyclization of unprotected peptides. *Journal of the American Chemical Society*, **121**, 3311–3320.

111 (a) Ösapay, G., Profit, A., and Taylor, J.W. (1990) Synthesis of tyrocidine A: use of oxime resin for peptide chain assembly and cyclization. *Tetrahedron Letters*, **31**,

6121–6124; (b) Ösapay, G. and Taylor, J.W. (1990) Multicyclic polypeptide model compounds. 1. Synthesis of a tricyclic amphiphilic α-helical peptide using an oxime resin, segment-condensation approach. *Journal of the American Chemical Society*, **112**, 6046–6051; (c) Xu, M., Nishino, N., Mihara, H., Fujimoto, T., and Izumiya, N. (1992) Synthesis of [D-Pyrenylalanine 4,4′]gramicidin S by solid-phase-synthesis and cyclization-cleavage method with oxime resin. *Chemistry Letters*, 191–194; (d) Nishino, N., Xu, M., Mihara, H., Fujimoto, T., Ohba, M., Ueno, Y., and Kumagai, H. (1992) Facile synthesis of cyclic peptides containing α-aminosuberic acid with oxime resin. *Journal of the Chemical Society, Chemical Communications*, 180–181; (e) Nishino, N., Xu, M., Mihara, H., Fujimoto, T., Ueno, Y., and Kumagai, H. (1992) Sequence dependence in solid-phase-synthesis-cyclization-cleavage for *cyclo* (-arginyl-glycyl-aspartyl-phenylglycyl-). *Tetrahedron Letters*, **33**, 1479–1482.

**112** (a) Clark, T.D., Sastry, M., Browna, C., and Wagner, G. (2006) Solid-phase synthesis of backbone-cyclized β-helical peptides. *Tetrahedron*, **62**, 9533–9540; (b) Ravn, J., Bourne, G.T., and Smythe, M.L. (2005) A safety catch linker for Fmoc-based assembly of constrained cyclic peptides. *Journal of Peptide Science*, **11**, 572–578.

**113** (a) Alsina, J., Rabanal, F., Chiva, C., Giralt, E., and Albericio, F. (1998) Active carbonate resins: application to the solid-phase synthesis of alcohol, carbamate and cyclic peptides. *Tetrahedron*, **54**, 10125–10152; (b) Tulla-Puche, J. and Barany, G. (2004) On-resin native chemical ligation for cyclic peptide synthesis. *The Journal of Organic Chemistry*, **69**, 4101–4107; (c) Alsina, J., Yokum, T.S., Albericio, F., and Barany, G. (1999) Backbone amide linker (BAL) strategy for $N^{\alpha}$-9-fluorenylmethoxycarbonyl (Fmoc) solid-phase synthesis of unprotected peptide *p*-nitroanilides and thioesters. *The Journal of Organic Chemistry*, **64**, 8761–8769; (d) Sabatino, G., Chelli, M., Mazzucco, S., Ginanneschi, M., and Papini, A.M. (1999) Cyclization of histidine containing peptides in the solid-phase by anchoring the imidazole ring to trityl resins. *Tetrahedron Letters*, **40**, 809–812; (e) Chan, W.C., Bycroft, B.W., Evans, D.J., and White, P.D. (1995) A novel 4-aminobenzyl ester based carboxy-protecting group for synthesis of atypical peptides by Fmoc-But solid phase chemistry. *Journal of the Chemical Society, Chemical Communications*, 2209–2210; (f) Spatola, A.F., Darlak, K., and Romanovskis, P. (1996) An approach to cyclic peptide libraries: reducing epimerization in medium sized rings during solid phase synthesis. *Tetrahedron Letters*, **37**, 591–594; (g) Trzeciak, A. and Bannwarth, W. (1992) Synthesis of "head-to-tail" cyclized peptides on solid support by Fmoc-chemistry. *Tetrahedron Letters*, **33**, 4557–4560; (h) Delforge, D., Art, M., Gillon, B., Dieu, M., Delaive, E., Raes, M., and Remacle, J. (1996) Automated solid-phase synthesis of cyclic peptides bearing a side-chain tail designed for subsequent chemical grafting. *Analytical Biochemistry*, **242**, 180–186; (i) Kates, S.A., Sole, N.A., Johnson, C.R., Hudson, D., Barany, G., and Albericio, F. (1993) A novel, convenient, three-dimensional orthogonal strategy for solid-phase synthesis of cyclic peptides. *Tetrahedron Letters*, **34**, 1549–1552.

**114** (a) Cudic, M., Wade, J.D., and Otvos, L. (2000) Convenient synthesis of a head-to-tail cyclic peptide containing an expanded ring. *Tetrahedron Letters*, **41**, 4527–4531; (b) Dixon, M.J., Nathubhai, A., Andersen, O.A., van Aalten, D.M.F., and Eggleston, I.M. (2009) Solid-phase synthesis of cyclic peptide chitinase inhibitors: SAR of the argifin scaffold. *Organic & Biomolecular Chemistry*, **7**, 259–268.

**115** (a) Monroc, S., Badosa, E., Feliu, L., Planas, M., Montesinos, E., and Bardajl, E. (2006) De *novo* designed cyclic cationic peptides as inhibitors of plant pathogenic bacteria. *Peptides*, **27**, 2567–2574; (b) Salvati, M., Cordero, F.M., Pisaneschi, F., Melani, F., Gratteri, P., Cini, N., Bottoncetti, A., and Brandia, A. (2008) Synthesis, SAR and *in vitro* evaluation of new cyclic Arg-Gly-Asp pseudopentapeptides containing a s-*cis* peptide bond as integrin $\alpha_v\beta_3$ and $\alpha_v\beta_5$ ligands. *Bioorganic & Medicinal Chemistry*, **16**, 4262–4271.

**116** Berthelot, T., Goncalves, M., Läin, G., Estieu-Gionnet, K., and Deleris, G. (2006) New strategy towards the efficient solid phase synthesis of cyclopeptides. *Tetrahedron*, **62**, 1124–1130.

**117** Eleftheriou, S., Gatos, D., Panagopoulos, A., Stathopoulos, S., and Barlos, K. (1999) Attachment of histidine, histamine and urocanic acid to resins of the trityl-type. *Tetrahedron Letters*, **40**, 2825–2828.

**118** Bernhardt, A., Drewello, M., and Schutkowski, M. (1997) The solid-phase synthesis of side-chain-phosphorylated peptide-4-nitroanilides. *The Journal of Peptide Research*, **50**, 143–152.

**119** (a) Cabrele, C., Langer, M., and Beck-Sickinger, A.G. (1999) Amino acid side-chain attachment approach and its application to the synthesis of tyrosine-containing cyclic peptides. *The Journal of Organic Chemistry*, **64**, 4353–4361; (b) Basso, A. and Ernst, B. (2001) Solid-phase synthesis of hydroxyproline-based cyclic hexapeptides. *Tetrahedron Letters*, **42**, 6687–6690.

**120** Scott, L.T., Rebek, J., Ovsyanko, L., and Sims, C.L. (1977) Organic chemistry on solid-phase. Site–site interactions on functionized polystyrene. *Journal of the American Chemical Society*, **99**, 625–626.

**121** (a) Malesevic, M., Strijowski, U., Bächle, D., and Sewald, N. (2004) An improved method for the solution cyclization of peptides under pseudo-high dilution conditions. *Journal of Biotechnology*, **112**, 73–77; (b) Minta, E., Kafarski, P., Martinez, J., and Rolland, V. (2008) Synthesis of cyclooctapeptides: constraints analogues of the peptidic neurotoxin, omega-agatoxine IVB – an experimental point of view. *Journal of Peptide Science*, **14**, 267–277.

**122** (a) Davies, J.S. (2003) The cyclization of peptides and depsipeptides. *Journal of Peptide Science*, **9**, 471–501; (b) Jiang, S., Li, Z., Ding, K., and Roller, P.P. (2008) Recent progress of synthetic studies to peptide and peptidomimetic cyclization. *Current Organic Chemistry*, **12**, 1502–1542; (c) Hargittai, B., Sole, N.A., Groebe, D.R., Abramson, S.N., and Barany, G. (2000) Chemical syntheses and biological activities of lactam analogues of α-conotoxin SI. *Journal of Medicinal Chemistry*, **43**, 4787–4792; (d) Cavallaro, V., Thompson, P., and Hearn, M. (1998) Solid phase synthesis of cyclic peptides: model studies involving $i-(i+4)$ side-chain-to-side-chain cyclization. *Journal of Peptide Science*, **4**, 335–343; (e) Tang, Y.-C., Xie, H.-B., Tian, G.-L., and Ye, Y.-H. (2002) Synthesis of cyclopentapeptides and cycloheptapeptides by DEPBT and the influence of some factors on cyclization. *The Journal of Peptide Research*, **60**, 95–103.

**123** (a) Haubner, R., Gratias, R., Diefenbach, B., Goodman, S.L., Jonczyk, A., and Kessler, H. (1996) Structural and functional aspects of RGD-containing cyclic pentapeptides as highly potent and selective integrin $\alpha_V\beta_3$ antagonists. *Journal of the American Chemical Society*, **118**, 7461–7472; (b) Fletcher, J.M. and Hughes, R.A. (2009) Modified low molecular weight cyclic peptides as mimetics of BDNF with improved potency, proteolytic stability and transmembrane passage *in vitro*. *Bioorganic and Medicinal Chemistry*, **17**, 2695–2702.

**124** Gobbo, M., Biondi, L., Filira, F., Gennaro, R., Benincasa, M., Scolaro, B., and Rocchi, R. (2002) Antimicrobial peptides: synthesis and antibacterial activity of linear and cyclic drosocin and apidaecin 1b analogues. *Journal of Medicinal Chemistry*, **45**, 4494–4504.

**125** (a) Napolitano, A., Rodriquez, M., Bruno, I., Marzocco, S., Autore, G., Riccio, R., and Gomez-Paloma, L. (2003) Synthesis, structural aspects and cytotoxicity of the natural cyclopeptides yunnanins A, C and phakellistatins 1,10. *Tetrahedron*, **59**, 10203–10211; (b) Avrutina, O., Schmoldt, H.-U., Kolmar, H., and Diederichsen, U. (2004) Fmoc-assisted synthesis of a 29-Residue cystine-knot trypsin inhibitor containing a guaninyl amino acid at the P1-position. *European Journal of Organic Chemistry*, 4931–4935; (c) Robinson, J.A.,

Shankaramma, S.C., Jetter, P., Kienzl, U., Schwendener, R.A., Vrijbloed, J.W., and Obrecht, D. (2005) Properties and structure–activity studies of cyclic β-hairpin peptidomimetics based on the cationic antimicrobial peptide protegrin I. *Bioorganic & Medicinal Chemistry*, **13**, 2055–2064; (d) Schabbert, S., Pierschbacher, M.D., Mattern, R.-H., and Goodman, M. (2002) Incorporation of (2*S*,3*S*) and (2*S*,3*R*)-methyl aspartic acid into RGD-containing peptides. *Bioorganic & Medicinal Chemistry*, **10**, 3331–3337; (e) Prusis, P., Muceniece, R., Mutule, I., Mutulis, F., and Wikberg, J.E.S. (2001) Design of new small cyclic melanocortin receptor-binding peptides using molecular modelling: role of the His residue in the melanocortin peptide core. *European Journal of Medicinal Chemistry*, **36**, 137–146; (f) Caba, J.M., Rodriguez, I.M., Manzanares, I., Giralt, E., and Albericio, F. (2001) Solid-phase total synthesis of trunkamide A. *The Journal of Organic Chemistry*, **66**, 7568–7574; (g) De Luca, S., De Capua, A., Saviano, M., Della Moglie, R., Aloj, L., Tarallo, L., Pedone, C., and Morelli, G. (2007) Synthesis and biological evaluation of cyclic and branched peptide analogues as ligands for cholecystokinin type 1 receptor. *Bioorganic & Medicinal Chemistry*, **15**, 5845–5853.

**126** (a) Thongyoo, P., Roqué-Rosell, N., Leatherbarrow, R.J., and Tate, E.W. (2008) Chemical and biomimetic total syntheses of natural and engineered MCoTI cyclotides. *Organic & Biomolecular Chemistry*, **6**, 1462–1470; (b) Lin, H., Thayer, D.A., Wong, C.-H., and Walsh, C.T. (2004) Macrolactamization of glycosylated peptide thioesters by the thioesterase domain of tyrocidine synthetase. *Chemistry & Biology*, **11**, 1635–1642; (c) Kohli, R.M., Takagi, J., and Walsh, C.T. (2002) The thioesterase domain from a nonribosomal peptide synthetase as a cyclization catalyst for integrin binding peptides. *Proceedings of the National Academy of Sciences of United States of America*, **99**, 1247–1252; (d) Trauger, J.W., Kohli, R.M., Mootz, H.D., Marahiel, M.A., and Walsh, C.T., Peptide cyclization catalysed by the thioesterase domain of tyrocidine synthetase. *Nature*, (2000) **407**, 215–218; (e) Tseng, C.C., Bruner, S.D., Kohli, R.M., Marahiel, M.A., Walsh, C.T., and Sieber, S.A. (2002) Characterization of the surfactin synthetase C-terminal thioesterase domain as a cyclic depsipeptide synthase. *Biochemistry*, **41**, 13350–13359.

**127** Simonsen, S.M., Sando, L., Rosengren, K.J., Wang, C.K., Colgrave, M.L., Daly, N.L., and Craik, D.J. (2008) Alanine scanning mutagenesis of the prototypic cyclotide reveals a cluster of residues essential for bioactivity. *The Journal of Biological Chemistry*, **283**, 9805–9813.

**128** Lecaillon, J., Gilles, P., Subra, G., Martinez, J., and Amblard, M. (2008) Synthesis of cyclic peptides via *O–N*-acyl migration. *Tetrahedron Letters*, **49**, 4674–4676.

**129** (a) Guibe, F. (1997) Allylic protecting groups and their use in a complex environment. Part I: allylic protection of alcohols. *Tetrahedron*, **53**, 13509–13556; (b) Guibe, F. (1998) Allylic protecting groups and their use in a complex environment. Part II: allylic protecting groups and their removal through catalytic palladium π-allyl methodology. *Tetrahedron*, **54**, 2967–3042.

**130** Isidro-Llobet, A., Alvarez, M., and Albericio, F. (2005) Semipermanent *p*-nitrobenzyloxycarbonyl (*p*NZ) protection of Orn and Lys side-chains: prevention of undesired α-Fmoc removal and application to the synthesis of cyclic peptides. *Tetrahedron Letters*, **46**, 7733–7736.

**131** Zhu, J., Tang, C., Kottke-Marchant, K., and Marchant, R.E. (2009) Design and synthesis of biomimetic hydrogel scaffolds with controlled organization of cyclic RGD peptides. *Bioconjugate Chemistry*, **20**, 333–339.

**132** Sanclimens, G., Crespo, L., Pons, M., Giralt, E., Albericio, F., and Royo, M. (2003) Saturated resins or stress of the resin. *Tetrahedron Letters*, **44**, 1751–1754.

# 12
# Peptide-Coupling Reagents

*Ayman El- Faham and Fernando Albericio*

## 12.1
## Introduction

In recent years, peptide-coupling reactions have significantly advanced in accord with the development of new peptide-coupling reagents, which has been covered in a number of valuable reviews [1].

Success in the chemical synthesis of peptides and also, in the construction of peptide libraries relies on an efficient combination of protecting groups and coupling reagents (Scheme 12.1).

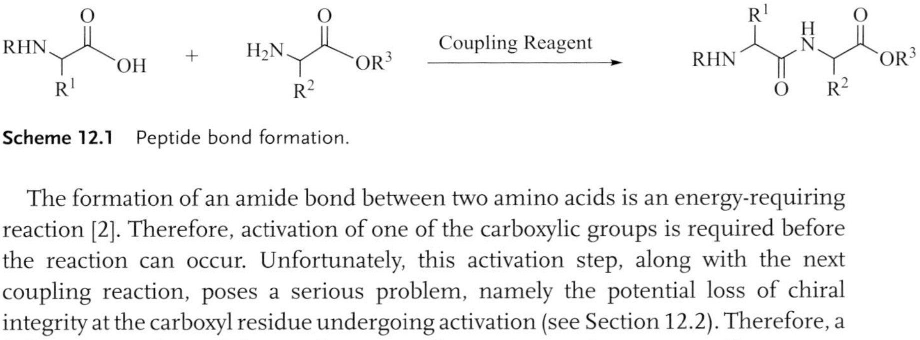

RHN–CH($R^1$)–C(O)OH + $H_2N$–CH($R^2$)–C(O)$OR^3$ —Coupling Reagent→ RHN–CH($R^1$)–C(O)–NH–CH($R^2$)–C(O)$OR^3$

**Scheme 12.1** Peptide bond formation.

The formation of an amide bond between two amino acids is an energy-requiring reaction [2]. Therefore, activation of one of the carboxylic groups is required before the reaction can occur. Unfortunately, this activation step, along with the next coupling reaction, poses a serious problem, namely the potential loss of chiral integrity at the carboxyl residue undergoing activation (see Section 12.2). Therefore, a full understanding of the mechanisms of racemization is necessary if we are to overcome this problem. Two major pathways for the loss of configuration, both base-catalyzed, have been recognized: (i) direct enolization (Path A) and (ii) 5(4*H*)-oxazolone (**1**) formation (Path B) (Scheme 12.2) [2–4].

Several parameters have been used to deal with such side-reactions during peptide-coupling reactions. A key issue is the use of appropriate *N*-protecting groups, such as the carbamate (*tert*-butyloxycarbonyl (Boc, **2**) [5], benzyloxycarbonyl (Cbz, Z, **3**) [6], 9-fluorenylmethyloxcarbonyl (Fmoc, **4**) [7], 2,7-di-*tert*-butyl-9-fluorenylmethyloxycarbonyl (Dtb-Fmoc, **5**) [8], and 2,7-bis(trimethylsilyl)-9-fluorenylmethyloxycarbonyl (Bts-Fmoc, **6**) [9, 10], or the recent type of base-sensitive amino-protecting groups 1,1-dioxobenzo[*b*]thiophene-2-ylmethyloxycarbonyl (Bsmoc, **7**) [11], 2-(*tert*-butylsul-

*Amino Acids, Peptides and Proteins in Organic Chemistry.*
*Vol.3 – Building Blocks, Catalysis and Coupling Chemistry.* Edited by Andrew B. Hughes

ISBN: 978-3-527-32102-5

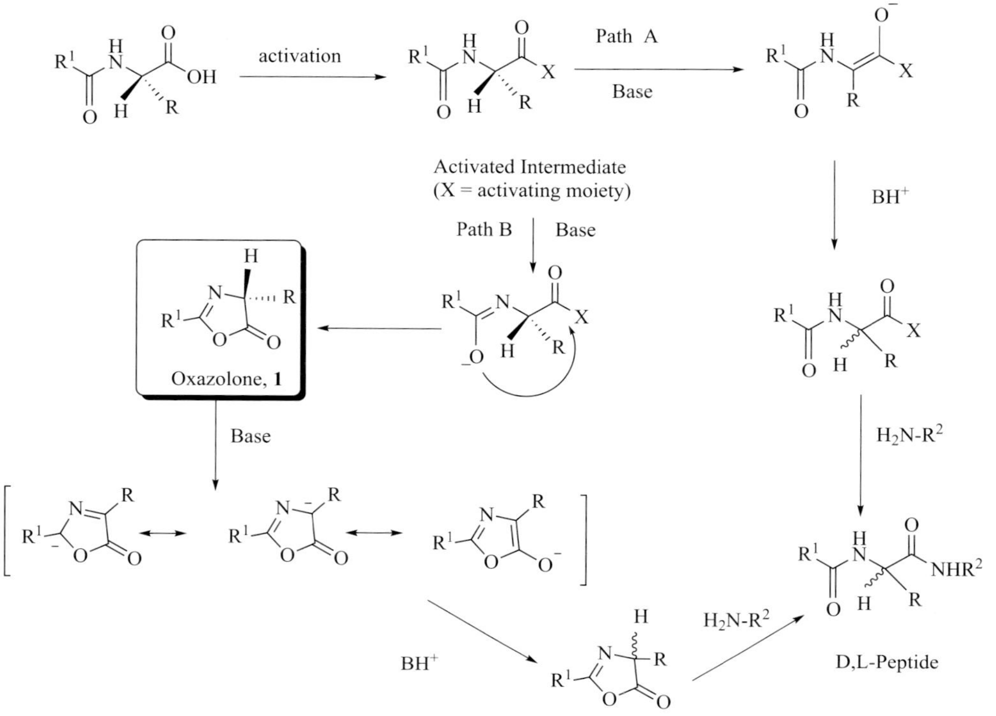

**Scheme 12.2** Racemization mechanisms.

fonyl)-2-propyloxycarbonyl (Bspoc, **8**) [12], and 2-methylsulfonyl-3-phenyl-1-prop-2-enyloxycarbonyl (Mspoc, **9**) [13] (Figure 12.1), because the electron-withdrawing effect of the "O" reduces the risk of oxazolone formation. Furthermore, the hindrance of the trityl group (**10**) prevents racemization, thereby protecting the α-H against removal. A further issue is the basicity and purity of the tertiary amines commonly used during the coupling reaction. Thus, diethylisopropylamine (DIEA) and *N*-methylmorpholine (NMM), which are considered practical bases because of their non-nucleophilic property, are often contaminated by secondary and primary amines. Recently, Carpino *et al.* [14] recommended the use of collidine (2,4,6-trimethylpyridine (TMP)) **11**, 2,3,5,6-tetramethylpyridine (TEMP) **12**, and 2,6-di-*tert*-butyl-4-(dimethylamino)pyridine (DBDMAP) **13** (Figure 12.1), which are greatly hindered, thereby jeopardizing H abstraction, and considerably less basic than DIEA and NMM, for assistance during the coupling.

In addition, more side-reactions occur during coupling. Namely, the formation of *N*-carboxyanhydrides **14**, when the protection of the α-amino is a carbamate, and diketopiperazines (DKPs) **15**, when at least one dipeptide is present [15] (Scheme 12.3). These two reactions are strongly favored by the presence of the leaving group in the carboxyl group (the C-terminal in the case of DKP formation).

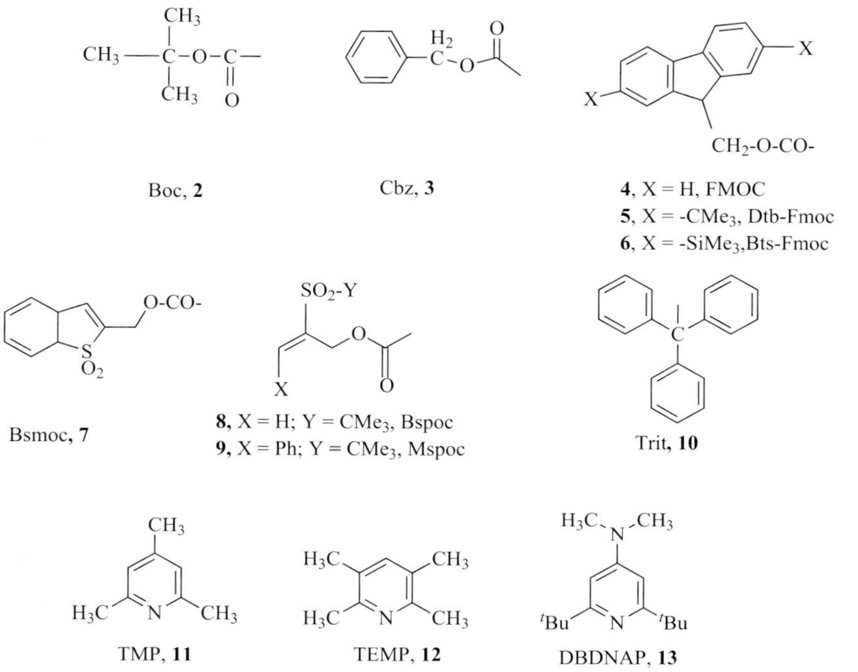

Boc, **2**

Cbz, **3**

**4**, X = H, FMOC
**5**, X = -$CMe_3$, Dtb-Fmoc
**6**, X = -$SiMe_3$,Bts-Fmoc

Bsmoc, **7**

**8,** X = H; Y = $CMe_3$, Bspoc
**9,** X = Ph; Y = $CMe_3$, Mspoc

Trit, **10**

TMP, **11**

TEMP, **12**

DBDNAP, **13**

**Figure 12.1** Structure of *N*-protecting amino acids and sterically hindered bases.

Furthermore, the formation of DKP is also facilitated by the presence of *N*-methylamino acids (favoring the *cis*-amide bond conformation) and/or amino acids of L-and D-configuration (more stability than the six-member ring DKP). When an uronium salt is used as coupling reagent (see Section 12.4), a guanidine side product (**16**) is produced when the uronium coupling reagent reacts directly with the amine moiety of the amino acid residue [16] (Scheme 12.3). This result is often due to the slow preactivation of the carboxylic acid or the use of excess uronium reagent.

## 12.2 Carbodiimides

The potential of carbodiimides, such as DCC (**17**) [17] (Figure 12.2), as peptide-coupling reagents in the synthesis of complex molecules has particularly attracted the attention of organic chemists. In this regard, the use of carbodiimides as peptide-coupling reagents was fairly common until 1985 and continues to be a useful synthetic procedure. DCC has been applied mainly in solid-phase peptide synthesis (SPPS) for the Boc/Bzl strategy and *N,N'*-diisopropylcarbodiimide (DIPCDI, DIC, **18**) for the Fmoc/*t*Bu strategy. Solution-phase couplings have involved mainly 1-ethyl-3-(3'-dimethylaminopropyl)carbodiimide (EDC or water-soluble carbodiimide

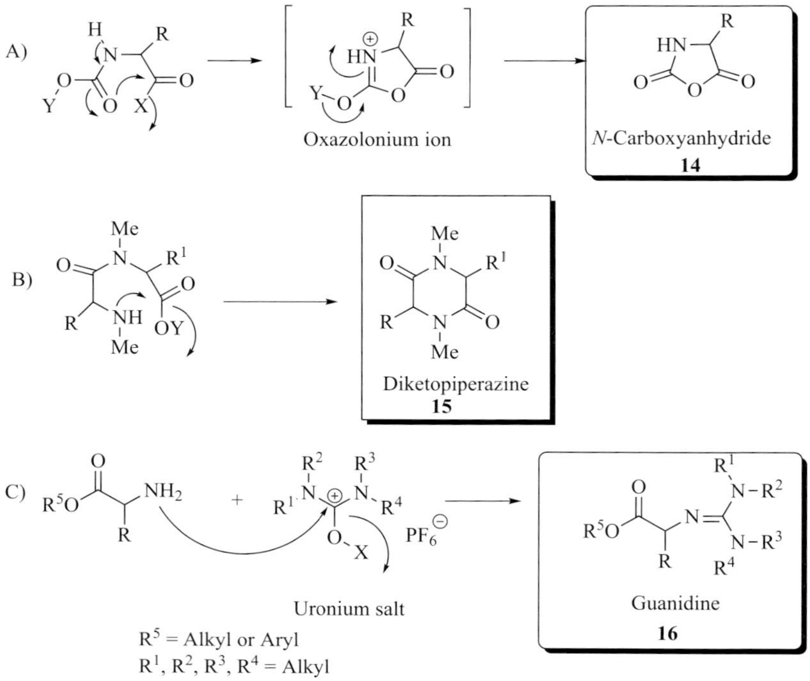

A) Oxazolonium ion → *N*-Carboxyanhydride **14**

B) Diketopiperazine **15**

C) Uronium salt → Guanidine **16**

$R^5$ = Alkyl or Aryl
$R^1, R^2, R^3, R^4$ = Alkyl

**Scheme 12.3** Side-reactions that may occur during coupling.

(WSCI), **19**) [18]. The more recent 1,3-*bis*(2,2-dimethyl-1,3-dioxolan-4-ylmethyl)carbodiimide (BDDC, **20**) has also become commercially available and has been used for solution-phase peptide couplings, achieving a maximum of 1.3% epimerization [19].

Nowadays, carbodiimides are used with a HOX additive as a trapping agent of the *O*-acylisourea (**21**) intermediate to form the corresponding active esters **24**, thus decreasing the degree of racemization in numerous cases. The highly reactive *O*-acylisourea can lead to oxazolone formation, which facilitates the loss of chiral integrity (Scheme 12.2), or to an unreactive *N*-acylurea (**22**). 1-Hydroxybenzotriazole (HOBt, **25**) [20] or aza-1-hydroxybenzotriazole (HOAt, **26**) [21] and other HOX (**27–39**) additives give the corresponding active esters after reaction with the

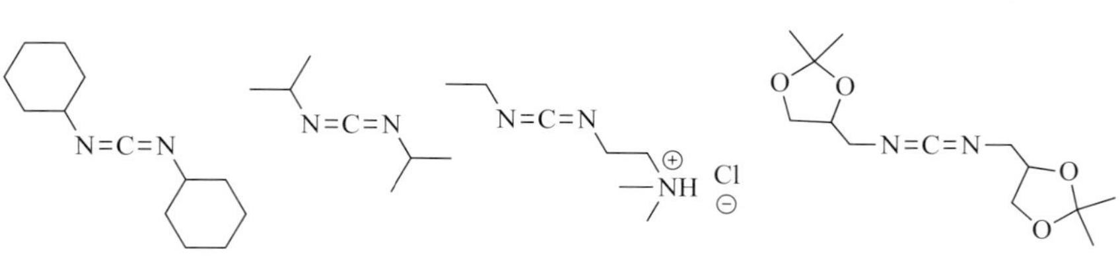

DCC, **17** DIPCDI, DIC, **18** EDC, WSC, **19** BDDC, **20**

**Figure 12.2** Structures of carbodiimides.

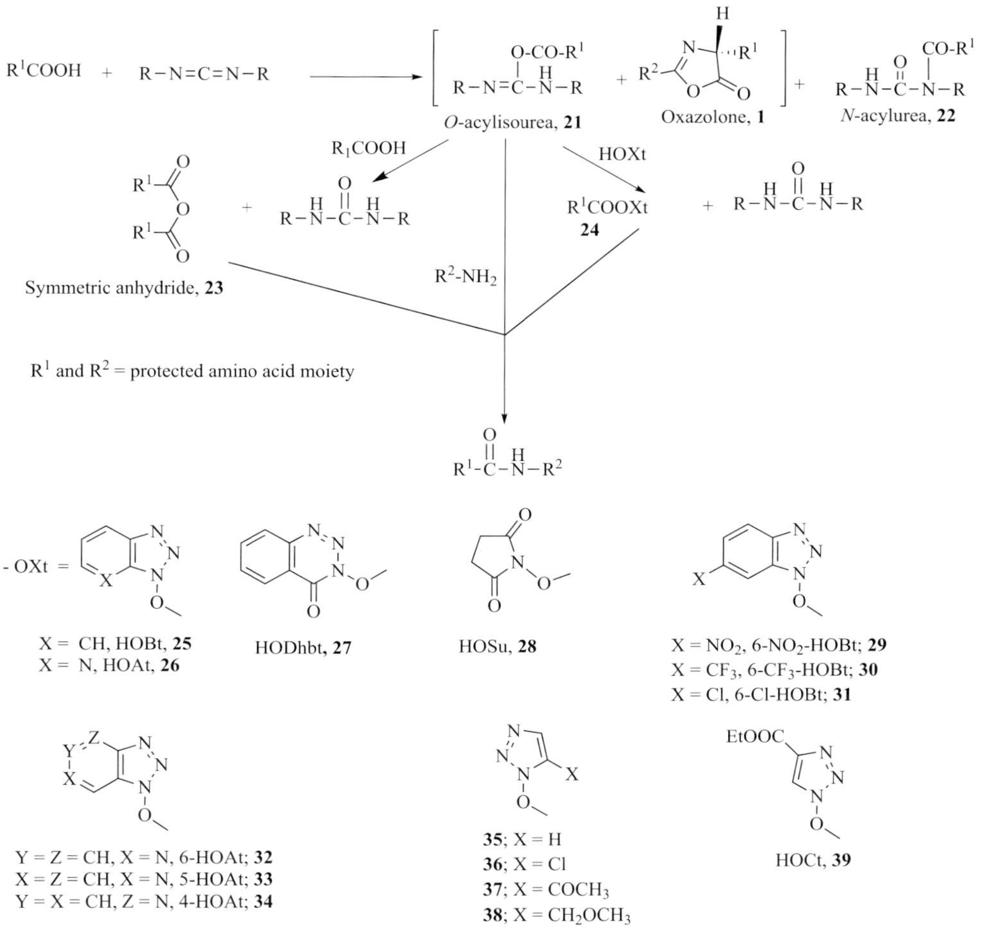

$R^1COOH$ + R−N=C=N−R → [ *O*-acylisourea, **21** + Oxazolone, **1** ] + *N*-acylurea, **22**

$R_1COOH$; HOXt; $R^1COOXt$ **24**; $R^2$-$NH_2$

Symmetric anhydride, **23**

$R^1$ and $R^2$ = protected amino acid moiety

- OXt =

X = CH, HOBt, **25**
X = N, HOAt, **26**

HODhbt, **27**

HOSu, **28**

X = $NO_2$, 6-$NO_2$-HOBt; **29**
X = $CF_3$, 6-$CF_3$-HOBt; **30**
X = Cl, 6-Cl-HOBt; **31**

Y = Z = CH, X = N, 6-HOAt; **32**
X = Z = CH, X = N, 5-HOAt; **33**
Y = X = CH, Z = N, 4-HOAt; **34**

**35**; X = H
**36**; X = Cl
**37**; X = $COCH_3$
**38**; X = $CH_2OCH_3$

EtOOC

HOCt, **39**

**Scheme 12.4** Mechanism of peptide bond formation through carbodiimide and the most common additives used to trap the *O*-acylisourea.

*O*-acylisourea (Scheme 12.4). The presence of a tertiary amine favors formation of the active ester [22]. Alternatively, the symmetrical anhydride (**23**), which is formed when 2 equiv. of *N*-protected amino acid are used with 1 equiv. of carbodiimide, can be employed as the active species.

Compared with other additives, HOAt (**26**) forms superior active esters in terms of yield and degree of racemization in both solution and solid-phase synthesis [23], even when the coupling takes place with the hindered α-aminoisobutyric acid (Aib) [24]. The key behind the outstanding behavior of HOAt is the nitrogen atom located at position 7 of the benzotiazole, which provides a double effect [21, 24]. (i) The electron-withdrawing influence of a nitrogen atom (regardless of its position) improves the quality of the leaving group, thereby leading to greater reactivity. (i) Placement of this

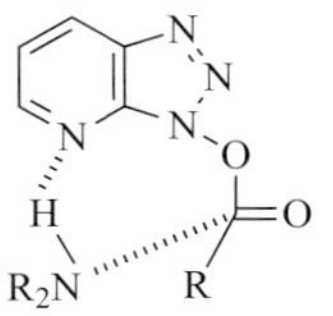

Figure 12.3 Neighboring group effect for HOAt.

nitrogen atom specifically at position 7 makes it feasible to achieve a classic neighboring group effect (Figure 12.3), which can both increase reactivity and reduce the loss of configurational integrity [22, 24]. Compared to HOBt, the corresponding 6-HOAt (**32**), 5-HOAt (**33**), and 4-HOAt (**34**) all lack the capacity to participate in such a neighboring group effect, and have little influence on the extent of stereomutation during the segment coupling reaction [25].

1-Oxo-2-hydroxydihydrobenzotriazine (HODhbt, **27**) [22] gives highly reactive esters but their formation is accompanied by 3-(2-azidobenzyloxy)-4-oxo-3,4-dihydro-1,2,3-benzotriazine (**40**) as a byproduct, which can then react with the amino group to terminate chain growth [20] (Scheme 12.5).

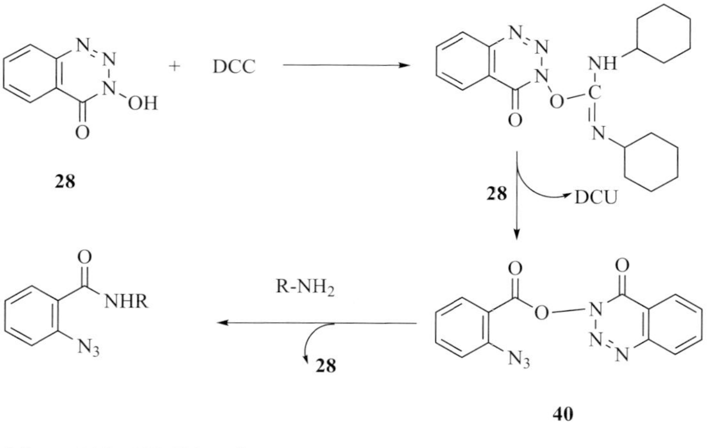

Scheme 12.5 HODhbt side-reaction.

Several coupling additives with triazole and tetrazole structures (**35–39**) (Scheme 12.4) in the presence of DIPCDI (**18**) have been evaluated in solid-phase Fmoc-based peptide synthesis [26]. As an acylating agent, the performance of 5-chloro-1-hydroxytriazole (**36**) is comparable to that of HOAt and HOBt; however, it shows less capacity to suppress racemization. Ethyl 1-hydroxy-1*H*-1,2,3-triazole-4-carboxylate (HOCt, **39**) has been applied in SPPS and is more efficient than HOBt with DIPCDI (**18**), presenting almost no racemization [27]. These reagents show an additional advantage because they do not have absorption in the UV at 302 nm, thus allowing the monitoring of the coupling process, a feature incompatible with Fmoc methodology in the case of HOBt or HOAt.

Recently, 6-Cl-HOBt (**31**) has been introduced into solid-phase synthesis. This additive is a good compromise between HOAt and HOBt in terms of reactivity and price [28]. More recently, Carpino *et al.* [29] described the aza derivative of HODhbt (HODhat, **41** and HODhad, **42**). Active esters of this additive (**41**) are slightly more reactive than OAt ones, which are considered the most reactive derivatives among these esters; however, the additive **41** gives the side-product **43**, as occurs with HODhbt (**28**) [29].

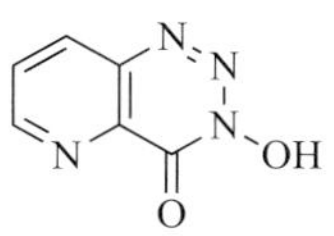

HODhat, **41** 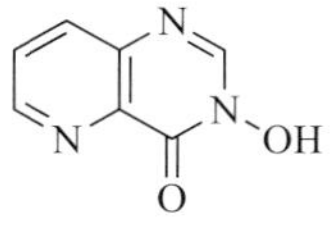 HODhad, **42** 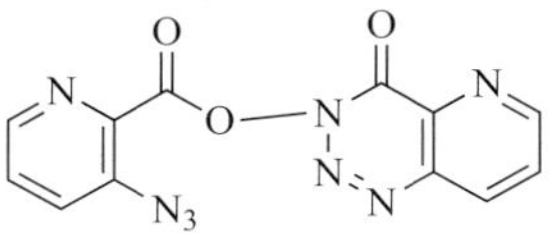 **43** 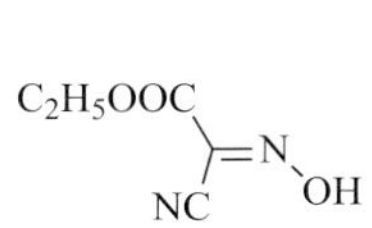 Oxyma, **44**

El-Faham and Albericio [30] recently reported a safe and highly efficient additive, ethyl 2-cyano-2-(hydroxyimino)acetate (Oxyma, **44**) to be used mainly in the carbodiimide approach for forming the peptide bond. Oxyma (**44**) displays a remarkable capacity to suppress racemization and an impressive coupling efficiency in both automated and manual synthesis. These effects are superior to those shown by HOBt and comparable to HOAt. Stability assays show that there is no risk of capping the resin in standard coupling conditions. Finally, calorimetry assays (differential scanning calorimetry and accelerating rate calorimetry) confirm the explosive potential of the benzotriazole-based additives and demonstrate the lower risk of explosion induced by Oxyma [30]. This point is highly relevant because all benzotriazole derivatives, such as HOBt and HOAt, exhibit explosive properties [31].

## 12.2.1
## General Procedure for Coupling Using Carbodiimide and HOXt; Solution Phase [22]

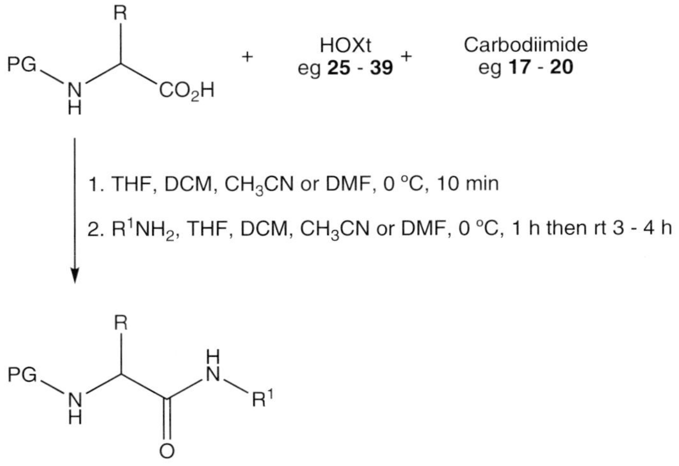

+ HOXt eg **25** - **39** + Carbodiimide eg **17** - **20**

1. THF, DCM, $CH_3CN$ or DMF, 0 °C, 10 min
2. $R^1NH_2$, THF, DCM, $CH_3CN$ or DMF, 0 °C, 1 h then rt 3 - 4 h

*N*-Protected amino acid (1 mmol), carbodiimide (1 mmol), and HOXt (1 mmol) in solvent (tetrahydrofuran (THF), dichloromethane (DCM), $CH_3CN$ or *N,N*-dimethylformamide (DMF)) (5 ml) is stirred at 0 °C for 10 min and then the amino component (1 mmol) in the same solvent (5 ml) is added. The reaction mixture is stirred at 0 °C for 1 h and then at room temperature for 3–4 h. The solvent is then removed under reduced pressure and the residue is dissolved in EtOAc (50 ml). The EtOAc solution is washed with 1 N HCl (2 × 10 ml), saturated $Na_2CO_3$ (2 × 10 ml), saturated NaCl (2 × 10 ml), dried ($MgSO_4$), and filtered. The filtrate is removed under vacuum and the residue is recrystallized from the appropriate solvent.

#### 12.2.1.1 General Procedure for Solid-Phase Coupling via Carbodiimide Activation

Carbodiimide couplings can be carried out either via automation or manually. Manual synthesis can be performed in a polypropylene syringe fitted with a polyethylene disk and a stopcock with occasional stirring or in a mechanically shaken silanized screw-cap reaction vessel with a Teflon-lined cap, a sintered glass frit, and a stopcock [32]. All the procedures presented here are limited to protocols for manual synthesis, because those for automatic synthesizers are either stipulated by the manufacturer or adapted rapidly from the manual procedure. The concentration of the activated species should be maintained at a maximum (0.6 M) based on the solubility of the reagents. *N*-Methylpyrrolidone (NMP) or other convenient solvents, as discussed earlier, can be used. Coupling times are dependent on the nature of the substrates and should be examined for each case. For standard solid-phase assembly of peptides, typically a 15–60 min cycle time provides quantitative coupling.

Carbodiimide-mediated couplings are usually performed with preactivation of the protected amino acid at either 4 or 25 °C using DCM as a solvent. For Fmoc-amino acids that are not totally soluble in DCM, mixtures with DMF may be used [33]. The use of DCM is optimal for the solid-phase acylation of isolated nucleophiles [34]. However, for linear assembly, where interchain aggregation may occur, the use of a more polar solvent to inhibit the formation of secondary structure is recommended. In these cases, if DCM is used as an initial solvent, a more effective procedure would involve filtration of the urea byproduct and evaporation followed by addition of DMF as the coupling medium [34–36].

For large-scale synthesis, preactivation at 4 °C is recommended because of the exothermic nature of the reaction. Carbodiimides, as well as other coupling reagents, are acute skin irritants and should be handled with great care. Thus, manipulation in a well-ventilated hood, using glasses, gloves, and if possible a face mask, is recommended. DCC, which has a low melting point (34 °C), can be handled as a liquid by gentle warming of the reagent container [37]. HOBt normally crystallizes with one molecule of water. Use of the hydrated form is highly satisfactory, but if anhydrous material is required, the dehydration should be carried out with care. Heating HOBt or HOAt above 180 °C can cause rapid exothermic decomposition.

The use of *N*-hydroxytetrazoles as trapping reagents should be precluded because of their explosive nature [31].

### 12.2.2 Loading of Wang Resin Using Carbodiimide [34]

Carbodiimides can be also used for the incorporation of *N*-protected amino acids on hydroxymethyl resins, such as the Wang resin.

Fmoc-amino acid (8 equiv.) is dissolved in DMF and Diisopropylcarbodiimide (DIC) (4 equiv.) is added. The resulting solution is allowed to stand for 15 min at room temperature before its addition to the Wang resin (preswollen in DCM and then DMF). 4-(*N,N*,-dimethylamino)pyridine (DMAP) (0.4 equiv.) is then added. The resulting mixture is then left under stirring for 4–5 h before isolating the resin by filtration through a sintered glass funnel. The resin is washed with DMF, DCM and ether (2 times each) and the loading is tested.

*Notes*

- Preferably fresh distilled solvent should be used.
- In the case of DCC as the carbodiimide, the byproduct dicyclohexylurea (DCU) should be filtered before work-up and then washed with the solvent. Removal of DCU is usually incomplete, and therefore a small amount remaining in solution can contaminate the product and should be removed by crystallization or chromatography. This difficulty can be circumvented by the application of water-soluble carbodiimides since they give rise to a water-soluble urea byproduct.
- In the case of using EDC · HCl (**19**), 1 equiv. of base, such as DIEA or NMM, should be used.
- 1 mmol of base, such as NMM or DIEA, can be used with carbodiimide/HOXt, but the preactivation time should be 5–7 min to prevent racemization.
- In the case of using an ester hydrochloride or amide hydrochloride, 1 mmol of base, such as NMM or DIEA, should be used.
- The reaction can be followed by TLC using EtOAc/hexane or DCM/MeOH or otherwise by IR spectroscopy following the disappearance of the active ester peak at 1824 $cm^{-1}$ [22].
- For large-scale syntheses (about 10 mmol of peptide), it is mandatory to pre-activate at 4 °C, because the exothermic nature of the reaction increases the risk of racemization, even when urethane-type protecting groups are used.
- Some protocols recommend a capping reaction for unreacted chains after each coupling. The capping step is usually carried out by acetylation with $Ac_2O$/DIEA (1: 1, 30 equiv.) in DMF. In some automated instruments $Ac_2O$ is replaced by a 0.3 M solution of *N*-acetylimidazole in DMF.
- In DMF, the activation is slow, thus a hindered base such as TMP can be used to enhance the step involving preactivation of the carboxylic acid residue. This contrasts with the normal situation, in which bases such as DIEA, NMM, or nonhindered pyridine base inhibit this step [22].

## 12.3
## Phosphonium Salts

Kenner *et al.* [38] were the first to describe the use of acylphosphonium salts as coupling reagents. These species were widely adopted only after the extensive studies of Castro and Coste [39], which introduced CloP [40] (**45**) and BroP [41] (**46**) as peptide-coupling reagents with noticeable racemization. After HOBt was discovered as a racemization suppressant, a new coupling reagent, known as BOP (benzotriazol-1-yloxy)tris(dimethylamino) phosphonium hexafluorophosphate) (**47**), was introduced in 1975 [39b,40]. BOP is a nonhygroscopic crystalline material that can be prepared effortlessly in large amounts, is easy to use and promotes rapid coupling.

Later, PyCloP (**48**), PyBroP (**49**), and PyBOP (**50**) (Figure 12.4) were introduced. In these compounds the dimethylamine moiety is replaced by pyrrolidine (Figure 12.4) [39, 43]. These reagents prevent the generation of poisonous hexamethylphosphoramide (HMPA, **51**) byproduct [44]. In a following study, Coste reported that halogenophosphonium reagents often give better results than other phosphonium-HOBt reagents for the coupling of *N*-methylated amino acid [45].

Later, 2-(benzotriazol-1-yloxy)-1,3-dimethyl-2-pyrrolidin-1-yl-1,3,2-diazaphospholidinium hexafluorophosphate (BOMP, **52**, Figure 12.4) was introduced as useful reagent for peptide coupling [46]. Scheme 12.6 depicts the hypothetical main mechanistic pathway involved in coupling reactions mediated by phosphonium salts, such as BOP (**47**) reagent.

The formation of OBt ester (**54**) is achieved in the presence of 1 equiv. of a tertiary base such as DIEA, NMM [47, 48], or TMP [14]. The presence of an extra equivalent of HOBt accelerates the coupling and also reduces the loss of configuration [44]. Some controversy has arisen regarding the possible intermediacy of an acylphosphonium salt **53** and its lifetime. Kim and Patel [43a] reported that such intermediates are present at $-20\,^{\circ}C$ in the absence of excess of HOBt. Furthermore, Coste and

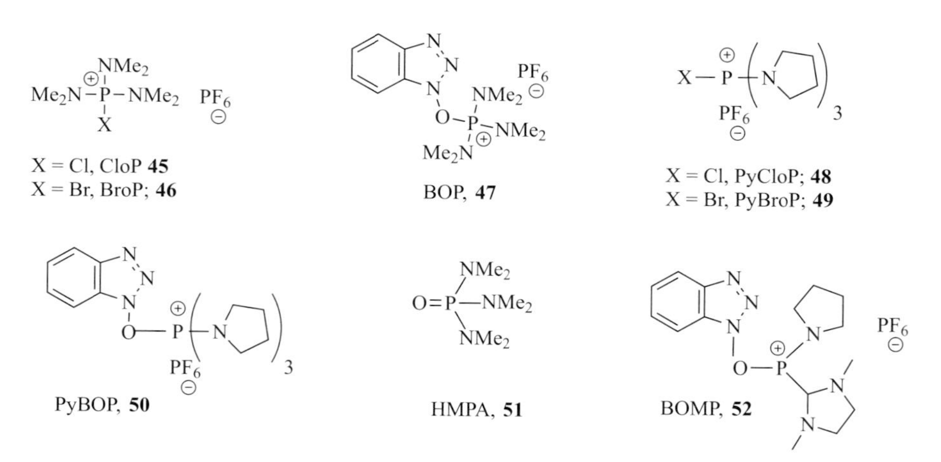


**Figure 12.4** Structure of phosphonium salts.

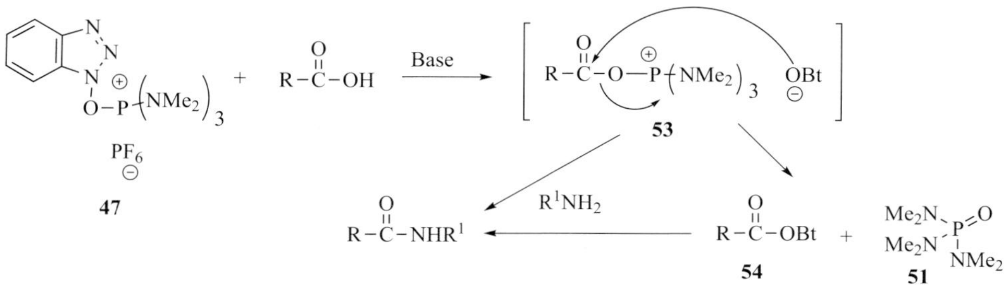

Scheme 12.6 Mechanism of BOP-mediated coupling reagent.

Campagne [48] proposed that this species is highly unstable and even at low temperature undergoes conversion to the OBt active ester **54** (Scheme 12.6).

Phosphonium salts derived from HOAt (**26**), such as (7-azabenzotriazol-1-yloxy) tris-(dimethylamino)phosphonium hexafluorophosphate (AOP, **55**) and (7-azabenzotriazol-1-yloxy)tris-(pyrrolidino)phosphonium hexafluorophosphate (PyAOP, **56**), have also been prepared and are generally more efficient than BOP (**47**) and PyBOP (**50**) as coupling reagents [23b, 49]. The pyrrolidino derivative PyAOP is slightly more reactive than the dimethylamino derivative AOP and does not release HMPA (**51**) in the activation step.

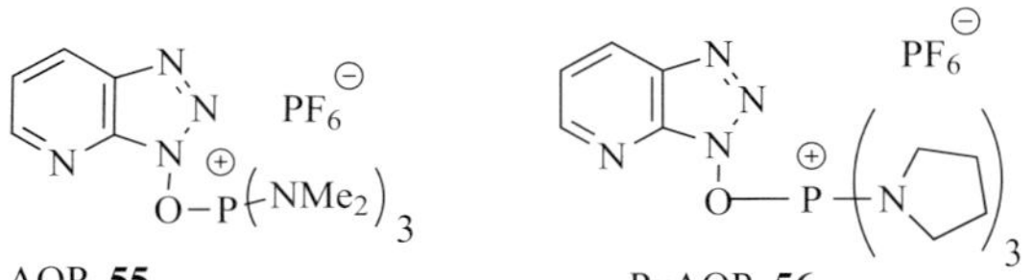

AOP, **55** PyAOP, **56**

PyAOP **56** is highly effective in the solution-phase cyclization of all-L-pentapeptides. This compound is a difficult case because of the restricted conformational flexibility of the linear precursor. For example, for the cyclization of H-Arg($NO_2$)–Lys (Z)–Asp(OBn)–Val–Tyr-OH, PyAOP gives 56% cyclomonomer with only 10.9% of the D-Tyr isomer after 1 h, whereas PyBOP leads to 52% cyclomonomer with 13% D-isomer and BOP only 38% of the desired product with 20.2% of C-terminal epimerization. In the course of the cyclization of H-Arg($H^+$)–Lys(Ac)–Ala–Val–Tyr-OH, only 8.7% of C-terminal epimerization is detected for PyAOP, whereas for PyBOP and BOP, epimerization reaches 16.8 and 20.7%, respectively.

Further research has revealed that PyAOP/HOAt is also an effective reagent for the cyclization of the highly hindered *N*-methyl amino acids [47e]. For the solid-phase synthesis of the cyclic peptide tachykinin antagonist, cyclo(Tyr–D-Trp–Val–D-Trp–D-Trp–Arg–Asp), the groups headed by Albericio and Carpino [49b] compared the cyclization of the monomer via PyAOP/HOAt. This strategy yields 74% of the cyclic product with 61% purity while PyBOP/HOBt gives only 69% of the cyclic product with 51% purity.

Since the discovery of HOBt-mediated coupling reagents, many racemization suppressants have been exploited as a part of the composition of new peptide-

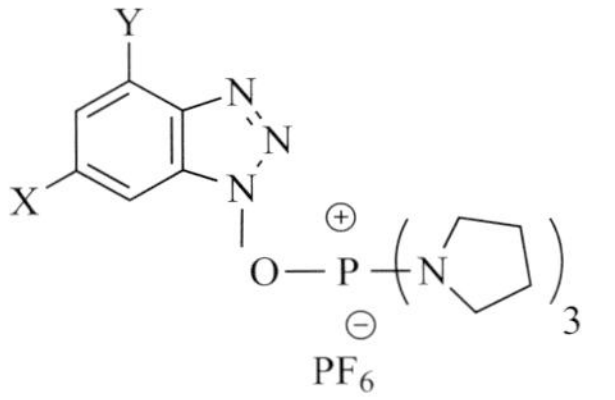

X = $NO_2$, Y = H, $NO_2$-PyBOP; **57**
X = $CF_3$, Y = H, $CF_3$-PyBOP; **58**
X = $CF_3$, Y = $NO_2$, $CF_3$-$NO_2$-PyBOP; **59**
X = Cl, Y = H, PyCloK; **60**

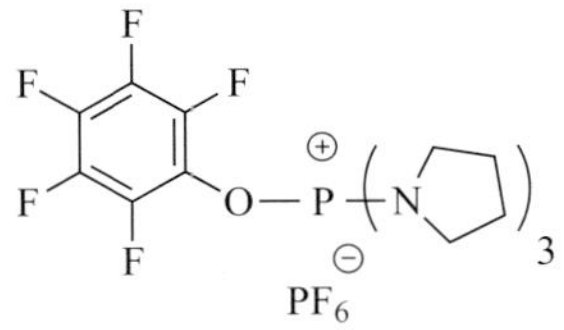

PyPOP, **61**

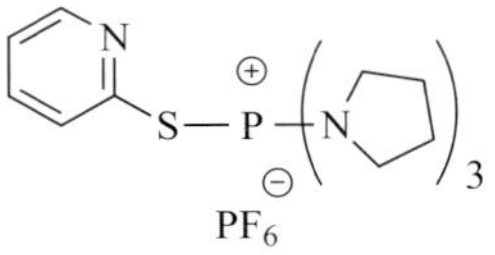

PyTOP, **62**

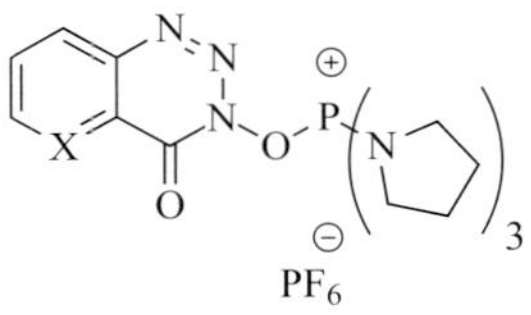

X = CH, PyDOP; **63**
X = N, PyDAOP; **64**

**Figure 12.5** Structure of phosphonium salts.

coupling reagents (Figure 12.5). For example, $NO_2$-PyBOP **57**, $CF_3$-PyBOP **58**, and $CF_3$-$NO_2$-PyBOP **59**, PyCloK **60**, PyPOP **61**, PyTOP **62**, PyDOP **63**, and PyDAOP **64** [29, 49, 50] were prepared in this regard and serve as efficient peptide-coupling reagents for the synthesis of dipeptides bearing *N*-methyl amino acids.

## 12.3.1 Preparation of Phosphonium Salts

The activation of carboxylic acids using phosphonium salts was described in the pioneering work by Kenner [38], Castro [51], Hruby [52], and Yamada [53]. Castro and Dormoy isolated the chlorotrisdimethylaminophosphonium cation in its perchlorate form (**65**) by reaction of tris(dimethylamino)phosphine with $CCl_4$ in ether followed by addition of an aqueous solution of ammonium perchlorate (Scheme 12.7) [40].

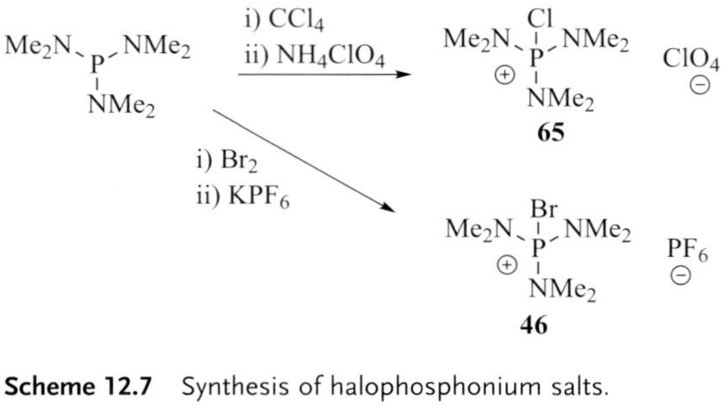

i) $CCl_4$ ii) $NH_4ClO_4$ → **65**
i) $Br_2$ ii) $KPF_6$ → **46**

**Scheme 12.7** Synthesis of halophosphonium salts.

The corresponding bromo derivative **46** (BroP) was prepared in a similar way in 85% yield by using $Br_2$ in ether at 0 °C followed by anion exchange with $KPF_6$ (Scheme 12.7) [41a, 43a]. The substitution of the tris(dimethylamino) group by the tripyrrolidino, affords the chlorotripyrrolidinophosphonium hexafluorophosphate (PyCloP, **48**) and bromotripyrrolidino phosphonium hexafluorophosphate (PyBroP, **49**) [54, 55], both of which are commercially available.

BOP (**47**), which was one of the first phosphonium salts to become commercially available, is prepared by reaction of tris(dimethylamino)phosphine with carbon tetrachloride in the presence of HOBt in THF at −30 °C followed by exchange of the chloride anion with the hexafluorophosphate anion (Scheme 12.8) [39b].

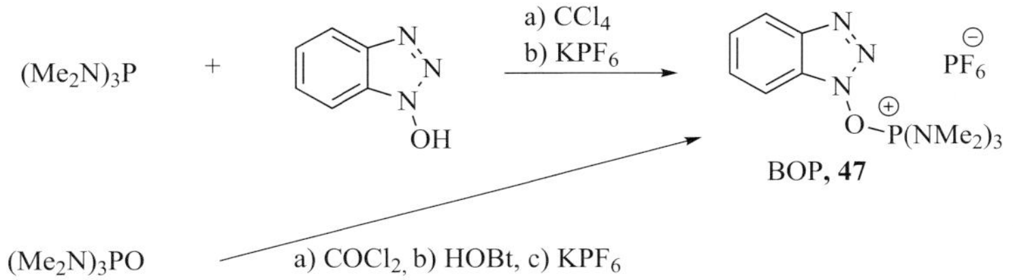

$(Me_2N)_3P$ + HOBt —— a) $CCl_4$ b) $KPF_6$ ——> BOP, **47**

$(Me_2N)_3PO$ —— a) $COCl_2$, b) HOBt, c) $KPF_6$ ——> BOP, **47**

**Scheme 12.8** Synthesis of BOP reagent.

A more economical method for preparation of this reagent was described by Castro [42a]. This involved the reaction of HMPA (**51**) with $COCl_2$ in toluene followed by reaction with HOBt in the presence of triethylamine (TEA) and then final anion exchange (Scheme 12.8). Phosphoryl chloride can also be used for the preparation of the chlorophosphonium cation intermediate, which can be isolated as hexafluorophosphate or perchlorate in almost quantitative yield [42b]. The pyrrolidine derivative **50** (PyBOP) has been prepared under similar reaction conditions and is also commercially available [42a]. Both BOP (**47**) and PyBOP (**50**), can be prepared by using triphosgene (BTC), which is less moisture-sensitive and easier to handle than $COCl_2$ [56].

Related HOBt-substituted reagents, such as $CF_3$-PyBOP (**58**), have been prepared [56] following the same protocol as for BOP (**47**) and PyBOP (**50**) [42a]. For the preparation of $NO_2$-PyOP (**57**) and $CF_3$-PyOP (**58**), the corresponding 6-nitro- or 6-trifluoromethyl-substituted benzotriazoles are allowed to react with PyCloP (**48**). The disubstituted benzotriazole derivative $CF_3$-$NO_2$-PyBOP (**59**) [4-nitro-6-(trifluoromethyl) benzotriazol-1-yl)oxy]tris-(pyrrolidino)phosphonium hexafluorophosphate [50a, 57] has been prepared by reaction of PyBrOP (**49**) with the corresponding disubstituted hydroxybenzotriazole. The dimethylamine and pyrrolidine phosphoramide derivatives AOP (**55**) and PyAOP (**56**) have been prepared in the same way [58], as well as the phosphonium salt PyDOP (**63**) and PyDAOP (**64**), by reaction of the corresponding additive with the phosphonium salt PyCloP (**48**) in the presence of TEA [44, 59]. Pentafluorophenol and -thiophenol derivatives of tris(pyrrolidino) phosphine oxide **61** (PyPOP) and **62** (PyPSP) [49] have also been obtained by reaction of the corresponding phenol or thiophenol with PyCloP. The thiophosphonium salt **62** (PyTOP) has been prepared by treating tris(pyrrolidino)phosphine with 2,2′-dipyridyl disulfide followed by precipitation of the phosphonium salt as its hexafluorophosphate (Scheme 12.9).

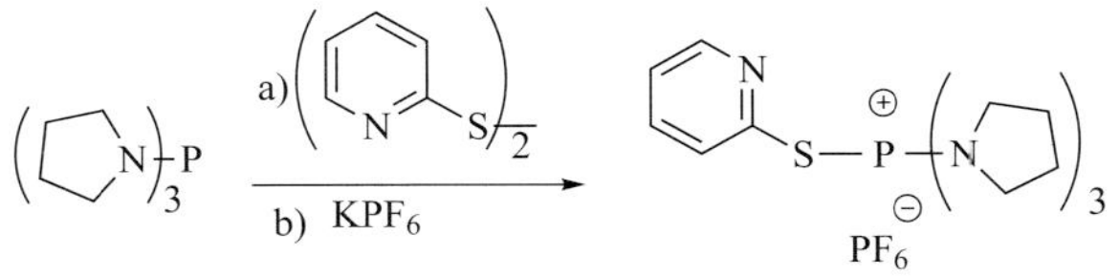

a) b) $KPF_6$

PyTOP, **62**

**Scheme 12.9** Synthesis of PyTOP.

### 12.3.2 General Method for the Synthesis of Phosphonium Salts [49]

Chlorophosphonium salt (CloP, TCloP, PyCloP, or PyBroP) (1 mmol), the additive (HOAt, HOBt, HODhbt, HODhat, HOxO, or HOBt derivatives) (1 mmol), and TEA (1 mmol) are dissolved in acetone (10 ml) and stirred for 1 h. The solution is filtered off and the solvent is removed under reduced pressure. The crude product is purified by precipitating from acetone by addition of ether, yielding white solids. Additive and TEA are substituted by the potassium or sodium salt of the additives.

*Notes*

- Halophosphonium salts have been used mainly in the solution-phase for the synthesis of peptides containing *N*-methyl and hindered α-amino acids.
- The pyrrolidine derivatives PyCloP (**48**) and PyBroP (**49**) give good results during the coupling of Z- or Fmoc-protected *N*-methyl amino acids [60]. However, in the case of Boc-protected systems, *N*-carboxyanhydrides (**15**) are formed when using these reagents [60].
- Some examples of the use of PyCloP (**48**) and PyBroP (**49**) in solution-phase as well as solid-phase have been reported [60–70].

## 12.4 Aminium/Uronium Salts

Initially, the product obtained by reaction of HOBt with tetramethylchlorouronium salt (TMUCl) was assigned to an uronium-type structure, presumably by analogy with the corresponding phosphonium salts, which bear a positive carbon instead of the phosphonium residue [71]. Recently [72], it has been shown by X-ray analysis that salts crystallize as aminium salts (guanidinium *N*-oxides) rather than the corresponding uronium salts. This occurs for *N*-[(1*H*-benzotriazol-1-yl)(dimethylamino)methylene]-*N*-methylmethanaminium hexafluorophosphate *N*-oxide (*N*-HBTU, **65**),

*N*-[(dimethylamino)-1*H*-1,2,3-triazolo[4,5-*b*]pyridin-1-ylmethylene]-*N*-methylmethanaminium hexafluorophosphate *N*-oxide (*N*-HATU, **66**), and 1-(1-pyrrolidinyl-1*H*-1,2,3-triazolo[4,5-*b*]pyridin-1-ylmethylene) pyrrolidinium hexafluorophosphate *N*-oxide (HAPyU, **68**) [73]. Nuclear magnetic resonance (NMR) studies in the case of HAPyU show that the same structure is found in solution (Scheme 12.10) [73].

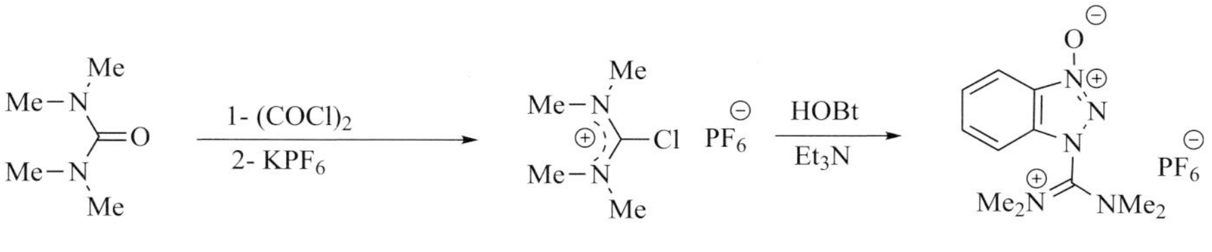

1- $(COCl)_2$
2- $KPF_6$
HOBt
$Et_3N$
**69** **70** **65**

**Scheme 12.10** Synthesis of *N*-HBTU **65**.

The preparation of these commercially available reagents is achieved by transformation of tetramethylurea (TMU, **69**) into the corresponding chlorouronium salt (TMUCl), by treatment with $COCl_2$ in toluene followed by exchange with $NH_4PF_6$ or $KPF_6$ and then reaction of **70** with HOBt to afford *N*-HBTU **65** (Figure 12.6) [23, 74]. Chlorotetramethyluronium chloride **70** has also been prepared by replacement of the extremely toxic $COCl_2$ by oxalyl chloride [75] or $POCl_3$ [76]. *N*-HBTU has been obtained using a one-pot procedure in organic solvents, and also the analogous tetrafluoroborate reagent, (TBTU, **71**), which could not be prepared by the previous procedure. These two reagents are commercially available.

This one-pot method has also been applied to the preparation of the HODhbt derivative 2-(3,4-dihydro-4-oxo-1,2,3-benzotriazin-3-yl)-1,1,3,3-tetramethyluronium tetrafluoroborate (TDBTU, **72**), the pyridone derivative 2-(2-oxo-1(2*H*)-pyridyl-1,1,3,3-tetramethyluronium tetrafluoroborate (TPTU, **73**), and the hydroxysuccinimide derivatives 2-succinimido-1,1,3,3-tetramethyluronium tetrafluoroborate (TSTU, **74a**) [74a], which are also commercially available. The hexafluorophosphate **74b** has also been prepared following the same strategy [75]. Other HOX derivatives

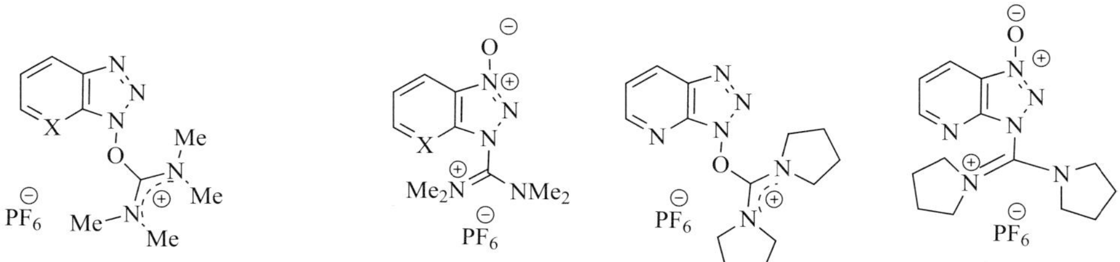

*O*-HBTU; **63,** X = CH,
*O*-HATU; **64,** X = N
*N*-HBTU; **65,** X = CH,
*N*-HATU; **66,** X = N
*O*-HAPyU, **67**
*N*-HAPyU,**68**

**Figure 12.6** Structures of aminium/uronium salts.

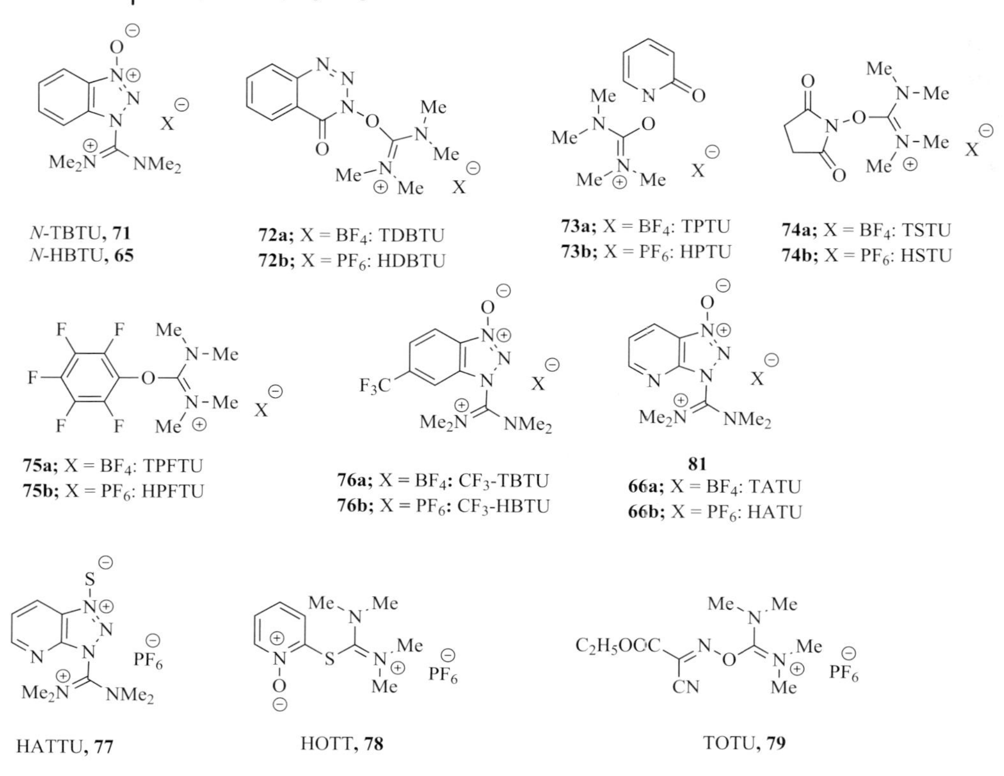


**Figure 12.7** Structures of aminium/uronium salts.

attached to the tetramethyluronium cation include the pentafluorophenol derivatives **75** (TPFTU) and HPFTU [77, 78] (Figure 12.7).

In the case of 1-hydroxybenzotriazole derivatives containing electron-withdrawing groups, the 6-trifluoromethyl derivative ($CF_3$-HBTU and $CF_3$-TBTU, **76**) has been prepared from tetrafluoromethylchloroformamidinium hexafluorophosphate [56]. The corresponding HBTU and TBTU analogs, containing the HOAt structure instead of HOBt, have been prepared from the TMU-Cl salts to give the corresponding reagents *N*-HATU **66** and tetrafluoroborate (TATU) [75], which have been shown to be *N*-oxides [73] with aminium structures. Two tetramethylurea-derived thiouronium reagents, the HOAt derivative **77** [79] (HATTU) and the *N*-hydroxy-2-pyridinethione derivatives **78** [80, 81] (HOTT), have been prepared, both following Knorr's strategy [75a, 81, 82]. The *O*-[(ethoxycarbonyl)cyanomethyleneamino]-*N,N,N′,N′*-tetramethyluronium tetrafluoroborate (TOTU, **79**) has been developed by a Hoechst group [80]. Further investigation has led to other aminium-type reagents such as HBPyU **80**, HAPyU **68**, HBMDU **81**, HAMDU **82**, HBPipU **83**, HAPipU **84**, and HOPyPfp **85** (Figure 12.8).

Recently, Xu and Li reported [73b] an alternative pathway to enhance coupling efficiency by modifying the carbon skeleton of uronium salts by replacing one of the

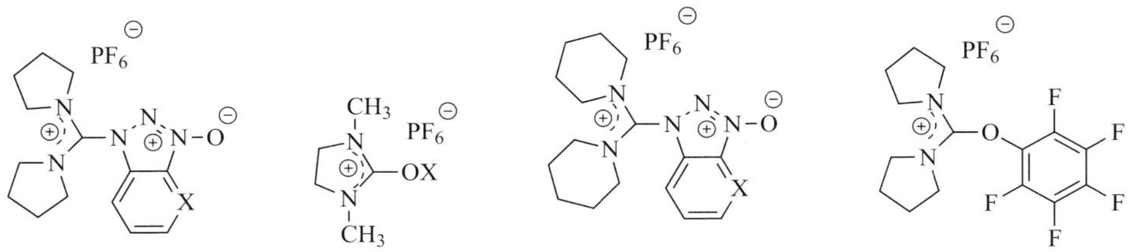

$X = CH$; HBPyU, **80**
$X = N$; HAPyU, **68**

$X = OBt$; HBMDU, **81**
$X = OAt$; HAMDU, **82**

$X = CH$; HBPipU, **83**
$X = N$; HAPipU, **84**

HOPyPfp, **85**

**Figure 12.8** Structure of aminium/uronium salts.

substituted amino groups with a hydrogen, alkyl, or aryl group (Scheme 12.11). The authors described that the use of these types of imonium reagents to synthesize peptides not only enhances the coupling efficiency, but also substantially suppresses the extent of racemization under relatively mild conditions.

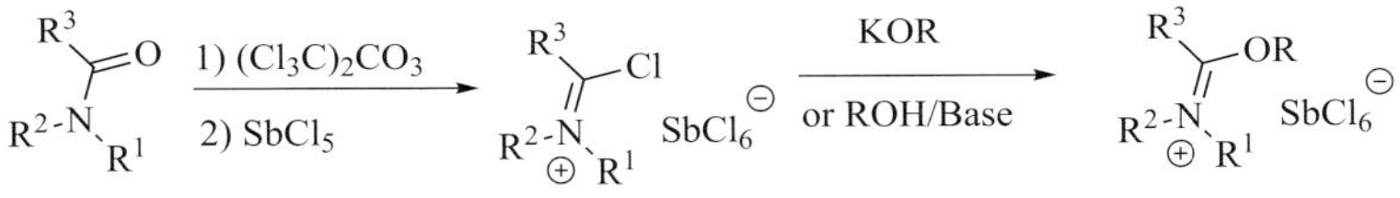

1) $(Cl_3C)_2CO_3$
2) $SbCl_5$

KOR or ROH/Base

$R^1, R^2$ = alkyl, aryl
$R^3$ = H, alkyl, aryl

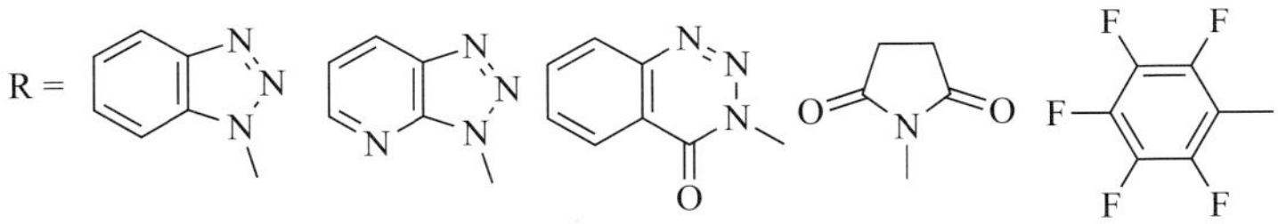

**Scheme 12.11** Synthesis of aminium/uronium salts.

More recently, Carpino *et al.* [29] reported the new coupling reagents HDATU **86** and HDAPyU **87** (Figure 12.9). These were prepared by a method analogous to that used for the preparation of HDTU **72** [75b], which is the HODhbt uronium salt reagent. As expected, **86** and **87** are more reactive than the analogous **72**.

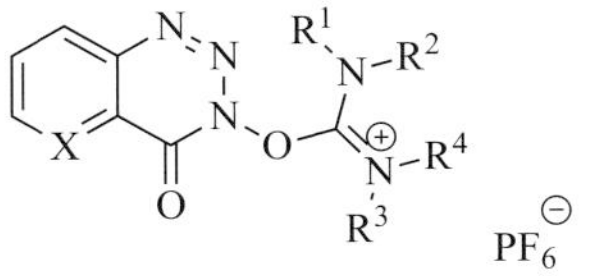

$X = N; R^1 = R^2 = R^3 = R^4 = CH_3$, HDATU; **86**
$X = N; R^1 = R^2 = R^3 = R^4 = (CH_2)_4$, HDAPyU; **87**
$X = CH; R^1 = R^2 = R^3 = R^4 = CH_3$, HDTU; **72**

**Figure 12.9** Structures of HDTU analogs.

El-Faham and Albericio described a new family of immonium-type coupling reagents based on the differences in the carbocation skeletons of coupling reagents which correlated with differences in stability and reactivity [83, 84]. The dihydroimidazole derivatives are highly unstable to air, while the salts derived from dimethyl morpholino are the most stable, and the pyrrolidino derivatives are of intermediate stability. Regarding both coupling yield and retention of configuration, derivatives of Oxyma (COMU, **91m**) have been confirmed to show superior performance to those of HOBt in all cases and the same performance as HOAt or sometimes better. The recent uronium-type reagents can be readily prepared by treating *N,N*-dialkylcarbamoyl chloride **88** with secondary amines such as diethylamine, pyrrolidine, piperidine or morpholine to give the corresponding urea derivatives **89** (Scheme 12.12). The urea derivatives then react with oxalyl chloride to yield the corresponding chlorosalts **90** (Scheme 12.12), which is stabilized by the formation of a $PF_6$ salt. Subsequent reaction with HOXt (A = B or A) in the presence of a tertiary amine such as $Et_3N$ affords the desired compound **91** as crystalline and shelf-stable solids (Scheme 12.12).

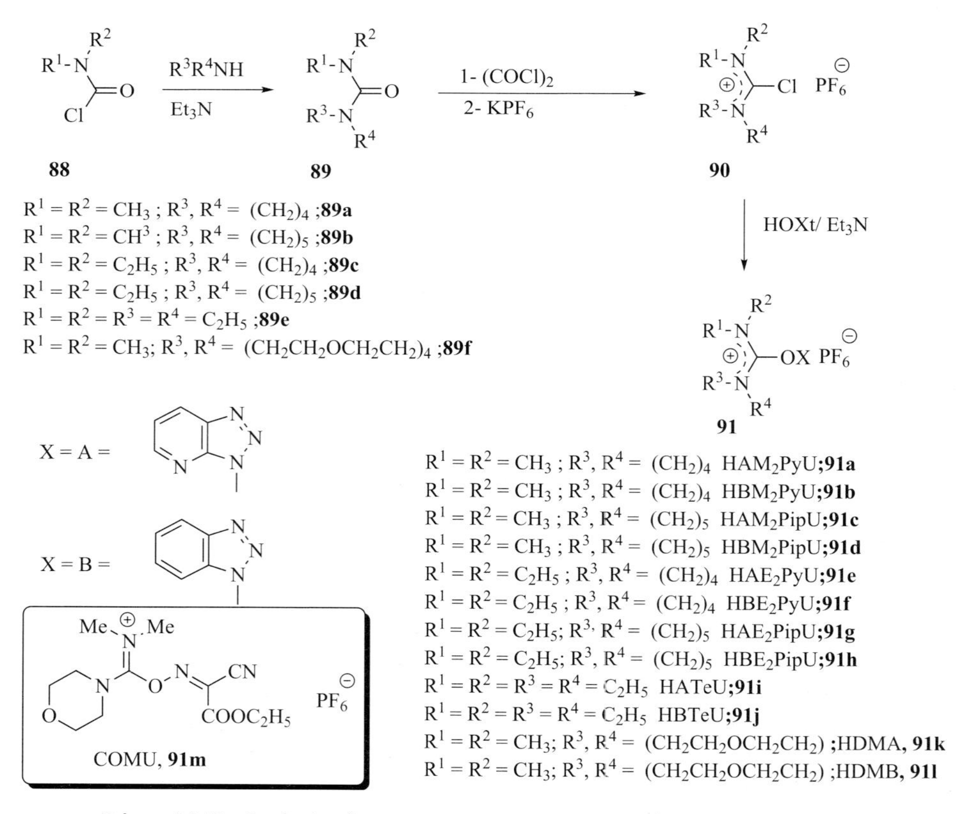


**Scheme 12.12** Synthesis of nonsymmetric uronium-type coupling reagents.

Mechanistically, aminium/uronium salts are thought to function in a manner similar to the phosphonium analogs (Scheme 12.13). Formation of carboxyl uronium salts which generate an active ester is achieved in the presence of 1 equiv. of tertiary base such as DIEA, NMM [48], or TMP [49]. The presence of an extra equivalent of HOXt accelerates coupling and reduces the loss of configuration [49].

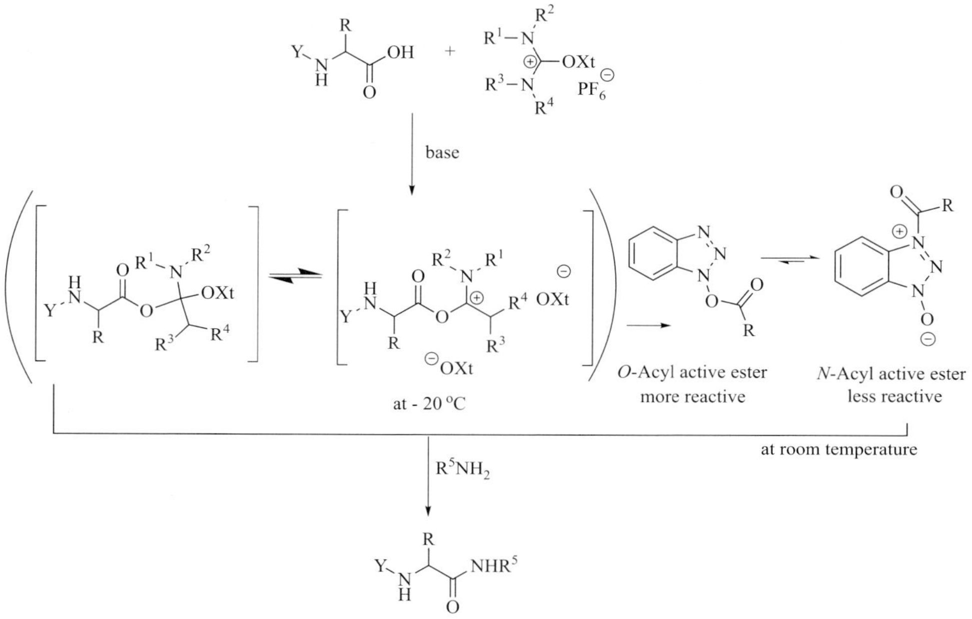

**Scheme 12.13** Proposed activation mechanism of aminium/uronium salts.

### 12.4.1
### Stability of Onium Salts

In order to determine the compatibility of benzotriazole-based coupling reagents with automated peptide synthesizers, the stability of these reagents in solution and in the solid state has been examined via high-performance liquid chromatography (HPLC) and $^{1}$H-NMR analysis (looking for the urea peak due to hydrolysis) [16]. Furthermore, evaluation of the stability of a given compound in solution can provide information about its reactivity. When stored under $N_2$, all the salts are stable at 25 °C for 3–4 weeks. DMF solutions (0.6 M) of reagents exposed to the atmosphere are stable for 1–2 weeks. From this information it is deduced that aza derivatives are less stable, and therefore more reactive, than the benzotriazole analogs [16]. The oxygen in the iminium structure of COMU (**91m**) increases the stability of the coupling reagent compared with the tetramethyl derivatives (HBTU, **65**, HATU, **66**, and TOTU, **79**). Furthermore, COMU (**91m**) has higher stability than the benzotriazole derivatives HATU (**66**) and HBTU (**65**) [84]. All these reagents show stability greater than 95% in a closed vial. These observations are of relevance for both solid-phase and solution strategies. Thus, when the activation of a carboxylic acid is slow and the coupling reagent is not stable, it is degraded and no longer able to activate the carboxylic function. This feature is crucial for cyclization steps and for segment coupling steps in convergent strategies, where excess of the carboxylic function is not present (cyclization) or is low (segment coupling) and therefore couplings are very slow. A further characteristic of COMU (**91m**) is that the course of reaction can be followed as a result of change of color, which depends on the type of base used. Thus, 2 min after the addition of the coupling reagent, the solution turns orange-red when DIEA is used as a base and pink in case of TMP. Once the reaction is complete, the solution becomes colorless and yellow, respectively. These results should be taken into account mainly when coupling reagents are placed in open vessels, such as in some automatic synthesizers. Furthermore, the nature of the carbon skeleton of a compound is of marked relevance to the stability of the compound. Both dihydroimidazole (HBMDU, **81**, and HAMDU, **82**) derivatives are highly unstable, whereas their corresponding morpholino and dimethylamine salts are the most stable, and the pyrrolidino derivatives are of intermediate stability. These results should be considered mainly for those syntheses performed in automatic synthesizers in which the coupling reagents are dispensed in open vessels. As expected, all the coupling reagents are more stable when the DMF solutions are stored under $N_2$ atmosphere [16, 84], conditions typically used in some automatic synthesizers.

Given that peptide bond formation is usually carried out in the presence of at least one extra equivalent of base, the stability of onium salts has also been examined in the presence of DIEA. Analysis of these results confirms that the various coupling reagents rapidly degrade in the absence of a carboxylic acid function. This observation has practical consequences for both solid-phase and solution strategies. Thus, if activation of a carboxylic acid is slow, the coupling reagents will be degraded and will no longer have the capacity to activate the carboxyl function. Under these conditions, aza derivatives are more labile than the benzotriazole derivatives, and pyrrolidino derivatives are more labile than morpholino, dimethylamino, and diethylamino derivatives. This observation is important for cyclization steps or in convergent strategies during fragment coupling because the yields tend to be lower than for other couplings.

### 12.4.2
### General Procedure for the Preparation of Chloroformamidinium Salts

Oxalyl chloride (100 mmol) is added dropwise to a solution of urea derivative (100 mmol) in dry DCM (300 ml) at room temperature over 5 min. The reaction mixture is stirred under reflux for 3 h, and the solvent is removed under reduced pressure. The residue is washed with anhydrous ether (2 × 100 ml), then bubbled with $N_2$ to remove excess of ether. The residue obtained is dissolved in DCM, and a saturated aqueous $KPF_6$ solution is added at room temperature and then the reaction mixture is stirred vigorously for 10–15 min. The organic layer is collected, washed once with water (100 ml), dried over anhydrous $MgSO_4$ and filtered, and the solvent is removed under reduced pressure. The crude product is recrystallized from DCM/ether. HOXt and DIEA are replaced using KOXt in $CH_3CN$.

### 12.4.3
### Synthesis of Aminium/Uronium Salts

Chloroformamidinium salt (10 mmol) is added to a stirring solution of HOXt (10 mmol) and TEA (10 mmol) in dry DCM (100 ml). The reaction mixture is stirred at room temperature overnight, filtered, and washed with DCM ($2 \times 10$ ml). The residue is then recrystallized from $CH_3CN$/ether.

### 12.4.4
### General Procedure for Coupling Using Onium Salts (Phosphonium and Uronium) in Solution Phase

The *N*-Protected amino acid (0.25 mmol), the amino component (0.25 mmol), and base (0.50 mmol or 0.75 mmol in case of ester hydrochloride) in DMF (2 ml) is treated with the coupling reagent (0.25 mmol) at 0 °C and the reaction mixture is stirred at 0 °C for 1 h and at room temperature for 2–3 h. The mixture is diluted with EtOAc (25 ml), and extracted with 1 N HC1 ($2 \times 5$ ml), 1 N $NaHCO_3$ ($2 \times 5$ ml), and saturated NaCl ($2 \times 5$ ml). The EtOAc is then dried with $MgSO_4$, the solvent is removed, and the crude peptide is directly analyzed by HPLC.

### 12.4.5
### General Procedure for Coupling Reaction in Solid-Phase Using Onium Salts (Phosphonium and Uronium)

*N*-Protected amino acid (3 equiv.), base (6 equiv.), and onium salt (3 equiv.) are preactivated in DMF (0.3 M) for 1 min and then added to the amino-resin with manual stirring for 2–5 min and allowed to stand at room temperature for 30 min (4 h for hindered residues). The resin is filtered and washed with DMF.

### 12.4.6
### General Procedures for Coupling Reaction in Solid-Phase Using Onium Salts (Phosphonium and Uronium) Boc-, Fmoc-Amino Acids via Phosphonium and Uronium Salts

*N*-Protected amino acid (4 equiv.) and the phosphonium or aminium salt (4 equiv.) are dissolved in DMF or the mixture is added to the resin bearing the free amino group, and finally the base (4–8 equiv.) is added. After 30 min the resin is filtered and washed with NMP. For Fmoc-Asn(Trt) coupling, HOXt (4 equiv.) should be added. Some researchers prefer to carry out a preactivation step involving the protected amino acid, the coupling reagent, the base, and HOXt for 10 min. This protocol is preferred in the case of aminium salt couplings, where chain termination via formation of the guanidino species can intervene. Alternatively, it is possible to reverse the addition of base and the other reagents in order to prevent contact of the aminium salt with the free amino group, because the activation step is faster than the formation of the chain-terminating guanidino derivative.

*General notes for onium salts (phosphonium and uronium)*

- Owing to of discrepancies in racemization levels when using different samples of commercial tertiary amines, DIEA (Aldrich, 99%) and NMM (Aldrich, 99%) are distilled first from ninhydrin and then from $CaH_2$ (boiling point 126 and 114–116 °C, respectively), and stored over molecular sieves. TMP (Eastman Kodak, 97%) is distilled from $CaH_2$ (boiling point 170–172 °C) and stored over molecular sieves. Other bases are treated similarly.
- Untreated DIEA, which may contain various amounts of primary or secondary amines (positive ninhydrin test), leads to enhanced racemization (2–3%). In contrast, TMP (Aldrich, 99%), taken directly from the bottle, gives racemization levels comparable to those obtained with material distilled over $CaH_2$.
- DMF (Fisher HPLC grade) is aspirated with a stream of $N_2$ for 15 min and stored over molecular sieves.
- In solution-phase coupling, preactivation is not required so the coupling reagent is added last at 0 °C
- In solid-phase coupling, to prevent guanidinium formation, preactivation for a maximum of 2 min is required, especially with uronium salt. However, this does not apply to phosphonium salts.
- To keep the racemization low and to improve the yield, long preactivation is not recommended in onium salts because long preactivation gives the chance for formation of oxazolone and hydrolysis of the active intermediates as well as shift the active ester from the *O*-acyl to the *N*-acyl which is less reactive (Scheme 12.13).
- For cyclization, uronium salts are not recommended because of the formation of guanidine side products.
- The use of these various onium salts requires careful attention to the tertiary base used and the preactivation time. However, for some synthesizers the preactivation time is dictated by the instrument, while in others and for manual syntheses it can be modulated.
- For onium salts incorporating HOAt, the activation of ordinary amino acids gives the corresponding OAt esters almost instantly. Thus, in such cases, the preactivation time should be kept to a minimum, since on standing alone the activated species can give rise to several side-reactions, including racemization and the formation of δ-lactam (Arg), cyano derivatives (Asn or Gln), or α-aminocrotonic acid (Thr). The same consideration applies to coupling reagents that incorporate HOBt.
- Regarding the use of a base during coupling, for those reactions involving amino acids that are not likely to lose their configuration (all except Cys and His), the reactions are carried out in the presence of 1.5–2.0 equivalents of a tertiary amine, such as DIEA or NMM.
- For the coupling of protected peptides, where resistance to conversion to oxazolone does not apply, and for the coupling of Cys and His, the use of only

1 equivalent of a weaker or more hindered base is recommended: for such systems TMP (**12**) and the more basic DBDMAP (**13**) are very promising [14].

## 12.5
## Fluoroformamidinium Coupling Reagents

A notable advance in the field was the development of fluoroformamidinium salts. Carpino and El-Faham reported that the air-stable, nonhygroscopic solid tetramethylfluoroformamidinium hexafluorophosphate (TFFH, **92**) is a convenient *in situ* reagent for amino acid fluoride formation during peptide synthesis (Scheme 12.14) [85]. TFFH is especially useful for His and Arg since the corresponding amino acid fluorides are not stable towards isolation or storage.

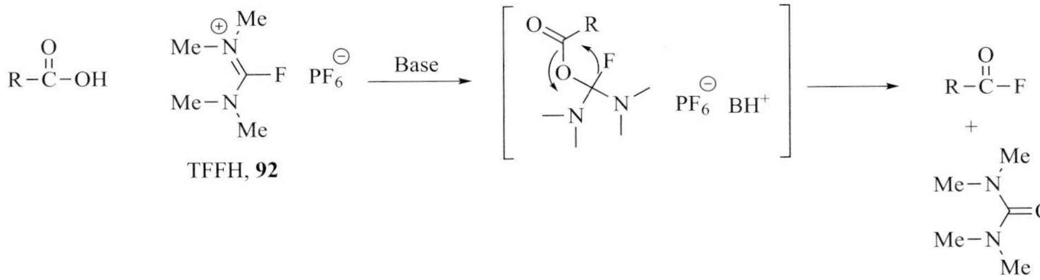

**Scheme 12.14** Synthesis of amino acid fluorides using TFFH.

IR examination shows that in the presence of DIEA, Fmoc-amino acids are converted to the acid fluorides by means of TFFH [85]. In DCM solution at room temperature, IR absorption characteristic of the carbonyl fluoride moiety (1842 $cm^{-1}$) appears after about 3 min, with complete conversion to the acid fluoride occurring after 8–15 min. For hindered amino acids (e.g., Aib), complete conversion may require 1–2 h [85–87]. If desired, the acid fluorides can be isolated and purified, thereby making TFFH a benign substitute for the corrosive cyanuric fluoride.

Other analogous reagents have also been synthesized (Figure 12.10). Bis(tetramethylene)fluoroformamidinium hexafluorophosphate (BTFFH, **93**) has an advantage over TFFH since in the work-up the reaction mixture of the latter generates toxic byproducts [87].

Fluorinating reagents **93**, **95**, **96**, **97** [83, 87], and **98** [84] show the same behavior as **93** in their capacity to provide routes to amino acid fluorides for both solution and solid-phase reactions [84, 85, 87]. In contrast **94**, which is more reactive but more sensitive to moisture, never gives complete conversion to the acid fluoride. The fluoro reagent **98** gives convenient results with 1 equiv. of base as a result of the morpholino moiety, which acts as proton acceptor [84]. Except for **94**, all of these reagents can be

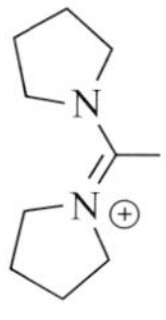

BTFFH, **93** 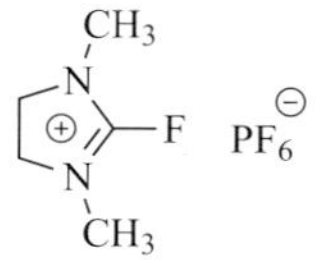 FIP, **94** 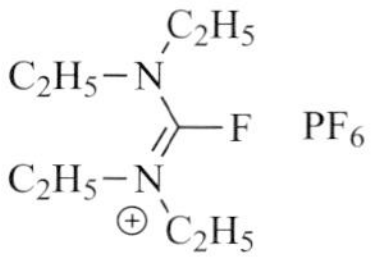 TEFFH, **95**

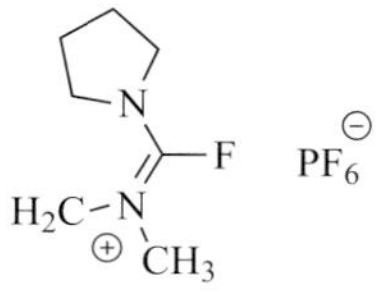

DMFFH, **96** 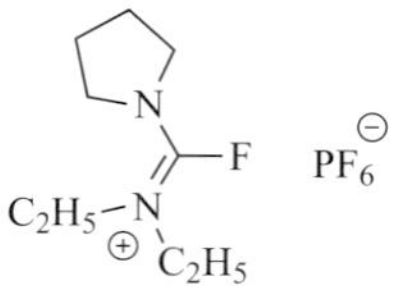 DEFFH, **97** 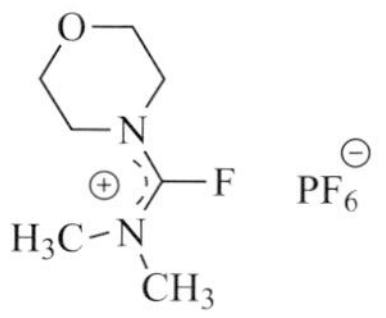 DMFH, **98**

**Figure 12.10** Structures of fluorinating reagents.

handled in air in the same way as common onium reagents [88], such as *N*-HBTU **65** and *N*-HATU **66**.

For some amino acids (e.g., Fmoc-Aib-OH) the use of TFFH alone gives results that are less satisfactory than those obtained with isolated amino acid fluorides. The deficiency has been traced to inefficient conversion to the acid fluoride, which under the conditions used (2 equiv. of DIEA) is accompanied by the corresponding symmetric anhydride and oxazolone formation [88, 89]. Furthermore, it has now been shown that when a fluoride additive, such as benzyltriphenylphosphonium dihydrogen trifluoride (PTF, **99**) or pyridine-hydrogen fluoride **100** [90, 91], is present during the activation step, the formation of the latter two products (symmetric anhydride and oxazolone) is prevented and a maximum yield of acid fluoride is obtained. Assembly of the difficult pentapeptide H-Tyr–Aib–Aib–Phe–Leu-$NH_2$ via TFFH coupling in the presence of **99** (PTF) gives a product with a quality similar to that obtained via isolated acid fluorides.

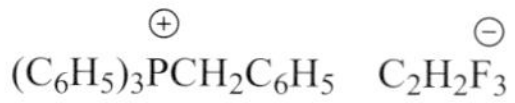

$(C_6H_5)_3\overset{\oplus}{P}CH_2C_6H_5 \quad \overset{\ominus}{C_2H_2F_3}$ **99**

$(HF)_n$ **100**

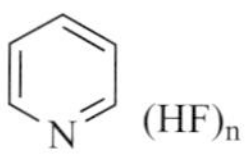

As the fluoride additive binds excess hydrogen fluoride as part of the complex, an accompanying acidic buffering effect might prove to be of useful in the case of coupling reactions in which considerable loss of configuration occurs at the activated carboxylic acid residue. In fact, such a protective effect has been observed for the sensitive histidine derivative (Fmoc-His(Trt)-OH) upon reaction with proline amide, which, with TFFH/DIEA and under ordinary conditions, gives the desired dipeptide in good yield with 7.4% stereomutation. In the presence of additive, stereomutation drops to 1.8% [91].

The related chloroformamidinium salt CIP (**101**) in the presence of HOAt as additive [87, 92] mediates the solution-phase Aib coupling in excellent yield (82–90% for Z-Aib-Aib-OMe) [92]. Two peptaibols, Ac-Aib-Pro-Aib-Ala-Aib-Ala-Gln-Aib-Val-Aib-Gly-Leu-Aib-Pro-Val-Aib-Aib-Glu-Gln-Pheol (alamethicin F-30) and Ac-Aib-Asn-Leu-Aib-Pro-Ala-Val-Aib-Pro-Aib-Leu-Aib-Pro-Leuol (trichovirin 14A) have been successfully synthesized using CIP/HOAt coupling techniques [92]. For Fmoc-Aib-OH (**102**) activation, the CIP-mediated reaction proceeds first through the oxazolone (**103**) and then through the HOAt active ester (**104**) (Scheme 12.15) [89].

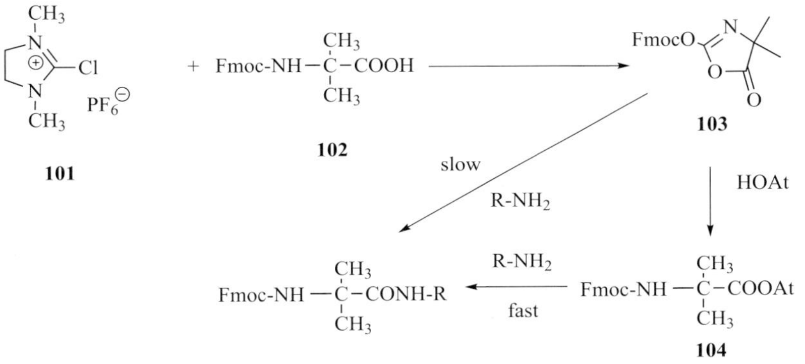

**Scheme 12.15** CIP-mediated reaction.

### 12.5.1
### General Method for the Synthesis of Fluoroformamidinium Salts [85,87]

A typical experiment for the preparation of TFFH (Scheme 12.16): In a 2-l three-necked round flask equipped with a mechanical stirrer, addition funnel and reflux condenser, oxalyl chloride (0.80 mol) is added in one portion to a solution of TMU (0.60 mol) in toluene (1 l) with vigorous stirring. The mixture is heated at 60 °C for 2 h and then cooled to room temperature. The addition funnel is replaced with a fritted adapter and the supernatant liquid is expelled using a positive pressure of nitrogen.

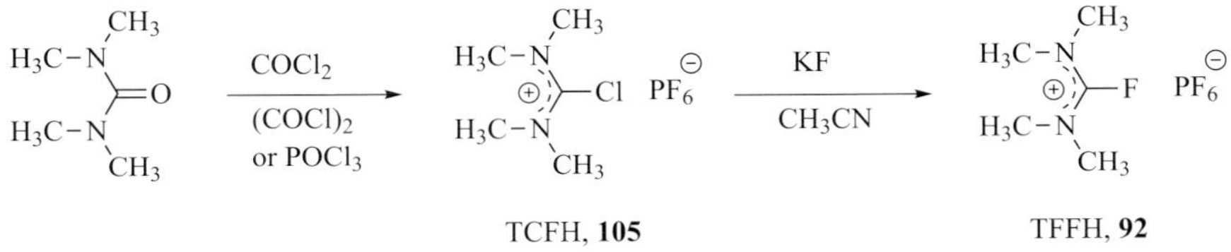

**Scheme 12.16** Synthesis of TFFH.

The precipitate is collected and washed with toluene and then with dry ether. The dichloro salts are collected and dissolved quickly in DCM (1 l) and treated with a saturated solution of $KPF_6$ (0.6 mol) in water. The reaction mixture is stirred vigorously at room temperature for 10–15 min and then the DCM is collected, dried ($MgSO_4$), and then removed under vacuum to give the chloro salt, TCFH **105**. To a solution of **105** (0.5 mol) in 300 ml of dry $CH_3CN$, 1.5 mol of oven-dried anhydrous KF is added and the mixture is stirred at room temperature for 3 h ($^1H$-NMR monitoring). Longer times are required for large-scale preparations. Following filtration of KCl, the filtrate is evaporated and the residue recrystallized from $CH_3CN$-ether to give TFFH in a yield 92% as nonhygroscopic white crystals (Scheme 12.16).

### 12.5.2
### Solution- and Solid-Phase Couplings via TFFH [92]

Not only does the acid fluoride methodology work well with acid-sensitive groups (Boc and *t*Bu side-chain protecting groups), but it is the acyl fluoride functionality itself that is likely to assure the widespread applicability of this general class of reagents. Due to the nature of the C—F bond, acyl fluorides show greater stability than the corresponding chlorides toward neutral oxygen nucleophiles, such as water or methanol, yet appear to be of equal or nearly equal reactivity toward anionic nucleophiles and amines [93].

Fluoroformamidinium salts TFFH and BTFFH are effective as isolated acid fluorides in both solution and solid-phase peptide assembly. Arginine represents a special case; reaction between Fmoc-Arg(Pbf)-OH and TFFH or BTFFH in the presence of DIEA (1/1/2) in DMF has been monitored by IR analysis. The acid fluoride (IR: 1845 $cm^{-1}$) is generated within 2 min and although it cyclizes slowly to the corresponding lactam (IR: 1794 $cm^1$), a significant amount of the acid fluoride remains unreacted even after 60 min.

### 12.5.3
### General Method for Solid-Phase Coupling via TFFH

*N*-Protected amino acid (4 equiv.), TFFH (4 equiv.), and the base (8 equiv.) are dissolved and preactivated for 5–7 min, and then added to the resin. After 30 min, the resin is filtered and washed with DMF or NMP. For Fmoc-Asn(Trt) coupling, HOXt (4 equiv.) should be added.

## 12.6
## Organophosphorus Reagents

Since Yamada introduced the mixed carboxylic-phosphoric anhydride method using diphenylphosphoryl azide (DPP, **106**) from diphenylphosphorochloridate and sodium azide to peptide chemistry in 1972 [94], various organophosphorus compounds

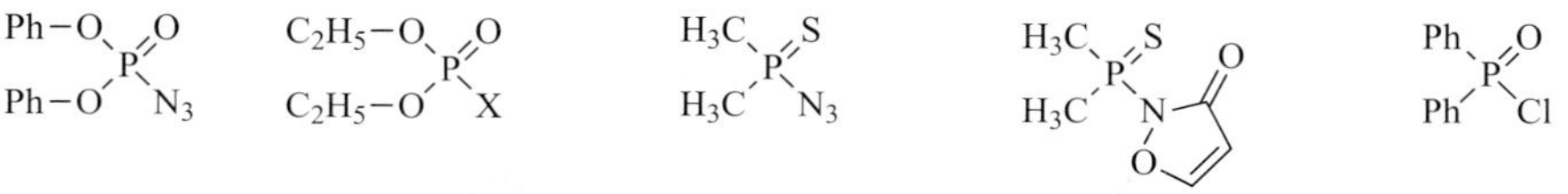

DPPA, **106** X = CN; DECP, **107** X = Br; DEPB, **108** X = Cl; DEPC, **109** MPTA, **110** MPTO, **111** DPPCl, **112**

**Figure 12.11** Structures of organophosphorus reagents.

have been developed as peptide-coupling reagents (Figure 12.11). This method usually gives higher regioselectivity towards nucleophilic attack by the amine component than a mixed carbonic anhydride method [95].

Modification of DPPA (**106**) has led to the development of thiophosphinic-type coupling reagents such as dimethylphosphinothioyl azide (MPTA, **110**) and 3-dimethylphosphinothioyl-2(3*H*)-oxazolone (MPTO, **111**) (Figure 12.11) [96].

These reagents are crystalline and stable for long-term storage. Since MPTA (**110**) generated a carbamoyl azide or urea derivative as the by-product, Ueki introduced MPTO (**111**), in which the azide group of MTPA is replaced by a 2-oxazolone group. On the basis of the earlier development of organophosphorus reagents, a great amount of effort has been focused on developing various coupling reagents of a similar kind. For example, norborn-5-ene-2,3-dicarboximidodiphenylphosphate (NDPP, **113**) [97], Cpt-Cl (**114**) [98], *N,N'*-bismorpholinophosphinic chloride (BMP-Cl, **115**) [99], diethyl phosphorobromidate (DEBP, **116**) [100], benzotriazol-1-yl diethylphosphate (BDP, **117**) [101], bis(*o*-nitrophenyl)phenyl phosphonate [102], (5-nitro-pyridyl)-diphenyl phosphinate [103], diphenyl 2-oxo-3-oxazolinyl phosphonate [104], and 1,2-benzisoxazol-3-yl diphenyl phosphate [105] have been prepared by several research groups (Figure 12.12).

More recently, Ye developed *N*-diethoxyphosphoryl benzoxazolone (DEPBO, **118**), *N*-(2-oxo-1,3,2-dioxaphosphorinanyl)-benzoxazolone (DOPBO, **119**), 3-[*O*-(2-oxo-

NDPP, **113** Cpt-Cl, **114** BMP-Cl, **115** DEBP, **116** BDP, **117**

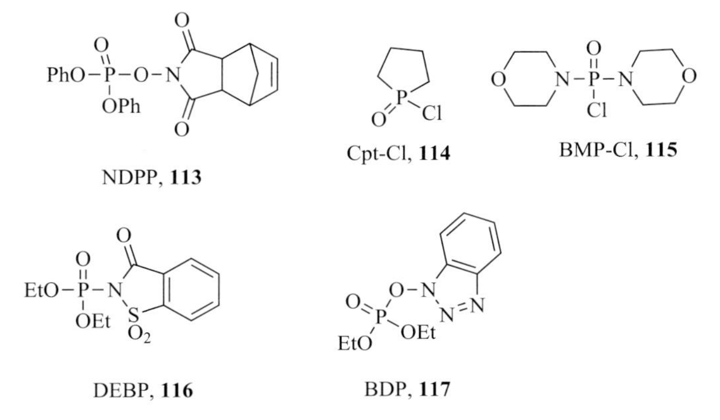

**Figure 12.12** Structures of organophosphorus reagents.

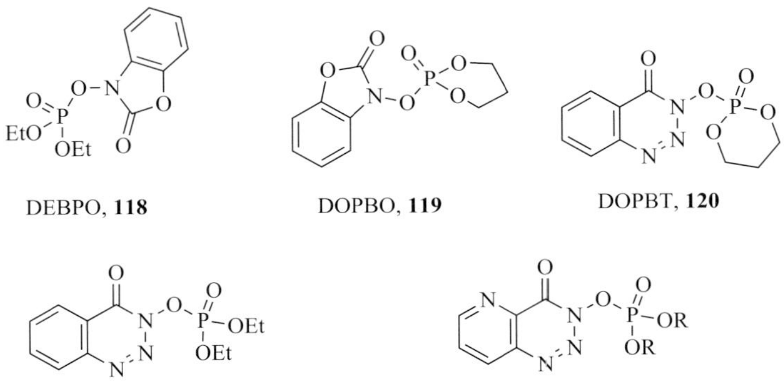

DEBPO, **118** DOPBO, **119** DOPBT, **120**

DEPBT, **121** R = Et or Ph, **122**

**Figure 12.13** Structures of organophosphorus reagents.

1,3,2-dioxaphosphorinanyl)-oxy]-1,2,3-benzotriazin-4(3*H*)-one (DOPBT, **120**), and 3-(diethoxyphosphoryloxy)-1,2,3-benzotriazin-4(3*H*)-one (DEPBT, **121**) (Figure 12.13) [106]. DEPBT derived from diphenylphosphorochloridate and HODhbt has been evaluated against other peptide-coupling reagents and gave good results in segment coupling reactions. Although the racemization-suppressing capacity of HODhbt is greater than that of HOBt, its utility was limited due to side-reactions.

Later, Carpino *et al.* [29] introduced new organophosphorus reagents (**122**, Figure 12.13). In this case the neighboring group effects believed to be relevant to the properties of HOAt are superimposed on the effects that enhance the efficiency of the phosphorus moiety. On the basis of the results described, these effects are related to the greater speed with which protected amino acids are converted to their active esters by the phosphorus derivatives. Long-term storage, unfortunately, requires great care because of the hydrolytic sensitivity of these materials. Given that these new phosphate esters are clearly superior to the older uronium/guanidinium reagents for segment coupling and, under certain conditions, for solid-phase peptide assembly, it was considered essential to search for reagents with greater shelf stability.

Recently [107], phosphoric acid diethyl ester 2-phenylbenzimidazol-1-yl ester, diphenylphosphinic acid 2-phenylbenzimidazol-1-yl ester (Scheme 12.17), and phosphoric acid diphenyl ester and 2-phenylbenzimidazol-1-yl ester have been reported as highly efficient coupling reagents. Their efficiency was evaluated through the synthesis of a range of amides and peptides, and the extent of racemization was found to be negligible. 1-Hydroxy-2-phenylbenzimidazole (**123**) was synthesized by coupling *ortho*-nitroaniline with benzyl bromide using sodium hydride as base, followed by benzyl deprotection (Scheme 12.17) [107]. The coupling reagents phosphoric acid diethyl ester 2-phenylbenzimidazol-1-yl ester (**124a**), diphenylphosphinic acid 2-phenylbenzimidazol-1-yl ester (**124b**), and phosphoric acid diphenyl ester 2-phenylbenzoimidazol-1-yl ester (**124c**) were synthesized by reaction of **123**

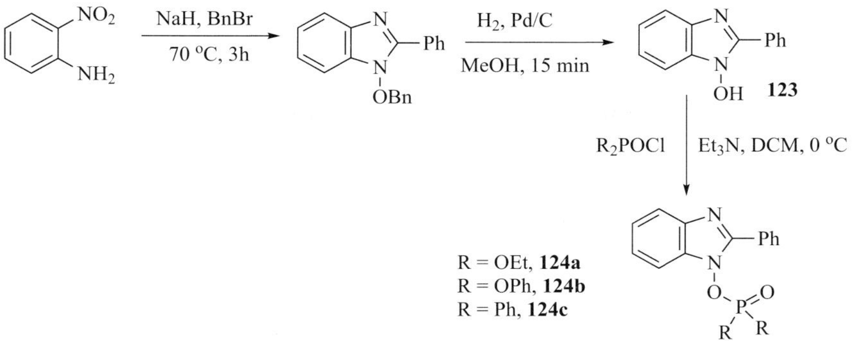

Scheme 12.17 Synthesis of phosphorus reagents of 1-hydroxy-2-phenylbenzimidazole.

with diethyl chlorophosphate, diphenylphosphorochloridate, or diphenylphosphinic chloride, respectively, using TEA base in DCM.

### 12.6.1 General Method for Synthesis of the Diphenylphosphoryl Derivatives

To a solution of diphenylchlorophosphate (0.01 mmol) and the corresponding additive HOXt (0.01 mmol) in anhydrous THF (50 ml), 0.01 mmol TEA is added under argon atmosphere (instead of using the additive (HOXt) and TEA for the synthesis, KOXt was used in some cases). After stirring for 3 h, the reaction mixture is filtered off and the solvent is removed under reduced pressure. The resulting yellow oils are washed with DCM (10 ml) and *n*-pentane (10 ml).

## 12.7 Triazine Coupling Reagents

1,3,5-Triazines have also been used as coupling reagents. Thus, 2-chloro-4,6-dimethoxy-1,3,5-triazine (CDMT, **127**) is a stable commercially available crystalline compound that is readily accessible from cyanuric chloride (CC) Scheme 12.18 [108].

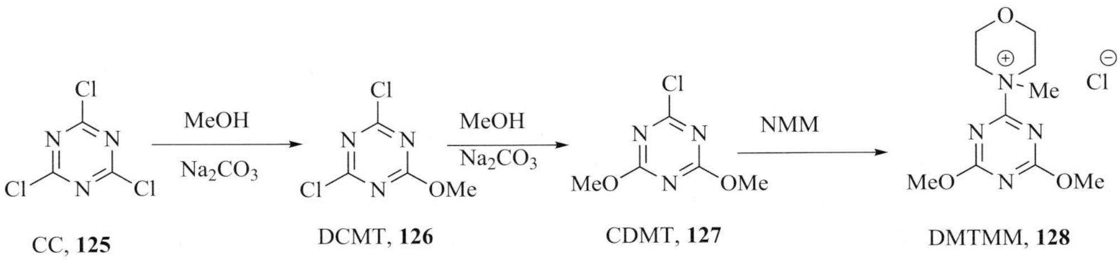

Scheme 12.18 Synthesis of DMTMM.

CC (**125**), 2,4-dichloro-6-methoxy-1,3,5-triazine (DCMT, **126**), and 2-chloro-4,6-dimethoxy-1,3,5-triazine (CDMT, **127**) are commercially available. In the laboratory, these triazine-based coupling reagents and also numerous analogs can be readily prepared from CC by substitution of one or two chlorine atoms with the appropriate nucleophile. Substitution usually proceeds stepwise. The chlorine atom can be substituted at 0 °C, the second one at room temperature, and the third one at 100 °C. DCMT (single chlorine substitution product) can be easily obtained by treating CC with methanol in the presence of $NaHCO_3$ [108–110], *N,N*-dimethylaniline [111], or TMP [112, 113]. However, DCMT is stable only when the reagent has been highly purified. The presence of impurities causes DCMT decomposition, which is accelerated further by the HCl released. Therefore, addition of HCl acceptors to DCMT, such as sodium phosphate, has been used to stabilize the compound [114].

However, it should be noted that this simple remedy carries a risk of side-reactions involving the interaction of phosphate anion with DCMT or CDMT. (For reaction of DCMT and CDMT with phosphates, see [115, 116]).

Synthesis of the bromo analog of CDMT, perdeuterated CDMT and transformation of DCMT into CDMT [117, 118]. Supplement the arsenal of synthetic methods available.

The related triazine, 4-(4,6-dimethoxy-1,3,5-triazin-2-yl)-4 methylmorpholinium chloride (DMTMM, **128**) has been prepared from CDMT (**127**) by a simple reaction with NMM [119] (Scheme 12.18).

The activation of carboxylic acids by CDMT (**127**) requires the presence of a tertiary amine in the reaction medium. This reaction can be considered erratic because only a few of the amines, such as NMM and NMP, have the capacity to react. In addition, the capacity of the amines to participate in the reaction does not correlate with the basicity of the amines in polar solvents. This observation suggests the existence of an intermediate involving the amine as a part of a multistep process [119].

The rate of formation of this intermediate, a triazinylammonium salt such as **128** (Scheme 12.19), will depend strongly on the steric hindrance of the amine. Thus, the concourse of hindered amines, such as triethylamine, causes a loss of reactivity of **127**. Only amines prone to the formation of salts, such as NMM when treated with **127**, are useful in the activation of carboxylic functions. TEA, which does not form a quaternary ammonium salt at low temperature in the reaction with **127**, does not have the capacity to activate benzoic acid [120]. Thus, the activation of carboxylic acids by **127** is comprised of two subsequent substitution reactions in the triazine ring. The first one involves substitution of the chlorine atom by the amine with the formation of a quaternary ammonium salt. This step is extremely sensitive to steric hindrance of amine substituents. The second step, which is highly tolerant of the steric hindrance of the carboxylic acid, involves substitution of the amine leaving group by the carboxylate ion to afford the triazine "superactive esters" **129**. In this regard, the DMTMM **128** prepared by reaction of **127** with NMM has been successfully applied to the synthesis of amides and esters [120, 121].

The monitoring of the quaternization of NMM at low temperature provides evidence of the formation of the zwitterionic addition product **128a**, the key

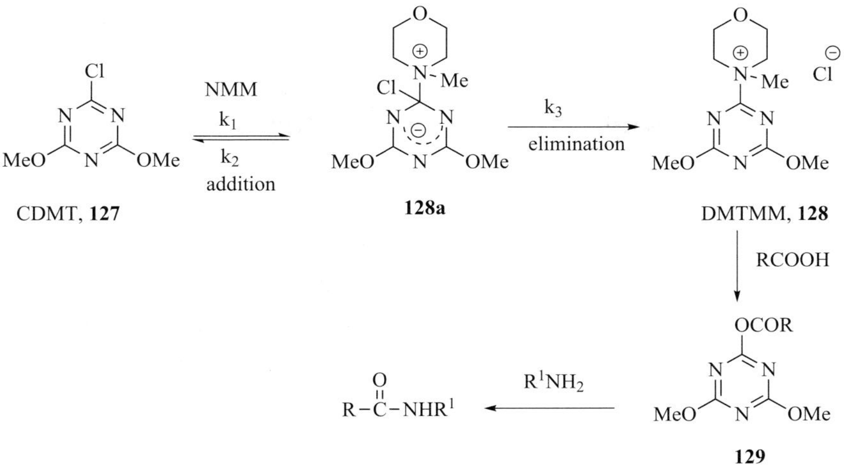

CDMT, **127** → (NMM, $k_1$ / $k_2$, addition) → **128a** → ($k_3$, elimination) → DMTMM, **128** → (RCOOH) → **129** → ($R^1NH_2$) → $R{-}C({=}O){-}NHR^1$

Scheme 12.19 Coupling using DMTMM.

intermediate in the classic two-step process AN + DN. Semiempirical modeling of the reaction, as well as measurements of nitrogen and chlorine kinetic isotope effects, also support this mechanism. Further data confirming this mechanism have been obtained during the enantioselective activation of carboxylic acids [122].

### 12.7.1 Formation of the Peptide Bond Using DMTMM (128) [122]

*N*-Protected amino acid (10 mmol) and NMM (10 mmol) are added to a vigorously stirred solution of **127** and NMM (10 mmol each) in $CH_3CN$ (20 ml), cooled to 0 °C. The stirring is continued for an additional 60 min, after which time the amino component (10 mmol) is added, and the mixture is stirred for an additional 2 h at 0 °C and overnight at room temperature. The solvent is evaporated under reduced pressure, and the residue is dissolved in $CHCl_3$ (30 ml). The solution is washed successively with water, 0.5 M aqueous $KHSO_4$, water, 0.5 M aqueous $NaHCO_3$, and water again. The organic layer is dried with $MgSO_4$, filtered, and concentrated to dryness.

## 12.8 Mukaiyama's Reagent

Mukaiyama's reagent, 2-chloro-1-methylpyridinium iodide **130**, in the presence of a carboxylic acid and a tertiary amine, gives an activated pyridinium ester **131** that reacts with a range of nucleophiles (Scheme 12.20) [123].

This reagent is not often used in peptide synthesis owing to the poor solubility of the pyridinium iodides in conventional solvents. The reaction has therefore to be

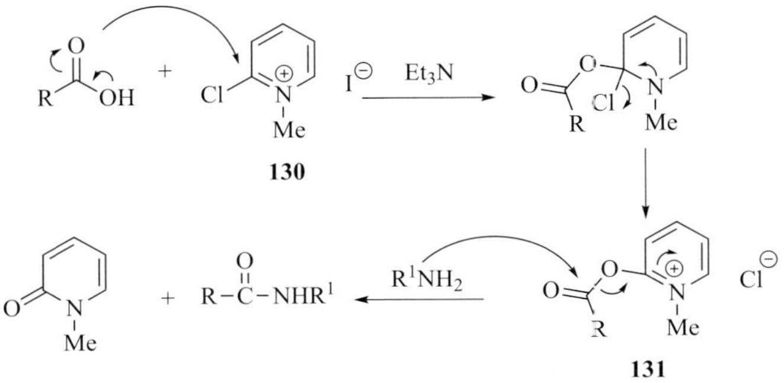

Scheme 12.20 One-pot coupling using Mukaiyama's reagents.

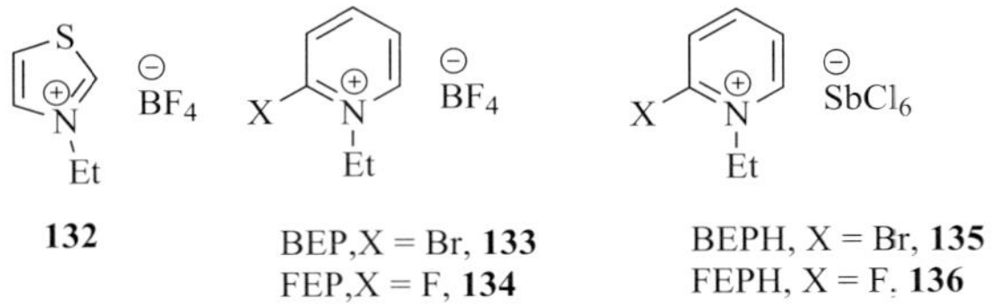

Figure 12.14 Alternative Mukaiyama-type reagents.

performed under reflux in DCM. Xu *et al.* [124] have recently published alternatives to Mukaiyama's reagent **131**. In order to improve the solubility of the pyridinium compounds, the tetrafluoroborate and hexachloroantimonate counter ions are adopted (Figure 12.14). 2-Bromo-3-ethyl-4-methylthiazolium tetrafluoroborate (BEMT, **132**) has been successfully applied to the synthesis of peptides containing *N*-alkyl or $C^{\alpha}$-dialkyl amino acids [124], and, later, these authors developed other 2-halopyridinium salts, such as 2-bromo-1-ethylpyridinium tetrafluoroborate (BEP, **133**), 2-fluoro-1-ethylpyridinium tetrafluoroborate (FEP, **134**), 2-bromo-1-ethylpyridinium hexachloroantimonate (BEPH, **135**), and 2-fluoro-1-ethylpyridinium hexachloroantimonate (FEPH, **13**) [125]. These α-halopyridinium-type coupling reagents have also been used in SPPS.

## 12.9 Conclusions

From the mid twentieth century, carbodiimides, azides, active esters, anhydrides, and stand-alone reagents, such as phosphonium and iminium salts, have been extensively used as coupling reagents both in solution- and SPPS. The application of the most effective coupling reagents has allowed for a rapid and marked expansion of peptide chemistry, which has consequently promoted advancements in other fields of organic chemistry which address amide bond formation. Thus, not only is the synthesis of small linear peptides containing natural amino acids now routinely possible, but also the laborious synthesis of large peptides containing 30–50 amino acid residues. Likewise, the synthesis of cyclic peptides containing non-natural and hindered amino acids and the synthesis of peptidomimetic building blocks has also been facilitated through the use of routine solid-phase techniques.

Despite these advancements several challenges still lie ahead in this field, such as the stepwise solid-phase synthesis of small proteins containing up to 100 coded residues and the synthesis of peptides containing extremely hindered building blocks, such as α,α-dialkyl amino acids, *N*-alkyl amino acids and the even more challenging *N*-aryl amino acids. These challenges will not be overcome through the use of new and more effective coupling methods alone. Rather, a proper combination of coupling reagent, α-amino protecting group, solid support, solvent, temperature, and other experimental conditions will be mandatory.

The existing coupling methods, together with a new generation of reagents, which are likely be developed in the very near future, in combination with improvements in other reagents and experimental conditions should allow for the facile and routine preparation of any peptide. Further improvements in peptide-coupling methodologies will make an enormous contribution to the introduction of peptides and peptidomimetics as drugs to treat a broad range of diseases.

## Acknowledgments

This work was partially supported by the Riyadh Techno Valley/King Saud University–Barcelona Science Park Alliance, AECYD (A/9846/07), CICYT (CTQ2006-03794/BQU), the "Generalitat de Catalunya" (2005SGR 00662), and the Institute for Research in Biomedicine.

## References

1 (a) Albericio, F., Chinchilla, R., Dodsworth, D.J., and Najera, C. (2001) *Organic Preparations and Procedures International*, **33**, 203; (b) Li, P. and Xu, J.C. (2001) *The Journal of Peptide Research*, **58**, 129; (c) Elmore, D.T. (2002) *Amino Acids, Peptides, Proteins*, **33**, 83; (d) Carpino, L.A., Beyermann, M., Wenschuh, H., and Bienert, M. (1996) *Accounts of Chemical Research*, **29**, 268; (e) Humphrey, J.M. and Chamberlin, A.R. (1997) *Chemical Reviews*, **97**, 2243; (f) Klausner, Y.S. and Bodanszky, M. (1972) *Synthesis*, 453; (g) Bailey, P.D. (1990) *An Introduction to Peptide Chemistry*, John Wiley & Sons, Ltd, Chichester; (h) Bodanszky, M. (1984) *Principles of Peptide Synthesis*, Springer, Berlin; (i) Hruby, V.J. and Schwyzer, R. (eds) (1998) *Peptide Chemistry, Design and Synthesis of Peptides, Conformational Analysis and Biological Functions*, Pergamon, Oxford; (j) Han S-.Y. and Kim, Y.-A. (2004) *Tetrahedron*, **60**, 2447; (k) Montalbetti, C.A.G.N. and Falque, V. (2005) *Tetrahedron*, **61**, 10827.

2 Bodanszky, M. (1993) *Principles of Peptide Synthesis*, 2nd edn, Springer, Berlin.

3 Lloyd-Williams, P., Albericio, F., and Giralt, E. (1997) *Chemical Approaches to the Synthesis of Peptide and Proteins*, CRC Press, Boca Raton, FL.

4 (a) Antonovics, I. and Young, G.T. (1967) *Journal of the Chemical Society (C)*, 595; (b) Carpino, L.A. (1988) *The Journal of Organic Chemistry*, **53**, 875.

5 McKay, F.C. and Albertson, N.F. (1957) *Journal of the American Chemical Society*, **79**, 4686.

6 Bergmann, M. and Zervas, L. (1932) *Berichte der Deutschen Chemischen Gesellschaft*, **65**, 1192.

7 Carpino, L.A. (1987) *Accounts of Chemical Research*, **20**, 401.

8 Stigers, K.D., Koutroulis, M.R., Chung, D.M., and Nowick, J.S. (2000) *The Journal of Organic Chemistry*, **65**, 3858.

9 Carpino, L.A. and Wu, A.-C. (2000) *The Journal of Organic Chemistry*, **65**, 9238.

10 Romoff, T.T. and Goodman, M. (1997) *The Journal of Peptide Research*, **49**, 281.

11 Carpino, L.A. and Philbin, M. (1999) *The Journal of Organic Chemistry*, **64**, 4315.

12 (a) Carpino, L.A., Philbin, M., Ismail, M., Truran, G.A., Mansour, E.M.E., Iguchi, S., Ionescu, D., El-Faham, A., Riemer, C., Warrass, R., and Weiss, M.S. (1999) *Journal of the American Chemical Society*, **119**, 9915; (b) Carpino, L.A., Ismail, M., Truran, G.A., Mansour, E.M.E., Iguchi, S., Ionescu, D., El-Faham, A., Riemer, C., and Warrass, R. (1999) *The Journal of Organic Chemistry*, **64**, 4324.

13 Carpino, L.A. and Mansour, E.M.E. (1999) *The Journal of Organic Chemistry*, **64**, 8399.

14 Carpino, L.A., Ionescu, D., and El-Faham, A. (1996) *The Journal of Organic Chemistry*, **61**, 2460.

15 (a) Ward, D.E., Lazny, R., and Pedras, M.S.C. (1997) *Tetrahedron Letters*, **38**, 339; (b) Bodanszky, M. and Martinez, J. (1981) *Synthesis*, 333.

16 Albericio, F., Bofill, J.M., El-Faham, A., and Kates, S.A. (1998) *The Journal of Organic Chemistry*, **63**, 9678.

17 Sheehan, J.C. and Hess, G.P. (1955) *Journal of the American Chemical Society*, **77**, 1067.

18 Williams, A. and Ibrahim, I.T. (1981) *Chemical Reviews*, **81**, 589.

19 Gibson, F.S., Park, M.S., and Rapoport, H. (1994) *The Journal of Organic Chemistry*, **59**, 7503.

20 König, W. and Geiger, R. (1970) *Chemische Berichte*, **103**, 788.

21 Carpino, L.A. (1993) *Journal of the American Chemical Society*, **115**, 4397.

22 Carpino, L.A. and El-Faham, A. (1999) *Tetrahedron*, **55**, 6813.

23 Carpino, L.A., El-Faham, A., Minor, C.A., and Albericio, F. (1994) *Journal of the Chemical Society, Chemical Communications*, 201.

24 Carpino, L.A., El-Faham, A., and Albericio, F. (1994) *Tetrahedron Letters*, **35**, 2279.

25 (a) Xu, Y. and Miller, M.J. (1998) *The Journal of Organic Chemistry*, **63**, 4314; (b) Carpino, L.A., Imazumi, H., Foxman, B.M., Vela, M.J., Henklein, P., El-Faham, A., Klose, J., and Bienert, M. (2000) *Organic Letters*, **2**, 2253.

26 Spetzler, J.C., Mendal, M., Felding, J., Vedsø, P., and Begtrup, M. (1998) *Journal of the Chemical Society, Perkin Transactions 1*, 1727.

27 (a) Jiang, L., Davison, A., Tennant, G., and Ramage, R. (1998) *Tetrahedron*, **54**, 14233; (b) Robertson, N., Jiang, L., and Ramage, R. (1999) *Tetrahedron*, **55**, 2713.

28 Sabatino, G., Mulinacci, B., Alcaro, M.C., Chelli, M., Rovero, P., and Papini, A.M. (2002) *Letters in Peptide Science*, **9**, 119.

29 Carpino, L.A., Xia, J., and El-Faham, A. (2004) *The Journal of Organic Chemistry*, **69**, 54.

30 Subirós-Funosas, R., Prohens, R., Barbas, R., El-Faham, A., and Albericio, F. (2009) *Chemistry – A European Journal*, **15**, 9394.

31 Wehrstedt, K.D., Wandrey, P.A., and Heitkamp, D. (2005) *Journal of Hazardous Materials*, **A126**, 1.

32 Barany, G. and Merrifield, R.B. (1979) In *The Peptides: Analysis, Synthesis, Biology*, vol **2** (eds E. Gross and J. Meinhofer), Academic Press, New York, pp. 1–284.

33 Jensen K.J., Alsina, J., Songster, M.F., Vagner, J.V., Albericio, F., and Barany, G. (1998) *Journal of the American Chemical Society*, **120**, 5441.

34 Kent, S.B.H. and Merrifield, R.B. (1981) In *Peptides 1980: Proceedings of the 16th European Peptide Symposium* (ed. K. Brunfeldt), Scriptor, Copenhagen, pp. 328–333.

35 Live, D.H. and Kent, S.B.H. (1983) In *Peptides – Structure Function* (eds V.J. Hruby and D.H. Rich), Pierce, Rockford, IL, pp. 65–68.
36 Kent, S.B.H. (1985) In *Peptides – Structure Function* (eds C.M. Deber and V.J. Hruby), Pierce, Rockford, IL, pp. 407–414.
37 Albert J.S. and Hamilton, A.D. (1995) In *Encyclopedia of Reagents for Organic Synthesis*, vol **3** (ed. L.A. Paquette), John Wiley & Sons, Ltd, Chichester, pp. 1751–1754.
38 Gawne G., Kenner, G., and Sheppard, R.C. (1969) *Journal of the American Chemical Society*, **91**, 5669.
39 (a) Castro, B. and Dormoy, J.R. (1973) *Bulletin de la Societe Chimique de France*, 3359; (b) Castro, B., Dormoy, J.R., Evin, G., and Selve, C. (1975) *Tetrahedron Letters*, **16**, 1219; (c) Castro, B., Dormoy, J.R., Evin, G., and Selve, C. (1977) *Journal of Chemical Research (S)*, 182.
40 Castro, B. and Dormoy, J.R. (1972) *Tetrahedron Letters*, 4747.
41 (a) Castro, B. and Dormoy, J.R. (1973) *Tetrahedron Letters*, 3243; (b) Coste, J., Dufour, M.-N., Pantaloni, A., and Castro, B. (1990) *Tetrahedron Letters*, **31**, 669.
42 (a) Castro, B., Dormoy, J.-R., Dourtoglou, B., Evin, G., Selve, C., and Ziebler, J.-C. (1976) *Synthesis*, 751; (b) Dormoy, J.-R. and Castro, B. (1979) *Tetrahedron Letters*, 3321; (c) Le-Nguyen, D., Heitz, A., and Castro, B. (1987) *Journal of the Chemical Society, Perkin Transactions 1*, 1915.
43 (a) Høeg-Jensen, T., Jakobsen, M.H., and Holm, A. (1991) *Tetrahedron Letters*, **32**, 6387; (b) Frérot, E., Coste, J., Pantaloni, A., Dufour, M.-N., and Jouin, P. (1991) *Tetrahedron*, **47**, 259.
44 Hudson, D. (1988) *The Journal of Organic Chemistry*, **53**, 617.
45 (a) Coste, J., Frérot, E., Jouin, P., and Castro, B. (1991) *Tetrahedron Letters*, **32**, 1967; (b) Coste, J., Frérot, F., and Jouin, P. (1994) *The Journal of Organic Chemistry*, **59**, 2437.
46 Wada, T., Sato, Y., Honda, F., Kawahara, S.-I., and Sekine, M. (1997) *Journal of the American Chemical Society*, **119**, 12710.
47 (a) Kim, M.H. and Patel, D.V. (1994) *Tetrahedron Letters*, **35**, 5603; (b) Coste, J. and Campagne, J.M. (1995) *Tetrahedron Letters*, **36**, 4253.
48 Campagne, J.M., Coste, J., and Joiun, P. (1995) *The Journal of Organic Chemistry*, **60**, 5214.
49 (a) Kates, S.A., Diekmann, E., El-Faham, A., Herman, L.W., Ionescu, D., McGuinness, B.F., Triolo, S.A., Albericio, F., and Carpino, L.A. (1996) In *Techniques in Protein Chemistry VII* (ed. D.R. Marshak), Academic Press, New York, p. 515; (b) Ehrlich A., Heyne, H.-U., Winter, R., Beyermann, M., Haber, H., Carpino, L.A., and Bienert, M. (1996) *The Journal of Organic Chemistry*, **61**, 8831; (c) Jou, G., Gonzalez, I., Albericio, F., Lloyd, P.W., and Giralt, E. (1997) *The Journal of Organic Chemistry*, **62**, 354; (d) Han, Y., Albericio, F., and Barany, G. (1997) *The Journal of Organic Chemistry*, **62**, 4307; (e) Albericio, F., Cases, M., Alsina, J., Triolo, S.A., Carpino, L.A., and Kates, S.A. (1997) *Tetrahedron Letters*, **38** 4851.
50 (a) Wijkmans, J.C.H.M., Blok, F.A.A., van der Marel, G.A., van Boom, J.H., and Bloemhoff, W. (1995) *Tetrahedron Letters*, **36**, 4643; (b) Høeg-Jensen, T., Olsen, C.E., and Holm, A. (1994) *The Journal of Organic Chemistry*, **59**, 1257.
51 Castro, B. and Dormoy, J.R. (1971) *Bulletin de la Societe Chimique de France*, 3034.
52 Barstow, L.E. and Hruby, V.J. (1971) *The Journal of Organic Chemistry*, **36**, 1305.
53 Yamada, S. and Takeuchi, Y. (1971) *Tetrahedron Letters*, **39**, 3595.
54 Coste, J., Le-Nguyen, D., Evin, G., and Castro, B. (1990) *Tetrahedron Letters*, **31**, 205.
55 Bates, A.J., Galpin, I.J., Hallet, A., Hudson, D., Kenner, G.W., Ramage, R., and Sheppard, R.C. (1975) *Helvetica Chimica Acta*, **58**, 688.
56 Wijkmans, J.C.H.M., Kruijtzer, J.A.W., van der Marel, G.A., van Boom, J.H., and Bloemhoff, W. (1994) *Recueil des Travaux Chimiques des Pays-Bas*, **113**, 394.
57 Reese, C.B. and Rei-Zhuo, Z. (1993) *Journal of the Chemical Society, Perkin Transactions 1*, 2291.
58 Ehrlich, A., Rothemund, S., Brudel, M., Beyermann, M., and Carpino, L.A. (1993) Bienert, M., *Tetrahedron Letters*, **34**, 4781.

**59** Frérot, F., Coste, J., Poncet, J., Jouin, P., and Castro, B. (1992) *Tetrahedron Letters*, **33**, 2815.

**60** Arendt, A. and Kolodziejczyk, A.M. (1978) *Tetrahedron Letters*, **40**, 3867.

**61** (a) Patino, N., Frérot, E., Galeotti, N., Poncet, J., Coste, J., Dufour, M.N., and Jouin, P. (1992) *Tetrahedron*, **48**, 4115; (b) Roux, F., Maugras, I., Poncet, J., Niel, G., and Jouin, P. (1994) *Tetrahedron*, **50**, 5345; (c) Poncet, J., Busquet, M., Roux, F., Pierré, A., Atassi, G., and Jouin, P. (1998) *Journal of Medicinal Chemistry*, **41**, 1524.

**62** Mattern, R.-H., Gunasekera, S., and McConnell, O. (1996) *Tetrahedron*, **52**, 425.

**63** Reissmann, S., Schwuchow, C., Seyfarth, L., Pineda De Castro, L.F., Liebmann, C., Paegelow, I., Werner, H., and Stewart, J.M. (1996) *Journal of Medicinal Chemistry*, **39**, 929.

**64** (a) Kurome, T., Inami, K., Inoue, T., Ikai, K., Takesako, K., Kato, I., and Shiba, T. (1993) *Chemistry Letters*, 1873; (b) Kurome, T., Inami, K., Inoue, T., Ikai, K., Takesako, T., Kato, I., and Shiba, T. (1996) *Tetrahedron*, **52**, 4327.

**65** Marsh, I.R., Bradley, M., and Teague, S.J. (1997) *The Journal of Organic Chemistry*, **62**, 6199.

**66** Frérot, E., Coste, J., Pantaloni, A., Dufour, M.-N., and Jouin, P. (1991) *Tetrahedron*, **47**, 259.

**67** Auvin-Guette, C., Frérot, E., Coste, J., Rebuffat, S., Jouin, P., and Bodo, B. (1993) *Tetrahedron Letters*, **34**, 2481.

**68** Walker, M.A. and Heathcock, C.H. (1992) *The Journal of Organic Chemistry*, **57**, 5566.

**69** Pop, I.E., Déprez, B.P., and Tartar, A.L. (1997) *The Journal of Organic Chemistry*, **62**, 2594.

**70** (a) Fournier, A., Wang, C.T., and Felix, A.M. (1988) *International Journal of Peptide and Protein Research*, **31**, 86; (b) Fournier, A., Danho, W., and Felix, A.M. (1989) *International Journal of Peptide and Protein Research*, **33**, 133.

**71** Gairí, M., Lloyd-Williams, P., Albericio, F., and Giralt, E. (1990) *Tetrahedron Letters*, **31**, 7363.

**72** (a) Abdelmoty, I., Albericio, F., Carpino, L.A., Foxman, B.M., and Kates, S.A. (1994) *Letters in Peptide Science*, **1**, 57; (b) Bofill, J.M. and Albericio, F. (1996) *Journal of Chemical Research (S)*, 302; (c) Carpino, L.A., Henklein, P., Foxman, B.M., Abdelmoty, I., Costisella, B., Wray, V., Domke, T., El-Faham, A., and Mugge, C. (2001) *The Journal of Organic Chemistry*, **66**, 5245.

**73** (a) del Fresno, M., El-Faham, A., Carpino, L.A., Royo, M., and Albericio, F. (2000) *Organic Letters*, **2**, 3539; (b) Li, P. and Xu, J.C. (2000) *Tetrahedron*, **56**, 4437.

**74** (a) Dourtoglou, V., Ziegler, J.-C., and Gross, B. (1978) *Tetrahedron Letters*, 1269; (b) Dourtoglou, V., Gross, B., Lambropoulou, V., and Zioudrou, C. (1984) *Synthesis*, 572.

**75** (a) Knorr, R., Trzeciak, A., Bannwarth, W., and Gillessen, D. (1989) *Tetrahedron Letters*, **30**, 1927; (b) Carpino, L.A., El-Faham, A., and Albericio, F. (1995) *The Journal of Organic Chemistry*, **50**, 3561.

**76** El-Faham, A. (1998) *Organic Preparations and Procedures International*, **30**, 477.

**77** Wessig, P. (1999) *Tetrahedron Letters*, **40**, 5987.

**78** Habermann, J. and Kunz, H. (1998) *Journal Fur Praktische Chemie*, **340**, 233.

**79** Klose, J., Henklein, P., El-Faham, A., Carpino, L.A., and Bienert, M. (1999) In *Peptides 1998: Proceedings of the 25th European Peptide Symposium* (eds S. Bajusz and F. Hudecz), Akadémiai Kiadó, Budapest, p. 204.

**80** Garner P., Anderson, J.T., Dey, S., Youngs, W.J., and Gabt, K. (1998) *The Journal of Organic Chemistry*, **63**, 5732.

**81** (a) Bailén, M.A., Chinchilla, R., Dodsworth, D.J., Nájera, C., Soriano, J.M., and Yus, M. (1999) In *Peptides 1998: Proceedings of the 25th European Peptide Symposium* (eds S. Bajusz and F. Hudecz), Akadémiai Kiadó, Budapest, p. 172; (b) Bailén M.A., Chinchilla, R., Dodsworth, D.J., and Nájera, C. (1999) *The Journal of Organic Chemistry*, **64**, 8936.

**82** Carpino, L.A., El-Faham, A., and Albericio, F. (1995) *The Journal of Organic Chemistry*, **50**, 3561.

**83** El-Faham, A., Khattab, S.N., Abdul-Ghani, M., and Albericio, F. (2006)

European Journal of Organic Chemistry, 1563.

**84** (a) El-Faham, A. and Albericio, F. (2007) *Organic Letters*, **9**, 4475; (b) El-Faham, A. and Albericio, F. (2008) *The Journal of Organic Chemistry*, **73**, 2731; (c) El-Faham, A., Subirós-Funosas, R., Prohens, R., and Albericio, F. (2009) *European Journal of Organic Chemistry*, **15**, 9416.

**85** (a) Carpino, L.A. and El-Faham, A. (1995) *Journal of the American Chemical Society*, **117**, 5401; (b) Boas, U., Pedersen, B., and Christensen, J.B. (1998) *Synthetic Communications*, **28**, 1223; (c) Vojkovsky, T. and Drake, B. (1997) *Organic Preparations and Procedures International*, **29**, 497; (d) El-Faham, A. and Khattab, S.N. (2009) *Synlett*, 886.

**86** (a) Yarovenko, N.N. and Radsha, M.A. (1959) *Journal of General Chemistry USSR (English Translation)* **29**, 2125; (b) Olah, G.A., Nojima, M., and Kerekes, I., *Journal of the American Chemical Society* (1974) **69**, 925.

**87** (a) El-Faham, A. (1998) *Chemistry Letters*, 671; (b) Akaji, K., Kuriyama, N., and Kiso, Y. (1994) *Tetrahedron Letters*, **35**, 3315.

**88** Carpino, L.A., Imazumi, H., El-Faham, A., Ferrer, F.J., Zhang, C., Lee, Y., Foxman, B.M., Henklein, P., Hanay, C., Mugge, C., Wenschuh, H., Klose, J., Beyermann, M., and Bienert, M. (2002) *Angewandte Chemie (International Edition in English)*, **41**, 441.

**89** Fiammengo, R., Licini, G., Nicotra, A., Modena, G., Pasquato, L., Scrimin, P., Broxterman, Q.B., and Kaptein, B. (2001) *The Journal of Organic Chemistry*, **66**, 5905.

**90** Carpino, L.A., Ionescu, D., El-Faham, A., Beyerman, M., Henklein, P., Hanay, C., Wenschuh, H., and Bienert, M. (2003) *Organic Letters*, **5**, 975.

**91** El-Faham, A., Kattab, S.N., and Abdul-Ghani, M. (2006) *Arkivoc*, **xiii**, 57.

**92** Redemann, T. and Jung, G. (1998) In *Peptide 1996: Proceedings of the 24th European Peptide Symposium* (eds R. Ramage and R. Epton), Mayflower Scientific, Kingswinford, p. 749.

**93** (a) Carpino L.A., Sadat-Aalaee, D., Chao, H.G., and DeSelms, R.H. (1990) *Journal of the American Chemical Society*, **112**, 9651; (b) Carpino, L.A., Mansour, E.M.E., and El-Faham, A. (1993) *The Journal of Organic Chemistry*, **58**, 4162.

**94** (a) Shioiri, T., Ninomiya, K., and Yamada, S. (1972) *Journal of the American Chemical Society*, **94**, 6203; (b) Shioiri, T. and Yamada, S.-I. (1974) *Chemical & Pharmaceutical Bulletin*, **22**, 849.

**95** (a) Takeuchi, Y. and Yamada, S.-I. (1974) *Chemical & Pharmaceutical Bulletin*, **22**, 832; (b) Jackson, A.G., Kenner, G.W., Moore, G.A., Ramage, R., and Thorpe, W.D. (1976) *Tetrahedron Letters*, **17**, 3627.

**96** (a) Katoh, T. and Ueki, M. (1993) *International Journal of Peptide and Protein Research*, **42**, 264; (b) Ueki, M., Inazu, T., and Ikeda, S. (1979) *Bulletin of the Chemical Society of Japan*, **52**, 2424; (c) Ueki, M. and Inazu, T. (1982) *Chemistry Letters*, 45.

**97** Kiso, Y., Miyazaki, T., Satomi, M., Hiraiwa, H., and Akita, T. (1980) *Journal of the Chemical Society, Chemical Communications*, 1029.

**98** (a) Ramage, R., Ashton, C.P., Hopton, D., and Parrott, M.J. (1984) *Tetrahedron Letters*, **25**, 4825; (b) Poulos, C., Ashton, C.P., Green, J., Ogunjobi, O.M., Ramage, R., and Tsegenidis, T. (1992) *International Journal of Peptide and Protein Research*, **40**, 315.

**99** (a) Panse, G.T. and Kamat, S.K. (1989) *Indian Journal of Chemistry*, 793; (b) Miyake, M., Kirisawa, M., and Tokutake, N. (1985) *Chemistry Letters*, 123.

**100** Kim, S., Chang, H., and Ko, Y.-K. (1985) *Tetrahedron Letters*, **26**, 1341.

**101** Watanabe, Y. and Mukaiyama, T. (1981) *Chemistry Letters*, 285.

**102** Mukaiyama, T., Kamekawa, K., and Watanabe, Y. (1981) *Chemistry Letters*, 1367.

**103** Kunieda, T., Abe, Y., Higuchi, T., and Hirobe, M. (1981) *Tetrahedron Letters*, **22**, 1257.

**104** Ueda, M. and Oikawa, H. (1985) *The Journal of Organic Chemistry*, **50**, 760.

**105** Fan, C.-X., Hao, X.-L., and Ye, Y.-H. (1996) *Synthetic Communications*, **26**, 1455.

**106** (a) Li, H., Jiang, X., Ye, Y.-H., Fan, C., Romoff, T., and Goodman, M. (1999)

*Organic Letters*, **1**, 91; (b) Xie, H.-B., Tian, G.-L., and Ye, Y.-H. (2000) *Synthetic Communications*, **30**, 4233; (c) Tang, Y.-C., Xie, H.-B., Tian, G.-L., and Ye, Y.-H. (2002) *The Journal of Peptide Research*, **60**, 95.

**107** Kokare, N.D., Nagawade, R.R., Rane, V.P., and Shinde, D.B. (2007) *Synthesis*, 766.

**108** (a) Gardiner, J.M. and Procter, J. (2001) *Tetrahedron Letters*, **42**, 5109; (b) Dudley, J.R., Thurston, J.T., Schaefer, F.C., Holm-Hansen, D., Hull, C.J., and Adams, P. (1951) *Journal of the American Chemical Society*, **73**, 2986.

**109** Ortega, F. and Bastide, J. (1997) *Journal of Bioorganic Chemistry*, **25**, 261.

**110** Schuldt, W. (1956) *Contributions from Boyce Thompson Institute*, **18**, 377.

**111** Beech, W.F. (1967) *Journal of the Chemical Society (C)*, 466.

**112** Audebert, R. and Neel, J. (1970) *Bulletin de la Societe Chimique de France*, 606.

**113** Koopman, H., Uhlenbroek, J.H., Haeck, H.H., Daams, J., and Koopmans, M.J. (1959) *Recueil des Travaux Chimiques des Pays-Bas*, **78**, 967.

**114** Siegel, E. (1978) *The Chemistry of Synthetic Dyes*, vol. **VI**, Academic Press, New York, p. 86.

**115** Zhizhin, M.T., Fomakhin, E.W., Goldfarb, E.I., and Pudovik, A.N. (1986) *Zhurnal General Khimii*, **56**, 553.

**116** Zimin, M.G., Fomakhin, E.V., Zheleznova, L.V., Islamov, R.G., and Pudovik, A.N. (1987) *Khimiya i Tekhnol Elementoorgan Soed i Polimerov, Kazan*, 4.

**117** Sochacki, M. and Kaminski, Z. (1994) *Journal of Organic Mass Spectrometry*, **29**, 102.

**118** Davies, A.G. and Sutcliffe, R. (1981) *Journal of the Chemical Society, Perkin Transactions 2*, 1512.

**119** Kaminski, Z.J., Paneth, P., and Rudzinski, J. (1998) *The Journal of Organic Chemistry*, **63**, 4248.

**120** Kunishima, M., Kawachi, C., Iwasaki, F., Terao, K., and Tani, S. (1999) *Tetrahedron Letters*, **40**, 5327.

**121** Kunishima, M., Morita, J., Kawachi, C., Iwasaki, F., Terao, K., and Tani, S. (1999) *Synlett*, **8**, 1255.

**122** (a) Kaminski, Z.J. and Kolesinska, B. (1999) Presented at XV Polish Peptide Symposium, Olstyn; (b) Kaminski, Z.J., Kolesinska, B., Kolesinska, J., Sabatino, G., Chelli, M., Rovero, P., Bøaszczyk, M., Gøowka, M.L., and Papini, A.M. (2005) *Journal of the American Chemical Society*, **127**, 16912.

**123** (a) Bald, E., Saigo, K., and Mukaiyama, T. (1975) *Chemistry Letters*, 1163; (b) Mukaiyama, T. (1979) *Angewandte Chemie (International Edition in English)*, **18**, 707; (c) Huang, H., Iwasawa, N., and Mukaiyama, T. (1984) *Chemistry Letters*, 1465.

**124** Li, P. and Xu, J.C. (1999) *Tetrahedron Letters*, **40**, 8301.

**125** Li, P. and Xu, J.-C. (2000) *Tetrahedron*, **56**, 8119.

# 13
# Chemoselective Peptide Ligation: A Privileged Tool for Protein Synthesis

*Christian P.R. Hackenberger, Jeffrey W. Bode, and Dirk Schwarzer*

## 13.1
## Introduction

The recognition that proteins are the machinery and mediators of living processes engendered the need to isolate, characterize, manipulate, and modify them for understanding and advancing biology and medicine. An era in which the only method to obtain proteins was extraction from living sources gave way to the age of genetic engineering, whereby proteins could be expressed in foreign host organisms, making possible the production of significant quantities of protein targets for biochemical studies, structural characterization, and even their use as pharmaceuticals. This recombinant approach to protein synthesis is limited by the restriction that the proteins thus prepared may contain only the 20 canonical amino acids. The introduction of unnatural amino acids or most post-translational modifications such as glycosylations or phosphorylations can be accomplished only with great difficulty or time-consuming techniques [1]. It is with these goals in mind that scientists have recently returned to the long-held dream of producing proteins by purely synthetic methods. Remarkable advances in chemical methods over the last 15 years have made chemical protein synthesis increasingly common, and emerging technologies promise even more versatile and practical methods in the coming years.

Within this chapter, we seek to provide the reader with an overview of the state of the art of peptide and protein synthesis using the concept of "chemoselective peptide ligation" – the selective coupling of two or more peptide fragments by an amide-bond-forming reaction without the need for chemical coupling reagents or the protection of side-chain functionalities (Scheme 13.1). In general, this is accomplished by the placement of uniquely reactive functional groups at the C-terminal end of one peptide fragment and the N-terminal end of the other; these functional groups or auxiliaries react exclusively with one another give a native (i.e., backbone) amide bond [2]. This chapter includes a brief history and explanation of its various forms and their development, an overview of the prevailing methods as they are currently practiced, and a survey of emerging chemical reactions for chemoselective ligation that will impact the field in the years to come. Additionally, we will limit ourselves

*Amino Acids, Peptides and Proteins in Organic Chemistry.*
*Vol.3 – Building Blocks, Catalysis and Coupling Chemistry.* Edited by Andrew B. Hughes

ISBN: 978-3-527-32102-5

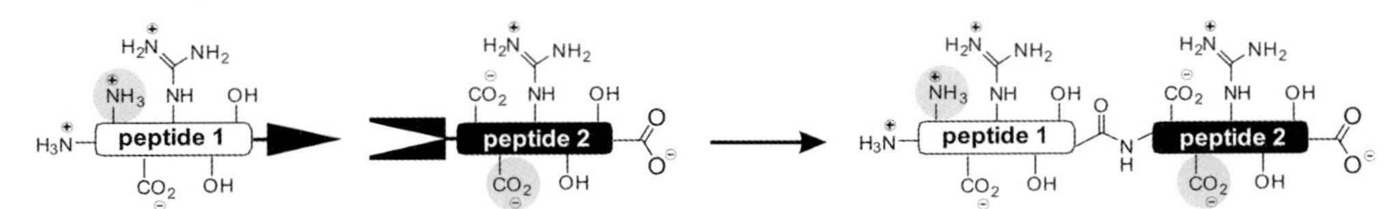


**Scheme 13.1** Concept of chemoselective amide ligation for peptide and protein synthesis. The cartoon elements represent unique functional groups that undergo a chemoselective reaction that produces a native (backbone) amide bond in the ligated product.

exclusively to chemical methods that allow for the formation of polypeptide conjugates in which the new linkage is an amide bond.

There are numerous aspects of chemical protein synthesis that cannot be covered by this chapter and have been the subject of other reviews [3]. These include, for instance, the synthesis of starting materials, namely thioesters and other activated *N*-peptides, enzymatic ligations and other techniques from molecular biology, and ligations which yield covalent linkages other than amide bonds and protein modification strategies. The interested reader is kindly referred to other sources for these issues [4].

It is important to distinguish chemoselective peptide ligations from the related and powerful fragment condensation strategies (Scheme 13.2) [5], such as the valuable Ag(I)-mediated coupling of C-terminal *N*-peptide thioesters with N-terminally unprotected *C*-peptides [6]. These methods have been essential to the history and development of chemical protein synthesis, including a starring role in the first chemical synthesis of intact proteins such as ribonuclease S [7] and the Green Fluorescent Protein (GFP) [8]. Interest in these methods for the coupling of partially protected glycopeptide fragments attest to the value of these approaches to chemical protein synthesis [9]. Despite the established utility of such methods, this general strategy suffers from several limitations that prevent its adoption as a unified approach to the preparation of pure, homogeneous peptides and proteins. These include a high probability of epimerization during the activation at the C-terminus of the *N*-peptide fragment, aggregation and poor solubility of the long, side-chain protected fragments, the usual need for an excess of one of the precious peptide fragments, and the

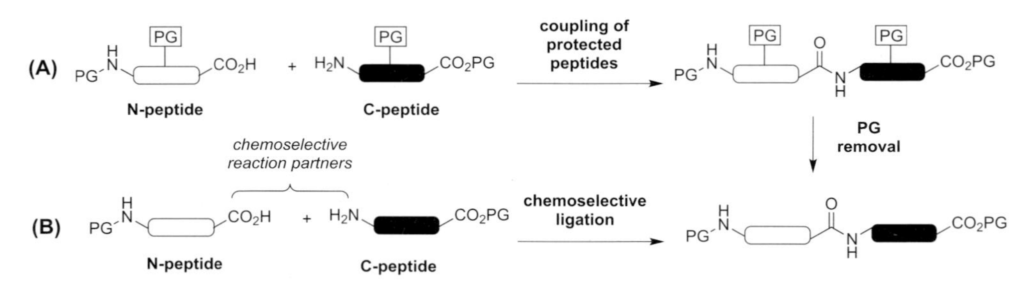


**Scheme 13.2** Fragment-coupling strategies: (A) coupling of protected peptides and (B) chemoselective ligation of unprotected peptides.

requirement for a global deprotection step that compromises the purity and yield of the final peptide [10].

The chemoselective ligation reactions covered in this chapter are built on a long tradition of synthetic peptide chemistry that is outside the scope of this chapter. Much of the chemistry discussed here relies on the powerful solid-phase methods for peptide synthesis pioneered by Merrifield [11]. Fifty years of advances in coupling reagents, linkers, resins, and protecting group strategies have made the synthesis of peptide fragments in the range of 30–40 amino acids routine and commonplace. The necessary protected amino acids and other materials are widely available, and advances in chromatographic purification render the synthesis of research quantities of linear peptides a reliable process. In fact, one of the major challenges in the development of new chemoselective ligation techniques is ensuring that the synthesis of the requisite peptide fragments containing orthogonally reactive functional groups or auxiliaries is compatible with the entrenched chemical methods for peptide synthesis.

The motivation for the discovery and development of new chemoselective ligation reactions for peptide and protein synthesis stems therefore from needs that are not met by the traditional methods of either linear peptide synthesis or protein production with recombinant expression techniques. These include (i) the large-scale manufacture of therapeutic peptides in which hundreds of kilograms, rather than the usual milligram quantities, of large peptides must be produced in high purity, (ii) the chemical syntheses of peptides containing unnatural amino acids including post-translational modifications, glycosylations, isotopically labeled segments, or D-amino acids, and (iii) protein semisynthesis, in which an expressed protein fragment is ligated to a synthetically produced peptide.

Each of these applications demands innovative chemical methods that allow selective backbone amide bond formation in the presence of the diverse, unprotected functional groups common to proteogenic side-chains. All of these applications have in common the synthesis of longer peptides in homogeneous form – a goal that is most effectively met by chemoselective ligation strategies that avoid the accumulation of difficult to remove addition or deletion sequences or impurities arising from late-stage global deprotection. Even for low-yielding ligation reactions it is usually trivial to separate the desired ligation product from unreacted starting materials (Scheme 13.3).

Intertwined in the clear need for chemical protein synthesis is the long history of peptide synthesis as a driving force for the development and discovery of new methods and concepts for organic synthesis in general. From the earliest syntheses of dipeptides [12] by Fischer to the modern implementation of native chemical ligation (NCL) for the synthesis of proteins by Kent and others, the overwhelming desire for chemical methods for protein and peptide synthesis has foreshadowed the development and direction of modern synthetic methods. Peptide chemists were among the first to demand and implement protecting groups for reactive functionalities – a concept of enormous importance and impact on the field of organic chemistry. Likewise, the recognition that truly chemoselective reactions for amide bond formation were the only way that protein synthesis would become routine presaged

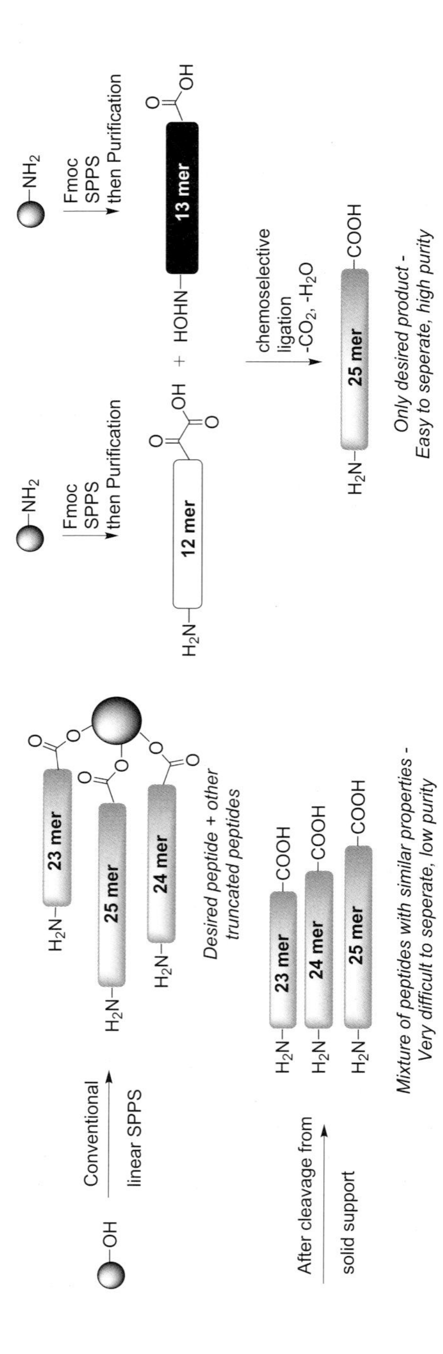


**Scheme 13.3** Cartoon illustrating the advantages of chemoselective ligation of unprotected peptide fragments for the synthesis of difficult peptide sequences. The purity of the final product is greatly enhanced by the use of chemoselective ligation.

the modern emphasis on bioorthogonal reactions for the synthesis, detection, and modification of biological materials. These concepts have recently ignited a new demand by the synthetic organic chemists for processes and strategic designs that dispense with the waste and added steps required by protecting groups and oxidation state adjustments [2]. It is within this spirit that we catalog, in Section 13.5, some of the most promising emerging methods for new chemoselective amide bond formation.

Before proceeding with our survey, it is useful to establish the graphical nomenclature employed in this chapter as defined in Figure 13.1.

## 13.2 Chemoselective Peptide Ligations Following a Capture/Rearrangement Strategy

### 13.2.1 Basic Concepts and Early Experiments

Most ligation techniques applied today are based on a "capture/rearrangement" strategy. The basic concept relies on two reactions that in combination mediate chemoselective peptide bond formation. The first reaction (capture) forms a selective but reversible linkage of two ligation partners. The second reaction (rearrangement) converts this flexible linkage into a stabile peptide bond in an irreversible manner. Although the terms reversible and irreversible are not a precise description of the reaction processes, which are primarily driven by the reaction kinetics, they are commonly used in this regard to illustrate the concept.

Among the earliest capture/rearrangement ligation schemes is a strategy called "prior thiol capture," which was developed by Kemp *et al.* [13, 14]. The procedure begins with attaching the peptide that will end up in the N-terminal region of the ligation product (i.e., the *N*-peptide, see Figure 13.1) to an auxiliary template (6-hydroxy-4-mercaptodibenzofuran) through an ester bond in molecule **1** (Scheme 13.4A) [15]. Alternatively, optimized solid-phase peptide synthesis (SPPS) protocols allow the direct assembly of the *N*-peptide on an immobilized template **1**. The capture reaction is a disulfide exchange between a Cys residue located at the N-terminus of the *C*-peptide **2** and the free thiol of the template **1** (Scheme 13.4A). The following rearrangement reaction is an intramolecular $O \rightarrow N$ transfer (**3**) to the peptide bond between both ligation partners in **4**, which is supported by the activated nature of the ester in **3**. After ligation product **5** is formed it can be released by treatment with reducing agents [16]. Although prior thiol capture has been optimized [16] and applied to polypeptides with up to 39 residues [17], nowadays the strategy of NCL is more routinely used for peptide ligations.

Important early experiments towards the later development of NCL were published by Wieland *et al.* in 1953. They investigated the chemical properties of amino acid thioesters [18] and reported that thiophenol-thioesters undergo an intermolecular aminolysis in the presence of amines. Most importantly, a glycine thioester of cysteamine **6** could not be synthesized and isolated. Instead the thioester rearranged rapidly through an intramolecular $S \rightarrow N$ shift to the corresponding amide **7**, which

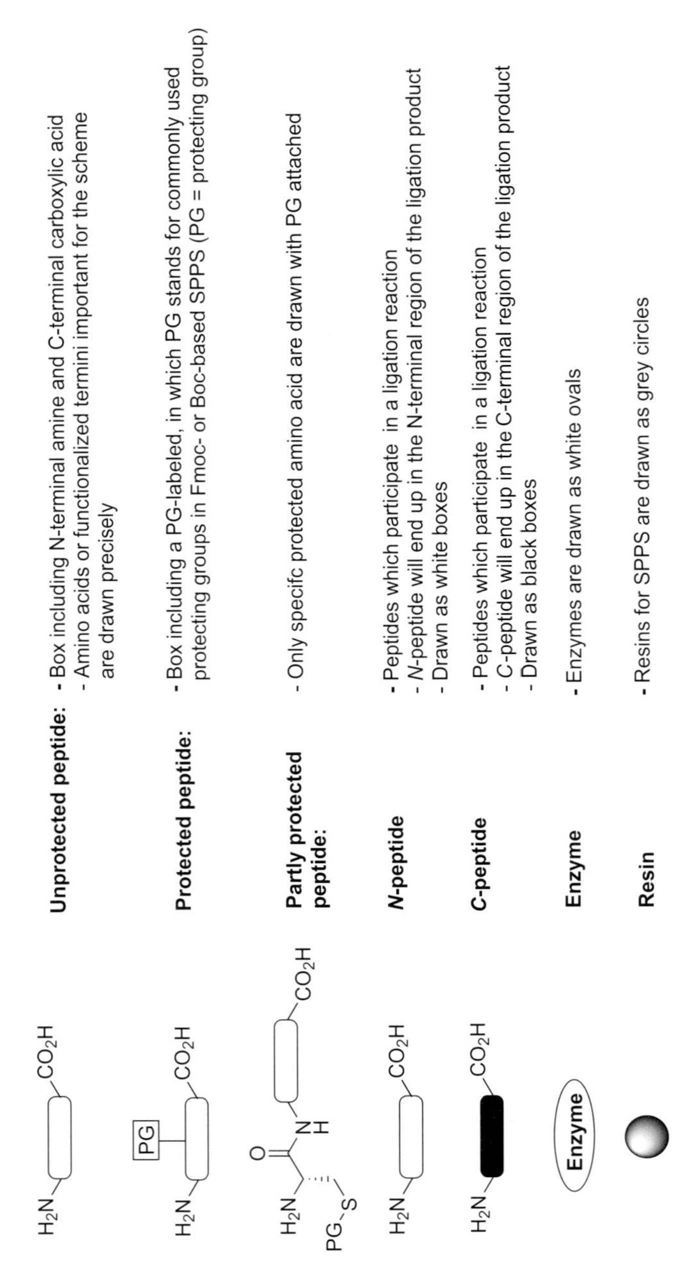


**Figure 13.1** Graphic toolbox for the different peptide fragments used in this chapter.

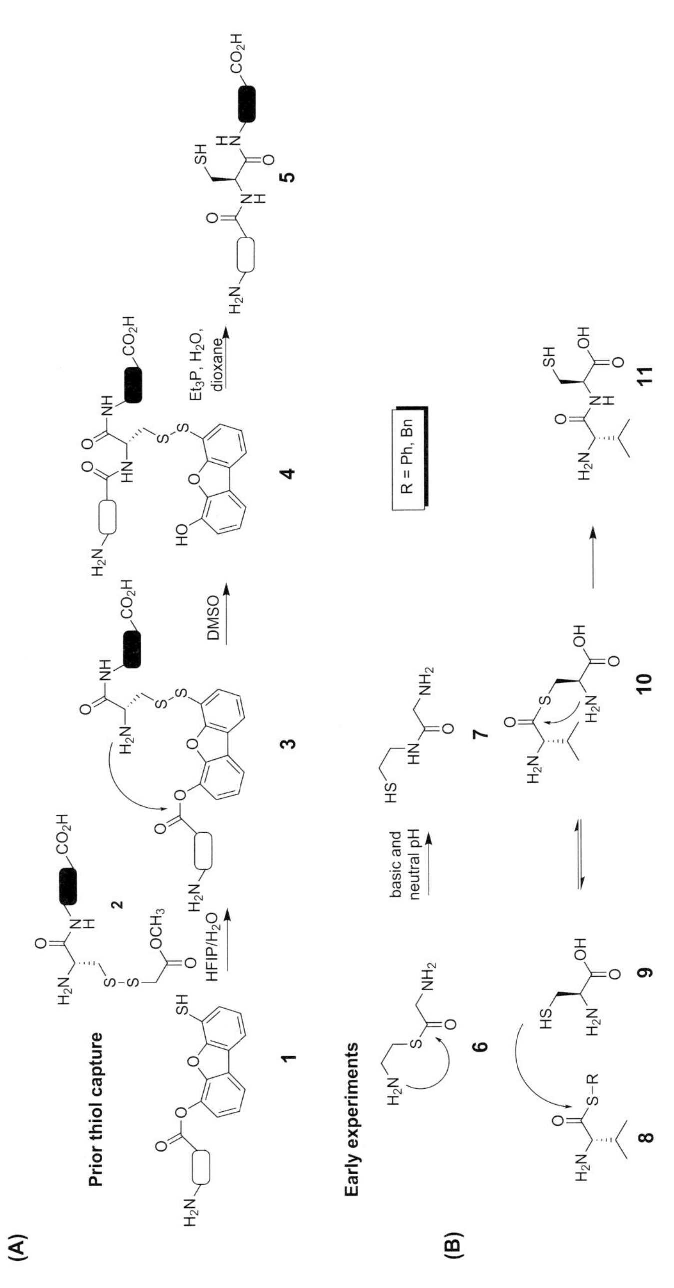


**Scheme 13.4** Aminolysis of thioesters by Wieland (A) and Kemp's prior thiol capture strategy (B). HFIP: Hexafluoroisopropanol.

was attributed to the proximity of the additional amino group of cysteamine to the thioester. Since this rearrangement occurred under mild conditions the authors concluded that this reaction could be used for the formation of peptide bonds. In today's nomenclature of a capture/rearrangement processess the latter reaction represents the rearrangement step. The preceding capture step is an intermolecular thiol–thioester exchange. Based on this observation, a Val-thiophenol thioester **8** was synthesized and treated with Cys (**9**) (Scheme 13.4B). Due to the reactive nature of the aryl thioester a rapid exchange with the thiol moiety of Cys occurred. The formed Val-*S*-Cys thioester **10** rearranged instantaneously to the Val–Cys dipeptide (**11**) linked with a native peptide bond [18].

### 13.2.2
### NCL

Today, NCL is the most widely used ligation technique. The key feature that makes NCL so successful is the chemoselectivity of this reaction, which was first fully exploited by Kent *et al.* in 1994 [19]. In their contribution, which also introduced the name "native chemical ligation" for the first time, the authors ligated two unprotected peptides to synthesize the human interleukin (IL)-8 (Scheme 13.5A). This primary publication has been cited nearly 1000 times, illustrating the importance of this finding for the chemical community. The *capture* step in NCL is mediated by a reversible thiol–thioester exchange between an electrophilic (aryl) thioester **12** at the C-terminus of the *N*-peptide and the nucleophilic thiol of a Cys residue located at the N-terminus of the *C*-peptide **13** (Scheme 13.5A). In the *rearrangement* reaction, the Cys-thioester **14** undergoes a rapid intramolecular $S \rightarrow N$ transfer which proceeds through a favorable five-membered transition state and finally forms a native peptide bond between the *C*- and the *N*-peptide in **5** [20]. Full chemoselectivity is only achieved when all unprotected amino acids, including other Cys residues, are tolerated and do not interfere with the overall reaction pathway. This is indeed the case for NCL because the irreversible intramolecular $S \rightarrow N$ shift can only occur at the unique N-terminal Cys residue. Therefore, Cys residues within the *N*- or *C*-peptide can form thioesters **15**, but due to the lack of a nearby amino group **15** cannot undergo a *rearrangement* reaction and instead rapidly exchange backwards to **12** or **14** (Scheme 13.5B).

Several further examples of NCL-based peptide ligations for various purposes and applications have followed the initial synthesis of human IL-8. These have been summarized in various excellent overviews and will therefore not be further discussed here.

NCL reactions are generally performed in buffered aqueous solutions at neutral pH. These physiological conditions are important in many aspects. Strongly basic conditions would make other residues such as Lys amenable to reactions with thioesters and, furthermore, thioesters can hydrolyze under basic conditions. Acidic conditions would reduce the reactivity of the Cys thiol and the N-terminal amine of the *C*-peptide is reduced [9g,20].

Another key feature determining the efficiency of the NCL reaction is the chemical composition of the C-terminal thioester. Dawson *et al.* have conducted

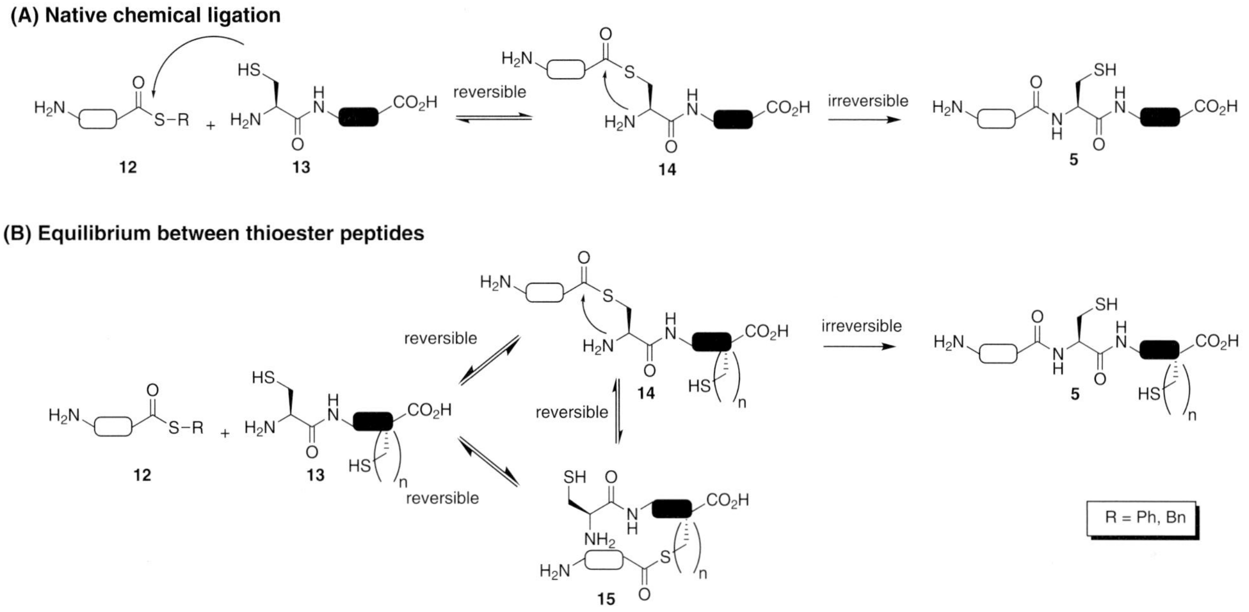

(A) Native chemical ligation

(B) Equilibrium between thioester peptides

R = Ph, Bn

**Scheme 13.5** Native chemical ligation. (A) Mechanistic pathway. (B) Equilibrium between thioester peptides. (C) Influence of residues at the C-terminus of the *N*-peptide. (D) Thioester exchange with aryl thiols.

**(C) Influence of the C-terminal thioester on the NCL reaction**

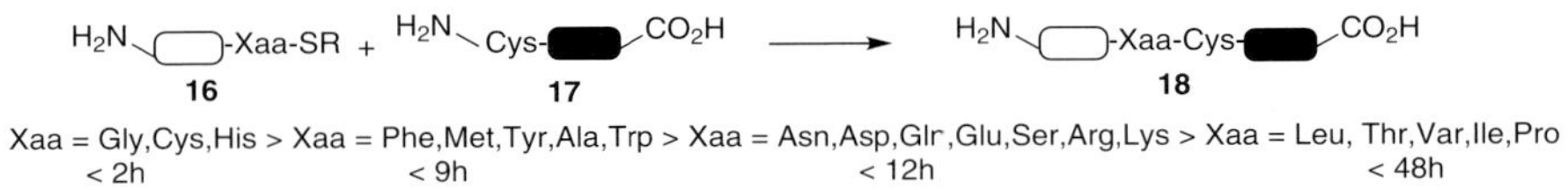

$H_2N$-Xaa-SR + $H_2N$-Cys-$CO_2H$ → $H_2N$-Xaa-Cys-$CO_2H$

16 17 18

Xaa = Gly,Cys,His > Xaa = Phe,Met,Tyr,Ala,Trp > Xaa = Asn,Asp,Glr,Glu,Ser,Arg,Lys > Xaa = Leu, Thr,Var,Ile,Pro

< 2h < 9h < 12h < 48h

**(D) Thioester exchange**

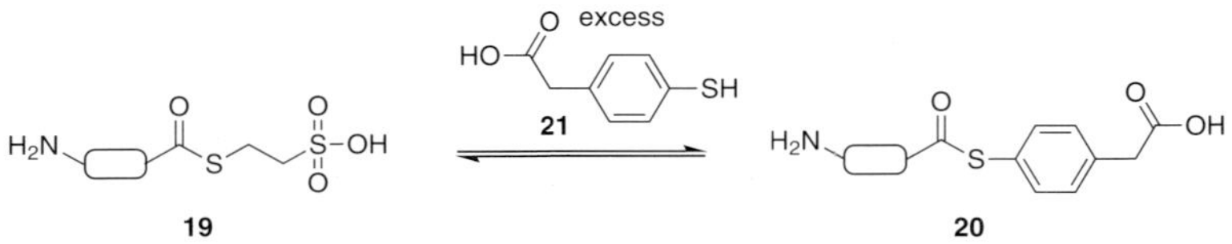

excess

HO SH

21

$H_2N$ S S-OH $H_2N$ S OH

19 20

**Scheme 13.5** (*Continued*)

a detailed analysis by synthesizing a set of model peptides containing each of the 20 proteinogenic amino acids as the C-terminal thioester **16** [21]. The reaction with a *C*-peptide **17** yielded the product **18** and the progress of the reactions were monitored over time by high-performance liquid chromatography (HPLC) (Scheme 13.5C). The fastest reaction rate was observed for the Gly-thioester, which reacted quantitatively in less than 4 h. In contrast, thioesters containing β-branched amino acids or Pro were not quantitatively converted into the ligation product **18** after even 2 days [21].

Finally, the nature of the thioester is important for the ligation efficiency [22]. In general, alkyl thioesters are less reactive and easier to synthesize than their aryl counterparts. Therefore, a common *modus operandi* is the synthesis of alkyl derivatives **19**, for example, with sodium-2-mercaptoethanesulfonate (MESNA), followed by an *in situ* conversion into the more reactive aryl thioesters **20** by the addition of an access of aryl thiols, like the water-soluble (4-carboxymethyl)thiophenol (MPAA, **21**, Scheme 13.5D) [23]. In addition to the mechanistic aspects, the addition of thiols to the ligation mixtures also prohibits the oxidation of the Cys thiol, which is another important aspect of an efficient ligation reaction. (For a recent application in kinetically controlled segment ligations, see Chapter 13.6. Several NCL-based ligation protocols also include tris(2-carboxyethyl) phosphine hydrochloride (TCEP) as a thiol free reducing agent.).

### 13.2.3
### Protein Semisynthesis with NCL

The term “protein semisynthesis” describes ligation reactions between both synthetic and recombinantly-produced starting materials. Chemoselective ligation techniques are ideally suited for such approaches because polypeptides from biological sources are unprotected. Protein semisynthesis combines the advantages of organic chemistry with biochemical techniques and allows incorporation of various unnatural functionalities into biopolymers without the size limitations of SPPS. The groundwork for modern protein semisynthesis was laid in the early 1970s by Offord *et al.* [24, 25]. However, these early semisynthesis schemes relied on the installation of protection groups into the isolated peptides and were therefore not chemoselective. Today, chemoselective ligation techniques that do not require complex protecting group strategies are used preferentially to link recombinant and synthetic peptides.

NCL has proven to be especially useful for protein semisynthesis because the capture/rearrangement process is mediated by Cys, which is genetically encoded (Scheme 13.6) [3, 19, 26]. Still, some manipulations of the starting material are required as all recombinant peptides and proteins are initially produced with an N-terminal Met. The essential N-terminal Cys in the *C*-peptide can be installed by genetic methods like site-directed mutagenesis and controlled cleavage with proteases. With this approach, a Cys residue can be introduced into the recognition site for a specific protease **23**, such as factor Xa [26, 27] or the TEV protease [28, 29] in a fusion protein **22** on the genetic level [65–67]. When the expressed protein is treated with the corresponding protease the N-terminal Cys is released in the recombinant peptide **12** which can be ligated with a synthetic peptide thioester **13** (Scheme 13.7). Alternative approaches use intein fusion proteins [30] or cyanogen bromide (CNBr)-

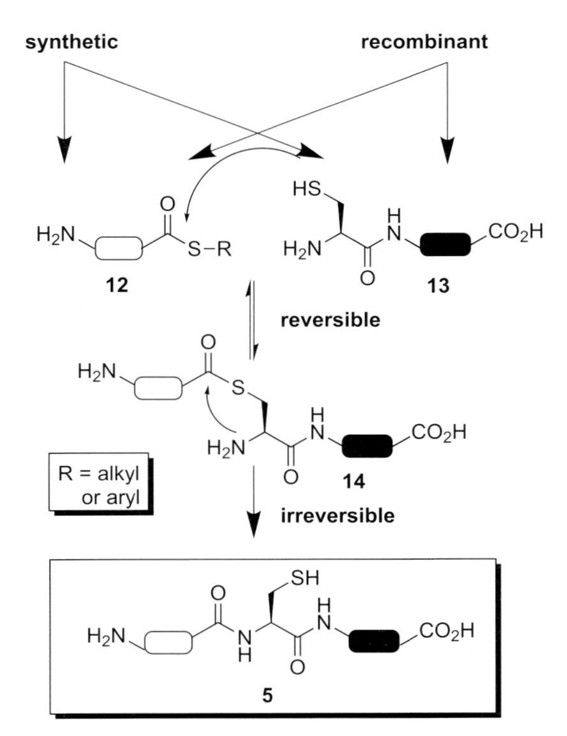


**Scheme 13.6** Concept of protein semisynthesis.

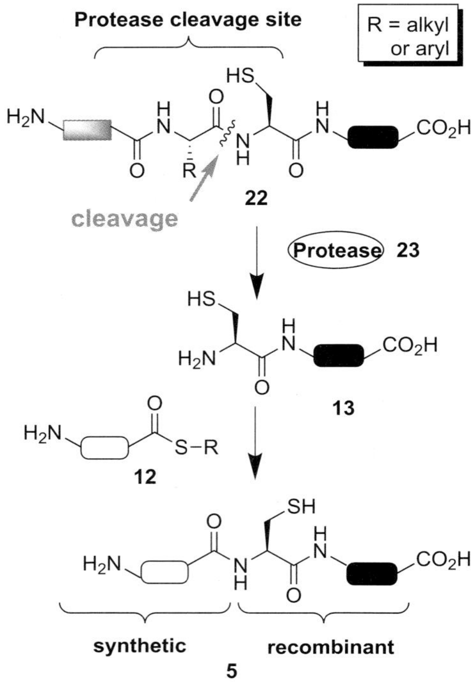

Scheme 13.7 Generating recombinant peptides with an N-terminal Cys.

mediated cleavage between a Met–Cys motif in a protein [31]. A construct containing a Cys following the initial Met can sometimes be obtained from bacteria with an N-terminal Cys without further manipulations. In these cases the endogenous methionine aminopeptidase of the host organism removes the initial Met residue and liberates the downstream Cys moiety [32].

## 13.2.4
## Protein Semisynthesis with Expressed Protein Ligation

The semisynthetic scheme described above is suitable for ligating synthetic peptides to the N-terminus of recombinant proteins. However, several biologically interesting proteins require a ligation method in which the C-terminus of a recombinant protein can be synthetically manipulated. In these cases a recombinant protein thioester is required, which cannot be synthesized with chemical tools. To address this problem, Muir and Cole introduced "expressed protein ligation" (EPL) [33]. EPL is based on an enzymatic approach to generate protein thioesters. The exploited process is referred to as "protein splicing" – a naturally occurring reaction where an internal protein domain excises itself out of a precursor polypeptide **24** and links the flanking N- and C-fragments with a native peptide bond [34–37]. The internal protein domain is called the "intein" and the flanking regions (N- and C-) "exteins". The intein is the sole catalyst possessing all enzymatic activities for the splicing process which is illustrated in Scheme 13.8A. Protein splicing is initiated by a reversible $N \rightarrow S$ acyl shift of the peptide backbone onto the side-chain of a catalytic Cys to form the thioester **25** at the N-terminus of the intein [38]. In the following step, the Cys-thioester migrates to a downstream Cys in **26**, which is the first amino acid of the C-extein. The final rearrangement involves a catalytic Asn residue at the C-terminus of the intein and releases the intein with a C-terminal succinimide **27** and the splice product **28** linked with a native peptide bond [39].

EPL uses a mutated intein that can only catalyze the initial $N \rightarrow S$ acyl shift. A protein of interest (POI) is fused N-terminally to this mutated intein on the genetic level representing the N-extein in **29** (Scheme 13.8B). The intein is usually extended by an affinity tag like a chitin-binding domain (CBD), which remains covalently attached to the intein and can be used for purification purposes. A reversible $N \rightarrow S$ shift yields the thioester **30**, which is intercepted by the addition of access thiols like MESNA **31** through a thiol–thioester exchange. The POI is delivered as a C-terminal thioester **32** that can be isolated by simple elution off the resin and used for a subsequent NCL with a synthetic *C*-peptide **13**. Alternatively, the synthetic peptide can be added directly to the thiol cleavage reaction and the ligation product **33** is formed *in situ* (Scheme 13.8B).

## 13.2.5
## Protein Trans-Splicing

Protein splicing can also be employed directly for protein semisynthesis. This process, "semisynthetic protein trans-splicing," requires a special kind of intein which is split into two fragments. These so-called "split inteins" occur in nature and reassociate spontaneously to a noncovalent, but fully functional complex that links the N- and C-exteins by the splicing process described above (Scheme 13.9A).

Split inteins were first discovered in the cyanobacterium *Synechocystis* sp. strain PCC6803. In this organism the catalytic subunit of the DNA polymerase III is expressed as two separate DnaE fragments (Scheme 13.9A), which are extended by N- and C-terminal parts of the intein. These intein pieces later rejoin and deliver the full-length DnaE [40].

For a semisynthetic application of protein trans-splicing a synthetic peptide including the short $Int^C$ fragment can be extended by an $Ex^C$ sequence and used for trans-splicing with a recombinant $Ex^N$–$Int^N$ fusion protein. This method has been applied successfully for ligating a synthetic FLAG epitope to a recombinant GFP (Scheme 13.9B) [41]. This method was also performed in live cells, as demonstrated for the ligation of construct **34** and the synthetic $Int^C$-FLAG peptide **35**, in which the cellular import of the synthetic peptide **35** was achieved by adding a protein transduction domain (PTD) for shuttling into the cell.

Split inteins can be also generated artificially [42–45] as, for example, reported by Mootz *et al.* [46]. Their artificially split Ssp DnaB intein system is split only 11 amino acids downstream of the N-terminus of the intein and was capable of ligating a

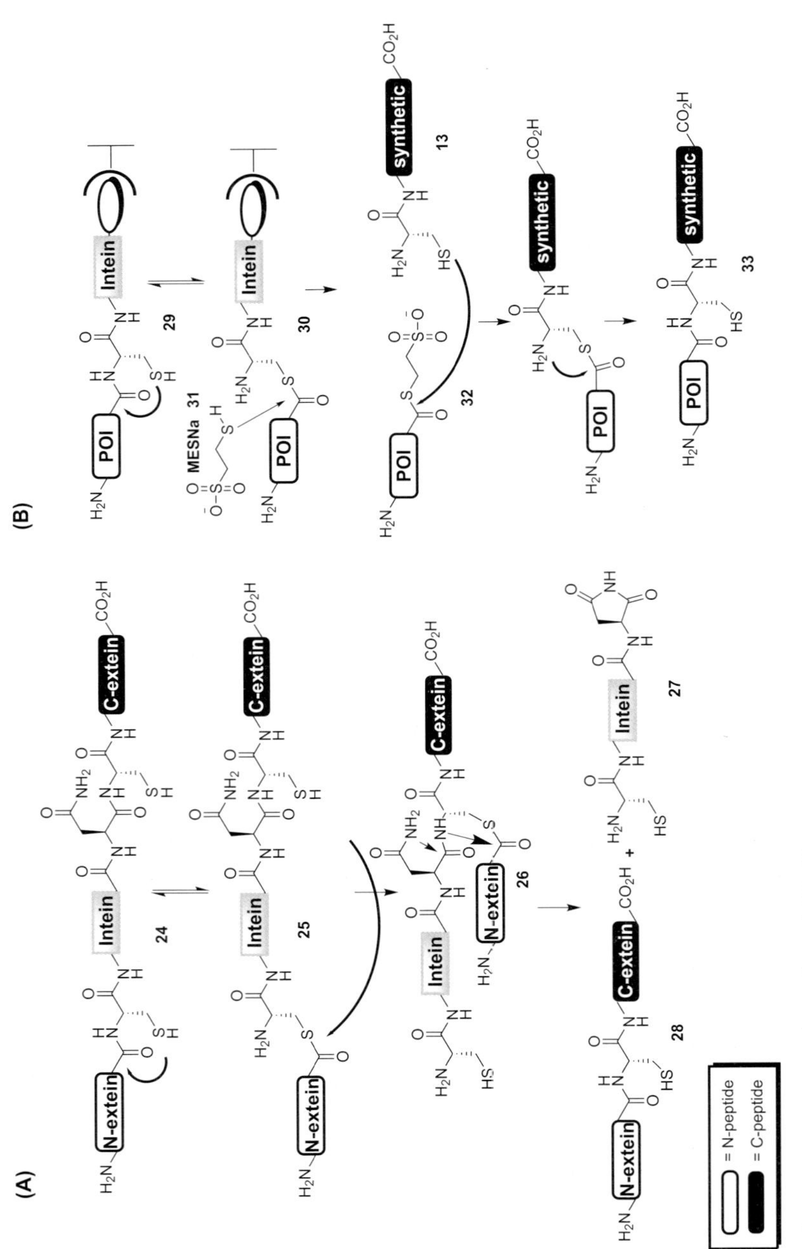


Scheme 13.8 Protein splicing (A) and EPL (B).

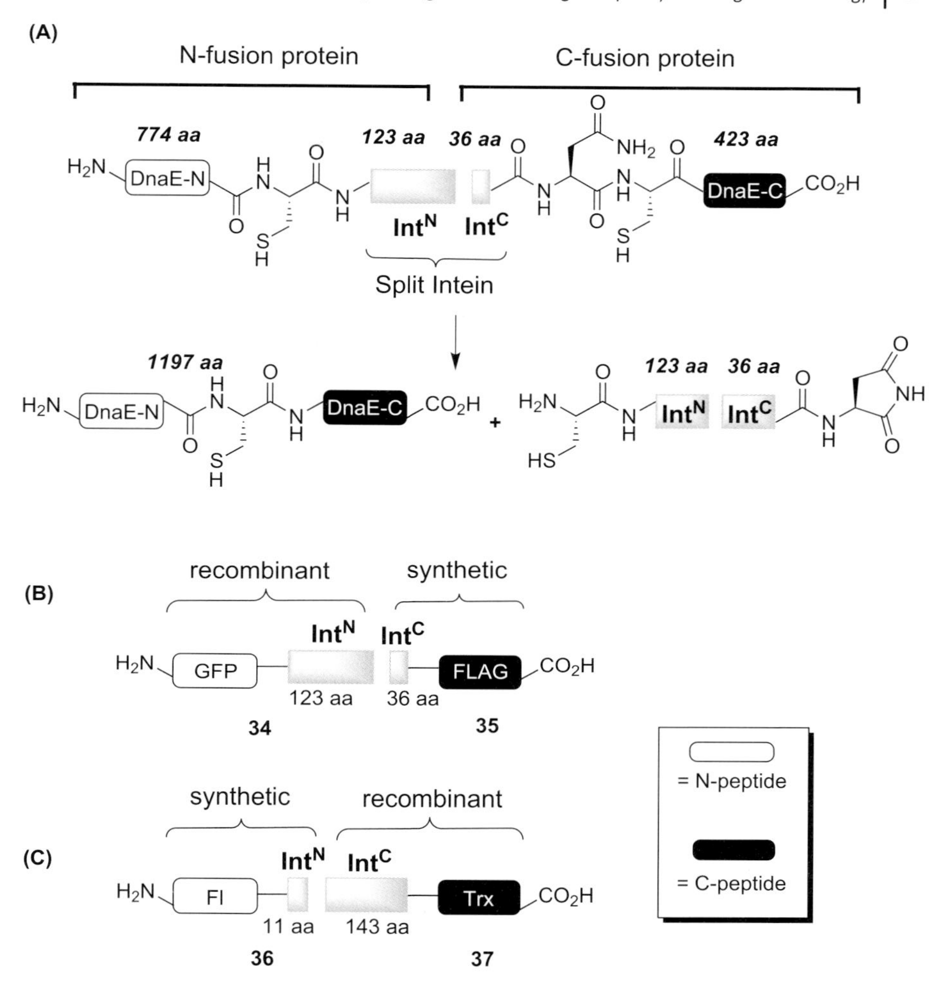


**Scheme 13.9** Protein trans-splicing. (A) Trans-splicing reaction of the split DnaE intein. (B) C-terminal semisynthetic protein trans-splicing, (C) N-terminal semisynthetic protein trans-splicing.

fluorescein (Fl)-conjugated peptide **36** to recombinant thioredoxin (Trx) **37**. The location of the split side decreased the length of the synthetic intein sequence, which will not end up in the ligation product, to one third of that of the naturally split Ssp DnaE intein. In addition, this system can be used to ligate synthetic peptides to the N-terminus of recombinant proteins, rather than the C-terminus as in the case of the Ssp DnaE intein. In a recently reported artificially split intein system a new C-terminal split site was identified and used for semisynthetic protein trans-splicing. This site is located only six amino acids upstream of C-terminus of the Ssp GyrB intein and is therefore very promising for ligating synthetic peptides to the C-terminus of recombinant proteins [47].

## 13.3
## Chemical Transformations for Cys-Free Ligations in Peptides and Proteins

NCL has emerged as the most powerful technique for the ligation of two unprotected peptide or protein fragments to yield a native amide bond. Nevertheless, the requirement of a Cys residue at the ligation site still motivates researchers to search for alternative ligation methods, because Cys residues are incorporated in proteins only to an extent of 1.7% [48]. This often excludes a direct retrosynthetic disconnection for peptide ligation via NCL. Additionally, Cys residues cannot necessarily be regarded as small "non-native" building blocks for the replacement of the naturally occurring residue at the ligation site, as they can oxidize to disulfide bonds and thereby influence the overall folding pathway.

As a possible solution to this problem, several methods have been developed to allow a chemoselective connection of amide bonds at non-Cys containing peptide junctions (**38**). All methods described in this chapter rely on a two-step protocol, in which either NCL (Chapter 13.3.1) or an auxiliary mediated capture/rearrangement strategy (Chapter 13.3.2) is followed by a final chemical transformation (Scheme 13.10). These approaches yield native peptide bonds with non-Cys or furnish amino acid analogs at the ligation site while maintaining the $\alpha$-amino acid backbone.

In this context it is important to note that ligation procedures have been reported in which nucleophilic amino acids other than Cys participate in the previously discussed capture/rearrangement steps when located at the N-terminus of the *C*-peptide. Examples include peptide ligations at a His residue [49] or ligations involving a -SeH nucleophile, thus leading to a ligation product with the rare amino acid selenocysteine (Sec) [50]. In particular the latter approach has been used in the synthesis of various peptides [51] and also proteins by EPL [52], which allowed important mechanistic investigations of proteins [53].

### 13.3.1
### Chemical Modification of NCL Products

The protein ligation strategies in this section employ a selective chemical transformation of a Cys (Scheme 13.11) [54, 55]. The advantage of a chemical modification strategy in combination with NCL stems from the fact that these transformations can address the unique chemical properties of Cys at the ligation site. Nevertheless, additional Cys residues in the ligation product limit the generality of this strategy, as selective transformations are often impossible to perform without further protecting group manipulations.

A critical aspect of this approach is the fact that chemical transformations have to be performed on protein material. Apart from chemoselectivity considerations, harsh reaction conditions could harm the valuable protein and thus significantly decrease the overall ligation yield or, even worse, influence the overall protein architecture by aggregation.

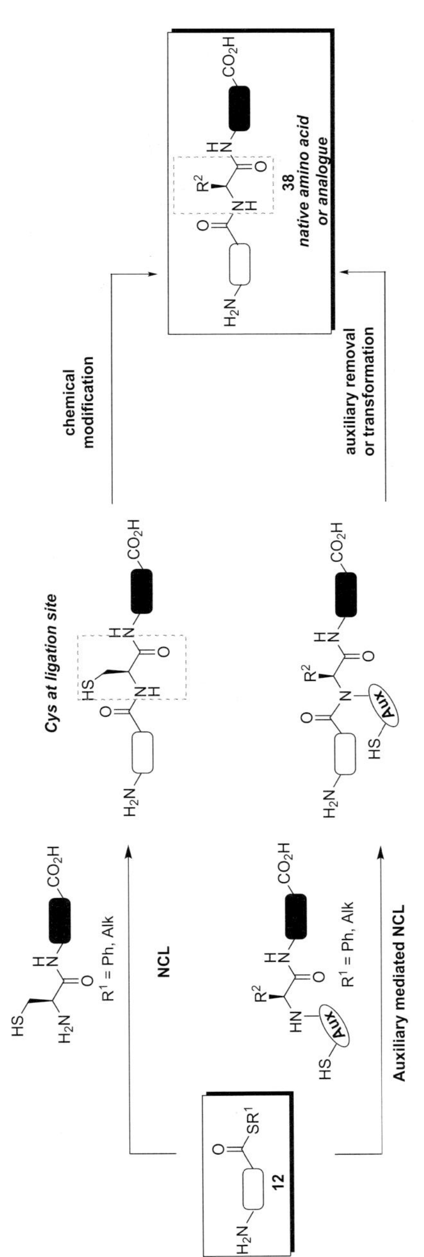


**Scheme 13.10** Synthetic pathways for non-Cys-containing peptides and proteins.

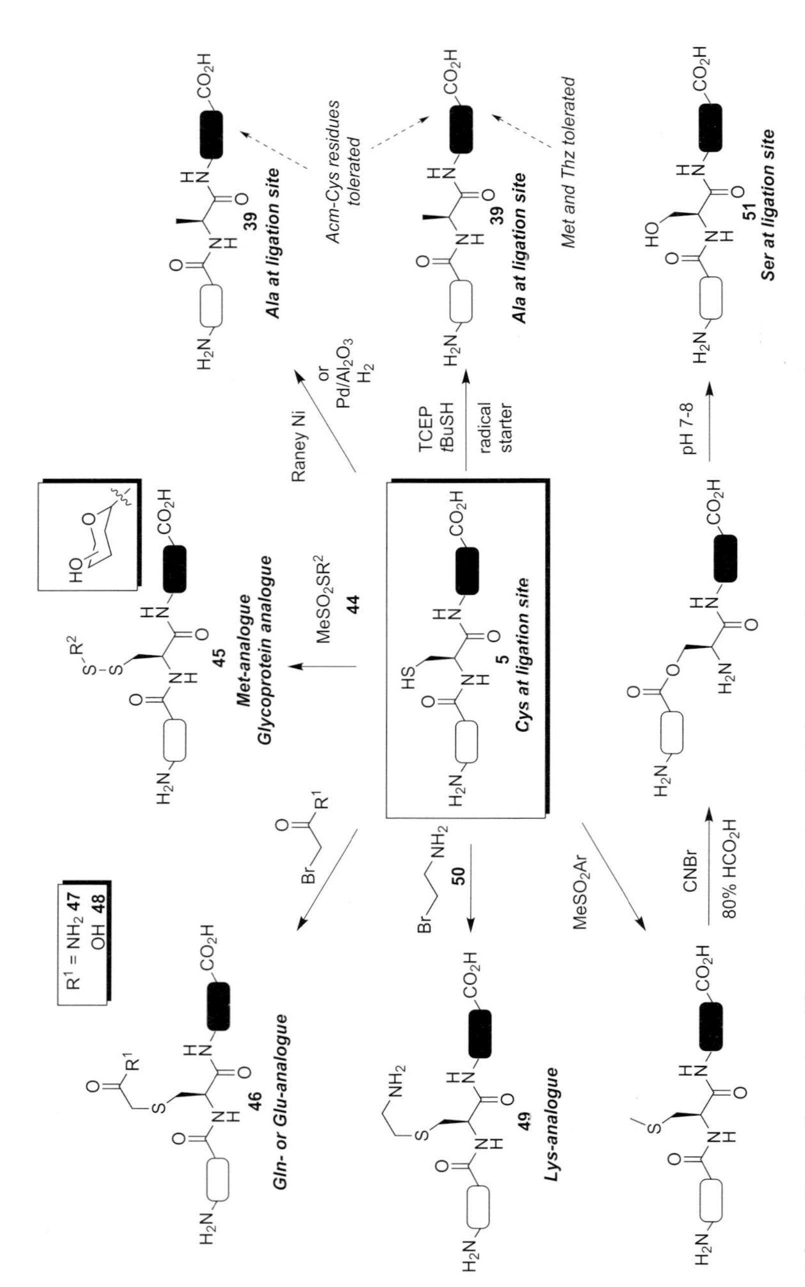


**Scheme 13.11** Chemical modification of NCL products: alkylation and desulfurization strategies.

#### 13.3.1.1 Desulfurization

**Ala Ligation: Desulfurization of Cys** Most frequently, Cys-containing ligation products are converted into other polypeptides by desulfurization techniques using Raney-Nickel or Ni-boride (Scheme 13.11). This strategy, which delivers ligation products with Ala (**39**) residues at the ligation site, was introduced by Perlstein [56] and has been applied in various challenging polypeptide systems [57, 58]. (For the desulfurization in the glycopeptide assisted ligation as introduced by Wong *et al.*, see Chapter 13.3.2.4.) Furthermore, it was extended to the selective desulfurization of unprotected Cys over acetamidomethyl (Acm)-protected Cys residues [59], thus allowing the ligation of Cys-containing polypeptides at Ala junctions after a final Acm-deprotection step. However, the heterogeneous nature of the nickel catalyst can lead to protein aggregation [60] and other side reactions including desulfurization of thioethers of Met or thiazolidines and the epimerization of secondary alcohols [61]. This disadvantage has been addressed by Danishefsky, who reported a metal-free radical based desulfurization to convert Cys residues in the presence of residues like Met, Acm-protected Cys and thiazolidines in high yields (Scheme 13.11) [62]. This protocol was recently applied to the desulfurization of phosphorylated and acetylated histone H2B proteins that were obtained by NCL (Scheme 13.12). The radical desulfurization conditions gave considerably higher yields than the heterogeneous Raney-Nickel catalyst (68–71 versus 30–35%) [63].

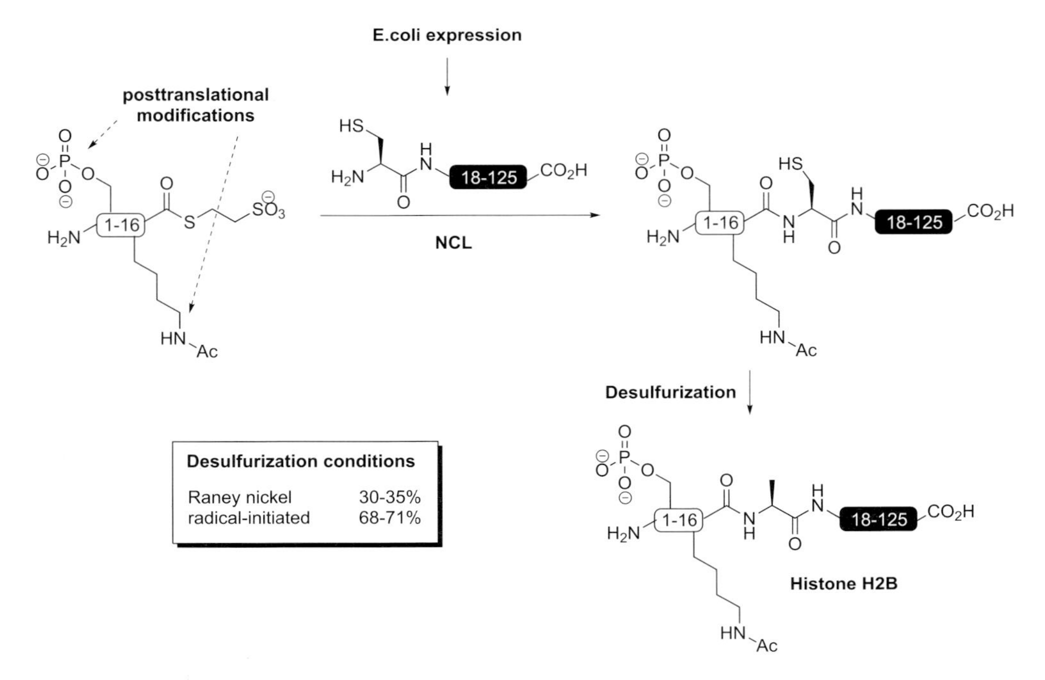


**Scheme 13.12** Desulfurization of histone H2B proteins.

**Phe Ligation and Val Ligation: Desulfurization of β-Mercapto -Building Blocks** In recent publications, NCL reactions were performed with β-branched Cys derivatives instead of Cys. Subsequent desulfurization of these unnatural thiol-containing amino acids introduced the natural residues at the ligation site. (For a previous discussion on the use of thiol-containing analogs of natural amino acids at ligation junctions see [58]). Although these protocols have not been used in the ligation of proteins they represent impressive examples in modern ligation techniques.

In the first example, β-mercaptophenylalanine was used as a template for ligations at Phe junctions (Scheme 13.13, top) [64]. This derivative can be attached to the N-terminus of the *C*-peptide as a disulfide building block **40**, which, under NCL conditions, generates **41** *in situ*. The benzylic thiol in the decapeptide **42** was removed by heterogeneous desulfurization in good overall yields.

In a related approach Val derivatives with an added thiol are used for the ligation at a hydrophobic ligation site. For example, a penicillamine building block **43a** is employed, in which the β,β-dimethylcysteine functionality reacted surprisingly fast and in high yields with an *N*-peptide thioester to penicillamine-containing model peptides with up to 22 amino acids (Scheme 13.13, bottom) [65]. These were transformed into the corresponding Val peptides by applying an optimized version of Danishefsky's homogeneous desulfurization technique.

Another system for the ligation at Val residues employs a primary thiol group at the γ-position of the Val building block **43b**. This auxiliary delivered, in an analogous two-step protocol, ligated peptides in very good yields. These two procedures have allowed ligations at junctions between various sterically demanding amino acids including Leu–Val and Thr–Val.

#### 13.3.1.2 Alkylation and Thioalkylation Protocols

The nucleophilic properties of Cys at the NCL site can also be employed in alkylation or thioalkylation reactions. Although the examples mentioned here have been applied mostly to expressed proteins and not to NCL products, they will be briefly discussed here because they demonstrate the scope of this concept. For example, Cys residues in proteins can be transformed via thioalkylation with methyl methanethiosulfonate **44** to Met analogs **45** [66, 67] or other important proteins carrying post-translation modification mimics such as disulfide-linked glycoconjugates [68]. The latter bioconjugates can be converted into stable derivatives with glycosyl thioether linkages by a chemoselective reaction with P(III) reagents [69]. Other alkylation protocols give Gln or Glu derivatives **46** by reaction with α-iodoacetamide (**47**) [70] or α-iodoacetic acid (**48**) [71]. Lys analogs **49** can be derived by reaction with aziridines or bromoethylamines (**50**) (Scheme 13.11) [72]. Although all of these transformations yield analogs of natural occurring amino acids, several studies have shown comparable protein function in comparison to their natural analogs [71].

In a recent publication the Cys residue was converted into a Ser in peptides **51** in high yields by a sophisticated procedure. First, the Cys was methylated with methyl-4-nitrobenzenesulfonate, which was followed by an intramolecular rearrangement with CNBr in formic acid and a subsequent $O \rightarrow N$ acyl shift under slightly basic conditions (pH 7–8) [73, 74]. This protocol has great potential in the synthesis of

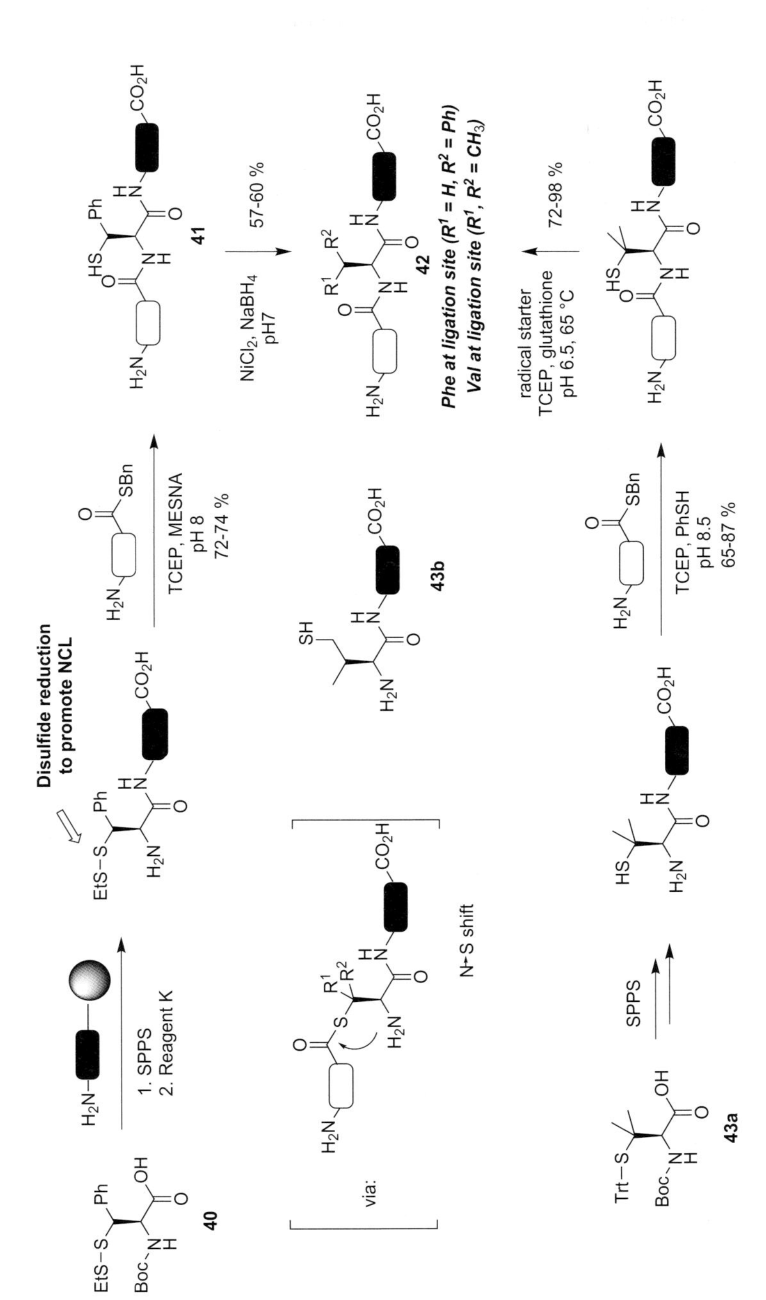


**Scheme 13.13** Native chemical ligation at a Phe (top) or Val junction (bottom) via a two-step ligation and desulfurization protocol.

*N*-linked glycopeptides as they contain an Asn–Xaa–Ser or Asn–Xaa–Thr consensus sequence in which Xaa resembles any amino acid except proline [75].

### 13.3.2
### Auxiliary Methods

Since the above discussed methods affect all Cys residues, auxiliary groups have been developed which facilitate the ligation reaction and can be removed without modifying any other amino acid residue of the ligation product [76].

Here, a thiol-containing auxiliary **52** is placed in a peptide, which subsequently reacts with an *N*-peptide thioester **12** in a rearrangement similar to the NCL (Scheme 13.14). The auxiliary is removed in a final synthetic transformation to furnish the desired native peptide or protein sequence **53**. Obviously, these chemical transformations have to meet the same requirements concerning mild reaction conditions and chemoselectivity as discussed previously.

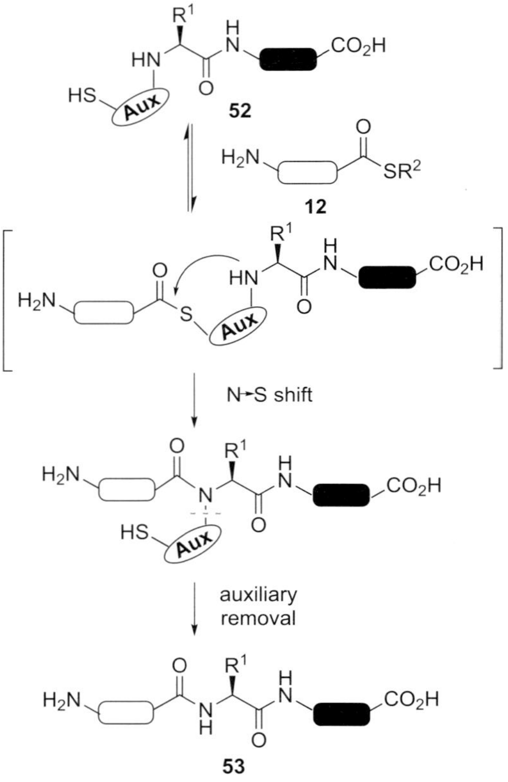

**Scheme 13.14** Mechanism and scope of the auxiliary mediated ligation.

#### 13.3.2.1 (Oxy-)Ethanethiol Auxiliary

A very simple auxiliary system employed an ethanethiol (X = $CH_2$, **54**) or oxyethanethiol (X = $OCH_2$, **55**) substituent (Table 13.1, entry 1), in which the ethanethiol

**Table 13.1** Auxiliary systems with possible ligation junctions and auxiliary removal conditions.

| | Structure | Ligation junction | Auxiliary removal conditions |
|---|---|---|---|
| | 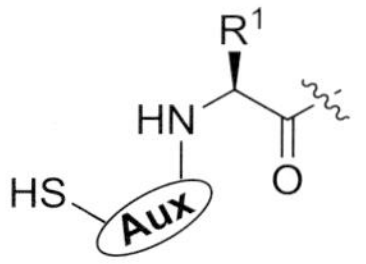 $R^1$, HN, HS, Aux, O | | |
| 1 | 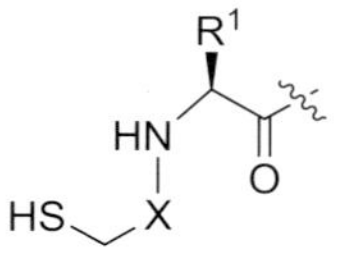 X = $CH_2$ **54**<br>X = $OCH_2$ **55** | Gly–Gly (for **54**)<br>Gly–Gly<br>Phe–Gly<br>Ala–Gly (for **55**) | None (for **54**)<br>Zn, acetic acid (for **55**) |
| 2 | 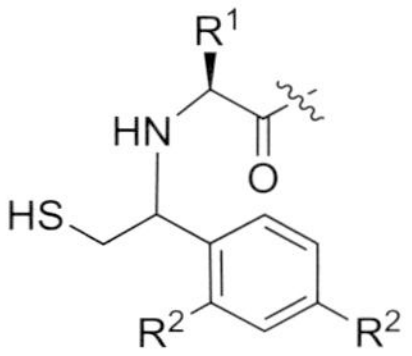 $R^2$ = H, OMe **56** | Gly–Gly<br>Ala–Gly<br>His–Gly<br>Lys–Gly | HF (for $R^2$ = H)<br>TFA (for $R^2$ = OMe) |
| 3 | 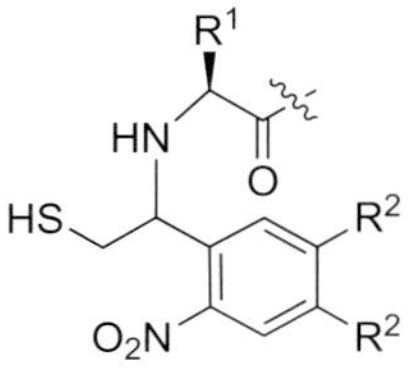 $R^2$ = H, OMe **57** | Gly–Gly<br>Ala–Gly | hν (310–365 nm) |
| 4 | 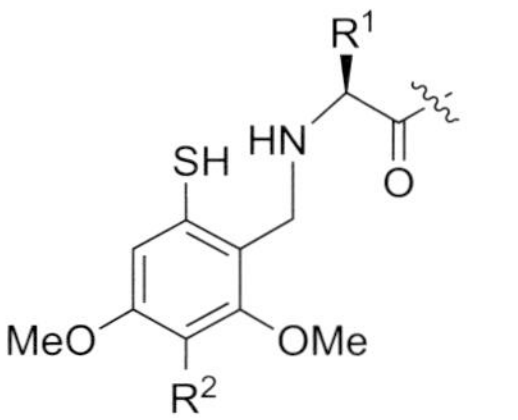 $R^2$ = H **58a** (Dmb)<br>OMe **58b** (Tmb) | Gly–Gly<br>Ala–Gly<br>Lys–Gly | HF (for **58a** (Dmb))<br>TFA ($Hg^{2+}$) (for **58b** (Tmb)) |

Dmb: 4,6-dimethoxy-2-mercaptobenzylamine, Tmb: 4,5,6-trimethoxy-2-mercaptobenzylamine.

auxiliary delivered better ligation yields at Gly junctions (up to 90%) in comparison to the oxyethanethiol analog. However, in contrast to the ethanethiol auxiliary, the oxyethanethiol analog can be removed from the protein material by reductive cleavage of the N–O bond with zinc [77]. Both systems have been applied to advanced peptide and protein synthesis, including peptide cyclization as well as the acquisition of disulfide-bridged protein analogs [78, 79].

#### 13.3.2.2 Photoremovable $N^{\alpha}$-1-Aryl-2-Mercaptoethyl Auxiliary

In auxiliary system **56** the N-terminus of the *C*-peptide is converted into a secondary amine with an aryl in the 1-position and a thiol functionality in the 2-position (Table 13.1, entry 2) [80]. This auxiliary can be incorporated into the peptide either by reaction of the parent amine with an α-bromo peptide or by a building block approach of the Boc- or Fmoc-protected $N^{\alpha}$-(1-aryl-2-mercaptoethyl) amino acid derivatives [81]. The auxiliary system enables sterically nondemanding peptide ligations at Gly and Ala junctions in good overall yields in peptides; however, the auxiliary has to be removed under strong acidic conditions with HF or trifluoroacetic acid (TFA)/bromotrimethylsilane (TMSBr), which limits its applicability in advanced protein examples.

Therefore, the development of a mild light-induced removal procedure for this auxiliary by the groups of Aimoto [82] and Dawson [83] is particularly important. They have identified light-sensitive auxiliary **57** with an *o*-nitrobenzene function, which efficiently ligated two nonapeptides, again at sterically nondemanding peptide ligations at Gly–Gly or Ala–Gly junctions, in high yields prior to photolytic removal (Table 13.1, entry 3). This improvement has recently found an important application in the semisynthesis of ubiquitylated histone peptides and proteins, which represented the first entry of an EPL at a non-Cys ligation site [81]. The combination with a second EPL will be discussed further in the final part of this chapter, since it involves an impressive example of a three-segment ligation approach of a biologically most relevant post-translationally modified histone protein [85].

#### 13.3.2.3 4,5,6-Trimethoxy-2-Mercaptobenzylamine Auxiliary

A 2-mercaptobenzyl based auxiliary system **58** with an electron-rich arylthiol can also undergo the thioester exchange with a C-terminal peptide thioester (Table 13.1, entry 4). After the subsequent $S \rightarrow N$ shift of the secondary amine, a tertiary amide is formed, which can be removed with TFA [86]. The most potent auxiliary is the 4,5,6-trimethoxy-2-mercaptobenzylamine (Tmb) **58b** with increased nucleophilicity of the arylthiol and increased acid lability of the tertiary amide. This auxiliary has been employed in peptide ligations that contained at least one Gly at the ligation site, with short coupling times of 0.5 (Gly–Gly), 2 (Lys–Gly) and 5 h (Ala–Gly) [87]. Furthermore, it has been used in the synthesis of cytochrome $b_{562}$ – a 106-residue protein [88]. In another study, the Tmb auxiliary was combined with a C-terminal peptide phenol ester **59** with an *o*-disulfide for sterically demanding ligations of two glycopeptide fragments **59** and **60**, in which each contained an *N*-linked chitobiose residue at a Gly–Gln junction [89]. In this investigation it was found that the acidic conditions applied for the auxiliary removal lead to an accumulation of the thioester

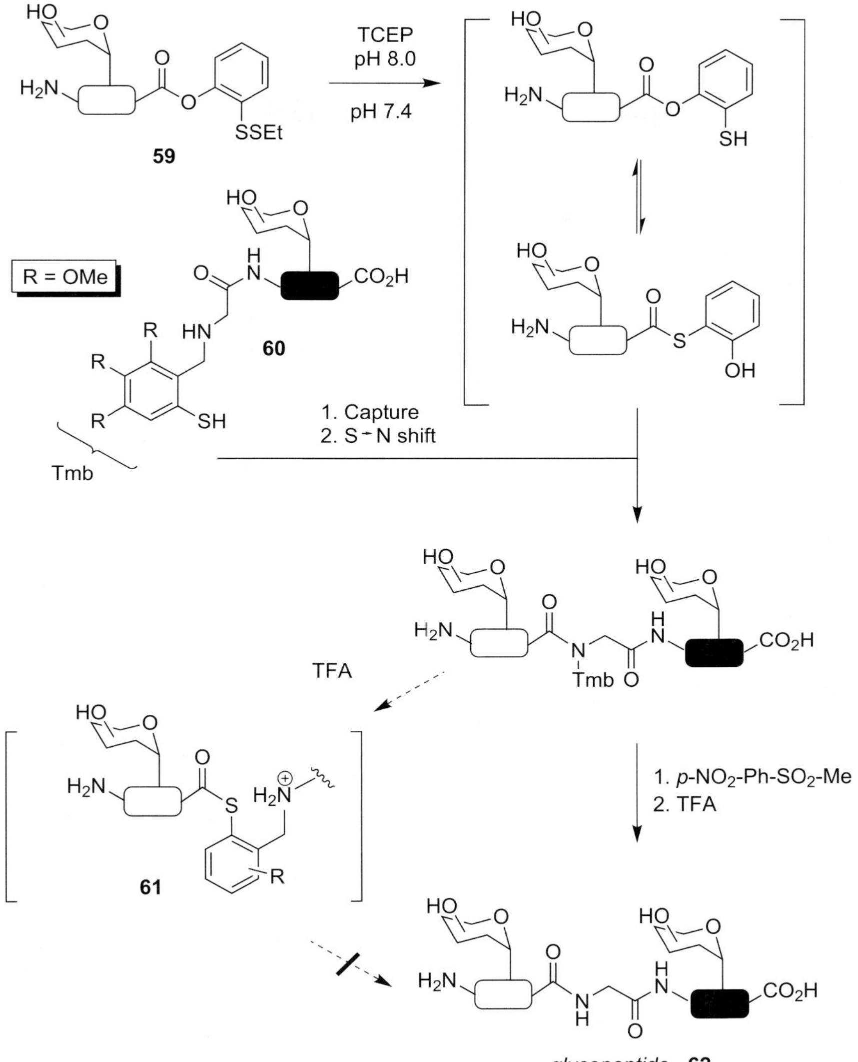

Scheme 13.15 Cys-free glycopeptide couplings using the Tmb auxiliary approach.

intermediate **61** via protonation of the secondary amine, thereby resulting in an $N \rightarrow S$ shift [90]. Methylation of the arylthiol increased the removal efficiency to give glycopeptide **62** (Scheme 13.15).

#### 13.3.2.4 Sugar-Assisted Glycopeptide Ligations

Recently, a ligation strategy was published by Wong *et al.* in which the auxiliary for the capture step is incorporated into the glycan called "sugar-assisted glycopeptide

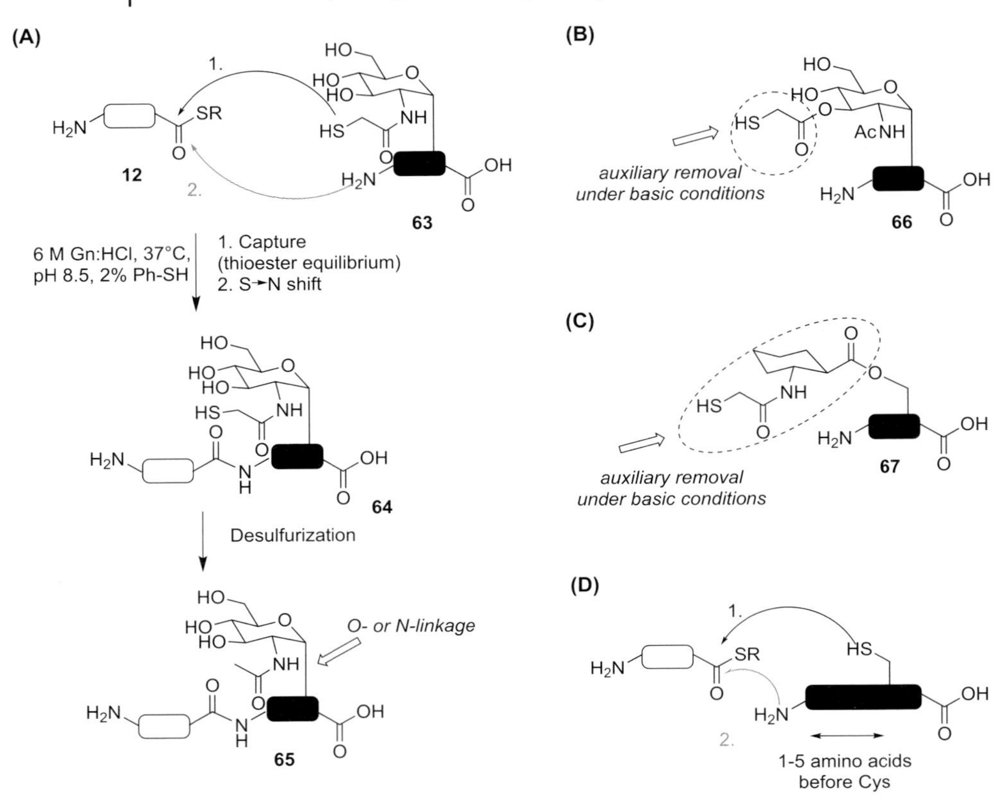


**Scheme 13.16** SAL.

ligation" (SAL, Scheme 13.16A) [91]. The thiol function can be introduced as a β-thioacetamido substituent at the C-2 position, either in a GalNAc for *O*-linked [92] or in a GlcNAc residue for *N*-linked glycopeptides **63** [93]. The thiol is assumed to initially form a macrocyclic thioester with a C-terminal peptide thioester before the free amine at the N-terminus of the glycopeptide fragment performs the $S \rightarrow N$ shift to generate the native amide bond in **64** under slightly basic conditions. Final desulfurization conditions (see Section 13.3.3.1) yield the native glycopeptide **65**. It was demonstrated that glycopeptides with up to six additional amino acids N-terminal to the glycosylation site were able to undergo SAL [94]. In further studies, a β-thioacetic ester in the C-3 position of a GalNAc residue as the auxiliary **66** can be used as well, which allows a final auxiliary removal with hydrazine (Scheme 13.16B) [95]. Additionally, a cyclohexyl-based auxiliary **67** was introduced [96a] and applied to the ligation of the HIV-1 Tat protein (Scheme 13.16C) [96b]. This system delivers, after auxiliary removal, Ser at the ligations junction; however, the auxiliary cleavage under basic conditions (pH 10–11) only proceeded well with a peptide, whereas attempts for the corresponding protein failed.

Based on these findings another investigation probed whether internal Cys residues within the *N*-peptide can accelerate the aminolysis of the N-terminal amine via the formation of macrocyclic thioester intermediates (Scheme 13.16D) [97]. It was found that appropriately positioned Cys residues can induce rate enhancements of up to 25-fold in comparison to a Cys-lacking control peptide.

In light of these investigations, another report from the Wong group was particularly important, because it demonstrated the fine balance between a direct aminolysis reaction and the macrocyclic thioester participation. In this report, a (glyco-)peptide thioester and another (glyco-)peptide at a non-Cys or sugar auxiliary junction were coupled [98]. In this nonchemoselective conjugation strategy, which is based upon early observations by Wieland, a carefully optimized *N*-methylpyrrolidone (NMP)/HEPES buffer at neutral to slightly basic pH makes the N-terminus of the *C*-peptide sufficiently nucleophilic to perform an aminolysis with the *N*-peptide thioester and simultaneously suppresses thioester hydrolysis. However, this method cannot be employed as a general chemoselective ligation technique, since additional amino groups in Lys side-chains have to be protected.

## 13.4 Other Chemoselective Capture Strategies

As discussed in Section 13.2, NCL is based on a thioester equilibrium resembling the capture step before the amide bond is formed via an $S \rightarrow N$ shift. The following two methods utilize different chemoselective capture steps before a rearrangement to an amide bond occurs.

### 13.4.1 Traceless Staudinger Ligation

This ligation reaction was introduced in 2000 and represents a traceless variant of the (nontraceless) Staudinger ligation for the synthesis of amide bonds [99, 100]. Both reactions originate from the classical Staudinger reaction, in which an azide reacts with a phosphine to form an iminophosphorane [101]. This reaction was identified by Bertozzi as chemoselective and further extended to a chemoselective functionalization of metabolically engineered azido-glycans in cellular environment. This could be achieved by placing an electrophilic ester in proximity to the iminophosphorane to prevent undesired hydrolysis to the phosphine oxide and amine [102]. The traceless Staudinger ligation utilizes this concept; however, the electrophilic ester moiety is changed. Here, the iminophosphorane **70**, formed between an α-azido-amino acid or -peptide **68** and a phosphinothioester **69**, reacts with an internal thioester to yield a native amide bond in high yields even in dipeptides at an Ala–Ala junction (**71**) (Scheme 13.17A) [103]. In this regard the formation of **70** in principle resembles the chemoselective capture step.

The traceless Staudinger ligation has received considerable attention as a very potent, universal strategy for peptide ligations, since it does not, like NCL, require a

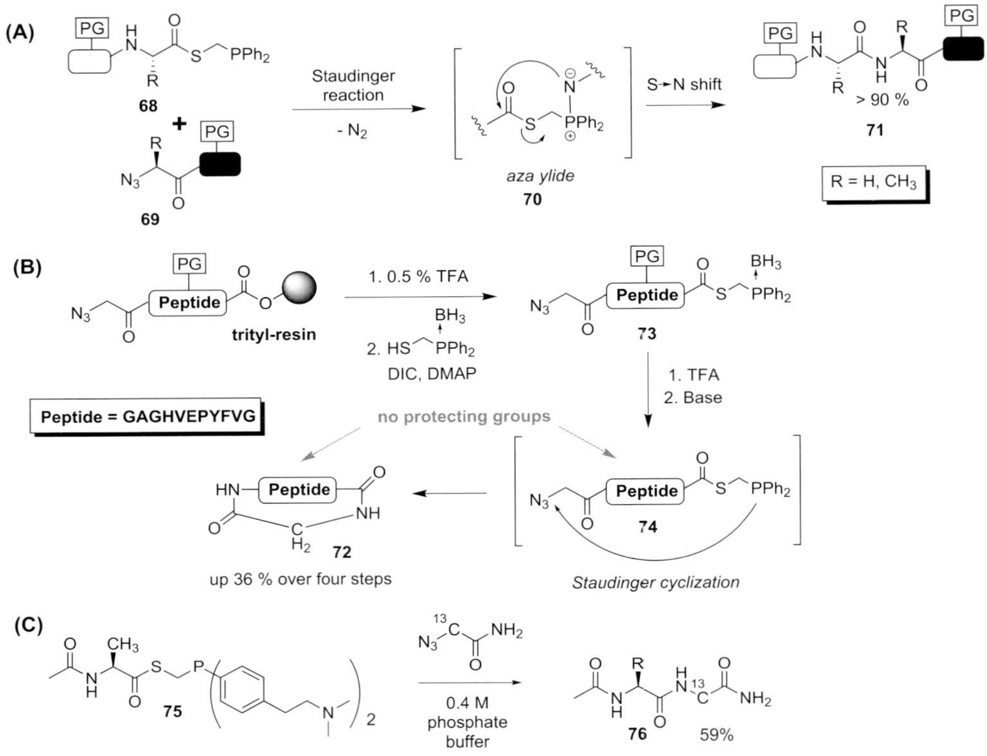

Scheme 13.17 (A) Scope and mechanism of the traceless Staudinger ligation. (B) Staudinger cyclization after acidic deprotection of borane-protected phosphinothiols. DIC: diisopropylcarbodiimide. (C) Water-soluble Staudinger ligation.

Cys residue at the ligation site. In recent years considerable improvements and mechanistic investigations [104] have been performed to reach this goal. These included various applications in the area of bioorganic chemistry, such as the site-specific immobilization of peptides on surfaces [105], the ligation of *protected* (glyco-) peptide fragments [106], the conjugation of unprotected sugar derivatives on protein carriers [107], or the cyclization of unfunctionalized lactams [108]. Recently, the chemoselectivity of the traceless Staudinger ligation was probed in the cyclization of unprotected peptide sequences. This was realized by the finding that borane-protected peptide azidophosphinothioesters **73** can be deprotected under acidic conditions, which simultaneously removed the protecting groups of the peptide side-chains as well as the borane protecting group to yield phosphonium salts **74**. After the addition of a base the cyclization to medium sized peptides **72** with up to 11 amino acids proceeded in good yields (Scheme 13.17B) [109]. Another important and very promising study for future intermolecular peptide ligations was the development of a water-soluble version of the traceless Staudinger ligation. Here, water soluble phosphinothiol linkers **75** with a tertiary amine were utilized, which resulted in the formation of an amide bond between a dipeptide **76** at pH 8 in good yields (Scheme 13.17C) [110].

#### 13.4.1.1 Imine Ligations with Subsequent Pseudo-Pro Formation

In another ligation the formation of an imine is used as the capture step. This procedure by Tam *et al.* delivers a pseudo-Pro by reaction of a C-terminal glycolaldehyde peptide **77** with another peptide **78** containing a hydroxyl or thiol function in a Cys, Thr or Ser residue at the N-terminus (Scheme 13.18) [111]. After imine condensation to **79** the nucleophile of the N-terminal amino acid can react with the imine to form an oxazolidine (Oxz, **80a**) or thiazolidine (Thz, **80b**). Then an irreversible $O \rightarrow N$ shift via a bicyclic five-membered intermediate delivers a hydroxylmethylene substituted pseudo-Pro **81** at the ligation junction with a new stereogenic center at the C-2 position [112]. In this process, the thiazolidine can be formed in aqueous solution whereas the oxazolidine synthesis requires anhydrous conditions.

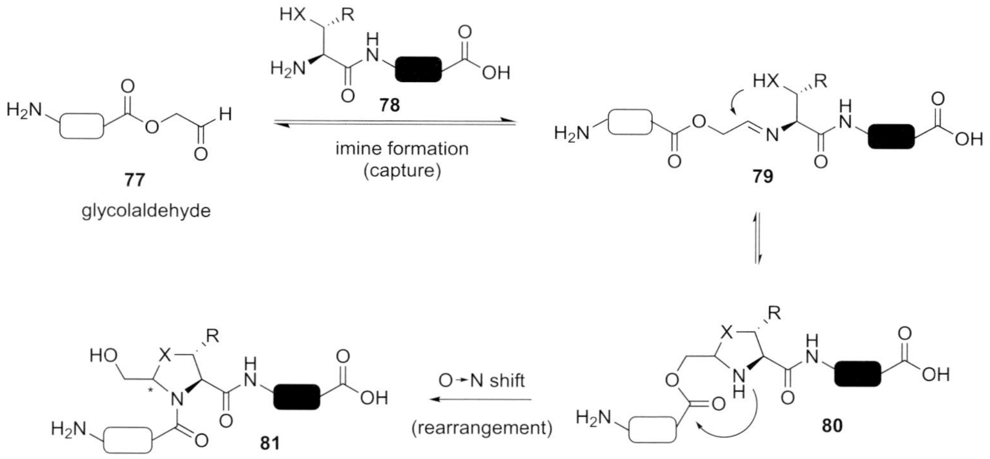

77
glycolaldehyde
78
imine formation
(capture)
79
80
O→N shift
(rearrangement)
81

| | |
|---|---|
| **Thiaproline-ligation:** X = S, R = H (Cys) | |
| **80a** | |
| **Oxaproline-ligation**: X = O, R = $CH_3$ (Thr) | |
| **80b** | X = O, R = H (Ser) |

**Scheme 13.18** Pseudo-Pro ligation of unprotected *N*-peptides with C-terminal glycoaldehyde *C*-peptides.

This reaction was applied in the assembly of proline-rich analogs of the 59-residue helical antibacterial peptide bactenecin 7, which showed comparable biological activities to the natural product [113]. In this aspect it is important to note that a three-segment condensation was used, in which the first junction was constructed by the formation of a thiazolidine of the Cys peptide – a faster step than the formation of the corresponding oxazolidine of the Thr peptide. Simultaneously, an oxidative conversion of a C-terminal glycerol ester of a peptide dimer into the corresponding glycolaldehyde and subsequent formation of the second pseudo-Pro was reported as another segment ligation protocol.

## 13.5 Peptide Ligations by Chemoselective Amide-Bond-Forming Reactions

The development of chemoselective ligation strategies allows the selective formation of a covalent bond between highly complex biological molecules without the requirement of protecting group transformations. Apart from the two-step "capture/rearrangement" strategy presented in the previous two sections, several organic reactions have been recently identified that allow a "direct" chemoselective bond formation in **84** (Scheme 13.19A). In an ideal scenario, these chemoselective reactions proceed under physiologically benign conditions (neutral pH in aqueous media and at room temperature) to yield a native amide bond. This bond is constructed from two "bioorthogonal" [114] functionalities in **82** and **83** that do not show any cross-reactivity with other functional groups present in a biological environment.

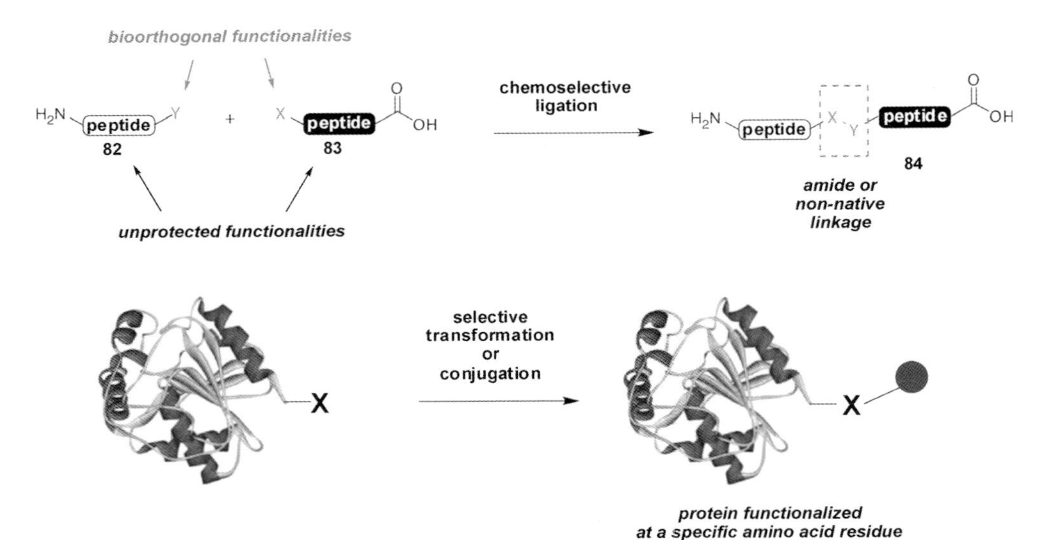


**Scheme 13.19** Chemoselective reaction for a native or non-native conjugation of polypeptides and selective transformation of natural amino acids.

Within this section we will give an overview of examples of alternative amide bond forming strategies with the potential for impact as chemoselective reactions in peptide and protein synthesis. A distinguishing feature of these reactions is their departure from the canonical mechanistic paradigm of acyl substitution in the formation of the amide bond. A hallmark of all of these processes is the use of two *unnatural* functional groups as the amide precursors. This has the advantage of offering outstanding chemoselectivity, but brings with it the disadvantage of requiring modifications of the C- and N-termini of the ligation partners that must be made either synthetically or by challenging site-specific modifications of expressed protein fragments.

In addition to the examples in this chapter it is important to note that the emerging generation of chemoselective peptide ligation reactions has grown in concert with the recognition that highly chemoselective reactions enable researchers to (site-)selectively functionalize or modify complex biomolecules even in cells or living organisms (Scheme 13.19B) [115]. These applications have significantly advanced study of the function of post- and cotranslationally modified proteins, such as by site-selective incorporation of biophysical probes into a biopolymer, commonly referred to as the "(bioorthogonal) chemical reporter strategy" [116], or by chemoselective transformations that resemble natural protein modifications, such as a chemoselective phosphorylation strategy [117].

## 13.5.1 Thio Acid/Azide Amidation

The coupling of thio acids with azides, first reported by Rosen in 1988 [118], provides the basis for a promising new approach for amide ligation. In 2003, Williams *et al.* explored this reaction for the synthesis of simple peptides and *N*-linked carbohydrates from glycosyl azides **85** [119]. These authors postulated that the reaction does not occur via an azide conversion to amine **86** but via the formation of a thiatriazoline derivative **87** as a result of [3 + 2] cycloaddition (Scheme 13.20) [120]. Additionally, a $RuCl_3$-promoted variant of the reaction was demonstrated to be high-yielding, even with sterically hindered azides [121]. Although not applied to the ligation of longer peptide substrates yet, this reaction is a promising new approach to chemoselective amide formation, and may have other important applications in the synthesis and modification of peptides and glycoproteins [122].

## 13.5.2 Thio Acid/N-Arylsulfonamide Ligations

A variant of the former reaction, first reported by Tomkinson [123], employing *N*-arylsulfonamides as the electrophilic amidation partner has been advanced as an exciting route to the synthesis of peptides (Scheme 13.21). Crich *et al.* exploited this process, which proceeds via nucleophilic addition of the thio acid to the highly electrophilic sulfonamide to give Meisenheimer complex **88**, to the synthesis of small peptide fragments including the couplings of sterically demanding substrates [124]. Alternatively, the amide component can be activated by the Mukaiyama or Sanger reagents or as an isocyanate [125]. More recently, the ability of these reactions to be sequenced for the formation of triblock peptides by iterative

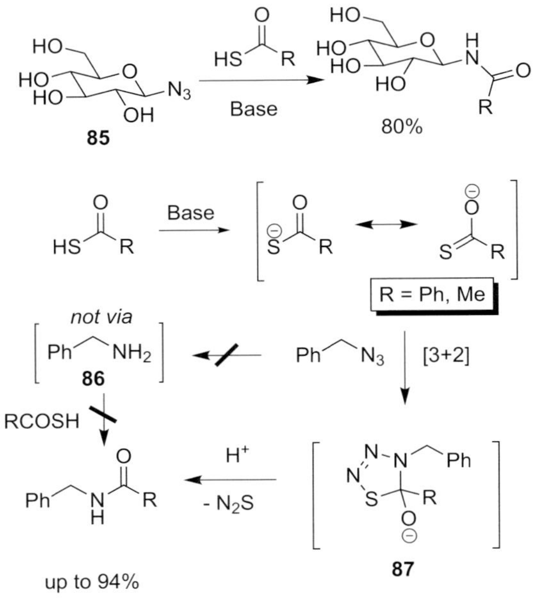

Scheme 13.20 Mechanism of the thio acid/azide amidation.

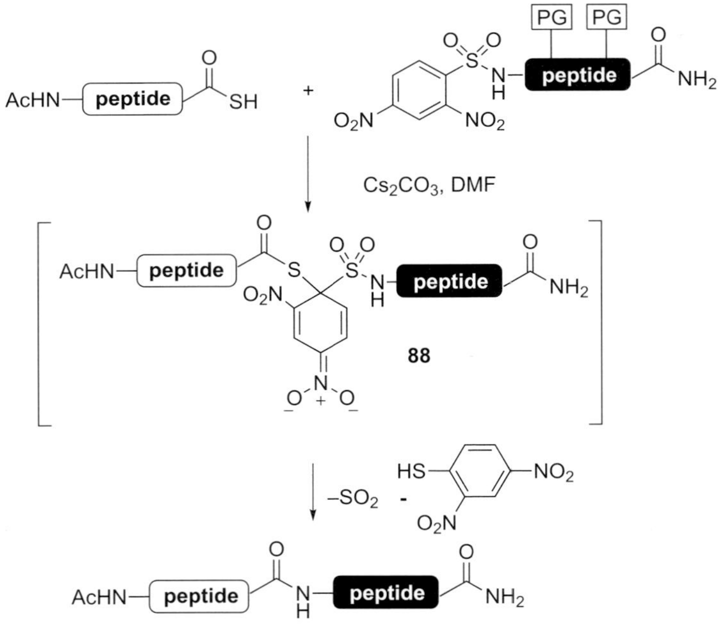

Scheme 13.21 Amide-forming ligation of peptide thio acids with electron-deficient sulfonamides.

couplings has been demonstrated [126]. If it can be established these reactions proceed in the presence of unprotected peptide side-chains containing amine and carboxylic acid functional groups, this process has the potential to be a general solution to peptide ligation.

### 13.5.3
### Chemoselective Decarboxylative Amide Ligation

In 2006, Bode *et al.* reported the decarboxylative condensation of *N*-alkylhydroxylamines **89** with α-ketoacids **90** to yield chemoselectively amide bonds **91** (Scheme 13.22) [127]. This reaction does not involve, in contrast to almost all other amide bond forming strategies, an addition/elimination reaction of an activated carboxylic acid derivative but the decarboxylation and elimination of water from a hemiaminal. In preliminary studies of this amide ligation, Bode established that the reaction proceeds without additional reagents or catalysts in polar solvents at 37–60 °C. The reaction is highly chemoselective, and readily tolerates the presence of unprotected amines, acids, alcohols, thiols, and nitrogen-based functional groups without any observed cross reactivity. The more convenient salts of the hydroxylamines can be employed directly in the reaction. There is potential for epimerization at the C-terminal α-ketoacid fragment but the stereochemical integrity is maintained throughout the ligation reaction, although care must be taken in the preparation of the α-ketoacid. Preliminary investigations established the potential utility of this process as a general approach to peptide ligation, with ligation reactions at sites including Phe–Ala, Pro–Ala, Val–Gly, Ala–Ala, and many others already demonstrated.

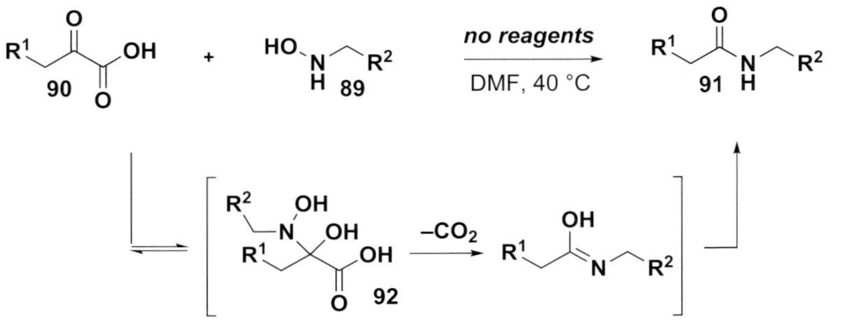

Scheme 13.22 Decarboxylative amide ligation ($R^1$, $R^2$ = alkyl, aryl, heteroatoms, heteroaryl).

The Bode group has recently reported novel linkers and reagents for preparing the requisite C-terminal α-ketoacids [128] and N-terminal hydroxylamines [129] by Fmoc-based SPPS. As these methods continue to improve and develop, they should make access to the requisite C- and N-terminal-modified peptides a robust and straightforward process.

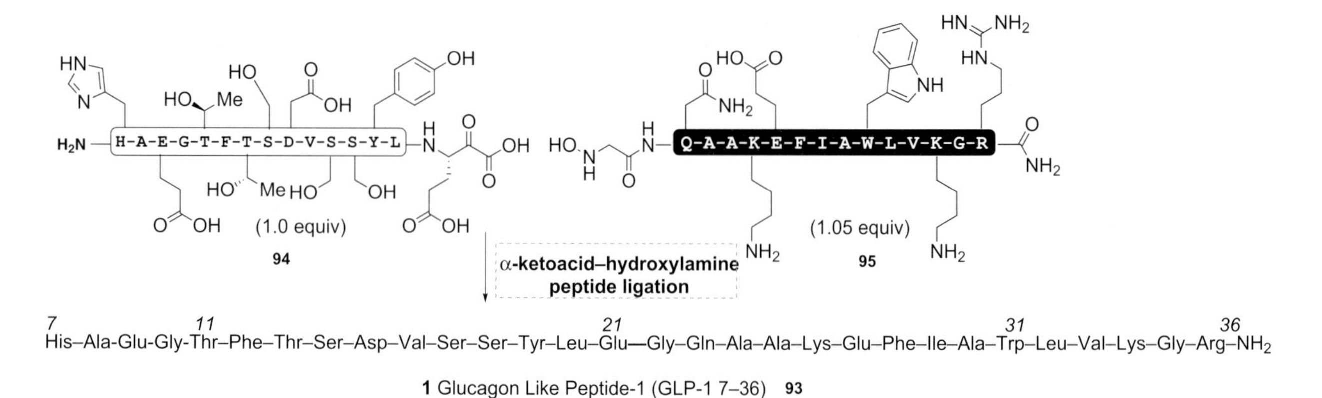


**Scheme 13.23** Synthesis of human GLP-1 (7–36) by chemoselective ligation of unprotected peptide fragments.

While the application of this reaction to the chemical synthesis of proteins is yet to be demonstrated, recent work on its application to the synthesis of human glucagon-like peptide (GLP)-1 [7–35] **93** reveals the remarkable chemoselectivity and potential of this reaction (Scheme 13.23). The 15mer N-terminal fragment **94** was prepared by Fmoc-based solid-phase synthesis from a side-chain linked sulfur ylide. Following resin cleavage and deprotection, this fully unprotected sulfur ylide was converted, in the presence of Oxone, to α-ketoacid **94**. The hydroxylamine fragment (C-terminus of the assembled peptide) **95** was prepared by displacement of a commercially synthesized N-terminal bromoacetate with a protected hydroxylamine. For solubility reasons, the key ligation reaction was performed in a 3 : 1 mixture of dimethylacetamide (DMA)/dimethylsulfoxide (DMSO) at 60 °C for 12 h, leading to a 70% isolated yield of the ligated peptide. Importantly, the reaction was conducted using a 1 : 1 stoichiometry of the ligation partners at only 10 mM.

A promising feature of this ligation reaction for the future development of new protocols for peptide synthesis is the still unexplored scope of the reaction partners for chemoselective amide formation. For example, the reaction is not limited to hydroxylamines of the structure RNHOH. *O*-Alkyl substituted hydroxylamines can undergo the ligation reaction, a finding exploited in an elegant approach to the iterative coupling of enantiopure isoxazolidine monomers for an iterative synthesis of $\beta^3$-oligopeptides [130]. For these later reactions, the preferred solvent is water or mixtures of water and *t*BuOH. Acylated hydroxylamines (e.g., RNH–OBz) are also excellent substrates for ligations in either *N*,*N*-dimethylformamide (DMF) or water, but their use for α-peptides is currently limited by the propensity of these groups to undergo elimination.

## 13.6 Strategies for the Ligation of Multiple Fragments

The size limitation of SPPS has for a long time prohibited the total chemical synthesis of proteins. Chemoselective ligation techniques represent an important contribution to extending this limit. However, the size of ligation products that can be obtained from two synthetic peptides is still smaller than most natural proteins [131, 132]. Therefore, multiple segment ligation schemes must be developed to access fully synthetic proteins, in which ligation procedures can be used iteratively to construct larger structures. For NCL-type ligation schemes an additional protection group manipulation is required to assemble the fragments in the desired manner, since the central segment in ligations with three (or more) ligation partners has to contain an N-terminal Cys *and* a C-terminal thioester. To prohibit undesired polymerization and/or cyclization, either the N-terminal Cys or the C-terminal thioester of the central segment must be protected. In contrast to thioesters, orthogonal protecting groups are available for Cys moieties, which can be installed during the synthesis and removed after a ligation reaction.

### 13.6.1
### Synthetic Erythropoietin

A prominent example for such a ligation strategy is the four-segment ligation of the glycoprotein hormone erythropoietin (EPO), which stimulates the proliferation of erythroid cells (Scheme 13.24) [133, 134]. In order to mimic the natural glycosylation patterns of EPO, branched, negatively charged precision-length polymers were installed in the protein by oxime ligation. The N-terminal Cys residues of both central fragments were orthogonally protected by an Acm group, which can be removed by treatment with Hg(II) [135]. The process started with the ligation of the polymer modified C-terminal fragment **96** with the first central fragment followed by removal of the Acm group. Repetition of this process with the two remaining fragments yielded synthetic full-length EPO **97**. Since two of the three Cys residues installed at ligation sites represented mutations, they were masked by alkylation with bromoacetic acid (**48**), thus reinstalling analogs of the naturally occurring glutamic acids (see Section 13.3.1.2). The biological activity of the synthetic EPO **97** was similar to wild-type EPO; however, the synthetic hormone maintained a 2- to 3-fold higher plasma level for several hours when injected into rats, which could be contributed by the stabilizing polymers.

### 13.6.2
### Convergent Strategies for Multiple Fragment Ligations

The example mentioned above requires the stepwise assembly of individual fragments from the C- to N-terminus. In these, the N-terminus of one peptide segment has to be protected to avoid undesired cyclization by an intramolecular NCL. A kinetically controlled ligation strategy was introduced by Kent *et al.*, which is convergent and does not require the strict C $\rightarrow$ N assembly [136]. In principle, this approach exploits the different reactivities of aryl and alkyl thioesters in a convergent protein syntheses [137]. Alkyl thioesters are sufficiently unreactive and do not participate in ligation reactions when competing aryl thioesters are present nor do they undergo an intramolecular NCL with an unprotected N-terminal Cys residue. Nevertheless, they can be converted into reactive aryl thioesters by thiol–thioester exchange upon adding an excess of aryl thiols. These features are employed in kinetically controlled ligation procedures: A central peptide fragment **99** containing an N-terminal Cys residue and a C-terminal alkyl thioester is ligated at the N-terminal Cys residue with an aryl thioester **98**. The ligation product **100** is then converted into an aryl thioester by thioester exchange for a ligation reaction (Scheme 13.25), thereby allowing the ligation of multiple fragments in parallel. This approach was applied to the convergent synthesis of crambin and lysozyme (Scheme 13.25) [138]. For lysozyme, the 130-amino-acid enzyme was synthesized from four fragments, resulting into a protein with full enzymatic activity and an identical X-ray structure as compared to biologically produced lysozyme.

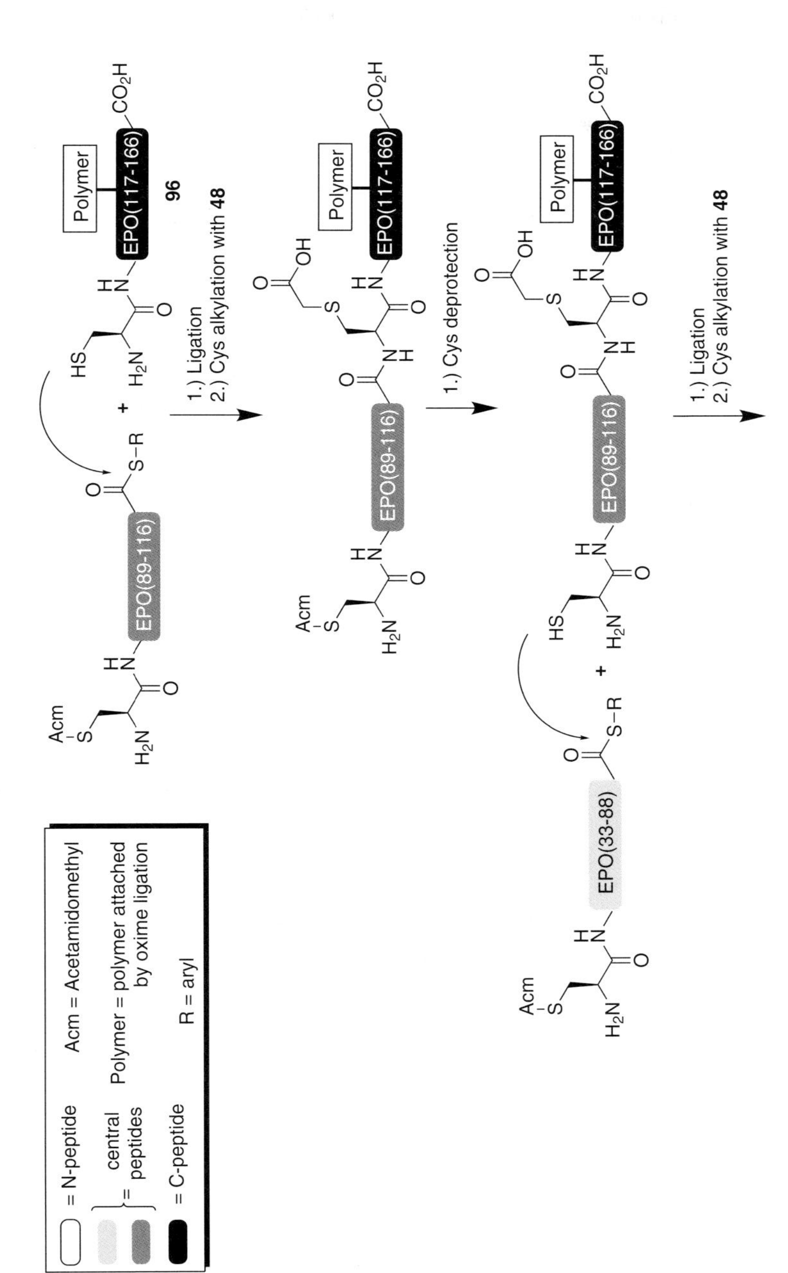


**Scheme 13.24** Total chemical synthesis of synthetic EPO.

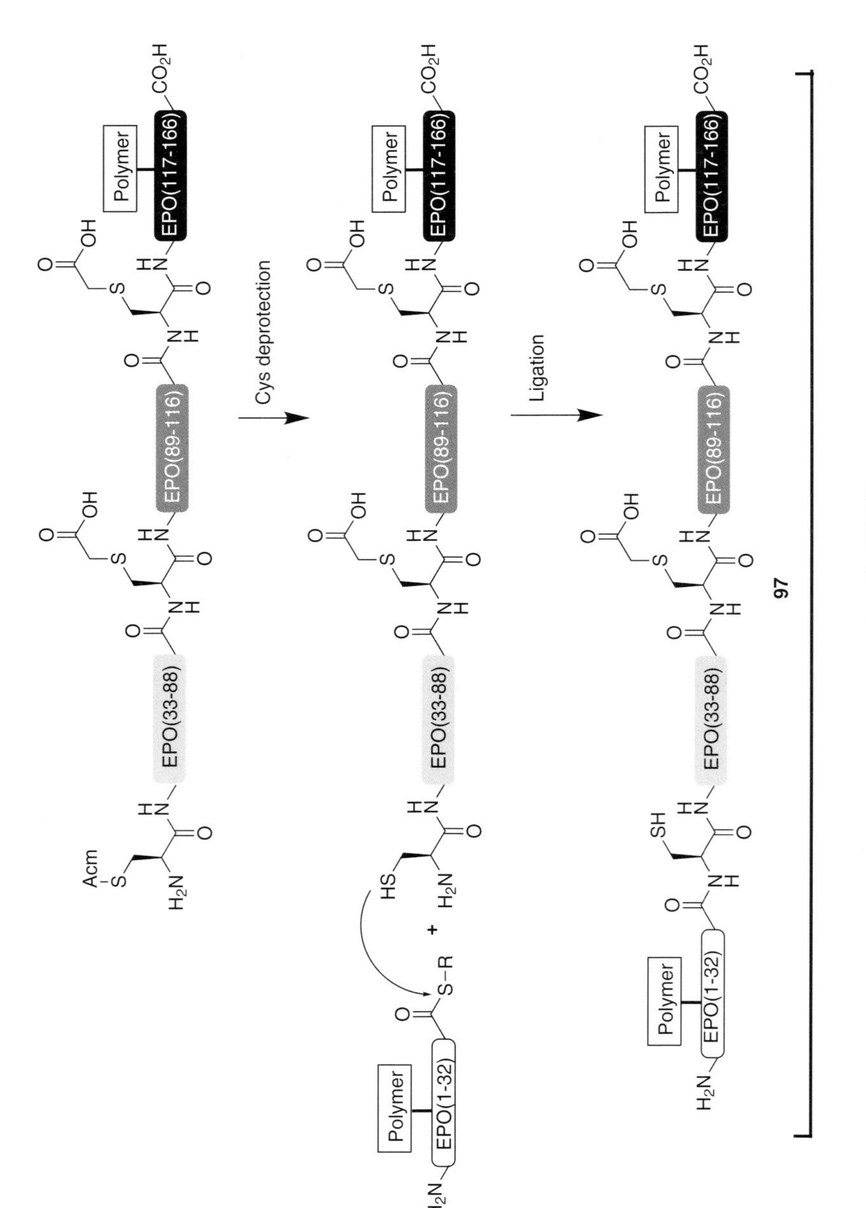


Scheme 13.24 *(Continued)*

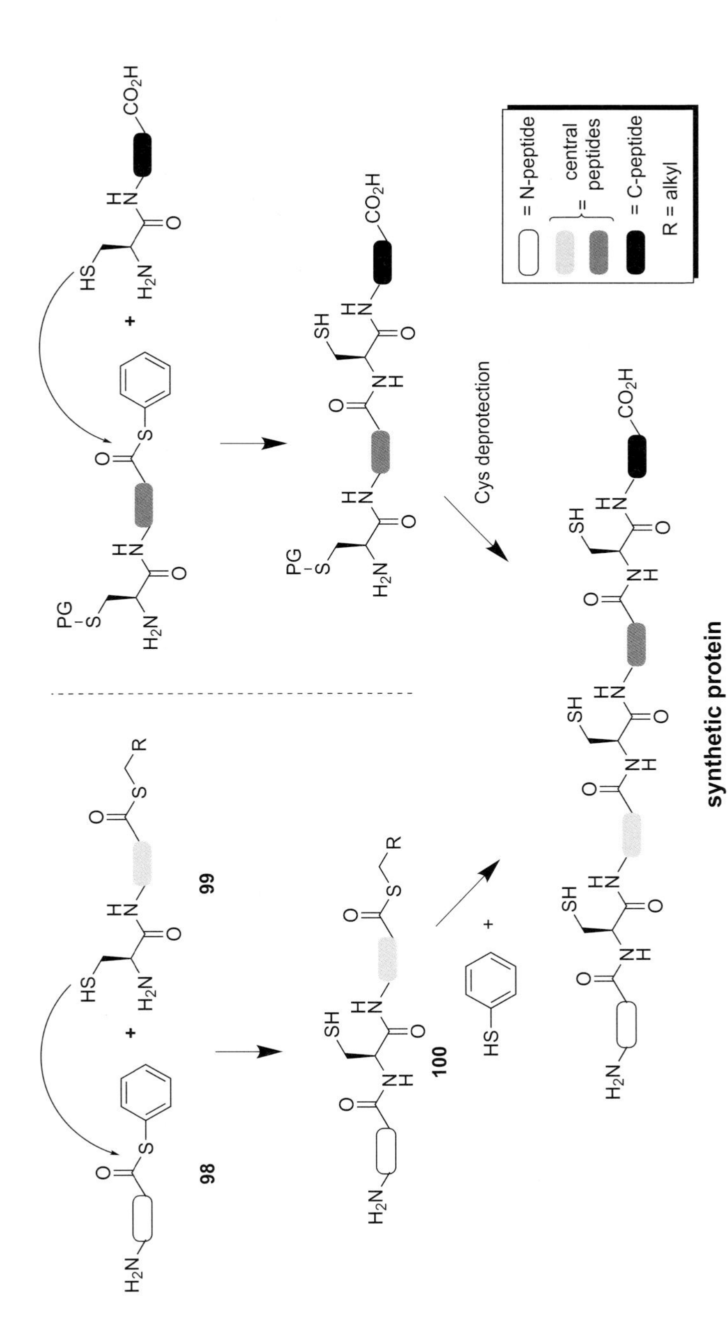


**Scheme 13.25** Convergent multiple fragment ligations via kinetically controlled ligation.

#### 13.6.2.1 Ubiquitylated Histone Proteins

Another impressive three-segment ligation strategy was developed by Muir *et al.* for the synthesis of ubiquitylated proteins [139–142]. The universal 76-amino-acid protein ubiquitin is attached to the ε-$NH_2$ groups of lysine residues via an isopeptide bond. The functions of this post-translational modification are manifold, including the triggering of proteosomal degradation or regulation of gene activity. In the latter case histone proteins are of particular interest, since they are essential for all eukaryotic organisms; they form a complex with DNA called chromatin, which packs DNA molecules of enormous size into the limited space of the cellular nuclei. The basic structural unit of chromatin is called nucleosome and consists of four pairs of the core histones H2A, H2B, H3, and H4, which form the protein scaffold around which the DNA is wrapped. In addition, histones also play an important role in the regulation of adjacent genes. These regulatory effects are mediated by post-translation modifications, including acetylation, methylation, and ubiquitylation, which can dynamically interact with each other. One example is the ubiquitylation of Lys120 of histone H2B, which triggers the methylation of Lys79 on histone H3. To uncover the mechanistic aspects for this modification cross-talk an EPL-based scheme for the synthesis of ubiquitylated peptides was developed which was later extended to homogeneously ubiquitylated recombinant histones proteins [139–141].

The first step in this process is the ligation of ubiquitin to the ε-$NH_2$ group of Lys120 in a synthetic C-terminal peptide of histone H2B (H2B-C, **101**, Scheme 13.26). Since Lys lacks a thiol adjacent to the ε-$NH_2$, which could perform the capture step, an auxiliary based strategy was employed. Therefore, a retrosynthetic cut was made between Gly76 and Gly75 of ubiquitin. The synthesis starts with the C-terminal H2B with an orthogonally trityl-protected K120. After selective deprotection, the K120 ε-$NH_2$ group was coupled with bromoacetic acid **48** to afford **102** before the disulfide protected auxiliary **103** (see Section 13.3.2.2) was reacted with the protected peptide to yield an auxiliary peptide **104**. After cleavage, peptide **105** was ligated to the ubiquitin thioester **106**, which was generated by EPL and contained all native ubiquitin residues except the C-terminal Gly76. This ligation yielded the ubiquitylated peptide **107**, in which the ligation auxiliary and the Cys protection group were simultaneously removed by UV irradiation. Afterwards, the resulting H2B-C peptide **108** can participate in a second ligation with a recombinant H2B-N thioester **109**, which was generated by thiolysis of an intein fusion protein. The ligated semisynthetic H2B **110** was desulfurized (see Section 13.3.1.1) to render the native Ala at the ligation site in the homogeneously ubiquitylated uH2B **111**.

In the following uH2B **111** was used with the other core histones to reconstitute nucleosomes and probe the cross-talk between the ubiquitylation of H2B at K120 and the methylation of H3 at K79. The authors could show that the methyl transferase Dot1, which is responsible for catalyzing the H3 K79 methylation, is actively stimulated by the ubiquitin moiety attached to the neighboring uH2B. These and further observations lead to the intriguing conclusion that all nucleosomes methylated at H3 K79 must at some point have been ubiquitylated on K120 of H2B.

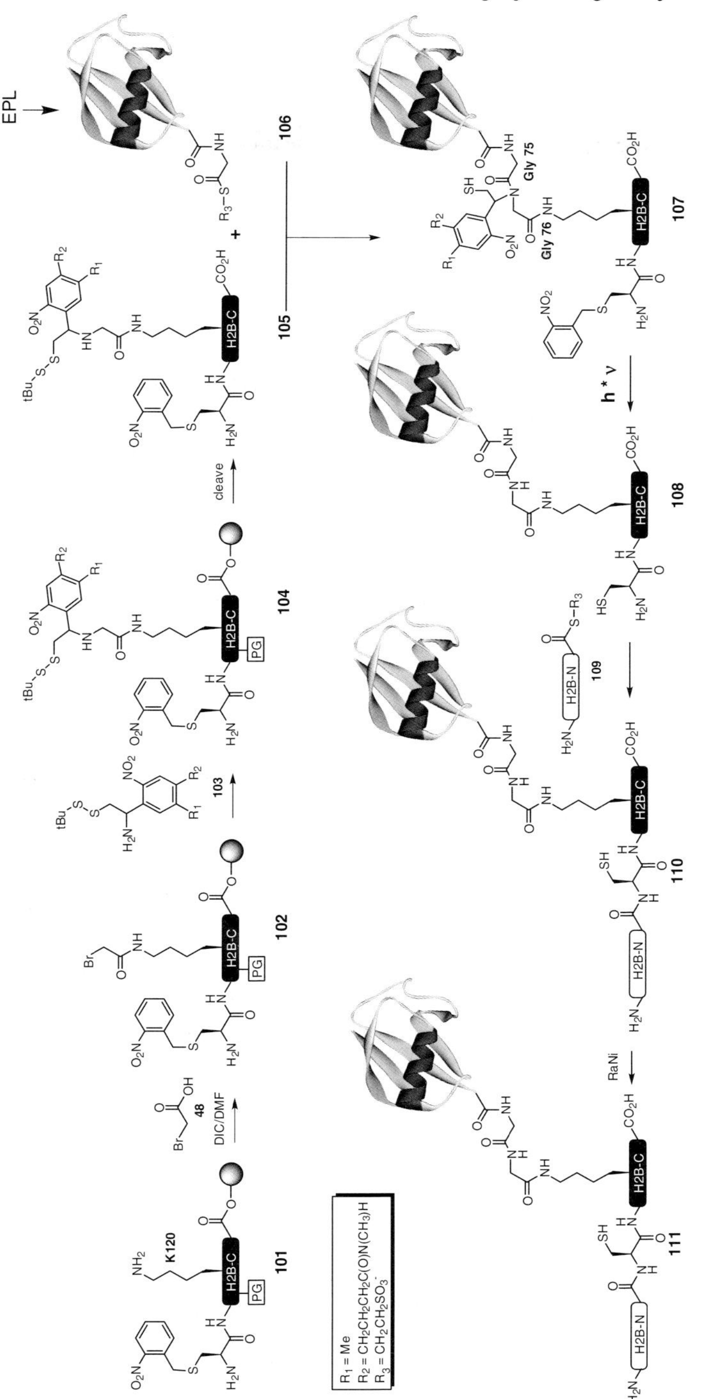


**Scheme 13.26** Semisynthesis of homogenously ubiquitylated histone H2B.

## References


1 (a) Walsh, C.T. (2003) In *Posttranslational Modifications of Proteins*, Roberts, Greenwood Village, CO; (b) Davis, B.G. (2004) *Science*, **303**, 480–482; (c) Wold, F. (1981) *Annual Review of Biochemistry*, **50**, 783–814; (d) Walsh, C.T., Garneau-Tsodikova, S., and Gatto Jr, G.J. (2005) *Angewandte Chemie International Edition*, **44**, 7342–7372; (2005) *Angewandte Chemie*, **117**, 7508–7539; (e) Gamblin, D.P., van Kasteren, S.I., Chalker, J.M., and Davis, B.G. (2008) *FEBS Journal*, **275**, 1949–1959; (f) Pellois, J.-P. and Muir, T.W. (2006) *Current Opinion in Chemical Biology*, **10**, 487–491.

2 For a recent reviews on protecting-group-free synthesis in the total synthesis of natural products, see: (a) Hoffmann, R.W. (2006) *Synthesis*, **21**, 3531–3541; (b) Young, I.S. and Baran, P.S. (2009) *Nature Chemistry*, **1**, 193–205; (c) Shenvi, R.A., O'Malley, D.P. and Baran, P.S. (2009) *Accounts of Chemical Research*, **42**, 530–541.

3 (a) Hackenberger, C.P.R. and Schwarzer, D. (2008) *Angewandte Chemie*, **120**, 10182–10228;(2008) *Angewandte Chemie International Edition*, **47**, 10030–10074; (b) Muir, T.W. (2003) *Annual Review of Biochemistry*, **72**, 249–289; (c) Pellois, J.-P. and Muir, T.W. (2006) *Current Opinion in Chemical Biology*, **10**, 487–491; (d) Durek, T. and Becker, C.F.W. (2005) *Biomolecular Engineering*, **22**, 153–172; (e) Nilsson, B.L., Soellner, M.B., and Raines, R.T. (2005) *Annual Review of Biophysics and Biomolecular Structure*, **34**, 91–118; (f) Seitz, O. (2003) *Organic Synthesis Highlights V*, 368–383; (g) Rauh, D. and Waldmann, H. (2007) *Angewandte Chemie*, **119**, 840–844;(2007) *Angewandte Chemie International Edition*, **46**, 826–829; (h) Haase, C. and Seitz, O. (2008) *Angewandte Chemie International Edition*, **47**, 1553–1556;(2008) *Angewandte Chemie*, **120**, 1575–1579; (i) Rademann, J. (2004) *Angewandte Chemie International Edition*, **43**, 4554–4556;(2004) *Angewandte Chemie*, **116**, 4654–4656; (j) Kent, S.B.H. (2009) *Chemical Society Reviews*, **38**, 338–351; (k) Hackenberger, C.P.R., Arndt, H.-D., and Schwarzer, D. (2010) *Chemie in unserer Zeit*, **44**, 198–206; (l) Sletten, E.M. and Bertozzi, C.R. (2009) *Angewandte Chemie International Edition*, **48**, 6974–6998;(2009) *Angewandte Chemie*, **121**, 7108–7133.

4 For a recent review concerning new strategies for the formation of amide bonds, see: Bode, J.W. (2006) *Current Opinion in Drug Discovery & Development*, **9**, 765–775.

5 (a) Bodanszky, M. (1984) In *Principles of Peptide Synthesis*, Springer, Berlin; (b) Sewald, N. and Jakubke, H.-D. (2002) In *Peptides: Chemistry and Biology*, Wiley-VCH Verlag GmbH, Weinheim.

6 Hojo, H., Matsumoto, Y., Nakahara, Y., Ito, E., Suzuki, Y., Suzuki, M., Suzuki, A., and Nakahara, Y. (2005) *Journal of the American Chemical Society*, **127**, 13720–13725.

7 Hirschmann, R., Nutt, R.F., Veber, D.F., Vitali, R.A., Varga, S.L., Jacob, T.A., Holly, F.W., and Denkewalter, R.G. (1969) *Journal of the American Chemical Society*, **91**, 507–508.

8 Nishiuchi, Y., Inui, T., Nishio, H., Bódi, J., Kimura, T., Tsuji, F.I., and Sakakibara, S. (1998) *Proceedings of the National Academy of Sciences of the United States of America*, **95**, 13549–13554.

9 Chen, G., Wan, Q., Tan, Z., Kan, C., Hua, Z., Ranganathan, K., and Danishefsky, S.J. (2007) *Angewandte Chemie International Edition*, **46**, 7383–7387.

10 The term C- and *N*-peptide is used in accordance with the terminology introduced by Tam, see also: Tam, J.P., Xu, J., and Eom, K.D. (2001) *Biopolymers*, **60**, 194–205.

11 (a) Merrifield, R.B. (1963) *Journal of the American Chemical Society*, **85**, 2149–2154;(b) Nobel lecture: Merrifield, R.B. (1985) *Angewandte Chemie (International Edition in English)*, **24**, 799–810;(1985) *Angewandte Chemie*, **97**, 801–812.

12 (a) Fischer, E. and Fourneau, E. (1901) *Berichte der Deutschen chemischen*

*Gesellschaft*, **34**, 2868–2877; (b) for an excellent historical overview on peptide synthesis and ligation strategies, see: Kimmerlin, T. and Seebach, D. (2005) *The Journal of Peptide Research*, **65**, 229–260.

13 (a) Kemp, D.S., Leung, S.-L. and Kerkman, D.J. (1981) *Tetrahedron Letters*, **22**, 181–184; (b) Kemp, D.S. and Kerkman, D.J. (1981) *Tetrahedron Letters*, **22**, 185–186; (c) Kemp, D.S. and Galakatos, N.G. (1986) *The Journal of Organic Chemistry*, **51**, 1821–1829; (d) Coltart, D.M. (2000) *Tetrahedron*, **56** 3449–3491.

14 (a) This method represents a extension of the previously reported "prior amine capture:" Kemp, D.S., Grattan, J.A., and Reczek, J. (1975) *The Journal of Organic Chemistry*, **40**, 3465–3466; (b) Kemp, D.S., Kerkman, D.J., Leung, S.-L., and Hanson, G. (1981) *The Journal of Organic Chemistry*, **46**, 490–498; (c) Kemp, D.S. (1981) *Biopolymers*, **20**, 1793–1804;(d) for a more recent extension of the amine capture strategy via quinolinium thioester salts, see: Leleu, S., Penhoat, M., Bouet, A., Dupas, G., Papamicaël, C., Marsais, F., and Levacher, V. (2005) *Journal of the American Chemical Society*, **127**, 15668–15669.

15 Kemp, D.S., Galakatos, N.G., Bowen, B., and Tan, K. (1986) *The Journal of Organic Chemistry*, **51**, 1829–1838.

16 (a) Fotouhi, N., Galakatos, N.G. and Kemp, D.S. (1989) *The Journal of Organic Chemistry*, **54**, 2803–2817; (b) Kemp, D.S. and Carey, R.I. (1991) *Tetrahedron Letters*, **32**, 2845–2848.

17 Kemp, D.S. and Carey, R.I. (1993) *The Journal of Organic Chemistry*, **58**, 2216–2222.

18 Wieland, T., Bokelmann, E., Bauer, L., Lang, H.U., Lau, H., and Schafer, W. (1953) *Liebigs Annalen der Chemie*, **583**, 129–149.

19 Dawson, P.E., Muir, T.W., Clark-Lewis, I., and Kent, S.B. (1994) *Science*, **266**, 776–779.

20 $N \rightarrow S$ shifts were reported under acidic conditions for the synthesis of (glyco-) peptide thioesters: Nagaike, F., Onuma, Y., Kanazawa, C., Hojo, H., Ueki, A., Nakahara, Y., and Nakahara, Y. (2006) *Organic Letters*, **8**, 4465–4468. These findings indicate that the $S \rightarrow N$ shift in the NCL leading to the amide bond cannot be generally regarded as fully irreversible. However, the rapid kinetics of the rearrangement result in a fast adjustment of the equilibrium and thereby formation of the thermodynamically favoured amide. In the literature the $S \rightarrow N$ acyl transfer is commonly regarded as irreversible which we will continue to do for consistency.

21 Hackeng, T.M., Griffin, J.H., and Dawson, P.E. (1999) *Proceedings of the National Academy of Sciences of the United States of America*, **96**, 10068–10073.

22 Dawson, P.E., Churchill, M.J., Ghadiri, M.R., and Kent, S.B.H. (1997) *Journal of the American Chemical Society*, **119**, 4325–4329.

23 For a systematic study, see: Johnson, E.C.B. and Kent, S.B.H. (2006) *Journal of the American Chemical Society*, **128**, 6640–6646.

24 Offord, R.E. (1969) *Nature*, **221**, 37–40.

25 Borras, F. and Offord, R.E. (1970) *Nature*, **227**, 716–718.

26 Erlanson, D.A., Chytil, M., and Verdine, G.L. (1996) *Chemistry & Biology*, **3**, 981–991.

27 Shogren-Knaak, M.A. and Peterson, C.L. (2004) *Methods in Enzymology*, **375**, 62–76.

28 Tolbert, T.J., Franke, D., and Wong, C.-H. (2005) *Bioorganic and Medicinal Chemistry*, **13**, 909–915.

29 Tolbert, T.J. and Wong, C.-H. (2002) *Angewandte Chemie International Edition*, **41**, 2171–2174;(2002) *Angewandte Chemie*, **114**, 2275–2278.

30 (a) Hackenberger, C.P.R., Chen, M.M., and Imperiali, B. (2006) *Bioorganic and Medicinal Chemistry*, **14**, 5043–5048; (b) Mathys, S., EvansJr, T.C., Chute, I.C., Wu, H., Chong, S., Benner, J., Liu, X.-Q., and Xu, M.-Q. (1999) *Gene*, **231**, 1–13.

31 Macmillan, D. and Arham, L. (2004) *Journal of the American Chemical Society*, **126**, 9530–9531.

32 Gentle, I.E., De Souza, D.P., and Baca, M. (2004) *Bioconjugate Chemistry*, **15**, 658–663.

**33** Muir, T.W., Sondhi, D., and Cole, P.A. (1998) *Proceedings of the National Academy of Sciences of the United States of America*, **95**, 6705–6710.

**34** Evans, T.C., Benner, J., and Xu, M.Q. (1998) *Protein Science*, **7**, 2256–2264.

**35** Muir, T.W. (2003) *Annual Review of Biochemistry*, **72**, 249–289.

**36** David, R., Richter, M.P.O., and Beck-Sickinger, A.G. (2004) *European Journal of Biochemistry*, **271**, 663–677.

**37** Noren, C.J., Wang, J., and Perler, F.B. (2000) *Angewandte Chemie International Edition*, **39**, 450–466;(2000) *Angewandte Chemie*, **112**, 458–476.

**38** Saleh, L. and Perler, F.B. (2006) *Chemical Record*, **6**, 183–193.

**39** Inteins are evolutionarily related to the autoprocessing domain of hedgehog proteins that utilize this process in an intermolecular fashion to link a cholesterol moiety to the C-terminus of the hedgehog signaling domain; reviewed in: Mann, R.K. and Beachy, P.A. (2000) *Biochimica et Biophysica Acta*, **1529**, 188–202.

**40** Wu, H., Hu, Z., and Liu, X.-Q. (1998) *Proceedings of the National Academy of Sciences of the United States of America*, **95**, 9226–9231.

**41** Giriat, I. and Muir, T.W. (2003) *Journal of the American Chemical Society*, **125**, 7180–7181.

**42** Sun, W., Yang, J., and Liu, X.-Q. (2004) *The Journal of Biological Chemistry*, **279**, 35281–35286.

**43** Brenzel, S., Kurpiers, T., and Mootz, H.D. (2006) *Biochemistry*, **45**, 1571–1578.

**44** Southworth, M.W., Adam, E., Panne, D., Byer, R., Kautz, R., and Perler, F.B. (1998) *The EMBO Journal*, **17**, 918–926.

**45** Mills, K.V., Lew, B.M., Jiang, S.-Q., and Paulus, H. (1998) *Proceedings of the National Academy of Sciences of the United States of America*, **95**, 3543–3548.

**46** (a) Ludwig, C., Pfeiff, M., Linne, U., and Mootz, H.D. (2006) *Angewandte Chemie International Edition*, **45**, 5218–5221; (2006) *Angewandte Chemie*, **118**, 5343–5347; (b) Mootz, H.D. (2009) *ChemBioChem*, **10**, 2579–2589.

**47** Appleby, J.H., Zhou, K., Volkmann, G., and Liu, X.-Q. (2009) *The Journal of Biological Chemistry*, **284**, 6194–6199.

**48** McCaldon, P. and Argos, P. (1988) *Proteins*, **4**, 99–122.

**49** Lianshan, Z. and Tam, J.P. (1997) *Tetrahedron Letters*, **38**, 3–6.

**50** For a review on the incorporation of Sec into proteins, see: Hondal, R.J. (2005) *Protein and Peptide Letters*, **12**, 757–764.

**51** (a) Quaderer, R., Sewing, A. and Hilvert, D. (2001) *Helvetica Chimica Acta*, **84**, 1197–1206; (b) Gieselman, M.D., Xie, L. and van der Donk, W.A. (2001) *Organic Letters*, **3**, 1331–1334.

**52** (a) Gieselman, M.D., Zhu, Y., Zhou, H., Galonic, D., and van der Donk, W.A. (2002) *ChemBioChem*, **3**, 709–716; (b) Roelfes, G. and Hilvert, D. (2003) *Angewandte Chemie International Edition*, **42**, 2275–2277; (2003) *Angewandte Chemie*, **115**, 2377–2379; (c) Hondal, R.J., Nilsson, B.L., and Raines, R.T. (2001) *Journal of the American Chemical Society*, **123**, 5140–5141.

**53** (a) Berry, S.M., Gieselman, M.D., Nilges, M.J., van der Donk, W.A., and Lu, Y. (2002) *Journal of the American Chemical Society*, **124**, 2084–2085; (b) Ralle, M., Berry, S.M., Nilges, M.J., Gieselman, M.D., van der Donk, W.A., Lu, Y., and Blackburn, N.J. (2004) *Journal of the American Chemical Society*, **126**, 7244–7256.

**54** (a) For reviews concerning the chemical modification of proteins, see: Qi, D., Tann, C.-M., Haring, D., and Distefano, M.D. (2001) *Chemical Reviews*, **101**, 3081–3111; (b) Tann, C.-M., Qi, D. and Distefano, M.D. (2001) *Current Opinion in Chemical Biology*, **5**, 696–704; (c) Hodgson, D.R.W. and Sanderson, J.M. (2004) *Chemical Society Reviews*, **33**, 422–430; (d) Davis, B.G. (2003) *Current Opinion in Biotechnology*, **14**, 379–386; for the first examples of chemical modifications of amino acids in active sites, see: (e) Neet, K.E. and Koshland, D.E. (1966) *Proceedings of the National Academy of Sciences of the United States of America*, **56**, 1606–1611; (f) Polgar, L. and Bender, M.L. (1966) *Journal of the American Chemical Society*, **88**, 3153–3154; (g) Clark, P.I. and Lowe, G.

(1977) *Journal of the Chemical Society. Chemical Communications*, 923–924.

55 Besides NCL products also Cys in simple expressed proteins can be converted with the following reactions into other amino acids, thereby generating enzymes with improved enzymatic function or with natural protein modifications. For the use of site-specific modifications in the directed molecular evolution of enzyme catalysts, see the following reviews: (a) Reetz, M.T. (2002) *Tetrahedron*, **58**, 6595–6602; (b) Reetz, M.T. (2004) *Proceedings of the National Academy of Sciences of the United States of America*, **101**, 5716–5722. For a recent example on a site-specific glycosylation using this approach, see: van Kasteren, S.I., Kramer, H.B., Jensen, H.H., Campbell, S.J., Kirkpatrick, J., Oldham, N.J., Anthony, D.C., and Davis, B.G. (2007) *Nature*, **446**, 1105–1109.

56 Perlstein, M.T., Atassi, M.Z., and Cheng, S.H. (1971) *Biochimica et Biophysica Acta*, **236**, 174–182.

57 Yan, L.Z. and Dawson, P.E. (2001) *Journal of the American Chemical Society*, **123**, 526–533.

58 (a) Brik, A., Yang, Y.-Y., Ficht, S., and Wong, C.-H. (2006) *Journal of the American Chemical Society*, **128**, 5626–5627; (b) Bang, D., Makhatadze, G.I., Tereshko, V., Kossiakoff, A.A. and Kent, S.B. (2005) *Angewandte Chemie International Edition*, **44**, 3852–3856; (2005) *Angewandte Chemie*, **117**, 3920–3924; (c) Bang, D., Gribenko, A.V., Tereshko, V., Kossiakoff, A.A., Kent, S.B., and Makhatadze, G.I. (2006) *Nature Chemical Biology*, **2**, 139–143; (d) Bayro, M.J., Mukhopadhyay, J., Swapna, G.V.T., Huang, J.Y., Ma, L.-C., Sineva, E., Dawson, P.E., Montelione, G.T. and Ebright, R.H. (2003) *Journal of the American Chemical Society*, **125**, 12382–12383.

59 Pentelute, B.L. and Kent, S.B.H. (2007) *Organic Letters*, **9**, 687–690.

60 He, S., Bauman, D., Davis, J.S., Loyola, A., Nishioka, K., Gronlund, J.L., Reinberg, D., Meng, F., Kelleher, N., and McCafferty, D.G. (2003) *Proceedings of the National Academy of Sciences of the United States of America*, **100**, 12033–12038.

61 (a) Nishide, K., Shigeta, Y., Obata, K., Inoue, T. and Node, M. (1996) *Tetrahedron Letters*, **37**, 2271–2274; (b) Node, M., Nishide, K., Shigeta, Y., Obata, K., Shiraki, H., and Kunishige, H. (1997) *Tetrahedron*, **53**, 12883–12894.

62 Wan, Q. and Danishefsky, S.J. (2007) *Angewandte Chemie International Edition*, **46**, 9248–9252;(2007) *Angewandte Chemie*, **119**, 9408–9412.

63 Chiang, K.P., Jensen, M.S., McGinty, R.K., and Muir, T.W. (2009) *ChemBioChem*, **10**, 2182–2187.

64 Crich, D. and Banerjee, A. (2007) *Journal of the American Chemical Society*, **129**, 10064–10065.

65 (a) Haase, C., Rohde, H. and Seitz, O. (2008) *Angewandte Chemie International Edition*, **47**, 6807–6810;(2008) *Angewandte Chemie*, **120**, 6912–6915; (b) Chen, J., Wan, Q., Yuan, Y., Zhu, J. and Danishefsky, S.J. (2008) *Angewandte Chemie International Edition*, **47**, 8521–8524;(2008) *Angewandte Chemie*, **120**, 8649–8652.

66 (a) Kenyon, G.L. and Bruice, T.W. (1977) *Methods in Enzymology*, **47**, 407–430; (b) Wynn, R., and Richards, F.M. (1995) *Methods in Enzymology*, **251**, 351–356; (c) DeSantis, G., Berglund, P., Stabile, M.R., Gold, M., and Jones, J.B. (1998) *Biochemistry*, **37**, 5968–5973; (d) DeSantis, G., Shang, X., and Jones, J.B. (1999) *Biochemistry*, **38**, 13391–13397.

67 For the alkylation of homocysteine derivatives (Hcy) to Met residues in analogous NCL ligations, see: Tam, J.P. and Yu, Q. (1998) *Biopolymers*, **46**, 319–327.

68 (a) For a chemoselective functionalization of Cys residues to disulfide-linked glycoconjugates with glycomethane- or phenylthiosulfonates, see: Davis, B.G., Maughan, M.A.T., Green, M.P., Ullman, A. and Jones, J.B. (2000) *Tetrahedron: Asymmetry*, **11**, 245–262; (b) Gamblin, D.P., Garnier, P., Ward, S.J., Oldham, N.J., Fairbanks, A.J., and Davis, B.G. (2003) *Organic & Biomolecular Chemistry*, **1**, 3642–3644; for a strategy employing

glycoselenenylsulfides, see: (c) Gamblin, D.P., Garnier, P., van Kasteren, S., Oldham, N.J., Fairbanks, A.J., and Davis, B.G. (2004) *Angewandte Chemie*, **116**, 846–851;(2004) *Angewandte Chemie International Edition*, **43**, 828–833; (d) Bernardes, G.J.L., Gamblin, D.P. and Davis, B.G. (2006) *Angewandte Chemie*, **118**, 4111–4115;(2006) *Angewandte Chemie International Edition*, **45**, 4007–4011.

**69** Bernardes, G.J.L., Grayson, E.L.J., Thompson, S., Chalker, J.M., Errey, J.C., Oualid, F.E., Claridge, T.D.W., and Davis, B.G. (2008) *Angewandte Chemie*, **120**, 2276–2279;(2008) *Angewandte Chemie International Edition*, **47**, 2244–2247.

**70** Bochar, D.A., Tabernero, L., Stauffacher, C.V., and Rodwell, V.W. (1999) *Biochemistry*, **38**, 8879–8883.

**71** Kochendoerfer, G.G., Chen, S.-Y., Mao, F., Cressman, S., Traviglia, S., Shao, H., Hunter, C.L., Low, D.W., Cagle, E.N., Carnevali, M., Gueriguian, V., Keogh, P.J., Porter, H., Stratton, S.M., Wiedeke, M.C., Wilken, J., Tang, J., Levy, J.J., Miranda, L.P., Crnogorac, M.M., Kalbag, S., Botti, P., Schindler-Horvat, J., Savatski, L., Adamson, J.W., Kung, A., Kent, S.B., and Bradburne, J.A. (2003) *Science*, **299** 884–887.

**72** (a) Smith, H.B. and Hartman, F.C. (1988) *The Journal of Biological Chemistry*, **263**, 4921–4925; (b) Hopkins, C.E., O'Connor, P.B., Allen, K.N., Costello, C.E. and Tolan, D.R. (2002) *Prot Sci*, **11**, 1591–1599.

**73** Okamoto, R. and Kajihara, Y. (2008) *Angewandte Chemie*, **120**, 5482–5486; (2008) *Angewandte Chemie International Edition*, **47**, 5402–5406.

**74** (a) For an ester to amide migration under slightly basic conditions in the synthesis of difficult peptide sequences, see: Coin, I., Dölling, R., Krause, E., Bienert, M., Beyermann, M., Sferdean, C.D. and Carpino, L.A. (2006) *The Journal of Organic Chemistry*, **71**, 6171–6177; (b) Carpino, L.A., Krause, E., Sferdean, C.D., Schümann, M., Fabian, H., Bienert, M. and Beyermann, M. (2004) *Tetrahedron Letters*, **45**, 7519–7523; (c) Sohma, Y., Sasaki, M., Hayashi, Y., Kimura, T. and Kiso, Y. (2004) *Chemical Communications*, 124–125; (d) Mutter, M., Chandravarkar, A., Boyat, C., Lopez, J., Dos Santos, S., Mandal, B., Mimna, R., Murat, K., Patiny, L., Saucède, L. and Tuchscherer, G. (2004) *Angewandte Chemie*, **116**, 4267–4273;(2004) *Angewandte Chemie International Edition*, **43**, 4172–4178.

**75** Naturally occurring Met residues which would interfere with this strategy can be protected as their sulfoxides and reconverted into the thioethers. See also: Hackenberger, C.P.R. (2006) *Organic & Biomolecular Chemistry*, **4**, 2291–2295.

**76** For a recent highlight on non-Cys-containing amide bond ligations, see: Macmillan, D. (2006) *Angewandte Chemie International Edition*, **45**, 7668–7672; (2006) *Angewandte Chemie*, **118**, 7830–7834.

**77** Canne, L.E., Bark, S.J., and Kent, S.B.H. (1996) *Journal of the American Chemical Society*, **118**, 5891–5896.

**78** (a) Meutermans, W.D.F., Golding, S.W., Bourne, G.T., Miranda, L.P., Dooley, M.J., Alewood, P.F. and Smythe, M.L. (1999) *Journal of the American Chemical Society*, **121**, 9790–9796; (b) Shao, Y., Lu, W. and Kent, S.B.H. (1998) *Tetrahedron Letters*, **39**, 3911–3914.

**79** Bark, S.J. and Kent, S.B.H. (1999) *FEBS Letters*, **460**, 67–76.

**80** (a) Botti, P., Carrasco, M.R. and Kent, S.B.H. (2001) *Tetrahedron Letters*, **42**, 1831–1833; (b) Marinzi, C., Bark, S.J., Offer, J., and Dawson, P.E. (2001) *Bioorganic and Medicinal Chemistry*, **9**, 2323–2328.

**81** (a) Clive, D.L.J., Hisaindee, S., and Coltart, D.M. (2003) *The Journal of Organic Chemistry*, **68**, 9247–9254; (b) Tchertchian, S., Hartley, O., and Botti, P. (2004) *The Journal of Organic Chemistry*, **69**, 9208–9214; (c) Macmillan, D. and Anderson, D.W. (2004) *Organic Letters*, **6**, 4659–4662.

**82** Kawakami, T. and Aimoto, S. (2003) *Tetrahedron Letters*, **44**, 6059–6061.

**83** Marinzi, C., Offer, J., Longhi, R., and Dawson, P.E. (2004) *Bioorganic and Medicinal Chemistry*, **12**, 2749–2757.

**84** (a) Chatterjee, C., McGinty, R.K., Pellois, J.-P., and Muir, T.W. (2007) *Angewandte Chemie International Edition*, **46**, 2814–2818;(2007) *Angewandte Chemie*, **119**, 2872–2876; for a highlight, including the relevance of ubiquilated proteins, see: (b) Hackenberger, C.P.R. (2007) *ChemBioChem*, **8**, 1221–1223.

**85** McGinty, R.K., Kim, J., Chatterjee, C., Roeder, R.G., and Muir, T.W. (2008) *Nature*, **453**, 812–816.

**86** Offer, J. and Dawson, P.E. (2000) *Organic Letters*, **2**, 23–26.

**87** Offer, J., Boddy, C.N.C., and Dawson, P.E. (2002) *Journal of the American Chemical Society*, **124**, 4642–4646.

**88** Low, D.W., Hill, M.G., Carrasco, M.R., Kent, S.B.H., and Botti, P. (2001) *Proceedings of the National Academy of Sciences of the United States of America*, **98**, 6554–6559.

**89** Wu, B., Chen, J., Warren, J.D., Chen, G., Hua, Z., and Danishefsky, S.J. (2006) *Angewandte Chemie*, **118**, 4222–4231; (2006) *Angewandte Chemie International Edition*, **45**, 4116–4125.

**90** This rearrangement was previously observed and utilized, see: Vizzavona, J., Dick, F., and Vorherr, T. (2002) *Bioorganic & Medicinal Chemistry Letters*, **12**, 1963–1965.For an application of this reaction in the synthesis of thioesters, see: Kawakami, T., Sumida, M., Nakamura, K., Vorherr, T., and Aimoto, S. (2005) *Tetrahedron Letters*, **46**, 8805–8807.

**91** Brik, A. and Wong, C.-H. (2007) *Chemistry – A European Journal*, **13**, 5670–5675.

**92** (a) Brik, A., Yang, Y.-Y., Ficht, S., and Wong, C.-H (2006). *Journal of the American Chemical Society*, **128**, 5626–5627; (b) Yang, Y.-Y., Ficht, S., Brik, A., and Wong, C.-H. (2007) *Journal of the American Chemical Society*, **129**, 7690–7701.

**93** Brik, A., Ficht, S., Yang, Y.-Y., Bennett, C.S., and Wong, C.-H. (2006) *Journal of the American Chemical Society*, **128**, 15026–15033.

**94** Payne, R.J., Ficht, S., Tang, S., Brik, A., Yang, Y.-Y., Case, D.A., and Wong, C.-H. (2007) *Journal of the American Chemical Society*, **129**, 13527–13536.

**95** Ficht, S., Payne, R.J., Brik, A., and Wong, C.-H. (2007) *Angewandte Chemie International Edition*, **46**, 5975–5979; (2007) *Angewandte Chemie*, **119**, 6079–6083.

**96** (a) Lutsky, M.-Y., Nepomniaschiy, N. and Brik, A. (2008) *Chem Commun*, 1229–1231; (b) Kumar, K.S.A., Harpaz, Z., Haj-Yahya, M., and Brik, A. (2009) *Bioorganic & Medicinal Chemistry Letters*, **19**, 3870–3874.

**97** Haase, C. and Seitz, O. (2009) *European Journal of Organic Chemistry*, 2096–2101.

**98** Payne, R.J., Ficht, S., Greenberg, W.A., and Wong, C.-H. (2008) *Angewandte Chemie International Edition*, **47**, 4411–4415;(2008) *Angewandte Chemie*, **120**, 4483–4487.

**99** (a) Nilsson, B.L., Kiessling, L.L. and Raines, R.T. (2000) *Organic Letters*, **2**, 1939–1941; (b) Saxon, E., Armstrong, J.I. and Bertozzi, C.R. (2000) *Organic Letters*, **2**, 2141–2143; (c) Nilsson, B.L., Kiesling, L.L. and Raines, R.T. (2001) *Organic Letters*, **3**, 9–12.

**100** For a review, see: Köhn, M. and Breinbauer, R. (2004) *Angewandte Chemie*, **116**, 3168–3178;(2004) *Angewandte Chemie International Edition*, **43**, 3106–3116.

**101** Staudinger, H. and Meyer, J. (1919) *Helvetica Chimica Acta*, **2**, 635–646.

**102** (a) Saxon, E. and Bertozzi, C.R. (2000) *Science*, **287**, 2007–2010; (b) Agard, N.J., Baskin, J.M., Prescher, J.A., Lo, A. and Bertozzi, C.R. (2006) *Chemical Biology*, **1**, 644–648.

**103** Soellner, M.B., Tam, A., and Raines, R.T. (2006) *The Journal of Organic Chemistry*, **71**, 9824–9830.

**104** Soellner, M.B., Nilsson, B.L., and Raines, R.T. (2006) *Journal of the American Chemical Society*, **128**, 8820–8828.

**105** (a) Soellner, M.B., Dickson, K.A., Nilsson, B.L., and Raines, R.T. (2003) *Journal of the American Chemical Society*, **125**, 11790–11791; (b) Köhn, M., Wacker, R., Peters, C., Schröder, H., Soulère, L., Breinbauer, R., Niemeyer, C.M., and Waldmann, H. (2003) *Angewandte Chemie*, **115**, 6010–6014;(2003) *Angewandte Chemie International Edition*,

**42**, 5830–5834; (c) Watzke, A., Gutierrez-Rodriguez, M., Köhn, M., Wacker, R., Schroeder, H., Breinbauer, R., Kuhlmann, J., Alexandrov, K., Niemeyer, C.M., Goody, R.S. and Waldmann, H. (2006) *Bioorganic and Medicinal Chemistry*, **14**, 6288–6306; (d) Watzke, A., Köhn, M., Gutierrez-Rodriguez, M., Wacker, R., Schröder, H., Breinbauer, R., Kuhlmann, J., Alexandrov, K., Niemeyer, C.M., Goody, R.S. and Waldmann, H. (2006) *Angewandte Chemie*, **118**, 1436–1440;Watzke, A. (2006) *Angewandte Chemie International Edition*, **45** 1408–1412.

**106** (a) Nilsson, B.L., Hondal, R.J., Soellner, M.B., and Raines, R.T. (2003) *Journal of the American Chemical Society*, **125**, 5268–5269; (b) Liu, L., Hong, Z.-Y., and Wong, C.-H. (2006) *ChemBioChem*, **7**, 429–432; (c) Merkx, R., Rijkers, D.T.S., Kemmink, J., and Liskamp, R.M.J. (2003) *Tetrahedron Letters*, **44**, 4515–4518.

**107** Grandjean, C., Boutonnier, A., Guerreiro, C., Fournier, J.-M., and Mulard, L.A. (2005) *The Journal of Organic Chemistry*, **70**, 7123–7132.

**108** (a) David, O., Meester, W.J.N., Bieräugel, H., Schoemaker, H.E., Hiemstra, H., and van Maarseveen, J.H. (2003) *Angewandte Chemie*, **115**, 4509–4511;(2003) *Angewandte Chemie International Edition*, **42**, 4373–4375; (b) Masson, G., den Hartog, T., Schoemaker, H.E., Hiemstra, H. and van Maarseveen, J.H. (2006) *Synlett*, 865–868.

**109** Kleineweischede, R. and Hackenberger, C.P.R. (2008) *Angewandte Chemie*, **120**, 6073–6077;(2008) *Angewandte Chemie International Edition*, **47**, 5984–5988.

**110** Tam, A., Soellner, M.B., and Raines, R.T. (2007) *Journal of the American Chemical Society*, **129**, 11421–11430.

**111** (a) Liu, C.-F. and Tam, J.P. (1994) *Journal of the American Chemical Society*, **116**, 4149–4153; (b) Liu, C.F. and Tam, J.P. (1994) *Proceedings of the National Academy of Sciences of the United States of America*, **91**, 6584–6588.

**112** Tam, J.P. and Miao, Z. (1999) *Journal of the American Chemical Society*, **121**, 9013–9022.

**113** Miao, Z. and Tam, J.P. (2000) *Journal of the American Chemical Society*, **122**, 4253–4260.

**114** The term "bioorthogonality" was introduced by C. R. Bertozzi in the following paper: Lemieux, G.A., de Graffenried, C.L., and Bertozzi, C.R. (2003) *Journal of the American Chemical Society*, **125**, 4708–4709. For a subsequent review, see: Dube, D.H. and Bertozzi, C.R. (2003) *Current Opinion in Chemical Biology*, **7**, 616–625.

**115** Sletten, E.M. and Bertozzi, C.R. (2009) *Angewandte Chemie*, **121**, 7108–7133; (2009) *Angewandte Chemie International Edition*, **48**, 6974–6998.

**116** (a) Prescher, J.A. and Bertozzi, C.R. (2005) *Nature Chemical Biology*, **1**, 13–21; (b) Hang, H.C. and Bertozzi, C.R. (2001) *Accounts of Chemical Research*, **34**, 727–736.

**117** Serwa, R., Wilkening, I., del Signore, G., Mühlberg, M., Claußnitzer, I., Weise, C., Gerrits, M., and Hackenberger, C.P.R. (2009) *Angewandte Chemie*, **121**, 8382–8387;(2009) *Angewandte Chemie International Edition*, **48**, 8234–8239.

**118** Rosen, T., Lico, I.M., and Chu, D.T.W. (1988) *The Journal of Organic Chemistry*, **53**, 1580–1582.

**119** Shangguan, N., Katukojvala, S., Greenberg, R., and Williams, L.J. (2003) *Journal of the American Chemical Society*, **125**, 7754–7755.For a similar process involving selenocarboxylates instead of thio acids, see: (a) Wu, X. and Hu, L. (2005) *Tetrahedron Letters*, **46**, 8401–8405; (b) Wu, X. and Hu, L. (2007) *The Journal of Organic Chemistry*, **72**, 765–774.

**120** Electron-poor azides can form thiatriazolines via a stepwise process. For a detailed mechanistic discussion, see: Kolakowski, R.V., Shangguan, N., Sauers, R.R., and Williams, L.J. (2006) *Journal of the American Chemical Society*, **128**, 5695–5702.

**121** Fazio, F. and Wong, C.-H. (2003) *Tetrahedron Letters*, **44**, 9083–9085.

**122** (a) Merkx, R., Brouwer, A.J., Rijkers, D.T.S., and Liskamp, R.M.J. (2005) *Organic Letters*, **7**, 1125–1128; (b) Barlett, K.N., Kolakowski, R.V., Katukojvala, S., and Williams, L.J. (2006) *Organic Letters*, **8**, 823–826.

**123** (a) Messeri, T., Sternbach, D.D., and Tomkinson, N.C.O. (1998) *Tetrahedron Letters*, **39**, 1669–1672; (b) Messeri, T., Sternbach, D.D., and Tomkinson, N.C.O. (1998) *Tetrahedron Letters*, **39**, 1673–1676.

**124** Crich, D., Sana, K., and Guo, S. (2007) *Organic Letters*, **9**, 4423–4426.

**125** Crich, D. and Sharma, I. (2009) *Angewandte Chemie International Edition*, **48**, 2355–2358.

**126** Crich, D. and Sharma, I. (2009) *Angewandte Chemie International Edition*, **48**, 7591–7594.

**127** Bode, J.W., Fox, R.M., and Baucom, K.D. (2006) *Angewandte Chemie International Edition*, **45**, 1248–1252;(2006) *Angewandte Chemie*, **118**, 1270–1274.

**128** (a) Ju, L. and Bode, J.W. (2009) *Organic & Biomolecular Chemistry*, **7**, 2259–2264; (b) Ju, L., Lippert, A.R. and Bode, J.W. (2008) *Journal of the American Chemical Society*, **130**, 4253–4255.

**129** Fukuzumi, T. and Bode, J.W. (2009) *Journal of the American Chemical Society*, **131**, 3864–3865.

**130** Carrillo, N., Davalos, E.A., Russak, J.A., and Bode, J.W. (2006) *Journal of the American Chemical Society*, **128**, 1452–1453.

**131** Muir, T.W., Dawson, P.E., and Kent, S.B.H. (1997) *Methods in Enzymology*, **289**, 266–298.

**132** Beligere, G.S. and Dawson, P.E. (1999) *Biopolymers*, **51**, 363–369.

**133** Kochendoerfer, G.G., Chen, S.Y., Mao, F., Cressman, S., Traviglia, S., Shao, H., Hunter, C.L., Low, D.W., Cagle, E.N., Carnevali, M., Gueriguian, V., Keogh, P.J., Porter, H., Stratton, S.M., Wiedeke, M.C., Wilken, J., Tang, J., Levy, J.J., Miranda, L.P., Crnogorac, M.M., Kalbag, S., Botti, P., Schindler-Horvat, J., Savatski, L., Adamson, J.W., Kung, A., Kent, S.B., and Bradburne, J.A. (2003) *Science*, **299** 884–887.

**134** Jacobs, K., Shoemaker, C., Rudersdorf, R., Neill, S.D., Kaufman, R.J., Mufson, A., Seehra, J., Jones, S.S., Hewick, R., Fritsch, E.F., Kawakita, M., Shimizu, T., and Miyake, T. (1985) *Nature*, **313**, 806–810.

**135** Veber, D.F., Milkowski, J.D., Varga, S.L., Denkewalter, R.G., and Hirschmann, R. (1972) *Journal of the American Chemical Society*, **94**, 5456–5461.

**136** Bang, D., Pentelute, B.L., and Kent, S.B.H. (2006) *Angewandte Chemie International Edition*, **45**, 3985–3988; (2006) *Angewandte Chemie*, **118**, 4089–4092.

**137** For a direct on-resin synthesis of peptide phenylthioesters, see: Bang, D., Pentelute, B.L., Gates, Z.P., and Kent, S.B. (2006) *Organic Letters*, **8**, 1049–1052.

**138** Durek, T., Torbeev, V.Y., and Kent, S.B.H. (2007) *Proceedings of the National Academy of Sciences of the United States of America*, **104**, 4846–4851.

**139** Chatterjee, C., McGinty, R.K., Pellois, J.-P., and Muir, T.W. (2007) *Angewandte Chemie International Edition*, **46**, 2814–2818;(2007) *Angewandte Chemie*, **119**, 2872–2876.

**140** McGinty, R.K., Chatterjee, C., and Muir, T.W. (2009) *Methods in Enzymology*, **462**, 225–243.

**141** McGinty, R.K., Kim, J., Chatterjee, C., Roeder, R.G., and Muir, T.W. (2008) *Nature*, **453**, 812–816.

**142** (a) For other synthetic routes to access ubiquitylated proteins, see: Ajish Kumar, K.S., Haj-Yahya, M., Olschewski, D., Lashuel, H.A. and Brik, A. (2009) *Angewandte Chemie*, **121**, 8234–8238; (2009) *Angewandte Chemie International Edition*, **48**, 8090–8094; (b) Yang, R., Pasunooti, K., Li, F., Liu, X.-W. and Liu, C.-F. (2009) *Journal of the American Chemical Society*, **131**, 13592–13593.

# 14
# Automation of Peptide Synthesis

*Carlo Di Bello, Andrea Bagno, and Monica Dettin*

## 14.1
## Introduction

Peptide synthesis was traditionally performed in homogeneous phase (solution), with intermediate separation and purification steps to isolate the desired product; this approach is expensive and time/labor-consuming. As is well acknowledged, the greatest breakthrough in peptide synthesis was proposed and developed in the early 1960s by Robert Bruce Merrifield who introduced the idea of an insoluble polymeric matrix to keep the growing peptide chain on a solid support [1–3]. As Merrifield declared in his Nobel lecture (8 December 1984) [4]:

The plan was to assemble a peptide chain in a stepwise manner while it was attached at one end to a solid support. With the growing chain covalently anchored to an insoluble matrix at all stages of the synthesis the peptide would also be completely insoluble and, furthermore, would be in a suitable physical form to permit rapid filtration and washing.

Actually, the main advantage of Merrifield's technique, termed solid-phase peptide synthesis (SPPS), is the replacement of traditional separations and purifications by simple washings and filtrations: this is noteworthy in the perspective of a fully "mechanized" process [5, 6]. It is important to state that the synthetic cycle originally conceived as the way for synthesizing peptides is based on a general scheme not dependent on the nature of the monomer units; thus, the solid-phase technique was soon applied to units other than amino acids, extending it to the synthesis of depsipeptides, polyamides, polynucleotides, and polysaccharides [4].

Indeed, Merrifield's fundamental idea was to attach the C-terminal amino acid of the growing chain to a solid support through a stable linkage; then, the subsequent addition of "protected" amino acids is performed stepwise. During peptide chain assembly, it is necessary to block (protect) potentially reactive functionalities to prevent undesired reactions, while others, which have to be directly involved in

*Amino Acids, Peptides and Proteins in Organic Chemistry.*
*Vol.3 – Building Blocks, Catalysis and Coupling Chemistry.* Edited by Andrew B. Hughes

ISBN: 978-3-527-32102-5

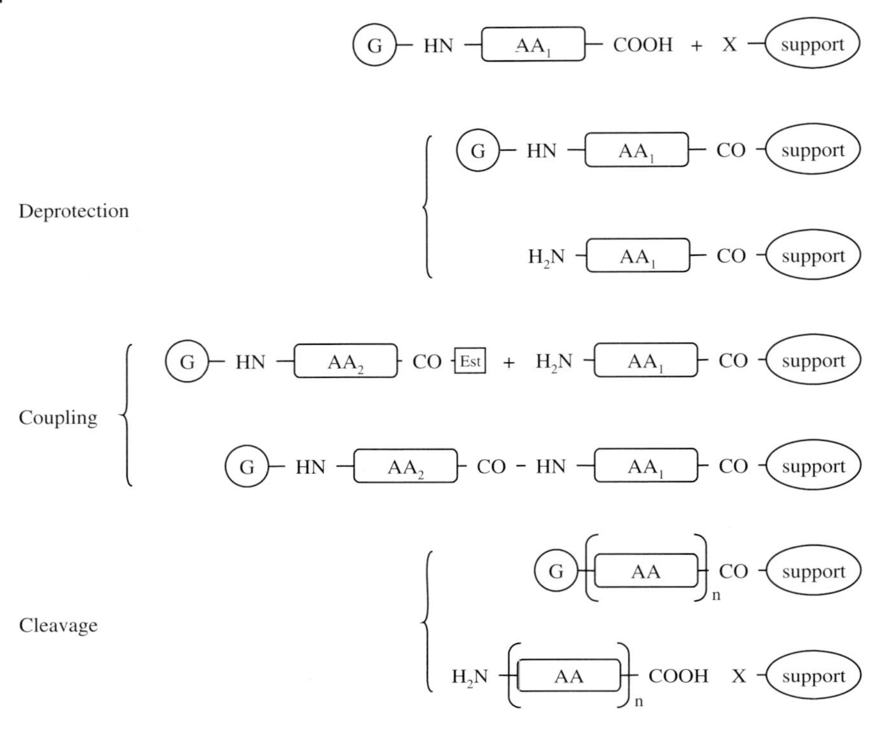


**Figure 14.1** General scheme for the solid-phase synthesis of peptides. G is the protecting group; AA indicates the amino acid; Est is an activated ester.

peptide bond formation, are "activated". Two types of protections are usually required: (i) temporary protection, to be removed from the α-amino group after each peptide bond formation, and (ii) reversible side-chain protection, which should be stable throughout the synthesis and be removed after the entire peptide chain is assembled [7–9].

Two of the main chemical strategies for the reversible protection of the α-amino groups are the *tert*-butyloxycarbonyl (Boc) and the fluorenylmethyloxycarbonyl (Fmoc) chemistries [8, 10, 11]. The key steps of both Boc and Fmoc procedures are deprotection and coupling (Figure 14.1). Deprotection provides for the removal of temporary protecting group (G in Figure 14.1) from the α-amino terminal; coupling allows the stepwise addition of activated amino acids (one *per* cycle) to the growing chain.

In SPPS, peptide bond formation can be ensured by three methods:

i) The addition of *in situ* reagents such as (a) *N,N'*-dicyclohexylcarbodiimide (DCC) with 1-hydroxybenzotriazole (HOBt), which accelerates coupling and suppresses racemization, (b) 2-(1*H*-benzotriazol-1-yl)-1,1,3,3-tetramethyluronium hexafluorophosphate (HBTU), or (c) (2-(7-aza-1*H*-benzotriazole-1-yl)-1,1,3,3-tetramethyluronium hexafluorophosphate (HATU).

ii) The use of active esters such as 2- or 4-dinitrophenyl esters, *N*-hydroxysuccinimide (OSu) esters, pentafluorophenyl (OPfp) esters, or HOBt esters,
iii) The use of preformed symmetrical anhydrides known for being highly reactive.

After each coupling step, all residual unreacted α-amino functions can be quantitatively converted into nonreactive species by reaction with acetic anhydride [5]; this operation is usually called "capping." In the Boc chemistry, the addition of dimethylsulfoxide (DMSO) improves resin–peptide solvation by avoiding hydrogen bond formation [12, 13].

Solid-phase procedures have allowed obtaining peptides up to around 50 residues with a high degree of purity [14] to be formed; longer peptides or the so-called "difficult sequences" are still complicated to synthesize due to aggregation [15, 16] and/or poor solvation of the growing peptide chain [17]. From a general point of view, these problems can be faced whenever incomplete deprotection and/or coupling may result in the formation of deletion sequences; these latter are very difficult to be separated due to similarities with the desired product. This is one of the reasons for suggesting the need for a truly automated procedure – reaction yield optimization has to be achieved automatically by monitoring and controlling the extension of both deprotection and coupling.

More recently, procedures have been introduced for the generation of large numbers of (usually related) peptide sequences in a reasonably short time. A collection of peptides, frequently consisting of all possible combinations of amino acids making up an *n*-amino acid peptide, named peptide libraries, might be used to assay protein functions or to individuate antigenic sequences that can be used as vaccines. In fact, the screening of thousands of parallel binding events is now possible through microarray techniques consisting of captured molecules or probes, immobilized on a planar surface.

## 14.2 SPPS: From Mechanization to Automation

As Merrifield stated, "the ability to purify after each reaction by simple filtration and washing and the fact that all reactions could be conducted within a single reaction vessel appeared to lend themselves ideally to a mechanized and automated process" [4]. At first, a very simple and manually operated instrument was assembled, just to work out the methodology and to synthesize small peptides. Moreover, the original Merrifield's "mechanical" synthesizer [18] was characterized by few essential features: the reaction vessel, containing the resin with the growing peptide chain, and the necessary plumbing to enable the appropriate solvents and reagents to be pumped in, mixed, and removed in the proper sequence. Interestingly, all mechanical events were performed by a simple stepping drum programmer and a set of timers.

Over the years, many synthesizers have been developed and many of them were made commercially available, too. Basically, they operate under computer control in the sense that the operator is allowed to select specific procedures from a database of preprogrammed routes: then, all the operations required by the synthetic cycle (e.g., reagents and solvents delivery, mixing, washings, filtrations) are performed in the correct sequence by a software program.

Two main hydraulic configurations have been proposed, depending on the reaction vessel: batch and flow. Batch reactors, like Merrifield's one, contain the resin–peptide system and provide agitation by stirring, inert gas bubbling, vortexing, shaking, or liquid recirculation. Synthesizers based on liquid-phase recirculation through a stationary bed are quite rare: however, this principle is used in industrial-scale synthesizers (Dan-Process); synthesizers can be single-channel or multiple-channel. Historically, the first commercially available automated multiple synthesizer was brought to the market by Zinsser Analytical in 1988. The machine provided the parallel synthesis of 144 peptides. Removal of the liquid from the solid-phase slurry was performed by a needle. Actually, one key issue in multiple solid-phase synthesis is the parallel removal of excess reagents and washing solutions from the solid support in all the synthetic compartments; in most of the currently available synthesizers the removal is achieved through the porous bottoms of the synthetic compartments (e.g., using plastic syringes with tightly fitting frit material at the bottom) or by vacuum filtration (e.g., Applied Biosystems), or by pressure application from the top of the compartments (e.g., Advantage ChemTech). These methods bear the inherent risk of clogging of one or more compartments, resulting in insufficient liquid removal from the clogged compartments, overflow, and consequently contamination of neighboring compartments.

Three strategies for parallel synthesis (i.e., the SPOT approach, the multipin synthesis, and the teabag synthesis) solve this problem. The SPOT synthesizer [19] avoids the clogging problem because the filtration support becomes the synthetic substrate. The size of the individual spot on which the synthesis is performed determines the scale of the synthesis. Usually only nanomolar amounts of peptides are produced. In the multipin approach the synthesis of peptide arrays is carried out on polyethylene "pins;" this technique [20] utilizes a solid support in the form of rods functionalized with a layer of swellable polymer, on which the repetitive coupling reaction is performed. These pins are arranged in a grid mapping the microtiter plate format and coupling is realized by dipping this grid into the plate preloaded with appropriate activated amino acids. Washings and deprotection reaction do not require segregation of individual pins and can be realized by simple dipping of the array into the common container. In the teabag method [21], polyethylene bags with fine holes are filled with resin and each bag is put in different reaction vessels to carry out amino acid coupling reaction. After reaction, all the bags are collected and processed together for protecting group removal and washing. About 100 different peptides in 500 micromolar quantity can be synthesized by this method.

CEM has employed microwave irradiation to shorten both coupling and deprotection duration: their synthesizer performs one Fmoc-based cycle of peptide synthesis in 10 min. The main advantages of microwave-assisted chemistry, originating from a combination of thermal and nonthermal factors, are shorter reaction times, higher yields, and milder reaction conditions [22]. Centrifugation is a powerful technique allowing the parallel processing of an unlimited number of reaction compartments (the parallel synthesis of 3072 different compounds was reported!) [23]. The key feature of the centrifugation synthetic technology is a new method for the separation of the solid support from reagent solution termed "tilted plate centrifugation." The great improvement of earlier centrifugation methods was the use of the wells of microtiter plates as synthetic compartments, thus enabling the parallel synthesis of much larger compound arrays. The plates are mounted on a centrifugal plate and tilted slightly down toward the center of centrifugation, thus generating a pocket in each well, in which the solid support is collected during centrifugation while the supernatant solutions are expelled. An essential feature of this approach is that well-to-well cross-contamination with reagent solution or resin is avoided since the plates are tilted while the direction of centrifugation is horizontal. Consequently, any liquid or resin expelled from the wells is either captured in the inter-well space of the plate or on the wall of the centrifuge.

Over the years, other technological improvements have been proposed to enhance synthetic procedure effectiveness by boosting instrument "mechanization" (e.g., "robotic" synthesizers able to assure high-throughput peptide synthesis have been created). In particular, robotic arms and dosing pumps have been developed for the precise and fast delivery of reagents and solvents.

Anyway, simply operating the synthetic steps on the basis of a preprogrammed list of actions is not sufficient for real process automation; doing so, it is only possible to select the duration of the chemical reactions before starting synthetic cycles on the basis of previous experience, with the idea of allowing the reactions enough time to go to completion. Thus, it is only possible to check the reaction yield *a posteriori* (i.e., when the machine is already running the succeeding step of the program). Under these constraints, it is easy to understand that in-line monitoring, by stopping the reactions only after reaching a satisfactory yield, would permit the optimization of chemical conversion, the reduction in byproduct formation, and/or improvement in the synthesis of "difficult sequences" by the addition of appropriate solvents/reagents. Indeed, a truly automatic synthesizer should be able to perform all the operations without depending on the operator's control. In addition, it has to provide a warning if difficulties arise during the different steps and, most important, it should automatically act on the chemical procedures by extending their duration, adding/changing reagents, or stopping the process if necessary.

The first step in going from simply "mechanized" towards "fully automated" synthesis is represented by the development of monitoring devices for detecting difficult reactions (e.g., reactions not going, for any reason, to completion). Afterwards, an "intelligent" system able to take decisions on the basis of the monitored data has to be implemented.

Over the years many efforts have been devoted to develop opportune devices to monitor the extent of both deprotection and coupling, thus controlling the overall synthetic process in a truly automatic way [14]: a brief review on the most interesting contributions in this field is herewith provided.

## 14.3
## Deprotection Step: Monitoring and Control

Two methods have been developed and used for evaluating the extent of the deprotection step in solid-phase peptide synthesis: the first is electrochemically based (conductivity), the second depends on photometric detection (absorbance).

The feasibility of on-line conductivity monitoring in a flow synthesizer during removal of Fmoc groups has been demonstrated [24, 25]. In fact, deprotection via Fmoc chemistry requires piperidine as the deprotecting reagent, and it results in the formation of carbon dioxide and a carbamic acid salt. The salt, reacting with the base, generates a conductance signal linearly related to the amount of Fmoc-piperidine adduct; the precision of this method is not satisfactory because conductance values of the piperidinyl carbamate are low. In another paper, the application of this method to Boc deprotection was discussed [26]. The protecting group is removed by excess trifluoroacetic acid (TFA); the corresponding trifluoroacetate salt is treated with a base forming ion pairs that generate measurable conductivity changes. When conductivity remains steady for at least 30 s, the deprotection is completed. It is worth noting that deprotection reaction time was claimed to be indicative of the duration of the subsequent coupling reaction. This approach to the "control" of the synthetic cycles has been applied to a commercial synthesizer [27, 28].

In the case of batch-type synthesizers, the extent of the deprotection step via Fmoc chemistry was assessed by comparing the conductivity of two samples withdrawn from the deprotecting mixture at different time-points [29]. The conductance signal is due to the formation of a carbamate salt by piperidine addition; then, since the signal is proportional to the amount of Fmoc groups removed, the deprotection time is prolonged until the conductivity signal is lower than a preset threshold. Consequently, problems in the deprotection progress are directly related to the height and the number of peaks before the conductivity signal is reduced under the preset threshold.

The conductivity based monitoring is suitable for the batch configuration [30]; in fact, the batch process is not affected by changes in conductivity due to varying reagent concentrations. The method proposed by Fox *et al.* controls the extent of the deprotection by evaluating the rate of Fmoc group removal. The system automatically extends the deprotection time until the rate of change falls within preset limits taking into account the duration of the slowest step detected.

Alternatively, the extent of deprotection step can be measured by the photometric detection of a byproduct released during Fmoc group removal. Fmoc groups are cleaved by excess organic base (e.g., piperidine), which results in the formation of a dibenzofulvene-piperidine adduct [31, 32]. Samples of the deprotecting mixture are forced to flow through a UV cell and the absorbance signal is measured. Since the signal is proportional to the amount of adduct generated, by evaluating the height of the absorbance peak, it is possible to estimate the amount of Fmoc group removal. Dryland and Sheppard [33] applied this strategy in a flow SPPS system arguing that

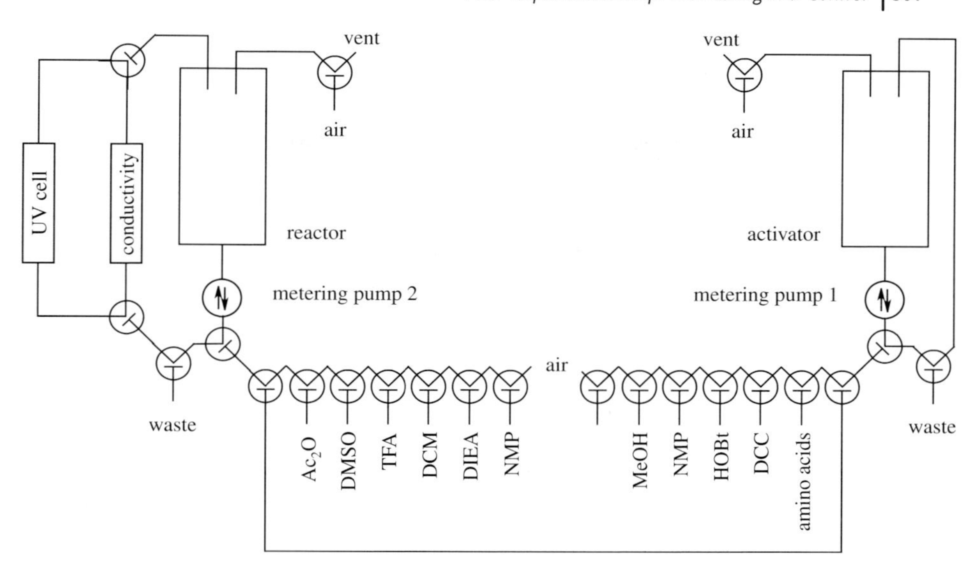


**Figure 14.2** Hydraulic scheme and delivery system of the SPPS apparatus. Two cells are located in parallel with the reactor to allow absorbance and conductivity measurements.

the rapidity by which the absorbance peak at 304 nm rises and falls is an indication of the deprotection reaction rate [34]. Initially, the UV measuring method was applied by Sheppard to Fmoc chemistry making use of flow systems, which were considered "particularly easily and simply mechanized" [33], but it was also used on batch commercial synthesizers [35, 36].

A comparison between conductimetric and photometric techniques applied to deprotection monitoring was performed in our lab [37]. For this purpose, we designed and assembled an original batch-type synthesizer equipped with devices to quantitatively determine the extent of deprotection by both conductimetric and photometric techniques [38]. In this system (Figure 14.2), the reaction vessel and the activator are located within two separate hydraulic loops served by independent metering pumps and delivery lines; the activator is required by the active esters chemical approach. The conductimetric and spectrophotometric cells are also set within the hydraulic loop of the reaction vessel; the reaction mixture is pumped through these cells at intervals. The signals coming from the conductimeter and the spectrophotometer are transformed by an A/D converter, and then sent to the I/O port of a PC. Original software collects, memorizes, and displays all data on the screen; data are then elaborated by dedicated software, which also plots the conductivity and absorbance trends. By integrating the different peaks, it is possible to correlate the curves to the deprotection yield. Once verified that the maximum value has been achieved, a suitable mathematical function would decide whether to stop and/or repeat the reaction, or proceed to the following chemical operation. This system was tested by synthesizing the H-Lys-Ala-Ala-Ala-Ala-Ala-Ala-Arg-Ala-OH peptide, which was shown to present chemical problems both in the coupling and deprotection steps [39], via Fmoc chemistry. While standard procedures were used for the coupling steps, three different conditions were examined for deprotection: (#1) 20% piperidine in *N*-methylpyrrolidone, (NMP) (#2) 20% piperidine in (15% DMSO in NMP), and (#3) 2% piperidine + 2% 1,8-diazabicyclo[5.4.0]undec-7-ene (DBU) in NMP. Deprotection extent was monitored in-line by both spectrophotometric and conductimetric methods. The signal trends relative to a sample residue (Ala9) are shown in Figure 14.3. As discussed above, the conductimetric signal is produced by formation of an ionic couple between piperidine and carbamic acid, which is generated in turn by the $CO_2$ released by degradation of the Fmoc group and excess piperidine. Therefore, conductivity is proportional to the amount of the Fmoc group removed from the α-amino terminal. On the other hand, UV monitoring is based on dibenzofulvene determination at 300 nm.

Data show that the most effective deprotection conditions were those used in experiment (#3) (Figure 14.3e): in fact, absorbance signal is largely reduced within the third addition of fresh deprotecting solution. Then, by integrating absorbance peaks (Figure 14.4), we correlated the curves to the deprotection yield: thus, a mathematical function decides to stop and/or repeat the reaction, or proceed to the following chemical operation.

On the other hand, conductimetric signal variations do not allow precise evaluation of changes in conductivity. In fact, signal variations of a few microsiemans (Figure 14.3b, d and f) are too close to the limit of instrumental sensitivity.

The proposed system was extensively utilized to collect data during the synthesis of "difficult sequences" [14], allowing us to conclude that monitoring devices based on absorbance measurements represent an effective tool to evaluate the extent of the deprotection step in SPPS [40].

More recently, a photometric approach for deprotection monitoring in continuous flow reactors utilizing Boc chemistry was proposed [41]. In this case the key step is the formation of carbon dioxide bubbles as a consequence of Boc removal with TFA in dichloromethane (DCM). Monitoring of gas release, at 313 nm, indicates the completion of the deprotection reaction. What the UV detector really measures is the light absorption due to bubbles passing through the cell, which, in turn, depends on several parameters (e.g., mean bubble size, number of bubbles and light source). If these parameters are not changed, the absorbance peak area is proportional to the amount of gas produced, while the peak shape is determined by bubble formation dynamics. It was found that complete Boc removal with TFA/DCM takes place in 10–30 min.

It is worthwhile mentioning that a completely different approach to deprotection monitoring in SPPS was presented by Due Larsen *et al.* [42] who proposed the near IR-Fourier transform Raman spectroscopy to follow deprotection steps in Fmoc chemistry and monitor peptide secondary structure, too. The extent of the deprotection is evaluated by quantifying the amount of residual Fmoc group (i.e., monitoring the appearance/disappearance of a characteristic band in the spectrum). As a matter of fact, at present, Fmoc quantification is still affected by the overlapping between the fluorenyl system bands and signals from other chemical groups (i.e., linkers) [42].

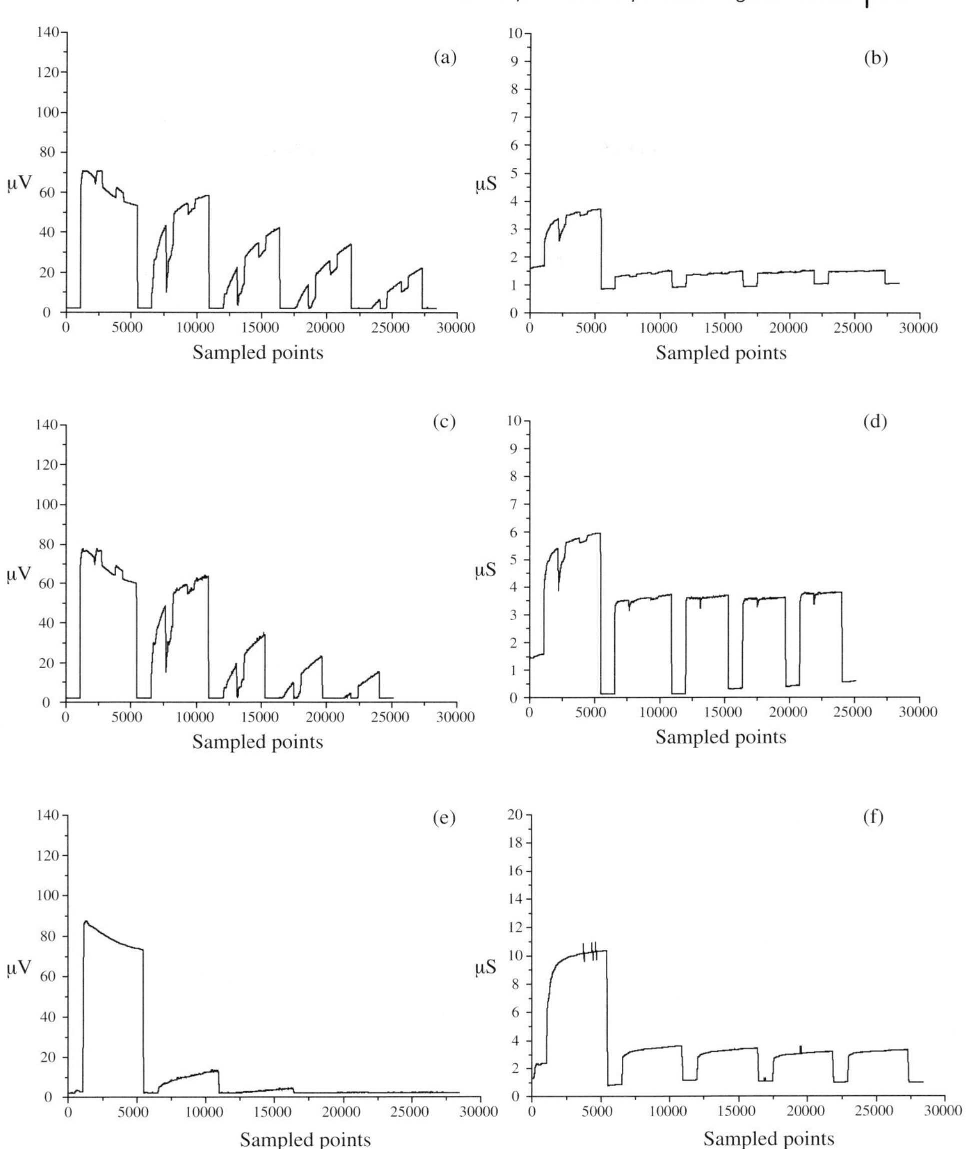


**Figure 14.3** Absorbance (μV, on the left column) and conductivity (μS, on the right column) trends monitored during the deprotection step of the Ala9 residue under conditions #1 (a and b), #2 (c and d), and #3 (e and f).

Considerable improvements are expected by the application of computational methods. Advances are also expected from the applications of spectroscopic devices that have been introduced in the field of solid-phase organic synthesis (SPOS) for the "in-real-time monitoring and probing" of reaction kinetics [43–45].

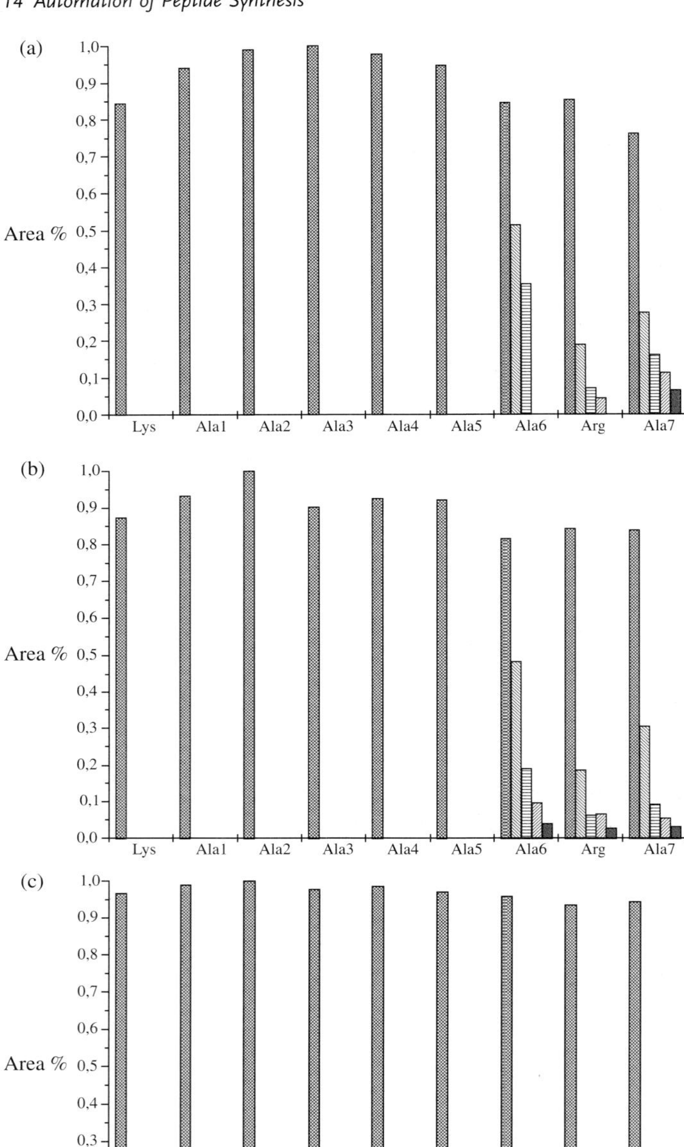


**Figure 14.4** Integrated values of absorbance during the deprotection step under conditions #1 (a), #2 (b), and #3 (c) for all the amino acid residues of the H-Lys-Ala-Ala-Ala-Ala-Ala-Ala-Arg-Ala-OH peptide sequence. Different backgrounds indicate different stepwise addition of fresh deprotecting solution.

## 14.4
## Coupling Step: Monitoring and Control

The yield of the coupling reaction is usually measured by off-line tests based on ninhydrin [46–48] and fluorescamine. These assays [49] generally allow quantification of the free residual (unreacted) amino groups on the resin. Even though time-consuming, invasive, and discontinuous, these methods represent the most widely used techniques for the quantitative off-line estimation of the coupling completion.

Other methods have been proposed for monitoring the coupling step by the means of photometric devices [25]: as the reaction proceeds, the number of cationic sites due to protonated N-terminal groups decreases. If an inert dye is introduced into the reaction mixture, it will distribute between all available cations and it will progressively be displaced from the resin as the reaction goes to completion. Therefore, the absorbance of the solution will reach a maximum when the reaction is complete [50]. The technique was generally termed counterion distribution monitoring (CDM). However, all methods based on this approach require the addition of chemical reagents (e.g., picric acid, 4,4′-dimethoxytrityl chloride, bromophenol blue dye, quinoline yellow dye) that can interfere with the synthetic process [51–54]. Definitely, these methods are not suitable for on-line applications; furthermore, a critical appraisal indicated that CDM is effective in monitoring the coupling rate, but less suitable at measuring the extent of the reaction [55].

Dryland and Sheppard tried to extend the absorbance approach to the coupling reaction [33]. In a flow reactor, coupling completion is qualitatively indicated by absorbance signal plateauing at 304 nm; thus, it is theoretically possible to monitor both deprotection and coupling reactions by means of the same photometric device.

In the wake of Sheppard's contributions, it is worthwhile mentioning a paper presented by Hellstern and Wirth [56] which, suggests following coupling reaction by absorbance changes in the liquid phase in a batch-type system. At 300 nm the Fmoc group, covalently linked to the amino function of amino acids, exhibits a very high extinction coefficient; therefore, it should be possible to follow variations of Fmoc protected amino acids in the liquid phase. Unfortunately, the method is not suitable for practical applications since it requires very low amino acid concentrations, close to the stoichiometric amount, while the most widely used methodologies use high amino acid excess to reduce reaction time and improve yields.

Another significant contribution regarding colorimetric monitoring has been proposed by Atherton *et al.* [57, 58]. By utilizing 3,4-dihydro-3-hydroxy-4-oxo-1,2,3-benzotriazine as activating agent, it is possible to detect the coupling extent by measuring absorbance changes in a range around 440 nm: an intense yellow coloration is immediately produced, fading as the reaction goes to completion. Once again, the principal drawback is represented by the necessity to use low amino acid concentrations. Furthermore, none of the above mentioned techniques is compatible with base-catalyzed reactions and on-line applications.

If the photometric approach seems to be unsuitable for the on-line monitoring of the coupling step in SPPS, great efforts have been made in order to apply conductivity measurements. A simple and reliable method based on continuous measurement of conductivity in the reaction vessel was described by Schafer-Nielsen *et al.* [59]. This method was applied to the Fmoc strategy in a flow reactor using *N*,*N*-dimethylformamide (DMF) as general solvent with pentafluorophenol derivatives as activated esters but it is also suitable for other chemistries (e.g., dihydroxybenzotriazole esters and symmetric anhydrides). Addition of small amounts of tertiary amine (e.g., *N*,*N*-diisopropylethylamine (DIEA)) results in the formation of counter-ions to the carboxyl groups released during the coupling reaction. The ion pairs between carboxyl groups and tertiary amine produce a conductivity increase in the reaction mixture, being proportional to the amount of ester groups reacted and thus to the extent of the coupling reaction. To obtain a stable signal, electrodes have to be introduced into the reaction vessel and the liquid phase has to be stopped flowing through the resin bed during measurements. Each amino acid shows a specific conductivity profile and for each amino acid the absolute conductivity value is affected by the composition of the coupling mixture. The extent of the reaction is monitored as the rate of conductivity change; when this value is lower than 2% per hour, the reaction is complete. This method allows determining both the reaction rate and the amount of ester consumed.

Other similar conductivity-based monitoring systems have been presented [24, 26, 60]; in all coupling reactions, using either symmetrical anhydrides or activated esters, acids or alcohols are released during peptide bond formation. The addition of a base generates an ionic pair responsible for a readily detectable conductivity signal allowing estimation of the reaction progress.

An alternative approach, avoiding any addition of base to the coupling mixture, was developed and applied to a batch reactor by Fox *et al.* [30, 61]. The HOBt used as a catalyst forms an ion pair with the amino group present on the resin support. During the coupling step, the incoming amino acid is coupled with the growing peptide chain and the basic amino group on the resin is thus removed. It produces a decrease of conductivity related to the number of unreacted amino acids on the resin at any time. According to Fox *et al.*, the conductivity detection system was applied to an automatic instrument and syntheses were carried out under computer control. The computer monitors the reaction in real-time and operates by automatically extending the reaction time. The coupling reaction is complete when a time corresponding to at least 12 half-lives has expired. The half-life of the reaction is calculated on the basis of the curve decay.

Another interesting method was introduced by Baru *et al.* [62]. The strategy is based on a noninvasive pressure monitoring for both Boc- and Fmoc-based synthesis. Pressure measurements are carried out by means of a resistance strain gage connected to the inlet of a continuous flow reactor. The basic idea is to correlate changes in pressure drop through the reactor to variations in peptide–resin structure. It was demonstrated elsewhere [42] that the structure of the growing peptide chains may influence the yields in peptide synthesis. The results obtained in a study utilizing a polyalanine model sequence are in line with its tendency to aggregation. Baru *et al.* also applied the pressure-based monitoring approach to the deprotection step, and concluded that it can detect onset, development, and termination of aggregation in the coupling step of Boc chemistry, and in both deprotection and coupling steps of

Fmoc chemistry. As a matter of fact, the paper by Baru *et al.* does not indicate which strategy should be applied to deprotection and coupling for improving the reaction yield; indeed, the coupling duration was set at 2 h while the proposed method has proved useful for reducing washing time.

With the aim of moving from a simple mechanized process towards a truly automated synthesis, we developed an original algorithm specifically designed for process yield prediction to control the coupling step [14] utilizing the chemistry proposed by Schafer-Nielsen *et al.* [59, 61]. In more detail, a feed-forward artificial neural network (FANN) [63, 64] was used to analyze the conductivity signals acquired during the initial stage of the coupling reaction (i.e., the period in which the conductivity signal usually shows its widest and most rapid variation). A relationship between the conductivity gradient and the final yield of the reaction was postulated, and the FANN was trained to handle the experimental results and to predict the corresponding final yield values. The experimental apparatus is described in [38]: it is equipped with a batch reactor and photometric and conductimetric devices. Fmoc-amino acids were activated as HOBt esters and NMP was used as main solvent. DIEA in NMP was added during the coupling step, whose duration was initially set at 60 min. After data acquisition each conductivity profile (Figure 14.5) was filtered, differentiated, normalized, and sampled (e.g., 25 conductivity values representing the initial velocity of the signal were identified). Among all the monitored coupling reactions, 34 observations were selected and for each observation a vector with 26 components, including 25 conductivity values and the final yield, determined by

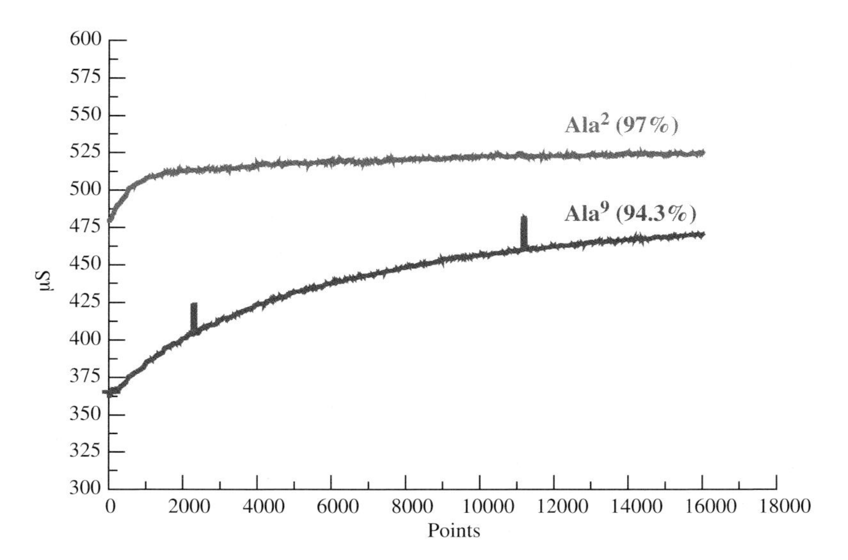


**Figure 14.5** Typical shape of conductimetric signals acquired during the coupling step: the highest curve corresponds to a reaction yield of 97% for the Ala2 residue of a polyalanine model sequence; the lowest curve corresponds to a reaction yield of 94.3% for the Ala9 residue of the same polyalanine model sequence.

an off-line assay, was created [46, 47]. The observation matrix was randomly split into two subsets, one for the training phase, the other for a cross-validation procedure. Comparison between the yield values predicted by the FANN and the ones determined through the off-line ninhydrin test, confirmed the reliability of the model to predict the yield of previously unseen couplings with an average accuracy of ±1.5%.

It is worthwhile to underline that the method allows prediction of the final yield within the first 5 min of the reaction. Then, in the on-line controlled system, the algorithm allows the reaction going to completion if the predicted yield is higher than the preset threshold (e.g., 97%). If the predicted value is lower than the threshold, the reaction is automatically stopped and a second coupling with fresh reagents is immediately activated. This method avoids the extension of the reaction time when the concentration of the acidic component in the liquid phase has reached a steady state. This prevents waste of time and, more importantly, since the steady state does not necessarily indicate a 100% yield, the occurrence of secondary reactions.

A further improvement was developed by our group [65] utilizing a statistical process control model. All the conductivity profiles monitored during the syntheses of several model sequences were collected in a historical database. The database was then split into two classes on the basis of the final yield: class 1 or *high-yield class* whose members have a yield equal to or higher than 97%, and class 2 or *low-yield class* with samples having a yield lower than 97%. The first step of the database analysis consists in the application of two statistical algorithms, Mean Hypothesis Testing (MHT) and Cluster Analysis (CA), to access the information content and the consistency of the data (Figure 14.6). The MHT evaluates the discriminating power of the data and allows verification if and where the measured variables are useful to describe process behaviors. The MHT was applied to different versions of the data (e.g., raw, filtered, derivative) but only derivatives calculated over the signals acquired in the first 10 min of the reaction were able to differentiate between high- and low-yield couplings. Afterwards, CA was used to explore data structure by grouping together data sharing similar properties (e.g., values, shapes, profiles). This method allowed verifying the arbitrary definition of the classes based on the final coupling yield and it was found to be consistent with the data. At this point, the database has been utilized to design and train a model able to identify a high yield coupling analyzing the derivative of the conductivity profiles monitored within the first minutes of the reactions. A principal component analysis (PCA)-based model has been formulated. PCA captures the variance in the data and partitions it among new variables, called principal components. PCA has been used to transform the conductivity profiles into two components vectors so that each coupling sample is represented by a point in the plane of the first two components (Figure 14.7). By defining statistical confidence limits in the two-dimensional plane, points (couplings) falling outside the limits are to be considered "not high-yield" couplings, while points within the limits are high-yield couplings.

Interestingly, the control scheme described above has been applied during the synthesis of the RGD sequence [66] performed under two different conditions: synthesis 1 was done with our prototype synthesizer applying the control procedure; synthesis 2 was carried out by means of a commercial synthesizer (control synthesis).

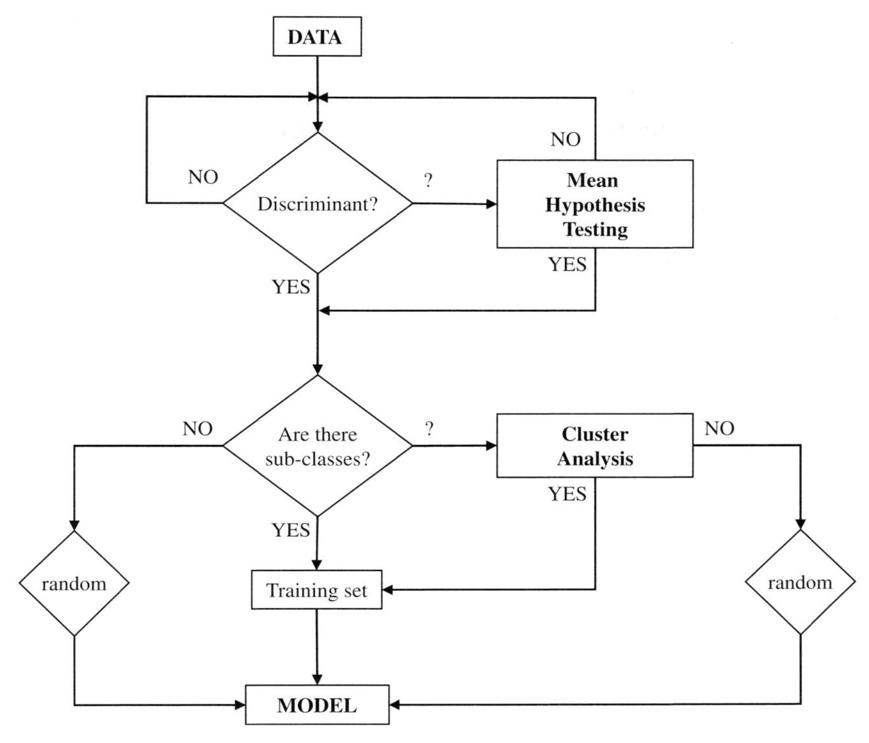


**Figure 14.6** To check the consistency of the data collected in the database containing the conductivity profiles monitored during the synthesis of the model peptides, two algorithms were applied: MHT and CA.

Chemical conditions applied during the syntheses of the RGD sample sequence are described in Table 14.1. The high-performance liquid chromatography (HPLC) chromatograms corresponding to the two crude products are illustrated in Figure 14.8: the results obtained confirmed the effectiveness of the control procedure (Table 14.2).

## 14.5 Integrated Deprotection and Coupling Control

As discussed, no single device is suitable for monitoring both deprotection and coupling reactions in SPPS. Thus, the synthetic procedure can be only optimized by integrating different approaches. As it has been previously described, two devices for monitoring deprotection and coupling reactions, through photometric and conductimetric systems, respectively, have been utilized in our laboratories [66, 67]; two algorithms have been formulated to control the syntheses using the Fmoc chemistry and the HOBt activated esters. In particular, the deprotection step is monitored by

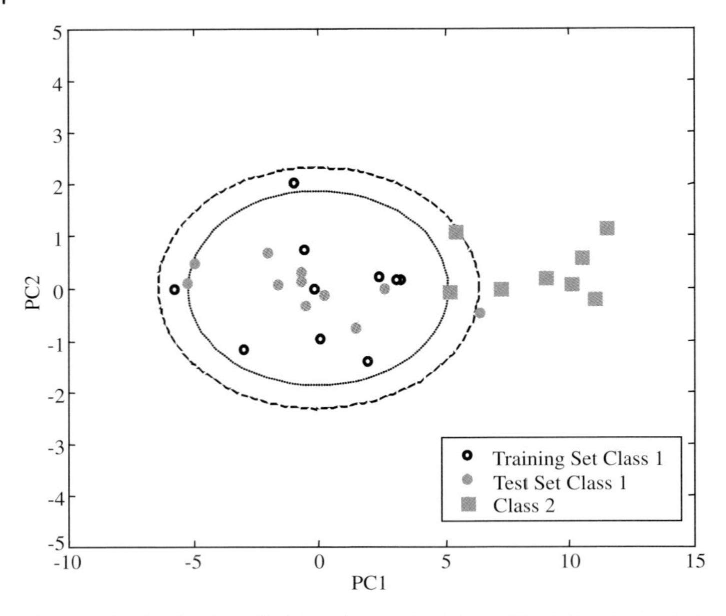


**Figure 14.7** Application of PCA on the conductivity profiles belonging to the historical database.

**Table 14.1** Chemical conditions applied during the synthesis of the RGD sample sequence.

| Conditions | Synthesis 1 | Syntheses 2 |
|---|---|---|
| Synthesizer | prototype | commercial |
| Resin | Fmoc-Lys(Boc)SAC TentaGel | Fmoc-Lys(Boc)SAC TentaGel |
| Scale | 0.25 mml | 0.25 mml |
| Deprotection step | | |
| Control | yes | no |
| Reagent | 20% piperidine/NMP | 20% piperidine/NMP |
| Time | 5 + 5 min (5 + 5 min) | 3 + 10 min |
| Coupling step | | |
| Control | yes | no |
| Amino acid excess | 4-fold | 4-fold |
| Reagent | HOBt/DCC<br>(HBTU/HOBt second coupling) | HBTU/HOBt |
| Time | 60 min (yield prediction > 97%)<br>30 min (yield prediction < 97%)<br>30 min (second coupling) | 30 min |
| Double coupling | yes (Arg) | no |
| Cleavage | 0.75 g phenol<br>TFA/thioanisole/<br>1,2-ethanedithiol/$H_2O$<br>10: 0.5: 0.25: 0.5 | 0.75 g phenol<br>TFA/thioanisole/<br>1,2-ethanedithiol/$H_2O$<br>10: 0.5: 0.25: 0.5 |

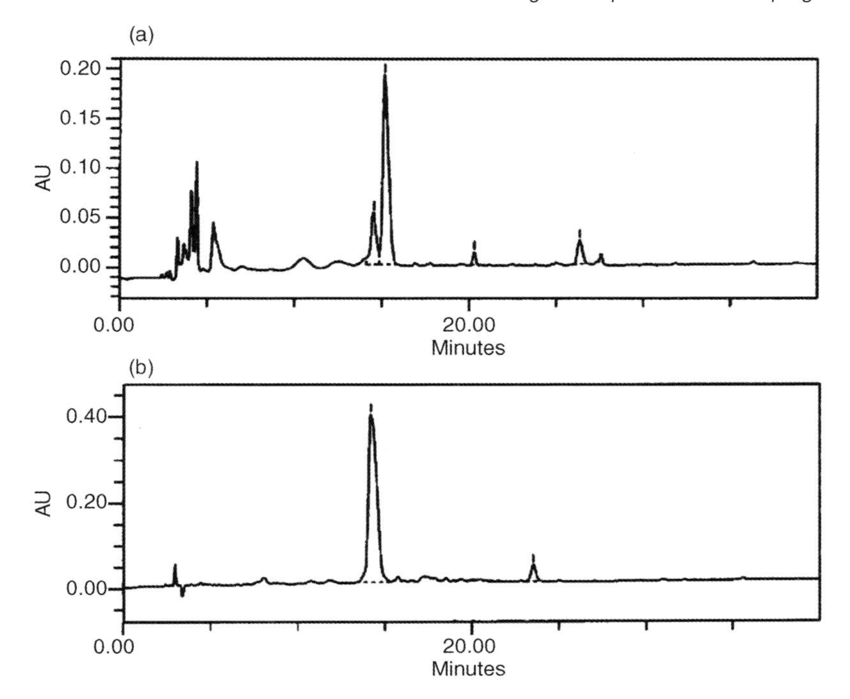


**Figure 14.8** HPLC chromatograms of the crude RGD product obtained by synthesis 1 (a) and synthesis 2 (b). Conditions used: column, Vydac $C_{18}$ (5 μm, 300 Å, 0.46 × 25 mm); eluent A, 0.05% TFA in $H_2O$; eluent B, 0.05% TFA in $CH_3CN$; gradient, 0–20% B over 40 min, 20–30% B over 1 min, 30–50% B over 20 min; flow rate, 1 ml/min; detector, 214 nm.

absorbance measurement at 301 nm: the deprotection reaction is thus carried out by a maximum of five subsequent introductions of fresh reagents for 5 min each. The absorbance peak is integrated and the area is proportional to the release of Fmoc groups. Deprotection is finally stopped when absorbance is nullified (i.e., is lower than a preset threshold being a function of the signal-to-noise ratio). If the signal due to the first introduction of piperidine in NMP is not appreciable (i.e., no Fmoc groups have been removed), a warning is automatically sent to the operator, otherwise the controlling system operates further introductions of fresh reagents

**Table 14.2** Yield values corresponding to the integration of the HPLC chromatograms.

| | Peak<br>Synthesis 1 (%) | Synthesis 2 (%) |
|---|---|---|
| Byproducts | — | 18.8 |
| Target sequence | 94.7 | 71 |
| Fmoc-protected sequence | 5.2 | 7.8 |

until the absorbance peak disappears. To improve reaction effectiveness, from the third introduction up to the fifth, different reagents are delivered (i.e., DBU/piperidine in NMP); the controlling system also increases the threshold. If the signal is still appreciable even after the last reagent introduction, a warning is sent to

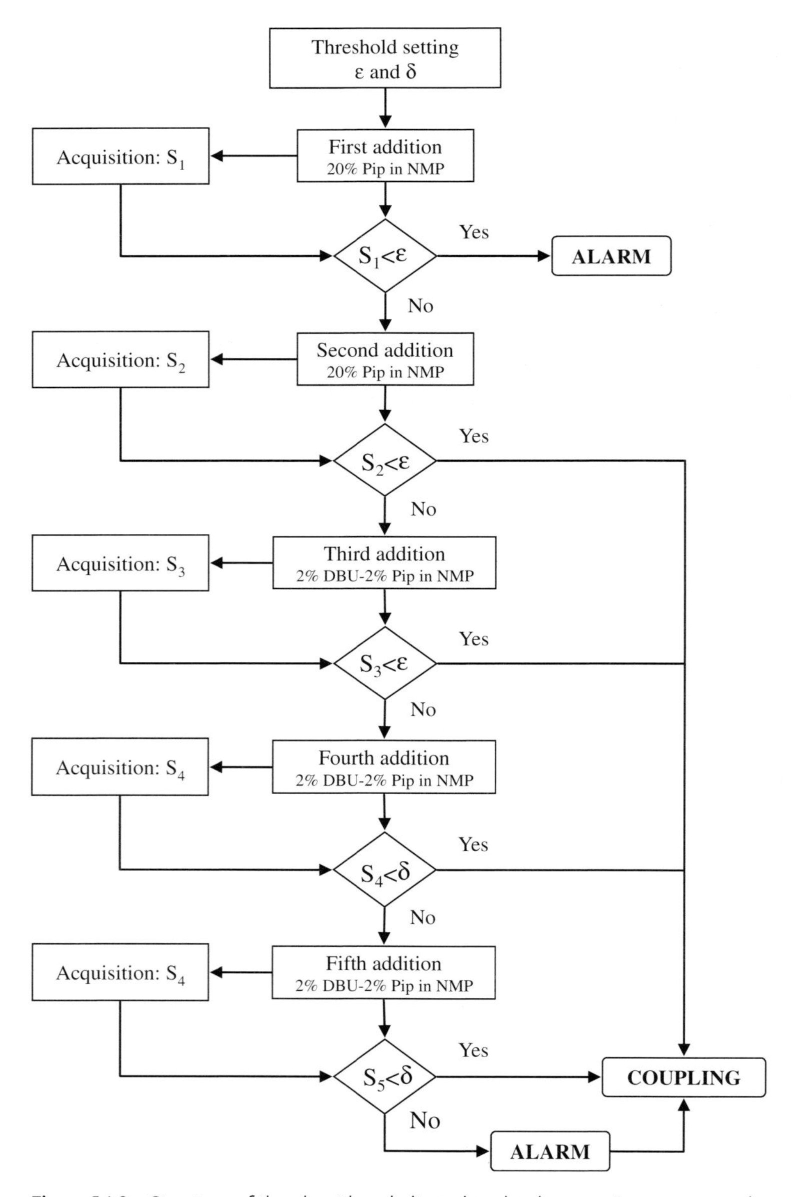


**Figure 14.9** Structure of the algorithm dedicated to the deprotection step control.

the operator and the system automatically switches to the coupling step. At any rate, if more than two introductions of fresh reagents are required, a warning for the operator is recorded since the presence of a difficult deprotection normally implies a difficult coupling. The control scheme used for optimizing the deprotection step is summarized in Figure 14.9.

During the coupling reaction, the conductivity of the liquid phase is measured [65]. The data collected during the first minutes are managed and analyzed in order to predict the final yield. If the value is higher than a convenient threshold, the control system allows the reaction to go to completion and then switches to the subsequent deprotection step. If the predicted yield is lower than the threshold, the first coupling via active esters is stopped and the system automatically provides a second coupling via a more effective HBTU strategy, which, however, cannot be monitored.

The control scheme used for optimizing the coupling step is summarized in Figure 14.10.

This monitoring and control system has been thoroughly tested during the synthesis of several model sequences. The quality of the products obtained has been successfully compared to the one of the same peptides synthesized with a commercial instrument in order to verify the reliability and the advantages offered by the new system.

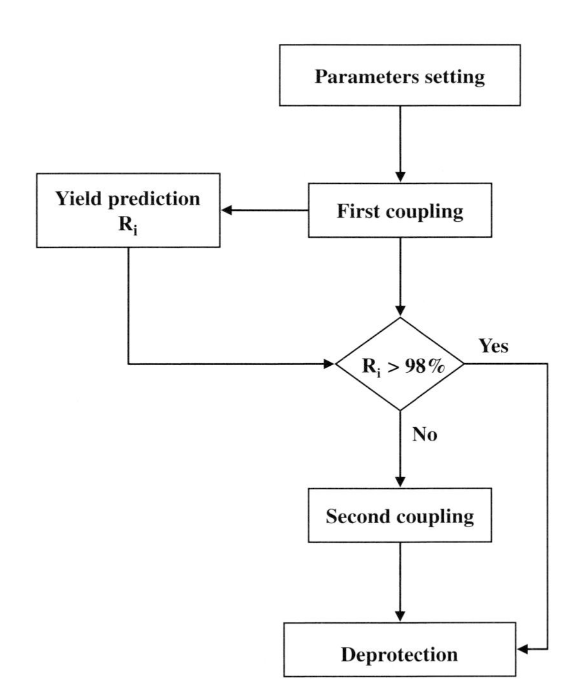


**Figure 14.10** Structure of the algorithm dedicated to the coupling step control.

## References


1 Merrifield, R.B. (1963) Solid phase peptide synthesis. I. The synthesis of a tetrapeptide. *Journal of the American Chemical Society*, **85**, 2149–2154.

2 Merrifield, B. (1968) The automatic synthesis of proteins. *Scientific American*, **218**, 56–62 and 67–72.

3 Merrifield, B. (1986) Solid phase synthesis. *Science*, **232**, 341–347.

4 Merrifield, B. (1985) Solid phase synthesis. Nobel lecture, 8 December 1984. *Bioscience Reports*, **5**, 353–376.

5 Barany, G., Kneib-Cordonier, N., and Mullen, D.G. (1987) Solid-phase peptide synthesis: a silver anniversary report. *International Journal of Peptide and Protein Research*, **30**, 705–739.

6 Stewart, J. and Young, J.D. (1984) *Solid Phase Peptide Synthesis*, Pierce Chemical, Rockford, IL.

7 Bagno, A., Bicciato, S., Buso, O., Dettin, M., and Di Bello, C. (1993) Development of a prototype for the automated solid-phase synthesis of peptides. *Chemical and Biochemical Engineering Quarterly*, **7**, 209–214.

8 Kent, S.B.H. (1988) Chemical synthesis of peptides and proteins. *Annual Review of Biochemistry*, **57**, 957–989.

9 Bailey, P.D. (1992) *An Introduction to Peptide Chemistry*, John Wiley & Sons, Inc., New York.

10 Atherton, E. and Sheppard, R.C. (1989) *Solid Phase Peptide Synthesis: A Practical Approach*, IRL Press, Oxford.

11 Grant, G.A. (1992) *Synthetic Peptides: A User's Guide*, Freeman, New York.

12 Dugas, H. and Penny, C. (1981) *Bioorganic Chemistry*, Springer, New York.

13 Applied Biosystems (1992) *Peptide Synthesizer 431A User Manual*, Version 1.12. Applied Biosystems, Carlsbad, CA.

14 Bagno, A., Bicciato, S., Dettin, M., and Di Bello, C. (1997) A novel algorithm for the coupling control in solid-phase peptide synthesis. *Journal of Peptide Research*, **50**, 231–237.

15 Milton, R.C. de L., Milton, S.C.F., and Adams, P.A. (1990) Prediction of difficult sequences in solid-phase peptide synthesis. *Journal of the American Chemical Society*, **112**, 6039–6046.

16 Krchnák, V., Flegelová, Z., and Vágner, J. (1993) Aggregation of resin-bound peptides during solid-phase peptide synthesis. Prediction of difficult sequences. *International Journal of Peptide and Protein Research*, **42**, 450–454.

17 Sheppard, R.C. (1973) Solid phase peptide synthesis. An assessment of the present position, in *Peptides 1971* (ed. H. Nesvadba), North Holland, Amsterdam.

18 Merrifield, R.B., Stewart, J.M., and Jernberg, N. (1966) Instrument for automated synthesis of peptides. *Analytical Chemistry*, **38**, 1905–1914.

19 Frank, R., Güler, S., Krause, S., and Lindenmaier, W. (1991) Facile and rapid "spot synthesis" of large numbers of synthetic peptides on membrane sheets, in *Peptides 1990* (eds E. Gerald and D. Andreu), ESCOM, Leiden.

20 Geysen H.M., Meloen, R.H., and Barteling, S.J. (1984) Use of peptide synthesis to probe viral antigens for epitopes to a resolution of a single amino acid. *Proceedings of the National Academy of Sciences of the United States of America*, **81**, 3998–4002.

21 Houghten, R.A. (1985) General method for the rapid solid-phase synthesis of large numbers of peptides: specificity of antigen–antibody interaction at the level of individual amino acids. *Proceedings of the National Academy of Sciences of the United States of America*, **82**, 5131–5135.

22 de la Hoz, A., Diaz-Ortiz, A., and Moreno, A. (2005) Microwaves in organic synthesis. Thermal and non-thermal microwave effects. *Chemical Society Reviews*, **34**, 164–178.

23 Lebl, M. (1999) New technique for high-throughput synthesis. *Bioorganic and Medicinal Chemistry Letters*, **9**, 1305–1310.

24 McFerran, N.V., Walker, B., McGurk, C.D., and Scott, F.C. (1991) Conductance measurements in solid phase peptide synthesis. Monitoring coupling and deprotection in Fmoc chemistry. *International Journal of Peptide and Protein Research*, **37**, 382–387.

25 Bagno, A., Bicciato, S., Dettin, M., and Di Bello, C. (2002) Automation and control of solid-phase peptide synthesis: methods and technology, in *Recent Research Development in Peptides*, vol. **1**, Research Signpost, Trivandrum.

26 McFerran, N. and Walker, B. (1990) Potentiated conductance as a general on-line monitoring system in peptide synthesis, in *Solid Phase Synthesis 1990* (ed. R. Epton), SPCC, Birmingham.

27 Applied Biosystems (1993) *Biosystems Reporter*, 5 May.

28 McCandless, J., Noble, R.L., Otteson, K.M., and Rawer, S. (1994) Applications of the new conductance monitoring technology in feedback control of SPPS, in *Solid Phase Synthesis 1994* (ed. R. Epton), Mayflower Worldwide, Birmingham.

29 Pipkorn, R. and Rawer, S. (1994) Conductivity monitoring of the Fmoc deprotection using different resins in batch synthesis of HTLV II ENV 174–209, in *Solid Phase Synthesis 1994* (ed. R. Epton), Mayflower Worldwide, Birmingham.

30 Fox, J.E., Newton, R., and Stroud, C.H. (1991) A new detector for fully automatic peptide synthesis. *International Journal of Peptide and Protein Research*, **38**, 62–65.

31 Hoffmann, R., Rawer, S., and Zeppezauer, M. (1994) On-line monitoring of the fulvene-piperidine adduct in a 430A batch synthesiser by UV absorption detection at 301 nm, in *Solid Phase Synthesis 1994* (ed. R. Epton), Mayflower Worldwide, Birmingham.

32 Meienhofer, J., Waki, M., Heimer, E.P., Lambros, T.J., Makofske, R.C., and Chang, C.-D. (1979) Solid phase synthesis without repetitive acidolysis. Preparation of leucyl-alanyl-glycyl-valine using 9-fluorenyl-methyloxycarbonylamino acids. *International Journal of Peptide and Protein Research*, **13**, 35–42.

33 Dryland, A. and Sheppard, R.C. (1986) Peptide synthesis. Part 8. A system for solid-phase synthesis under low pressure continuous flow conditions. *Journal of the Chemical Society, Perkin Transactions 1*, 125–137.

34 Dryland, A. and Sheppard, R.C. (1988) Peptide synthesis. Part 11. A system for continuous flow solid phase peptide synthesis using fluorenylmethoxy-carbonyl-amino acid pentafluorophenyl esters. *Tetrahedron*, **44**, 859–876.

35 Otteson, K.M., MacDonald, R.L., Noble, R.L., and Hoeprich, P.D. (1991) Research News, Applied Biosystems, Carlsbad, CA.

36 Nalis, D. and Jacob, R. (1993) UV-monitoring of Fmoc deprotection in batch peptide synthesis, in *Peptides 1992* (eds C.H. Schneider and A.N. Eberle), ESCOM, Leiden.

37 Bicciato, S., Bagno, A., Dettin, M., and Di Bello, C. (1995) Presented at 6th International Conference on Computer Applications in Biotechnology, Garmish-Patenkirchen, May 14–17.

38 Bicciato, S., Bagno, A., Dettin, M., Buso, O., and Di Bello, C. (1995) An improved system for automated peptides synthesis. *Chemical Engineering and Technology*, **18**, 210–215.

39 Pegoraro, S., Vigano., S., Rovero, P., Rivoltella, R., Bocciato, S., Bagno, A., Dettin, M., and Di Bello, C. (1994) Presented at 23rd European Peptide Symposium, Braga, September 4–10.

40 Dettin, M., Pegoraro, S., Rovero, P., Bicciato, S., Bagno, A., and Di Bello, C. (1997) SPPS of difficult sequences. A comparison of chemical conditions, synthetic strategies and on-line monitoring. *Journal of Peptide Research*, **49**, 103–111.

41 Baru, M.B., Mustaeva, L.G., Gorbunova, E.Y., Vagenina, I.V., Kitaeva, M., and Cherskii, V.V. (1999) Spectrophotometric monitoring in continuous-flow Boc-based solid-phase peptide synthesis. *Journal of Peptide Research*, **54**, 263–269.

42 Due Larsen, B., Christensen, D.H., Holm, A., Zillmer, R., and Nielsen, O.F. (1993) The Merrifield peptide synthesis studied by near-infrared Fourier-transform Raman spectroscopy. *Journal of the American Chemical Society*, **115**, 6247–6253.

43 Henkel, B. and Bayer, E. (1998) Monitoring of solid phase peptide synthesis by FT-IR spectroscopy. *Journal of Peptide Science*, **4**, 461–470.

44 Gremlich, H.-U. (1998/1999) The use of optical spectroscopy in combinatorial

chemistry. *Biotechnology and Bioengineering*, **61**, 179–187.

45 De Miguel, Y.R. and Shearer, A.S. (2000) Infrared spectroscopy in solid-phase synthesis. *Biotechnology and Bioengineering*, **71**, 119–129.

46 Kaiser, E., Colescott, R.L., Bossinger, C.D., and Cook, P.I. (1970) Color test for detection of free terminal amino groups in the solid-phase synthesis of peptides. *Analytical Biochemistry*, **34**, 595–598.

47 Kaiser, E., Bossinger, C.D., Colescott, R.L., and Olsen, D.B. (1980) Color test for terminal prolyl residues in the solid-phase synthesis of peptides. *Analytica Chimica Acta*, **118**, 149–151.

48 Sarin, V.K., Kent, S.B.H., Tam, J.P., and Merrifield, R.B. (1981) Quantitative monitoring of solid-phase peptide synthesis by the ninhydrin reaction. *Analytical Biochemistry*, **117**, 147–157.

49 Felix, A.M. and Jimenez, M.H. (1973) Rapid fluorometric detection for completeness in solid phase coupling reactions. *Analytical Biochemistry*, **52**, 377–381.

50 Salisbury, S.A., Tremeer, E.J., Davies, J.W., and Owen, D.E.I.A. (1990) Acylation monitoring in solid phase peptide synthesis by the equilibrium distribution of coloured ions. *Journal of the Chemical Society, Chemical Communications*, 538–540.

51 Hodges, R.S. and Merrifield, R.B. (1975) Monitoring of solid phase peptide synthesis by an automated spectrophotometric picrate method. *Analytical Biochemistry*, **65**, 241–272.

52 Arad, O. and Houghten, R.A. (1990) An evaluation of the advantages and effectiveness of picric acid monitoring during solid phase peptide synthesis. *Peptide Research*, **3**, 42–50.

53 Krchnák, V., Vágner, J., and Lebl, M. (1988) Noninvasive continuous monitoring of solid-phase peptide synthesis by acid–base indicator. *International Journal of Peptide and Protein Research*, **32**, 415–416.

54 Flegel, M. and Sheppard, R.C. (1990) A sensitive, general method for quantitative monitoring of continuous flow solid phase peptide synthesis. *Journal of the Chemical Society, Chemical Communications*, 536–538.

55 Kinsmann, R.G. and Olibvier, G.W.J. (1994) Counter-ion distribution monitoring: a critical appraisal, in *Solid Phase Synthesis 1994* (ed. R. Epton), Mayflower Worldwide, Birmingham.

56 Hellstern, H. and Wirth, W. (1993) A new large-scale peptide synthesiser with on-line monitoring, in *Peptides 1992* (eds C.H. Schneider and A.N. Eberle), ESCOM, Leiden.

57 Atherton, E., Cameron, L., Meldal, M., and Sheppard, R.C. (1986) Self-indicating activated esters for use in solid phase peptide synthesis. Fluorenylmethoxy-carbonylamino acid derivatives of 3-hydroxy-4-oxodihydrobenzotriazine. *Journal of the Chemical Society, Chemical Communications*, 1763–1765.

58 Cameron, L., Meldal, M., and Sheppard, R.C. (1987) Feedback control in organic synthesis. A system for solid phase peptide synthesis with true automation. *Journal of the Chemical Society, Chemical Communications*, 270–272.

59 Schafer-Nielsen, C., Hansen, P.H., Lihme, A., and Heegaard, P.M.H. (1989) Real time monitoring of acylations during solid phase peptide synthesis: a method based on electrochemical detection. *Journal of Biochemical and Biophysical Methods*, **20**, 69–79.

60 McFerran, N., Scott, F., and Walker, B. (1993) Amino acid activation: solvent composition for monitoring reaction progress, in *Peptides 1992* (eds C.H. Schneider and A.N. Eberle), ESCOM, Leiden.

61 Fox, J., Newton, R., Heegard, P., and Schafer-Nielsen, C. (1990) A novel method of monitoring the coupling reaction in solid phase synthesis, in *Solid Phase Synthesis 1990* (ed. R. Epton), SPCC, Birmingham.

62 Baru, M.B., Mustaeva, L.G., Vagenina, I.V., Gorbunova, E.Y., and Cherskii, V.V. (2001) Pressure monitoring of continuous-flow solid-phase peptide synthesis. *Journal of Peptide Research*, **57**, 193–202.

63 Rumelhart, D.E. and McClelland, J.L. (1986) *Parallel Distributed Processing:*

*Exploration in the Microstructure of Cognition*, vol. **1**, *Foundations*, MIT Press, Cambridge, MA.

64 Lippmann, R.P. (1987) An introduction to computing with neural nets. *IEEE ASSP Magazine* **35**, 4–22.

65 Bicciato, S., Bagno, A., Dettin, M., and Di Bello, C. (2001) Application of statistical process control to solid phase peptide synthesis, in *Solid Phase Synthesis 2000* (ed. R. Epton), Eaton Press, Wallasey.

66 Bicciato, S., Bagno, A., Dettin, M., and Di Bello, C. (1995) Presented at IcheaP-2, Florence, May 15–17.

67 Bicciato, S., Bagno, A., and Di Bello, C. (1995) Presented at AIChE Annual Meeting, Miami Beach, FL, November 12–17.

# 15
# Peptide Purification by Reversed-Phase Chromatography

*Ulrike Kusebauch, Joshua McBee, Julie Bletz, Richard J. Simpson, and Robert L. Moritz*

## 15.1
## RP-HPLC of Peptides

Reversed-phase high-performance liquid chromatography (RP-HPLC) has dominated peptide purification since the late 1970s due to its efficient separation and ease of use with volatile compatible buffer systems for numerous applications [1, 2]. RP-HPLC separation of chains of amino acids (peptides and proteins) is based predominantly on reversible hydrophobic interactions between the amino acid side chains with the hydrophobic surface of the chromatographic stationary phase (solid particles or packings) in competition with the flowing mobile phase (chromatographic buffer or solvent). Separation is performed by binding peptides onto the hydrophobic RP stationary phase in polar conditions (e.g., water) and eluting in nonpolar solvents (e.g., organic solvent such as acetonitrile) in the presence of a pH modifying agent. As peptides bind to the stationary phase, the amount of hydrophobic area on the surface of the peptide(s) exposed to the solvent is minimized. Thus, the degree of organized water is decreased with a concomitant favorable increase in entropy of the system. For this reason, it is advantageous, under these solvent conditions, for peptides to associate with the stationary phase upon loading onto the RP-HPLC column. Mobile phase composition is then subsequently modified so that bound peptides are differentially eluted (desorbed) back into the mobile phase. The order of peptide desorption is simply based on their relative hydrophobicity (i.e., least hydrophobic proteins elute first, followed by peptides in increasing order of heir surface hydrophobicity). Peptide elution is usually performed by increasing the organic solvent concentration (either in a stepwise fashion or in a gradient manner). By these means peptides, which are concentrated or trace enriched during the binding and separation process, are eluted in a purified and concentrated form ready for collection or subsequent analysis using hyphenated techniques such as mass spectrometry.

*Amino Acids, Peptides and Proteins in Organic Chemistry.*
*Vol.3 – Building Blocks, Catalysis and Coupling Chemistry.* Edited by Andrew B. Hughes

ISBN: 978-3-527-32102-5

## 15.2
## Peptide Properties

Peptides are covalently bonded strings of amino acids. The α-carboxyl group of one amino acid is bonded to the α-amino group of another amino acid and, through the loss of a water molecule, forms a peptide bond or amide bond. Many amino acids joined in this fashion are called polypeptides or peptides for short. The peptide chain consists of a regularly repeating backbone and a variable part consisting of 20 different side chains of common amino acids bonded to the α-carbon of the amino acid chain. Amino acids are dipolar ions and can vary their ionization state depending on pH. The $pK_a$ can vary from as low as 4 for the acidic amino acids (e.g., aspartic acid) to as high as 12 for the basic amino acids (e.g., arginine). RP-HPLC takes advantage of these properties by using specific pH buffer systems and ion-paring reagents to manipulate the elution pattern of peptides (i.e., selectivity) to enable the separation of hundreds of different peptides in a single chromatographic elution process.

## 15.3
## Chromatographic Principles

### 15.3.1
### Choice of Mobile Phase

For most general RP-HPLC for the purification of peptides, the mobile phase is generally composed of only three components: (i) an aqueous "buffer", (ii) , ion-pairing agent or pH modifier to optimize selectivity (retention), and (iii) organic solvent.

#### 15.3.1.1 Mobile-Phase Aqueous Buffer pH

Peptide separations can be manipulated by the eluent pH. Shifts in retention time can be induced through the protonation or deprotonation of acidic or basic side-chains and may cause peptides to separate differently as the pH is varied depending on the side-chains that are present. In the example in Figure 15.1, the peptides elute differently at pH 2 than at pH 10 because of the dissociation of acidic side-chains. Since most carboxylic acids have $pK_a$ values near 4, these groups would be completely ionized. Manipulation of aqueous buffer pH needs to be performed with caution as both the HPLC and the stationary phase used for peptide separations can be severely affected by extremes in pH and may cause irreparable damage. Conventional silica-based RP packings suffer from two disadvantages. (i) They have a limited useable pH range, typically pH 2–8. Below pH 2, the bonded phase is susceptible to hydrolysis; above pH 8, hydroxide ion can attack and dissolve the silica, which causes the collapse of the packed bed (resulting in decreased column lifetime) and loss of efficiency. (ii) Depending on the efficiency of the bonding chemistry, basic analytes interact strongly with residual silanols and cause tailing peaks that are detrimental to resolution. Silica-based RP-HPLC packing materials

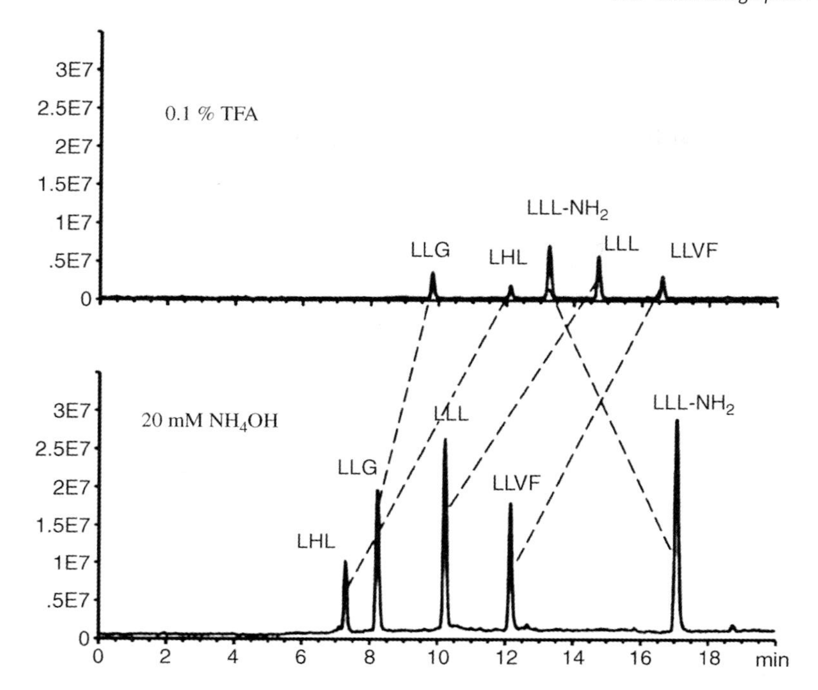


**Figure 15.1** Effect of pH mobile phase on peptide selectivity. TFA conditions: A, 0.1% TFA in water, B, 0.085% TFA in 80% acetonitrile; $NH_4OH$ conditions: A, 20 mM $NH_4OH$ in water, B, 20 mM $NH_4OH$ in 80% acetonitrile. Column: Zorbax Extend C18, 2.1 × 150 mm; flow rate: 0.25 ml/min; temperature: 25 °C; gradient: 5–60% B in 20 min; LC/MS: positive ion ESI, $V_f$ 70 V, $V_{cap}$ 4.5 kV, $N_2$, 35 psi, 12 l/min, 300 °C, 4 μl (50 ng each peptide). (Figure courtesy of Agilent Technologies, Inc.)

are thus most often performed at low pH values, generally between pH 2 and 4. Further advantages of using a low pH include good sample solubility, ion suppression of both acidic side-chains on the sample, and residual unbonded silanol groups on the silica support.

Peptides with the same charge are resolved very similarly independent of the ion-pairing reagent used, although the overall retention times of the peptides increase with increasing hydrophobicity of the anion (Figure 15.2). Peptides of differing charge elute at differing rates relative to each other depending on concentration of ion-pairing reagents. Ion-pairing reagents increase peptide retention time with increasing concentration, albeit to different extents, again based on hydrophobicity of the anion, that is, the more hydrophobic the anion, the greater the increase in peptide retention time at the same reagent concentration. Commonly used acids in RP-HPLC include trifluoroacetic acid (TFA), heptafluorobutryric acid (HFBA), formic acid (FA), acetic acid (AC), and *o*-phosphoric acid in the concentration range of 0.05–0.1% (w/v) or 10–100 mM. Mobile phases containing unbuffered 1% (w/v) NaCl, ammonium acetate (pH 6–7), ammonium bicarbonate (pH 7), or phosphate salts (e.g., triethylammonium phosphate, pH 6) are suitable for use at pH values

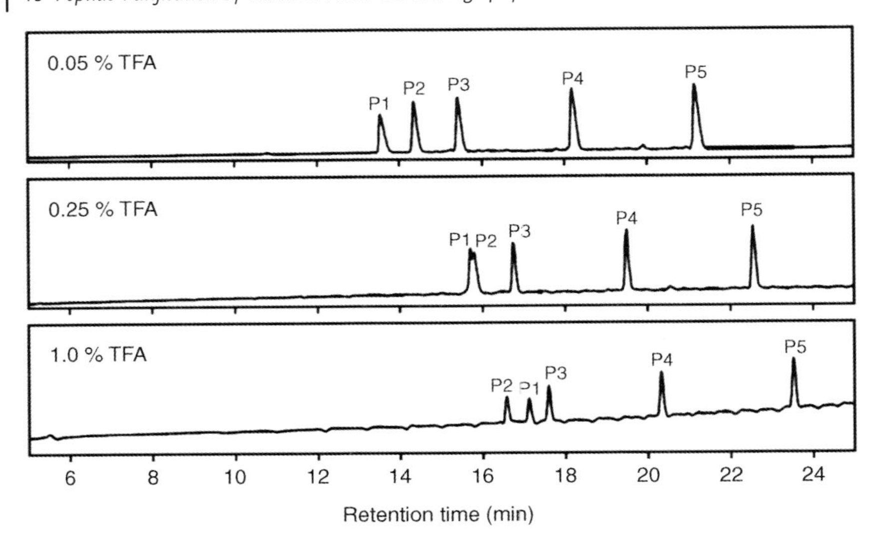


**Figure 15.2** Effect of ion-pairing concentration of mobile phase on peptide selectivity. Column: Zorbax 300SB C8, 4.6 × 150 mm; mobile phase: A = $H_2O$ and TFA, B = acetonitrile and 0.05–1.0 % TFA; gradient: 0–30% B in 30 min; flow rate: 1 ml/min; temperature: 40 °C; detection: UV 254 nm; sample: 6 μl injection volume; sample: peptide decapeptide standards differing slightly in hydrophobicity. (Figure courtesy of Agilent Technologies, Inc.)

closer to neutrality but, these salts are not volatile and are generally used for first-dimension separations prior to a desalting and second-dimension separation. For the use of buffers at higher pH (above pH 7) such as ammonium formate (pH 10), stable bonded silica packings or polystyrene-based packings have proven excellent alternatives for the separation of peptides.

#### 15.3.1.2 **Organic Solvent**

When injected onto the column, peptides bind to the surface of the adsorbent and desorb only when the organic solvent reaches a specific and unique concentration. Once desorbed, they interact only slightly with the adsorbent surface as they elute down the column. Peptides may be thought of as "sitting" on the stationary phase, with most of the molecule exposed to the mobile phase and only a part of the molecule termed the "hydrophobic foot" in contact with the RP surface. For details on chromatographic theory, see [3–5]. For elution of peptides from RP-HPLC packings, the mobile phase must be manipulated to provide an environment of reduced polarity. The most common organic solvents used in RP-HPLC of peptides, and their order of elutropic strength, are: 1-propanol > 2-propanol > ethanol > acetonitrile > methanol [1]. In practice, the most widely used organic modifiers are 2-propanol, acetonitrile, and methanol, the most popular choice being acetonitrile. Although isopropanol (2-propanol) is often used because of its strong eluting properties, it is limited because of its high viscosity, which results in lower column efficiencies and high back-pressures. Acetonitrile, methanol, and isopropanol are

UV-transparent to low wavelengths (e.g., 210 nm), which is an essential requirement for RP-HPLC because column elution is typically monitored by UV absorbance. UV transparency is particularly important because most RP-HPLC separations (especially peptides) are monitored below 220 nm for optimal detection sensitivity of peptides (e.g., proteins and peptides lacking the aromatic amino acids tryptophan, tyrosine, and phenylalanine can only be detected by UV transparency using wavelengths below 225 nm). For use in other detection techniques such as MS, the most popular organic solvent is acetonitrile due to its stability and good chromatographic performance. The use of alcohol-based solvents is problematic due to the formation of esters once acid is added to the alcohol-based organic solvent and builds up over time.

With conventional RP-HPLC, an organic solvent is added to the aqueous mobile phase (using either isocratic concentration or commonly using a gradient of 0–80% organic solvent) to lower its polarity, thereby causing peptides to elute from the column. The lower the polarity of the mobile phase, the greater its eluting strength in RP chromatography. As there is strong dependence on the relative retention (or capacity factor, $k'$) of individual peptides and the concentration of organic modifier required for their elution, the separation of a mixture of peptides of slightly different relative hydrophobicities requires gradient elution. The use of nonvolatile buffers such as phosphate and NaCl are problematic, and care must be taken to ensure that the concentration of organic modifier does not exceed 50%, otherwise these salts will precipitate and block the pump, solvent tubing, column, or detector flow cell if using UV to analyze the column eluent.

## 15.3.2 Stationary Phase

### 15.3.2.1 Surface Bonding

RP-HPLC packings are formed by bonding a hydrocarbon phase to the silica matrix by means of chlorosilanes, silicon-based molecules with chlorine as the reactive group and to which a hydrocarbon group is attached. The hydrocarbon group forming the hydrophobic phase is usually a linear aliphatic hydrocarbon of 4 (C4), 8 (C8) or 18 (C18), carbons (Figure 15.3). It is commonly accepted that the length of the hydrocarbon chain contributes to the hydrophobicity of the stationary phase and can affect the overall retention of peptides during separations. Peptides (less than 5000 Da) are best separated on small-pore C18 columns because of higher surface area than larger-pore adsorbents. C8 columns are similar to C18 columns in their application, but they sometimes offer a different selectivity or ability to separate particular peptides such as very hydrophobic ones that may be refractive to elution from C18 type packings.

### 15.3.2.2 Pore Diameter

RP-HPLC packings, typically based on silica gel, are porous particles and the majority of the interactive surface is inside the pores. Consequently, peptides must enter a pore in order to be adsorbed and separated. RP-HPLC has generally been

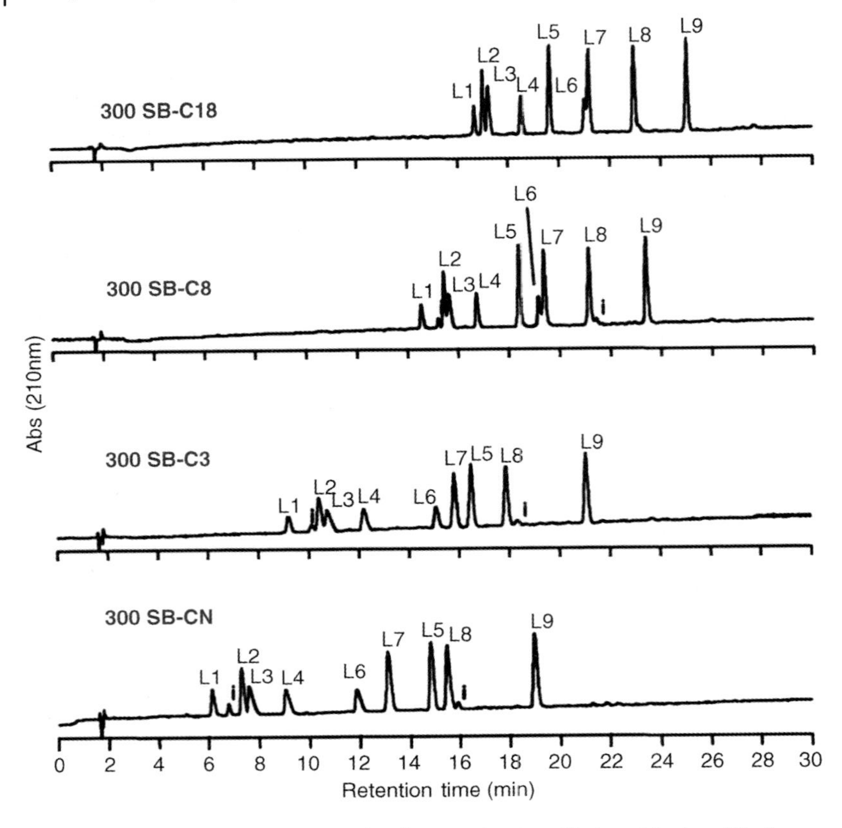


**Figure 15.3** Effect of stationary-phase bonding chemistry on peptide selectivity. Columns: Zorbax 300SB with C18, C8, and C3 carbon chain lengths, and CN (cyanopropyl) bonded phase, 4.6 × 150 mm; mobile phase: gradient, 0–26% B in 30 min, A = 0.1% TFA in water, B = 0.1% TFA in acetonitrile; temperature: 40 °C sample: 2 μg in 20 μl of each peptide (L1 = Leu–Gly–Leu; L2 = Leu–His–Leu; L3 = Leu–Arg–Leu; L4 = Leu–Leu–Leu-$NH_2$; L5 = Leu–Leu–Val–Tyr; L6 = Leu–Leu–Leu; L7 = Leu–Leu–Phe-$NH_2$; L8 = Leu–Leu–Phe; L9 = Leu–Leu–Val–Phe); flow rate: 1.0 ml/min; detection: UV 210 nm. (Figure reproduced with permission from [6].)

performed with particles having pores around 100 Å in diameter. Proteins and larger peptides usually chromatograph poorly on these materials, in part because they are too large to enter pores of this diameter. Consequently, the RP-HPLC separation of most peptides is performed on columns with particles having pores 120 Å in diameter.

#### 15.3.2.3 **Particle Size**

The size of the packing in the column affects the efficiency of the desorption of the eluting peaks due to the amount of diffusion within the particle. The use of larger particles increases peak widths and reduces resolution. However, these materials are much less expensive and produce lower back-pressure, both beneficial when purifying larger quantities of peptides. Smaller-diameter particles produce sharper peaks and better resolution; 3- to 5-μm materials are recommended for micro, analytical, and small-scale preparative separations (columns from 50 μm to 10 mm internal diameter). A number of commercial vendors are now offering sub-3-μm particles for RP-HPLC applications, with 2.4 and 1.7 μm being commonplace [7]. One driving force behind the reduction of particle size is the ability to dramatically shorten run times for analytical or narrowbore RP-HPLC analyses, but these smaller particles also increase resolution by reducing column dead volume and diffusion as the analyte passes over the particles in the column. With higher resolution, shorter columns can be utilized, along with higher flow rates, very short run times, and comparable back-pressures. Similarly for nanobore RP-HPLC, smaller particle diameters increase surface area-to-volume ratio, allowing not only for better packing but better resolution.

#### 15.3.2.4 Ultra-High-Pressure Liquid Chromatography

As particle size has been reduced for RP-HPLC separations, the direct effect has been the concomitant increase in back-pressure from these smaller packings. These effects can easily overwhelm the back-pressure limitations of conventional HPLC hardware which typically have an upper limit of 400 bar (6000 psi). In recent years, the development of hardware for handling excessive back-pressures (up to 1400 bar, 20 000 psi) has created a new subclass of HPLC called ultra-high-pressure liquid chromatography (UHPLC). This new hardware is specifically designed to operate at very high back-pressures specifically using sub-2-μm particle sizes [8]. Several vendors are specifically designing UHPLC systems that produce stable flow rates as low as 1–10 nl/min without preflow splitting to take advantage of the high-separating power of micron sized RP-HPLC packings.

#### 15.3.2.5 Synthetic Polymer Packings

Although silica-based packings perform very well under moderate operating conditions of pH and temperature, there is sometimes a need to operate at higher than normal pH or temperature or in the presence of high concentrations of chaotropic agents such as guanidine–HCl, without the problem of degrading silica columns in rapid fashion. A robust synthetic polymer matrix, such as polystyrene–divinylbenzene, is stable under these harsh conditions and thus offers a practical alternative to silica [9].

#### 15.3.2.6 Monolithic Stationary Phase

In the quest for new column technologies to provide faster and/or higher-resolution separations, many researchers have reported on work with "monolithic" columns where the first monolithic silica columns were reported in 1996 [10–12]. Rather than the stationary phase being comprised of individual discrete beads, monolithic columns create a single monolithic block or "rod" made up of cross-linked porous silica or polymers [13]. Monoliths have several advantages over conventional bead-packed columns. Namely, an interconnected skeleton with large flow channels and a large surface area with high porosity [14]. These large channels allow increased linear velocity of the mobile phase and significantly reduced back-pressures without sacrificing sensitivity or peak capacity. Also, the low mass transfer resistance makes them rather suitable for separation of high-molecular-weight molecules, particularly peptides and proteins. The final result is generally much faster separations with resolution comparable to that obtained with conventional particle columns. Several monolithic columns are now commercially available, and a few of these have large porous channels suitable for peptides and small proteins. Indeed, silica monolithic columns have been manufactured in 10–20 μm diameter fused silica columns up to 70 cm long [15, 16]. Luo *et al.* demonstrated that a 25 cm × 10 μm internal diameter silica monolithic column could be integrated with a MS electrospray emitter to eliminate dead volume to the mass spectrometer detector and achieve high separations without the deleterious diffuse effects of post-column dead volume. They were able to identify 5510 unique peptides covering 1443 proteins from a 300-ng *Shewanella oneidensis* tryptic digest when their monolithic electrospray ionization (ESI) emitter was coupled to a Thermo LTQ linear ion trap mass spectrometer [17].

#### 15.3.2.7 **Packed Bed (Column) Length**

The adsorption/desorption responsible for the separation of peptides takes place primarily at the head of the column. Therefore, column length can be reduced without overly affecting separation and resolution of peptides, and relatively short columns (5–15 cm) can often be used [18]. However, the advantage of longer columns provides for more uniform packed beds and can enhance the overall efficiency of the column. Peptides, such as those from protease digests, are more affected by column length and are best separated on columns 15 or 25 cm in length. The larger the number of peptides in the mixture, the longer the column required. Stone and Williams found that more peptide fragments from a tryptic digest of carboxymethylated transferrin were separated on a column of 250 mm length (104 peaks) than on a column of 150 mm (80 peaks) or a column of 50 mm (65 peaks) [19]. Column length may affect other aspects of the separation, such as sample capacity, which has been shown to be a function of column volume. Thus, for columns of equal diameter, this means longer columns. Consequently, to maximize sample capacity, the longest available column in a given diameter should be selected. Column back-pressure is a function of column length and longer columns do result in higher back-pressures and may limit the overall flow rate due to the pressure limit of the packed bed.

One major difficulty in utilizing longer column lengths and smaller particle sizes is the dramatic increase in back-pressure. Doubling the length of a column doubles the back-pressure required to maintain the same linear velocity. Similarly, reducing particle size diameter in half increases back-pressure 4-fold for columns of equal length [14]. This is also compounded by the fact that increased linear velocities are needed to optimize the resolving capabilities of smaller particle sizes [7]. Exceedingly high back-pressures also generate heat caused by friction from the solvent being pushed through the column which requires careful temperature control of the column during separations [20]. Elevated temperatures can reduce back-pressures by reducing the viscosity of the mobile phase. Nanobore columns are particularly suited for this application due to low radial temperature gradients [21].

Typical lengths for analytical to nanobore applications run from 50 to 250 mm. Increasing column lengths has been demonstrated to improve peak capacity. For example, Gilar *et al.* demonstrated that doubling the column length increased peak capacity by 40% [22]. Shen *et al.* have even demonstrated nanobore columns with packing lengths as long as 200 cm [23]. Longer column lengths also offer increased loading capacities and greater surface area for separations, which can contribute to increased sensitivity for MS analysis.

#### 15.3.2.8 **Gradient Effect**

The organic solvent in the mobile phase of the RP-HPLC run helps solubilize the polypeptide and desorb it from the hydrophobic packing surface. A practical consequence of the mechanism of interaction is that peptides are very sensitive to organic modifier concentration. The sensitivity of peptide retention to subtle changes in the modifier concentration makes isocratic elution (where the solvent modifier concentration is kept constant) difficult because the organic modifier concentration must be maintained precisely. Slowly raising the concentration of organic solvent as the peptide elutes (solvent gradient) results in the sharpest peaks and best resolution. Consequently, gradient elution is preferred for RP-HPLC polypeptide separations. Typical changes in organic solvent concentration (gradient slope) are of the order of 0.5–2% change/min. However, shallow gradients with slopes of below 0.5% have proven to be very effective in separating complex mixtures of peptides. When fractionating very complex mixtures of peptides, such as those resulting from the digestion of cellular proteins, long gradient times of 120–180 min with slopes of 0.1–0.5% are recommended. Optimizing the gradient slope is an important aspect of method development in polypeptide separations. Longer gradient times and lower gradient slopes nearly always result in better resolution, especially for complex mixtures of peptides (Figure 15.4). Minor selective effects can sometimes be obtained with different organic solvents and, in some cases, a combination of different solvents has been used to great effect [1]. Additional optimization of protein and peptide separations can be accomplished by careful manipulation of the flow rate and gradient steepness [24–26].

#### 15.3.2.9 **Temperature**

Column temperature not only affects solvent viscosity and column back-pressure, but can also affect peptide elution selectivity [27]. The effect of temperature is illustrated in Figure 15.5 by the separation of model peptides. At 35 °C, peptides 1 and 2 coelute. As the temperature is raised to 60 °C, peptide 1 decreases retention, whereas peptide 2 increases retention, resulting in good resolution between the two peptides without affecting the resolution of peptide 3. This illustrates the significant impact that temperature may have on peptide selectivity.

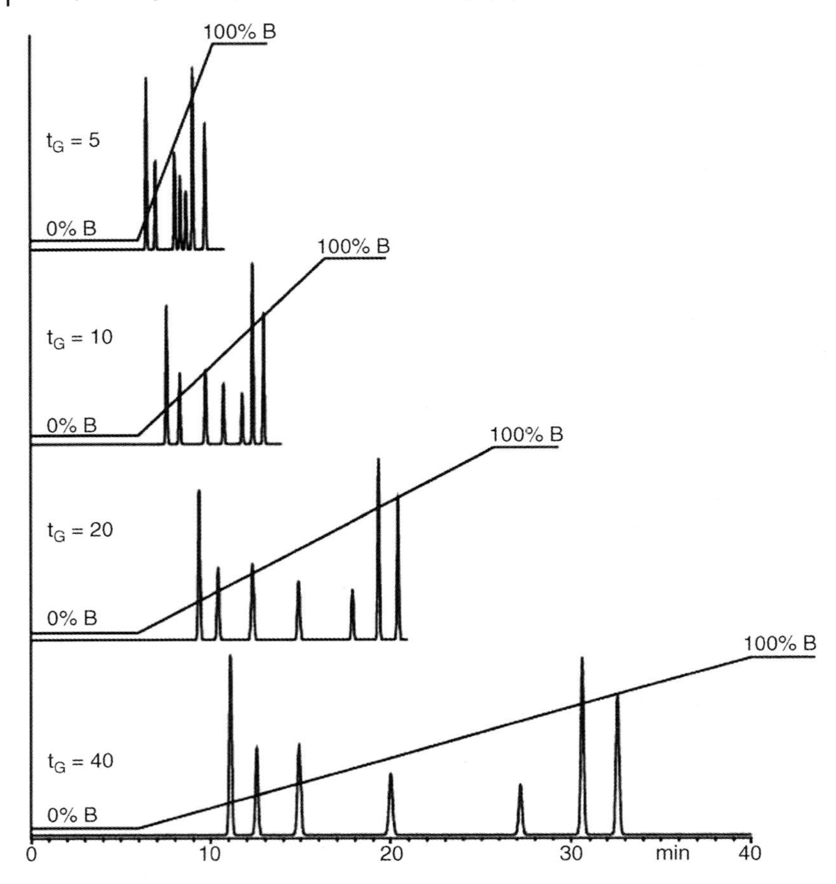


**Figure 15.4** Effect of gradient concentration of mobile phase on peptide separation. Columns: Zorbax 300SB C18, 4.6 × 150 mm; mobile phase: gradient, 0–26% B in 30 min, A = 0.1% TFA in water, B = 0.1% TFA in acetonitrile; flow rate: 1.0 ml/min; temperature: 40 °C; detection: UV 210 nm; $t_G$, gradient time; sample: 2 μg of each peptide standard. (Figure courtesy of Agilent Technologies, Inc.)

## 15.4 Prediction of Peptide Retention Times

Predicting peptide retention and elution conditions from RP-HPLC columns is of great importance for peptide purification, characterization (and quantitation). For example, the current use of MS targeted analysis within complex mixtures (e.g., multiple reaction monitoring (MRM)) for identification and quantitation of proteotypic peptides, (i.e., signature peptides that can identify a specific protein from all other proteins in an organism) relies heavily on knowing where a particular peptide will elute in RP-HPLC packing as well as its signature ions. Usage of peptide

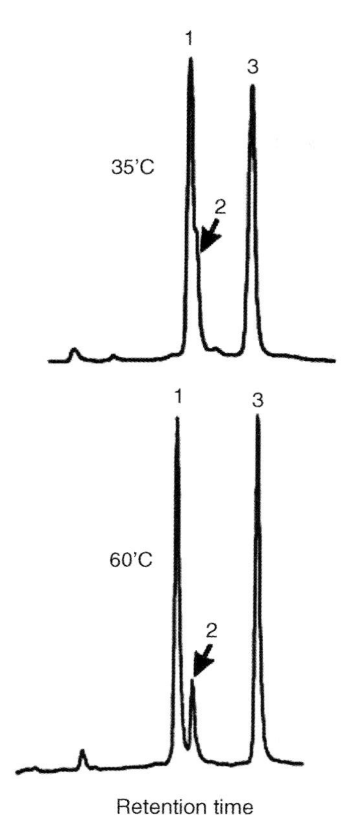


**Figure 15.5** Effect of temperature on polypeptide selectivity. Columns: Zorbax 300SB, 4.6 × 150 mm; mobile phase: gradient, 0–26% B in 30 min, A = 0.1% TFA in water, B = 0.1% TFA in acetonitrile; flow rate: 1.0 ml/min; temperature: 40 °C; detection: UV 210 nm; sample: 2 μg of each peptide. (Figure courtesy of Agilent Technologies, Inc.)

retention knowledge can greatly assist alignment of LC-MS data using differential instrument set-up, comparative liquid chromatography (LC) analysis (with or without MS setup) or MRM analysis by allowing for highly multiplexed MS quantitation in an extremely complex mixture. MRM analysis performs targeted selection of peptides by discarding all other peptide fragments at a particular unit of time and then allowing the instrument to move on to the next peptide parameters without over-taxing the computational system. Several approaches to predict peptide elution times and calibrate RP-HPLC columns have been proposed [28–35]. The most commonly used predictor is the Sequence Specific Retention Calculator (SSRCalc) developed by Krohkin *et al.* [36–40]. The SSRCalc program calculates peptide hydrophobicities and correlates these with retention times. The model algorithm relies on the summation of retention coefficients of individual amino acids and takes in account several other criteria such as retention coefficients for N- and C-terminal amino acids, residue position within a peptide chain, peptide length, additional length correction for smaller peptides, overall hydrophobicity, p*I*, nearest-neighbor effect of basic amino acids (K, R, H), and propensity to form helical structure. SSRCalc was developed for 100- and 300-Å pore size RP C18 sorbents, linear water/acetonitrile gradient systems with a constant flow rate, and 0.1% TFA as ion-pairing modifier. Formic acid and acetic acid can be used as alternative ion-pairing reagents without invalidating the approach. Separations at pH 10 in combination with small pore size sorbents of 100 Å have also been investigated. Changing the ion-pairing modifier influences the algorithms predictability, thus SSRCalc models developed for different eluent conditions cannot be compared directly with each other and need to be standardized for each system [39]. Cysteine-containing peptides need to be reduced and alkylated with iodoacetamide for correct predictions with SSRCalc as other cysteine protection groups or free cysteines result in different retention coefficients. The SSRCalc model can distinguish peptides with close mass over charge ($m/z$) values derived from mass spectrometers that could be advantageous for detailed peptide mapping applications of selected peptides.

To calibrate the RP-HPLC system, the gradient delay time ($A$) has first to be determined, and then the hydrophobicity values ($H$) have to be calculated and plotted versus the retention time ($RT$). The resulting linear function $RT = A + B^* H$ allows determination of the constant ($B$) – a value that is related to the slope of the acetonitrile gradient. These parameters can be applied to the analysis of unknown samples where reproducible LC is required. Constant $B$ can also be transferred to different LC systems as long as the same gradient is used. The algorithm has been improved over time and currently Version 3 of the program is available at the Manitoba Center for Proteomics web site (see http://hs2.proteome.ca/SSRCalc/SSRCalc.html).

The current version of hydrophobicity index (HI)-linked SSRCalc applies additional calculation to determine the HI so that the actual acetonitrile concentrations of peptide elution over multiple separation systems can be determined. In addition to the calculation of relative hydrophobicity, the percentage of acetonitrile (or molar fraction of organic solvent, $\varphi$) as a dimension unit to express peptide hydrophobicity was proposed and further developed to express the HI. A peptide HI indicates the organic solvent concentration which provides a fixed retention factor measured under isocratic conditions and is characteristic of peptide–sorbent interaction depending only on the type of stationary phase and ion-pairing modifier [40]. Formulae for the conversion of the SSRCalc hydrophobicity into HI for different sorbent-ion-pairing modifiers are described by Krokhin *et al.* [40]. Expression of hydrophobicity in acetonitrile percentage scale is beneficial for both comparison of peptide retention and method development as a facile transfer from isocratic to gradient methods.

SSRCalc was developed on a small database of 346 tryptic peptides. Recent studies by other groups investigated the influence of post-translational modifications [41, 42]. Phosphorylation events on 33 peptides were examined and were shown to generally decrease retention time on RP-HPLC when compared to their nonphosphorylated forms. In contrast, some peptides (five phosphorylated peptides from the dataset) showed increased retention. Other possible peptide modifications such as acetyl and amide protecting groups indicated that the acetyl group increases retention on RP-HPLC, whereas amide protecting groups decrease retention time.

In addition to the SSRCalc procedure for predicting peptide elution, approaches based on (i) sequence-dependant retention time prediction called "liquid chromatography at critical conditions" (LCCC), was developed to take into consideration the location of amino acids in the primary structure [43], (ii) prediction models based on the investigation of physicochemical descriptors (e.g., the logarithm of the Van der Waals volume of the peptide and the logarithm of the peptide calculated *n*-octanol/water partition coefficient) using quantitative structure–retention relationships (QSRRs) [44, 45], (iii) techniques relying on statistical learning [46, 47], and (iv) prediction of RP-HPLC retention time with the use of artificial neural networks (ANNs), introduced and further developed by Petritis *et al.* [48, 49]. The use of ANNs is of particular interest as this takes into consideration very large peptide databases of approximately 345 000 non-redundant peptides derived from more than 12 000 LC-MS/MS experiments of more than 20 different organisms. Initially, the predictive capability was based solely on an amino acid composition-based model, but was significantly improved in later studies by incorporation of peptide sequence information including peptide length, sequence, hydrophobicity and hydrophobic moment, nearest-neighbor amino acid, as well as peptide predicted structural configurations such as helix, sheet or coil. During the development of the algorithm, the ANN system was tested with a set of 1300 peptides that were identified with highest confidence and excluded from the 345 000 peptide training set beforehand. The model displayed an average elution time precision of about 1.5%, allowing for accurate retention time prediction of both isomers and isobaric peptides, which has resulted in more confidence in peptide retention prediction and identifications.

To standardize the elution pattern of a RP-HPLC system by using a standard protein digest or synthetic peptide mixture for calibration of the LC system, peptides should be selected to cover the entire hydrophobicity range to resemble complex biological samples. The peptides should not contain amino acids that are subject to chemical modifications such as oxidation or N-terminal cyclization reactions (e.g., pyroglutamate). Further, standard peptides should have appropriate molecular weights and ionization properties enabling the detection with common MS instruments [40].

## 15.5 Advantages of Reduced Scale

Frequently, biological samples to be analyzed by peptide RP-HPLC-ESI-MS are precious and as such it is essential to maximize the signal-to-noise ratio of the chromatographic eluate into the mass spectrometer. By reducing eluate volume and peak widths the peptide is introduced into the mass spectrometer as a sharp, concentrated signal that improves the opportunity for detection and analysis. One of the most common and simple ways to improve sensitivity is reducing the column diameter and flow rate. This offers several benefits, including less solvent consumption, less sample required for loading, and smaller elution volume for increased

**Table 15.1** Sensitivity as a function of column diameter.

| Column: internal diameter (mm) | Flow rate | Detection sensitivity |
|---|---|---|
| Analytical: 4.6 | 1.0 ml/min | 1.0 |
| Narrowbore: 2.1 | 200 μl/min | 5 |
| Microbore: 1.0 | 50 μl/min | 20 |
| Capillary: 0.3 | 5 μl/min | 200 |
| Nanobore: 0.075 | 300 nl/min | 3300 |
| Open tubular: 0.01 | 20 nl/min | 50 000 |

peptide concentration. The increase in sensitivity achieved by this is inversely proportional to the square of the column diameter (Eq. (15.1)) [50]. Table 15.1 shows the increase in sensitivity achieved by reducing the column diameter compared to a standard 4.6-mm HPLC column.

$$\text{Flow } a = \left(\frac{r_a}{r_b}\right)^2 \times \text{Flow } b. \tag{15.1}$$

Today, a typical RP-HPLC-ESI-MS experiment will employ a nanobore column bearing an internal diameter of 75 μm running at a flow rate of approximately 300 nl/min. While column diameters as small as 15 μm have been used with flow rates down to 20 nl/min columns become progressively more difficult to maintain and pack as column diameter decreases [50]. The 75-μm diameter column has become generally accepted as a balance between increased sensitivity and ease of use.

Nanobore RP-HPLC columns often double as ESI emitters for MS applications. Unlike analytical and narrowbore columns, which are typically packed as stainless steel columns, RP-HPLC packings for nanobore are slurry packed into a length of fused silica of appropriate internal diameter in which one end is pulled to a fine tip less than 5 μm across [51]. Voltage applied at a liquid junction before the column initiates spray and charge transfer to the eluate. Coupling RP-HPLC and ESI eliminates postcolumn peak broadening due to dead volumes allowing the eluate to spray directly into the MS.

## 15.6
## Two-Dimensional Chromatographic Methods

The use of separations orthogonal to RP-HPLC has grown dramatically in recent years. Chromatography by differing retention mechanisms offers the ability to separate peptide species that otherwise would coelute, risking the potential inability to detect one of the species, particularly if the concentration of the two species differ greatly. In a perfect orthogonal chromatographic system, the peak capacity of the RP-HPLC separation is multiplied by the peak capacity of each additional chromatographic dimension [52]. A number of separation mechanisms can be employed as a first dimension to complement RP-HPLC. These include strong-cation exchange (SCX) [53], isoelectric focusing [54], sodium dodecyl sulfate–polyacrylamide gel electrophoresis (SDS–PAGE) [55], and even RP-HPLC conducted in basic pH versus acidic pH [56]. While methods such as isoelectric focusing or SDS–PAGE are considered "off-line" methods for first dimension separations [57], some orthogonal techniques may be placed "in-line" by packing the first-dimension particles behind the RP particles in the same column or linking two columns in series. Fractions eluted from the first dimension are caught on the second, RP dimension which can then be separated by standard RP-HPLC gradients.

One of the most common in-line multidimensional techniques is to combine SCX with RP-HPLC. Peptides are eluted via salt step gradients off the SCX media and captured on the RP-HPLC media. Washburn *et al.* demonstrated this technique in a 100-μm fused silica integrated ESI-tip column which was first packed with 10 cm of 5-μm C18 media, followed by 4 cm of 5-μm SCX media. By introducing step gradients of ammonium acetate, they were able to fractionate a tryptic digest of yeast into 15 separate RP-HPLC runs. This experiment yielded the identification of 5540 unique peptides and 1484 proteins from yeast, and demonstrated that in-line multidimensional separations interface quite well with MS analysis.

A number of specialized multidimensional separations have been implemented coupled to RP-HPLC. For example, the use of inverse-gradient chromatography [58, 59] (or more recently known as hydrophilic interaction liquid chromatography (HILIC)) [60] for the enrichment of peptides and proteins at high organic solvent strength. This has proven to be an excellent fractionation scheme for phosphopeptides [61, 62], and the immunoaffinity depletion of highly abundant proteins in blood serum samples to reduce complexity prior to proteolytic digestion and RP-HPLC-ESI-MS analysis [63].

As mass spectrometers become faster and more sensitive, chromatography must keep up in order to provide suitable separations for robust analysis. New techniques such as nanobore RP-HPLC, UHPLC, and multidimensional chromatography have provided ways of improving peak capacities and generate increasingly efficient peptide separations. With the recent efforts to attempt to identify every protein in an organism's proteome [64, 65] high peak capacity, high sample capacity RP-HPLC is required to overcome the extremely complex nature of biological samples such as cell lysates or, worse yet, blood plasma fractions.

## 15.7 Peptide Analysis in Complex Biological Matrices

The analysis of protein composition in biological samples has been a subject of interest and considerable challenge in recent years. Biological samples such as blood serum can be phenomenally complex, having as many as 25 000 individual proteins, with 100 000 or more variations of these, covering a concentration range of at least 10 orders of magnitude [66]. Recently, advances in MS have provided a rapid high-throughput method for the analysis of peptides generated by proteolytic digestion from complex protein samples. However, given the above example of the complexity of blood serum, it has been estimated that a proteolytic digest of blood serum will generate as many as 600 000 peptides for potential analysis [52]. As technology in the development of mass spectrometers has advanced, the ability to analyze increasingly complex mixtures of peptides has become available. This, in turn, has renewed interest in the ability to separate complex peptide samples via chromatography as a way of managing sample complexity and increase sensitivity for MS analysis. Many of the chromatographic methodologies currently used are merely improvements of traditionally used techniques such as reduction in scale, column lengths, particle size, and adjustments in pore size and compositions. Additionally, efforts to make chromatography more compatible with MS analysis have been quite successful. Also, several novel technologies are just beginning to be implemented that will offer significant improvements in peak capacities for the nanoscale separations which have become commonplace for MS analysis.

## 15.8 Standard Methods for Peptide Separations for Analysis by Hyphenated Techniques

The gold standard for peptide separation and purification well before its application to MS has been RP-HPLC. RP-HPLC offers high resolution and sensitivity, often with the ability to separate polypeptides with as few as one amino acid difference [67]. In addition, RP-HPLC offers several advantages to peptide separation for analysis by MS including direct interfacing of the HPLC eluate to an ESI source (ESI-MS) (see [68] for a more thorough review of interfacing RP-HPLC to MS), a MS-compatible solvent composition (i.e., no salts), and good orthogonality to other peptide and protein separation techniques such as SCX, isoelectric focusing, and hydrophobic interaction chromatography (HIC) [52, 53]. Separations orthogonal to RP-HPLC prior to ESI-MS analysis can dramatically increase the number of identified peptides from a complex biological sample by MS analysis in a single experiment, albeit at the expense of run-time. Despite the many advantages RP-HPLC has for performing peptide separations for MS analysis, numerous adaptations and modifications have been required and implemented in order to improve RP-HPLC compatibility with ESI-MS. Optimizing both RP-HPLC separations and electrospray conditions is a delicate balance that can return tremendous gains in sensitivity while maintaining optimal peak capacity in separations.

## 15.9 Emerging Methods for Peptide Separations for Analysis by Hyphenated Techniques

Emerging technologies particularly suited to RP-HPLC separation of peptides is the development of microfabricated fluidic "chips" based on polyimide structures. These novel chromatographic systems are suitable for Lab-on-a-Chip applications that can be either packed with silica or monolithic phases or manufactured with arrays of ordered pillars of silica rods with the earliest being published in 1995 [69–71]. The advantages of this type of column structure are numerous, both in ease of use, separation standardization and transferability, and increases in chromatographic performance. The structure of columns from microfabricated chips decreases band dispersion. The major cause of the efficiency limits in modern LC is the band dispersion occurring because of the poor structural homogeneity of the stationary phase support (i.e., the random arrangement of particles in a packed bed inside a column or the random structure of monoliths) [72]. With the design and structure of microfabricated chips, dead volumes and defined connection volumes are kept to the low nanoliter volumes, extremely hard to achieve with the manual construction of traditional fused silica columns and finger-tight nano fittings. Recent commercial introduction of HPLC chip constructions with various configurations and RP-HPLC stationary phases have been made by Agilent [73] (see Figure 15.6) and Eksigent Technologies.

Numerous applications have emerged using RP-HPLC and the focus on hyphenated techniques is particularly suited to Lab-on-a-Chip applications. Peptide separation for proteomics discovery and applications to targeted quantitation through MRM technology will increase the use of this design in peptide separation [74].

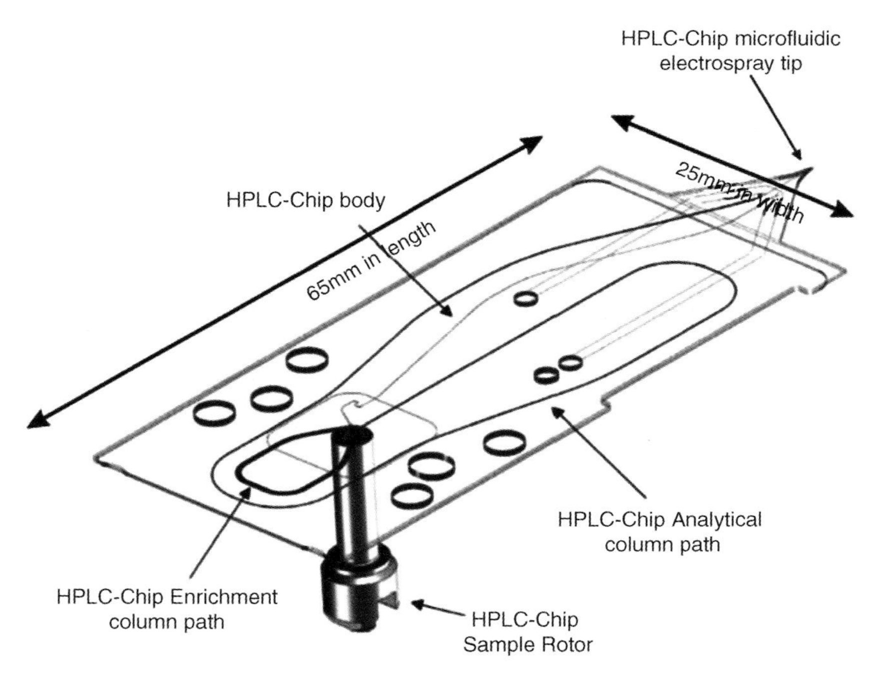


**Figure 15.6** Construction of "Lab-on-a-Chip" HPLC column for RP-HPLC of peptides and MS detection. Construction: polyimide chip construction 65 × 24 mm. Columns: Zorbax 300SB separation column, 50–150 mm (HPLC-Chip has a micromachined "square" LC channel of 50 μm (D) × 75 μm (W) × 50–150 mm (L) looped 3 times the length of the chip) and 40 nl enrichment column; flow rate: 200–600 nl/min. (Figure courtesy of Agilent Technologies, Inc.)

**Table 15.2** Current HPLC silica and column manufacturers.

| Manufacturer | URL | RP | Silica | Polymer | Packed | Bulk | Notes |
|---|---|---|---|---|---|---|---|
| ACE | www.ace-hplc.com | yes | yes | | yes | yes | |
| Agela Tech | www.agela.com | yes | yes | yes | yes | | |
| Agilent Technologies | www.chem.agilent.com | yes | yes | | yes | | |
| Benson Polymeric | www.bensonpolymeric.com | yes | | yes | yes | yes | |
| BIA Separations | www.biaseparations.com | | | yes | yes | | manufactures monolithic columns |
| Bischoff Chromatography | www.bischoff-chrom.com | yes | yes | | yes | yes | refill columns |
| Capital HPLC | www.capital-hplc.co.uk | yes | yes | | yes | | |
| Chromatographic Specialties | www.chromspec.com | yes | yes | | yes | | |
| Chromatography Research Supplies | www.chromres.com | yes | | yes | yes | | |
| Cohesive Technologies/ Thermo Scientific | www.cohesivetech.com | | | | | | manufactures turbulent flow columns |
| Column Engineering | www.column-engineering.com | yes | yes | | yes | yes | |
| Dionex | www.dionex.com | yes | yes | | yes | | |
| Dr. Maisch | www.dr-maisch.com | yes | yes | | yes | yes | |
| ESA | www.esainc.com | yes | yes | | yes | | |
| Fortis Technologies | www.fortis-technologies.com | yes | yes | | yes | | |
| GL Sciences | www.glsciences.com | yes | yes | yes | yes | | |
| Grace Davison Discovery Sciences | www.discoverysciences.com | yes | yes | yes | yes | | |
| GS-Tek | www.gs-tek.com | | yes | | yes | | distributor of Agela Tech products |
| Hamilton | www.hamiltoncompany.com | yes | yes | yes | yes | yes | |

| | | | | | | | |
|---|---|---|---|---|---|---|---|
| HiChrom | www.hichrom.co.uk | yes | yes | | yes | | |
| Higgins Analytical | www.higanalyt.com | yes | yes | yes | yes | | |
| Honeywell | http://www51.honeywell.com/sm/rlss/bandj/products-applications/chromato/hplc_columns.html?c=243 | yes | yes | | yes | | |
| IDEX Health and Science | www.idex-hs.com | | | | | | manufactures HPLC hardware and accessories |
| Jordi Labs | www.jordilabs.com | yes | | yes | yes | yes | |
| Knauer | www.knauer.net | yes | yes | | yes | | |
| Kromasil | www.kromasil.com | | yes | | yes | | Manufactures HPLC columns for distribution |
| Macherey-Nagel | www.mn-net.com | yes | yes | | yes | yes | |
| Michrom BioResources | www.michrom.com | yes | yes | yes | yes | yes | |
| MZ-Analysentechnik | www.mz-at.de | yes | yes | | yes | yes | |
| Nacalai USA | www.nacalaiusa.com | yes | yes | | yes | | |
| Nomura Chemical | http://www.develosil.net/english/index.html | yes | yes | | yes | | |
| Orachrom | www.orachrom.com | yes | | yes | yes | | |
| Phenomenex | www.phenomenex.com | yes | yes | yes | yes | yes | |
| Pickering Labs | www.pickeringlabs.com | yes | | | | | specialty columns available |
| PolyLC | www.polylc.com | yes | yes | | yes | yes | |
| Restek | www.restek.com/ | yes | yes | yes | yes | | |
| Separation Methods Technologies | www.separationmethods.com | yes | yes | | yes | yes | |
| SeQuant-Merck | www.sequant.com | | yes | yes | yes | | manufacturer of HILIC columns |
| Shiseido HPLC | http://www.shiseido.co.jp/e/hplc/ | yes | yes | yes | yes | | |
| Shodex | www.shodex.com/ | yes | yes | yes | yes | | |
| Sielc Technologies | www.sielc.com/ | yes | yes | | yes | | |

*(Continued)*

**Table 15.2** (*Continued*)

| Manufacturer | URL | RP | Silica | Polymer | Packed | Bulk | Notes |
|---|---|---|---|---|---|---|---|
| Thomson Instruments | http://www.hplc.com/index.htm | yes | yes | | yes | | |
| Tosoh Bioscience | www.separations.eu.tosohbioscience.com | yes | yes | yes | yes | | |
| Unimicro Technologies | www.unimicrotech.com | yes | yes | | yes | | |
| Varian | www.varianinc.com | yes | yes | yes | yes | yes | |
| Waters | www.waters.com | yes | yes | yes | yes | | |
| YMC America | www.ymcamerica.com | yes | yes | yes | yes | yes | |
| ZirChrom Separations | www.zirchrom.com | yes | yes | | yes | | zirconia-based packing |

## 15.10
## Practical use of RP-HPLC for Purifying Peptides (Analytical and Preparative Scale)

This protocol will provide both practical and simple tips to set-up the HPLC system for successful RP-HPLC analysis of peptides. To design efficient separation schemes using RP-HPLC, it is extremely important to evaluate the entire chromatographic system, HPLC hardware, column and buffers, before it is applied to the desired separation. This is especially true when a new column enters the laboratory and applies to both beginners and more experienced chromatographers, who wish to obtain reproducible peptide maps. Many styles of stationary phase and column configurations are available to the modern chromatographer and a brief list of current manufacturers of hardware for RP-HPLC of peptides is shown in Table 15.2. Upon purchasing a new column or stationary phase for packing into a column and before using a new column, familiarize yourself with the manufacturer's instructions and recommendations for proper conditioning, usage, and storage of the column ensuring adherence to the proper use of buffers and pressure. Often, subtle operating conditions that are crucial for correct column usage are overlooked, resulting not only in irreversible column damage, but also the possible loss of important samples before one realizes that the damage is irrevocable. Once the operating procedures are understood, a test mixture separation should be performed using a set of well understood and characterized standards (e.g., peptide or protein mixtures or even a simple tryptic digest of a model protein such as myoglobin) to evaluate column performance. For this exercise, it is essential to reproduce the standard set of operating conditions described in this protocol before proceeding with an analysis of the test sample. If an equivalent elution profile cannot be obtained on the new column using this set of standards and the standard operating conditions described here, then any new or old column should be considered suspect and replaced by the manufacturer or simply repurchased.

For standardization within the laboratory as well as providing a set of simple test procedures for all laboratory personnel to carry out and results agreed upon, these standard operating procedures should be carried out on a regular basis (and carefully logged) to monitor column performance and project the life-time of the column. Because stationary phases differ widely from one manufacturer to another (see Table 15.2) in terms of their physical characteristics, the functional ligand used, ligand density, base support (silica or polystyrene–divinylbenzene), particle size, pore size, as well as column dimensions, standard peptide separations such as the one described here are a useful diagnostic of these differences (and can provide useful information as to the separation power of that particular chromatographic phase and can be exploited in a purification strategy). If samples are to be collected for further characterization, careful attention should be given to the "dead volume" of the plumbing between the flow cell and outlet in order to accurately collect purified peptides into correct fractions and avoid peak mixing.

For illustration purposes, the purification of Glu-fibrinogen peptide from crude chemical syntheses is shown in Figure 15.7. Prior to purification, an analytical RP-HPLC 4.6-mm internal diameter column was prepared and validated according to

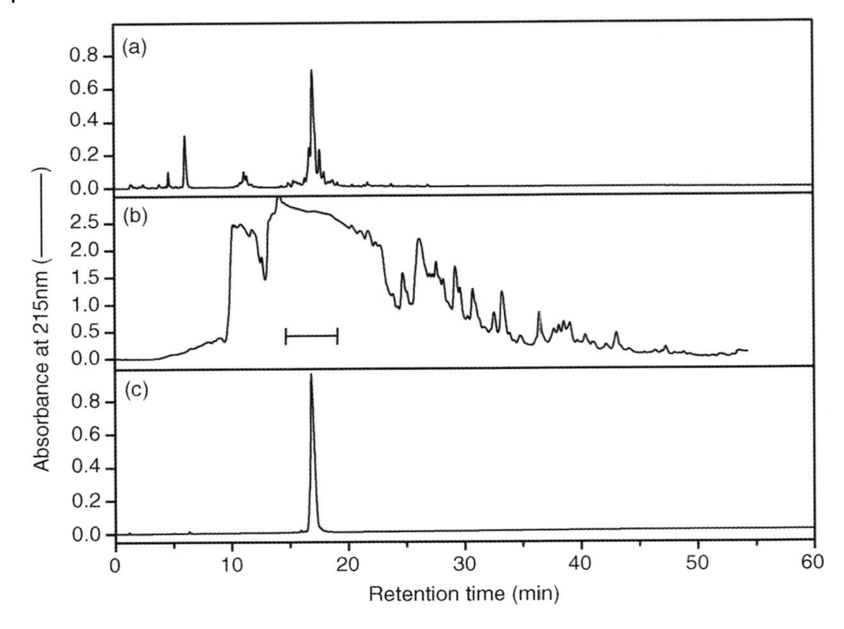


**Figure 15.7** Example of RP-HPLC purification of crude peptide Glu-fibrinogen peptide. Columns: Vydac C4, 4.6 × 150 mm (panels A and C) and 10 × 150 mm (panel B) columns; mobile phase: gradient, 0–100% B in 60 min, A = 0.1% TFA in water, B = 0.0935% TFA/60% acetonitrile; flow rate: 5.0 ml/min (10 mm internal diameter) and 1.0 ml/min (4.6 mm internal diameter); temperature: 45 °C; detection: UV 210 nm; sample: 50 mg of Glu-fibrinogen peptide (EGVNDNEEGFFSAR). (A) Analytical separation of 50 μg crude synthetic peptide, (B) 50 mg crude synthetic peptide purification by preparative RP-HPLC, and (C) analytical separation of 50 μg purified peptide from pool (indicated by bar in B) from preparative RP-HPLC.

the following procedure and an analysis of the crude preparation was performed (Figure 15.7A). A preparative RP-HPLC column (10 mm internal diameter) with the same stationary phase was also prepared and validated the same way and a separation of 50 mg of crude peptide was performed (Figure 15.7B). Pooled fractions from the preparative separation was evaluated for purity on the same analytical RP-HPLC column used to evaluate the crude peptide preparation (Figure 15.7C).

## 15.10.1 Simple Protocol for Successful RP-HPLC

### 15.10.1.1 Buffer Preparation

1) Acetonitrile (HPLC grade or higher).
2) Deionized and polished water (18 MΩ, Milli-Q, Millipore, or bottled HPLC grade equivalent).
3) TFA (HPLC grade, Thermo or equivalent) for UV detection.
4) FA (HPLC grade, Aldrich *puriss* or equivalent) for MS.
5) 1 ml Gilson pipette or equivalent.

Solvent A aqueous 0.1% v/v TFA. Make by adding 1.0 ml of neat TFA to 1 l of $H_2O$ in a glass-stoppered measuring cylinder, mix thoroughly end over end, and then pour into a clean HPLC reservoir bottle (extensively washed with Milli-Q $H_2O$ beforehand). Substitute TFA for FA if using MS.

Solvent B (60% v/v acetonitrile/40% $H_20$ containing 0.1% v/v TFA). Make by adding 600 ml of acetonitrile to a 1-l glass-stoppered cylinder and adjust to 1 l with $H_2O$ (400 ml). (*NOTE*: Mixing of acetonitrile and water will reduce the volume by around 4% and be endothermic. For consistent results, always add water to the acetonitrile.) Add 1 ml of neat TFA (final concentration 0.1% v/v TFA) and mix the solution thoroughly. Substitute TFA for FA if using MS.

*Solvent preparation tip:*

- If both Solvent A and Solvent B both contain 0.1% (v/v) TFA, then the absorbance at 214 nm of Solvent A will be slightly higher – 20-40 milli-absorption units (mAU) – than that of Solvent B, due to the contributing absorbance of acetonitrile. This will result in a rising baseline during the development of a gradient from 0 to 60% acetonitrile due to the additional absorbance of the acetonitrile. Although this does not present a problem when working with 50–100 μg of material (using a 4.6-mm internal diameter column), the rising baseline can become a serious problem (i.e., the baseline will be off-scale if using high-sensitivity setting on the UV detector and late-eluting peaks cannot be detected without adjusting the baseline, which is extremely difficult to accomplish during the course of a chromatographic run). To overcome this potential problem, it is recommended that the amount of TFA in Solvent B be slightly less than that in Solvent A (e.g., 0.09–0.095% or 900–950 μl of TFA/l) compared with 1.0% (1000 μl of TFA/l). By these means, the amount of TFA is adjusted carefully to accomplish a flatter baseline.
- For MS applications, replace TFA with FA to ensure good ionization of the eluting peptides as using TFA will compete for the charge with the peptide and give poor-quality results. For the mass spectrometer, there is no need for a UV detector, therefore the concentration of FA should be kept constant for both Solvent A and B.
- If solvents are not degassed, then bubble formation can occur during a chromatographic run as a result of out-gassing. Such bubbles can lodge in the flow cell, resulting in an erratic detector signal (i.e., baseline). To minimize this risk, Solvent B should be a "percentage buffer" (i.e., containing $H_2O$ (e.g., 60% acetonitrile, 40% $H_2O$ containing 0.1% TFA)). Additionally, solvents should be degassed every 1–2 days with high-purity helium or constant vacuum over a period of 30 min. Most modern HPLC systems include a degassing device so there would be no need to add additional degassing and the instrument manufacturer's recommendations should be followed.
- For specific applications, the researcher may want to use 2-propanol or methanol as a gradient elution system. With alcohol-type gradients, a slow build up of esters occurs at room temperature, resulting in an increased baseline absorbance. For this reason, solvents must be prepared every 1–2 days and then discarded properly according to local environment laws.

15.10.1.2 **HPLC Chromatographic System**

The HPLC system should be equipped with programmed gradient elution, UV detection to at least 210 nm, software for acquisition of chromatographic data and integration of peak areas, and be capable of accurate and reproducible solvent delivery and gradient formation at a range of flow rates required for the particular separation size. Flow rates from 50 μl to 5 ml/min for microbore (1 mm internal diameter column) to semi-preparative (10 mm internal diameter) range.

- RP column (C4 or C18, 5 μm particle size, 120 or 300 Å pore size).
- Semipreparative column 150 × 10-mm internal diameter (flow rate 5 ml/min).
- Analytical column 150 × 4.6 mm internal diameter or cartridge of 100 × 4.6 mm internal diameter (flow rate 1 ml/min).
- Narrow bore column 150 × 2.1 mm internal diameter or cartridge of 100 × 2.1 mm internal diameter (flow rate 0.2–0.1 ml/min).
- Microbore column 150 × 1.0 mm internal diameter (flow rate 50 μl/min).
- Capillary column 150 × 0.3 mm internal diameter (flow rate 4 μl/min).
- Nanobore column 150 × 0.075–0.05 mm internal diameter (flow rate 0.6–0.3 μl/min),
- Routine for new column usage. Prime both solvent lines with Solvents A and B, respectively, to ensure that there are no bubbles in the system. Flush the HPLC system completely before attaching the column by making sure a steady flow of liquid is flowing out of the final liquid tubing prior to attaching the column. Attachment of the solvent line to the column with air present can irreparably damage the column bed. Connect the solvent delivery tubing from the injector to the top end of the column and let liquid flow out of the end of the column while monitoring the back-pressure to ensure the column is not over-pressured. Then connect the tubing from the column outlet to the flow cell and flow 30 column volumes of Solvent B though the column, followed by 30 volumes of Solvent A to completely wet the packed bed. The column is now ready to be evaluated using the procedure described below.

*Connection of columns to HPLC plumbing tip:*

- Before connecting a column to the HPLC plumbing, make sure that column end-fittings, ferrules, and tubing depths are correctly matched. Incorrect use of column end-fittings can seriously damage a column, resulting in solvent leakage and/or high dead volumes at the connections (and hence poor chromatographic performance). For capillary and nanobore columns this is especially important as the slightest dead volume could equal the volume of the entire elution gradient, resulting in extremely poor performance, resolution, and sensitivity.

15.10.1.3 **Test Sample**

Dissolve the peptide standards (1 mg of each peptide/ml) in 80% $H_2O$, 20% acetonitrile, and 0.1% TFA (can be stored frozen for several weeks). It is recommended that the 1 mg/ml solution of peptide standards be stored in small aliquots (e.g., 100 μl) and not freeze-thawed more than 5 times.

1) Run a blank or "control chromatogram" (i.e., no sample loaded). Typically, this is the first chromatographic run of the day. The following are the recommended standard chromatographic conditions for the RP-HPLC system evaluation:
   - Linear gradient: From 0% to 100% B, where Solvent A is aqueous and Solvent B is aqueous 60% acetonitrile.
   - Gradient rate: typically 1% acetonitrile/min (i.e., 0–100% Solvent B (60% acetonitrile/40% $H_2O$) in 60 min).
   - Flow rate: matched to the column internal diameter (e.g., 1 ml/min for a 4.6-mm internal diameter column, 5 ml/min for a 10-mm internal diameter column).
   - Temperature: Ambient or 45 °C.

   Before commencing the blank run, pump 20 column volumes of Solvent B through the column, followed by an equal volume of Solvent A (ensure that baseline is stabilized, i.e., remains flat). Then adjust the flow rate to an operating flow rate of 1 ml/min (or 5 ml/min for a 10-mm internal diameter column or 0.1 ml/min for a 2.1-mm internal diameter column) and wait until the baseline stabilizes (around 5 min). Column volumes are calculated by measuring column length and internal diameter and applying the following equation $V = \pi r^2 h$, where $V$ is the column volume, $r$ is the internal radius of the column, and $h$ is the length of the column.
2) Run the peptide standard (or mixed peptide standards) (control run) as described in Step 1, but inject 100-μl of peptide standard(s) (0.1 of mg stock standards/ml) for the 4.6-mm internal diameter column (i.e., 10 μg of each peptide standard; 10 μg equates around 100 mAU at a flow rate of 1 ml/min). For a 2.1-mm internal diameter column, load 10 μl of 0.1 mg peptide standards/ml (i.e., 1 μg of each peptide standard (1 μg equates around 100 mAU at a flow rate of 0.1 ml/min)).
3) Run the peptide digest or the crude synthetic peptide mixture, using the same conditions described in Step 2.
4) For sample collection, collect eluting peptide peaks in polypropylene tubes (microfuge or 96-well plates) fitted with tight-fitting caps and store them at −20 °C for further analysis. *DO NOT USE* polystyrene-based plastics (e.g., clear type used for ELISA assays, etc.) for collection as these tubes or plates will absorb protein and peptide.

   If sample collection is to be performed manually, the dead volume of the plumbing from the detector flow cell to the outlet tubing must be determined accurately.
5) For long-term RP-HPLC column storage, flush the columns and store them with an aqueous organic solvent (e.g., 50% v/v acetonitrile or methanol/50% $H_2O$). For overnight storage, pump the columns with Solvent B at a very slow flow rate (e.g., 50 μl/min) to keep the system free of air and at the ready for analysis the next morning without having to repeat priming the system and can be used to immediately start at Step 1.

## References


1 Mahoney, W.C. and Hermodson, M.A. (1980) Separation of large denatured peptides by reverse phase high performance liquid chromatography. Trifluoroacetic acid as a peptide solvent. *The Journal of Biological Chemistry*, **255**, 11199–11203.

2 O'Hare, M.J. and Nice, E.C. (1979) Hydrophobic high-performance liquid chromatography of hormonal polypeptides and proteins on alkylsilane-bonded silica. *Journal of Chromatography*, **171**, 209–226.

3 Snyder, L.R. and Kirkland, J.J. (1979) *Introduction to Modern Liquid Chromatography*, Wiley-Interscience, New York.

4 Mant, C.T. and Hodges, R.S. (1991) *High-Performance Liquid Chromatography of Peptides and Proteins: Separation, Analysis and Conformation*, CRC Press, Boca Raton, FL.

5 Janson, J. and Ryden, L. (1998) *Protein Purification*, 2nd edn, Wiley-VCH Verlag GmbH, Weinheim.

6 Boyes, B.E. and Walker, D.G. (1995) Selectivity optimization of reversed-phase high-performance liquid chromatographic peptide and protein separations by varying bonded-phase functionality. *Journal of Chromatography A*, **691**, 337–347.

7 Mazzeo, J.R., Neue, U.D., Kele, M., and Plumb, R.S. (2005) Advancing LC performance with smaller particles and higher pressures. *Analytical Chemistry*, **77**, 460A–467A.

8 Cunliffe, J.M., Adams-Hall, S.B., and Maloney, T.D. (2007) Evaluation and comparison of very high pressure liquid chromatography systems for the separation and validation of pharmaceutical compounds. *Journal of Separation Science*, **30**, 1214–1223.

9 Fulton, S., Meys, M., Protentis, J., Afeyan, N.B., Carlton, J., and Haycock, J. (1992) Preparative peptide purification by cation-exchange and reversed-phase perfusion chromatography. *BioTechniques*, **12**, 742–747.

10 Petro, M., Svec, F., Gitsov, I., and Frechet, J.M.J. (1996) Molded monolithic rod of macroporous poly(styrene-*co*-divinylbenzene) as a separation medium for HPLC of synthetic polymers: "on-column" precipitation-redissolution chromatography as an alternative to size exclusion chromatography of styrene oligomers and polymers. *Analytical Chemistry*, **68**, 315–321.

11 Minakuchi, H., Nakanishi, K., Soga, N., Ishizuka, N., and Tanaka, N. (1996) Octadecylsilylated porous silica rods as separation media for reversed-phase liquid chromatography. *Analytical Chemistry*, **68**, 3498–3501.

12 Svec, F. and Frechet, J.M.J. (1996) Molded separation media. An inexpensive, efficient, and versatile alternative to packed columns for the fast HPLC separation of peptides, proteins, and synthetic oligomers and polymers. *Macromolecular Symposia*, **110**, 203–216.

13 Unger, K.K., Skudas, R., and Schulte, M.M. (2008) Particle packed columns and monolithic columns in high-performance liquid chromatography-comparison and critical appraisal. *Journal of Chromatography A*, **1184**, 393–415.

14 Sandra, K., Moshir, M., D'hondt, F., Verleysen, K., Kas, K., and Sandra, P. (2008) Highly efficient peptide separations in proteomics: Part 1. Unidimensional high performance liquid chromatography. *Journal of Chromatography B*, **866**, 48–63.

15 Luo, Q., Shen, Y., Hixson, K.K., Zhao, R., Yang, F., Moore, R.J., Mottaz, H.M., and Smith, R.D. (2005) Preparation of 20-microm-i.d. silica-based monolithic columns and their performance for proteomics analyses. *Analytical Chemistry*, **77**, 5028–5035.

16 Luo, Q., Tang, K., Yang, F., Elias, A., Shen, Y., Moore, R.J., Zhao, R., Hixson, K.K., Rossie, S.S., and Smith, R.D. (2006) More sensitive and quantitative proteomic measurements using very low flow rate

porous silica monolithic LC columns with electrospray ionization-mass spectrometry. *Journal of Proteome Research*, **5**, 1091–1097.

17 Luo, Q., Page, J.S., Tang, K., and Smith, R.D. (2007) MicroSPE-nanoLC-ESI-MS/MS using 10-μm-i.d. silica-based monolithic columns for proteomics. *Analytical Chemistry*, **79**, 540–545.

18 Moritz, R.L. and Simpson, R.J. (1992) Purification of proteins and peptides for sequence analysis using microcolumn liquid chromatography. *Journal of Microcolumn Separations*, **4**, 485–489.

19 Stone, K.L., Elliott, J.I., Peterson, G., McMurray, W., and Williams, K.R. (1990) Reversed-phase high-performance liquid chromatography for fractionation of enzymatic digests and chemical cleavage products of proteins. *Methods in Enzymology*, **193**, 389–412.

20 de Villiers, A., Lauer, H., Szucs, R., Goodall, S., and Sandra, P. (2006) Influence of frictional heating on temperature gradients in ultra-high-pressure liquid chromatography on 2.1mm I.D. columns. *Journal of Chromatograph A*, **1113**, 84–91.

21 Xiang, Y., Liu, Y., and Lee, M.L. (2006) Ultrahigh pressure liquid chromatography using elevated temperature. *Journal of Chromatography A*, **1104**, 198–202.

22 Gilar, M., Daly, A.E., Kele, M., Neue, U.D., and Gebler, J.C. (2004) Implications of column peak capacity on the separation of complex peptide mixtures in single- and two-dimensional high-performance liquid chromatography. *Journal of Chromatography A*, **1061**, 183–192.

23 Shen, Y., Tolic, N., Masselon, C., Pasa-Tolic, L., Camp, D.G., 2nd; Hixson, K.K., Zhao, R., Anderson, G.A., and Smith, R.D. (2004) Ultrasensitive proteomics using high-efficiency on-line micro-SPE-nanoLC-nanoESI MS and MS/MS. *Analytical Chemistry*, **76**, 144–154.

24 Glajch, J.L. (1986) Analytical challenges in biotechnology. *Analytical Chemistry*, **58**, 385A–394A.

25 Stadalius, M.A., Gold, H.S., and Snyder, L.R. (1985) Optimization model for the gradient elution separation of peptide mixtures by reversed-phase high-performance liquid chromatography: verification of band width relationships for acetonitrile–water mobile phases. *Journal of Chromatography A*, **327**, 27–45.

26 Stadalius, M.A., Gold, H.S., and Snyder, L.R. (1984) Optimization model for the gradient elution separation of peptide mixtures by reversed-phase high-performance liquid chromatography: verification of retention relationships. *Journal of Chromatography A*, **296**, 31–59.

27 Hancock, W.S., Chloupek, R.C., Kirkland, J.J., and Snyder, L.R. (1994) Temperature as a variable in reversed-phase high-performance liquid chromatographic separations of peptide and protein samples. I. Optimizing the separation of a growth hormone tryptic digest. *Journal of Chromatography A*, **686**, 31–43.

28 Meek, J.L. (1980) Prediction of peptide retention times in high-pressure liquid chromatography on the basis of amino acid composition. *Proceedings of the National Academy of Sciences of the United States of America*, **77**, 1632–1636.

29 Meek, J.L. and Rossetti, Z.L. (1981) Factors affecting retention and resolution of peptides in high-performance liquid chromatography. *Journal of Chromatography*, **211**, 15–28.

30 Browne, C.A., Bennett, H.P.J., and Solomon, S. (1982) The isolation of peptides by high-performance liquid chromatography using predicted elution positions. *Analytical Biochemistry*, **124**, 201–208.

31 Casal, V., Martin-Alvarez, P.J., and Herraiz, T. (1996) Comparative prediction of the retention behaviour of small peptides in several reversed-phase high-performance liquid chromatography columns by using partial least squares and multiple linear regression. *Analytica Chimica Acta*, **326**, 77–84.

32 Guo, D., Mant, C.T., Taneja, A.K., Parker, J.M.R., and Hodges, R.S. (1986) Prediction of peptide retention times in reversed-phase high-performance liquid

chromatography. I. Determination of retention coefficients of amino acid residues of model synthetic peptides. *Journal of Chromatography*, **359**, 499–518.

33 Guo, D., Mant, C.T., Taneja, A.K., and Hodges, R.S. (1986) Prediction of peptide retention times in reversed-phase high-performance liquid chromatography. II. Correlation of observed and predicted peptide retention times and factors influencing the retention times of peptides. *Journal of Chromatography*, **359**, 519–532.

34 Palmblad, M., Ramström, M., Bailey, C.G., McCutchen-Maloney, S.L., Bergquist, J., and Zeller, L.C. (2004) Protein identification by liquid chromatography-mass spectrometry using retention time prediction. *Journal of Chromatography B*, **803**, 131–135.

35 Palmblad, M., Ramström, M., Markides, K.E., Hakansson, P., and Bergquist, J. (2002) Prediction of chromatographic retention and protein identification in liquid chromatography/mass spectrometry. *Analytical Chemistry*, **74**, 5826–5830.

36 Krokhin, O.V., Craig, R., Spicer, V., Ens, W., Standing, K.G., Beavis, R.C., and Wilkins, J.A. (2004) An improved model for prediction of retention times of tryptic peptides in ion pair reversed-phase HPLC: its application to protein peptide mapping by off-line HPLC-MALDI MS. *Molecular & Cellular Proteomics*, **3**, 908–919.

37 Krokhin, O.V. (2006) Sequence-specific retention calculator. Algorithm for peptide retention prediction in ion-pair RP-HPLC: application to 300- and 100-A pore size C18 sorbents. *Analytical Chemistry*, **78**, 7785–7795.

38 Krokhin, O.V., Ying, S., Cortens, J.P., Ghosh, D., Spicer, V., Ens, W., Standing, K.G., Beavis, R.C., and Wilkins, J.A. (2006) Use of peptide retention time prediction for protein identification by off-line reversed-phase HPLC-MALDI MS/MS. *Analytical Chemistry*, **78**, 6265–6269.

39 Spicer, V., Yamchuk, A., Cortens, J., Sousa, S., Ens, W., Standing, K.G., Wilkins, J.A., and Krokhin, O.V. (2007) Sequence-specific retention calculator. A family of peptide retention time prediction algorithms in reversed-phase HPLC: applicability to various chromatographic conditions and columns. *Analytical Chemistry*, **79**, 8762–8768.

40 Krokhin, O.V. and Spicer, V. (2009) Peptide retention standards and hydrophobicity indexes in reversed-phase high-performance liquid chromatography of peptides. *Analytical Chemistry*, **81**, 9522–9530.

41 Kim, J., Petritis, K., Shen, Y., Camp, D.G. 2nd, Moore, R.J., and Smith, R.D. (2007) Phosphopeptide elution times in reversed-phase liquid chromatography. *Journal of Chromatography A*, **1172**, 9–18.

42 Baczek, T. and Sieradzka, M. (2008) Influence of acetyl and amide groups on peptides RP-LC retention behaviour. *Journal of Liquid Chromatography and Related Technologies*, **16**, 2417–2428.

43 Gorshkov, A.V., Tarasova, I.A., Evreinov, V.V., Savitski, M.M., Nielsen, M.L., Zubarev, R.A., and Gorshkov, M.V. (2006) Liquid chromatography at critical conditions: comprehensive approach to sequence-dependent retention time prediction. *Analytical Chemistry*, **78**, 7770–7777.

44 Baczek, T., Wiczling, P., Marszall, M., Heyden, Y.V., and Kaliszan, R. (2005) Prediction of peptide retention at different HPLC conditions from multiple linear regression models. *Journal of Proteome Research*, **4**, 555–563.

45 Kaliszan, R., Baczek, T., Cimochowska, A., Juszczyk, P., Wiśniewska, K., and Grzonka, Z. (2005) Prediction of high-performance liquid chromatography retention of peptides with the use of quantitative structure-retention relationships. *Proteomics*, **5**, 409–415.

46 Pfeifer, N., Leinenbach, A., Huber, C.G., and Kohlbacher, O. (2007) Statistical learning of peptide retention behavior in chromatographic separations: a new kernel-based approach for computational proteomics. *BMC Bioinformatics*, **8**, 468.

47 Pfeifer, N., Leinenbach, A., Huber, C.G., and Kohlbacher, O. (2009) Improving peptide identification in proteome analysis by a two-dimensional retention time filtering approach. *Journal of Proteome Research*, **8**, 4109–4115.

48 Petritis, K., Kangas, L.J., Ferguson, P.L., Anderson, G.A., Pasa-Tolic, L., Lipton, M.S., Auberry, K.J., Strittmatter, E.F., Shen, Y., Zhao, R., and Smith, R.D. (2003) Use of artificial neural networks for the accurate prediction of peptide liquid chromatography elution times in proteome analyses. *Analytical Chemistry*, **75**, 1039–1048.

49 Petritis, K., Kangas, L.J., Yan, B., Monroe, M.E., Strittmatter, E.F., Qian, W.-J., Adkins, J.N., Moore, R.J., Xu, Y., Lipton, M.S., Camp, D.G., and Smith, R.D. (2006) Improved peptide elution time prediction for reversed-phase liquid chromatography-MS by incorporating peptide sequence information. *Analytical Chemistry*, **78** 5026–5039.

50 Scott, R.P.W. and Kucera, P. (1979) Use of microbore columns for the separation of substances of biological origin. *Journal of Chromatography*, **185**, 27–41.

51 Gatlin, C.L., Kleemann, G.R., Hays, L.G., Link, A.J., and Yates, J.R. 3rd (1998) Protein identification at the low femtomole level from silver-stained gels using a new fritless electrospray interface for liquid chromatography-microspray and nanospray mass spectrometry. *Analytical Biochemistry*, **263**, 93–101.

52 Motoyama, A. and Yates, J.R. 3rd (2008) Multidimensional LC separations in shotgun proteomics. *Analytical Chemistry*, **80**, 7187–7193.

53 Washburn, M.P., Wolters, D., and Yates, J.R. (2001) Large-scale analysis of the yeast proteome by multidimensional protein identification technology. *Nature Biotechnology*, **19**, 242–247.

54 Hörth, P., Miller, C.A., Preckel, T., and Wenz, C. (2006) Efficient fractionation and improved protein identification by peptide OFFGEL electrophoresis. *Molecular & Cellular Proteomics*, **5**, 1968–1974.

55 Schirle, M., Heurtier, M.-A., and Kuster, B. (2003) Profiling core proteomes of human cell lines by one-dimensional PAGE and liquid chromatography-tandem mass spectrometry. *Molecular & Cellular Proteomics*, **2**, 1297–1305.

56 Gilar, M., Olivova, P., Daly, A.E., and Gebler, J.C. (2005) Two-dimensional separation of peptides using RP-RP-HPLC system with different pH in first and second separation dimensions. *Journal of Separation Science*, **28**, 1694–1703.

57 Moritz, R.L., Ji, H., Schütz, F., Connolly, L.M., Kapp, E.A., Speed, T.P., and Simpson, R.J. (2004) A proteome strategy for fractionating proteins and peptides using continuous free-flow electrophoresis coupled off-line to reversed-phase high-performance liquid chromatography. *Analytical Chemistry*, **76**, 4811–4824.

58 Simpson, R.J., Moritz, R.L., Nice, E.E., and Grego, B. (1987) A high-performance liquid-chromatography procedure for recovering subnanomole amounts of protein from SDS-gel electroeluates for gas-phase sequence-analysis. *European Journal of Biochemistry*, **165**, 21–29.

59 Simpson, R.J. and Moritz, R.L. (1989) Chromatographic fractionation of proteins at high organic-solvent modifier concentrations. *Journal of Chromatography*, **474**, 418–423.

60 Alpert, A.J. (1990) Hydrophilic-interaction chromatography for the separation of peptides, nucleic acids and other polar compounds. *Journal of Chromatography A*, **499**, 177–196.

61 McNulty, D.E. and Annan, R.S. (2008) Hydrophilic interaction chromatography reduces the complexity of the phosphoproteome and improves global phosphopeptide isolation and detection. *Molecular & Cellular Proteomics*, **7**, 971–980.

62 Boersema, P.J., Mohammed, S., and Heck, A.J.R. (2008) Hydrophilic interaction liquid chromatography (HILIC) in proteomics. *Analytical and Bioanalytical Chemistry*, **391**, 151–159.

63 Martosella, J. and Zolotarjova, N. (2008) Multi-component immunoaffinity subtraction and reversed-phase

chromatography of human serum. *Methods in Molecular Biology*, **425**, 27–39.
64 Picotti, P., Bodenmiller, B., Mueller, L.N., Domon, B., and Aebersold, R. (2009) Full dynamic range proteome analysis of *S. cerevisiae* by targeted proteomics. *Cell*, **138**, 795–806.
65 Malmström, J., Beck, M., Schmidt, A., Lange, V., Deutsch, E.W., and Aebersold, R. (2009) Proteome-wide cellular protein concentrations of the human pathogen *Leptospira interrogans*. *Nature*, **460**, 762–765.
66 Anderson, N.L. and Anderson, N.G. (2002) The human plasma proteome: history, character, and diagnostic prospects. *Molecular & Cellular Proteomics*, **1**, 845–867.
67 Rivier, J. and McClintock, R. (1983) Reversed-phase high-performance liquid chromatography of insulins from different species. *Journal of Chromatography*, **268**, 112–119.
68 Aebersold, R. (2003) Constellations in a cellular universe. *Nature*, **422**, 115–116.
69 Ocvirk, G., Verpoorte, E., Manz, A., Grasserbauer, M., and Widmer, H.M. (1995) High-performance liquid-chromatography partially integrated onto a silicon chip. *Analytical Methods and Instrumentation*, **2**, 74–82.
70 Hasselbrink, E.F., Shepodd, T.J., and Rehm, J.E. (2002) High-pressure microfluidic control in lab-on-a-chip devices using mobile polymer monoliths. *Analytical Chemistry*, **74**, 4913–4918.
71 De Pra, M., De Malsche, W., Desmet, G., Schoenmakers, P.J., and Kok, W.T. (2007) Pillar-structured microchannels for on-chip liquid chromatography: Evaluation of the permeability and separation performance. *Journal of Separation Science*, **30**, 1453–1460.
72 Knox, J.H. (2002) Band dispersion in chromatography – a universal expression for the contribution from the mobile zone. *Journal of Chromatography A*, **960**, 7–18.
73 Yin, H. and Killeen, K. (2007) The fundamental aspects and applications of Agilent HPLC-Chip. *Journal of Separation Science*, **30**, 1427–1434.
74 Staes, A., Timmerman, E., Van Damme, J., Helsens, K., Vandekerckhove, J., Vollmer, M., and Gevaert, K. (2007) Assessing a novel microfluidic interface for shotgun proteome analyses. *Journal of Separation Science*, **30**, 1468–1476.

# 16
# Difficult Peptides

*M. Terêsa Machini Miranda, Cleber W. Liria, and Cesar Remuzgo*

## 16.1
## Importance of Peptide Synthesis

As thousands of biologically active peptides have been discovered in the last 50 years, to date no one contests that all living organisms – from bacteria to mammals and complex plant forms – produce such compounds for vital purposes. A few examples are given in Table 16.1.

In association with the first discoveries of biologically active peptides, it was noted that natural sources could provide quite restricted amounts of them. At that time, the composition and structure of proteins were not sufficiently understood and the ribosomal machinery remained unknown. Hence, many first-rate chemists started searching for methods, procedures, and protocols that could allow for the preparation of peptides in a variable range of scales and purities. Such circumstances turned peptide synthesis into an important research topic [1, 2].

In the following decades, peptide synthesis reached a high level of efficacy, and also became a fundamental tool for research in the chemistry and biology of these macromolecules [1, 3]. Indeed, most of the current knowledge on peptide hormones, antibiotics, cytotoxins, opioids, and hormone-releasing factors has been generated with the help of synthetics.

Peptide synthesis has also been critical for providing a wide range of information in biochemistry, medicine, biology, and chemistry as its applications include: (i) therapeutic purposes [4, 5]; (ii) development of techniques for peptide detection, separation, identification, quantification, and analysis [6, 7]; (iii) characterization of proteinase–substrate or inhibitor interactions [8, 9]; (iv) study of protein–protein interactions and the basis of protein folding [10, 11]; (v) mapping of protein epitopes [12, 13]; (vi) mimicking of enzyme activity [14, 15]; (vii) engineering of new peptide-based drugs with therapeutic potential [16, 17]; (viii) design of new biomaterials [18, 19]; (ix) synthesis of prodrugs [20, 21]; and (x) chemical synthesis of proteins [22, 23].

Despite these advances, peptide synthesis remains a research topic that attracts an impressive number of scientists worldwide. Most current studies focus

*Amino Acids, Peptides and Proteins in Organic Chemistry.*
*Vol.3 – Building Blocks, Catalysis and Coupling Chemistry.* Edited by Andrew B. Hughes

ISBN: 978-3-527-32102-5

**Table 16.1** Biological functions played by peptides in mammals.

| Peptide | Structure | Function | Reference |
|---|---|---|---|
| Oxytocin | (cyclo 1–6)CYIQNCPLG-$NH_2$ | uterine contraction; milk ejection | [24] |
| Bradykinin | RPPGFSPFR | vasodilation | [25] |
| Cholecystokinin-33 | KAPSGRMSIVKNLQNLDPSHRIS DRDY($SO_3H$)MGWMDF-$NH_2$ | gallbladder contraction and relaxation; pancreatic enzyme secretion | [26] |
| Glutathione | γECG | strong antioxidant effect | [27] |
| Vasopressin | (cyclo 1–6)CYFQNCP(K or R)G | vasoconstriction and anti-diuresis | [28] |
| Glucagon | HSQGTFTSDYSKYLDSRRAQDFV QWLMNT | response to changes in blood glucose concentration | [29, 30] |
| Histatin-5 | DSHAKRHHGYKRKFHEKHHSHR GY | fungicidal effect | [31] |
| Protegrins | (cyclo 6–15, 8–13)RGGRLCYCRR RFCVCVGR<br>(cyclo 6–15, 8–13)RGGRLCYCRR RFCICV<br>(cyclo 6–15, 8–13)RGGGLCYCRRR FCVCVGR | antimicrobial effect | [32] |
| Met- or | YGGFM | analgesia | [33] |
| Leu-enkephalin | YGGFL | | |
| Transforming growth factor-β | (cyclo 8–21, cyclo 16–32, cyclo 34–43) VVSHFNKCPDSHTQYCFHGTCRFL VQEEKPACVCHSGYVGVRCEH ADLLA | collagen production stimulation | [34] |

on: (i) methods, procedures, and protocols that may furnish comparatively higher quality products (free or surface-bound) with higher yields in less time or, at least, overcome problems that are inherent in existing technologies [35, 36]; and (ii) new uses [37, 38].

In summary, for decades peptide synthesis has been a fascinating target for research as well as an essential tool for research and production.

## 16.2 Methods for Peptide Synthesis

The formation of a peptide bond involves the α-carboxyl group of an amino acid derivative (acyl donor) and the α-amine of another (acyl acceptor). As an amine is usually a good nucleophile at high pH, but the hydroxyl of a carboxyl group is a poor leaving group, the α-carboxyl group of the amino acid that acts as the acyl donor ($R^1$-COOH) must be activated ($R^1$-COX; where X is an electron-withdrawing group) to allow for the nucleophilic attack by the α-amino group of the amino acid that acts as acyl acceptor ($H_2N$-$R^2$):

$$R^1\text{-COOH} + \text{activating agent} \rightarrow R^1\text{-COX} + H_2N\text{-}R^2 \rightarrow R^1\text{-CONH-}R^2$$

The means of such activation determines the method used to make peptides in the laboratory. The main methods are the following:

**Chemical** – employs a wide range of chemical reagents, such as carbodiimides and uronium salts [1, 22, 39].
**Enzymatic** – employs isolated proteases or esterases (free or immobilized), most being enzymes whose mechanisms of action comprise the formation acyl-enzyme intermediates with the substrate [40, 41].
**Recombinant DNA** – employs a transporter RNA for each of the DNA-encoded amino acids as it uses the ribosomal machinery of bacteria, fungi, or phages [42, 43].

All these methods furnish a crude peptide that usually contains the desired peptide and impurities. Thus, it must be purified by liquid chromatography. An exception is made for phage display technology (a modality of recombinant DNA technology) since, in this case, the peptides are produced on the surface of bacteriophages and remain attached to it while used in binding assays [44, 45].

Compared with the others, the chemical method of peptide synthesis has been studied far more, and is considered the first option for routinely obtaining short and medium-sized peptides. In fact, such a method has the potential of being able to furnish the synthetic form of naturally occurring peptides protein fragments or model peptides, as well as their analogs containing unusual amino acids, D-amino acids, chemical modifications, and conformational constraints or N- and/or C-truncated analogs. On the other hand, chemical peptide synthesis requires full or, at least, partial protection of the reactive groups, produces large quantities of waste, and leads to amino acid enantiomerization in extents that are largely dependent on the amino acid nature, type of protecting groups, and synthesis conditions [1, 2, 39].

The enzyme-mediated and the recombinant DNA methods are enantioselective, friendlier to the environment, and, once properly adapted, quite useful for large-scale production in reactors [46, 47]. Nevertheless, they are not as general as the chemical methods.

## 16.3 Chemical Peptide Synthesis

It is well-known that this method can be performed classically (in solution) or on a solid support. Figure 16.1 summarizes the chemical strategies and all existing approaches for both [39, 48–50].

In the classical method (i) all reagents are fully soluble in the reaction media, (ii) the α-carboxyl group of the acyl acceptor is usually esterified or amidated, (iii) peptide elongation might occur in the N → C or C → N direction, and (iv) the product of

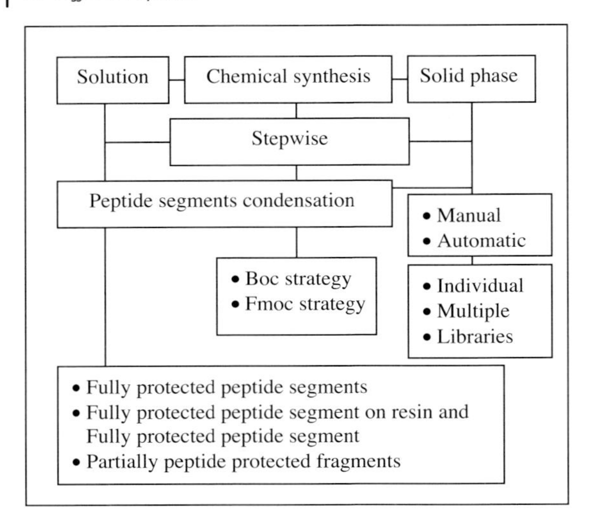


**Figure 16.1** Chemical strategies and practical variations of chemical peptide synthesis.

each step of the synthetic process is isolated, purified, quantified, and characterized. Therefore, it furnishes crude products of high purity. In addition, it can be performed in any scale to provide milligram to kilogram amounts of peptide. On the other hand, this is very time-consuming – a feature that inspired the creation of the solid-phase method.

In fact, in the early 1960s, R.B. Merrifield made a major contribution to science by demonstrating that peptide synthesis could be significantly sped-up if it was carried out on a resin (solid-phase peptide synthesis (SPPS)) [51]. Being covalently bound to the resin, the products of each step of peptide elongation (aminoacyl-resin or growing peptide-resin resulting from (i) resin aminoacylation, (ii) removal of the α-amino group blocker, (iii) α-amine group, neutralization and (iv) amino acid coupling to the growing peptide-resin; Figure 16.2) could be separated from the reaction media by simple filtration. Then, the aminoacyl-resin or growing peptide-resin could be submitted to alternate washings with solvents (that swell or shrink it) for efficient elimination of the reagents and soluble byproducts [1, 39]. More importantly, it could be submitted to reaction with ninhydrin or picrate, allowing the monitoring of each synthetic step [52, 53].

SPPS was initially designed for stepwise incorporation of the amino acids to a functionalized polystyrene-based resin by the Boc/Bzl chemistry, but in the 1980s a new chemical strategy was established (Fmoc strategy [54]). This significant event unleashed the study and development of numerous other polymeric supports and linkers [1, 55], and made possible the preparation of carboxyl-free or C-terminal-modified peptides with high quality and in good yields. The combination of the Fmoc strategy and the new resins/linkers also expanded the use of SPPS to the synthesis of side-chain-modified (e.g., sulfated, phosphorylated, and glycosylated) and head-to-tail cyclized peptides. Moreover, it allowed for the preparation of peptide

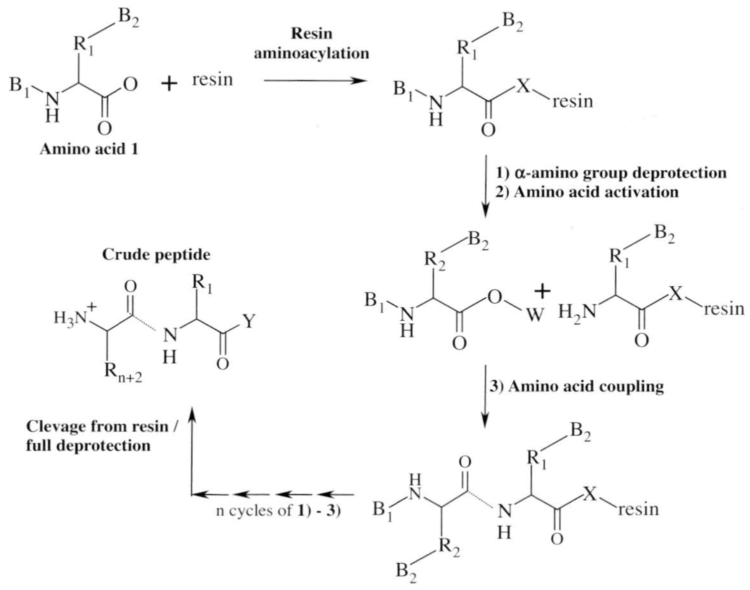

**Figure 16.2** Stepwise SPPS ($B_1$: α-amino group blocker; $B_2$: amino acid side-chain blocker; R: amino acid side-chain; resin: functionalized resin; X: O or NH; W: activating group; Y: OH or $NH_2$).

libraries [56, 57], peptide arrays for screening purposes [58, 59], multiple antigenic peptides [60, 61], and peptide fragments for convergent chemical synthesis of protein analogs in solution by a variety of approaches that include native chemical ligation [22], template-assisted coupling, and domain ligation [62].

The interest in using peptide segments instead of amino acids as acyl donors in SPPS also made fragment condensation on resin achievable in recent decades. In principle, convergent solid-phase synthesis (CSPS) is a fastest approach for making large peptides whose syntheses are not practicable by stepwise SPPS [3, 4]. Although it has been successfully used by many authors, CSPS of trifunctional amino acid-rich peptides is not always so straightforward. The difficulties arise from the limited solubility of fully protected peptide segments in the organic solvents commonly used in stepwise SPPS [2, 63] as well as from steric hindrance effects that may hamper their coupling to the growing peptide-resin.

The recent idea of using elevated temperatures in all steps of SPPS made it even faster and, consequently, a method even more attractive for preparing milligram to gram quantities of a peptide. Although it needs further studies before being used routinely, SPPS at high temperature has been effectively used for the preparation of peptides, glycopeptides, and peptoids under a variable range of conditions that employ conventional heating [35, 64] or microwave irradiation [36, 65]. In the last case, automatic peptide synthesizers for single synthesis have already been designed and tested [66, 67].

It is essential to stress that the classical method and SPPS are complementary tools for peptide production. Many factors must be considered when deciding which one must be used.

## 16.4 "Difficult Peptide Sequences"

It is well known that, despite the progress of chemical peptide synthesis, certain peptides are extremely difficult or, even, impractical to synthesize by the methods, strategies, and approaches shown in Figure 16.1. Synthetic difficulties may arise from peptide features such as the following: (i) presence of chemical bonds that are labile in conditions typically employed for peptide assembly [2, 68, 69]; (ii) amino acid sequence highly prone to undergo a large number of the side-reactions typical of chemical peptide synthesis [1, 70]; (iii) presence of a single amino acid (homopolymers) [71, 72]; (iv) elevated content of repetitive motifs rich in a specific amino acid [35, 73]; (v) high content of peptide bonds involving amino acids with bulky side-chains or protecting groups [74, 75]; (vi) high proportion of amino acids that induce peptide folding [76]; and (vii) combination of amino acid sequence with side-chain protecting groups highly prone to favor intra- or intermolecular association that leads to peptide aggregation [77]. Although all of these peptides could be called "difficult peptides" or "difficult peptide sequences," the expression has been reserved to describe peptides with a high tendency for aggregation during their syntheses [78, 79]. Coincidently, when released from resin and fully deprotected, many so-called "difficult peptides" are also difficult to dissolve in aqueous solution and exhibit a tendency to adopt secondary structure [80]. Table 16.2 contains examples of a few peptides whose chemical syntheses are difficult to perform (problematic) as well as some typical "difficult peptides."

Peptide aggregation occurs in the classical method and in SPPS. It may be more severe when the peptide is attached to the resin. The phenomenon is explained by the formation of hydrogen bonds involving the –NH and the –C=O groups of the peptide bonds of a growing peptide chain, leading to β-sheet-like structures (Figure 16.3) that make the α-amine group of the growing peptide less accessible to the acyl donor or other reagents used for its extension. As a result, the rates of $N^{\alpha}$-deprotection and incorporation of a new amino acid to the growing peptide chain are significantly reduced, and are not completed under standard conditions even after prolonged reaction times, recouplings, substitution of the coupling reagents, and change of the chemical strategy. Actually, such synthetic difficulty can be mastered only if the aggregates formed are disrupted, which requires unusual experimental conditions and/or introduction of modified amino acids. However, even when peptide aggregation is overcome and peptide assembly on resin is concluded by the Boc or Fmoc strategy, the resulting crude peptide is a complex mixture of the "difficult peptide sequence" with many structurally related byproducts (deleted analogs). As fractioning of such components by liquid chromatography, the separation technique mostly used for such purpose [81], is not an easy task, the

**Table 16.2** Examples of peptides difficult to synthesize and of typical "difficult peptides.".

| Peptides | Amino acid sequence | Reference |
|---|---|---|
| **Peptides difficult to synthesize** | | |
| small cyclic peptide | cyclo(VTVTVT) | [84] |
| human cholecystokinin-58 | VSQRTDGESRAHLGALLARYIQQARKAPSG RMSIVKNLQNLDPSHRISDRDY($SO_3H$) MGWMDF-$NH_2$ | [85] |
| N-terminal domain of γ-zein | $(VHLPPP)_8$ | [76] |
| branched poly(proline) peptide | Fmoc-PPPPP-Amp(Fmoc-PPPPP)-PPPPPP | [86] |
| RANTES (Regulated upon Activation Normal T-cell Expressed and Secreted) (1–68) | SPYSSDTTPCCFAYIARPLPRAHIKEY FYTSGKCSNPAVVFVTRKNRQVCANP EKKWVREYINSLEMS-$NH_2$ | [87] |
| **Typical "difficult peptides** | | |
| acyl carrier protein (65–74) | VNVNVQVQVD | [88] |
| poly(Ala) ($Ala_{10}$) | AAAAAAAAAA | [89] |
| amyloid β peptide ($Aβ_{42}$) | DAEFRHDSGYEVHHQKLVFFAEDVGSNKG AIIGLMVGGVVIA | [90] |
| amyloidogenic neurotoxin prion peptide (106–126) | KTNMKHMAGAAAAGAVVGGLG | [91] |
| acanthoscurrin (101–132) | GGGLGGGRGGGYGGGGGYGGGYGGGYG GGKYK-$NH_2$ | [35] |

purification process invariably leads to significant loss of the desired peptide and, thus, to extremely low final peptide yields.

Chemists have tried to predict "difficult peptide sequences" by using different approaches. Based on previous observations of Narita *et al.* that protected peptide segments in a random coil conformation are more soluble in organic solvent suitable

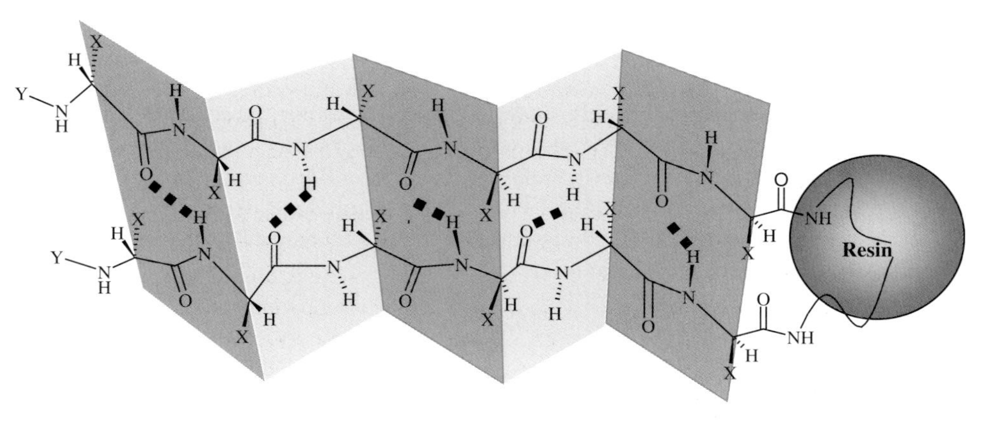


**Figure 16.3** Schematic representation of peptide intermolecular aggregation during solid-phase synthesis (X: R or R-P, where R is the amino acid side-chain and P is its protecting group; Y: Boc or Fmoc).

for aminoacylation conditions [82, 83], Milton *et al.* proposed a method that uses the Chou–Fasman conformational parameter $P_c$ (coil parameter for each amino acid) as follows: the average, $<P_c>$, of a peptide segment reveals its tendency to assume a random coil conformation instead of a α-helix or β-sheet structure; consequently, $<P_c^*>$ ($P_c^*$ can be obtained from the linear regression of the function $1/P_c = 0.739P_\alpha + 0.345P_\beta$) values greater than 1.0 indicated easy coupling of the subsequent residue in an acceptable time; values in the range of 0.9–1.0 indicated a longer reaction time and need for "recoupling"; values lower than 0.9 indicated persistent difficult coupling [79]. Van Woerkon and Van Nispen analyzed the results of 696 couplings using a computer program that considered the Fmoc-amino acid used, its side-chain protecting group, the amino acid to be acylated, the length of the growing peptide, and the result of the Kaiser tests performed [92]. In 1993, Krchnak *et al.* followed the volume changes of swollen peptide-resins during SPPS and derived a $P_a$ for each amino acid coupled ($P_a$: aggregation potential, which reflects the propensity of the amino acid to contribute to peptide aggregation). The use of $P_a$ allowed for the preparation of aggregation profiles, such as those of acyl carrier protein (63–74), $(Ala)_{10}$, cytochrome *c* (66–104), HIV-1 aspartyl protease (86–99), and growth hormone releasing factor (1–44), and for predicting potentially difficult couplings [93].

Finally, still in the early 1990s, many reputable researchers recognized that (i) typical "difficult sequences" present reproducible repetitive incomplete aminoacylations whose improvement by recoupling or capping is limited; (ii) such synthetic difficulty is sequence-dependent, occurring irrespective of resin type or chemical strategy, and aggravated by high resin loadings and sterically hindered amino acids in the sequence; (iii) there is weak or no correlation with the N-terminal amino acid of the growing peptide-resin; and (iv) these symptoms are accompanied by reduced swelling of the growing peptide-resin [78, 79, 94]. Afterwards, the identification of new "difficult peptide sequences" became quite easy.

## 16.5 Means to Overcome Peptide Aggregation in SPPS

It has been proposed that the following factors determine the occurrence of peptide aggregation during SPPS: type of resin, its degree of substitution, degree of swelling of the growing peptide-resin in the solvent used for the incorporation of new amino acid derivatives, and nature of the incorporating amino acid derivatives [80, 90]. Such a proposal is based on a huge number of results obtained in the earlier studies focused on the following areas.

### 16.5.1 *In Situ* Neutralization

Until the early 1980s, SPPS was mostly performed by Boc chemistry [96] and aggregation on resin during the chain assembly was thought to occur when the deprotected α-amine group of the growing peptide resin (resulting from Boc group removal in trifluoroacetic acid (TFA)) was neutralized with an organic base. In the late 1980s and early 1990s, this was confirmed to be true because amino acid sequences found to aggregate during their synthesis by this chemical strategy also aggregated when a new chemical strategy, Fmoc [54], was used [79, 92]. Based on such knowledge, in the early 1990s, a few authors reported *in situ* neutralization protocols for minimizing the occurrence of the phenomenon during stepwise SPPS by the Boc strategy [96, 97]. Among the syntheses successfully carried out using such protocols was that of HIV-1 protease (89–99) – a typical "difficult peptide" [96].

### 16.5.2 Solvents for Peptide Chain Assembly

For many years after the introduction of the solid-phase method, all its steps were performed almost exclusively in dichloromethane (DCM) because this solvent was able to swell the resins based on polystyrene–divinylbenzene (PS-DVB) copolymers (an important factor in peptide build-up on resin), and dissolve Boc-amino acids, carbodiimides, and 1-hydroxybenzotriazole (HOBt) [95]. Later, it was found that the solvation of the growing peptide-resin in DCM was insufficient. Hence, peptide chemists searched for new options by mixing it or replacing it with highly polar solvents such as dimethylformamide (DMF), *N*-methypyrrolidone (NMP), and dimethylsulfoxide (DMSO) as these aprotic solvents and their mixtures with DCM were capable of efficiently swelling peptide-resins and/or reducing the rigidity of β-sheet structures [98, 99]. The mixture of 20% hexafluoro-2-propanol (HIFP)/DCM was also demonstrated to be more suitable than DCM for the incorporation of a Boc-amino acid in a peptide-resin due to swelling enhancement [100].

That being so, a large number of studies have been performed with the aim of understanding the correlation between solvent properties and solvation of peptide-resins derived from the so-called "difficult peptide sequences." As a result, aggregation in stepwise SPPS using resin based on 1% DVB-PS copolymer, for instance, was found to be minimized in 20% TFE/DCM, 35% tetrahydrofuran (THF)/NMP, HFIP/DCM, NMP, DCM/DMF/NMP (1 : 1 : 1, containing 1% Triton X-100 and 2 M ethylenecarbonate) at 55 °C and in 25% DMSO/toluene at 60 °C [94, 101–105].

### 16.5.3 Type and Substitution Degree of Resins for Peptide Chain Assembly

The PS-DVB-based resins were almost exclusively used for at least two decades (in fact, until now, SPPS has relied on such nonpolar supports [106]). Nevertheless, with the elucidation of their swelling limitations, discovery of new coupling reagents, and conception of the Fmoc chemistry [54], this situation has changed. In fact, resins based on comparatively more polar polymers were used in combination with DMF, being the solvent of choice (i) for Fmoc removal in the presence of an organic base, and (ii) for coupling reactions involving Fmoc-amino acid derivatives and coupling reagents such as 2-(1*H*-benzotriazol-1-yl)1-1,3,3-tetramethyluronium hexafluorophosphate (TBTU) or *O*-(7-azabenzotriazol-1-yl)-1,1,3-3,-tetramethyluroniumhexafluorophosphate (HATU) [107].

Based on successful chemical syntheses of aggregating peptide sequences in solution using poly(ethylene glycol) (PEG) as a blocker of the α-carboxyl group, in 1995, Merrifield reported its incorporation into polystyrene resin to give PEG-PS. Due to its more polar character compared to traditional polystyrene-based resins, the resin PEG-PS was able to swell in protic solvents [106].

In the following years, various laboratories developed new resins with the aim of increasing peptide-resin solvation and, consequently, SPPS efficiency. Among the new resins reported are TentaGel (contains polystyrene in its structure [108]), CLEAR (does not contain polystyrene in its structure [109]), ChemMatrix (a 100% PEG support [105]), and 2% 1,4-butanediol dimethacrylate cross-linked polystyrene (BDDMA-PS [110]). All of them have been successfully employed for the synthesis of typical "difficult peptides."

It is important to point out that the higher the resin loading, the higher the peptide concentration on resin and the greater the chance of interaction between the growing peptide chains. In other words, it is highly recommended to use resins with low loading degrees when attempting the synthesis of a peptide with a chronic tendency to aggregate.

### 16.5.4
### Use of Chaotropic Salts During Peptide Chain Assembly

The positive effect of chaotropic salts in peptide solubility in anhydrous organic solvents and in peptide conformation (due to complexation) was demonstrated in the late 1980s [111]. Thus, in the early 1990s, 2 M LiBr anhydrous THF, quite effective in disrupting the β-sheet structure in peptide-resins, was used for disaggregation of a peptidyl-Kaiser oxime resin [112]. Subsequently, anhydrous solvents such as DMF, NMP, and THF, or the binary mixture DMF/DCM (1 : 1), containing or not LiCl, LiBr, KSCN, or $NaClO_4$, were employed in solid-phase coupling reactions starting from Fmoc-$(Ala)_5$Phe, a well-known "difficult peptide sequence," on polystyrene, poly(*N*,*N*-dimethylacrylamide) "kieselguhr" (PDMAA-KG), and poly(ethylene oxide) (PEO)-PS resins. The results obtained indicated that the reaction yields were mainly enhanced in the presence of the lithium salts due to swelling enhancement of the peptide-resins [113].

### 16.5.5
### Use of Amide Backbone Protection

The 2,4-dimethoxybenzyl (Dmb) group as amide backbone blocker was first employed during chemical synthesis in solution to prevent the decrease of peptide solubility due to acquisition of secondary structures [114]. In the early 1980s, the idea of using it (the so-called "tertiary peptide bond") was rescued and employed in classical convergent synthesis of large peptides and small proteins: briefly, Narita *et al.* observed that the solubility of oligoleucine fragments in polar solvents (DMSO, THF, and MeOH) was significantly reduced owing to formation of β-sheet structures, and that the phenomenon was reversed when Dmb and Pro were incorporated in the sequence (due to acquisition of an unordered structure) [115, 116]. On the other hand, Eckert and Seigel employed the chromophoric ferrocenylmethyl (FEM) group to protect the NH of peptide bonds involving Asn and/or Gly and disrupt secondary structures [117], achieving the synthesis of a human β-endorphin fragment (24–31) in solution [118].

In 1992, Bartl and Frank applied the same approach to the synthesis of $Ala_{13}$ on cellulose paper by substituting the NH of peptide bonds involving Ala with methyloxymethyl (Mom), benzyloxymethyl (Bom), ethyloxymethyl (Etm), or phenylthiomethyl (Ptm) groups and found that Ptm was the most efficient for disrupting peptide aggregation [119]. In the same year, Bedford *et al.* synthesized the model peptide poly (Ala)–Val and the dodecapeptide derived from the C-terminal fragment of human KDEL receptor by the solid-phase method, having observed that the insertion of Pro and Sar (*N*-methylglycine) precluded aggregation and had a long range effect (four to six residues) [71].

With the introduction of the reversibly protected tertiary peptide bond 2-hydroxy-4-methoxybenzyl (Hmb) group in Ala, stepwise solid-phase synthesis of the acyl carrier protein fragment (65–74) and $(Ala)_{17}$–Val were achieved, confirming the long range and structure dissociating effects of the NH protection [120]. Although, it became obvious that the acylation of the *N,O*-bis-Fmoc derivatives was not an easy task, it must, therefore, be performed with symmetric anhydrides or pentafluorophenyl (Pfp) esters; the syntheses of a few peptides by such an approach furnished crude peptides with good quality [121]. The Hmb protecting group was also successfully used for the synthesis of difficult sequences, such as aggrecan fragment (368–379), human $\alpha_{1E\text{-}3}$ calcium channel subunit fragment (985–1004) [122], Aβ amyloid fragment (1–40) [123], and human prion protein fragment (106–126) [124].

Based on these studies, Quibell *et al.* tried to extend the use of backbone protecting groups to the synthesis of difficult sequences using a Boc/Bzl strategy by modifying Hmb. Since acetylated Hmb (AcHmb) as the amide backbone protecting group led to a substantial improvement in the solubility of the synthetic βA4 amyloid fragment (34–42) [125], they concluded that the 2-hydroxy moiety was Hmb's essential feature and proposed the 2-hydroxybenzyl (Hbz) group, TFA-stable and trifluoromethanesulfonic acid (TFMSA) labile, for SPPS by the Boc/Bzl strategy. This new amide backbone blocker was successfully used for the synthesis of an acyl carrier protein fragment (65–74) [126].

Other NH peptide bond protecting groups were developed and tested. The groups 5-methylfurfuryl, 5-methylthienylmethyl, 5-methoxythienylmethyl, 3-methoxythienylmethyl, Dmb, and 2,4,6-trimethoxybenzyl (Tmb) were employed in the syntheses of Fmoc-(X)Ala–Gly–Val-resin or Fmoc-Ala–(X)Gly–Val–resin (where X is one of the groups cited above). The syntheses using the Tmb provided the highest coupling yields (90%) [127]. Miranda *et al.* introduced the 2-hydroxy-6-nitrobenzyl (Hnb) group and demonstrated its effectiveness through solid-phase synthesis of a known hindered difficult peptide sequence, STAT-91 [128]. The Dmb group was also employed by means of the backbone protected dipeptide Fmoc-Gly–(Dmb)Gly-OH allowing for the synthesis of a 61-amino-acid residue peptide [73].

### 16.5.6
### The Use of Pseudo-Prolines

In the early 1980s, Toniolo *et al.* demonstrated that in solution or in solid-state, the α-helical and β-sheet secondary structures of linear oligopeptides could be disrupted if a Pro residue was placed in their central positions [129, 130]. The same effect was observed when a proline was inserted in the sequence of poly(Ala) on a solid-phase support as peptide aggregation was not only prevented, but also facilitated the incorporation of six additional Ala residues [71]. In this context, Haack and Mutter reported the use of oxazolidines and thiazolidines derived from Ser, Thr, and Cys as temporary protecting groups to be used in peptide synthesis (Figure 16.4) [131, 132]. In fact, it has been shown that such structures, called "pseudo-Pro" because they are Pro structure analogs, disturb the onset of regular structure by forming a *cis*-amide bond and, consequently, prevent peptide aggregation [133].

A comparative study of the efficiencies of the Hmb protection and pseudo-Pro dipeptides insertion in avoiding peptide aggregation during the synthesis of "difficult peptide sequences," such as the nonphosphorylated peptide MEDSTYYKASKGC-$NH_2$, its corresponding *O*-phosphotyrosine analog, and a fragment of influenza virus hemagglutinin, has demonstrated that both approaches improve the assembly of these peptide sequences. Nevertheless, the acylation of the growing peptide-resin containing the amino acid protected with Hmb at the N-terminal position may be difficult to complete and lead to deleted byproducts. On the other hand, the use of pseudo-Pro dipeptides is restricted to the synthesis of peptides containing Ser, Thr, or Cys, whereas that of the Hmb group is not [134].

In recent years, the use of pseudo-Pro dipeptides in the peptide synthesis resins expanded. The following peptides have been prepared and, therefore, illustrate the usefulness of the approach for such purpose: human amylin [135], [$Pro^{215}$]-myelin proteolipid protein (PLP) fragment (181–230) and its analog [$Ser^{215}$]-PLP (181–230) [136], RANTES fragment (1–68) [105], Pan DR epitope [137], insulin-like peptide-3 [138], cecropin (1–7)–melittin (2–9) hybrid, gp41 ectodomain peptide, extended calcipressin fragment with N-terminal extension, CCL4-L1 and CCL4-L2 chemokines [139], and the D2 domain of human vascular endothelial growth factor receptor-1 [140].

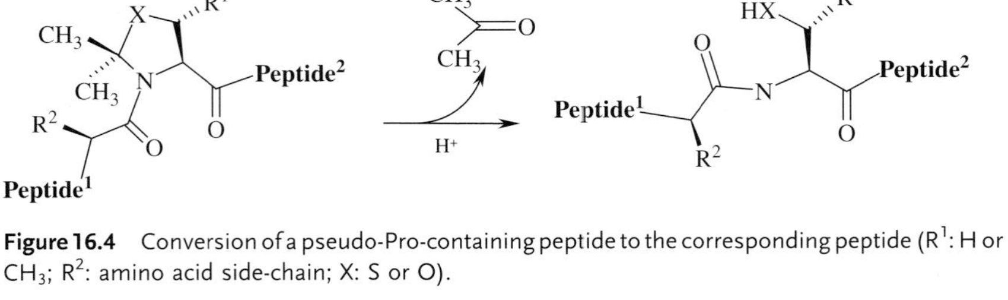

**Figure 16.4** Conversion of a pseudo-Pro-containing peptide to the corresponding peptide ($R^1$: H or $CH_3$; $R^2$: amino acid side-chain; X: S or O).

## 16.5.7
## O-Acyl Isopeptide Approach

In 1998, Horikawa *et al.* reported the synthesis of [Leu]-enkephalin analogs containing α-substituted serines to replace $Gly^2$ via an *O,N* migration method [141]. In 2003, the synthetic tool was first applied in syntheses of "difficult peptides" [142]. In the following years, the research groups of Carpino, Mutter, and Kiso confirmed the suitability of the *O*-acyl isopeptide approach for the same purpose [142–145].

The *O*-acyl isopeptide approach, also called "depsipeptide methodology," is based on the *O*-acylation of Ser or Thr by another amino acid for prevention of secondary structure formation stabilized by hydrogen bonds: the amide bond in X–Ser or X–Thr sequences (where X is an amino acid) is replaced by an ester bond formed due to reaction of the α-carboxyl group of X with the hydroxyl group of Ser or Thr side-chain (Figure 16.5). Once the peptide chain assembly is concluded, the amide bond is restructured through an *O,N*-acyl shift mediated by base – a known procedure [143–145].

As high enantiomerization levels of Ser and Thr were observed (the highest was 21% for Thr), the approach was improved with the introduction by Kiso's and Carpino's group of the "*O*-acyl isodipeptide" or "depsidipeptide unit." Its use in SPPS of Alzheimer's disease-related amyloid β peptides (1–42) and analogs was enantiomerization-free, making it suitable for use in automated synthesis of long peptides [145, 146].

The depsipeptide methodology has been also used for the synthesis of: (i) other "difficult peptide sequences" such as the extremely difficult WW domain FBP28 [147], the globular protein Crambin-46 fragment (16–46) [146], the pentapeptide Ac-VVTVV-$NH_2$ [145], and corticotropin releasing factor [148]; (ii) the "Click or Switch peptides," which have the potential to be used in molecular recognition, prodrug design, biosensor technology, self-association, and irreversible aggregation studies in degenerative cerebral diseases [142, 144, 149]. These peptides are folding precursor *O*-acyl peptides, and have the capacity of being quickly and easily converted to the native peptide via an *O,N*-acyl migration reaction at pH 7.4 [142, 145, 149] or by means of two steps: photoirradiation when the photocleavable 6- nitroveratryloxycarbonyl (Nvoc) protecting group protects the α-amino group of Ser followed by *O, N*-acyl migration reaction at pH 7.4 [150].

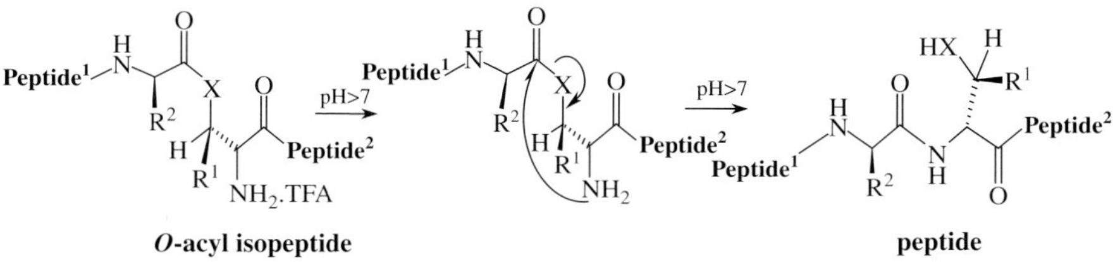

**Figure 16.5** Schematic representation of the *O*-acyl isopeptide approach ($R^1$: H or $CH_3$; $R^2$: amino acid side-chain; X: S or O).

### 16.5.8
### Use of Elevated Temperatures

In traditional stepwise SPPS, all steps are performed at room temperature [2, 39, 51, 54, 95]. However, peptide chemists have been trying to carry them out at elevated temperatures (SPPS-ET). Although their main goal is to significantly speed-up peptide chain assembly, many of these scientists are also interested in minimizing difficulties associated with peptide aggregation on resin.

Until the mid-1990s, despite important contributions made by different research groups [151–154], little information was available in the literature on SPPS-ET. For instance, potential to minimize the tendency of the growing peptide-resin to establish intra- and interchain interactions, extent of swelling in commercially available resins in solvents suitable for use at temperatures higher than 40 °C, and potential of amino acid enantiomerization had not been satisfactorily studied. Thus, Varanda and Miranda investigated such aspects and reported that, under the experimental conditions employed (conventional heating, 60 °C and 25% DMSO/toluene as the basic solvent), the solid-phase syntheses of a "difficult peptide" (acyl carrier protein (65–74)) and of other peptides (fragments of unsulfated human cholecystokinin) were efficient, fast, and quite simple [103], which was revised by Rivier and Miranda [64]. Later, Souza and coworkers demonstrated that, under such conditions, SPPS-ET does not enhance enantiomerization of Ile, Tyr, Lys, and Phe compared to traditional SPPS [155]. In 2007, they disclosed that their protocols using conventional heating also led to low enantiomerization levels (below 2% for Pro, Ser, Arg, Ala, Tyr, Asp, Met, Phe, His and Trp; the levels for Cys were equivalent to those obtained for couplings at room temperature) [156]. Therefore, quite recently, this group reported it in more detail [36] as well as successfully synthesized fragment 101–132 of acanthoscurrin, a glycine-rich "difficult peptide sequence," by combining their protocols for SPPS at 60 °C using conventional heating with (i) CLEAR amide of low substitution degree, (ii) Hmb amide backbone protection, and (iii) presence of LiCl in the coupling reaction media [35].

Simultaneously, many other research groups studied microwave-assisted SPPS [74, 157, 158] inspired by Wang *et al.*'s pioneering work [159]. Some groups have successfully attempted the syntheses of peptoids, glycopeptides, or peptides difficult to synthesize by the traditional method [67, 160], but only a few have studied and achieved the solid-phase synthesis of typical "difficult peptides": Collins and Leadbeater [161], Kappe *et al.* [65], and Chung *et al.* [75]. All these authors reported acceptable peptide purities and synthesis yields in quite short times.

## 16.6
## Monitoring the Synthesis of a "Difficult Peptide"

Even though there are many techniques suitable for the investigation of peptide conformation in solution, only a few of them allow for directly accessing the structural features of a peptide bound to a resin.

Since $^{13}$C-nuclear magnetic resonance (NMR) had been successfully employed in conformational studies of alamethicin derived peptide-polyoxyethylene supports [162] and was shown to be suitable for accessing the structure of the peptide in solution [163], Giralt *et al.* employed it to monitor every synthesis step of Asn-(N-Me)-Thr–Ala-OH on polystyrene resins [164]. Later, quadrupole echo deuterium NMR was also used to study the chain association of the growing polyglycine-benzydrylamino (BHA) resin, having revealed that while some polyglycine chains form hydrogen bonds with neighboring chains, others form cross-links with the polymeric support [165]. In the last decade, the application of NMR techniques for the study of peptidyl-resins was expanded by the introduction of magic angle spinning (MAS). MAS-NMR has been used successfully to monitor the solid-phase synthesis of "difficult sequences" such as aggrecan (370–379) on Wang resin [166] and acyl carrier protein (65–74) on BHA resin [167].

After the demonstration of the possibility of introducing the paramagnetic amino acid 2,2,6,6-tetramethylpiperidine-*N*-oxyl-4-amino-4-carboxylic acid (TOAC) at any position of a peptide sequence attached to a resin [168], electron paramagnetic resonance (EPR) became a useful tool for examining mobility and microenvironment of growing peptide-resins and, consequently, for monitoring the synthesis of difficult sequences in DCM, DMF, DMSO, and NMP or in mixtures able to disaggregate peptide chains such as the magic one (DCM/DMF/NMP; 1: 1: 1, 1% Triton, 4 N ethylene carbonate) and polyfluorinated alcohols (10% HFIP/DCM and 50% TFE/DCM) [169].

Fourier transform (FT)-IR spectroscopy is another spectroscopic technique used for detecting the formation of β-sheet secondary structure in peptides in solution [83, 170] and also monitoring every step of SPPS [171]. In fact, FT-IR spectra can be used for revealing aggregation in peptides attached to the resin as they may present bands in the amide I region (C=O stretching mode; 1685 (weak) and 1630 $cm^{-1}$ (strong)) and in the amide II region (N−H bending mode, 1525–1530 $cm^{-1}$ (strong)) characteristic of β-sheet secondary structure [171, 172].

By the introduction of the Nd: YAG laser in the near-IR (NIR) at 1064 $cm^{-1}$, Raman spectroscopy entered the Fourier domain and allowed for proper detection of the secondary structure in peptides [173]. NIR-FT-Raman spectroscopy was shown to be suited for monitoring SPPS of "difficult peptides" such as poly(Ala) on PepSyn Gel [174] or TentaGel resin [72] and the acanthoscurrin fragment (101–132) on CLEAR amide resin [35] due to the presence of the following bands in the spectra collected for the growing peptide-resins: at 1665–1680 (the amide I region; C=O stretching mode) and 1230–1240 $cm^{-1}$ (amide III region; C−N stretching and N−H bending modes) – characteristics of β-sheet secondary structure [175].

Of the four spectroscopic techniques cited above for monitoring SPPS of the so-called "difficult peptide sequences," the most advantageous are NIR-FT-Raman and solid-state NMR as they are not destructive. However, the second requires more sophisticated and quite expensive equipment as well as special preparation or labeling of samples.

## 16.7 Conclusions

The so-called "difficult sequences" are of major importance in peptide chemistry and biology as they have constantly served as model peptides in studies in which the ultimate aim is (i) to understand intra- and intermolecular peptide interactions and, consequently, peptide folding in solution or bound to a resin; and (ii) the creation of alternative chemical strategies, protocols, coupling reagents, side-chain protecting groups, amide backbone protecting groups, resins, and linkers suited for efficient stepwise and convergent synthesis of peptides and small proteins.

## References


1 Sewald, N. and Jakubke, H.-D. (2002) *Peptide: Chemistry and Biology*, Wiley-VCH Verlag GmbH, Weinheim.

2 Lloyd-Williams, P., Albericio, F., and Giralt, E. (1997) *Chemical Approaches to the Synthesis of Peptides and Proteins*, CRC Press, Boca Raton, FL.

3 Albericio, F. (2004) *Current Opinion in Chemical Biology*, **8**, 211–221.

4 Bray, B.L. (2003) *Nature Reviews Drug Discovery*, **2**, 587–593.

5 Molinski, T.F., Dalisay, D.S., Lievens, S.L., and Saludes, J.P. (2009) *Nature Reviews Drug Discovery*, **8**, 69–85.

6 Lee, H.-J., Na, K., Kwon, M.-S., Kim, H., Kim, K.S., and Paik, Y.-K. (2009) *Proteomics*, **9**, 3395–3408.

7 van den Broek, I., Sparidans, R.W., Schellens, J.H.M., and Beijnen, J.H. (2008) *Journal of Chromatography B*, **872**, 1–22.

8 Diamond, S.L. (2007) *Current Opinion in Chemical Biology*, **11**, 46–51.

9 Zablotna, E., Jáskiewicz, A., Legowska, A., Miecznikowska, H., Lesner, A., and Rolka, K. (2007) *Journal of Peptide Science*, **13**, 749–755.

10 Rigter, A., Langeveld, J.P.M., Timmers-Parohi, D., Jacobs, J.G., Moonen, P.L.J.M., and Bossers, A. (2007) *BMC Biochemistry*, **8**, 6.

11 Best, R.B. and Hummer, G. (2009) *The Journal of Physical Chemistry. B*, **113**, 9004–9015.

12 Fournel, S. and Muller, S. (2003) *Current Protein & Peptide Science*, **4**, 261–276.

13 Andresen, H., Zarse, K., Grötzinger, C., Hollidt, J.-M., Ehrentreich-Förster, E., Bier, F.F., and Kreuzer, O.J. (2006) *Journal of Immunological Methods*, **315**, 11–18.

14 Soulère, L. and Bernard, J. (2009) *Bioorganic & Medicinal Chemistry Letters*, **19**, 1173–1176.

15 Brandt, E.G., Hellgren, M., Brinck, T., Bergman, T., and Edholm, O. (2009) *Physical Chemistry Chemical Physics*, **11**, 975–983.

16 Loffet, A. (2002) *Journal of Peptide Science*, **8**, 1–7.

17 Nicholson, B., Lloyd, G.K., Miller, B.R., Palladino, M.A., Kiso, Y., Hayashi, Y., and Neuteboom, S.T.C. (2006) *Anti-Cancer Drugs*, **17**, 25–31.

18 Holmes, T.C. (2002) *Trends in Biotechnology*, **20**, 16–21.

19 Gelain, F., Horii, A., and Zhang, S. (2007) *Macromolecular Bioscience*, **7**, 544–551.

20 Chung, M.C., Gonçalves, M.F., Colli, W., Ferreira, E.I., and Miranda, M.T.M. (1997) *Journal of Pharmaceutical Sciences*, **86**, 1127–1131.

21 Santos, C., Mateus, M.L., dos Santos, A.P., Moreira, R., de Oliveira, E., and Gomes, P. (2005) *Bioorganic & Medicinal Chemistry Letters*, **15**, 1595–1598.

22 Dawson, P.E. and Kent, S.B.H. (2000) *Annual Review of Biochemistry*, **69**, 923–960.

23 Kent, S.B.H. (2009) *Chemical Society Reviews*, **38**, 338–351.

24 du Vigneaud, V., Ressler, C., and Trippett, S. (1953) *The Journal of Biological Chemistry*, **205**, 949–957.

25 Rocha, M. and Silva, E. (1962) *Biochemical Pharmacology*, **10**, 3–21.

26 Jensen, R.T., Wank, S.A., Rowley, W.H., Sato, S., and Gardner, J.D. (1989) *Trends in Pharmacological Sciences*, **10**, 418–423.

27 Hopkins, F.G. and Harris, L.J. (1929) *The Journal of Biological Chemistry*, **84**, 269–320.

28 du Vigneaud, V., Gish, D.T., and Katsoyannis, P.G. (1954) *Journal of the American Chemical Society*, **76**, 4751–4752.

29 Burger, M. and Brandt, W. (1935) *Zeitschrift fur die Gesamte Experimentelle Medizin*, **96**, 375–397.

30 Farah, A.E. (1983) *Pharmacological Reviews*, **35**, 181–217.

31 Oppenheim, F.G., Xu, T., McMillian, F.M., Levitz, S.M., Diamond, R.D., Offner, G.D., and Troxler, R.F. (1988) *The Journal of Biological Chemistry*, **263**, 7472–7477.

32 Kokryakov, V.N., Harwig, S.S.L., Panyutich, E.A., Shevchenko, A.A., Aleshina, G.M., Shamova, O.V., Korneva, H.A., and Lehrer, R.I. (1993) *FEBS Letters*, **327**, 231–236.

33 Hughes, J., Smith, T.W., Kosterlitz, H.W., Fothergill, L.A., Morgan, B.A., and Morris, H.R. (1975) *Nature*, **258**, 577–579.

34 Roberts, A.B., Lamb, L.C., Newton, D.L., Sporn, M.B., De Larco, J.E., and Todaro, G.J. (1980) *Proceedings of the National Academy of Sciences of the United States of America*, **77**, 3494–3498.

35 Remuzgo, C., Andrade, G.F.S., Temperini, M.L.A., and Miranda, M.T.M. (2009) *Biopolymers*, **92**, 65–75.

36 Loffredo, C., Assunção, N.A., Gerhardt, J., and Miranda, M.T.M. (2009) *Journal of Peptide Science*, **15**, 808–817.

37 Apostolopoulos, V. (2009) *Expert Review of Vaccines*, **8**, 259–260.

38 Pandey, V.N., Upadhyay, A., and Chaubey, B. (2009) *Expert Opinion on Biological Therapy*, **9**, 975–989.

39 Benoiton, N.L. (2005) *Chemistry of Peptide Synthesis*, Taylor & Francis, Boca Raton, FL.

40 Liria, C.W., Romagna, C.D., Rodovalho, N.N., Marana, S.R., and Miranda, M.T.M. (2008) *Journal of the Brazilian Chemical Society*, **19**, 1574–1581.

41 Wehofsky, N., Pech, A., Liebscher, S., Schmidt, S., Komeda, H., Asano, Y., and Bordusa, F. (2008) *Angewandte Chemie International Edition*, **47**, 5456–5460.

42 Thie, H., Meyer, T., Schirrmann, T., Hust, M., and Dubel, S. (2008) *Current Pharmaceutical Biotechnology*, **9**, 439–446.

43 Yasuda, H., Tada, Y., Hayashi, Y., Jomori, T., and Takaiwa, F. (2005) *Transgenic Research*, **14**, 677–684.

44 Szardenings, M. (2003) *Journal of Receptor and Signal Transduction Research*, **23**, 307–349.

45 Arap, M.A. (2005) *Genetics and Molecular Biology*, **28**, 1–9.

46 Ager, D.J., Pantaleone, D.P., Henderson, S.A., Katritzky, A.R., Prakash, I., and Walters, D.E. (1998) *Angewandte Chemie (International Edition in English)*, **37**, 1802–1817.

47 Walsh, G. (2005) *Applied Microbiology and Biotechnology*, **67**, 151–159.

48 Machado, A., Liria, C.W., Proti, P.B., Remuzo, C., and Miranda, M.T.M. (2004) *Quimica Nova*, **27**, 781–789.

49 Sakakibara, S. (1999) *Biopolymers*, **51**, 279–296.

50 Okada, Y. (2004) In *Houben-Weyl: Synthesis of Peptides and Peptidomimetics*, vol. E22a (eds A. Felix, L. Moroder, C. Toniolo, and M. Goodman), Thieme, Stuttgart, p. 591.

51 Merrifield, R.B. (1963) *Journal of the American Chemical Society*, **85**, 2149–2154.

52 Kaiser, E., Colescott, R.L., Bossinger, C.D., and Cook, P.I. (1970) *Analytical Biochemistry*, **34**, 595–598.

53 Gisin, B.F. (1972) *Analytica Chimica Acta*, **58**, 248–249.

54 Atherton, E. and Sheppard, R.C. (1989) *Solid Phase Peptide Synthesis: A Practical Approach*, IRL Press, Oxford.

55 White, P.D. and Chan, W.C. (2000) In *Fmoc Solid Phase Peptide Synthesis: A Practical Approach* (eds W.C. Chan and P.D. White), Oxford University Press, Oxford, p. 9.

56 Marasco, D., Perretta, G., Sabatella, M., and Ruvo, M. (2008) *Current Protein & Peptide Science*, **9**, 447–467.

57 Boschetti, E., Bindschedler, L.V., Tang, C., Fasoli, E., and Righetti, P.G. (2009) *Journal of Chromatography. A*, **1216**, 1215–1222.

58 Uttamchandani, M. and Yao, S.Q. (2008) *Current Pharmaceutical Design*, **14**, 2428–2438.

59 Breitling, F., Nesterov, A., Stadler, V., Felgenhauer, T., and Bischoff, F.R. (2009) *Molecular Biosystems*, **5**, 224–234.

60 Pini, A., Falciani, C., and Bracci, L. (2008) *Current Protein & Peptide Science*, **9**, 468–477.

61 Tam, J.P. (2006) In *Handbook of Biologically Active Peptides* (ed. A.J. Kastin), Academic Press, Burlington, MA, p. 541.

62 Tam, J.P., Xu, J., and Eom, K.D. (2001) *Biopolymers*, **60**, 194–205.

63 Benz, H. (1994) *Synthesis*, 337–358.

64 Rivier, J.E. and Miranda, M.T.M. (2002) In *Houben-Weyl: Synthesis of Peptide and Peptidomimetics*, vol. E22a (eds M. Goodman, A. Felix, L. Moroder, and C. Toniolo), Thieme, Stuttgart, p. 806.

65 Bacsa, B., Horváti, K., Bosze, S., Andreae, F., and Kappe, C.O. (2008) *The Journal of Organic Chemistry*, **73**, 7532–7542.

66 Ferguson, J.D. (2003) *Molecular Diversity*, **7**, 281–286.

67 Rizzolo, F., Sabatino, G., Chelli, M., Rovero, P., and Papini, A.M. (2007) *International Journal of Peptide Research and Therapeutics*, **13**, 203–208.

68 Wang, S.S., Yang, C.C., Kulesha, I.D., Sonenberg, M., and Merrifield, R.B. (1974) *International Journal of Peptide and Protein Research*, **6**, 103–109.

69 Blake, J. (1979) *International Journal of Peptide and Protein Research*, **13**, 418–425.

70 Mergler, M. and Dick, F. (2005) *Journal of Peptide Science*, **11**, 650–657.

71 Bedford, J., Hyde, C., Johnson, T., Jun, W., Owen, D., Quibell, M., and Sheppard, R.C. (1992) *International Journal of Peptide and Protein Research*, **40**, 300–307.

72 Ryttersgaard, J., Larsen, B.D., Holm, A., Christensen, D.H., and Nielsen, O.F. (1997) *Spectrochimica Acta A*, **53**, 91–98.

73 Zahariev, S., Guarnaccia, C., Zanuttin, F., Pintar, A., Esposito, G., Maravic, G., Krust, B., Hovanessian, A.G., and Pongor, S. (2005) *Journal of Peptide Science*, **11**, 17–28.

74 Erdélyi, M. and Gogoll, A. (2002) *Synthesis*, 1592–1596.

75 Katritzky, A.R., Haase, D.N., Johnson, J.V., and Chung, A. (2009) *The Journal of Organic Chemistry*, **74**, 2028–2032.

76 Dalcol, I., Rabanal, F., Ludevid, M.-D., Albericio, F., and Giralt, E. (1995) *The Journal of Organic Chemistry*, **60**, 7575–7581.

77 Atherton, E., Woolley, V., and Sheppard, R.C. (1980) *Journal of the Chemical Society, Chemical Communications*, **20**, 970–971.

78 Kent, S.B.H. (1988) *Annual Review of Biochemistry*, **57**, 957–989.

79 Milton, R.C. de L., Milton, S.C.F., and Adams, P.A. (1990) *Journal of the American Chemical Society*, **112**, 6039–6046.

80 Tickler, A.K. and Wade, J.D. (2007) *Current Protocols in Protein Science*, **50**, 18.8.1.

81 Ambulos, N.P. Jr, Bibbs, L., Bonewald, L.F., Kates, S.A., Khatri, A., Medzihradszky, K.F., and Weintraub, S.T. (2000) In *Solid-Phase Synthesis* (eds S.A. Kates and F. Albericio), Dekker, New York, p. 751.

82 Narita, M., Ishikawa, K., Chen, J.-Y., and Kim, Y. (1984) *International Journal of Peptide and Protein Research*, **24**, 580–587.

83 Narita, M., Honda, S., Umeyama, H., and Obana, S. (1988) *Bulletin of the Chemical Society of Japan*, **61**, 281–284.

84 Skropeta, D., Jolliffe, K.A., and Turner, P. (2004) *The Journal of Organic Chemistry*, **69**, 8804–8809.

85 Kitagawa, K., Adachi, H., Sekigawa, Y., Yagami, T., Futaki, S., Gu, Y.J., and Inoue, K. (2004) *Tetrahedron*, **60**, 907–918.

86 Crespo, L., Sanclimens, G., Royo, M., Giralt, E., and Albericio, F. (2002) *European Journal of Organic Chemistry*, 1756–1762.

87 McNamara, J.F., Lombardo, H., Pillai, S.K., Jensen, I., Albericio, F., and Kates, S.A. (2000) *Journal of Peptide Science*, **6**, 512–518.

88 Arunan, C. and Pillai, V.N.R. (2000) *Tetrahedron*, **56**, 3005–3011.

89 Larsen, B.D. and Holm, A. (1998) *The Journal of Peptide Research*, **52**, 470–476.

**90** Tickler, A.K., Clippingdale, A.B., and Wade, J.D. (2004) *Protein and Peptide Letters*, **11**, 377–384.

**91** Cardona, V., Eberle, I., Barthélémy, S., Beythien, J., Doerner, B., Schneeberger, P., Keyte, J., and White, P.D. (2008) *International Journal of Peptide Research and Therapeutics*, **14**, 285–292.

**92** van Woerkom, W.J. and van Nispen, J.W. (1991) *International Journal of Peptide and Protein Research*, **38**, 103–113.

**93** Krchnak, V., Flegelova, Z., and Vagner, J. (1993) *International Journal of Peptide and Protein Research*, **42**, 450–454.

**94** Milton, S.C.F. and Milton, R.C. de L. (1990) *International Journal of Peptide and Protein Research*, **36**, 193–196.

**95** Steward, J.M. and Young, J.D. (1984) *Solid Phase Peptide Synthesis*, Pierce, Rockford, IL.

**96** Schnölzer, M., Alewood, P., Jones, A., Alewood, D., and Kent, S.B.H. (1992) *International Journal of Peptide and Protein Research*, **40**, 180–193.

**97** Alewood, P., Alewood, D., Miranda, L., Love, S., Meutermans, W., Wilson, D., and Gregg, B.F. (1997) In *Methods in Enzymology: Solid-Phase Peptide Synthesis*, vol. 289 (ed G.B. Fields), Academic Press, New York, pp. 14–29.

**98** Narita, M., Kojima, Y., and Isokawa, S. (1989) *Bulletin of the Chemical Society of Japan*, **62**, 1976–1981.

**99** Bagley, C.J., Otteson, K.M., May, B.L., McCurdy, S.N., Pierce, L., Ballard, F.J., and Wallace, J.C. (1990) *International Journal of Peptide and Protein Research*, **36**, 356–361.

**100** Yamashiro, D., Blake, J., and Li, C.H. (1976) *Tetrahedron Letters*, **17**, 1469–1472.

**101** Fields, G.B. and Fields, C.G. (1991) *Journal of the American Chemical Society*, **113**, 4202–4207.

**102** Tam, J.P. and Lu, Y.-A. (1995) *Journal of the American Chemical Society*, **117**, 12058–12063.

**103** Varanda, L.M. and Miranda, M.T.M. (1997) *The Journal of Peptide Research*, **50**, 102–108.

**104** Cilli, E.M., Marchetto, R., Schreier, S., and Nakaie, C.R. (1999) *The Journal of Organic Chemistry*, **64**, 9118–9123.

**105** Garcia-Martin, F., Quintanar-Audelo, M., Garcia-Ramos, Y., Cruz, L.J., Gravel, C., Furic, R., Cote, S., Tulla-Puche, J., and Albericio, F. (2006) *Journal of Combinatorial Chemistry*, **8**, 213–220.

**106** Merrifield, B. (1995) In *Peptides: Synthesis, Structures and Applications* (ed. B. Gutte), Academic Press, San Diego, CA, p. 93.

**107** Fields, G.B. and Noble, R.L. (1990) *International Journal of Peptide and Protein Research*, **35**, 161–214.

**108** Bayer, E., Hemmasi, B., Albert, K., Rapp, W., and Dengler, M. (1983) Presented at 8th American Peptide Symposium, Tucson.

**109** Kempe, M. and Barany, G. (1996) *Journal of the American Chemical Society*, **118**, 7083–7093.

**110** Ajikumar, P.K. and Devaky, K.S. (2000) *Letters in Peptide Science*, **7**, 207–215.

**111** Seebach, D., Thaler, A., and Beck, A.K. (1989) *Helvetica Chimica Acta*, **72**, 857–867.

**112** Hendrix, J.C., Halverson, K.J., Jarrett, J.T., and Lansbury, P.T. (2002) *The Journal of Organic Chemistry*, **55**, 4517–4518.

**113** Thaler, A., Seebach, D., and Cardinaux, F. (1991) *Helvetica Chimica Acta*, **74**, 628–643.

**114** Weygand, F., Steglich, W., and Bjarnason, J. (1968) *Chemische Berichte*, **101**, 3642–3648.

**115** Narita, M., Fukunaga, T., Wakabayashi, A., Ishikawa, K., and Nakano, H. (1984) *International Journal of Peptide and Protein Research*, **23**, 306–314.

**116** Narita, M., Ishikawa, K., Nakano, H., and Isokawa, S. (1984) *International Journal of Peptide and Protein Research*, **24**, 14–24.

**117** Eckert, H., and Seidel, C. (1986) *Angewandte Chemie (International Edition in English)*, **25**, 159–160.

**118** Eckert, H., Kiesel, Y., Seidel, C., Kaulberg, C., and Brinkmann, H. (1986) *Chemistry for Peptide and Protein*, **3**, 19–28.

**119** Bartl, R., Klöppel, K.-D., and Frank, R. (1992) In *Peptides: Chemistry and Biology (Proceedings of the American Peptide Symposium)* (eds J.A. Smith and J.E. Rivier), ESCOM, Leiden, p. 505.

**120** Johnson, T., Quibell, M., Owen, D., and Sheppard, R.C. (1993) *Journal of the*

*Chemical Society, Chemical Communications*, 369–372.

121 Hyde, C., Johnson, T., Owen, D., Quibell, M., and Sheppard, R.C. (1994) *International Journal of Peptide and Protein Research*, **43**, 431–440.

122 Simmonds, R.G. (1996) *International Journal of Peptide and Protein Research*, **47**, 36–41.

123 Clippingdale, A.B., Macris, M., Wade, J.D., and Barrow, C.J. (1999) *The Journal of Peptide Research*, **53**, 665–672.

124 Jobling, M.F., Barrow, C.J., White, A.R., Masters, C.L., Collins, S.J., and Cappai, R. (1999) *Letters in Peptide Science*, **6**, 129–134.

125 Quibell, M., Turnell, W.G., and Johnson, T. (1994) *Tetrahedron Letters*, **35**, 2237–2238.

126 Johnson, T. and Quibell, M. (1994) *Tetrahedron Letters*, **35**, 463–466.

127 Johnson, T., Quibell, M., and Sheppard, R.C. (1995) *Journal of Peptide Science*, **1**, 11–25.

128 Miranda, L.P., Meutermans, W.D.F., Smythe, M.L., and Alewood, P.F. (2000) *The Journal of Organic Chemistry*, **65**, 5460–5468.

129 Toniolo, C., Bonora, G.M., Mutter, M., and Pillai, V.N.R. (1981) *Makromolekulare Chemie*, **182**, 2007–2014.

130 Toniolo, C., Bonora, G.M., Mutter, M., and Pillai, V.N.R. (1981) *Makromolekulare Chemie*, **182**, 1997–2005.

131 Haack, T. and Mutter, M. (1992) *Tetrahedron Letters*, **33**, 1589–1592.

132 Wöhr, T. and Mutter, M. (1995) *Tetrahedron Letters*, **36**, 3847–3848.

133 Dumy, P., Keller, M., Ryan, D.E., Rohwedder, B., Wohr, T., and Mutter, M. (1997) *Journal of the American Chemical Society*, **119**, 918–925.

134 Sampson, W.R., Patsiouras, H., and Ede, N.J. (1999) *Journal of Peptide Science*, **5**, 403–409.

135 Abedini, A. and Raleigh, D.P. (2005) *Organic Letters*, **7**, 693–696.

136 Trifilieff, E. (2005) *The Journal of Peptide Research*, **66**, 101–110.

137 Cremer, G.-A., Tariq, H., and Delmas, A.F. (2006) *Journal of Peptide Science*, **12**, 437–442.

138 Shabanpoor, F., Bathgate, R.A.D., Hossain, M.A., Giannakis, E., Wade, J.D., and Hughes, R.A. (2007) *Journal of Peptide Science*, **13**, 113–120.

139 de la Torre, B.G., Jakab, A., and Andreu, D. (2007) *International Journal of Peptide Research and Therapeutics*, **13**, 265.

140 Goncalves, V., Gautier, B., Huguenot, F., Leproux, P., Garbay, C., Vidal, M., and Inguimbert, N. (2009) *Journal of Peptide Science*, **15**, 417–422.

141 Horikawa, M., Shigeri, Y., Yumoto, N., Yoshikawa, S., Nakajima, T., and Ohfune, Y. (1998) *Bioorganic & Medicinal Chemistry Letters*, **8**, 2027–2032.

142 Sohma, Y., Taniguchi, A., Yoshiya, T., Chiyomori, Y., Fukao, F., Nakamura, S., Skwarczynski, M., Okada, T., Ikeda, K., Hayashi, Y., Kimura, T., Hirota, S., Matsuzaki, K., and Kiso, Y. (2006) *Journal of Peptide Science*, **12**, 823–828.

143 Carpino, L.A., Krause, E., Sferdean, C.D., Schümann, M., Fabian, H., Bienert, M., and Beyermann, M. (2004) *Tetrahedron Letters*, **45**, 7519–7523.

144 Mutter, M., Chandravarkar, A., Boyat, C., Lopez, J., Dos Santos, S., Mandal, B., Mimna, R., Murat, K., Patiny, L., Saucède, L., and Tuchscherer, G. (2004) *Angewandte Chemie International Edition*, **43**, 4172–4178.

145 Sohma, Y., Taniguchi, A., Skwarczynski, M., Yoshiya, T., Fukao, F., Kimura, T., Hayashi, Y., and Kiso, Y. (2006) *Tetrahedron Letters*, **47**, 3013–3017.

146 Coin, I., Dolling, R., Krause, E., Bienert, M., Beyermann, M., Sferdean, C.D., and Carpino, L.A. (2006) *The Journal of Organic Chemistry*, **71**, 6171–6177.

147 Bang, D., Chopra, N., and Kent, S.B.H. (2004) *Journal of the American Chemical Society*, **126**, 1377–1383.

148 Coin, I., Beyermann, M., and Bienert, M. (2007) *Nature Protocols*, **2**, 3247–3256.

149 Dos Santos, S., Chandravarkar, A., Mandal, B., Mimna, R., Murat, K., Saucede, L., Tella, P., Tuchscherer, G., and Mutter, M. (2005) *Journal of the American Chemical Society*, **127**, 11888–11889.

150 Taniguchi, A., Sohma, Y., Kimura, M., Okada, T., Ikeda, K., Hayashi, Y., Kimura, T., Hirota, S., Matsuzaki, K., and Kiso, Y. (2006) *Journal of the American Chemical Society*, **128**, 696–697.

151 Barlos, K., Papaioannou, D., Patrianakou, S., and Tsegenidis, T. (1986) *Liebigs Annalen der Chemie*, **11**, 1950–1955.

152 Tam, J.P. (1987) *International Journal of Peptide and Protein Research*, **29**, 421–431.

153 Wang, S. and Foutch, G.L. (1991) *Chemical Engineering Science*, **46**, 2373–2376.

154 Rabinovich, A.K. and Rivier, J.E. (1994) *American Biotechnology Laboratory*, **12**, 48–51.

155 Souza, M.P., Tavares, M.F.M., and Miranda, M.T.M. (2004) *Tetrahedron*, **60**, 4671–4681.

156 Loffredo, C., Assunção, N.A., and Miranda, M.T.M. (2009) In *Peptides for Youth (Proceedings of the 20th. American Peptide Symposium)*, vol. **61** (eds S. Del Valle, E., Escher and W.D. Lubell), Springer, New York, p. 165.

157 Murray, J.K. and Gellman, S.H. (2005) *Organic Letters*, **7**, 1517–1520.

158 Palasek, S.A., Cox, Z.J., and Collins, J.M. (2007) *Journal of Peptide Science*, **13**, 143–148.

159 Yu, H.M., Chen, S.T., and Wang, K.T. (2002) *The Journal of Organic Chemistry*, **57**, 4781–4784.

160 Fara, M.A., Díaz-Mochón, J.J., and Bradley, M. (2006) *Tetrahedron Letters*, **47**, 1011–1014.

161 Collins, J.M. and Leadbeater, N.E. (2007) *Organic & Biomolecular Chemistry*, **5**, 1141–1150.

162 Leibfritz, D., Mayr, W., Oekonomopulos, R., and Jung, G. (1978) *Tetrahedron*, **34**, 2045–2050.

163 Epton, R., Goddard, P., and Ivin, K.J. (1980) *Polymer*, **21**, 1367–1371.

164 Giralt, E., Rizo, J., and Pedroso, E. (1984) *Tetrahedron*, **40**, 4141–4152.

165 Ludwick, A.G., Jelinski, L.W., Live, D., Kintanar, A., and Dumais, J.J. (1986) *Journal of the American Chemical Society*, **108**, 6493–6496.

166 Dhalluin, C., Boutillon, C., Tartar, A., and Lippens, G. (1997) *Journal of the American Chemical Society*, **119**, 10494–10500.

167 Valente, A.P., Almeida, F.C.L., Nakaie, C.R., Schreier, S., Crusca, E. Jr, and Cilli, E.M. (2005) *Journal of Peptide Science*, **11**, 556–563.

168 Marchetto, R., Schreier, S., and Nakaie, C.R. (1993) *Journal of the American Chemical Society*, **115**, 11042–11043.

169 Cilli, E.M., Vicente, E.F., Crusca, E. Jr, and Nakaie, C.R. (2007) *Tetrahedron Letters*, **48**, 5521–5524.

170 Oh-uchi, S.-k., Lee, J.-S., and Narita, M. (1996) *Bulletin of the Chemical Society of Japan*, **69**, 1303–1307.

171 Henkel, B. and Bayer, E. (1998) *Journal of Peptide Science*, **4**, 461–470.

172 Kumar, I.M.K., Pillai, V.N.R., and Mathew, B. (2002) *Journal of Peptide Science*, **8**, 183–191.

173 Hallmark, V. and Rabolt, J.F. (1989) *Macromolecules*, **22**, 500.

174 Larsen, B.D., Christensen, D.H., Holm, A., Zillmer, R., and Nielsen, O.F. (1993) *Journal of the American Chemical Society*, **115**, 6247–6253.

175 Colthup, N.B., Daly, H.L., and Wiberley, S.E. (1990) *Introduction to Infrared and Raman Spectroscopy*, 3rd edn, Academic Press, Boston, MA.

# Index



*Amino Acids, Peptides and Proteins in Organic Chemistry.*
*Vol.3 – Building Blocks, Catalysis and Coupling Chemistry.* Edited by Andrew B. Hughes

ISBN: 978-3-527-32102-5

***m***



***n***



***o***

***q***



***r***



***s***